Herbal Medicine in Depression

Clara Grosso
Editor

# Herbal Medicine in Depression

## Traditional Medicine to Innovative Drug Delivery

Springer

*Editor*
Clara Grosso
REQUIMTE/LAQV, Laboratory of Pharmacognosy
Department of Chemistry, Faculty of Pharmacy
University of Porto
Porto, Portugal

ISBN 978-3-319-14020-9 ISBN 978-3-319-14021-6 (eBook)
DOI 10.1007/978-3-319-14021-6

Library of Congress Control Number: 2016937643



Printed on acid-free paper

This Springer imprint is published by Springer Nature
The registered company is Springer International Publishing AG Switzerland

# Preface

This book was conceived by professionals and researchers from different backgrounds, combining ethnopharmacology, phytochemistry, pharmacology and neuroscience topics, for a better understanding of depression. According to the World Health Organization, more than 350 million people of all ages suffer from depression, and the existing drug therapies are only suitable to treat ca. 60–70 % of patients. Therefore, this book intends to debate the current knowledge on the aetiology of depression, the mechanism of action of the available antidepressants and what routes should be followed to pursue more efficient tools for treatment-resistant patients.

This book starts with a general chapter (Chap. 1) dealing with aspects of prevalence and costs of the disease worldwide as well as the theories postulated since the 1950s to explain the physiopathology of depression. The reader will understand that in the light of nowadays knowledge a unified hypothesis of depression is not sustained due to its multifactorial features.

Chapter 2 summarizes the mechanism of action of the antidepressant drugs available to treat depression, as well as possible new targets to take into account in new drug discovery programs.

Chapters 3 and 4 highlight the importance that cellular and animal models have had in our understanding of the pathophysiology of depression. A critical analysis of the advantages and disadvantages of each model is also presented.

Since ancient times, nature has been a rich source of drugs which display higher structural diversity than the synthetic ones. With this in mind and being aware that sooner or later new antidepressant drugs have to be tested, approved and released, big pharma companies are encouraged to create new programs of drug discoveries based on natural products. Therefore, several medicinal plants with antidepressant activity were compiled in Chaps. 5–10, aiming at contributing (a) to gather the traditional knowledge of communities transferred orally from generation to generation or already established in traditional medicines and (b) to arouse the interest of the most skeptical persons. These chapters will stimulate readers to appreciate the benefits of natural products on human health.

In a more alternative and futuristic view to treat depression, new drug delivery systems for antidepressant drugs are described in detail in Chap. 11.

Finally, Chap. 12 takes into account all the issues discussed over the previous chapters and proposes new strategies that can be followed to reduce the burden of depression in societies around the world.

I sincerely hope the readers will enjoy and learn from the contents of this book.

Porto, Portugal Clara Grosso

# Acknowledgements

First of all, I would like to express my gratitude to all the authors of this book for the effort and dedication demonstrated and knowledge shared. I also would like to acknowledge the photographers and institutions who provided us some of the photos of the plant species covered in this book.

This book was financed through project UID/QUI/50006/2013, receiving financial support from FCT/MEC through national funds, and co-financed by FEDER, under the Partnership Agreement PT2020. I am also indebted to FCT for the FCT Investigator (IF/01332/2014).

# Contents

# Contributors

**Paula B. Andrade** REQUIMTE/LAQV, Laboratório de Farmacognosia, Departamento de Química, Faculdade de Farmácia, Universidade do Porto, Porto, Portugal

**Luís Catarino** Faculdade de Ciências, Centre for Ecology, Evolution and Environmental Changes, (Ce3C), Universidade de Lisboa, Lisbon, Portugal

**Matthew J. Cheesman** School of Biomedical Sciences, University of Queensland, St Lucia, QLD, Australia

**Oana Cioanca** Department of Pharmacognosy-Phytotherapy, Faculty of Pharmacy, University of Medicine and Pharmacy "Gr. T. Popa", Iasi, Romania

**Ian Edwin Cock** Environmental Futures Research Institute, Griffith University, Nathan Campus, Nathan, QLD, Australia

School of Natural Sciences, Griffith University, Nathan Campus, Nathan, QLD, Australia

**Fabiola Domínguez** Centro de Investigación Biomédica de Oriente, Instituto Mexicano del Seguro Social, Metepec, Mexico

**Rong Fan** Institute of Integrated Traditional Chinese and Western Medicine, Xiangya Hospital, Central South University, Changsha, China

**Dandan Feng** Institute of Integrated Traditional Chinese and Western Medicine, Xiangya Hospital, Central South University, Changsha, China

**Juan Manuel Gallardo** Unidad de Investigación Médica en Enfermedades Nefrológicas, Hospital de Especialidades, Centro Médico Nacional Siglo XXI, Instituto Mexicano del Seguro Social, Ciudad de, Mexico

**Eva González-Trujano** Dirección de Investigaciones en Neurociencias, Instituto Nacional de Psiquiatría Ramón de la Fuente Muñiz, México, DF, Mexico

**Clara Grosso** REQUIMTE/LAQV, Laboratory of Pharmacognosy, Department of Chemistry, Faculty of Pharmacy, University of Porto, Porto, Portugal

**Christian Guerra-Araiza** Unidad de Investigación Médica en Farmacología. Centro Médico Nacional Siglo XXI, Instituto Mexicano del Seguro Social, México, DF, Mexico

**Lucian Hritcu** Department of Biology, Alexandru Ioan Cuza University, Iasi, Romania

**Danuta Jantas** Department of Experimental Neuroendocrinology, Institute of Pharmacology, Polish Academy of Sciences, Krakow, Poland

**Vandita Kakkar** University Institute of Pharmaceutical Sciences, Panjab University, Chandigarh, India

**Manoj Kumar** Advanced Drug Delivery Research, R&D, Pancea Biotec Limited, Lalru, Punjab, India

**Zhiqing Li** Institute of Integrated Traditional Chinese and Western Medicine, Xiangya Hospital, Central South University, Changsha, China

**Xiangping Lin** Institute of Integrated Traditional Chinese and Western Medicine, Xiangya Hospital, Central South University, Changsha, China

**Nilufar Mamadalieva** Institute of the Chemistry of Plant Substances of Academy of Sciences of Uzbekistan, Tashkent, Uzbekistan

**Nazim Mamedov** Stockbridge School of Agriculture, University of Massachusetts at Amherst, Amherst, MA, USA

**Eurico S. Martins** JBT/IICT, Jardim Botânico Tropical, Instituto de Investigação Científica Tropical, Lisbon, Portugal

**Nancy Modgill** University Institute of Pharmaceutical Sciences, Panjab University, Chandigarh, India

**Sandra Orozco-Suárez** Unidad de Investigación Médica en Enfermedades, Hospital de Especialidades, Centro Médico Nacional Siglo XXI, Instituto Mexicano del Seguro Social, México, DF., Mexico

**Eleni Palazidou** Barts & London School of Medicine, Queen Mary University, London, UK

**Julia J. Segura-Uribe** Unidad de Investigación Médica en Enfermedades, Hospital de Especialidades, Centro Médico Nacional Siglo XXI, Instituto Mexicano del Seguro Social, México, Mexico

**Tao Tang** Institute of Integrated Traditional Chinese and Western Medicine, Xiangya Hospital, Central South University, Changsha, China

**Generosa Teixeira** Faculdade de Farmácia, Centre for Ecology, Evolution and Environmental Changes, (Ce3C), Universidade de Lisboa, Lisbon, Portugal

**Patrícia Valentão** REQUIMTE/LAQV, Laboratório de Farmacognosia, Departamento de Química, Faculdade de Farmácia, Universidade do Porto, Porto, Portugal

**Yang Wang** Institute of Integrated Traditional Chinese and Western Medicine, Xiangya Hospital, Central South University, Changsha, China

**Zian Xia** Institute of Integrated Traditional Chinese and Western Medicine, Xiangya Hospital, Central South University, Changsha, China

**Zhaoyu Yang** Institute of Integrated Traditional Chinese and Western Medicine, Xiangya Hospital, Central South University, Changsha, China

**Shu Yang** Institute of Integrated Traditional Chinese and Western Medicine, Xiangya Hospital, Central South University, Changsha, China

**Chunhu Zhang** Institute of Integrated Traditional Chinese and Western Medicine, Xiangya Hospital, Central South University, Changsha, China

# Chapter 1
# Depressive Disorders: Prevalence, Costs, and Theories

**Clara Grosso, Patrícia Valentão, and Paula B. Andrade**

**Abstract** The estimated global cost of mental health conditions, including depressive disorders, was US$2.5 trillion in 2010. According to the World Health Organization (WHO), more than 350 million people of all ages suffer from depression, 10–40 % of them not improving their condition with the current drug therapies, thus contributing to the increased burden of mental disorders. Indeed, prognostics are not encouraging since it is predicted that unipolar depressive disorders will be the first cause of disability-adjusted life years (DALYs) by the year 2030. For these reasons, it is urgent to find new antidepressant drugs that act on other targets rather than the conventional one (monoamine transmission). Having this in mind, this chapter provides a critical review of the theories available to explain the pathogenesis of depressive disorders, namely, those focusing on disturbances of monoamine, glutamate and GABA transmission, changes in the hypothalamic-pituitary-adrenal (HPA) axis, neuroinflammation, neurogenesis and neurotrophic factors, glial pathology, epigenetic mechanisms, and disturbance of the circadian rhythm.

**Keywords** Neurotransmitters • Neuroinflammation • Hypothalamic-pituitary-adrenal axis • Neuroplasticity • Epigenetics • Glial cells • Circadian rhythm

## 1.1 Introduction

The fifth edition of the *Diagnostic and Statistical Manual of Mental Disorders (DSM-5)* considers two separate branches on the mood disorder diagnostic tree: (1) depressive disorders, comprising disruptive mood dysregulation disorder, major

---

C. Grosso
REQUIMTE/LAQV, Laboratory of Pharmacognosy, Department of Chemistry, Faculty of Pharmacy, University of Porto, Porto, Portugal

P. Valentão • P.B. Andrade (✉)
REQUIMTE/LAQV, Laboratório de Farmacognosia, Departamento de Química, Faculdade de Farmácia, Universidade do Porto, Rua de Jorge Viterbo Ferreira, n°. 228, 4050-313 Porto, Portugal
e-mail: pandrade@ff.up.pt



C. Grosso (ed.), *Herbal Medicine in Depression*,
DOI 10.1007/978-3-319-14021-6_1

depressive disorder, persistent depressive disorder, premenstrual dysphoric disorder, substance/medication-induced depressive disorder, depressive disorder due to another medical condition, other specified depressive disorder, and unspecified depressive disorder, and (2) bipolar and related disorders, including bipolar I disorder, bipolar II disorder, cyclothymic disorder, substance/medication-induced bipolar and related disorder, bipolar and related disorder due to another medical condition, other specified bipolar and related disorder, and unspecified bipolar and related disorder (American Psychiatric Association 2013).

An approach to these branches is provided below.

### *1.1.1 Depressive Disorders*

Disruptive mood dysregulation disorder (DMDD) is characterized by severe and recurrent temper outbursts that are grossly out of proportion in intensity or duration to the situation. On average, these occur three or more times per week for 1 year or more. Between outbursts, children with DMDD display persistently irritable or angry mood, most of the day and nearly every day. The diagnostic requires the above symptoms to be present in at least two settings (at home, at school, or with peers) for 12 or more months, and symptoms must be severe in at least one of these settings. The onset of symptoms must be before the age of 10, and DMDD diagnostic should not be made for the first time before the age of 6 or after the age of 18 (American Psychiatric Association 2013).

Major depressive disorder (MDD) is characterized by five or more of the following nine symptoms during 2 weeks: (1) depressed mood, (2) loss of interest or pleasure, (3) change of weight (5 % change over 1 month) or appetite, (4) insomnia or hypersomnia, (5) psychomotor retardation or agitation, (6) loss of energy or fatigue, (7) worthlessness or guilt, (8) impaired concentration or indecisiveness, and (9) thoughts of death or suicidal ideation or attempt. From these symptoms, at least (1) and (2) are always present (Uher et al. 2014; American Psychiatric Association 2013).

Persistent depressive disorder (dysthymia) is characterized by depressed mood for the most of the day, for more days than not, for 2 years or longer. The diagnostic requires the presence of two or more of the following symptoms during the same period and never without symptoms for more than 2 months: (1) poor appetite or overeating, (2) insomnia or hypersomnia, (3) low energy or fatigue, (4) low self-esteem, (5) impaired concentration or indecisiveness, and (6) hopelessness (Uher et al. 2014; American Psychiatric Association 2013).

Premenstrual dysphoric disorder is characterized by the presence of five (out of eleven) symptoms in the week before menstruation, the symptoms starting to disappear few days after the beginning of menstruation and vanishing completely in the week after. The presence of, at least, one of the following symptoms is characteristic: (1) marked affective lability (e.g., suddenly feeling sad or tearful or experiencing increased sensitivity to rejection); (2) persistent and marked anger or irritability or increased interpersonal conflicts; (3) marked depressed mood, feeling of hopelessness, or self-deprecating thoughts; and (4) marked anxiety, tension, feelings of being "keyed up" or "on the edge." Additionally, one or more of the following symptoms are also present:

1. Decreased interest in common activities (e.g., work, school, friends, and hobbies)
2. Subjective sense of difficulty in concentrating
3. Lethargy, easy fatigability, or marked lack of energy
4. Marked change in appetite, overeating, or specific food cravings
5. Hypersomnia or insomnia
6. Subjective sense of being overwhelmed or out of control
7. Other physical symptoms, such as breast tenderness or swelling, headaches, joint or muscle pain, sensation of bloating, or weight gain (American Psychiatric Association 2013)

Substance-/medication-induced depressive disorder is a prominent and persistent disturbance of mood or loss of interest or pleasure as a direct physiological consequence of a drug of abuse, a medication, another somatic treatment for depression, or toxin exposure (American Psychiatric Association 2013).

A mood disorder due to another medical condition is characterized by a prominent and persistent disturbance of mood or loss of interest or pleasure as a direct physiological consequence of another medical condition (American Psychiatric Association 2013).

Other specified depressive disorders and unspecified depressive disorders are those with features that do not meet criteria for specified depressive disorders. For the first group, the clinician chooses to communicate the specific reason why the presentation does not meet criteria for any specific depressive disorder, while in the second one, he does not specify. Other specified depressive disorder class includes recurrent brief depression, brief depressive episode (4–13 days), and depressive episode with insufficient symptoms (American Psychiatric Association 2013).

### *1.1.2 Bipolar and Related Disorders*

In bipolar I disorder, a manic episode can be preceded or followed by a hypomanic episode or a major depressive episode. The manic episode is characterized by a distinct period of abnormally and persistently elevated, expansive, or irritable mood, lasting, at least, 1 week (or any duration if hospitalization is necessary). During this period, three or more of the following symptoms are present (four if the mood is only irritable):

1. Inflated self-esteem or grandiosity
2. Decreased need for sleep (e.g., feels rested after only 3 h of sleep)
3. More talkative than usual or pressure to keep talking
4. Flight of ideas or subjective experience that thoughts are racing
5. Distractibility (i.e., attention too easily drawn to unimportant or irrelevant external stimuli)

6. Increase in goal-directed activity (either socially, at work or school, or sexually) or psychomotor agitation (nondirected activity)
7. Excessive involvement in pleasurable activities that have a high potential for painful consequences (e.g., engaging in unrestrained buying sprees, sexual indiscretions, or foolish business investments)

The hypomanic episode is characterized by the same symptoms, but differs in severity and duration (at least 4 days) (American Psychiatric Association 2013).

Bipolar II disorder is characterized by one or more major depressive episodes, accompanied by at least one hypomanic episode, the symptoms of both being those described above (American Psychiatric Association 2013).

A cyclothymic disorder is characterized by at least 2 years of numerous periods of hypomanic and depressive symptoms that do not meet criteria for a manic episode and for a major depressive episode, respectively (American Psychiatric Association 2013).

In substance/medication-induced bipolar and related disorder, a prominent and persistent disturbance of mood or loss of interest or pleasure as a direct physiological consequence of a drug of abuse, a medication, another somatic treatment for depression, or toxin exposure is observed (American Psychiatric Association 2013).

Bipolar and related disorder due to another medical condition consists of a prominent and persistent disturbance of mood and increased energy and activity as a direct physiological consequence of another medical condition (American Psychiatric Association 2013).

Other specified bipolar and related disorder and unspecified bipolar and related disorder are disorders with bipolar features that do not meet criteria for any of the specified bipolar disorders. In the first group, the clinician chooses to communicate the specific reason why the presentation does not meet criteria for any specified bipolar and related disorder, while in the second one, he does not specify. Other specified bipolar and related disorders include the following presentations: (1) short-duration hypomanic (2–3 days) and major depressive episodes, (2) hypomanic episodes with insufficient symptoms and major depressive episodes, (3) hypomanic episode without prior major depressive episode, and (4) short-duration cyclothymia (less than 24 months).

## 1.2 Prevalence

According to the World Health Organization (WHO), more than 350 million people of all ages suffer from depression (WHO 2015). Unfortunately, the condition of 10–40 % of them does not improve with the current drug therapies or shows a partial response combined with functional impairment, poor quality of life, suicide ideation and attempts, self-injurious behavior, and high relapse rate (Møldrup and Oestergaard 2011; Al-Harbi 2012).

Mental health conditions are a leading cause of disability-adjusted life years (DALYs) worldwide, neuropsychiatric disorders accounting to 199 million DALYs (Bloom et al. 2011). In particular, unipolar depressive disorders correspond to 65 million DALYs and bipolar affective disorder to 14 million DALYs (WHO 2008).

Among the top 25 causes of global years lived with disability (YLDs) in 2013, major depression and dysthymia are the 2nd and the 16th leading causes, respectively, while in 1990 were ranked in lower positions (3rd and the 20th positions, respectively). Bipolar disorder occupies the 17th position (19th in 1990) (Global Burden of Disease Study 2013 Collaborators 2015). Alarmingly, the report from WHO published in 2008 foresees that unipolar depressive disorders will be the first cause of DALYs by the year 2030 (WHO 2008).

The Global Burden of Disease (GBD) Study 2013 ranked the prevalence and YLDs of several diseases and the results for depressive disorders (comprising major depressive disorder and persistent depressive disorder) and bipolar disorder (BPD) arc shown in Tables 1.1 and 1.2.

**Table 1.1** Prevalence and YLDs 2013 (thousands) by age for depressive disorders and bipolar disorder (Global Burden of Disease Study 2013 Collaborators 2015)

| | Depressive disorders | | Bipolar disorder | |
|---|---|---|---|---|
| Age | Prevalent cases (thousands) | YLDs (thousands) | Prevalent cases (thousands) | YLDs (thousands) |
| 0–6 days | – | – | – | – |
| 7–27 days | – | – | – | – |
| 28–364 days | – | – | – | – |
| 1–4 years | 320.0 | 68.7 | – | – |
| 5–9 years | 5544.8 | 1169.4 | – | – |
| 10–14 years | 13709.7 | 2773.5 | 641.8 | 136.0 |
| 15–19 years | 24403.6 | 4709.3 | 2958.1 | 623.8 |
| 20–24 years | 33395.4 | 6187.4 | 5891.0 | 1232.1 |
| 25–29 years | 35658.5 | 6391.7 | 7037.7 | 1460.9 |
| 30–34 years | 34041.5 | 5972.3 | 6261.2 | 1290.2 |
| 35–39 years | 32901.7 | 5662.0 | 5392.9 | 1102.7 |
| 40–44 years | 32586.6 | 5516.2 | 4716.7 | 957.8 |
| 45–49 years | 30754.8 | 5143.9 | 4044.6 | 814.8 |
| 50–54 years | 27309.0 | 4525.2 | 3333.6 | 665.1 |
| 55–59 years | 23511.0 | 3849.7 | 2667.3 | 525.8 |
| 60–64 years | 20370.4 | 3296.4 | 2141.7 | 416.5 |
| 65–69 years | 13771.9 | 2196.6 | 1352.3 | 258.9 |
| 70–74 years | 11222.5 | 1762.4 | 1032.9 | 194.4 |
| 75–79 years | 7963.2 | 1221.9 | 691.1 | 127.3 |
| +80 years | 8135.7 | 1186.1 | 598.6 | 104.9 |

**Table 1.2** Global prevalence and YLDs 2013 (thousands) of depressive disorders and bipolar disorder (Global Burden of Disease Study 2013 Collaborators 2015)

| Country | Prevalent cases 2013 (thousands) | | YLDs 2013 (thousands) | |
|---|---|---|---|---|
| | Depressive disorders | Bipolar disorder | Depressive disorders | Bipolar disorder |
| Afghanistan | 976.6 | 168.3 | 172.3 | 34.0 |
| Albania | 152.0 | 22.9 | 26.0 | 4.6 |
| Algeria | 2097.2 | 310.1 | 381.1 | 63.2 |
| Andorra | 4.3 | 0.6 | 0.7 | 0.1 |
| Angola | 1041.7 | 111.7 | 193.1 | 22.7 |
| Antigua and Barbuda | 5.0 | 0.7 | 0.9 | 0.1 |
| Argentina | 1992.7 | 331.3 | 336.0 | 67.6 |
| Armenia | 158.6 | 22.5 | 27.4 | 4.6 |
| Australia | 1573.6 | 213.7 | 272.3 | 43.0 |
| Austria | 444.0 | 65.5 | 73.7 | 13.2 |
| Azerbaijan | 488.7 | 71.0 | 85.0 | 14.5 |
| Bahrain | 78.1 | 11.9 | 14.3 | 2.4 |
| Bangladesh | 8126.1 | 1097.9 | 1434.8 | 221.6 |
| Barbados | 16.8 | 2.4 | 2.9 | 0.5 |
| Belarus | 628.9 | 76.7 | 108.8 | 15.5 |
| Belgium | 500.6 | 82.7 | 79.8 | 16.6 |
| Belize | 15.9 | 2.4 | 2.9 | 0.5 |
| Benin | 531.9 | 58.7 | 97.6 | 11.9 |
| Bhutan | 38.2 | 5.4 | 6.8 | 1.1 |
| Bolivia | 503.0 | 74.2 | 90.5 | 15.2 |
| Bosnia and Herzegovina | 198.1 | 29.8 | 33.4 | 6.0 |
| Botswana | 112.5 | 13.6 | 20.5 | 2.8 |
| Brazil | 11154.4 | 1597.9 | 1961.1 | 323.9 |
| Brunei | 18.3 | 2.7 | 3.1 | 0.6 |
| Bulgaria | 388.0 | 54.0 | 64.3 | 10.8 |
| Burkina Faso | 788.9 | 91.3 | 145.9 | 18.7 |
| Burundi | 580.6 | 56.2 | 109.7 | 11.5 |
| Cambodia | 483.7 | 91.7 | 77.9 | 18.6 |
| Cameroon | 1011.1 | 124.7 | 184.2 | 25.3 |
| Canada | 1741.3 | 284.5 | 279.7 | 57.5 |
| Cape Verde | 30.6 | 3.4 | 5.6 | 0.7 |
| Central African Republic | 243.6 | 27.5 | 44.4 | 5.5 |
| Chad | 530.6 | 65.0 | 96.6 | 13.2 |
| Chile | 889.2 | 146.2 | 150.2 | 29.8 |
| China | 63932.0 | 8321.3 | 10691.0 | 1706.7 |
| Colombia | 3404.0 | 469.4 | 628.5 | 95.1 |
| Comoros | 45.7 | 4.3 | 8.6 | 0.9 |
| Congo | 234.4 | 25.5 | 43.1 | 5.2 |

(continued)

**Table 1.2** (continued)

| Country | Prevalent cases 2013 (thousands) | | YLDs 2013 (thousands) | |
|---|---|---|---|---|
| | Depressive disorders | Bipolar disorder | Depressive disorders | Bipolar disorder |
| Costa Rica | 307.7 | 43.9 | 56.2 | 8.9 |
| Cote d'Ivoire | 950.0 | 116.5 | 172.1 | 23.6 |
| Croatia | 225.4 | 33.8 | 37.6 | 6.8 |
| Cuba | 689.1 | 99.1 | 119.4 | 19.9 |
| Cyprus | 47.4 | 7.0 | 8.0 | 1.4 |
| Czech Republic | 556.6 | 85.5 | 92.2 | 17.0 |
| Democratic Republic of the Congo | 3387.8 | 367.2 | 619.6 | 73.7 |
| Denmark | 286.5 | 41.2 | 47.9 | 8.3 |
| Djibouti | 59.4 | 5.8 | 11.1 | 1.2 |
| Dominica | 3.9 | 0.6 | 0.7 | 0.1 |
| Dominican Republic | 529.6 | 78.2 | 95.0 | 16.0 |
| Ecuador | 791.9 | 116.7 | 141.7 | 23.9 |
| Egypt | 2701.1 | 603.5 | 447.9 | 122.3 |
| El Salvador | 323.9 | 50.4 | 58.0 | 10.2 |
| Equatorial Guinea | 41.2 | 4.5 | 7.6 | 0.9 |
| Eritrea | 377.4 | 36.7 | 70.6 | 7.4 |
| Estonia | 122.2 | 10.7 | 22.0 | 2.1 |
| Ethiopia | 5200.0 | 528.2 | 966.0 | 107.8 |
| Federated States of Micronesia | 3.0 | 0.5 | 0.5 | 0.1 |
| Fiji | 30.0 | 5.0 | 4.9 | 1.0 |
| Finland | 344.9 | 43.8 | 58.7 | 8.7 |
| France | 3608.4 | 465.9 | 611.9 | 93.7 |
| Gabon | 92.5 | 10.2 | 16.8 | 2.0 |
| Georgia | 244.8 | 33.5 | 41.8 | 6.8 |
| Germany | 4001.4 | 605.5 | 640.1 | 120.6 |
| Ghana | 1398.5 | 158.4 | 257.9 | 32.4 |
| Greece | 566.7 | 84.4 | 93.7 | 17.0 |
| Grenada | 5.6 | 0.8 | 1.0 | 0.2 |
| Guatemala | 638.5 | 104.7 | 114.1 | 21.2 |
| Guinea | 558.2 | 66.1 | 102.4 | 13.5 |
| Guinea-Bissau | 72.9 | 9.8 | 13.1 | 2.0 |
| Guyana | 36.9 | 5.5 | 6.6 | 1.1 |
| Haiti | 477.9 | 71.4 | 85.4 | 14.5 |
| Honduras | 499.8 | 60.4 | 93.6 | 12.3 |
| Hungary | 534.1 | 80.2 | 89.5 | 16.1 |
| Iceland | 13.4 | 2.2 | 2.2 | 0.5 |
| India | 65346.3 | 8730.7 | 11529.6 | 1774.8 |

(continued)

**Table 1.2** (continued)

| Country | Prevalent cases 2013 (thousands) | | YLDs 2013 (thousands) | |
|---|---|---|---|---|
| | Depressive disorders | Bipolar disorder | Depressive disorders | Bipolar disorder |
| Indonesia | 8569.5 | 1592.4 | 1407.7 | 327.9 |
| Iran | 4526.3 | 654.6 | 823.7 | 133.3 |
| Iraq | 1175.2 | 213.8 | 207.2 | 43.5 |
| Ireland | 262.6 | 33.5 | 45.0 | 6.8 |
| Israel | 352.0 | 49.7 | 59.5 | 10.1 |
| Italy | 3285.0 | 474.1 | 537.3 | 94.7 |
| Jamaica | 148.6 | 21.5 | 26.4 | 4.4 |
| Japan | 6087.4 | 854.7 | 985.3 | 173.6 |
| Jordan | 377.1 | 53.0 | 69.1 | 10.8 |
| Kazakhstan | 824.1 | 118.4 | 142.5 | 24.1 |
| Kenya | 2710.2 | 258.6 | 510.3 | 52.7 |
| Kiribati | 3.2 | 0.6 | 0.5 | 0.1 |
| Kuwait | 191.7 | 28.4 | 35.3 | 5.8 |
| Kyrgyzstan | 257.4 | 37.2 | 45.4 | 7.6 |
| Laos | 194.6 | 38.1 | 31.9 | 7.8 |
| Latvia | 97.4 | 15.9 | 15.6 | 3.2 |
| Lebanon | 242.0 | 40.5 | 42.3 | 8.2 |
| Lesotho | 123.5 | 13.0 | 22.6 | 2.6 |
| Liberia | 181.2 | 24.3 | 32.1 | 4.9 |
| Libya | 334.0 | 48.0 | 60.6 | 9.7 |
| Lithuania | 197.9 | 22.7 | 34.1 | 4.6 |
| Luxembourg | 27.5 | 4.1 | 4.6 | 0.8 |
| Macedonia | 106.6 | 16.5 | 18.0 | 3.3 |
| Madagascar | 1405.5 | 130.5 | 262.9 | 26.4 |
| Malawi | 910.8 | 88.0 | 170.5 | 17.9 |
| Malaysia | 1070.0 | 197.4 | 176.7 | 40.6 |
| Maldives | 11.5 | 2.2 | 1.9 | 0.5 |
| Mali | 639.5 | 78.9 | 117.0 | 16.1 |
| Malta | 22.7 | 3.3 | 3.8 | 0.7 |
| Marshall Islands | 2.1 | 0.3 | 0.3 | 0.1 |
| Mauritania | 205.5 | 23.0 | 37.8 | 4.7 |
| Mauritius | 50.7 | 8.9 | 8.3 | 1.8 |
| Mexico | 6900.0 | 1023.3 | 1259.2 | 209.5 |
| Moldova | 224.7 | 27.8 | 39.2 | 5.6 |
| Mongolia | 135.3 | 20.2 | 23.8 | 4.2 |
| Montenegro | 30.9 | 4.7 | 5.2 | 0.9 |
| Morocco | 1721.3 | 256.3 | 309.6 | 52.0 |
| Mozambique | 1433.5 | 138.9 | 265.2 | 27.9 |
| Myanmar | 1917.0 | 358.3 | 312.2 | 73.4 |

(continued)

**Table 1.2** (continued)

| Country | Prevalent cases 2013 (thousands) | | YLDs 2013 (thousands) | |
|---|---|---|---|---|
| | Depressive disorders | Bipolar disorder | Depressive disorders | Bipolar disorder |
| Namibia | 126.3 | 14.7 | 23.0 | 3.0 |
| Nepal | 977.7 | 178.4 | 163.3 | 36.4 |
| Netherlands | 769.6 | 134.0 | 123.2 | 26.8 |
| New Zealand | 240.7 | 47.0 | 40.1 | 9.5 |
| Nicaragua | 332.6 | 48.1 | 60.7 | 9.7 |
| Niger | 781.1 | 86.8 | 144.7 | 17.8 |
| Nigeria | 7706.8 | 957.2 | 1378.7 | 191.8 |
| North Korea | 1014.2 | 139.7 | 165.7 | 28.5 |
| Norway | 244.0 | 37.8 | 40.3 | 7.6 |
| Oman | 188.7 | 30.8 | 34.7 | 6.3 |
| Pakistan | 9356.6 | 1175.9 | 1679.5 | 239.8 |
| Palestine | 313.3 | 27.3 | 60.3 | 5.6 |
| Panama | 220.7 | 32.4 | 40.1 | 6.6 |
| Papua Nova Guinea | 209.6 | 36.2 | 34.8 | 7.4 |
| Paraguay | 323.8 | 48.2 | 57.9 | 9.8 |
| Peru | 1580.3 | 232.8 | 281.5 | 47.2 |
| Philippines | 3046.9 | 575.2 | 501.0 | 117.8 |
| Poland | 2031.6 | 312.3 | 340.6 | 62.8 |
| Portugal | 705.8 | 82.2 | 120.5 | 16.4 |
| Qatar | 130.3 | 21.1 | 24.0 | 4.3 |
| Romania | 1139.4 | 189.3 | 190.9 | 38.2 |
| Russia | 9400.4 | 1169.3 | 1619.0 | 235.8 |
| Rwanda | 736.6 | 67.7 | 137.4 | 13.6 |
| Saint Lucia | 10.2 | 1.5 | 1.8 | 0.3 |
| Saint Vincent and the Grenadines | 5.9 | 0.9 | 1.0 | 0.2 |
| Samoa | 5.6 | 0.9 | 0.9 | 0.2 |
| Sao Tome and Principe | 10.3 | 1.1 | 1.9 | 0.2 |
| Saudi Arabia | 1553.6 | 224.4 | 282.7 | 45.3 |
| Senegal | 728.5 | 80.0 | 134.4 | 16.3 |
| Serbia | 489.9 | 74.8 | 82.8 | 15.1 |
| Seychelles | 3.6 | 0.6 | 0.6 | 0.1 |
| Sierra Leone | 289.4 | 35.4 | 52.8 | 7.2 |
| Singapore | 276.3 | 27.5 | 49.2 | 5.6 |
| Slovakia | 287.0 | 44.6 | 48.3 | 9.0 |
| Slovenia | 109.5 | 16.7 | 18.2 | 3.3 |
| Solomon Islands | 15.6 | 2.7 | 2.6 | 0.6 |
| Somalia | 576.6 | 54.1 | 108.6 | 11.0 |
| South Africa | 2728.7 | 376.5 | 475.5 | 75.3 |

(continued)

**Table 1.2** (continued)

| Country | Prevalent cases 2013 (thousands) | | YLDs 2013 (thousands) | |
|---|---|---|---|---|
| | Depressive disorders | Bipolar disorder | Depressive disorders | Bipolar disorder |
| South Korea | 1926.2 | 344.4 | 296.9 | 70.0 |
| South Sudan | 627.8 | 64.4 | 115.0 | 12.9 |
| Spain | 2641.5 | 366.5 | 441.1 | 73.6 |
| Sri Lanka | 810.6 | 142.9 | 131.8 | 29.2 |
| Sudan | 1712.9 | 233.6 | 311.9 | 47.3 |
| Suriname | 28.3 | 4.2 | 5.0 | 0.9 |
| Swaziland | 60.6 | 7.7 | 10.9 | 1.5 |
| Sweden | 551.4 | 71.8 | 93.2 | 14.4 |
| Switzerland | 405.4 | 62.1 | 65.8 | 12.4 |
| Syria | 1106.0 | 152.6 | 202.7 | 31.0 |
| Taiwan | 722.6 | 146.7 | 102.1 | 29.9 |
| Tajikistan | 347.7 | 50.3 | 61.6 | 10.3 |
| Tanzania | 2858.5 | 269.7 | 536.7 | 54.8 |
| Thailand | 2816.9 | 489.3 | 455.8 | 100.3 |
| The Bahamas | 21.5 | 3.2 | 3.8 | 0.6 |
| The Gambia | 114.9 | 10.0 | 21.9 | 2.0 |
| Timor-Leste | 24.7 | 5.1 | 4.0 | 1.0 |
| Togo | 343.4 | 39.6 | 63.3 | 8.1 |
| Tonga | 3.2 | 0.5 | 0.5 | 0.1 |
| Trinidad and Tobago | 76.8 | 11.5 | 13.5 | 2.3 |
| Tunisia | 648.6 | 91.5 | 117.7 | 18.6 |
| Turkey | 3323.1 | 600.6 | 573.7 | 121.7 |
| Turkmenistan | 247.7 | 36.3 | 43.6 | 7.5 |
| Uganda | 2272.8 | 189.6 | 432.1 | 38.5 |
| Ukraine | 3065.7 | 375.7 | 523.3 | 75.7 |
| United Arab Emirates | 528.9 | 91.5 | 97.9 | 18.9 |
| United Kingdom | 3019.6 | 475.1 | 492.8 | 95.5 |
| United States | 19695.5 | 2471.8 | 3340.3 | 496.4 |
| Uruguay | 168.5 | 27.4 | 28.2 | 5.6 |
| Uzbekistan | 1365.7 | 198.5 | 240.9 | 40.8 |
| Vanuatu | 7.4 | 1.3 | 1.2 | 0.3 |
| Venezuela | 1684.7 | 252.1 | 305.9 | 51.4 |
| Vietnam | 3795.6 | 635.7 | 629.2 | 129.7 |
| Yemen | 857.2 | 151.5 | 152.1 | 30.9 |
| Zambia | 790.5 | 76.8 | 147.5 | 15.5 |
| Zimbabwe | 681.5 | 84.8 | 123.7 | 17.1 |

## 1.3 Economic Burden

Based on the data published by Hu (2006), plus new reports from the United States, China, Kenya, and Australia, the World Economic Forum, and the Harvard School of Public Health estimated that the global cost of mental health conditions in 2010 was US$2.5 trillion (US$1.6 and 0.9 trillion from high-income and low-income countries, respectively), with the costs projected to rise to US$6.0 trillion by 2030. Two-thirds of the total costs correspond to indirect costs (i.e., lost income due to mortality, disability, care seeking, or other and nonpersonal costs), and the remaining are from direct costs (i.e., personal medical care costs and personal nonmedical care costs) (Bloom et al. 2011).

Concerning only depressive disorders, the economic burden of depression, including MDD, BPD, and persistent depressive disorder, was estimated at US$83.1 billion in 2000 in the United States. This total comprised US$26.1 billion from direct medical costs, US$5.4 billion from suicide-related mortality costs, and US$51.5 billion from indirect workplace costs. The economic burden of MDD, in particular, was US$173.2 billion in 2005 and US$210.5 billion in 2010, direct costs accounting to US$77.5 billion in 2005 and US$98.9 billion in 2010 (Greenberg et al. 2015).

The scenario in Europe is also alarming. In a study comprising data from Austria, Belgium, Bulgaria, Cyprus, Czech Republic, Denmark, Estonia, Finland, France, Germany, Greece, Hungary, Iceland, Ireland, Italy, Latvia, Lithuania, Luxembourg, Malta, the Netherlands, Norway, Poland, Portugal, Romania, Slovakia, Slovenia, Spain, Sweden, Switzerland, and United Kingdom, the cost of brain disorders in 2010 was €798 billion, direct health care costing 37 %, direct nonmedical cost corresponding to 23 %, and indirect cost to 40 %. MDD and BPD caused direct costs of 37.9 and 3.5 billion euros and indirect costs of 54.0 and 18.0 billion euros, respectively (Olesen et al. 2012).

The total estimated cost of depression in China was US$6.2 billion in 2002. Direct and indirect costs were about 16 and 84 % of the total cost of depression, respectively (Hu et al. 2007).

In Australia, depression costs approximately US$12.6 billion per year (Manicavasagar 2012). Indirect cost per patient was around US$134.3 thousand for bipolar disorder I, US$76.8 thousand for bipolar disorder II, and US$68.3 thousand for unipolar depression. Direct costs for the same disorders were around US$26.3 thousand, US$17.6 thousand, and US$27.2 thousand per patient, respectively (Parker et al. 2013).

In 2001, South Africa spent US$88 per patient with BPD with medical management, while with medication the costs were almost ten times higher (US$751 per patient) (McLeod et al. 2002). On the other hand, in a study performed between 2002 and 2004, severe depression in South Africa was associated with indirect costs of US$4798 per patient/year (Lund et al. 2013).

In Colombia, the costs per patient (in 2007 US$) for antidepressant drug treatment were estimated at $1618 (for venlafaxine), $1207 (for fluoxetine), and $1068 (for amitriptyline) (Machado et al. 2008).

Unfortunately, studies dealing with the cost of illness are lacking for most of the countries from Africa, Asia, Eastern Europe, and Latin America.

## 1.4 Theories of Depression

### *1.4.1 Monoamine Theories*

Monoamine theories state that depression is associated to a decrease in monoamine levels in the synaptic cleft, namely, of the catecholamine noradrenaline and of the indoleamine serotonin. Several classes of antidepressants work by increasing the levels of monoamines in the synaptic cleft, namely, tricyclic antidepressants (TCAs), monoamine oxidase inhibitors (MAOIs), selective serotonin reuptake inhibitors (SSRI), noradrenaline reuptake inhibitors (NRIs), and serotonin and noradrenaline reuptake inhibitors (SNRIs) (Fig. 1.1).

The 1950s is frequently called "the golden decade of psychopharmacology." In this decade, several events occurred: the discovery of imipramine (a TCA), the use of iproniazid (an MAOI), the discovery of the antimanic action of lithium, and the clinical introduction of three drugs, chlorpromazine (a dopamine antagonist), meprobamate (an anxiolytic drug), and chlordiazepoxide (a benzodiazepine) (López-Muñoz and Alamo 2009).

Iproniazid, a hydrazine derivative first tested as an antituberculosis agent, started showing to induce greater vitality to tuberculosis patients; additionally, animals given iproniazid, and subsequently reserpine, instead of becoming sedated, became excited and overactive (Chessin et al. 1957). Later, the psychiatrists Kline, Loomer, and Saunders were the first to assess the efficacy of this MAOI in non-tuberculosis depressed patients and observed that 70 % of them had undergone a substantial improvement (Loomer et al. 1958). During the 1960s, iproniazid was replaced by more efficient MAOIs, such as tranylcypromine and phenelzine (López-Muñoz and Alamo 2009). All these MAOIs irreversibly inhibited both isoenzymes (MAO-A and MAO-B), leading to hypertensive crisis. This fact, encouraged, much later, the development of new inhibitors with two key characteristics: to be selective inhibitors only of MAO-A and to cause reversible and competitive inhibition. The first reversible and selective inhibitor of MAO-A (RIMA) reaching the market was moclobemide in 2000 (López-Muñoz and Alamo 2009).

During the 1960s and 1970s, there was almost a division between continents (Mulinari 2012) concerning the role of noradrenaline and of serotonin in the pathogenesis of depression. The American psychiatrist Joseph Schildkraut defended the role of noradrenaline in his paper entitled "The catecholamine hypothesis of affective disorder: A review of supporting evidence," published in 1965 (Schildkraut 1965). He took into account that (1) MOAIs and TCAs were both inferred to exert their

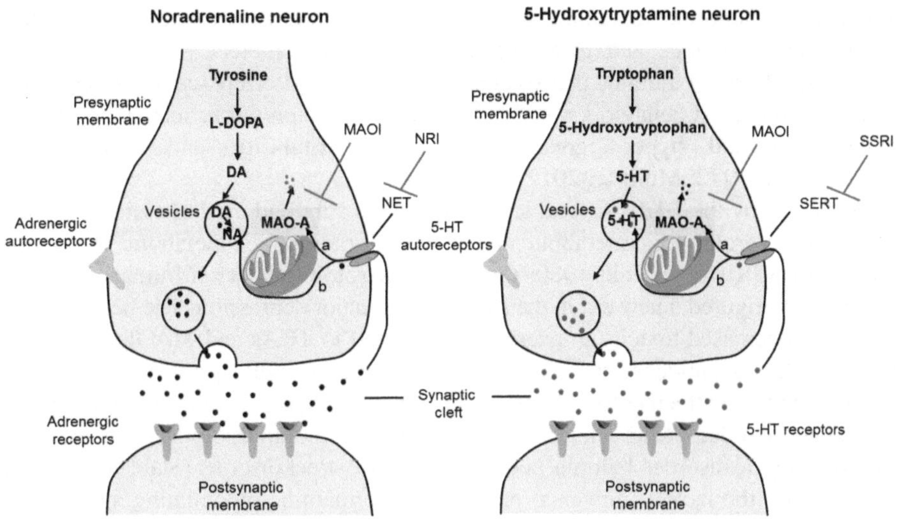


**Fig. 1.1** Schematic representation of noradrenaline (NA, *black dots*) and serotonin (5-hydroxytryptamine, 5-HT, *brown dots*) synaptic terminals. Monoamines are transported by vesicles along the axon and released to the synaptic cleft, where they interact with the postsynaptic receptors. Exocytosis may be $Ca^{2+}$ dependent or independent. The reuptake of noradrenaline and serotonin to the presynaptic neuron occurs through the noradrenaline reuptake transporter (NET) and serotonin reuptake transporter (SERT), respectively. Monoamines are catabolized by the enzyme MAO-A. Catecholamines, such as noradrenaline, are additionally catabolized by other enzyme, catechol-O-methyltransferase (COMT), but its location in controversial. After reuptake, monoamines can be deaminated by MAO-A (a) or incorporated into vesicles by vesicular monoamine transporter (b). *DA* dopamine, *MAOI* monoamine oxidase inhibitor, *NRI* noradrenaline reuptake inhibitor, *SSRI* selective serotonin reuptake inhibitor (Adapted from Torres et al. 2003)

antidepressant effect by elevating brain catecholamine levels; (2) when reserpine was given to laboratory animals, it induced sedation (or a "depression" behavior), which was associated with depletion of brain catecholamines; and (3) reserpine-induced sedation in animals could be prevented by administration of the catecholamine precursor L-DOPA or by administration of MAOIs or TCAs. On the other hand, the British psychiatrist Alec Coppen in his work "The biochemistry of affective disorders" published in 1967 (Coppen 1967) in the *British Journal of Psychiatry* reviewed evidence linking serotonin deficit to depression, since serotonin precursor tryptophan potentiated the antidepressant effect of an MAOI in patients with severe depressive illness, although the same effect was not observed with L-DOPA. The serotonin hypothesis gained additional support with the works by the Soviet researchers Lapin and Oxenkrug (1969) and by the Swedish scientist Carlsson (Carlsson et al. 1958), who showed that the reserpine-induced depression was mediated by depletion of the catecholamine dopamine, but not of noradrenaline. Carlsson showed that L-DOPA is converted into dopamine, but the subsequent step, conversion of dopamine into noradrenaline, was never completed because it is blocked by reserpine.

Nonetheless, Baldessarini (1972) showed that animal studies favored noradrenaline hypothesis, since interference with catecholamine metabolism in laboratory animals was reported to induce a decrease in locomotion, exploratory behavior, conditioned avoidance, and reward-seeking behavior parameters, which are features similar to those of human depression. By contrast, interference with serotonin metabolism caused behaviors not resembling human depression, such as insomnia, hypersexual and hyper-aggressive behaviors, irritability, and hyperactivity (Baldessarini 1972; Mulinari 2012).

Although by the 1970s several teams suggested subdividing depression into two classes, one serotonergic metabolic and another noradrenergic metabolic, the development of SSRIs in the mid-1980s favored the serotonin theory (Mulinari 2012).

SSRIs prefigured a new era in the treatment of major depression. The better tolerability and decreased toxicity in overdose, compared to TCAs and MAOIs, has made them the class of antidepressants of choice for the treatment of depression. SSRIs have also proven to be effective for other CNS disorders, such as generalized anxiety disorder, panic disorder, social anxiety disorder, obsessive-compulsive disorder, premenstrual dysphoric disorder, bulimia, and post-traumatic stress disorder (Stahl et al. 2013).

However, the lack of universal efficacy of monoamine potentiating antidepressants, as well as their 2–3 weeks of therapeutic latency, led to the search for other mechanisms that could better explain the pathogenesis of depression or that could complement the monoamine theory (Groves 2007).

### *1.4.2 Hypothalamic-Pituitary-Adrenal Axis and Depression*

Adaptive responses to stress involve short-term activation of the hypothalamic-pituitary-adrenal (HPA) axis. However, efficient regulation is required to return stress hormone to basal levels, as excessive stimulation or reactivity of the HPA axis can trigger a long-lasting state of distress, impairment of the limbic system, and increased risk for psychiatric disorders (Goel et al. 2014).

In the past 30 years, numerous researchers have noted HPA axis hyperactivity in a subset of patients with major depression (Varghese and Brown 2001). Besides this disorder, a panoply of other conditions may be associated with increased and prolonged activation of HPA axis, namely, anorexia nervosa, obsessive-compulsive disorder, panic anxiety, chronic active alcoholism, alcohol and narcotic withdrawal, excessive exercise, poorly controlled diabetes mellitus, childhood sexual abuse, and hyperthyroidism (Juruena et al. 2004).

The HPA axis consists of a feedback loop. It comprises the hypothalamus, pituitary gland, and adrenal glands (Varghese and Brown 2001), being activated during an emotional stress. The neurosecretory cells of the paraventricular nucleus (PVN) of the hypothalamus receive neuronal inputs from many brain regions, including the amygdala, hippocampus, and nuclei of the midbrain (McQuade and Young 2000). The hypothalamus secretes the corticotropin-releasing hormone (CRH, also known as corticotropin-releasing factor (CRF)) and arginine vasopressin (AVP). These two hormones act on the pituitary gland to increase the secretion of adrenocorticotropic hormone (ACTH). ACTH is then transported in the blood stream to the adrenal cortex and interacts with receptors on adrenocortical cells that stimulate the production and release of cortisol, which is the adrenal glucocorticoid stress hormone. Glucocorticoids, then, interact with mineralocorticoid receptor (MR) and the glucocorticoid receptor (GR) in HPA tissues, where they are responsible for the feedback inhibition of the secretion of CRH and ACTH (Juruena et al. 2003, 2004; Varghese and Brown 2001). These receptors differ in their affinity for glucocorticoids, with MR demonstrating the highest affinity and low capacity for cortisol and GR showing lower affinity, but high capacity for cortisol. Dexamethasone (DEX) only binds to GR (Juruena et al. 2003, 2004). Moreover, receptor distribution in the rodent brain is different, MR being predominantly found in limbic areas, particularly the hippocampus, and GR more extensively distributed across all brain regions; in primates, MR is also found in the cortex and subcortical structures (Young et al. 2003) (Fig. 1.2).

Two opposing mechanisms seems to contribute to HPA axis hyperactivity: for one side the impaired feedback inhibition is related to a decreased function of the GRs causing high levels of glucocorticoids, and for the other side, the harmful

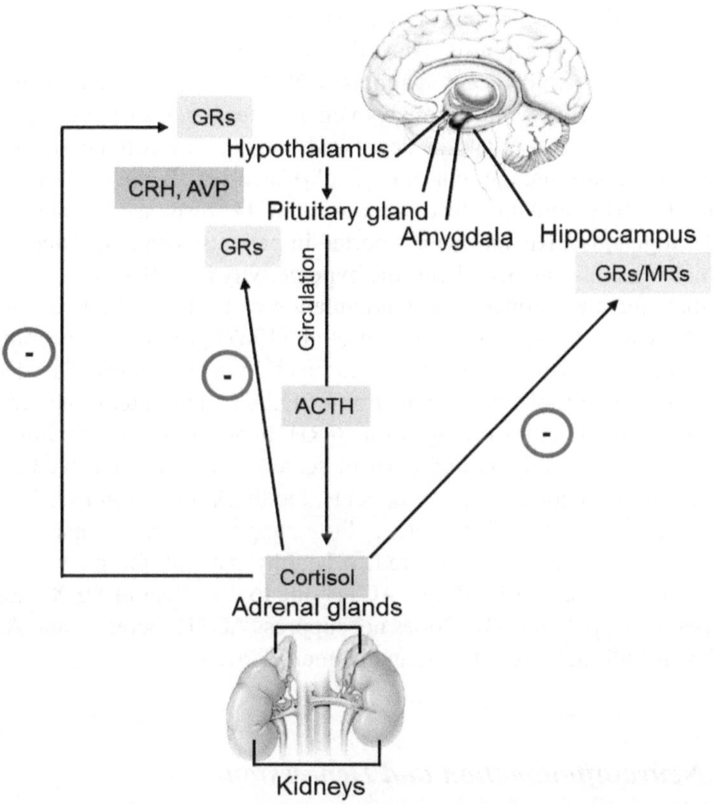


**Fig. 1.2** Schematic diagram of the hypothalamus-pituitary-adrenal (HPA) axis, describing negative feedback of cortisol. *CRH* corticotropin-releasing hormone, *AVP* arginine vasopressin, *ACTH* adrenocorticotropic hormone, *GRs* glucocorticoid receptors, and *MRs* mineralocorticoid receptors (Adapted from Juruena et al. 2004)

effects of excessive glucocorticoid signaling depend on functional GRs (Anacker et al. 2011). Antidepressants like clomipramine (TCA), fluoxetine (SSRI), desipramine (TCA), paroxetine (SSRI), citalopram (SSRI), and amitriptyline (TCA) increase GR expression, enhance GR function, and promote GR nuclear translocation in depressed patients, animals, and cellular models (Juruena et al. 2004; Barden 2004). Rats treated with amitriptyline, desipramine, or citalopram have showed elevated MR and GR mRNA expression in the hippocampus (Seckl and Fink 1992). Moreover, successful treatment of depression using electroconvulsive therapy (ECT) (Nemeroff et al. 1991) or fluoxetine (de Bellis et al. 1993) has been shown to result in the reduction of the high pretreatment concentrations of cerebrospinal fluid CRF.

Abnormalities of the HPA axis include increased concentrations of cortisol in plasma and urine, early escape of plasma cortisol from suppression following administration of DEX, and increased cerebrospinal fluid concentrations of CRH (Krishnan et al. 1993). Two tests have been employed to assess HPA axis dysregulation. Dexamethasone suppression test (DST) consists of administering synthetic DEX and monitoring the plasma cortisol concentrations to identify those who fail to maintain a suppressive response. Thus, this test aims to measure the ability of an exogenous glucocorticoid receptor agonist (DEX) to suppress endogenous stress hormone release. DEX, which provides negative feedback to HPA axis, does not suppress plasma cortisol levels in depressed subjects, as compared with normal ones, due to the abnormal amount of cortisol (Carroll et al. 1968, 1981). The other test is called CRH stimulation test. A reduced ACTH response following administration of exogenous CRH has been reported in patients with major depressive disorder, which is a consequence of chronic hyperactivity of CRH-secreting neurons in the hypothalamus that induces a downregulation of CRH receptors at the anterior pituitary (Newport et al. 2003). Holsboer et al. (1987) proposed a combined DEX/CRH test for the assessment of HPA system function in psychiatric disorders, which is the most sensitive tool for the detection of HPA system alterations. In this test, DEX is administered in the evening before CRH. In normal controls without depression, DEX suppresses ACTH and cortisol secretion via negative feedback to the pituitary gland. In major depression patients, feedback inhibition is reduced by the presence of elevated cortisol; therefore, little suppression takes place and ACTH levels remain high. CRH administered to healthy patients the next morning will show little or no increase of CRH or ACTH due to the effect of DEX suppression, while in depressed patients, DEX does not suppress ACTH secretion and ACTH and cortisol levels will increase (Ozbolt and Nemeroff 2013).

### *1.4.3 Neuroinflammation and Depression*

Neuroinflammation is the brain's first response to stress, injury, infection, or disease. It is primarily mediated by one of two cell systems: microglia of the central nervous system (CNS) and lymphocytes, monocytes, and macrophages of the hematopoietic system (Hurley and Tizabi 2013).

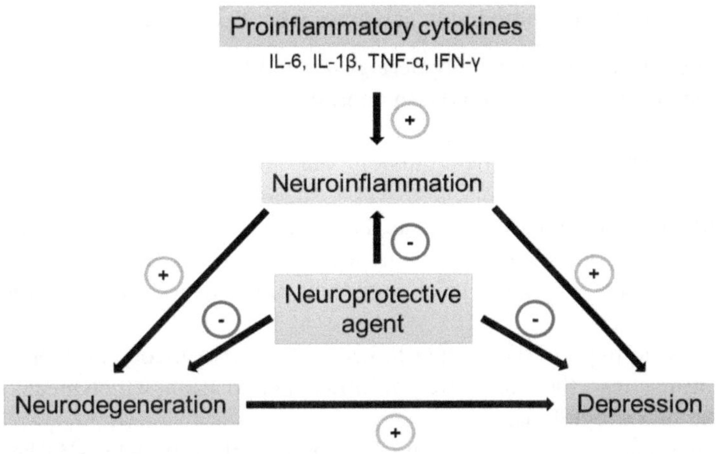


**Fig. 1.3** Relationship between neuroinflammation, neurodegeneration, and depression. (+) and (−) indicate a positive and negative influence, respectively (Adapted from Hurley and Tizabi 2013)

The possibility of stress, through activation of neuroendocrine system, neurotransmitter levels changes and pro-inflammatory cytokine release, contribute to the pathogenesis of depression led to the development of the cytokine hypothesis of depression. It is known that chronic stress may exacerbate the release of pro-inflammatory cytokines (tumor necrosis factor (TNF)-α, interleukin (IL)-1β, and IL-6) and then precipitate depressive episodes (Hurley and Tizabi 2013). Maes et al. (1998a) were the first researchers who showed that psychological stress in humans induced an inflammatory response via production of pro-inflammatory cytokines. Therefore, drugs that interfere with harmful consequences of stress on inflammatory pathways may offer novel treatments for mood disorders (Fig. 1.3; Hurley and Tizabi 2013).

Glial cells have a dual effect. While glial activation initially has a protective function (release of anti-inflammatory cytokines), its continuous activation can lead to exaggerated release of pro-inflammatory cytokines and hence to neuronal damage (Hurley and Tizabi 2013).

**The cytokine theory was based in several facts:**

1. Elevated circulating cytokines are frequently observed in depressed patients.
2. Administration of cytokines to humans can induce symptoms of depression.
3. Depression is frequent in peripheral inflammatory diseases.
4. Administration of cytokines to animals induces behavior that looks like the symptoms of depression, e.g., anhedonia; sleep, appetite, and activity level disturbance; cognitive deficits; and flu-like complains.
5. Elevated circulating concentrations of cortisol observed in many depressed patients reflect the well-established activation of the HPA axis by pro-inflammatory cytokines.
6. Chronic antidepressant treatments in animals can prevent behaviors induced by lipopolysaccharide (LPS) (Beaurepaire 2002; Wagner-Smith and Markou 2011).

**However, there are some inconsistencies with this theory:**

1. Cytokines have somnogenic, hyperalgesic, and thermogenic effects, which are the opposite symptoms observed in depression.
2. Cytokines (e.g., IL-1 and IL-2) induce cognitive impairments, but severe impairments in explicit or implicit memory are not common in typical forms of depression.
3. Pro-inflammatory cytokines activate the HPA axis and can inhibit GR function at multiple levels, but 30–50 % of patients do not show HPA axis hyperactivity.
4. In many cases, cytokine levels are not reduced following antidepressant administration.
5. The cytokine hypothesis is often linked to the stress theory of depression, but concerning stress, there are different forms of depression, which means different implications of cytokines.
6. Cytokines, such as IL-1, IFN-γ, and TNF-α, not only stimulate 5-HT neurotransmission but also reduce the production of 5-HT by stimulating the enzyme indoleamine 2,3-dioxygenase (IDO), which converts tryptophan, the precursor of 5-HT, into kynurenine.
7. Long-term administration of TCAs and SSRIs attenuates the sickness behavior induced by LPS or cytokines, but the administration of the TCA clomipramine led to increased production of IL-1 and IL-3 (Beaurepaire 2002; Wichers and Maes 2002; Schiepers et al. 2005).

Moreover, anti-inflammatory drugs like nonsteroidal anti-inflammatory drugs (NSAIDs), steroidal anti-inflammatory prednisone, COX-2 inhibitor celecoxib, and minocycline have a positive effect on mood (Levine et al. 1996; Bouwer et al. 2000; Pae et al. 2008; Akhondzadeh et al. 2009; Iyengar et al. 2013). On the other hand, in a pharmacovigilance study using data from a large health care system, authors confirmed a significant association between NSAID exposure and poor antidepressant treatment outcome in major depressive disorder. They concluded that salicylates and COX-2 inhibitors might be preferable to NSAIDs when indicated for antidepressant-treated patients (Gallagher et al. 2012).

### *1.4.4 Neurogenesis Theory*

The finding that new neurons are not created in the adult CNS began to be seriously confronted in the 1960s by the work of Altman and coworkers (1965, 1966; 1969) and later by the work of Kaplan and Hinds (1977) with rats. Nevertheless, only after the 1990s, adult neurogenesis became accepted (Drew and Hen 2007).

In adult mammals (rodents, primates), neurogenesis occurs in two germinal zones: the subgranular zone (SGZ) of the dentate gyrus (DG) and the subventricular zone (SVZ). In humans, SGZ neurogenesis has been conclusively demonstrated, but SVZ neurogenesis remains controversial (Eriksson et al. 1998; Sanai et al. 2011; Calabrese et al. 2014). Reduced hippocampus volume was observed in depressed patients and is thought to result from a reduction in SGZ neurogenesis (Bremner et al. 2000).

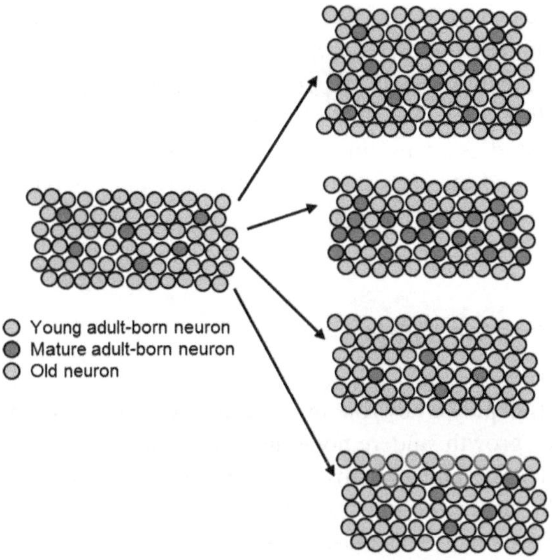


**Fig. 1.4** Hypotheses explaining how antidepressants affect neurogenesis. The right side of the image, from top to down: hypothesis 1 - antidepressants increase the total number of granule cells; hypothesis 2 - antidepressants increase the turnover of granule cells; hypothesis 3 - antidepressants increase the number of young neurons; hypothesis 4 - antidepressants change the behavior of young neurons (Adapted from Drew and Hen 2007)

The neurogenesis theory of depression was formally proposed in the last decade and is based on two postulates: (1) decreased adult neurogenesis results in depressive or anxious phenotypes, and (2) for effective treatment of these disorders, intact hippocampal neurogenesis is required (Santarelli et al. 2003; Petrik et al. 2012). This theory was very attractive because it helped to explain the delay of several weeks between the initiation of antidepressant treatment and the beginning of their effects, since the maturation of the hippocampal progenitors into mature DG granule neurons and their incorporation into circuitry also takes about 3–4 weeks (Kee et al. 2007; Lavergne and Jay 2010).

Several observations led to the establishment of this theory: stress potently suppressed adult neurogenesis and increased cell death due to increased glucocorticoid release (Gould et al. 1992; Gould et al. 1991), increasing brain levels of serotonin with chronic antidepressant treatment enhance the basal rate of DG neurogenesis (Fig. 1.4; Jacobs et al. 2000), and depletion of serotonin inhibits adult neurogenesis (Brezun and Daszuta 1999); some animal (rodents and nonhuman primates) stress models, including repeated restraint (Pharm et al. 2003), inescapable foot shock (Malberg and Duman 2003), defeat paradigm (Czéh et al. 2002), and social isolation (Stranahan et al. 2006; Cinini et al. 2014), show neurogenesis impairments, which agrees with the first postulate.

Conversely, several studies have demonstrated that decreasing neurogenesis by methylazoxymethanol acetate or γ-ray irradiation does not change depression behavior in non-stress and stressed animals (unpredictable chronic mild stress (UCMS), corticosterone treatment (Cort), social stress, or restraint stress). However, this discrepancy of results may be dependent on the method, animal strain, or extent of neurogenesis disruption (Petrik et al. 2012).

There are also studies supporting, as well as refuting, the second postulate, both with non-stressed and stressed animals (UCMS, Cort, social stress, or restraint stress) (Petrik et al. 2012).

### *1.4.5 Neurotrophic Theory*

Neurotrophic factors play a key role in neuroplasticity, because they modulate axonal and dendritic growth and remodeling, membrane receptor trafficking, neurotransmitter release, and synapse formation and function. Among these neurotrophins, the neurotrophin brain-derived neurotrophic factor (BDNF) is an essential mediator of neuronal plasticity, also mediating neurogenesis (Park and Poo 2013; Calabrese et al. 2013). BDNF is able to regulate neuronal survival via the phosphatidylinositol 3-kinase (PI3-kinase)/Akt pathway. Akt phosphorylates and controls the biological functions of several proteins important for modulating cell survival. Among the substrates of Akt is BAD, a Bcl-2 family member that promotes apoptosis by binding to the anti-apoptotic protein Bcl-xL, which in the absence of binding, inhibits the proapoptotic activity of Bax. Phosphorylation of BAD by Akt/protein kinase B results in its association with 14-3-3 proteins instead of BcL-xL, consequently preventing apoptosis. Chronic stress may disrupt the anti-apoptotic regulation of Bcl-2, thus reducing neurogenic cell survival (Groves 2007; Huang and Reichardt 2001).

The neurotrophin hypothesis postulates that a loss of BDNF plays a key role in the pathophysiology of depression and its restoration may be a critical mechanism behind antidepressant efficacy (Groves 2007).

It has been reported that the administration of pro-inflammatory cytokines or of the cytokine-inducer LPS causes a significant reduction of mRNA levels of BDNF and BDNF level (Lapchak et al. 1993; Guan and Fang 2006; Schnydrig et al. 2007), as well as of other neurotrophins (Guan and Fang 2006). On the other hand, antidepressant treatments increase neurotrophin levels within the hippocampus (Fig. 1.5).

For instance, chronic electroconvulsive seizure (ECS) treatment, a model of ECT, induces a high level of expression of several neurotrophins: BDNF and its receptor, tyrosine kinase receptor B (TrkB) (Nibuya et al. 1995; Dias et al. 2003), vascular-endothelial growth factor (VEGF) (Segi-Nishida et al. 2008) and fibroblast growth factor-2 (FGF2) (Gwinn et al. 2002). MDD patients subjected to ECT also showed increased serum levels of BDNF and insulin growth factor-1 (IGF-1) (Stelzhammer et al. 2013). Similar results have been obtained with animal models involving numerous antidepressant drugs, including TCAs, MAOs, and SSRIs (Dias et al. 2003; Saarelainen et al. 2003; Warner-Schmidt and Duman 2007; Bachis et al. 2008; Grunbaum-Novak et al. 2008; Watanabe et al. 2010).

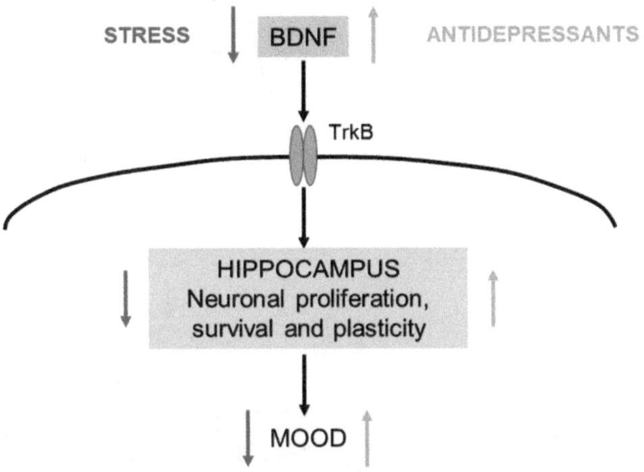


**Fig. 1.5** Schematic diagram showing the opposing roles of stress and antidepressant therapy on hippocampal BDNF expression. *BDNF* brain-derived neurotrophic factor, *TrkB* tyrosine kinase receptor B (Adapted from Groves 2007)

However, some animal studies do not support this hypothesis. Acute fluoxetine treatment increased the expression of BDNF in the cortex and the inferior olive, but decreased it in other areas, including the hippocampus (CA fields and dentate gyrus) (Miró et al. 2002). A decreased BDNF mRNA level in cortex and hippocampus was also observed after single treatment with sertraline (Nibuya et al. 1995). However, Dias et al. (2003) reported an increase in BDNF following chronic electroconvulsive shock, tranylcypromine, and desipramine treatment, suggesting that those negative results may be specific to serotonin modulators or to insufficient days of treatment.

Eisch et al. (2003) showed that 7 days of BDNF infusions into the ventral tegmental area (VTA) increased immobility in the forced swim test, indicative of a depressive behavior and that viral-mediated suppression of the receptor TrkB delivered significant antidepressive-like effects. Similar results were obtained by Berton et al. (2006), indicating that the effect of BDNF in VTA is the opposite of that reported for the hippocampus.

Finally, genetically manipulated mice (e.g., BDNF(+/−) mice) do not exhibit depressive behaviors in some studies (McQueen et al. 2001; Chourbaji et al. 2004), while in other works, a decreased response to antidepressants or depressive behaviors in FST and TST was observed (Chan et al. 2006; Monteggia et al. 2007; Govindarajan et al. 2006).

### *1.4.6 Glutamate Hypothesis*

Glutamate (Glu) is an abundant neurotransmitter that is essential for learning and memory, though at high concentrations can produce harmful effects, such as neuronal cell death (McCarthy et al. 2012). It acts via two classes of receptors, namely,

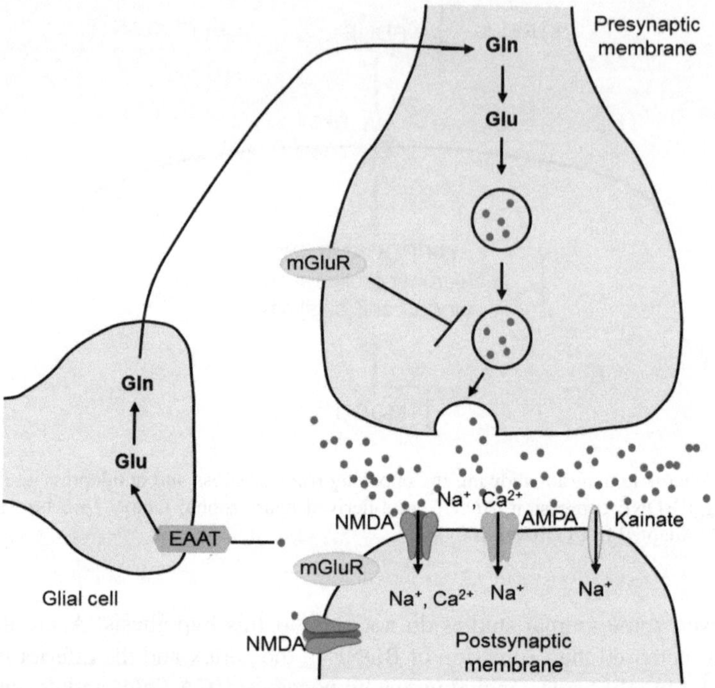


**Fig. 1.6** Glutamatergic neurotransmission. Glutamine (Gln) is converted to glutamate (Glu), though it can also be derived from the TCA cycle (not shown). Glu (*brown dots*) is packaged into presynaptic vesicles and released to the synaptic cleft. Exocytosis can be $Ca^{2+}$-dependent or independent. Glu is removed from the extracellular space by excitatory amino acid transporters (EAATs) predominantly present on glia. In glial cells Glu is converted to Gln. Several Glu receptors are found on pre- and postsynaptic neurons, as well as on glial cells. These include both ionotropic receptors (AMPA, NMDA, and kainate) and metabotropic receptors (mGluRs) (Adapted from Sanacora et al. 2008)

ligand-gated ion channels (ionotropic receptors) and G protein-coupled (metabotropic; mGluR) receptors. Ionotropic receptors are divided into α-amino-3-hydroxy-5-methyl-4-isoxazole-propionate (AMPA), *N*-methyl-D-aspartate (NMDA), and kainate receptors (Traynelis et al. 2010). The NMDA receptor complex is a heterooligomer with multiple, allosterically coupled recognition sites for glutamate, glycine, and use-dependent channel blockers (Paul et al. 1994).

In the brain, glutamate can be synthesized de novo from glucose via the Krebs cycle and the transamination of α-oxoglutarate or be recycled through the glutamate/glutamine cycle (Fig. 1.6). Glutamate is transported into synaptic vesicles by vesicular Glu transporters (VGLUTs), being released in a $Ca^{2+}$-dependent manner into the synaptic cleft. After releasing, glutamate binds to and activates specialized ionotropic and metabotropic receptors. The neurotransmitter is removed from the extracellular space via high-affinity excitatory amino acid transporters (EAATs) in neighboring glial cells, which convert it into glutamine, in a reaction catalyzed by glutamine synthetase. Glutamine is then transported back into the glutamatergic neuron where it is hydrolyzed by glutaminase back into glutamate. Failure of EAATs to effectively clear glutamate from the extracellular space results in various forms of cellular damage and is believed to be linked with neuropsychiatric disorders (Sanacora et al. 2008).

With the glutamate hypothesis of depression aroused the term "glutamate-based depression," being defined as a chronic depressive illness related to environmental stress and diseases associated with altered glutamate neurotransmission. McCarthy et al. (2012) hypothesized that glutamate-induced overactivation of extrasynaptic NMDA receptors in the subgenual cingulate area known as Brodmann's 25 plays an important role in the etiology of depression. The excessive activation of the extrasynaptic NMDA receptors leads to structural and functional changes in these regions, disrupting neurotransmission in distal regions associated with mood regulation, such as the prefrontal cortex.

However, the roots of a glutamate hypothesis of depression have almost 20 years and can be traced to the early 1990s, when it was found that NMDA receptor antagonists (the competitive NMDA antagonist 2-amino-7-phosphonoheptanoic acid, the noncompetitive NMDA antagonist dizocilpine, and the glycine partial agonist 1-aminocyclopropanecarboxylic acid) possess antidepressant-like action in mice by reducing the immobility in the FST (Trullas and Skolnick 1990). Moreover, NMDA antagonists are also active in the chronic mild stress model of anhedonia. Papp and Moryl (1993, 1994) examined the effects of NMDA antagonists in this paradigm and demonstrated that dizocilpine and the competitive NMDA antagonists CGP 37849 and CGP 40116 were as effective as the TCA imipramine in restoring stress-induced deficits in sucrose consumption. On the other hand, typical and atypical antidepressants, apart from modulating the monoaminergic system, can also modulate NMDA receptors. To assess if alterations in radioligand binding to NMDA receptors were a general consequence of chronic antidepressant treatment, groups of mice were administered with antidepressants (desipramine, amitriptyline, chlorgyline, tranylcypromine, citalopram, sertraline, alaproclate, and mianserin). These treatments produced significant reductions in the potency of glycine to inhibit [$^3$H]5,7-dichlorokynurenic acid binding to strychnine-insensitive glycine receptors in neocortical membranes (Paul et al. 1994). Furthermore, Nowak et al. (1993) and Paul et al. (1993) showed that chronic treatment of imipramine and of ECS altered the radioligand binding properties at several loci on the NMDA receptor complex. These treatments reduced the potency of glycine to inhibit [$^3$H]5,7-dichlorokynurenic acid binding to strychnine-insensitive glycine receptors, as well as decreased the potency of glycine to inhibit [$^3$H]CGP 39653 binding to NMDA receptors.

More recently, preclinical and clinical observations have strengthened this hypothesis since depressed patients show elevated levels of glutamate and compounds which reduce transmission at NMDA receptors have antidepressant properties. Hashimoto et al. (2007) observed increased levels of glutamate in *post-mortem* samples of frontal cortex from patients with major depressive disorder. Moreover, the plasma levels of glutamate in patients with major depression have been reported to be higher than those in normal control subjects (Mauri et al. 1998), and a 5-week treatment with antidepressants (trazodone, trazodone + pindolol, or trazodone + fluoxetine) significantly decreased the levels of glutamate in sera of depressed patients (Maes et al. 1998a, b). Nonetheless, there are some reports showing decreased glutamate levels in the brains of adults with major depression (Auer et al. 2000) or in pediatric major depression (Rosenberg et al. 2005).

At last, Berman et al. (2000) and Zarate et al. (2006) observed that patients receiving ketamine (noncompetitive NMDA receptor antagonist) showed significant improvement of depression compared with subjects receiving placebo.

### *1.4.7 The GABAergic Hypothesis*

γ-Aminobutyric acid (GABA) and glycine are two inhibitory neurotransmitters that are released separately or co-released from presynaptic terminals of inhibitory neurons. After their binding to $GABA_A$ receptors and glycine receptors, respectively, at the postsynaptic membrane, they elicit inhibitory postsynaptic currents (IPSCs). IPSCs provide phasic inhibition in the neuronal network and are important for information processing. In addition to its action at synaptic sites, low concentrations of ambient GABA can activate high-affinity $GABA_A$ receptors that are expressed at extrasynaptic sites to elicit a sustained inhibitory current called tonic inhibitory current. This tonic inhibitory currents have been identified in rat cerebellar granule cells, granule cells of the dentate gyrus, thalamocortical relay neurons of the ventral basal complex, layer V pyramidal neurons in the somatosensory cortex, and inhibitory interneurons in the CA1 region of the hippocampus (Ataka and Gu 2006).

A role for the neurotransmitter GABA in mood disorders was first proposed in the 1980s, based on the clinical observation that valproic acid, a GABA agonist, was effective in the treatment of BPD (Emrich et al. 1980). Since then, substantial preclinical and clinical data have been accumulated, supporting the involvement of GABA in mood disorders. The strongest evidence is based on reduced GABA levels in plasma and cerebrospinal fluid or resected cortical tissue of depressed patients (Fig. 1.7) (Gerner and Hare 1981; Petty and Schiesser 1981; Petty 1994; Sanacora et al. 1999). This neurochemical phenotype is consistent with a selective loss of calbindin-positive GABAergic interneurons observed in the dorsal prefrontal cortex of depressed patients (Rajkowska et al. 2007). However, Roy et al. (1991) did not observe significant differences in GABA level between depressed patients and the control group, but, among depressed patients, those with unipolar melancholia had lower cerebrospinal fluid (CSF) levels of GABA.

It has been shown that depressed patients have altered expression and subunit composition of $GABA_A$ receptors. For instance, comparison of *post-mortem* brains of depressed *vs*. nondepressed suicide victims revealed an increased expression of the α5, γ2, and δ subunit mRNAs in the dorsolateral prefrontal cortex (BA44/46) (Klempan et al. 2009). Additionally, there is growing evidence that genetic polymorphisms in $GABA_A$ receptor subunits genes, such as GSBRB1, GABRA4,

**Fig. 1.7** GABAergic neurotransmission. γ-Aminobutyric acid (GABA, *green dots*) is synthesized from glutamine/glutamate. GABA is transported through vesicles and released in the synaptic cleft by a $Ca^{2+}$-dependent or independent exocytosis. Then, GABA act on the postsynaptic $GABA_A$ receptors (Adapted from Hensch 2005; Hyland and Cryan 2010)

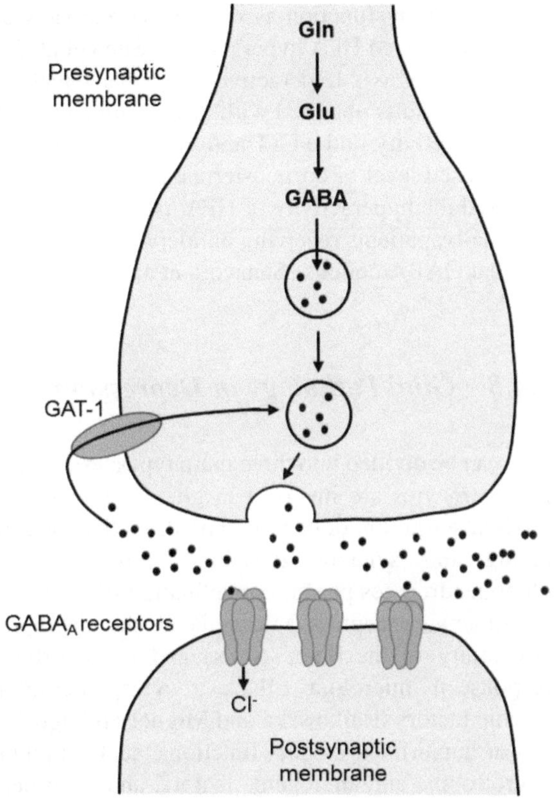


GABRB3, GABRA5, and GABRR1, are involved in BPD and GABRA5, GABRA1, GABRA6, and GABRG2 in MDD (Yamada et al. 2003; Craddock et al. 2010).

The vast majority of $GABA_A$ receptors found at postsynaptic sites are composed of α1, 2, 3, β1, 2, 3, and γ2 subunits. The γ2 one is of particular interest, as it appears to be part of all naturally occurring postsynaptic $GABA_A$ receptors. Indeed, genetically modified mice heterozygous for the γ2 subunit gene ($\gamma 2^{+/-}$) have been characterized as an animal model of anxious depression that displays both anxious- and depressive-like behaviors, such as heightened neophobia and behavioral inhibition to naturally aversive situations, reduced escape attempts under highly stressful conditions, anhedonia-like effects, and selective cognitive deficits (Luscher et al. 2011).

GABAergic inhibition regulates the activity of HPA axis. CRH neurons are innervated by a high density of GABAergic terminals and are regulated by robust GABAergic inhibition. Furthermore, numerous other brain regions exert control over the HPA axis via indirect GABAergic connections onto CRH neurons. The GABAergic control of the HPA axis is highly plastic and is altered by both acute and chronic stress (Maguire 2014). However, it is not known if GABAergic deficits independently contribute to HPA axis hyperactivity associated with depression or if

GABAergic dysfunction associated with anxiety and mood disorders is secondary to stress-induced HPA hyperactivity (Shen et al. 2010). The phenotype of $\gamma 2^{+/-}$ also induces HPA axis hyperactivity and has selective antidepressant drug responses, since the results obtained with desipramine but not with fluoxetine showed antidepressant activity in the FST and TST (Shen et al. 2010). The authors also observed a decreased level of corticosterone after desipramine, concluding that desipramine reversed the hyperactivity of HPA axis.

Finally, patients receiving antidepressant drug treatment and ECT show reduced cortical GABA deficits (Sanacora et al. 1998, 2000).

### *1.4.8 Glial Pathology in Depression*

Glia can be divided into three main types: astrocytes, oligodendrocytes, and microglia. Astrocytes are involved in glucose metabolism, neurotransmitter (glutamate, GABA, serotonin, dopamine, noradrenaline, and histamine) modulation, brain signaling and synaptic plasticity, immune response, and neurodegeneration. Oligodendrocytes promote myelination of axons and also possess several subtypes of glutamate receptors. Microglia are the macrophages of the brain and respond to any injury in the brain. Stress and glucocorticoids induce a pro-inflammatory response in microglia. Like astrocytes, microglia express cytokines and neurotrophic factors (Rajkowska and Miguel-Hidalgo 2007).

An impairment of glial functions, such as glutamate metabolism, deficiency in neurotrophic and angiogenic factors, and disturbed myelination, contributes to the pathophysiology of depression (Rajkowska and Miguel-Hidalgo 2007).

Glial cell reduction has been noticed in *post-mortem* brain tissues, more precisely at the anterior cingulate cortex, dorsolateral prefrontal cortex, and orbitofrontal cortex in MDD and/or BPD subjects (Ongür et al. 1998; Rajkowska et al. 1999; Cotter et al. 2001, 2002) and in amygdala of MDD patients (Bowley et al. 2002; Hamidi et al. 2004). On the other hand, an increased glial cell density was observed in the hippocampus CA subfield and in the granule cell layer of the DG in patients with MDD (Stockmeier et al. 2004). Moreover, glial size can be increased or unchanged in the brains of MDD and BPD patients (Rajkowska et al. 1999, 2001; Cotter et al. 2002, 2005; Chana et al. 2003; Stockmeier et al. 2004).

Considering these studies, it has been speculated that a reduction in glial density accompanied by an increase in glial nuclei size suggests that there are a decreased number of functional glial cells that may be forced to work harder to maintain the metabolic needs of neurons (Rajkowska and Miguel-Hidalgo 2007).

Furthermore, a reduction in the level of immunoreactive myelin basic protein, a marker for mature oligodendrocytes, was found in homogenates from the anterior frontal cortex of depressed subjects (Honer et al. 1999). On the other hand, myelin basic protein was unchanged but the level of cholesterol, a major component of myelin, was lower in the cortex of MDD and BPD patients (Beasley et al. 2005).

In addition, ultrastructural changes in oligodendrocytes were observed in the dorsolateral prefrontal and anterior frontal cortex in MDD and BPD patients (Uranova et al. 2004) and a reduction in the expression of key oligodendrocyte-related and myelin-related genes was recently reported in the temporal cortex in MDD and in the dorsolateral prefrontal cortex in BPD (Tkachev et al. 2003; Aston et al. 2004; Sokolov 2007).

The levels and expression of the astrocytic markers glial fibrillary acidic protein (GFAP), gap junction proteins, such as connexins 40 and 43, the water channel aquaporin-4 (AQP4), a calcium-binding protein S100B, of the glutamatergic markers including the excitatory amino acid transporters 1 and 2 (EAAT1, EAAT2), and of the glutamine synthetase are changed in MDD. There was a significant decrease in the density of GFAP-immunoreactive astrocytes and GFAP area fraction in gray matter of the dorsolateral prefrontal cortex in younger depressed patients (<60 years old), as compared to control subjects (Miguel-Hidalgo et al. 2000), while GFAP-immunoreactive area fraction is significantly decreased in the gray matter of the orbitofrontal cortex in younger and older patients with MDD (Miguel-Hidalgo et al. 2010). In contrast, older subjects with late-onset depression showed increases in GFAP-immunoreactive area fraction and cell density in the gray matter of dorsolateral prefrontal cortex (Miguel-Hidalgo et al. 2000; Davis et al. 2002).

Decreases in the gene and protein expression of connexins 30 and 43 have been observed in several brain regions of patients diagnosed with MDD (Ernst et al. 2011) and rats submitted to chronic unpredictable stress present the same pattern (Sun et al. 2012). This decrease may alter calcium wave propagation and communication between astrocytes.

A decrease in the expression of mRNA for AQP4 was identified in locus coeruleus in MDD (Bernard et al. 2011), and high levels of S100B in the cerebrospinal fluid and serum as a result of astrocyte leakage were observed in MDD patients (Schroeter et al. 2002).

Reduced expression of mRNA for EAAT1, EAAT2, and glutamine synthetase was noted in the anterior cingulate and dorsolateral prefrontal cortex in *post-mortem* brain samples from patients with MDD (Choudary et al. 2005).

For all the reasons presented above, glial cells may be seen as a possible new target for new antidepressant drugs.

### *1.4.9 Epigenetic Theory*

Epigenetics comprises mechanisms of gene expression modifications that do not alter the genomic code itself (Schroeder et al. 2010).

Stable alterations in gene expression that contribute to structural and functional changes in multiple brain regions are implicated in the heterogeneity and pathogenesis of MDD. It is known that transcriptional dysregulation may underlie the behavioral manifestation of many psychiatric disorders, including depression.

Several different types of epigenetic mechanisms have been implicated to date, including stress- or antidepressant-induced changes in DNA methylation, histone acetylation, methylation and phosphorylation, and noncoding RNA (e.g., miRNAs) (Vialou et al. 2013).

Methylation of DNA at cytosines adjacent to guanines (CpG) in promoter regions may inhibit local gene transcription by two different mechanisms, or by interfering with transcription factor binding or by recruiting methylated-DNA binding proteins that alter transcription efficiency. Thus, cytosine methylation is typically a silent mark while hypomethylation is usually associated with increased expression (Elliott et al. 2010; Hunter 2012). DNA methylation is catalyzed by DNA methyltransferases (DNMTs), while methyl-CpG-binding domain protein 2 (MBD2), growth arrest and DNA-damage-inducible protein alpha (GADD45a) and GADD45b are responsible for DNA demethylation. Moreover, ten-eleven translocation (TET) enzymes oxidize 5-methyl cytosine into 5-hydroxymethylcytosine, which can be further oxidized into 5-formolcytosine and 5-carboxylcytosine (Fig. 1.8) (Vialou et al. 2013).

For instance, changes in CpG methylation levels at gene promoters implicated in HPA axis reactivity have been shown. In animal models, social defeat stress-induced demethylation of *Crf* promoter coincides with a social avoidance behavior, which is reversed by chronic imipramine treatment (Elliott et al. 2010). On the other hand, animals subjected to stress showed increased methylation of the *Bdnf* promoter in hippocampus (Roth and Sweatt 2011).

Histone modifications include histone acetylation of lysine residues, methylation of lysine, arginine, and histidine residues and phosphorylation of serine and threonine (Huang et al. 2005; Sun et al. 2013). Acetylation of histones is also associated with a transcriptionally active state. Histone acetylation at lysine residues is controlled by histone acetyltransferases (HATs) and histone deacetylases (HDACs) (Hunter 2012; Vialou et al. 2013). On the other hand, histone methylation can be associated with both transcriptional activation and repression, depending on the residue modified by methylation (Fig. 1.8) (Vialou et al. 2013), the lysine one being the only form implicated in depression to date (Sun et al. 2013; Nestler et al. 2015). The methylation of lysine 4, 36, and 79 of histone H3 (H3K4, H3K36, and H3K79) is a gene activation mark, whereas H3K9, H3K27, and H4K20 methylation is a gene-repressive modification (Büttner et al. 2010; Gu and Lee 2013; Sun et al. 2013; Nestler et al. 2015). For instance, after chronic social defeat stress in mice, there is a transient decrease, followed by a persistent increase, of H3 acetylation of lysine 14 (acH3K14) in nucleus accumbens (NAc) and a decreased level of HDAC2. Similar effects were observed in the nucleus accumbens of depressed humans studied *post-mortem*. The authors also observed that HDAC inhibitors (MS-275 and SAHA) have antidepressant-like effects when infused into the NAc. MS-275 and SAHA, but not systemic fluoxetine, increase the levels of acH3K14 in NAc. Moreover, genes regulated only by MS-275, but not by fluoxetine, include gja5 (involved in gap junction formation) and dlgap1 (assembly of postsynaptic density complexes). On the other hand, neuropeptide cortistatin (involved in sleep regulation) is downregulated by stress in NAc, an effect reversed by MS-275 or fluoxetine (Covington et al. 2009).

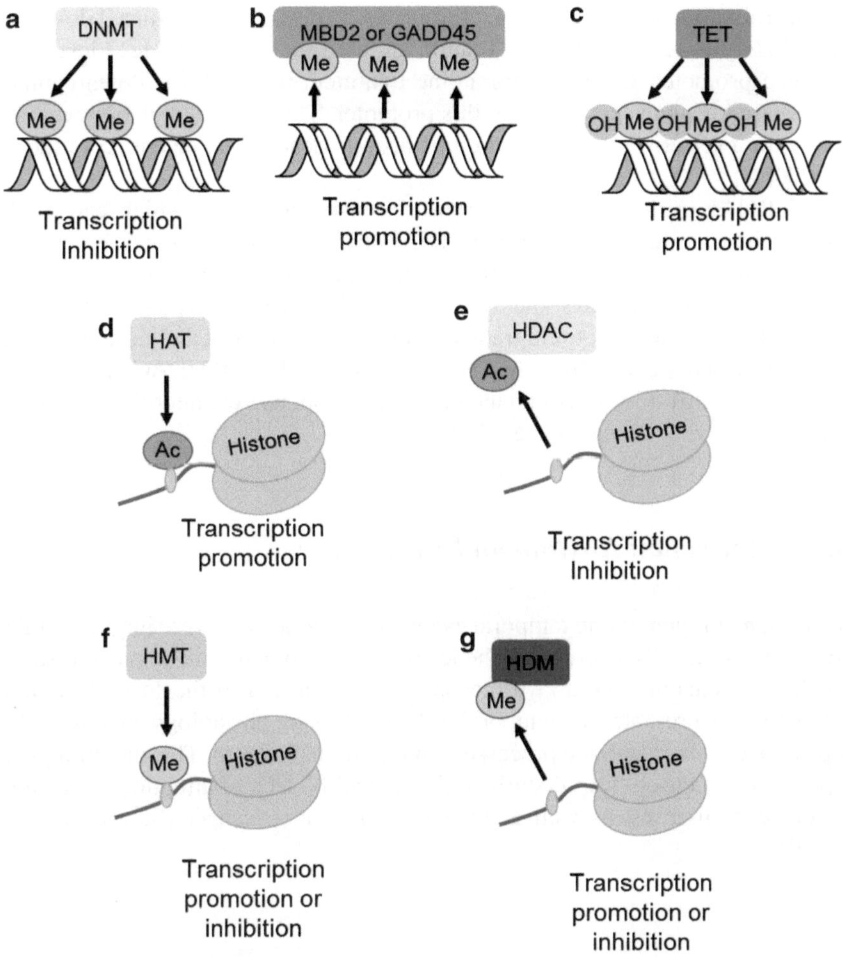


**Fig. 1.8** Overview of epigenetic regulatory mechanisms in depression. (**a**) DNA methyltransferases (DNMTs) add methyl groups to the cytosines of CpG islands and are generally associated with DNA silencing; (**b**) DNA demethylation promotes transcription and is mediated by methyl-CpG-binding domain protein 2 (MBD2) or by growth arrest and DNA-damage-inducible protein alpha (GADD45a) or GADD45b; (**c**) ten-eleven translocation (TET) enzymes oxidize 5mC into 5′-hydroxymethylcytosine which promotes transcription; (**d**) histone acetyltransferases (HATs) add acetyl groups to lysine residues on histone tails and are generally associated with transcription promotion; (**e**) histone deacetylases (HDACs) remove those acetyl groups and inhibit transcription; (**f**) histone methylation is mediated by histone methyltransferases (HMTs), which add methyl groups to lysine residues on histone tails. Methylation can promote or inhibit transcription; (**g**) demethylation by histone demethylases (HDMs) can promote or inhibit transcription (Adapted from Hedrich and Tsokos 2011; Niehrs and Schäfer 2012; Dias et al. 2015)

Concerning histone methylation, chronic social defeat stress induced downregulation of *Bdnf* transcripts III and IV and increased methylated H3K27 level at the *Bdnf* gene promoter. Chronic imipramine treatment reversed this downregulation and increased histone acetylation at this promoter. This hyperacetylation was associated with a selective downregulation of histone deacetylase (Hdac) 5 (Tsankova et al. 2006).

MicroRNAs (miRNAs) are posttranscriptional regulators, which can bind to complementary sequences on target mRNAs and induce translational repression or mRNA degradation (Vialou et al. 2013). It has been shown that chronic administration of fluoxetine increases the levels of miRNA-16 in serotonergic neurons. miRNA-16 targets the serotonin transporter (SERT) transcript and decreases SERT expression (Baudry et al. 2010). Moreover, SIRT 1, a class III HDAC, decreases the expression of miR-134, thus enhancing cAMP response element-binding protein (CREB) and BDNF levels (Gao et al. 2010).

### 1.4.10 Circadian Rhythm and Depression

The mechanisms generating temporal events in the organisms were ignored until the 1970s, when it was discovered that the lesion of the suprachiasmatic nucleus (SCN), a small hypothalamic site above the optical chiasm, led to the loss of circadian rhythm. SCN coordinates the temporal order of all body physiology, including daily sleep-wake cycles, metabolic processes, and hormonal release. The alteration of this temporal order causes sleep disturbances, irritability, lack of attention, gastrointestinal or heart diseases, as well as tendency to develop cancer (Salgado-Delgado et al. 2011).

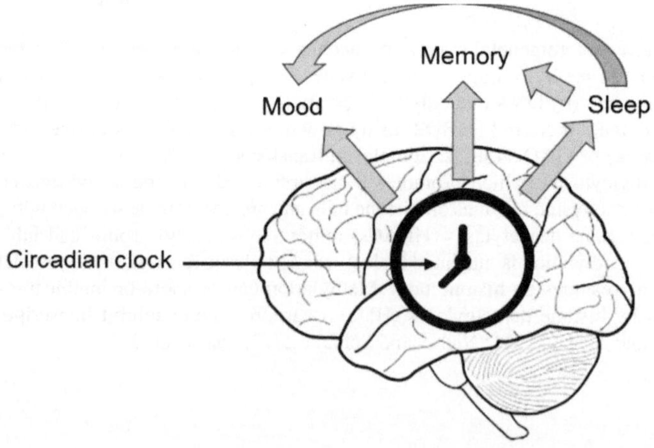


**Fig. 1.9** Circadian rhythm and depression (Adapted from Kondratova and Kondratov 2012)

Disruption of circadian rhythms is common in patients with unipolar and bipolar disorders (Fig. 1.9). The most obvious change is observed in mood, because commonly depressed individuals show an improvement of their mood state in the evening, while this pattern is inverted in bipolar depressive patients, which feel better in the morning (Salgado-Delgado et al. 2011). Moreover, the sleep/wake rhythm is affected by insomnia at night and hypersomnia during the day, with a consequent feeling of constant tiredness (Salgado-Delgado et al. 2011). Patients with MDD have nightmares at least twice a week and their frequency has been shown to be related to the risk of suicide (Ağargün et al. 1998). Concerning BPD, patients often experience a reduced need for sleep during manic episodes, while insomnia or hypersomnia can be experienced nearly every day during episodes of depression (Harvey et al. 2009).

Disturbances in the organization of the sleep/wake cycle in MDD and BPD patients are thought to be due to abnormalities in the timing of the rapid eye movement (REM) sleep and non-rapid eye movement (non-REM or NREM) sleep cycle (Wirz-Justice 2006). Increases in REM sleep density have also been found to be specific to affective disorders, both unipolar and bipolar (Gillin et al. 1979; Jernajczyk 1986; Hudson et al. 1988, 1992; Lauer et al. 1992; Wichniak et al. 2000). Nevertheless, in some studies reporting the nature and extent of sleep disturbance in children and teenagers with BPD, there was reduced REM sleep (Mehl et al. 2006). With the exception of trimipramine, all TCAs reduce or suppress REM sleep and increase REM latency, as well as the MAOIs phenelzine and tranylcypromine, the SNRI venlafaxine and the SSRIs paroxetine and sertraline (Vogel et al. 1990; Winokur et al. 2001).

In depressive patients, the nocturnal melatonin release is often diminished. On the other hand, depressive patients display increased levels of plasmatic cortisol, the stress hormone (Salgado-Delgado et al. 2011). Antidepressant medications have been shown to increase melatonin levels in rodents and humans (Murphy et al. 1986) and the hypnotic melatonin has also been successfully used for the treatment of sleep problems related to perturbations of the circadian time by acting on melatonin receptors in the hypothalamic suprachiasmatic nuclei (SCN) (Cardinali et al. 2012). In 2009, the use of a new antidepressant agomelatine was authorized in Europe. This new drug is a melatonergic agonist and complementary 5-$HT_{2C}$ antagonist (de Bodinat et al. 2010).

Besides MDD and BPD, disruption of circadian rhythms can also lead to seasonal affective disorder (SAD). In winter, when days are shorter and nights are longer than in spring and summer, SAD can be developed as a consequence of lower sunlight exposure. SAD is characterized by anhedonic mood symptoms, sleepiness in the day and insomnia at night, increased appetite (particularly for carbohydrates), and weight. This type of depression is characterized by a regular, annual episode of major depression during autumn and/or winter and remission or episode of mania/hypomania during spring and summer. Contrasting with MDD, in SAD patients, melatonin is released in higher proportions (Salgado-Delgado et al. 2011). Bright artificial light of at least 1500 lx has been used to treat some forms of depression, particularly SAD (Wirz-Justice 2008; Rastad et al. 2011; Popper 2013; Danilenko and Ivanova 2015).

## 1.5 Conclusions

Antidepressant drugs available on the market, such as MAOIs, SSRIs, TCAs, SNRIs, and NRIs, aim at increasing the synaptic level of monoamines, namely, serotonin and noradrenaline. However, the lack of efficacy of monoamine potentiating antidepressants, as well as their 2–3 weeks of therapeutic latency, led to the search for other mechanisms that could explain better the pathogenesis of depression or that could complement monoamine theory. Recent research points to the role of other targets for the development of new antidepressant drugs, which can help to decrease the number of patients with treatment-resistant depression. Therefore, there is a need to fully understand how drugs interact with the targets mentioned in this chapter and the effects they induce in order to develop a therapy tailored to individual patients and disease states.

**Acknowledgments** This work was financed through project UID/QUI/50006/2013, receiving financial support from FCT/MEC through national funds, and co-financed by FEDER, under the Partnership Agreement PT2020. To all financing sources the authors are greatly indebted. Clara Grosso thanks FCT for the FCT Investigator (IF/01332/2014).

## References

Ağargün MY, Cilli AS, Kara H, Tarhan N, Kincir F, Oz H. Repetitive and frightening dreams and suicidal behavior in patients with major depression. Compr Psychiatry. 1998;39(4):198–202.

Akhondzadeh S, Jafari S, Raisi F, Nasehi AA, Ghoreishi A, Salehi B, Mohebbi-Rasa S, Raznahan M, Kamalipour A. Clinical trial of adjunctive celecoxib treatment in patients with major depression: a double blind and placebo controlled trial. Depress Anxiety. 2009;26(7):607–11.

Al-Harbi KS. Treatment-resistant depression: therapeutic trends, challenges, and future directions. Patient Prefer Adherence. 2012;6:369–88.

Altman J. Autoradiographic and histological studies of postnatal neurogenesis. III. Dating the time of production and onset of differentiation of cerebellar microneurons in rats. J Comp Neurol. 1969;136(3):269–93.

Altman J, Das GD. Autoradiographic and histological evidence of postnatal hippocampal neurogenesis in rats. J Comp Neurol. 1965;124(3):319–35.

Altman J, Das GD. Autoradiographic and histological studies of postnatal neurogenesis. I. A longitudinal investigation of the kinetics, migration and transformation of cells incorporating tritiated thymidine in neonate rats, with special reference to postnatal neurogenesis in some brain regions. J Comp Neurol. 1966;126(3):337–89.

American Psychiatric Association. Diagnostic and statistical manual of mental disorders, fifth edition (DSM-5). Washington, DC: American Psychiatric Association; 2013.

Anacker C, Zunszain PA, Carvalho LA, Pariante CM. The glucocorticoid receptor: pivot of depression and of antidepressant treatment? Psychoneuroendocrinology. 2011;36(3):415–25.

Aston C, Jiang L, Sokolov BP. Microarray analysis of postmortem temporal cortex from patients with schizophrenia. J Neurosci Res. 2004;77(6):858–66.

Ataka T, Gu JG. Relationship between tonic inhibitory currents and phasic inhibitory activity in the spinal cord lamina II region of adult mice. Mol Pain. 2006;2:36.

Auer DP, Putz B, Kraft E, Lipinski B, Schill J, Holsboer F. Reduced glutamate in the anterior cingulate cortex in depression: an *in vivo* proton magnetic resonance spectroscopy study. Biol Psychiatry. 2000;47(4):305–13.

Bachis A, Mallei A, Cruz MI, Wellstein A, Mocchetti I. Chronic antidepressant treatments increase basic fibroblast growth factor and fibroblast growth factor-binding protein in neurons. Neuropharmacology. 2008;55(7):1114–20.

Baldessarini RJ. Biogenic amines and behavior. Annu Rev Med. 1972;23(5):343–54.

Barden N. Implication of the hypothalamic–pituitary–adrenal axis in the physiopathology of depression. J Psychiatry Neurosci. 2004;29(3):185–93.

Baudry A, Mouillet-Richard S, Schneider B, Launay JM, Kellermann O. MiR-16 targets the serotonin transporter: a new facet for adaptive responses to antidepressants. Science. 2010; 329(5998):1537–41.

Beasley CL, Honer WG, Bergmann K, Falkai P, Lütjohann D, Bayer TA. Reductions in cholesterol and synaptic markers in association cortex in mood disorders. Bipolar Disord. 2005; 7(5):449–55.

Beaurepaire R. Questions raised by the cytokine hypothesis of depression. Brain Behav Immun. 2002;16:610–7.

Berman RM, Cappiello A, Anand A, Oren DA, Heninger GR, Charney DS, Krystal JH. Antidepressant effects of ketamine in depressed patients. Biol Psychiatry. 2000;47(4): 351–4.

Bernard R, Kerman IA, Thompson RC, Jones EG, Bunney WE, Barchas JD, Schatzberg AF, Myers RM, Akil H, Watson SJ. Altered expression of glutamate signaling, growth factor, and glia genes in the locus coeruleus of patients with major depression. Mol Psychiatry. 2011; 16(6):634–46.

Berton O, McClung CA, Dileone RJ, Krishnan V, Renthal W, Russo SJ, Graham D, Tsankova NM, Bolanos CA, Rios M, Monteggia LM, Self DW, Nestler EJ. Essential role of BDNF in the mesolimbic dopamine pathway in social defeat stress. Science. 2006;311(5762):864–8.

Bloom DE, Cafiero ET, Jané-Llopis E, Abrahams-Gessel S, Bloom LR, Fathima S, Feigl AB, Gaziano T, Mowafi M, Pandya A, Prettner K, Rosenberg L, Seligman B, Stein AZ, Weinstein C. The global economic burden of noncommunicable diseases. Geneva: World Economic Forum; 2011.

Bouwer C, Claassen J, Dinan TG, Nemeroff CB. Prednisone augmentation in treatment-resistant depression with fatigue and hypocortisolaemia: a case series. Depress Anxiety. 2000; 12(1):44–50.

Bowley MP, Drevets WC, Ongür D, Price JL. Low glial numbers in the amygdala in major depressive disorder. Biol Psychiatry. 2002;52(5):404–12.

Bremner JD, Narayan M, Anderson ER, Staib LH, Miller HL, Charney DS. Hippocampal volume reduction in major depression. Am J Psychiatry. 2000;157(1):115–8.

Brezun JM, Daszuta A. Depletion in serotonin decreases neurogenesis in the dentate gyrus and the subventricular zone of adult rats. Neuroscience. 1999;89(4):999–1002.

Büttner N, Johnsen SA, Kügler S, Vogel T. Af9/Mllt3 interferes with Tbr1 expression through epigenetic modification of histone H3K79 during development of the cerebral cortex. Proc Natl Acad Sci U S A. 2010;107(15):7042–7.

Calabrese F, Guidotti G, Racagni G, Riva MA. Reduced neuroplasticity in aged rats: a role for the neurotrophin brain-derived neurotrophic factor. Neurobiol Aging. 2013;34(12):2768–76.

Calabrese F, Rossetti AC, Racagni G, Gass P, Riva MA, Molteni R. Brain-derived neurotrophic factor: a bridge between inflammation and neuroplasticity. Front Cell Neurosci. 2014;8:430.

Cardinali DP, Srinivasan V, Brzezinski A, Brown GM. Melatonin and its analogs in insomnia and depression. J Pineal Res. 2012;52(4):365–75.

Carlsson A, Lindqvist M, Magnusson T, Waldeck B. On the presence of 3-hydroxytyramine in brain. Science. 1958;127(3296):471.

Carroll BJ, Martin FIR, Davies B. Resistance to suppression by dexamethasone of plasma 11-O.H.C.S. levels in severe depressive illness. Br Med J. 1968;3:285.

Carroll BJ, Feinberg M, Greden JF, Tarika J, Albala AA, Haskett RF, James NM, Kronfol Z, Lohr N, Steiner M, de Vigne JP, Young E. A specific laboratory test for the diagnosis of melancholia. Standardization, validation, and clinical utility. Arch Gen Psychiatry. 1981;38(1):15–22.

Chan JP, Unger TJ, Byrnes J, Rios M. Examination of behavioral deficits triggered by targeting BDNF in fetal or postnatal brains of mice. Neuroscience. 2006;142:49–58.

Chana G, Landau S, Beasley C, Everall IP, Cotter D. Two-dimensional assessment of cytoarchitecture in the anterior cingulate cortex in major depressive disorder, bipolar disorder, and schizophrenia: evidence for decreased neuronal somal size and increased neuronal density. Biol Psychiatry. 2003;53(12):1086–98.

Chessin M, Dubnick B, Kramer ER, Scott CC. Modifications of pharmacology of reserpine and serotonin by iproniazid. J Pharmacol Exp Ther. 1957;119(4):453–60.

Choudary PV, Molnar M, Evans SJ, Tomita H, Li JZ, Vawter MP, Myers RM, Bunney Jr WE, Akil H, Watson SJ, Jones EG. Altered cortical glutamatergic and GABAergic signal transmission with glial involvement in depression. Proc Natl Acad Sci U S A. 2005;102(43):15653–8.

Chourbaji S, Hellweg R, Brandis D, Zörner B, Zacher C, Lang UE, Henn FA, Hörtnagl H, Gass P. Mice with reduced brain-derived neurotrophic factor expression show decreased choline acetyltransferase activity, but regular brain monoamine levels and unaltered emotional behavior. Mol Brain Res. 2004;121(1-2):28–36.

Cinini SM, Barnabe GF, Galvão-Coelho N, de Medeiros MA, Perez-Mendes P, Sousa MB, Covolan L, Mello LE. Social isolation disrupts hippocampal neurogenesis in young non-human primates. Front Neurosci. 2014;8:45.

Coppen A. The biochemistry of affective disorders. Br J Psychiatry. 1967;113(504):1237–64.

Cotter D, Mackay D, Landau S, Kerwin R, Everall I. Reduced glial cell density and neuronal size in the anterior cingulate cortex in major depressive disorder. Arch Gen Psychiatry. 2001;58(6):545–53.

Cotter D, Mackay D, Chana G, Beasley C, Landau S, Everall IP. Reduced neuronal size and glial cell density in area 9 of the dorsolateral prefrontal cortex in subjects with major depressive disorder. Cereb Cortex. 2002;12(4):386–94.

Cotter D, Hudson L, Landau S. Evidence for orbitofrontal pathology in bipolar disorder and major depression, but not in schizophrenia. Bipolar Disord. 2005;7(4):358–69.

Covington HE, Maze I, LaPlant QC, Vialou VF, Ohnishi YN, Berton O, Fass DM, Renthal W, Rush AJ, Wu EY, Ghose S, Krishnan V, Russo SJ, Tamminga C, Haggarty SJ, Nestler EJ. Antidepressant actions of histone deacetylase inhibitors. J Neurosci. 2009;29(37): 11451–60.

Craddock N, Jones L, Jones IR, Kirov G, Green EK, Grozeva D, Moskvina V, Nikolov I, Hamshere ML, Vukcevic D, Caesar S, Gordon-Smith K, Fraser C, Russell E, Norton N, Breen G, St Clair D, Collier DA, Young AH, Ferrier IN, Farmer A, McGuffin P, Holmans PA, Wellcome Trust Case Control Consortium (WTCCC), Donnelly P, Owen MJ, O'Donovan MC. Strong genetic evidence for a selective influence of GABA(A) receptors on a component of the bipolar disorder phenotype. Mol Psychiatry. 2010;15(2):146–53.

Czéh B, Welt T, Fischer AK, Erhardt A, Schmitt W, Müller MB, Toschi N, Fuchs E, Keck ME. Chronic psychosocial stress and concomitant repetitive transcranial magnetic stimulation: effects on stress hormone levels and adult hippocampal neurogenesis. Biol Psychiatry. 2002;52(11):1057–65.

Danilenko KV, Ivanova IA. Dawn simulation vs. bright light in seasonal affective disorder: Treatment effects and subjective preference. J Affect Disord. 2015;180:87–9.

Davis S, Thomas A, Perry R, Oakley A, Kalaria RN, O'Brien JT. Glial fibrillary acidic protein in late life major depressive disorder: an immunocytochemical study. J Neurol Neurosurg Psychiatry. 2002;73(5):556–60.

de Bellis MD, Gold PW, Geracioti Jr TD, Listwak SJ, Kling MA. Association of fluoxetine treatment with reductions in CSF concentrations of corticotropin-releasing hormone and arginine vasopressin in patients with major depression. Am J Psychiatry. 1993;150(4):656–7.

de Bodinat C, Guardiola-Lemaitre B, Mocaër E, Renard P, Muñoz C, Millan MJ. Agomelatine, the first melatonergic antidepressant: discovery, characterization and development. Nat Rev Drug Discov. 2010;9(8):628–42.

Dias BG, Banerjee SB, Duman RS, Vaidya VA. Differential regulation of brain derived neurotrophic factor transcripts by antidepressant treatments in the adult rat brain. Neuropharmacology. 2003;45:553–63.

Dias BG, Maddox SA, Klengel T, Ressler KJ. Epigenetic mechanisms underlying learning and the inheritance of learned behaviors. Trends Neurosci. 2015;38(2):96–107.

Drew MR, Hen R. Adult hippocampal neurogenesis as target for the treatment of depression. CNS Neurol Disord Drug Targets. 2007;6(3):205–18.

Eisch AJ, Bolaños CA, de Wit J, Simonak RD, Pudiak CM, Barrot M, Verhaagen J, Nestler EJ. Brain-derived neurotrophic factor in the ventral midbrain-nucleus accumbens pathway: a role in depression. Biol Psychiatry. 2003;54(10):994–1005.

Elliott E, Ezra-Nevo G, Regev L, Neufeld-Cohen A, Chen A. Resilience to social stress coincides with functional DNA methylation of the *Crf* gene in adult mice. Nat Neurosci. 2010;13(11):1351–3.

Emrich HM, Zerssen DV, Kissling W, Möller H-J, Windorfer A. Effect of sodium valproate on mania: the GABA-hypothesis of affective disorders. Arch Psychiatry Neurol Sci. 1980; 229(1):1–16.

Eriksson PS, Perfilieva E, Björk-Eriksson T, Alborn AM, Nordborg C, Peterson DA, Gage FH. Neurogenesis in the adult human hippocampus. Nat Med. 1998;4(11):1313–7.

Ernst C, Nagy C, Kim S, Yang JP, Deng X, Hellstrom IC, Choi KH, Gershenfeld H, Meaney MJ, Turecki G. Dysfunction of astrocyte connexins 30 and 43 in dorsal lateral prefrontal cortex of suicide completers. Biol Psychiatry. 2011;70(4):312–9.

Gallagher PJ, Castro V, Fava M, Weilburg JB, Murphy SN, Gainer VS, Churchill SE, Kohane IS, Iosifescu DV, Smoller JW, Perlis RH. Antidepressant response in patients with major depression exposed to NSAIDs: a pharmacovigilance study. Am J Psychiatry. 2012;169(10): 1065–72.

Gao J, Wang WY, Mao YW, Gräff J, Guan JS, Pan L, Mak G, Kim D, Su SC, Tsai LH. A novel pathway regulates memory and plasticity via SIRT1 and miR-134. Nature. 2010; 466(7310):1105–9.

Gerner RH, Hare TA. CSF GABA in normal subjects and patients with depression, schizophrenia, mania, and anorexia nervosa. Am J Psychiatry. 1981;138(8):1098–101.

Gillin JC, Duncan W, Pettigrew KD, Frankel BL, Snyder F. Successful separation of depressed, normal and insomniac subjects by EEG sleep data. Arch Gen Psychiatry. 1979;36(1):85–90.

Global Burden of Disease Study 2013 Collaborators. Global, regional, and national incidence, prevalence, and years lived with disability for 301 acute and chronic diseases and injuries in 188 countries, 1990–2013: a systematic analysis for the Global Burden of Disease Study 2013. Lancet. 2015;386(9995):743–800.

Goel N, Innala L, Viau V. Sex differences in serotonin (5-HT) 1A receptor regulation of HPA axis and dorsal raphe responses to acute restraint. Psychoneuroendocrinology. 2014;40:232–41.

Gould E, Woolley CS, McEwen BS. Adrenal steroids regulate postnatal development of the rat dentate gyrus: I. Effects of glucocorticoids on cell death. J Comp Neurol. 1991;313(3):479–85.

Gould E, Cameron HA, Daniels DC, Woolley CS, McEwen BS. Adrenal hormones suppress cell division in the adult rat dentate gyrus. J Neurosci. 1992;12(9):3642–50.

Govindarajan A, Rao BS, Nair D, Trinh M, Mawjee N, Tonegawa S, Chattarji S. Transgenic brain-derived neurotrophic factor expression causes both anxiogenic and antidepressant effects. Proc Natl Acad Sci U S A. 2006;103(35):13208–13.

Greenberg PE, Fournier A-A, Sisitsky T, Pike CT, Kessler RC. The economic burden of adults with major depressive disorder in the United States (2005 and 2010). J Clin Psychiatry. 2015;76(2):155–62.

Groves JO. Is it time to reassess the BDNF hypothesis of depression? Mol Psychiatry. 2007;12: 1079–88.

Grunbaum-Novak N, Taler M, Gil-Ad I, Weizman A, Cohen H, Weizman R. Relationship between antidepressants and IGF-1 system in the brain: possible role in cognition. Eur Neuropsychopharmacol. 2008;18(6):431–8.

Gu B, Lee MG. Histone H3 lysine 4 methyltransferases and demethylases in self-renewal and differentiation of stem cells. Cell Biosci. 2013;3(1):39.

Guan Z, Fang J. Peripheral immune activation by lipopolysaccharide decreases neurotrophins in the cortex and hippocampus in rats. Brain Behav Immun. 2006;20(1):64–71.

Gwinn RP, Kondratyev A, Gale K. Time-dependent increase in basic fibroblast growth factor protein in limbic regions following electroshock seizures. Neuroscience. 2002;114(2):403–9.
Hamidi M, Drevets WC, Price JL. Glial reduction in amygdala in major depressive disorder is due to oligodendrocytes. Biol Psychiatry. 2004;55(6):563–9.
Hashimoto K, Sawa A, Iyo M. Increased levels of glutamate in brains from patients with mood disorders. Biol Psychiatry. 2007;62(11):1310–6.
Hedrich CM, Tsokos GC. Epigenetic mechanisms in systemic lupus erythematosus and other autoimmune diseases. Trends Mol Med. 2011;17(12):714–24.
Hensch TK. Critical period plasticity in local cortical circuits. Nat Rev Neurosci. 2005;6:877–88.
Holsboer F, von Bardeleben U, Wiedemann K, Müller OA, Stalla GK. Serial assessment of corticotropin-releasing hormone response after dexamethasone in depression. Implications for pathophysiology of DST nonsuppression. Biol Psychiatry. 1987;22:228–34.
Honer WG, Falkai P, Chen C, Arango V, Mann JJ, Dwork AJ. Synaptic and plasticity-associated proteins in anterior frontal cortex in severe mental illness. Neuroscience. 1999;91(4): 1247–55.
Hu TW. Perspectives: an international review of the national cost estimates of mental illness, 1990–2003. J Ment Health Policy Econ. 2006;9(1):3–13.
Hu TW, He Y, Zhang M, Chen N. Economic costs of depression in China. Soc Psychiatry Psychiatr Epidemiol. 2007;42(2):110–6.
Huang EJ, Reichardt LF. Neurotrophins: roles in neuronal development and function. Annu Rev Neurosci. 2001;24:677–736.
Huang S, Litt M, Felsenfeld G. Methylation of histone H4 by arginine methyltransferase PRMT1 is essential *in vivo* for many subsequent histone modifications. Genes Dev. 2005;19(16): 1885–93.
Hudson JI, Lipinski JF, Frankenburg FR, Grochocinski VJ, Kupfer DJ. Electroencephalographic sleep in mania. Arch Gen Psychiatry. 1988;45(3):267–73.
Hudson JI, Lipinski JF, Keck Jr PE, Aizley HG, Lukas SE, Rothschild AJ, Waternaux CM, Kupfer DJ. Polysomnographic characteristics of young manic patients. Comparison with unipolar depressed patients and normal control subjects. Arch Gen Psychiatry. 1992;49(5):378–83.
Hunter RG. Epigenetic effects of stress and corticosteroids in the brain. Front Cell Neurosci. 2012;6:article 18.
Hurley LL, Tizabi Y. Neuroinflammation, neurodegeneration, and depression. Neurotox Res. 2013;23:131–44.
Hyland NP, Cryan JF. A gut feeling about GABA: focus on GABA(B) receptors. Front Pharmacol. 2010;1:124.
Iyengar RL, Gandhi S, Aneja A, Thorpe K, Razzouk L, Greenberg J, Mosovich S, Farkouh ME. NSAIDs are associated with lower depression scores in patients with osteoarthritis. Am J Med. 2013;126(11):1017.e11–18.
Jacobs BL, van Praag H, Gage FH. Adult brain neurogenesis and psychiatry: a novel theory of depression. Mol Psychiatry. 2000;5:262–9.
Jernajczyk W. Latency of eye movement and other REM sleep parameters in bipolar depression. Biol Psychiatry. 1986;21(5-6):465–72.
Juruena MF, Cleare AJ, Bauer ME, Pariente CM. Molecular mechanisms of glucocorticoid receptor sensitivity and relevance to affective disorders. Acta Neuropsychiatr. 2003;15:354–67.
Juruena MF, Cleare AJ, Pariante CM. The hypothalamic pituitary adrenal axis, glucocorticoid receptor function and relevance to depression. Rev Bras Psiquiatr. 2004;26(3):189–201.
Kaplan MS, Hinds JW. Neurogenesis in the adult rat: electron microscopic analysis of light radioautographs. Science. 1977;197(4308):1092–4.
Kee N, Teixeira CM, Wang AH, Frankland PW. Preferential incorporation of adult-generated granule cells into spatial memory networks in the dentate gyrus. Nat Neurosci. 2007;10(3): 355–62.
Klempan TA, Sequeira A, Canetti L, Lalovic A, Ernst C, Ffrench-Mullen J, Turecki G. Altered expression of genes involved in ATP biosynthesis and GABAergic neurotransmission in the

ventral prefrontal cortex of suicides with and without major depression. Mol Psychiatry. 2009;14(2):175–89.

Kondratova AA, Kondratov RV. The circadian clock and pathology of the ageing brain. Nat Rev Neurosci. 2012;13:325–35.

Krishnan KRR, Rayasam K, Reed DR, Smith M, Chappell P, Saunders WB, Ritchie JC, Carroll BJ, Nemeroff CB. The corticotropin releasing factor stimulation test in patients with major depression: Relationship to dexamethasone suppression test results. Depression. 1993;1(3):133–6.

Lapchak PA, Araujo DM, Beck KD, Finch CE, Johnson SA, Hefti F. BDNF and trkB mRNA expression in the hippocampal formation of aging rats. Neurobiol Aging. 1993;14(2):121–6.

Lapin IP, Oxenkrug GF. Intensification of the central serotoninergic processes as a possible determinant of the thymoleptic effect. Lancet. 1969;1(7586):132–6.

Lauer CJ, Wiegand M, Krieg JC. All-night electroencephalographic sleep and cranial computed tomography in depression. A study of unipolar and bipolar patients. Eur Arch Psychiatry Clin Neurosci. 1992;242(2-3):59–68.

Lavergne F, Jay TM. A new strategy for antidepressant prescription. Front Neurosci. 2010;4:192.

Levine J, Cholestoy A, Zimmerman J. Possible antidepressant effect of minocycline. Am J Psychiatry. 1996;153(4):582.

Loomer HP, Saunders IC, Kline NS. A clinical and pharmacodynamics evaluation of iproniazid as a psychic energizer. Psychiatr Res Rep Am Psychiatr Assoc. 1958;8:129–41.

López-Muñoz F, Alamo C. Monoaminergic neurotransmission: the history of the discovery of antidepressants from 1950s until today. Curr Pharm Des. 2009;15(14):1563–86.

Lund C, Myer L, Stein DJ, Williams DR, Flisher AJ. Mental illness and lost income among adult South Africans. Soc Psychiatry Psychiatr Epidemiol. 2013;48(5):845–51.

Luscher B, Shen Q, Sahir N. The GABAergic deficit hypothesis of major depressive disorder. Mol Psychiatry. 2011;16(4):383–406.

Machado, M., Lopera, M.M., Diaz-Rojas, J., Jaramillo, L.E., Einarson, T.R., The Universidad Nacional de Colombia Pharmacoeconomics Group. Pharmacoeconomics of antidepressants in moderate-to-severe depressive disorder in Colombia. Rev Panam Salud Publica. 2008; 24(4):233–9.

Maes M, Song C, Lin A, De Jongh R, Van Gastel A, Kenis G, Bosmans E, De Meester I, Benoy I, Neels H, Demedts P, Janca A, Scharpe S, Smith RS. The effects of psychological stress on humans: increased production of pro-inflammatory cytokines and a Th1-like response in stress induced anxiety. Cytokine. 1998a;10:313–8.

Maes M, Verkerk R, Vandoolaeghe E, Lin A, Scharpe S. Serum levels of excitatory amino acids, serine, glycine, histidine, threonine, taurine, alanine and arginine in treatment-resistant depression: modulation by treatment with antidepressants and prediction of clinical responsivity. Acta Psychiatr Scand. 1998b;97(4):302–8.

Maguire J. Stress-induced plasticity of GABAergic inhibition. Front Cell Neurosci. 2014;8:157.

Malberg JE, Duman RS. Cell proliferation in adult hippocampus is decreased by inescapable stress: reversal by fluoxetine treatment. Neuropsychopharmacology. 2003;28(9):1562–71.

Manicavasagar V. A review of depression diagnosis and management. Melbourne, VIC: The Australian Psychological Society; 2012.

Mauri MC, Ferrara A, Boscati L, Bravin S, Zamberlan F, Alecci M, Invernizzi G. Plasma and platelet amino acid concentrations in patients affected by major depression and under fluvoxamine treatment. Neuropsychobiology. 1998;37(3):124–9.

McCarthy DJ, Alexander R, Smith MA, Pathak S, Kanes S, Lee CM, Sanacora G. Glutamate-based depression GBD. Med Hypotheses. 2012;78(5):675–81.

McLeod H, Rothberg A, Pels L, Eekhout S, Mubangizi DB, Fish T. The costing of the proposed chronic disease list benefits in South African medical schemes in 2001. Cape Town: Centre for Actuarial Research, University of Cape Town; 2002.

McQuade R, Young AH. Future therapeutic targets in mood disorders: the glucocorticoid receptor. Br J Psychiatry. 2000;177:390–5.

McQueen GM, Ramakrishnan K, Croll SD, Siuciak JA, Yu G, Young LT, Fahnestock M. Performance of heterozygous brain-derived neurotrophic factor knockout mice on behavioral analogues of anxiety, nociception, and depression. Behav Neurosci. 2001;115:1145–53.

Mehl RC, O'Brien LM, Jones JH, Dreisbach JK, Mervis CB, Gozal D. Correlates of sleep and pediatric bipolar disorder. Sleep. 2006;29(2):193–7.

Miguel-Hidalgo JJ, Baucom C, Dilley G, Overholser JC, Meltzer HY, Stockmeier CA, Rajkowska G. Glial fibrillary acidic protein immunoreactivity in the prefrontal cortex distinguishes younger from older adults in major depressive disorder. Biol Psychiatry. 2000;48(8):861–73.

Miguel-Hidalgo JJ, Waltzer R, Whittom AA, Austin MC, Rajkowska G, Stockmeier CA. Glial and glutamatergic markers in depression, alcoholism, and their comorbidity. J Affect Disord. 2010;127(1-3):230–40.

Miró X, Pérez-Torres S, Artigas F, Puigdomènech P, Palacios JM, Mengod G. Regulation of cAMP phosphodiesterase mRNAs expression in rat brain by acute and chronic fluoxetine treatment. An in situ hybridization study. Neuropharmacology. 2002;43:1148–57.

Møldrup C, Oestergaard S. Improving outcomes for patients with depression by enhancing antidepressant therapy with non-pharmacological interventions: a systematic review of reviews. Public Health. 2011;125:357–67.

Monteggia LM, Luikart B, Barrot M, Theobold D, Malkovska I, Nef S, Parada LF, Nestler EJ. Brain-derived neurotrophic factor conditional knockouts show gender differences in depression-related behaviors. Biol Psychiatry. 2007;61(2):187–97.

Mulinari S. Monoamine theories of depression: historical impact on biomedical research. J Hist Neurosci. 2012;21:366–92.

Murphy DL, Garrick NA, Tamarkin L, Taylor PL, Markey SP. Effects of antidepressants and other psychotropic drugs on melatonin release and pineal gland function. J Neural Transm Suppl. 1986;21:291–309.

Nemeroff CB, Bissette G, Akil H, Fink M. Neuropeptide concentrations in the cerebrospinal fluid of depressed patients treated with electroconvulsive therapy: corticotrophin-releasing factor, beta-endorphin and somatostatin. Br J Psychiatry. 1991;158:59–63.

Nestler EJ, Peña CJ, Kundakovic M, Mitchell A, Akbarian S. Epigenetic basis of mental illness. Neuroscientist. 2015. doi:10.1177/1073858415608147.

Newport DJ, Heim C, Owens MJ, Ritchie JC, Ramsey CH, Bonsall R, Miller AH, Nemeroff CB. Cerebrospinal fluid corticotropin-releasing factor (CRF) and vasopressin concentrations predict pituitary response in the CRF stimulation test: a multiple regression analysis. Neuropsychopharmacology. 2003;28(3):569–76.

Nibuya M, Morinobu S, Duman RS. Regulation of BDNF and trkB mRNA in rat brain by chronic electroconvulsive seizure and antidepressant drug treatments. J Neurosci. 1995;15(11): 7539–47.

Niehrs C, Schäfer A. Active DNA demethylation by Gadd45 and DNA repair. Trends Cell Biol. 2012;22(4):220–7.

Nowak G, Trullas R, Layer RT, Skolnick P, Paul IA. Adaptive changes in the *N*-methyl-D-aspartate receptor complex after chronic treatment with imipramine and 1-aminocyclopropanecarboxylic acid. J Pharmacol Exp Ther. 1993;265(3):1380–6.

Olesen J, Gustavsson A, Svensson M, Wittchen H-U, Jönsson B, on behalf of the CDBE2010 study group and the European Brain Council. The economic cost of brain disorders in Europe. Eur J Neurol. 2012;19:155–62.

Ongür D, Drevets WC, Price JL. Glial reduction in the subgenual prefrontal cortex in mood disorders. Proc Natl Acad Sci U S A. 1998;95(22):13290–5.

Ozbolt LB, Nemeroff CB. Chapter 2: HPA axis modulation in the treatment of mood disorders. In: Schoepf D, editor. Psychiatric disorders – new frontiers in affective disorders. Morn Hill: InTech; 2013.

Pae CU, Marks DM, Han C, Patkar AA. Does minocycline have antidepressant effect? Biomed Pharmacother. 2008;62(5):308–11.

Papp M, Moryl E. Similar effects of chronic treatment with imipramine and the NMDA antagonists CGP 37849 and MK-801 in a chronic mild stress model of depression in rats. Eur J Pharmacol. 1993;3(3):348–9.

Papp M, Moryl E. Antidepressant activity of non-competitive and competitive NMDA receptor antagonists in a chronic mild stress model of depression. Eur J Pharmacol. 1994; 263(1-2):1–7.

Park H, Poo M-m. Neurotrophin regulation of neural circuit development and function. Nat Rev Neurosci. 2013;14:7–23.

Parker G, McCraw S, Hadzi-Pavlovic D, Fletcher K. Costs of the principal mood disorders: a study of comparative direct and indirect costs incurred by those with bipolar I, bipolar II and unipolar disorders. J Affect Disord. 2013;149(1-3):46–55.

Paul IA, Layer RT, Skolnick P, Nowak G. Adaptation of the *N*-methyl-D-aspartate receptor complex in rat front cortex following chronic treatment with electroconvulsive shock or imipramine. J Pharmacol Exp Ther. 1993;247(3):305–12.

Paul IA, Nowak G, Layer RT, Popik P, Skolnick P. Adaptation of the *N*-methyl-D-aspartate receptor complex following chronic antidepressant treatments. J Pharmacol Exp Ther. 1994;269(1): 95–102.

Petrik D, Lagace D, Eisch AJ. The neurogenesis hypothesis of affective and anxiety disorders: are we mistaking the scaffolding for the building? Neuropharmacology. 2012;62(1):21–34.

Petty F. Plasma concentrations of gamma-aminobutyric acid (GABA) and mood disorders: a blood test for manic depressive disease? Clin Chem. 1994;40(2):296–302.

Petty F, Schiesser MA. Plasma GABA in affective illness. A preliminary investigation. J Affect Disord. 1981;3(4):339–43.

Pharm K, Nacher J, Hof PR, McEwen BS. Repeated restraint stress suppresses neurogenesis and induces biphasic PSA-NCAM expression in the adult rat dentate gyrus. Eur J Neurosci. 2003;17:879–86.

Popper CW. Mood disorders in youth: exercise, light therapy, and pharmacologic complementary and integrative approaches. Child Adolesc Psychiatr Clin N Am. 2013;22(3):403–41.

Rajkowska G, Miguel-Hidalgo JJ. Gliogenesis and glial pathology in depression. CNS Neurol Disord Drug Targets. 2007;6(3):219–33.

Rajkowska G, Miguel-Hidalgo JJ, Wei J, Dilley G, Pittman SD, Meltzer HY, Overholser JC, Roth BL, Stockmeier CA. Morphometric evidence for neuronal and glial prefrontal cell pathology in major depression. Biol Psychiatry. 1999;45(9):1085–98.

Rajkowska G, Halaris A, Selemon LD. Reductions in neuronal and glial density characterize the dorsolateral prefrontal cortex in bipolar disorder. Biol Psychiatry. 2001;49(9):741–52.

Rajkowska G, O'Dwyer G, Teleki Z, Stockmeier CA, Miguel-Hidalgo JJ. GABAergic neurons immunoreactive for calcium binding proteins are reduced in the prefrontal cortex in major depression. Neuropsychopharmacology. 2007;32(2):471–82.

Rastad C, Ulfberg J, Lindberg P. Improvement in fatigue, sleepiness, and health-related quality of life with bright light treatment in persons with seasonal affective disorder and subsyndromal SAD. Depress Res Treat. 2011;2011:543906.

Rosenberg DR, Macmaster FP, Mirza Y, Smith JM, Easter PC, Banerjee SP, Bhandari R, Boyd C, Lynch M, Rose M, Ivey J, Villafuerte RA, Moore GJ, Renshaw P. Reduced anterior cingulate glutamate in pediatric major depression: a magnetic resonance spectroscopy study. Biol Psychiatry. 2005;58(9):700–4.

Roth TL, Sweatt JD. Epigenetic mechanisms and environmental shaping of the brain during sensitive periods of development. J Child Psychol Psychiatry. 2011;52(4):398–408.

Roy A, Dejong J, Ferraro T. CSF GABA in depressed patients and normal controls. Psychol Med. 1991;21(3):613–8.

Saarelainen T, Hendolin P, Lucas G, Koponen E, Sairanen M, MacDonald E, Agerman K, Haapasalo A, Nawa H, Aloyz R, Ernfors P, Castrén E. Activation of the TrkB neurotrophin receptor is induced by antidepressant drugs and is required for antidepressant-induced behavioral effects. J Neurosci. 2003;23(1):349–57.

Salgado-Delgado R, Osorio AT, Saderi N, Escobar C. Disruption of circadian rhythms: a crucial factor in the etiology of depression. Depress Res Treat. 2011;2011:839743.

Sanacora G, Mason GF, Rothman DL, Berman RM, Charney DS, Krystal JH. ECT effects on cortical GABA levels as determined by $^{1}$H-MRS. Scientific abstracts from the 37th annual meeting of the American College of Neuropsychopharmacology. San Juan: American College of Neuropsychopharmacology; 1998. p. 200.

Sanacora G, Mason GF, Rothman DL, Behar KL, Hyder F, Petroff OA, Berman RM, Charney DS, Krystal JH. Reduced cortical gamma-aminobutyric acid levels in depressed patients determined by proton magnetic resonance spectroscopy. Arch Gen Psychiatry. 1999; 56(11):1043–7.

Sanacora G, Mason GF, Rothman DL, Berman RM, Zimolo Z, Krystal JH. Cortical GABA concentrations are increased in depressed patients following treatment with selective serotonin reuptake inhibitors. Scientific abstracts from the 39th annual meeting of the American College of Neuropsychopharmacology. San Juan: American College of Neuropsychopharmacology; 2000. p. 200.

Sanacora G, Zarate CA, Krystal JH, Manji HK. Targeting the glutamatergic system to develop novel, improved therapeutics for mood disorders. Nat Rev Drug Discov. 2008;7(5):426–37.

Sanai N, Nguyen T, Ihrie RA, Mirzadeh Z, Tsai HH, Wong M, Gupta N, Berger MS, Huang E, Garcia-Verdugo JM, Rowitch DH, Alvarez-Buylla A. Corridors of migrating neurons in the human brain and their decline during infancy. Nature. 2011;478(7369):382–6.

Santarelli L, Saxe M, Gross C, Surget A, Battaglia F, Dulawa S, Weisstaub N, Lee J, Duman R, Arancio O, Belzung C, Hen R. Requirement of hippocampal neurogenesis for the behavioral effects of antidepressants. Science. 2003;301:805–9.

Schiepers OJG, Wichers MC, Maes M. Cytokines and major depression. Prog Neuropsychopharmacol Biol Psychiatry. 2005;29:201–17.

Schildkraut JJ. The catecholamine hypothesis of affective disorders: a review of supporting evidence. Am J Psychiatry. 1965;122(5):509–22.

Schnydrig S, Korner L, Landweer S, Ernst B, Walker G, Otten U, Kunz D. Peripheral lipopolysaccharide administration transiently affects expression of brain-derived neurotrophic factor, corticotropin and proopiomelanocortin in mouse brain. Neurosci Lett. 2007;429(1):69–73.

Schroeder M, Krebs MO, Bleich S, Frieling H. Epigenetics and depression: current challenges and new therapeutic options. Curr Opin Psychiatry. 2010;23(6):588–92.

Schroeter ML, Abdul-Khaliq H, Diefenbacher A, Blasig IE. S100B is increased in mood disorders and may be reduced by antidepressive treatment. NeuroReport. 2002;13(13):1675–8.

Seckl JR, Fink G. Antidepressants increase glucocorticoid and mineralocorticoid receptor mRNA expression in rat hippocampus *in vivo*. Neuroendocrinology. 1992;55(6):621–6.

Segi-Nishida E, Warner-Schmidt JL, Duman RS. Electroconvulsive seizure and VEGF increase the proliferation of neural stem-like cells in rat hippocampus. Proc Natl Acad Sci U S A. 2008;105(32):11352–7.

Shen Q, Lal R, Luellen BA, Earnheart JC, Andrews AM, Luscher B. Gamma-and antidepressant drug sensitivity reminiscent of melancholic forms of depression. Biol Psychiatry. 2010; 68(6):512–20.

Sokolov BP. Oligodendroglial abnormalities in schizophrenia, mood disorders and substance abuse. Comorbidity, shared traits, or molecular phenocopies? Int J Neuropsychopharmacol. 2007;10(4):547–55.

Stahl SM, Lee-Zimmerman C, Cartwright S, Morrissette DA. Serotonergic drugs for depression and beyond. Curr Drug Targets. 2013;14(5):578–85.

Stelzhammer V, Guest PC, Rothermundt M, Sondermann C, Michael N, Schwarz E, Rahmoune H, Bahn S. Electroconvulsive therapy exerts mainly acute molecular changes in serum of major depressive disorder patients. Eur Neuropsychopharmacol. 2013;23(10):1199–1207.

Stockmeier CA, Mahajan GJ, Konick LC, Overholser JC, Jurjus GJ, Meltzer HY, Uylings HB, Friedman L, Rajkowska G. Cellular changes in the postmortem hippocampus in major depression. Biol Psychiatry. 2004;56(9):640–50.

Stranahan AM, Khalil D, Gould E. Social isolation delays the positive effects of running on adult neurogenesis. Nat Neurosci. 2006;9(4):526–33.

Sun JD, Liu Y, Yuan YH, Li J, Chen NH. Gap junction dysfunction in the prefrontal cortex induces depressive-like behaviors in rats. Neuropsychopharmacology. 2012;37(5):1305–20.

Sun H, Kennedy PJ, Nestler EJ. Epigenetics of the depressed brain: role of histone acetylation and methylation. Neuropsychopharmacology. 2013;38(1):124–37.

Tkachev D, Mimmack ML, Ryan MM, Wayland M, Freeman T, Jones PB, Starkey M, Webster MJ, Yolken RH, Bahn S. Oligodendrocyte dysfunction in schizophrenia and bipolar disorder. Lancet. 2003;362(9386):798–805.

Torres GE, Gainetdinov RR, Caron MG. Plasma membrane monoamine transporters: structure, regulation and function. Nat Rev Neurosci. 2003;4(1):13–25.
Traynelis SF, Wollmuth LP, McBain CJ, Menniti FS, Vance KM, Ogden KK, Hansen KB, Yuan H, Myers SJ, Dingledine R. Glutamate receptor ion channels: structure, regulation, and function. Pharmacol Rev. 2010;62(3):405–96.
Trullas R, Skolnick P. Functional antagonists at the NMDA receptor complex exhibit antidepressant actions. Eur J Pharmacol. 1990;185(1):1–10.
Tsankova NM, Berton O, Renthal W, Kumar A, Neve RL, Nestler EJ. Sustained hippocampal chromatin regulation in a mouse model of depression and antidepressant action. Nat Neurosci. 2006;9(4):519–25.
Uher R, Payne JL, Pavlova B, Perlis RH. Major depressive disorder in DSM-5: implications for clinical practice and research of changes from DSM-IV. Depress Anxiety. 2014;31:459–71.
Uranova NA, Vostrikov VM, Orlovskaya DD, Rachmanova VI. Oligodendroglial density in the prefrontal cortex in schizophrenia and mood disorders: a study from the Stanley Neuropathology Consortium. Schizophr Res. 2004;67(2-3):269–75.
Varghese BAFP, Brown ES. The hypothalamic-pituitary-adrenal axis in major depressive disorder: a brief primer for primary care physicians. Prim Care Companion J Clin Psychiatry. 2001;3(4):151–5.
Vialou V, Feng J, Robison AJ, Nestler EJ. Epigenetic mechanisms of depression and antidepressant action. Annu Rev Pharmacol Toxicol. 2013;53:59–87.
Vogel GW, Buffenstein A, Minter K, Hennessey A. Drug effects on REM sleep and on endogenous depression. Neurosci Biobehav Rev. 1990;14(1):49–63.
Wagner-Smith K, Markou A. Depression: a repair response to stress-induced neuronal microdamage that can grade into chronic neuroinflammatory condition? Neurosci Biobehav Rev. 2011;35:742–64.
Warner-Schmidt JL, Duman RS. VEGF is an essential mediator of the neurogenic and behavioral actions of antidepressants. Proc Natl Acad Sci U S A. 2007;104(11):4647–52.
Watanabe K, Hashimoto E, Ukai W, Ishii T, Yoshinaga T, Ono T, Tateno M, Watanabe I, Shirasaka T, Saito S, Saito T. Effect of antidepressants on brain-derived neurotrophic factor (BDNF) release from platelets in the rats. Prog Neuropsychopharmacol Biol Psychiatry. 2010; 34(8):1450–4.
Wichers M, Maes M. The psychoneuroimmuno-pathophysiology of cytokine-induced depression in humans. Int J Neuropsychopharmacol. 2002;5:375–88.
Wichniak A, Riemann D, Kiemen A, Voderholzer U, Jernajczyk W. Comparison between eye movement latency and REM sleep parameters in major depression. Eur Arch Psychiatry Clin Neurosci. 2000;250(1):48–52.
Winokur A, Gary KA, Rodner S, Rae-Red C, Fernando AT, Szuba MP. Depression, sleep physiology, and antidepressant drugs. Depress Anxiety. 2001;14(1):19–28.
Wirz-Justice A. Biological rhythm disturbances in mood disorders. Int Clin Psychopharmacol. 2006;21 Suppl 1:S11–5.
Wirz-Justice A. Biological rhythms and depression: treatment opportunities. WPA Bull Depress. 2008;12:5–8.
World Health Organization. The global burden of disease. 2004 Update. Geneva: WHO; 2008.
World Health Organization. Depression. Fact sheet N°369; 2015.
Yamada K, Watanabe A, Iwayama-Shigeno Y, Yoshikawa T. Evidence of association between gamma-aminobutyric acid type A receptor genes located on 5q34 and female patients with mood disorders. Neurosci Lett. 2003;349(1):9–12.
Young EA, Lopez JF, Murphy-Weinberg V, Watson SJ, Akil H. Mineralocorticoid receptor function in major depression. Arch Gen Psychiatry. 2003;60(1):24–8.
Zarate Jr CA, Singh JB, Carlson PJ, Brutsche NE, Ameli R, Luckenbaugh DA, Charney DS, Manji HK. A randomized trial of an N-methyl-D-aspartate antagonist in treatment-resistant major depression. Arch Gen Psychiatry. 2006;63(8):856–64.

# Chapter 2
# Traditional and Novel Possible Targets for Antidepressant Drugs

**Eleni Palazidou**

**Abstract** Antidepressant drugs have been available since the late 1950s. A large number of such drugs have been developed over the years. Unraveling their mechanism of action paved the way to the development of the earlier hypotheses of depression such as the monoamine theory and beta-receptor downregulation hypothesis. This led to yet more antidepressant drugs being produced, all of which worked via the noradrenergic and serotonergic systems. They increased monoamine neurotransmitter system activity usually by blocking the presynaptic reuptake mechanism, binding the monoamine oxidase intraneuronally or targeting specific receptors on the postsynaptic membrane.

Although generally effective, current antidepressants are successful in less than two thirds of depressed patients. More recent developments in the understanding of the pathophysiology of depression have stimulated a rethinking in the approach to drug development. Novel drugs need to take into account the current knowledge on brain dysfunction in relation to depression and target available and future biomarkers of this condition. Various avenues are being explored in keeping with the new approach of "precision medicine" with the view to develop drugs based on "molecular diagnoses."

Depression is a multifactorial condition and not all abnormalities and biomarkers apply to all patients. The skill of future approaches to treatment would be to identify the specific abnormalities relevant to the individual patient and target these biomarkers with specific drugs.

**Keywords** Antidepressant drugs • Depression • Monoamine neurotransmitter systems • Brain neurotransmitter systems • Biomarkers

E. Palazidou, MD, PhD, MRCP, FRCPSYCH (✉)
Barts & London School of Medicine, Queen Mary University, London, UK
e-mail: eleni.palazidou@doctors.org.uk



C. Grosso (ed.), *Herbal Medicine in Depression*,
DOI 10.1007/978-3-319-14021-6_2

## 2.1 Introduction

Depressive disorder is a mental condition recognized as such from ancient times. Its clinical picture was described by Hippocrates, the father of medicine, who coined the term "melancholia," which is still in existence. This common and highly morbid and lethal condition was largely untreatable until the middle of the twentieth century.

The discovery of the neurotransmitter noradrenaline (NA) by Axelrod (1971) laid the foundations of the psychopharmacology of depression in the late 1950s and the 1960s. The serendipitous discovery of the first effective antidepressant drugs together with the growing understanding of the function of the monoamine neurotransmitter systems and skillful clinical observations paved the way to the first biochemical theories of depression and the development of a range of pharmacological treatments. It is now accepted that depression has a multifactorial etiopathogenesis with genetic diathesis and stress playing a major role. There are abnormalities in the activity of various neurotransmitter systems in particular the monoamines and possibly others too, brain neurotrophins and the immune system. A complex interaction between these factors is associated with functional and structural deficits in the brain (Palazidou 2012).

Depression is a relapsing/remitting progressive condition. Despite the leaps made in our understanding of its pathophysiology and the availability of a wealth of pharmacological and psychological treatments, three quarters of patients experience more than one episode of depression. The risk of recurrence is higher if the first episode occurred at a younger age and if there is a family history of depression (Hollon et al. 2006). Furthermore the risk of recurrence increases with each episode, and as the number of episodes increases, the influence of stress on recurrence wanes (Kendler et al. 2000). The longer the duration of illness, the poorer the response to treatment and the likelihood of remission will be (Okuda et al. 2010). Subsyndromal states encourage relapse and possibly chronicity (Pintor et al. 2003). It is possible that at some point in the course of the disorder, functional and perhaps also structural abnormalities in the brain become irreversible, and this may explain the requirement for ongoing long-term antidepressant drug treatment in many patients with recurrent episodes and the resistance to treatment in chronic depression. Early effective treatment with complete remission may prevent possible irreversible brain abnormalities in the long term and lead to better long-term outcomes.

These are important facts to bear in mind when treating depressive disorder and also very importantly when assessing the efficacy of antidepressant drugs. Both the short- and long-term outcomes need to be considered. So far the drug treatments available simply control symptoms and this aids the natural process of healing rather than cure the underlying pathology.

## 2.2 Brief Historical Overview of Antidepressant Drug Development

The first effective antidepressant drugs were *iproniazid* (a monoamine oxidase inhibitor), developed originally as an antituberculous agent and similar to isoniazid (Bloch et al. 1954) and *imipramine*, a tricyclic with a chemical structure very similar to chlorpromazine, which was originally developed as an antipsychotic in 1958 (Kuhn 1970).

Over the following 20–30 years, the emphasis had been on understanding brain monoamine neurotransmission and unraveling the mechanisms of action of the antidepressant drugs as a way of working out the pathophysiology of depression. Theories about the neurochemical causes of depression at different time points were based and restricted to the hitherto available knowledge of the neurochemistry and pharmacology of the brain.

### *2.2.1 The Monoamine Hypothesis of Depression: Catecholamine (Schildkraut 1965) and Indoleamine Hypothesis (Coppen 1967)*

The well-replicated findings that the concentrations of noradrenaline (NA) and serotonin (5-HT) in various body fluids (blood, urine, cerebrospinal fluid) are decreased in patients with depression compared to healthy subjects pointed to monoamine deficiency as the cause of depressive illness. This evidence together with the clinical observation that treatment with drugs such as tetrabenazine, which after long-term treatment deplete neurons from monoamines, can cause depression and that the recently discovered antidepressants, imipramine and iproniazid, increased NA and 5-HT concentrations formed the foundations of the Monoamine Hypothesis of Depression in the 1960s.

### *2.2.2 Beta-Receptor Downregulation Theory of the Mechanism of Action of Antidepressant Drugs*

The identification of neuronal monoamine receptors in the 1970s and the finding that changes occurred in their density in relation to ongoing treatment with antidepressant drugs led to the hypothesis that downregulation of the postsynaptic beta-1 adrenergic receptors (β1 adrenoreceptors) and reductions in their second messenger, adenylate cyclase, activity was the key mechanism of action of antidepressant drugs (Vetulani and Sulser 1975). Beta-receptor downregulation is a very consistent and well-replicated finding in animal research, which occurred after about 10–15 days of ongoing antidepressant drug treatment and coincided with the timing of observed clinical improvement in depressive symptoms. However, other receptors outside the noradrenergic system, such as 5-$HT_2$ serotonergic receptors, were also downregulated, although this finding was not as universal as that of beta-receptor downregulation, and other pre- and postsynaptic, as well as somatodendritic, receptors showed changes in density and activity after repeated exposure to antidepressant drugs.

The "beta-receptor downregulation hypothesis" of the mechanism of action of antidepressant drugs was not supported by research in humans, which showed that both acute and chronic (2 or more weeks) treatments with antidepressant drugs result in increased adrenergic neurotransmission in normal subjects and in depressed patients (Palazidou et al. 1992a, b). Further developments in molecular biology confirmed that, although adaptive changes occur at various receptor sites (pre- and postsynaptically as well as at somatodendritic sites), the net output of the adrenergic and serotonergic system in response to repeated exposure to antidepressant drugs is one of increased neurotransmission. A cascade of events within the activated neuronal cells result in the increased production of brain-derived neurotrophin (BDNF), a protein widespread in the brain, which has the most important role in maintaining the resilience of neurons, restoring and enhancing neuronal plasticity. There is a wealth of evidence that links the concentrations of BDNF with neo-neurogenesis in the hippocampus restoring its volume to the pre-illness state, as observed during treatment with a variety of antidepressant drugs (Shimizu et al. 2003; Piccinni et al. 2009).

The monoamine depletion studies confirmed that antidepressant drugs exert their effect via an increase in NA or 5-HT concentrations in the brain. It was demonstrated through a series of experiments that decreasing the brain availability of 5-HT (through tryptophan depletion) in patients successfully treated with a serotonergic agent caused a relapse of depression (Delgado et al. 1990). Similar findings were shown when reducing the concentrations of NA (using α-methyl-para-tyrosine) in patients successfully treated with an adrenergic antidepressant (Delgado et al. 1993).

In the 1970s–1980s, several tricyclic (TCA) and monoamine oxidase inhibitor (MAOI) drugs became available, as well as some other agents with a different pharmacological profile and mode of action, such as trazodone and mianserin. Unfortunately all the drugs developed up to this point, although effective, were associated with significant side effects that made them less acceptable to patients. This stimulated an interest in the production of drugs with a narrower profile of pharmacological action. It was believed that higher selectivity not only would improve tolerability but would also increase efficacy although there was no good reason to consider the latter to be true. There followed a rigorous development of drugs such as the selective serotonin reuptake inhibitors (SSRIs) and to a much lesser extent the selective NA reuptake inhibitors (NARIs) as well as receptor-selective drugs such as 5-$HT_{1A}$ agonists.

The selective NA reuptake inhibitors such as maprotiline and reboxetine failed to effectively compete with the SSRIs. The SSRIs became the mainstay of the treatment of depression and are currently recommended as the first-choice antidepressant group by various treatment guidelines for depression (e.g., the NICE Guidelines for Depression in the UK). This is based on their better tolerability, although the SSRIs failed to demonstrate any superiority in efficacy. On the contrary, in more severe depression, dual-action antidepressants such as the TCAs are more effective (Anderson and Tomenson 1994) than the SSRIs.

Disappointment with the lack of demonstrable superiority in efficacy by the SSRIs led to the return to dual-action drugs like the tricyclic and the MAOIs but with selectivity for noradrenergic and serotonergic pathways without the additional pharmacological effects of the older drugs, which were mostly responsible for their poor tolerability and potential for toxicity. The new groups of this second-generation dual-action antidepressants were the selective 5-HT and NA reuptake inhibitors (SNRIs), such as venlafaxine, duloxetine, and milnacipran, the reversible monoamine oxidase inhibitors such as moclobemide, and the NA and selective serotonergic agents (NASSa) such as mirtazapine.

A variety of other drugs such as bupropion, a selective NA and dopamine (DA) reuptake inhibitor; tianeptine, a selective serotonin reuptake enhancer; and agomelatine, a melatonin receptor agonist and 5-$HT_{2C}$ receptor antagonist were also introduced in the last decade or so. Drugs such as the 5-HT1A agonists (buspirone, gepirone) failed to establish themselves as antidepressant drugs; their clinical use is limited to the treatment of anxiety.

## 2.3 Current Antidepressant Drugs: Mechanisms of Action

Unfortunately the great achievements of neuroscience, which improved our understanding of the pathogenesis of depressive illness, have not as yet been matched with the production of more effective and innovative antidepressant drugs. Almost universally the currently available antidepressants work through the activation of one or more monoamine neurotransmitter systems. The targets of their action are centered on the neuronal synapse and located mostly at the neuronal terminal site. The classical targets are the monoamine reuptake/transporter sites, the monoamine oxidase enzyme, and the alpha-2 autoreceptors and 5-HT receptors (see Box 2.1). Irrespective of their specific site of action, the end result of the treatment with the current antidepressant drugs is that of an increase in the concentrations of the targeted neurotransmitters (NA, 5-HT, DA) in the synaptic cleft with increased binding at the postsynaptic receptors and downstream neuronal activation. This activation is believed to increase the concentrations of brain-derived neurotrophin (BDNF). BDNF is responsible for various physiological neuronal functions including survival, neurite outgrowth, and synaptic plasticity in the brain, and its role in depression is believed to be the restoration or improvement of neuronal function and activity.

**Box 2.1. Current Antidepressant Sites of Action**

**Monoamine Oxidase**

- Nonselective, nonreversible inhibitors (MAOIs)
- Selective, reversible inhibitors (RIMAs)

**Monoamine reuptake/transporter**

- Nonselective NA and 5-HT reuptake inhibitors (TCAs)
- Selective NA and 5-HT reuptake inhibitors (SNRIs)
- Selective 5-HT reuptake inhibitors (SSRIs)
- Selective 5-HT reuptake enhancer (tianeptine)
- Selective NA reuptake inhibitors (NARIs)
- Selective NA and DA reuptake inhibitors (bupropion)

**Receptor sites**

- α2 and 5-$HT_2$ receptor antagonists—mianserin, mirtazapine
- 5-$HT_2$ and other receptors antagonists—trazodone, nefazodone
- Melatonin receptor ($MT_1$ and $MT_2$) agonist and 5-$HT_{2C}$ antagonist
- agomelatin

**Selective 5-HT dual action (reuptake and receptor binding) or multimodal action**

Vilazodone
Vortioxetine

## *2.3.1 Monoamine Oxidase Inhibitors (MAOIs)*

The enzyme monoamine oxidase (MAO) is located on the outer membrane of the mitochondria, and it is widely distributed in human tissues, including the brain. It has a very important function, which is to regulate the monoamine concentrations of the nervous system by breaking down (via deamination) the endogenous monoamines (NA, 5-HT, and DA) released in the neuronal cytoplasm, as well as trace amines (phenylethylamine) and dietary amines (tyramine) in order to prevent excessive build up and noxious interactions. Monoamine oxidase exists in two isoenzyme forms, MAO-A, found in DA and NA, and serotonin neurons and MAO-B, found to a greater extent in serotonin neurons and also preferentially in striatal DA neurons. Therefore the main substrates for MAO-A are adrenaline, NA, and 5-HT and for MAO-B, phenylethylamine, phenylethanolamine, tyramine, and benzylamine, while DA and tyramine are metabolized by both isoenzymes (Krishnan 2004). In the brain, the two MAO isoenzymes are found within the biogenic neuronal terminals where they regulate the concentrations of the relevant neurotransmitters by metabolizing them upon their reuptake into the cytoplasm of the presynaptic neuron.

*Iproniazid*, the first effective antidepressant drug to be produced, was originally developed as a treatment for tuberculosis, alongside isoniazid, in the early 1950s. It was soon observed that unlike isoniazid, it increased patients' sense of well-being and in some cases it actually caused psychosis. In the following few years, it was demonstrated that iproniazid, but, interestingly, not isoniazid, was a MAO inhibitor and that it was effective in the treatment of depression (Loomer et al. 1957; Crane 1957). Iproniazid was later abandoned as an antidepressant because of hepatotoxicity, and other MAOIs were produced such as the hydrazines, phenelzine (Nardil), isocarboxazid (Marplan), nialamide (Niamid), and the non-hydrazine tranylcypromine (Parnate), which is chemically similar to amphetamine. Although hepatotoxicity is rare with these agents in 3–5 % of patients, they increase liver enzymes.

By blocking the actions of the monoamine oxidase, the MAOIs prevent the breakdown of the monoamine neurotransmitters, an action that results in an increase in their concentrations at the synaptic cleft and at postsynaptic receptors.

Unfortunately, although effective, the first-generation MAOIs have two major drawbacks, which are due to (a) their lack of selectivity (binding both MAO-A and MAO-B) and (b) their irreversible binding to these isoenzymes.

#### 2.3.1.1 Nonselective MAO Binding

Their nonselective binding to both isoenzymes, MAO-A and MAO-B, is responsible for the MAOIs' serious pharmacodynamic interactions with some other drugs such as the sympathomimetics, tricyclics, and bupropion and also with tyramine containing foodstuffs (Table 2.1). These interactions manifest with hypertensive reactions.

Tyramine, normally broken down by the MAO enzymes, has a pressor effect and has indirect as well as direct sympathomimetic actions. Increased concentrations of tyramine displace NA from the storage synaptic vesicles into the adrenergic neuronal cytoplasm increasing its extracellular concentrations and resulting in increased acti-

**Table 2.1** MAOIs—toxicity/interactions

| | |
|---|---|
| Adverse effects/interactions | Increased weight<br>Sexual dysfunction<br>Dry mouth, headache |
| Toxicity in overdose | Lethal (hypertensive crisis, stroke, myocardial infarction) |
| +Sympathomimetics<br>+Carbamazepine<br>+Opiates<br>+Tyramine | Hypertensive crisis<br>Hypertensive crisis (cheese reaction) |
| +β-Blockers | More marked hypotension and bradycardia |
| +Oral hypoglycemics | Increased hypoglycemia |
| +SSRIs | Serotonin syndrome |
| +Clomipramine | Serotonin syndrome |
| +Bupropion | Hypertensive crisis<br>Seizures |

**Box 2.2**

**"Cheese Reaction"**

Occurs usually 20–60 min after ingestion of food
Hypertension
Headache
Palpitations
Nausea/vomiting
Apprehension
Sweating
Restlessness/motor agitation
Mild hyperpyrexia
Mydriasis (dilated pupils)

*Very Severe*

Loss of consciousness
Hyperpyrexia
Cerebral hemorrhage
Death

vation of postsynaptic receptors and excessive adrenergic activation; it may also act as a neurotransmitter itself. MAOI administration through its binding of hepatic MAO prevents the metabolism of tyramine, causing a buildup in its concentrations in the blood circulation. The combination of increased tyramine and NA concentrations results in excessive adrenergic activation and hypertensive crises ("cheese reaction") with potentially serious cerebrovascular and cardiovascular complications (Box 2.2).

Tyramine is found in variable quantities in some foodstuffs, in particular various types of mature cheese, fava beans, other fermented foods, and wines. Such high-risk foods in large amounts contain 10–36 mg of tyramine (Da Prada et al. 1988). The tyramine effect of food is potentiated by the MAOIs some 10–20 times. Concentrations of 6 mg may cause a mild reaction, 10 mg can cause a moderate one, and 20–25 mg can cause a hypertensive reaction (Amrein et al. 1988).

For the duration of treatment with MAOIs, therefore, patients are required to be on a strict diet, which ensures they avoid foodstuffs known to be rich in tyramine. Although this diet has been relaxed in more recent years (Gardner et al. 1996), it remains a burden to patients, and as safer antidepressants were introduced, the use of the MAOIs declined and it is currently very limited.

Drugs with selectivity for one or the other MAO isoenzyme failed to show improvements in efficacy or safety. Clorgiline, although it selectively binds MAO-A only, allowing MAO-B to function as normal, unfortunately exhibited almost as marked "cheese reaction" as the nonselective MAOIs. Selegiline (Deprenyl), a selective MAO-B inhibitor, although successful in the treatment of Parkinson's disease, lacked powerful enough antidepressant properties.

**Box 2.3**

**Serotonin Syndrome**

Coincident with the addition of or increase in a known serotonergic agent to an established medication regimen, at least three of the following clinical features are present:
Mental status changes (confusion, hypomania)
Agitation
Myoclonus
Hyperreflexia
Diaphoresis
Shivering
Tremor
Diarrhea
Incoordination
Fever
Other etiologies (e.g., infectious, metabolic, substance abuse, or withdrawal) have been ruled out.
A neuroleptic had not been started or increased in dosage prior to the onset of the signs and symptoms listed above.

#### 2.3.1.2 Serotonergic Pharmacodynamics Interactions (Serotonin Syndrome)

Concomitant treatment of MAOIs with selective serotonin reuptake inhibitors (SSRIs) and some other serotonergic drugs should be avoided as the combination causes excessive serotonergic activity presenting with "serotonin syndrome" (see Box 2.3), which is a serious condition, and if left untreated, it can cause death (Boyer and Shannon 2005).

#### 2.3.1.3 Irreversible Binding to MAO

A further problem with the MAOIs is that their binding of MAO-A and MAO-B is irreversible. Although the drug is eliminated from the circulation fairly rapidly, on discontinuation of treatment, the MAO-bound fraction continues to exert its actions until such time that adequate amounts of new, non-bound MAO is synthesized and this requires on average 2 weeks. In practice this means that drugs and foodstuffs that are known to have adverse interactions with the MAOIs should be avoided during this period, and therefore, some antidepressants such as the SSRIs cannot be prescribed and the patient may have to stay off medication for 2 weeks.

### 2.3.2 *Reversible Selective Monoamine Oxidase Inhibitors (RIMAs)*

Despite their major drawbacks, the MAOIs were successful in the treatment of depression, and alongside the tricyclics, they were the mainstay of treatment for depression for over 30 years. Patients were willing to put up with the limitations that their use imposed, in particular the dietary requirements, because they derived therapeutic benefit. It was therefore understandable that efforts were made to produce new and safer MAOIs, which would not require dietary restrictions. Such drugs should bind reversibly the monoamine oxidase and furthermore they should be selective for MAO-A.

As a reversible and selective inhibitor of MAO-A with no affinity for MAO-B, moclobemide is a typical representative of the RIMAs. It is free of adverse interactions with tyramine containing foodstuffs. While 20 mg of tyramine can cause a hypertensive reaction during treatment with the nonselective MAOIs (Amrein et al. 1988), concentrations of 100 mg and above are needed to have the same effect during moclobemide treatment (Da Prada et al. 1988). On discontinuing drug treatment and as moclobemide is eliminated from the body, the MAO is promptly freed to function as normal. Other drugs with the potential for interactions with MAO inhibitors can be started promptly without the 2 weeks drug-free period required by the traditional MAOIs.

As with the older MAOIs, however, moclobemide can cause serotonin syndrome if given concomitantly with other serotonergic agents such as the SSRIs, and the combination should be avoided.

### 2.3.3 *Tricyclic Antidepressant Drugs (TCAs)*

Tricyclic antidepressants target the reuptake site on the presynaptic membrane of the neuronal terminal. The reuptake site, currently referred to also as the transporter, through an energy-driven mechanism, takes back into the terminal about 90 % of the neurotransmitter remained in the synaptic cleft after release from the neuronal terminal, binding to receptors on the postsynaptic membrane. By inhibiting the reuptake of both NA and 5-HT, the tricyclic antidepressants increase the availability of these neurotransmitters for binding to adrenergic and serotonergic postsynaptic receptors and enhancing neurotransmission.

The first TCA, imipramine, was developed in 1958 as an antipsychotic, having a similar tricyclic chemical structure to chlorpromazine. Once its efficacy was established in the treatment of depression (Kuhn 1970), a series of chemically similar, tricyclic drugs were developed including the tertiary amines such as amitriptyline, imipramine, clomipramine, doxepin, and trimipramine (which have a central three-ring structure with two methyl groups at the end of the side chain) and the demethylated secondary amines such as nortriptyline, desipramine, and protriptyline (which have a central three-ring structure with only one methyl group at the end of the side chain). All TCAs inhibit reuptake of 5-HT and NA but with differential affinity for the 5-HT and NA transporter which is determined by their chemical structure. The tertiary amine TCAs have higher affinity and are more powerful 5-HT reuptake blockers; clomipramine is the most powerful 5-HT transporter blocker. The secondary amines are more potent NA reuptake inhibitors. The tertiary amines, for example, amitriptyline, have also alpha-1 adrenergic receptor ($\alpha 1$ adrenoceptor) antagonist properties, and this is associated with orthostatic hypotension. In practice this may cause mild faintness on standing up due to rapid drop in blood pressure in the young, but in the elderly, this can cause actual fainting and a fall with more serious consequences such as hip or other bone fractures.

The tricyclics have additional pharmacological actions outside the monoamine neurotransmitter systems, which account for many of their adverse effects. Some of these are not only unacceptable to patients but may also cause serious health problems. They block histaminic receptors causing sedation and weight gain and they have troublesome anticholinergic properties. The latter are associated with blurring of vision, dry mouth, urinary hesitancy and possible urinary retention in men with prostatic hypertrophy, worsening of narrow angle glaucoma or memory in people with Alzheimer's dementia, and others (see Box 2.3).

The TCAs have direct cardiac effects, and through their actions in prolonging the QTc interval, they can cause ventricular arrhythmias and sudden death in individuals with preexisting cardiac abnormalities. Their narrow therapeutic index renders them a high risk of cardiotoxicity even in healthy individuals after a drug overdose or because of adverse pharmacokinetic interactions.

### *2.3.4 Selective Serotonin Reuptake Inhibitors (SSRIs)*

The selective serotonin reuptake inhibitors (SSRIs) were introduced in the 1980s as a breakthrough in antidepressant drug development. At the time higher pharmacological selectivity was believed to be the answer to better efficacy and better tolerability. The SSRIs have no uniform chemical structure but share a common mode of action. They selectively inhibit the reuptake of serotonin at serotonergic neuronal synaptic sites, increasing serotonergic neurotransmission, and they do not have significant direct adrenergic properties. Their selectivity made them a popular choice for the treatment of depression, and they rapidly replaced the TCAs and the MAOIs because of their comparatively favorable side effect profile. Their side effects are due to activation of serotonergic receptors and serotonergic activity in general as they lack pharmacological actions beyond this system.

The SSRIs have become the antidepressant group of choice in the treatment of the average episode of depression, recommended by guidelines in the treatment of depression such as the NICE guidelines in the UK. Although not devoid of side effects, their comparatively better tolerability, ease of use (unlike the TCAs, they do not require dose titration), and safety in overdose are superior to the TCAs and

MAOIs. However their efficacy is not as good as it was hoped it would be. Only 35 % of depressed patients achieve remission with SSRI antidepressant drug treatment compared to 45 % treated with the SNRI venlafaxine (Thase et al. 2001).

Although generally thought to be similar in terms of efficacy, in a meta-analysis of antidepressant efficacy and tolerability, the SSRIs escitalopram and sertraline were found superior compared to other SSRIs, such as fluoxetine, paroxetine, and fluvoxamine, the selective NA reuptake inhibitor (NARI) reboxetine, and also the SNRI duloxetine, in terms of efficacy, with fewer dropouts than duloxetine, venlafaxine, paroxetine, fluvoxamine, and reboxetine (Cipriani et al. 2009). As a group, the SSRIs benefit many patients with mild to moderate depression, but both the response rates and speed of action are inferior to those of the dual-action antidepressants mirtazapine and venlafaxine (Quitkin et al. 2001; Thase et al. 2001). In more severe forms of depression, they are not as effective as the dual-action TCAs (Anderson and Tomenson 1994).

In the absence of significant pharmacological actions outside the serotonergic system, the SSRIs' side effects are limited to their actions on serotonin receptors, for example, nausea related to $5\text{-HT}_3$ receptor activation and agitation, insomnia, and sexual dysfunction due to $5\text{-HT}_2$ receptor activation (see Box 2.4).

**Box 2.4. Side Effect Profile of SSRIs**

*Nausea/vomiting*[a]

Abdominal pain
Dry mouth
Constipation/diarrhea

*Headache*[a]
Asthenia

Dizziness
Insomnia/somnolence

*Sweating*[a]

Anorexia
Weight loss
Nervousness/agitation
Tremor
Convulsions
Dystonic reactions

*Sexual dysfunction*[a]

(reduced libido, anorgasmia)

[a]Most common side effects

### 2.3.5 Selective Serotonin Reuptake Enhancing Antidepressants

Tianeptine, the only representative of this mechanism of action, which appears to be the opposite of that of the SSRIs (as far as the primary target of action is concerned), has been used in the treatment of depression and anxiety for many years. In contrast to the SSRIs, it enhances 5-HT reuptake, and therefore, its action is expected to reduce rather than increase extracellular 5-HT concentrations. However, after long-term treatment, like the SSRIs, it increases serotonergic activity (Ortiz et al. 1993).

### 2.3.6 Selective Noradrenaline Reuptake Inhibitors (NARIs)

The NARIs selectively block reuptake at noradrenergic synaptic sites. The noradrenergic neurotransmitter system was the first and most widely studied in the research into the pathophysiology of depression, and the first influential theories, such as the "catecholamine hypothesis" and the "beta-receptor downregulation" theory of depression, were based on observations within the noradrenergic system. Nevertheless, drugs acting selectively on noradrenergic activity alone, in the brain, failed to demonstrate efficacy of the magnitude shown by the dual-action antidepressants or even the SSRIs and failed to establish themselves as a group of drugs comparative to other antidepressants.

The NARIs, unlike the SSRIs, did not gain much popularity with prescribers and this is due to a variety of reasons. Researcher and clinician's attention had shifted away from noradrenergic neurotransmission and focused on the serotonergic system as the key to the pharmacology of depression from the 1980s onwards. Selective NA uptake inhibitors licensed for other indications such as atomoxetine (licensed for treatment of attention deficit disorder) failed to show antidepressant properties, and some NARIs (such as edivoxetine) and others failed to demonstrate efficacy in licensing clinical trials.

Reboxetine is a typical NARI, being a highly selective and potent NA reuptake inhibitor. It was licensed at a time that the SSRIs were already well established and were recommended by guidelines (NICE guidelines) as the first choice for the treatment of depression. It has been shown to be less effective than other antidepressants including some SSRIs and to be associated with higher dropout rates (Cipriani et al. 2009). Reboxetine's positive benefit-risk balance was questioned in a meta-analysis carried out by the German Institute IQWiG, as part of a health technology assessment of three antidepressants which concluded, rather harshly, that reboxetine is "an ineffective and potentially harmful antidepressant" (Eyding et al. 2010).

These negative findings were disputed by the UK Medicines and Healthcare Products Regulatory Agency (2011) and the European Pharmacovigilance Working Party, in a UK- and Europe-wide review of all available efficacy and safety data, which concluded that "reboxetine has benefit over placebo in its authorized indication." Importantly, on reviewing the previous meta- analysis, it was also found that studies in an in-patient setting and in severe depression consistently showed better efficacy results than those in outpatients. This suggests that the noradrenergic system may have a more important role to play in more severe forms of depression.

## 2.4 Second-Generation Dual-Action Antidepressants

The failure of antidepressant drugs, which act selectively on a single neurotransmitter system (either serotonergic or noradrenergic), to demonstrate superiority in efficacy over the less selective TCAs and MAOIs encouraged the return to dual-action agents and the development of a series of drugs acting on both monoamine systems. These range from monoamine reuptake inhibitors such as the selective 5-HT and NA reuptake inhibitors (SNRIs) and the NA and DA reuptake inhibitor, bupropion, to the reversible monoamine oxidase inhibitors (RIMAs), the NA and 5-HT-specific agents (NASSAs) and other receptor-specific drugs.

### *2.4.1 Selective Serotonin and Noradrenaline Reuptake Inhibitors (SNRIs)*

The introduction of the first selective 5-HT and NA reuptake inhibitor (SNRI) venlafaxine was preceded by the concomitant use of an SSRI and a tricyclic or NARI by clinicians in an attempt to achieve better antidepressant efficacy. This approach was supported by evidence (Nelson 1998; Seth et al. 1992).

Like the TCAs, SNRIs venlafaxine, milnacipran, and duloxetine block 5-HT and NA reuptake from the serotonergic and adrenergic synaptic cleft. However, unlike the TCAs, they have no significant affinity for neurotransmitter transporters or receptor sites outside the monoamine neurotransmitter systems. There are some variations in their binding of the 5-HT and NA transporters, with duloxetine having the highest affinity for both transporters compared to the other two SNRIs (see Table 2.2) (Vaishnavi et al. 2004). Milnacipran has equivalent affinity for the two transporters, while duloxetine is 10 times and venlafaxine 30 times more potent at binding the 5-HT transporter than the NA transporter. These differences, however, are not reflected in their clinical efficacy (Table 2.3).

#### 2.4.1.1 Venlafaxine and Desmethylvenlafaxine

Venlafaxine was the first SNRI licensed for the treatment of depression. At lower doses it is essentially a potent 5-HT reuptake inhibitor with the adrenergic effects becoming apparent at doses of 150 mg and above. At the highest doses (maximum dose 375 mg a day), venlafaxine may also inhibit DA synaptic reuptake.

Desmethylvenlafaxine (*O*-desmethylvenlafaxine) is the major active metabolite of venlafaxine, and it has similar affinity for the 5-HT and NA transporters and has been licensed as an antidepressant in some countries such as the USA.

**Table 2.2** TCAs adverse effect profile

| | |
|---|---|
| Anticholinergic | Dry mouth<br>Blurred vision<br>Urinary hesitancy (retention)<br>Constipation<br>Memory impairment<br>Aggravation of narrow angle glaucoma |
| Antihistaminic | Sedation<br>Weight gain |
| Alpha-1 adrenoceptor blockade | Orthostatic hypotension<br>Cardiac effects |
| Cardiovascular effects | Sinus tachycardia<br>Conduction delay<br>Arrhythmias<br>Sudden death |
| Other adverse effects | Sexual dysfunction<br>Impaired cognitive and psychomotor skills<br>Lowered seizure threshold |

**Table 2.3** Affinity of antidepressants for transporter binding and inhibition of monoamine uptake (adapted from Vaishnavi et al. 2004)

| | Transporter binding | | | Uptake inhibition | | |
|---|---|---|---|---|---|---|
| Drug | SERT | NET | DAT | SERT | NET | DAT |
| Milnacipran | 8.44 | 22 | >100,000 | 151 | 68 | >100,000 |
| Duloxetine | 0.07 | 1.17 | 230 | 3.7 | 20 | 439 |
| Venlafaxine | 7.8 | 1920 | 6050 | 145 | 1420 | 3070 |
| Amitriptyline | 3.45 | 13.3 | 2580 | 67 | 63 | 7500 |
| Citalopram | 1.13 | 4870 | >100,000 | 19 | 8690 | >100,000 |

#### 2.4.1.2 Milnacipran

Milnacipran unlike the other two SNRIs has equipotent affinity for the 5-HT and NA transporter. Its levo-isomer, levomilnacipran, has been recently licensed as a separate antidepressant. Unlike milnacipran, duloxetine, and venlafaxine, it has higher affinity for the NA transporter and inhibits NA reuptake more powerfully than 5-HT reuptake. This preferential NA reuptake inhibition renders it pharmacologically (at least as far as monoamine reuptake inhibition is concerned) similar to the secondary TCA amines (Briley et al. 1996).

#### 2.4.1.3 Duloxetine

Duloxetine, despite its much higher affinity for both 5-HT and NA transporters compared to the other SNRIs, is not more effective in the treatment of depression, and it may be actually inferior in efficacy compared to the dual-action antidepressants, venlafaxine, or mirtazapine as well as the SSRIs escitalopram and sertraline (Cipriani et al. 2009).

### 2.4.2 *Noradrenaline and Dopamine Reuptake Inhibitors*

Bupropion is the antidepressant representative of this group. Although its mechanism of action is not fully understood, its antidepressant efficacy is believed to be related primarily to NA and less so to DA reuptake inhibition, despite its weak affinity for these transporters. The powerful NA reuptake effect is attributed mainly to its metabolite hydroxybupropion. Despite its weak affinity, bupropion increases the expression of DA transporter mRNA, increasing DA concentrations. A further relevant pharmacological action of bupropion is the inhibition of nicotinic receptors (Dell'Orso et al. 2011).

Bupropion's main adverse effect is the induction of seizures. Although lowering of seizure threshold is an undesirable property of most antidepressants irrespective of what group they belong, it is more common in the case of bupropion and it is dose dependent.

### 2.4.3 *Receptor-Selective Antidepressant Drugs*

A chemically and pharmacologically diverse group of antidepressant drugs has been placed under this heading. What they have in common is that they target specific receptor sites rather than the monoamine transporters or the enzyme, monoamine oxidase. They became available at different time points in the development of antidepressant drugs, over the last 30 years or so, and they act on a variety of receptor sites.

**Box 2.5. Side Effect Profile of SNRIs**

DO NOT have:
- Anticholinergic
- Antihistaminic
- Alpha1 receptor antagonism

BUT DO have:
1. Cardiovascular effects
   - Sinus tachycardia, BP↑
   - Arrhythmias
   - Conduction delay
   - Risk of death in overdose (lower than TCAs and MAOIs)
2. Serotonergic-specific side effects
   (see SSRIs' side effects list—Box 2.4)
3. Can lower seizure threshold (convulsions)

#### 2.4.3.1 Noradrenergic and Specific Serotonergic Antidepressants (NASSAs)

Mianserin

Mianserin is a tetracyclic antidepressant drug, a racemic mixture of *R* (–) and *S* (+) mianserin. Lacking in significant cardiovascular effects, it became the drug of choice in the treatment of depressed patients with cardiovascular problems. Unfortunately the relative risk for agranulocytosis somehow limited its use and it was eventually replaced by mirtazapine.

Mirtazapine

Mirtazapine is a tetracyclic piperazinoazepine analogue of mianserin. It has multiple pharmacological sites of action and it is primarily a potent antagonist of central α2-adrenergic autoreceptors and heteroreceptors.

The blockade of the presynaptic α2 autoreceptors allows increased release of noradrenaline from the neuronal terminal. Noradrenaline stimulates serotonin neuronal cell firing via α1-adrenergic receptors (which are not blocked by mirtazapine), while the blockade of presynaptic α2 heteroreceptors located on serotonergic neuronal terminals increases the release of 5-HT in the synaptic cleft. All these mechanisms of action contribute to the dual action (activation of both adrenergic and serotonergic systems in the brain).

Mirtazapine also has 5-HT receptor antagonist properties, which contribute both to its efficacy and its tolerability. By blocking the postsynaptic $5\text{-HT}_2$ and $5\text{-HT}_3$ receptors, mirtazapine increases the concentrations of serotonin available to bind and activate $5\text{-HT}_{1A}$ postsynaptic receptors, which are believed to mediate the antidepressant effects. On the other hand, the blockade of these receptors has a beneficial effect on the side effect profile. The blockade of $5\text{-HT}_3$ receptors minimizes the likelihood of nausea, a side effect commonly associated with serotonergic agents such as the SSRIs. By blocking $5\text{-HT}_2$ receptors, mirtazapine reduces the likelihood of agitation, insomnia, and sexual dysfunction, which are recognized side effects of SSRIs, due to activation of $5\text{-HT}_{2A}$ and $5\text{-HT}_{2C}$ receptors.

In addition to its adrenergic and serotonergic properties, mirtazapine also has high affinity for H1-histamine receptors, and this action is responsible for its known side effects of somnolence and weight gain.

Trazodone

Trazodone was synthesized in the late 1970s and its chemical structure is that of a triazolopyridine derivative with a mechanism of action very different from that of the existing antidepressants (the TCAs and the MAOIs) and has both adrenergic and serotonergic as well as antihistaminic actions. Unlike other antidepressants, it is inactive in classical antidepressant screening tests such as the reserpine model and the forced swim test.

It has complex actions on the serotonergic system. It is a selective serotonin reuptake inhibitor but this action is weak compared to the SSRIs. It has agonist activity at $5\text{-HT}_{2A/2C}$ receptors and it acts as an antagonist at $5\text{-HT}_{1A}$ and $5\text{-HT}_2$ receptor subtypes. Its metabolite, m-chlorophenylpiperazine (m-CPP), is a potent 5-HT receptor agonist.

Trazodone is a weak antagonist of α2-adrenergic autoreceptors and a fairly potent antagonist at α1-adrenergic postsynaptic receptors. It also has moderate antihistaminic (H1-receptor antagonist), but it is devoid of any anticholinergic properties.

Trazodone is a sedative antidepressant, presumably because of its antihistaminic and $5\text{-HT}_2$ receptor antagonist actions, a property desirable for patients with agitated depression, particularly the elderly; however, it needs to be prescribed with caution in this population because of its potential for orthostatic hypotension. Blockade of α1 receptors is considered to be responsible for orthostatic hypotension, which is observed during treatment with trazodone as well as with other α1 adrenoceptor antagonist drugs such as some TCAs, quetiapine, and others. Priapism is an uncommon but serious side effect, and the risk for this is higher in the first 4 weeks of treatment with trazodone.

Nefazodone

Nefazodone is a phenylpiperazine with a chemical structure similar to trazodone and it was developed in the 1990s as an alternative to trazodone. It is a $5\text{-HT}_2$ receptor antagonist with weak inhibition of 5-HT and NA reuptake and has a lower affinity for the α1-adrenergic receptor and less likely to cause orthostatic hypotension.

Nefazodone has been discontinued because of serious hepatotoxic effects.

#### 2.4.3.2 Dual-Action Selective Serotonergic Antidepressants

Studies in the 1990s focused on the therapeutic role of individual 5-HT receptors such as $5\text{-HT}_{1A}$ and $5\text{-HT}_{2A}$ (Celada et al. 2004). There was also particular interest in the role of somatodendritic as well as postsynaptic $5\text{-HT}_{1A}$ receptors. It was shown that blocking reuptake of 5-HT and at the same time inhibiting the action of somatodendritic $5\text{-HT}_{1A}$ receptors (with a selective antagonist) while simultaneously activating postsynaptic $5\text{-HT}_{1A}$ receptors is associated with an acceleration of the antidepressant effect (Blier et al. 1997). Such pieces of evidence influenced the development of newer serotonergic agents aiming to combine 5-HT transporter blockade and 5-HT receptor actions.

The recently licensed 5-HT receptor-selective antidepressants such as vilazodone and vortioxetine are examples of these mechanisms of action. They have been promoted as novel antidepressants with better side effect profile than the SSRIs (in particular reduced likelihood of sexual dysfunction). Further studies and clinical experience will show whether they have any superiority in terms of efficacy and speed of action compared to the SSRIs or the SNRIs.

Vilazodone

Vilazodone is a dual-action serotonergic agent being a 5-HT reuptake inhibitor and also a 5-$HT_{1A}$ receptor partial agonist. It is associated with gastrointestinal side effects (diarrhea, the most common side effect) but no sexual dysfunction (a common side effect of SSRIs).

Vortioxetine

Vortioxetine is a more complex "multimodal action" antidepressant with multiple sites of action within the serotonergic neurotransmitter system. It is a 5-$HT_3$, 5-$HT_7$, and 5-$HT_{1D}$ receptor antagonist; a 5-$HT_{1B}$ receptor partial agonist; and a 5-$HT_{1A}$ receptor agonist as well as a serotonin (5-HT) transporter (SERT) inhibitor. These direct 5-HT receptor and transporter actions are associated with wide-ranging indirect pharmacological effects on brain structures within the neurocircuit of depression, including increases in serotonergic and to a lesser extent noradrenergic, dopaminergic, cholinergic, histaminic, and glutamatergic neurotransmission, possibly via negative feedback mechanisms controlling neuronal activity. It is described as serotonin modulator and stimulator (SMS) with its main activity within the serotonergic system (Sanchez et al. 2015).

Vortioxetine's side effect profile is similar to that of SSRIs, with gastrointestinal symptoms being most common, but unlike the SSRIs, sexual dysfunction and sleep disruption are less of a problem.

#### 2.4.3.3 Melatonin Receptor-Selective Drugs

Disruptions of the human circadian rhythms generated by the biological clock (which is located in the suprachiasmatic nucleus (SCN) in the hypothalamus) have been implicated in the pathophysiology of depression. Melatonin secretion by the pineal gland is under the control of the SCN and is linked to the day-night cycle, being synthesized and released at night. Melatonin acting via melatonin receptors has physiological, including chronobiotic, actions in the brain.

Agomelatine

Agomelatine acts as an agonist at $MT_1$ and $MT_2$ brain receptors and is believed that this action may improve mood and sleep via the regulation and resetting of the biological clock in the SCN. It is also an antagonist at 5-$HT_{2C}$ receptors releasing DA and noradrenaline in the frontal cortical areas, an action considered to be a significant contributor to its antidepressant actions.

Agomelatine has failed to show superiority in efficacy compared to existing antidepressants including the SSRIs (Hickie and Rogers 2011). It is well tolerated but at higher doses it has the potential for hepatotoxicity.

## 2.5 Future Antidepressants: Targets of Action

The rapid growth of the neurosciences and, in particular, neuroimaging and molecular biology over the last couple of decades has been gradually unraveling the workings of the brain in relation to mental disorders. Structural and functional neuroimaging helped identify areas of the brain involved in the regulation of mood and their connections, the neurocircuit of depression. All these developments reveal a very complex and multifactorial pathophysiology of depression, where monoamine neurotransmitter function is only one of the many links in the chain of factors involved. The current consensus view on the pathophysiology of depression points to disruptions of communications within the neurocircuit of depression, caused by a variety of abnormalities, which involve vulnerability (as well as some protective) genes, dysregulation of the hypothalamo-pituitary-adrenal (HPA) axis, the influence of early life adverse experiences, immune mechanisms, biochemical abnormalities, and other factors.

As far as treatment of depression is concerned, irrespective of the possible cause/s, by increasing the activity of one or more of the monoamine neurotransmitter systems, the antidepressant drugs help reestablish effective communications within the neurocircuit of depression. This is achieved probably via the restoration of adequate concentrations of brain-derived neurotrophin (BDNF), the role of which is to protect the resilience and enhance the neuronal plasticity of the brain. Restoring and maintaining neuronal growth and function to an optimum level is believed to ensure the maintenance and stability of the euthymic status.

However, the success of the currently available antidepressant treatments is limited. Despite the wealth of antidepressant drugs available, one third of patients fail to achieve and maintain long-term remission. According to the STAR-D report (Rush et al. 2006), the outcome of the first step of acute antidepressant treatment is 30.8 %; the effectiveness decreases with each successive antidepressant drug step, dropping down to only 13 % response after the fourth acute treatment step. The overall cumulative remission rate is 67 % and those who required more treatment steps have higher relapse rates during the naturalistic follow-up phase. Such findings demonstrate the great need for more effective drugs.

Neurotransmitter systems other than the monoamines, in particular glutamate and GABA, also operate within the neurocircuit of depression and interact with the noradrenergic and serotonergic pathways, and these have been investigated as targets for possible novel drugs. Attention has also turned to the cholinergic neurotransmitter system and its possible role in depression and its treatment.

Any new or novel antidepressant drugs therefore have a wide range of sites to target, which include monoamine and also non-monoamine neuronal systems, such as the GABAergic, cholinergic, and glutamatergic systems as well as the glucocorticoids and the immune system.

## *2.5.1 Monoamine Targets*

### 2.5.1.1 Triple Monoamine Reuptake Inhibitors

As interest is turning to drugs with a wider spectrum of action, the addition of DA reuptake inhibition to that of NA and 5-HT reuptake blockade has stimulated the production of several agents with triple action, by the pharmaceutical industry. This is not a new idea as there are already drugs with dopaminergic actions such as the MAOIs, bupropion, and some others.

There is a fair amount of evidence to support the notion that underactivity in dopaminergic neurotransmission is relevant in the pathophysiology of depression and that increasing dopaminergic neurotransmission may be beneficial. Some symptoms of depression such as anhedonia, poor motivation, low energy, and disruption of reward behavior, core features of this condition, are considered to be clinical manifestations of dopaminergic hypoactivity (Dunlop and Nemeroff 2007; Naranjo et al. 2001). Patients with Parkinson's disease, a condition caused by brain DA deficiency (mainly in the basal ganglia), have high rates of depression (McDonald et al. 2003), and the plasma concentrations of the DA metabolite, homovanillic acid, are reduced in depressed patients (Mitani et al. 2006). The existence of antidepressants with proven efficacy, which have dopaminergic activity, such as bupropion and the MAOIs, lends further support to the importance of the dopaminergic system in the treatment of depression.

The monoamine neurotransmitter systems are functionally closely connected and in general they appear to enhance each other's activity. Central dopaminergic neurotransmission regulates the activity of the serotonergic and noradrenergic neurons at the dorsal raphe nuclei and the locus coeruleus, respectively, and the dopaminergic neurons at the ventral tegmental area respond to the changes in 5-HT and NA release (Guiard et al. 2009).

The increased activity of the dopaminergic neurons may further increase (in addition to that resulting from the adrenergic and serotonergic activation) the concentrations of BDNF in the hippocampus and other relevant sites within the neurocircuit of depression. Whether these positive interactions, as expected in the acute treatment will be maintained during long-term treatment, needs to be investigated further.

It has been suggested that the triple reuptake inhibitors could be not only more effective than existing antidepressants but could potentially also have a faster onset of action (Skolnick et al. 2003). It is also thought that the addition of dopaminergic activation may counteract the problem of sexual dysfunction caused by serotonergic activity. On the other hand, however, dopaminergic activation may bring a different set of problems, such as psychological or physical dependence via the dopaminergic activation within the brain reward system. Dopamine reuptake inhibitors have been divided into Type 1, which cause euphoria and have the potential for dependence, and Type 2, which may not (Rothman 1990). Any new triple agents need to take this into consideration.

#### 2.5.1.2 Atypical Antipsychotics

Antipsychotic drugs both first- and second-generation have always had a place in the treatment of psychotic depression but have been of little use in nonpsychotic depressive disorder. In recent years in the absence of any major leaps in the development of more effective antidepressants, there has been an interest in atypical antipsychotics, used mainly as augmentation in resistant to treatment depression (DeBattista and Hawkins 2009). Quetiapine has shown most promise demonstrating effectiveness in the treatment of bipolar depression as well as in unipolar depressive disorder either as monotherapy or in addition to an antidepressant (Keating and Robinson 2007; Baune 2007). It is the drug of choice in the treatment of bipolar depression, as the standard antidepressants can cause a switch to mania and cycle acceleration in patients with bipolar disorder (Palazidou 2009). Its metabolite, norquetiapine, has monoamine reuptake inhibitory properties, and this action may be the major contributor to its antidepressant efficacy, as quetiapine appears to be also effective, as monotherapy, in the treatment of unipolar nonpsychotic depression (Thase et al. 2013). However, the high dropout rates (Maneeton et al. 2012) suggest that quetiapine's side effect profile makes it a less desirable option compared to the better tolerated, existing antidepressants.

### *2.5.2 Non-monoamine Neurotransmitter Targets*

#### 2.5.2.1 Acetylcholine System

Anticholinergic, in particular antimuscarinic, drugs are known to have euphoriant properties, and scopolamine has shown some antidepressant potential in a small study (Furey and Drevets 2006). The response is rapid at 3–5 days after a single intravenous infusion in both treatment naïve and in patients with recurrent depression (Ellis et al. 2014). A lot more work is needed to demonstrate the possible usefulness of antimuscarinic drugs as antidepressants in the long term.

Drugs acting on the nicotinic receptor have not shown potential as antidepressants. *S-mecamylamine* (TC-5214), a nicotinic receptor modulator, failed to show significant antidepressant efficacy and was discontinued.

#### 2.5.2.2 Glutamate System

The brain is rich in glutamate neurons, which by far outnumber other neurotransmitter systems apart from the GABA system, and it is argued that it is excitatory glutamatergic neurotransmission, counterbalanced by GABAergic inhibition, which is responsible for changes in emotion and cognition. There is glutamate dysfunction in the limbic/prefrontal cortex of depressed patients and evidence suggests that drugs targeting the glutamate system may accelerate the onset of antidepressant action (Sanacora et al. 2012).

Ketamine, a nonselective *N*-methyl-D-aspartic acid (NMDA) receptor antagonist and an established anesthetic in veterinary and human (pediatric) practice, has been investigated as a potential antidepressant. In subanesthetic doses, it strengthens cortical synapses through neurotrophic mechanisms and may lead to clinical improvement as it rapidly reverses prefrontal cortical deficits (Moghaddam 2002).

Studies in major depressive disorder (Berman et al. 2000), in bipolar depression (Zarate et al. 2012) and in treatment-resistant depression, showed dramatic improvement after a single dose of ketamine which lasted for 1 week (Zarate et al. 2006). A review of the trials so far (Covvey et al. 2012) suggested a benign side effect profile. Nevertheless it is too early to draw any conclusions about the long-term efficacy and safety of ketamine. The need (at least so far) for administering ketamine intravenously is associated with practical difficulties for long-term treatment, and it is not known whether its antidepressant effect will be sustained. Ketamine is known to be a hallucinogenic drug and it remains to be seen what the effects of chronic use may be.

The evidence so far simply suggests there is scope for further research into drugs that target the NMDA receptor, as potential antidepressants. Further work is needed to assess the efficacy and safety of such drugs as well as the development of drugs with antiglutamatergic properties, which are easier to administer and are safe to use in the treatment of depressed patients.

### *2.5.3 Hypothalamo-Pituitary-Adrenal (HPA) Axis*

The role of psychosocial stress is considered most important in the pathogenesis of depression, and dysregulation of the HPA axis with persistent hypercortisolemia is a well-recognized abnormality present in over 50 % of depressed patients (Carroll et al. 1976). This has been associated with impairment of neo-neurogenesis and structural changes in the hippocampus in depressed patients. Persistently high cortisol concentrations have toxic effects on the hippocampus possibly operating directly or via a reduction in BDNF concentrations (Bremner et al. 2000). It makes sense therefore to consider the HPA axis as a target in the treatment of depression and in particular to investigate the use of antiglucocorticoid drugs as possible antidepressants.

Ketoconazole, a glucocorticoid receptor antagonist, and metyrapone, a cortisol synthesis inhibitor, showed some potential antidepressant efficacy (Murphy 1991; Thakore and Dinan 1995; Wolkowitz et al. 1999; O'Dwyer et al. 1995), and a systematic review suggested that antiglucocorticoid augmentation of antidepressant drugs can be effective in depressed patients with metyrapone showing the largest effect size (Gallagher et al. 2008). However, in a randomized controlled study in treatment-resistant depressed patients, metyrapone augmentation of SSRIs, SNRIs, or mirtazapine failed to show any benefit (Ferrier et al. 2015). Cortisol-releasing hormone receptor 1 (CRHR1) antagonists investigated as possible antidepressants (Holsboer and Ising 2008) were discontinued because of unacceptable toxicity (DeBattista and Belanoff 2006).

### 2.5.4 Immune System

Immunological mechanisms have attracted a lot of interest in recent years as playing an important part in the complex pathophysiology of depression. Severe depression is associated with immune activation and in particular with raised cytokine concentrations. Pro-inflammatory cytokines (the signaling molecules of the immune system) elicit sickness behavior (fatigue and lethargy).

The activation of the inflammatory system response (IRS), however, does not operate by itself as a separate mechanism of clinical symptom production, and it appears to closely interact with both the monoamine neurotransmitter systems and the HPA axis. Raised pro-inflammatory cytokines are associated with peripheral tryptophan (serotonin precursor) depletion and may influence noradrenergic activity. IRS activation also stimulates the HPA axis. It has been suggested that impaired glucocorticoid receptor function may be related to chronic exposure to inflammatory cytokines associated with chronic physical illness or chronic stress (Pace et al. 2007). Glucocorticoid resistance in turn may cause further increase in inflammation.

Depression is such a common adverse event in patients receiving treatment with interferon (a drug used in the treatment of some viral infection, multiple sclerosis, autoimmune disorders, and others), and that it is not an uncommon practice to prescribe antidepressants concurrently for prophylaxis. However, treatment with SSRIs, with a reduction of clinical symptom scores, showed only limited decrease in IL-6 and TNF-$\alpha$ cytokine concentrations, although IL-$\beta$ levels are significantly decreased (Hannestad et al. 2011).

Tackling the inflammatory system directly with the use of nonsteroidal anti-inflammatory drugs (NSAIDs) such as the selective cyclooxygenase (COX)-2 and nonselective COX inhibitors, celecoxib and aspirin, and other drugs believed to have anti-inflammatory properties, such as minocycline, omega-3 fatty acids, vitamin D, and others, has been considered as a possible new avenue in the treatment of depression. The results so far are inconclusive with reports of both success and failure. NSAIDs, celecoxib, and aspirin, used as an add-on therapy with the SSRIs fluoxetine or sertraline, showed a reduction in depressive symptoms (Müller et al.

2006; Akhondzadeh et al. 2009; Abbasi et al. 2012). However, when used as monotherapy, they failed to show any significant improvement in patients with major depressive disorder (Eyre et al. 2015).

It appears that the presence of more significant inflammatory abnormalities may have a positive predictive value in the antidepressant response to NSAIDs. Patients with higher C-reactive protein (CRP), TNF-α, and TNFR1 and TNFR2 concentrations were more likely to respond to infliximab (Raison et al. 2012).

The antibiotic minocycline, a tetracycline derivative, regulates neuroplasticity via a pleiotropic mechanism of action, which has effects on apoptosis, anti-inflammatory, antioxidant, as well as effects on neurotransmission (monoamines and glutamatergic systems) (Soczynska et al. 2012). It remains to be seen whether it indeed has a real potential in the treatment of depression.

### *2.5.5 Pharmacogenetics*

Pharmacogenetics is a growing area and it is hoped in the not so long future it will guide drug therapies of psychiatric conditions, including depression. So far, limited progress has been made in this direction and it still lags far behind other areas of medicine and particularly oncology and cancer treatments.

Although some work has been done in pharmacodynamics and response to treatment, more emphasis has been placed on pharmacokinetics, which is relevant to response to treatment as well as to tolerability and toxicity. Two methods have been used, namely, the candidate polymorphism approach and the genome-wide association method, both of which have major limitations. Most of the results in pharmacodynamics but also pharmacokinetics have failed to reach a level of robustness that would encourage application in clinical practice in depression.

Most studies on pharmacodynamics have focused on genes coding receptors, transporters, and second messenger systems, while pharmacokinetic studies targeted mainly the cytochrome P450 family of enzymes, in particular CYP2D6 (known to metabolize most psychotropic drugs) but also others such as CYP2C19 as well as P-glycoprotein (Porcelli et al. 2011). The CYP450 family of enzymes metabolize drugs mostly in the liver, and their genes determine their level of activity. This subdivides individuals into ultra rapid, rapid, medium, or slow metabolizers, resulting in drug plasma concentrations ranging from too low to too high or toxic levels in different individuals. This variability in drug concentrations, of course, influences both the efficacy of the relevant drug as well as its tolerability and safety.

*CYP2C19* allele status has been implicated in the metabolism and response to treatment of escitalopram and citalopram. There was a positive association in the European Consortium project Genome-Based Therapeutic Drugs for Depression (GENDEP), while the STAR-D study did not find a clear association with either CYP2C19 or *CYP2D6* in terms of response; however, it did not preclude a possible association in metabolism.

**P-glycoprotein** is a plasma membrane transporter protein, which is involved in the efflux function of the blood-brain barrier and it is encoded by the ABCB1 gene. It appears to be involved in the removal of most antidepressants from the brain, and variations in the ABCB1 gene affect the activity of P-glycoprotein, which may influence response to antidepressant drugs.

The evidence at least so far suggests that there may not be an association between metabolizer status and clinical response. In the USA, the FDA-approved AmpliChip CYP450 test has been incorporated into clinical practice, although its reliability in determining appropriate antidepressant drug choice is not as yet proven (see review by Narasimhan and Loboff 2012). In their review of relevant studies up to the year 2012, Gvozdic et al. (2012) are more positive, claiming promising findings for several variants in candidate genes (involved in both pharmacodynamics and pharmacokinetics) including the serotonin transporter, serotonin receptors, ABCB1, and genes involved in the HPA axis.

Efforts to identify gene influences on treatment response have not been fruitful so far, although there is evidence that gene variations do have an effect on vulnerability to depression. For example, BDNF met-allele carriers (val/met and met/met) have smaller hippocampal volumes in both depressed patients and healthy controls (Frodl et al. 2008), but this does not influence antidepressant drug response (Domschke et al. 2010).

The 5-hydroxytryptamine transporter-linked polymorphic promoter region (5-HTTLPR) has been extensively studied showing an association with bipolar disorder and suicidality. There is a disputed link with response to antidepressant drug treatment (Goltser-Dubner et al. 2010).

## 2.6 Conclusions

In order to develop effective treatments, it is essential to have identified specific biomarkers as treatment targets, and in depression these are multiple. The success achieved in cancer therapies, based on molecular diagnoses, is encouraging efforts toward such "precision medicine" approaches to the pharmacotherapy of mental disorders including depression. The National Institute of Mental Health (NIMH) has led the way by funding a study, the Establishing Moderators/Mediators for a Biosignature of Antidepressant Response in Clinical Care (EMBARC), aiming to discover biomarkers, which can be used to develop novel antidepressant drugs and to guide a personalized approach to pharmacotherapy. The NIMH has also introduced the Research Domain Criteria (RDoC) project, which aims to reclassify mental disorders based on evidence from neuroscience in addition to the clinical presentation. There are similar initiatives in Europe such as the Roadmap for Mental Health Research funded by the European Commission and a call from the European Union Innovative Medicines Initiative to link clinical neuropsychiatry with quantitative neurobiology (Insel and Bruce 2015).

Depression is a multifactorial condition and not all abnormalities and biomarkers apply to all patients. The skill of future approaches to treatment would be to identify the specific abnormalities relevant to the individual patient and target these biomarkers with specific drugs. For example, not all patients have immune abnormalities present, but those who do are more likely to respond to treatments targeting these.

The failure so far of any drugs other than those targeting the monoamine systems suggests that it may be necessary to use combination therapies, i.e., a drug that increases monoamine activity together with an anti-inflammatory agent, a glucocorticoid antagonist, a drug targeting the glutamate system, or others, depending on the individual patient profile.

Further developments in pharmacogenetics will add to our ability to tailor the antidepressant drug treatment plan to the individual based on his or her identified abnormalities, genetic characteristics (pharmacokinetic and pharmacodynamic), and the presenting clinical picture.

Future developments need to also take into consideration the relapsing remitting nature of depression as it presents in most patients and the paramount importance of early treatment with full remission. Drugs may need to be used long term or even lifelong in some patients and therefore issues of tolerability and safety are essential in addition to efficacy. Ideally we should aim to develop drugs that are curative and not just controlling symptoms and preventing relapse but this is a much harder and realistically not an achievable task given the present understanding of the pathophysiology of depression.

## References

Abbasi SH, Hosseini F, Modabbernia A, Ashrafi M, Akhondzadeh S. Effect of celecoxib add-on treatment on symptoms and serum IL-6 concentrations in patients with major depressive disorder: randomized double-blind placebo-controlled study. J Affect Disord. 2012;141:308–14.

Akhondzadeh S, Jafari S, Raisi F, Nasehi AA, Ghoreishi A, Salehi B, Mohebbi-Rasa S, Raznahan M, Kamalipour A. Clinical trial of adjunctive celecoxib treatment in patients with major depression: a double blind and placebo controlled trial. Depress Anxiety. 2009;26:607–11.

Amrein R, Allen SR, Vranesic D, Stahl M. Antidepressant drug therapy: associated risks. J Neural Transm. 1988;26(suppl):73–86.

Anderson IM, Tomenson BM. The efficacy of selective serotonin reuptake inhibitors in depression: a meta-analysis of studies against tricyclic antidepressants. J Psychopharmacol. 1994;8:238–49.

Axelrod J. Noradrenaline: fate and control of its biosynthesis. Les Prix Nobel 1970. Imprimerie Royale P. A. Norstedt & Soner, 1971; p. 189–208

Baune BT, Caliskan S, Todder D. Effects fo adjunctive antidepressant therapy with quetiapine on clinical outcome, quality of sleep and daytime motor activity in patients with resistant depression. Hum Psychopharmacol Clin Exp. 2007;22:1–9.

Berman RM, Cappiello A, Anand A, Oren DA, Heninger GR, Charney DS, Krystal JH. Antidepressant effects of ketamine in depressed patients. Biol Psychiatry. 2000;47:351–4.

Blier P, Bergeron R, de Montigny C. Selective activation of post-synaptic 5-HT1A receptors induces rapid antidepressant response. Neuropsychopharmacology. 1997;16:333–8.

Bloch RG, Doonief AS, Buchberg AS, Spellman S. The clinical effect of isoniazid and iproniazid in the treatment of pulmonary tuberculosis. Ann Intern Med. 1954;40:881–900.
Boyer EW, Shannon M. The serotonin syndrome. N Eng J Med. 2005;352:1112–20.
Bremner JD, Narayan M, Anderson ER. Hippocampal volume reduction in major depression. Am J Psychiatry. 2000;157:115–7.
Briley M, Prost JF, Moret C. Preclinical pharmacology of milnacipran. Int Clin Psychopharmacol. 1996;11 suppl 4:9–14.
Carroll BJ, Curtis GC, Davies BM, Mendels J, Sugerman AA. Urinary free-cortisol excretion in depression. Psychol Med. 1976;6:43–50.
Celada P, Puia MV, Amargós-Bosch M, Adell A, Artigas F. The therapeutic role of 5-HT1A and 5-HT2A receptors in depression. J Psychiatry Neurosci. 2004;29:252–65.
Cipriani A, Furukawa TA, Salanti G, Geddes JR, Higgins JP, Churchill R, Watanabe N, Nakagawa A, Omori IM, McGuire H, Tansella M, Barbui C. Comparative efficacy and acceptability of 12 new-generation antidepressants: a multiple-treatments meta-analysis. Lancet. 2009;373: 746–58.
Coppen A. The biochemistry of affective disorders. Br J Psychiatry. 1967;113:1237–64.
Covvey JR, Crawford AN, Lowe DK. Intravenous ketamine for treatment-resistant major depressive disorder. Ann Pharmacother. 2012;46:117–23.
Crane GE. Iproniazid (Marsilid) phosphate, a therapeutic agent for mental disorders and debilitating disease. Psychiatr Res Rep Am Psychiatr Assoc. 1957;8:142–52.
Da Prada M, Zurcher G, Wuthrich I, Haefely WE. On tyramine, food, beverages and the reversible MAO inhibitor moclobemide. J Neural Transm. 1988;26(suppl):31–56.
DeBattista C, Belanoff J. The use of mifepristone in the treatment of neuropsychiatric disorders. Trends Endocrinol Metab. 2006;17:117–21.
DeBattista C, Hawkins J. Utility of atypical antipsychotics in the treatment of resistant unipolar depression. CNS Drugs. 2009;23:369–77.
Delgado PL, Charney DS, Price LH, Aghajanian GK, Landis H, Heninger GR. Serotonin function and the mechanism of antidepressant action. Reversal of antidepressant-induced remission by rapid depletion of plasma tryptophan. Arch Gen Psychiatry. 1990;47:411–8.
Delgado PL, Miller HL, Salomon RM, Licinio J, Heninger GR, Gelenberg AJ, Charney DS. Monoamines and the mechanism of antidepressant action: effects of catecholamine depletion on mood of patients treated with antidepressants. Psychopharmacol Bull. 1993;29:389–96.
Dell'Orso B, Palazzo MC, Oldani L, Altamura AC. The noradrenergic action in antidepressant treatments: pharmacological and clinical aspects. CNS Neurosci Ther. 2011;17:723–32.
Domschke K, Lawford B, Laje G, Berger K, Young R, Morris P, Deckert J, Arolt V, McMahon FJ, Baune BT. Brain-derived neurotrophic factor (BDNF) gene: no major impact on antidepressant treatment response. Int J Neuropsychopharmacol. 2010;13:93–101.
Dunlop BW, Nemeroff CB. The role of dopamine in the patho- physiology of depression. Arch Gen Psychiatry. 2007;64:327–37.
Ellis JS, Zarate CA, Luckenbaugh DA, Furey ML. Antidepressant treatment history as a predictor of response to scopolamine:clinical implications. J Affect Disord. 2014;162:39–42.
Eyding D, Lelgemann M, Grouve U, Härter M, Kromp M, Kaiser T, Kerekes MF, Gerken M, Wieseler B. Reboxetine for acute treatment of major depression: systematic review and meta-analysis of published and unpublished placebo and selective serotonin reuptake inhibitor controlled trials. BMJ. 2010;341:4737.
Eyre HA, Air T, Proctor S, Rositano S, Baune BT. A critical review of the efficacy of non-steroidal anti-inflammatory drugs in depression. Prog Neuropsychopharmacol Biol Psychiatry. 2015;7:11–6.
Ferrier IN, Anderson IM, Barnes J, Gallagher P, Grunze HCR, Haddad PM, House AO, Hughes T, Lloyd AJ, Mamasoula C, McColl E, Pearce S, Siddiqi N, Sinha B, Speed C, Steen N, Wainwright J, Watson S, Winter FH, McAllister-Williams RH. Randomised controlled trial of Antiglucocorticoid augmentation (metyrapone) of anti-depressants in depression (ADD study). Efficacy Mech Eval. 2015;2(4).
Frodl T, Moller HJ, Meisenzahl E. Neuroimaging genetics: new perspectives in research on major depression? Acta Psychiatr Scand. 2008;118:363–72.

Furey ML, Drevets WC. Antidepressant efficacy of the antimuscarinic drug scopolamine: a randomized, placebo-controlled clinical trial. Arch Gen Psychiatry. 2006;63:1121–9.

Gallagher P, Malik N, Newham J, Young AH, Ferrier IN, Mackin P. Antiglucocorticoid treatments for mood disorders. Cochrane Database Syst Rev. 2008;1:CD005168.

Gardner DM, Shulman KI, Walker SE, Tailor SA. The making of a user friendly MAOI diet. J Clin Psychiatry. 1996;57:99–104.

Goltser-Dubner T, Galili-Weisstub E, Segman RH. Genetics of unipolar major depressive disorder. Isr J Psychiatry Relat Sci. 2010;47:72–82.

Guiard BP, Mansari ME, Blier M. Prospect of a dopamine contribution in the next generation of antidepressant drugs: the triple reuptake inhibitors. Curr Drug Targets. 2009;10(1069 -1084):1069.

Gvozdic K, Brandl EJ, Taylor DL, Müller DJ. Genetics and personalized medicine in antidepressant treatment. Curr Pharm Des. 2012;18:5853–78.

Hannestad J, DellaGioia N, Bloch M. The effect of antidepressant medication treatment on serum levels of inflammatory cytokines: a meta-analysis. Neuropsychopharmacology. 2011;36(12): 2452–9.

Hickie IB, Rogers NL. Novel melatonin-based therapies: potential advances in the treatment of major depression. Lancet. 2011;378:621–31.

Hollon SD, Shelton RC, Wisniewski S, Warden D, Biggs MM, Friedman ES, Husain M, Kupfer DJ, Nierenberg AA, Petersen TJ, Shores-Wilson K, Rush AJ. Presenting characteristics of depressed outpatients as a function of recurrence: preliminary findings from the STAR*D clinical trial. J Psychiatry Res. 2006;40:59–69.

Holsboer F, Ising M. Central CRH system in depression and anxiety- evidence from clinical studies with CRH1 receptor antagonists. Eur J Pharmacol. 2008;583:350–7.

Insel TR, Bruce N. Brain disorders? Precisely Cuthbert. Science. 2015;348:499.

Keating GM, Robinson DM. Quetiapine: a review of its use in the treatment of bipolar depression. Drugs. 2007;67:1077–95.

Kendler KS, Thornton LM, Gardner CO. Stressful life events and previous episodes in the etiology of major depression in women: an evaluation of the 'kindling' hypothesis. Am J Psychiatry. 2000;157:1243–51.

Krishnan KRR. Monoamine oxidase inhibitors. In: Schatzberg AF, Nemeroff CB, editors. Textbook of psychopharmacology. Arlington, VA: The American Psychiatric Publishing; 2004. p. 303–14.

Kuhn R. The imipramine story. In: Ayd FJ, editor. Discoveries in biological psychiatry. Philadelphia, PA: Blackwell; 1970. p. 205–17.

Loomer HP, Saunders JC, Kline NA. Clinical and pharmacological evaluation of iproniazid as a psychic energizer. In: Cleghorn RA, editor. Research in affects psychiatric research reports no 8 of the American Psychiatric Association. Washington, DC: American Psychiatric Association; 1957. p. 129–41.

Maneeton M, Maneeton B, Srisurapanont M, Martin D. Quetiapine monotherapy in acute phase for major depressive disorder: a meta-analysis of randomized, placebo-controlled trials. BMC Psychiatry. 2012;12:160.

McDonald WM, Richard IH, DeLong MR. Prevalence, etiology, and treatment of depression in Parkinson's disease. Biol Psychiatry. 2003;54:363–75.

MHRA. Medicines and healthcare products regulatory agency-reboxetine: benefit-risk balance reviewed. Drug Safety Update. 2011;5(2):H1.

Mitani H, Shirayama Y, Yamada T, Kawahara R. Plasma levels of homovanillic acid, 5-hydroxyindoleacetic acid and cortisol and serotonin turnover in depressed patients. Prog Neuropsychopharmacol Biol Psychiatry. 2006;30:531–4.

Moghaddam B. Stress activation of glutamate neurotransmission in the prefrontal cortex: implications for dopamine-associated psychiatric disorders. Biol Psychiatry. 2002;51:775–87.

Müller N, Schwarz MJ, Dehning S, Douhe A, Cerovecki A, Goldstein-Müller B, Spellmann I, Hetzel G, Maino K, Kleindienst N, Möller HJ, Arolt V, Riedel M. The cyclooxygenase-2 inhibitor celecoxib has therapeutic effects in major depression: results of a double-blind, ran-

domized, placebo controlled, add-on pilot study to reboxetine. Mol Psychiatry. 2006; 11:680–4.
Murphy BE. Treatment of major depression with steroid suppressive drugs. J Steroid Biochem Mol Biol. 1991;39:239–44.
Naranjo CA, Tremblay LK, Busto UE. The role of the brain reward system in depression. Prog Neuropsychopharmacol Biol Psychiatry. 2001;25:781–823.
Narasimhan S, Loboff FW. Pharmacogenetics of antidepressant drugs. Pharmacogenomics. 2012;13:441–64.
Nelson JC. Augmentation strategies with serotonergic-noradrenergic combinations. J Clin Psychiatry. 1998;59 suppl 5:65–8.
O'Dwyer AM, Lightman SL, Marks MN, Checkley SA. Treatment of major depression with metyrapone and hydrocortisone. J Affect Disord. 1995;33:123–8.
Okuda A, Suzuki T, Kishi T, Yamanouchi Y, Umeda K, Haitoh H, Hashimoto S, Ozaki N, Iwata N. Duration of untreated illness and antidepressant fluvoxamine response in major depressive disorder. Psychiatry Clin Neurosci. 2010;64:268–73.
Ortiz J, Mariscot C, Alvarez E, Artigas F. Effects of the antidepressant drug tianeptine on plasma and platelet serotonin of depressive patients and healthy controls. J Affect Disord. 1993;29: 227–34.
Pace WWT, Hu F, Miller A. Cytokine effects on glucocorticoid receptor function: relevance to glucocorticoid resistance and the pathophysiology and treatment of major depression. Brain Behav Immun. 2007;21:9–19.
Palazidou E. Quetiapine: a new option in bipolar depression. Future Prescriber. 2009;10(4):9–14.
Palazidou E. The neurobiology of depression. Br Med Bull. 2012;101(1):127–45.
Palazidou E, Skene S, Arendt J. The acute and chronic affects of (þ)-and (2-)-oxaprotiline upon melatonin secretion in normal subjects. Psychol Med. 1992a;22:61–7.
Palazidou E, Papadopoulos A, Radcliff H, Dawling S, Checkley SA. Noradrenaline uptake inhibition increases melatonin secretion, a measure of noradrenergic neurotransmission, in depressed patients. Psychol Med. 1992b;22:309–15.
Piccinni A, Del Debbio A, Medda P, Bianchi C, Roncaglia I, Veltri A, Zanello S, Massimetti E, Origlia N, Domenici L, Marazziti D, Dell'Osso L. Plasma brain-derived neurotrophic factor in treatment-resistant depressed patients receiving electroconvulsive therapy. Eur Neuropsychopharmacol. 2009;19:349–55.
Pintor L, Gasto C, Navarro V, Torres X, Fañanas L. Relapse of major depression after complete and partial remission during a 2-year follow-up. J Affect Disord. 2003;73:237–44.
Porcelli S, Fabbri C, Drago A, Gibiino S, De Ronchi D, Serretti A. Genetics and antidepressants; where we are. Clin Neuropsychiatry. 2011;2:99–150.
Quitkin FM, Taylor BP, Kremer C. Does mirtazapine have a more rapid onset than SSRIs? J Clin Psychiatry. 2001;62:358–61.
Raison CL, Rutherford RE, Woolwine BJ, Shuo C, Schettler P, Drake DF, Haroon E, Miller AH. A randomized controlled trial of the tumor necrosis factor antagonist infliximab for treatment-resistant depression: the role of baseline inflammatory biomarkers. Arch Gen Psychiatry. 2012;2012:1–11.
Rothman RB. High affinity dopamine reuptake inhibitors as potential cocaine antagonists: a strategy for drug development. Life Sci. 1990;46:17–21.
Rush AJ, Trivedi MH, Wisniewski SR, Nierenberg AA, Stewart JW, Warden D, Niederehe G, Thase ME, Lavori PW, Lebowitz BD, McGrath PJ, Rosenbaum JF, Sackeim HA, Kupfer DJ, Luther J, Fava M. Acute and longer-term outcomes in depressed outpatients requiring one or several treatment steps: a STAR*D report. Am J Psychiatry. 2006;163:1905–17.
Sanacora G, Treccani G, Popoli M. Towards a glutamate hypothesis of depression: an emerging frontier of neuropsychopharmacology of mood disorders. Neuropsychopharmacology. 2012; 62:63–77.
Sanchez C, Asin K, Artigas F. Vortioxetine, a novel antidepressant with multimodal activity: review of preclinical and clinical data. Pharmacol Ther. 2015;145:43–57.
Schildkraut JJ. The catecholamine hypothesis of affective disorders: a review of supporting evidence. Am J Psychiatry. 1965;122:509–22.

Seth R, Jennings AL, Bindman J, Phillips J, Bergman K. Combination treatment with noradrenaline and serotonin reuptake inhibitors in resistant depression. Br J Psychiatry. 1992;161: 562–5.

Shimizu E, Hashimoto K, Okamura N, Koike K, Komatsu N, Kumakiri C, Nakazato M, Watanabe H, Shinoda N, Okada S, Iyo M. Alterations of serum levels of brain-derived neurotrophic factor (BDNF) in depressed patients with or without antidepressants. Biol Psychiatry. 2003;54:70–5.

Skolnick P, Popic P, Janowsky A, Beer B, Lippa AS. "Broad spectrum" antidepressants: is more better for the treatment of depression? Life Sci. 2003;73:3175–9.

Soczynska JK, Mansur RB, Brietzke E, et al. Novel therapeutic targets in depression: minocycline as a candidate treatment. Behav Brain Res. 2012;235:302–17.

Thakore JH, Dinan T. Cortisol synthesis inhibition: a new treatment strategy for the clinical and endocrine manifestations of depression. Biol Psychiatry. 1995;37:364–8.

Thase ME, Entsuah AR, Rudolph RL. Remission rates during treatment with venlafaxine or selective serotonin reuptake inhibitors. Br J Psychiatry. 2001;178:234–41.

Thase ME, Montgomery S, Papakostas GI, Bauer M, Trivedi MH, Svedsäter H, Locklear JC, Gustafsson U, Datto C, Eriksson H. Quetiapine XR monotherapy in major depressive disorder: a pooled analysis to assess the influence of baseline severity on efficacy. Int Clin Psychopharmacol. 2013;28:113.

Vaishnavi SN, Nemeroff C, Plott SJ, Rao SG, Kranzler J, Owens MJ. Milnacipran: a comparative analysis of human monoamine uptake and transporter binding affinity. Biol Psychiatry. 2004;55(3):320–2.

Vetulani J, Sulser F. Action of various antidepressant treatments reduces reactivity of noradrenergic cyclic AMP-generating system in limbic forebrain. Nature. 1975;257:495–6.

Wolkowitz OM, Reus VI, Chan T, Manfredi F, Raum W, Johnson R, Canick J. Antiglucocorticoid treatment of depression: double-blind ketoconazole. Biol Psychiatry. 1999;45:1070–4.

Zarate Jr CA, Singh JB, Carlson PJ, Brutsche NE, Ameli R, Luckenbaugh DA, Charney DS, Manji HK. A randomized trial of an N-methyl-D-aspartate antagonist in treatment-resistant major depression. Arch Gen Psychiatry. 2006;63:856–64.

Zarate Jr CA, Brutsche NE, Ibrahim L, Franco-Chaves J, Diazgranados N, Cravchik A, Selter J, Marquardt CA, Liberty V, Luckenbaugh DA. Replication of ketamine's antidepressant efficacy in bipolar depression: a randomized controlled add-on trial. Biol Psychiatry. 2012;71: 939–46.

# Chapter 3
# Cell-Based Systems of Depression: An Overview

**Danuta Jantas**

**Abstract** From several years, cellular models have been used solely or in combination with animal studies to investigate the mechanisms of depression and antidepressant drug action. Moreover, such models based on known mechanism could be used as a first screening platform for searching new agents with putative antidepressant activity. For those purposes, primary neuronal cell models (cortical or hippocampal neurons), neuronal cell lines (rat clonal pheochromocytoma PC12, human neuroblastoma SH-SY5Y, mouse hippocampal HT-22), primary glial cell cultures (astrocytes, microglia), and glial cell lines (rat glioma C6, microglial BV2) could be engaged. In this book chapter an overview on the used cellular models in experimental studies of depression will be presented. There will be an attempt to standardize all available protocols for particular cell systems with the emphasis on advantages and limitations of each model.

**Keywords** Primary neuronal and glial cell cultures • SH-SY5Y • PC12 • HT-22 • BV2 • Antidepressants • Natural products • Herbs • Neuroplasticity • Neurogenesis • BDNF

## 3.1 Introduction

Over 60 years of studies on depression mechanisms in human subjects as well as in experimental models have not fully explained the pathogenesis of this chronic, recurring, and potentially life-threatening illness (due to the risk for suicide), which affects up to 20 % of the population across the globe (Berton and Nestler 2006). Although depression is a very heterogeneous disease, it is believed that about 40–50 % of the risk for this syndrome is genetic; however, the specific genes have not been identified yet. The remaining 50–60 % of the non-genetic risk is also poorly defined, with suggestions that various adverse life events

D. Jantas (✉)
Department of Experimental Neuroendocrinology, Institute of Pharmacology,
Polish Academy of Sciences, Smetna 12 Street, PL31 343 Krakow, Poland
e-mail: jantas@if-pan.krakow.pl



C. Grosso (ed.), *Herbal Medicine in Depression*,
DOI 10.1007/978-3-319-14021-6_3

(e.g., early childhood trauma, emotional stress, physical illness, viral infections) might be involved (Vialou et al. 2013; Lopizzo et al. 2015). Thus it is highly probable that the abovementioned detrimental factors interact with undefined yet genes to cause the illness. Since depression is a complex neuropsychiatric disorder affecting several neuronal circuits in the brain (Nestler et al. 2002; Berton and Nestler 2006), one could argue if using simple cellular models is relevant for searching for mechanisms of depression and testing the efficiency of antidepressant drugs. In reality, the cellular models in depression research are usually employed to confirm observations from human subjects and/or animal studies, and quite often it is only a way for the clear distinction of cell-type-specific response. When the mechanisms are known, it is possible to use cell models as a first screening platform for searching new putative antidepressants. However, such search is limited to the studies on known mechanisms of action of clinically used drugs, which is based mostly on acute action of available antidepressant drugs acting on monoaminergic transmission which is not, per se, therapeutic (Berton and Nestler 2006). Thus it is rather very unlikely that new-target compounds with antidepressant activity will be discovered using templates for monoaminergic compounds. Recently discovered, once again by serendipity, the fast antidepressant effect of ketamine directed research to test the efficiency of other *N*-methyl-D-aspartate (NMDA) receptor antagonists and study the mechanisms involved in its fast therapeutic effect (Lodge and Mercier 2015; Iadarola et al. 2015). In this book chapter, an overview on the cellular models employed for investigation of mechanisms of depression as well as in searching for new classes of antidepressants will be presented. There will be an attempt to standardize all available protocols for particular cell systems with the emphasis on advantages and limitations of each model. The detailed description of currently used cell models will be preceded by short description of actual hypothesis which are thought to contribute to depression etiopathogenesis.

## 3.2 Mechanisms of Depression: Hypothesis

The oldest hypothesis of depression (monoaminergic) was established in 1965 and assumes that deficiencies and/or imbalances in the serotonin (5-HT), the noradrenaline (NA), and presumably the dopamine (DA) systems underlie the depressive symptoms (Schildkraut 1995; Nestler 1998; Berton and Nestler 2006). In fact, the nowadays available antidepressant drugs, for example, SSRIs (selective serotonin reuptake inhibitors) fluoxetine and citalopram; NRIs (noradrenaline reuptake inhibitors) reboxetine and atomoxetine; SNRIs (serotonin and noradrenaline reuptake inhibitors) venlafaxine and duloxetine; MAOIs (monoamine oxidase inhibitors) tranylcypromine and phenelzine; and atypical antidepressants mirtazapine, tianeptine, and bupropion, were developed on the template of the first tricyclics (TCAs: imipramine, desipramine) which act on the reconstitution of normal synaptic 5-HT and NA levels in the brain (Berton and Nestler 2006).

Another early hypothesis of depression was the impact of stress and the abnormalities in the hypothalamic–pituitary–adrenal (HPA) axis that arised from the clinical observations and showed a hyperactive HPA axis in one to two thirds of depressed patients, which were normalized by various antidepressant drugs (Nestler et al. 2002; Berton and Nestler 2006). Thus much effort has been directed to the corticotropin releasing factor (CRF), glucocorticoid, substance P, vasopressin, and orexin systems as targets for the development of novel antidepressants (Berton and Nestler 2006; Paez-Pereda et al. 2011; Raedler 2011; Aubry 2013; Watson et al. 2012; Yeoh et al. 2014; Chen et al. 2015).

The neuroinflammatory hypothesis of depression has been originated from clinical observations of elevated pro-inflammatory cytokines (e.g., interleukin-1α (IL-1α) and -6 (IL-6), tumor necrosis factor α (TNF α), and interferon-γ (INF-γ)) and other inflammation-related proteins (e.g., C reactive protein—CRP) in plasma and cerebral spinal fluid (CSF) of depressive subjects which were normalized by antidepressant treatment (Miller et al. 2009; Hashioka 2011; Anderson et al. 2013; Abelaira et al. 2014; Hashimoto 2015). However, despite of many studies, it is still not clear whether inflammation is a primary or secondary process contributing to depression (Anderson et al. 2013; Bakunina et al. 2015; Cattaneo et al. 2015). On the other hand, there are some evidences that immune suppression can lead to depression which is exemplified by high prevalence of depression among patients with hepatitis C treated with INF-γ (Nestler et al. 2002; Berton and Nestler 2006).

The neurotrophic hypothesis of depression has been proposed on the basis of decreased brain-derived neurotrophic factor (BDNF) expression observed in animal models as well in the postmortem suicide patients' brain tissue (Duman 2004). Thus there has been great hope in finding small BDNF mimetics or small molecules affecting the cell signaling pathways activated by BDNF (e.g., tropomyosin receptor kinase B (TrkB) receptor, Ras/Raf/ERK, phosphatidylinositol 3-kinase/v-akt murine thymoma viral oncogene homologue (PI3-K/Akt), phospholipase C gamma (PLCγ)) or by agents inducing BDNF expression (e.g., activators of protein kinase A/cAMP/cAMP response element-binding protein (PKA/cAMP/CREB) signaling or phosphodiesterase (PDE) inhibitors) (Duman et al. 1997; Duman 2004; Berton and Nestler 2006). Among other neurotrophic factors suggested to play a role in depression are fibroblast growth factor 2 (FGF2), vascular endothelial growth factor (VEGF), and glia-derived neurotrophic factor (GDNF) (Duman 2004; Newton and Duman 2004).

The mood-elevating properties of anti-infectious agents with some NMDA glutamate receptor antagonist activity (D-cycloserine and amantadine) led to the establishment of glutamatergic hypothesis of depression (Paul and Skolnick 2003; Sanacora et al. 2003, 2012). A recently described rapid antidepressant effect of ketamine reawakened studies on NMDA receptor antagonists (memantine, AZD6765, MK-0657, dextromethorphan) and glycine-site partial agonists of NMDAR (e.g., GLYX-13 and D-cycloserine) as novel classes of antidepressants with fast onset of action and stable and lasting effects (Niciu et al. 2014; Brachman et al. 2015; Iadarola et al. 2015; Lodge and Mercier 2015; Pochwat et al. 2014; Zhang et al. 2015). Among other glutamatergic-acting compounds studied as putative antidepressants are ligands for metabotropic glutamate receptors (mGluRs) and AMPA receptor potentiators (Alt et al. 2006; Bleakman et al. 2007; Krystal et al. 2010; Niswender and Conn 2010; Palucha-Poniewiera et al. 2010, 2013, 2014; Freudenberg et al. 2015; Partin 2015).

The neuroplasticity hypothesis of depression assumes that stress-induced changes in hippocampus (decreased neurotrophin level, dendritic de-branching, reduced neurogenesis, induction of apoptosis) could lead to network disturbances in the medial prefrontal cortex (mPFC) causing cognitive impairments observed in depressed patients (Newton and Duman 2004; Anacker et al. 2013; Jett et al. 2015). In contrast, the increased dendritic abortization in emotional circuits (e.g., amygdala) has been shown after stress. Therapeutic effects of rapid-acting antidepressants (ketamine) as well as some of long-acting (classical) one have been linked with their ability to rapidly regulate neuronal excitability and thereby to increase synaptic translation and release of BDNF, to activate the TrkB-mTOR-p70S6k signaling pathway, and to increase synaptogenesis within the prefrontal cortex changes in TrkB signaling, synaptic plasticity, and neuronal excitability. These observation led to conclusion that these are shared neurobiological phenomena implicated in antidepressant responses produced by both gradually and rapidly acting antidepressant drugs (Rantamäki and Yalcin 2015; Thompson et al. 2015).

More than a decade ago, a theory about abnormal functioning of glial cells in mood disorders has been put forward based on the numerous postmortem histopathological studies that demonstrated the reduced glial cell numbers in various fronto-limbic areas of depressed patients (Rajkowska and Miguel-Hidalgo 2007; Drevets et al. 2008). Recent developments suggest that apart from their housekeeping functions astrocytes in dynamic way could regulate synaptogenesis, synaptic strength, and stability and control adult hippocampal neurogenesis (Czéh and Di Benedetto 2013). Moreover, there are data showing that antidepressants not only modify astroglial physiology, morphology, and gliogenesis but could also have influence on intracellular signaling pathways (e.g., MAPK/ERK1/2, PI3-K, NFκB), gene expression (glial fibrillary acid protein (GFAP), vimentin, aquaporin 4, or Ndrg2), receptor expression (vascular endothelial growth factor receptor (VEGFR), TrkB), and the release of various trophic factors (BDNF, VEGF, GDNF) (Sofroniew 2005; Hamby and Sofroniew 2010; Czéh and Di Benedetto 2013).

In recent years, the discovery of epigenetic mechanisms, which control chromatin structure and function and mediate changes in gene expression that occur in response to diverse stimuli, have been considered to be involved in etiology of affective disorders and therapeutic efficiency of used medications (Vialou et al. 2013). Epigenetic investigations promise to improve the understanding of the mechanisms by which individuals exhibit widely varying responses to adverse life events, both during development and in adulthood. Several different types of epigenetic mechanisms have been implicated to date, including stress- or antidepressant-induced changes in DNA methylation, histone acetylation and methylation, and miRNAs, and histone deacetylase (HDAC) inhibitors have shown antidepressant-like responses in diverse animal models (Berton and Nestler 2006; Vialou et al. 2013).

## 3.3 Neuronal Cell Models for Depression

Based on bibliographic search of cellular models used in depression research, it could be assumed that neuronal cells are the most frequently used. This is tightly correlated with the evolution of depression theories which, at the very beginning, have been mostly concentrated on the impact of stress on neurons and neuronal brain networks. Since frontal regions of the cortex and hippocampus, which are thought to mediate the cognitive aspects of depression, are among the most often studied brain regions in depressive-like animal models of depression, thus, rodent (mouse or rat) primary neuronal hippocampal and cortical cell cultures are most commonly employed. Other neuronal cellular models studied are PC12 rat pheochromocytoma, human neuroblastoma SH-SY5Y, and HT-22-immortalized mouse hippocampal cell line.

### *3.3.1 Primary Neuronal Cell Cultures*

Despite of all difficulties connected with preparation and long-term maintenance of cultures, rodent primary neuronal cells are widely used for in vitro studies of neuronal toxicology, pharmacology, electrophysiology, gene expression, development, and effects of growth factors and hormones. Their main advantage over the other cellular models used in neuro-field is the fact that they present all the characteristics of living neurons with maintenance of excitability features under cell culture conditions and the presence of the whole neurotransmitters' signaling machinery (their synthesis/release/reuptake systems and receptors). The most commonly used neuronal cell cultures are the embryonic ones (hippocampal, cortical) since the procedure of isolation of cells from the embryo brain tissue is more efficient and survival rate of neurons is better when compared to postnatal neuronal cell cultures.

#### 3.3.1.1 Methodology

Brain tissues (hippocampi or cortices) for the preparation of primary neuronal cell cultures are usually taken from mouse/rat embryos at 15/16 (cortical) and 17/18 (hippocampal) day of gestation (Brewer et al. 1993; Brewer 1995). After dissection, meninges are removed and tissue is collected in ice-cold buffer (phosphate buffer solution (PBS) or Hanks' balanced salt solution (HBSS)). Next, minced with a scalpel, tissue is dissociated by enzymatic digestion with trypsin (usually 0.1 % in PBS/HBSS for 20–30 min. at room temperature (RT)) and mechanically triturated in presence of 20 % fetal bovine serum (FBS) and DNAse I (150 Kunitz units). After gentle centrifugation (5 min at $100 \times g$), the isolated cells are suspended in neuronal cell culture medium, counted and seeded onto poly-ornithine (0.05–0.1 mg/ml)- or polylysine (0.025–0.5 mg/ml)-coated multi-well plates. Since neurons do not proliferate in culture, it is important to seed cells in the right density which will allow them to grow properly and make connections. Neuronal cell culture medium regularly used nowadays for culturing of embryonic neurons is a serum-free medium described by Brewer et al. (1993, 1995) which was optimized for survival of rat embryonic hippocampal (and also for other types of neurons) after 4 days in culture. This medium consists of Neurobasal medium, supplement B27 (2 %), glutamine (0.5–2 mM), and penicillin/streptomycin mixture (usually at 10 times lower concentrations than usually used for other cell types or omitted for some studies since it can be harmful for neurons) which is fully or partially changed (about half) to a fresh one each 2–3 day. In Neurobasal/B27 medium, glial growth is reduced to less than 0.5 % of the nearly pure neuronal population, as can be judged by immunocytochemistry for GFAP and neuron-specific markers (e.g., enolase, MAP-2, or NeuN) (Brewer et al. 1993). Despite of this, at some points of cell culture (usually day 3 or 4), some researchers additionally use cell proliferation inhibitors (e.g., doxifluridine, cytosine β-D-arabinoside (Ara-C), or adenine-9-β-D-arabinofranoside (Ara A)) to halt the growth of nonneuronal cells (Okugawa et al. 1999; Chu et al. 2010; Wuwongse et al. 2013). However, these agents can sometimes be detrimental for neurons, especially in a situation when experimental setting requires the use of reduced level of B27 or usage of B27 without antioxidants. Neurons after seeding are kept in a humidified incubator at 37 °C and 5 % $CO_2$ usually for 12–14 days after which they are thought to be of mature phenotype (differentiated). With the aim of studying some developmental (e.g., apoptotic processes) or aging aspects, younger (less than 7 days) and older (more than 19 days) cell cultures can be used. More rarely, postnatal tissue (taken from 1/2-day-old mouse/rat pubs) is used for the preparation of postnatal hippocampal neurons. This type of neuronal cell cultures is usually more useful for electrophysiological studies (Andjus et al. 1997; Amaral and Pozzo-Miller 2007; Wang et al. 2010). However, for other applications (e.g., cell viability assays, ELISA, or Western blot analysis), embryonic neuronal cell cultures are used since the number of isolated cells from the brains of embryos are much higher than those from postnatal ones. Moreover, since the postnatal tissue is much thicker than the embryonic one, the dissociation of neurons is also challenging and the survival rate of these neurons are relatively lower than the embryonic ones. The main modifications required for culturing of postnatal neurons are the digestion step, which is done usually for a longer time (even for 45 min) with papain at 37 °C, and later on, the cells are seeded on poly-D-lysine/laminin- or poly-ornithine/matrigel-coated multi-well plates (Andjus et al. 1997; Amaral and Pozzo-Miller 2007; Heiser et al. 2013). Postnatal neurons can be successfully cultured in serum-free Neurobasal/B27 medium (Amaral and Pozzo-Miller 2007; Wang et al. 2010). It is also possible to maintain such cultures in another type of neuronal media containing serum, for example, Eagle's minimum essential medium (EMEM) supplemented with 10 % FBS and, optionally, with some other factors (e.g., D-glucose, apo-transferrin, insulin, vitamin B12, biotin), but cell proliferation inhibitors should be used during culture period to inhibit glial growth (Andjus et al. 1997; Okugawa et al. 1999). Among other primary neuronal models which have been used to study depression are midbrain, hindbrain, and hypothalamic and amygdala neuronal cell cultures; however, they are used rather occasionally since they are characterized by low purity (contamination with glial cells) and/or low yield (Pepin et al. 1989; al-Damluji et al. 1993; Zhang et al. 2012b).

#### 3.3.1.2 Depression Aspects Studied in Primary Neuronal Cell Cultures

At the very beginning of the use of primary neuronal cell cultures to study mechanistic aspects of depression and/or antidepressant drug action, research was mainly concentrated on the impact of stress on monoamines uptake, on neurotransmitter receptor expression and function, and on the expression of stress-related factors (e.g., GR mRNA and protein level) and how these changes were influenced by acute or/and chronic antidepressants (Pepin et al. 1989; Hösli and Hösli 1995; Okugawa et al. 1999; Pariante et al. 2003). Since the hippocampal area contains a large number of GR, making it prone to chronic stress, hippocampal neuronal cell cultures have been used to test the mechanisms by which corticosterone (CORT) or dexamethasone (DEX) evoke cell damage (mainly via induction of apoptosis) and to evaluate the neuroprotective effects of various antidepressant drugs as well as new putative antidepressant compounds (Crochemore et al. 2005; Moosavi et al. 2008; Liu et al. 2011a, b). There are also data demonstrating that CORT evoke changes in cell viability in other types of primary neurons as shown in primary hypothalamic neurons (Zhang et al. 2012a, b). Moreover, by using various types of neuronal cell cultures, it is possible to show the brain-specific mechanisms responsible for protective effects of several compounds, as, for example, was shown for icariin, a flavonol from the Chinese plant *Epimedium sagittatum* Maxim. Icariin attenuated the CORT-induced hippocampal and hypothalamic cell apoptosis via activation of the PI3-K/Akt pathway and inhibition of p38MAPK, respectively (Liu et al. 2011a, b; Zhang et al. 2012a, b). In respect to the involvement of neurodegeneration in depression, apart from GR-acting agents, other cell death inducers (e.g., NMDA, hypoxia, staurosporine, doxorubicin) have been used for testing neuroprotective potency of antidepressants and new compounds in cortical and in hippocampal neuronal cell cultures (Drzyzga et al. 2009; McKernan et al. 2009; Jantas et al. 2014a). Importantly, some new mechanisms can be first observed in cellular models, e.g., induction of autophagy by antidepressants (Zschocke and Rein 2011; Zschocke et al. 2011), and later on, they could be confirmed by in vivo studies revealing the impaired mechanism (in this case autophagy) in depression as well as possible therapeutic intervention by influence on this process (Mattson 2012; Polajnar and Zerovnik 2014; Gassen et al. 2014). Moreover, using a proteomic approach and primary neuronal cell cultures, it is possible to detect new mechanisms able to explain the differential effect of some compounds as has been shown, for example, for tianeptine which changes expression of proteins involved in neurite growth, metabolism of neurotrophic substances, synaptogenesis, and synaptic activity homeostasis (McEwen and Chattarji 2004; Chu et al. 2010; McEwen et al. 2010).

A reduced neurotrophic signaling (e.g., BDNF) and impaired neuronal plasticity connected with attenuated neurogenesis, synaptogenesis, and neuritogenesis found in stress-based animal studies and in clinic observations could be used in cellular models as template for searching for new putative antidepressant compounds. For example, Musazzi et al. (2014) recently applied a rapidly detectable and highly sensitive reporter gene assay in primary cortical and hippocampal neurons to characterize the selective activation profile of BDNF and cAMP response element (CRE) promoters. This could be used as a platform for searching new compounds with antidepressant and/or neuroprotective potential. Moreover, new mechanisms could be discovered as has been revealed for transient receptor potential-canonical 3 (TRPC3) channel which activation mediated the BDNF-evoked dendritic remodeling in hippocampal neurons (Amaral and Pozzo-Miller 2007). Modern molecular approaches used nowadays could be beneficial for confirmation of older hypothesis in both in vitro and in vivo models. For example, the lentiviral shRNA silencing of BDNF (LV-shBDNF-3) added to primary hippocampal neurons and injected into the dentate gyrus of the hippocampus reduced the cell viability and evoked changes in animal behaviors associated with depression, effects prevented by the herbal medicine formula "Kai-Xin San" (KXS) (Hu et al. 2014b).

The decrease in neuronal plasticity markers in the prefrontal cortex and hippocampal regions observed in depression could be modeled in primary hippocampal neurons exposed to various harmful agents (e.g., CORT, exogenous neurotoxicants) which evoked decrease in synaptogenesis, neuritogenesis, and NMDA receptor-mediated postsynaptic currents in hippocampal neurons (Wuwongse et al. 2013; Pizzurro et al. 2014). These changes could be prevented, for example, by natural compounds with antidepressant activity (e.g., the naphthodianthrone, hypericin) (Wuwongse et al. 2013). Moreover, neurotrophic and neuroprotective effects of synthetic antidepressant and some natural compounds with putative antidepressant activity were observed during development of hippocampal neurons in cell culture, where promoting effects on synaptogenesis, neuritogenesis, and NMDA receptor-mediated postsynaptic currents were shown for herbal medicines: *Moringa oleifera* Lam., *Gelidium amansii* J.V. Lamouroux, and hypericin (one of the active compounds of *Hypericum perforatum* L. extract) (Wang et al. 2010; Leuner et al. 2013; Hannan et al. 2014a, b). Moreover, the detailed mechanisms responsible for the improvements in synaptic plasticity promoted by tested agents could be detected as was for hyperforin (another active compound from *Hypericum perforatum* L. extract) which has been shown to improve synaptic plasticity by activation of TRPC6 followed by stimulation of different intracellular pathways including Ras/MEK/ERK, PI3-K/Akt, and calcium/calmodulin-dependent protein kinase type IV (CAMKIV) (Heiser et al. 2013). It should be underlined that in the era of increased high-content cell-based screenings applied for neuronal plasticity and regeneration studies, more sophisticated analysis tools (e.g., MorphoNeuroNet) are implemented which will help in determining the effects of tested compounds on cultures with dense neurite networks (Pani et al. 2014). Some neuronal developmental mechanisms can also be explored by using primary neuronal cell cultures in combination with animal studies, as was exemplified by research on epigenetic modulators of neuronal development and demonstration of a regulation of axonal branching by microRNA-29 via targeting doublecortin protein expression (Li et al. 2014).

From large-scale experimental approaches, a novel technique for rapid and efficient screening of neuroactivity by using primary cortical neuronal networks grown on multielectrode neurochips has been also proposed (Gramowski et al. 2006). In such system, the multiparametric assessment of electrical activity changes caused by psychoactive herbal extracts from *Hypericum perforatum* L., *Passiflora incarnata* L., and *Valeriana officinalis* L. and various combinations thereof revealed a receptor-specific and concentration-dependent inhibition of the firing patterns (Gramowski et al. 2006). Among the advantages of this such high-content screening platform are its sensitivity, selectivity, and robustness of neurochip recordings which enable testing complex mixtures of neuroactive substances to assess their therapeutic potential (Gramowski et al. 2006).

The impact of stressful conditions on neurogenesis as well as on neurogenic potential of various antidepressant drugs or new compounds with antidepressant activity is usually studied in animals by measurement of newborn neurons in neurogenic niches (e.g., subventricular zone (SVZ), dentate gyrus of the hippocampus, or olfactory bulbs). From cellular models, cerebellar granular cell (CGC) cultures have been demonstrated to be efficient for the assessment of antidepressant drug-induced neurogenesis (Zusso et al. 2004). Moreover, recently developed models of the embryonic cortical progenitor cells (Wahane et al. 2014), hippocampal neural stem cells (Pao et al. 2012; Wang et al. 2014), fetal mesencephalic neural stem cells (fmNSCs) (Storch et al. 2003; Brunn et al. 2014) or primary cultures of the adult subependymal zone (SEZ) cells (Ortega et al. 2013), or subventricular zone (SVZ)-derived neurospheres (Schramm and Schulte 2014) could be employed in future studies as a screening platform for searching new compounds with neurogenic activity.

## 3.3.2 *Rat Pheochromocytoma PC12 Cell Line*

PC12 is a clonal cell line of neuroectodermal origin, and it is derived from rat adrenal medulla pheochromocytoma (Greene and Tischler 1976). It can be induced to differentiate by the nerve growth factor (NGF) to a sympathetic phenotype expressing neurites and excitability (Greene and Tischler 1976). PC12 cell line is extensively used as a tool for studying the function of neurons, neuronal differentiation, and neurotoxicity. Since it originates from adrenal glands, which are a component of the HPA axis—a pathway disturbed in depression (Berton and Nestler 2006)—it is the most often used among various cell lines for studies on depression and antidepressant drug action mechanisms.

### 3.3.2.1 Methodology

PC12 cells are usually cultured in Dulbecco's modified Eagle's medium (DMEM) supplemented with heat-inactivated (10 %) fetal calf serum (FCF), 5 % heat-inactivated horse serum (HS), and penicillin/streptomycin mixture (50–100 U/ml and 50–100 μM/ml, respectively) and kept under standard mammalian cell culture conditions (37 °C in a humidified incubator containing 5 % $CO_2$) (Leuner et al. 2007; Heiser et al. 2013; Ishima et al. 2014). The percentage used in horse and calf sera varies between various studies (Li et al. 2003b; Han and Lee 2009) as well as other cell culture media (e.g., RPMI) (Hindley et al. 1997; Han and Lee 2009; Paul et al. 2014) or nutrient mixtures (e.g., Ham F12 nutrient mixture) used (Terada et al. 2014). Under basal conditions, PC12 cells grow in suspension and are often used for studies on neurotoxicity/neuroprotection (Kritis et al. 2015) or can be easily and efficiently transfected with liposome/polymer reagents to study various biological mechanisms (Karkoulias et al. 2006; Leuner et al. 2007; Heiser et al. 2013; Ip et al. 2014). This latter feature is one of the main advantages of the use of this cell line for mechanistic studies, since primary neuronal cultures as well as other neuronal cell lines (e.g., SH-SY5Y cells) are generally difficult to transfect with liposome/polymer reagents, and then use of viral transduction is required.

Most often, PC12 cells are used to study the impact of various detrimental (e.g., CORT, DEX) and protective compounds (e.g., antidepressants, neuroprotectants) on the neurite outgrowth of PC12 cell. For this purpose, PC12 cell should be differentiated, which is usually done by seeding cells in low density on collagen-, polylysine-, or poly-D-lysine/laminin-coated plates and culturing them in cell differentiation medium. In these protocols, such medium is the same as that used for culture maintenance (DMEM or RPMI with antibiotics), but serum level is usually decreased (e.g., 3 % of HS and 1.5 % FCF) and the medium is supplemented with NGF at various ranges of concentrations which is usually experimentally set (50–300 ng/ml) (Leuner et al. 2007; Nishimura et al. 2008; Wang et al. 2013; Terada et al. 2014). Such differentiation is usually done for 7–10 days with 3-times medium exchange during the week. When the impact of other stimulating factors (e.g., synthetic drugs, natural compounds) on neurite outgrowth is tested, NGF level is lowered to 0–2.5 ng/ml (Matsui et al. 2012; Xu et al. 2012; Heiser et al. 2013; Ishima et al. 2014, 2015). In literature, there are examples of differentiation of PC12 cells by glioma-conditioned medium (GCM) and retinoic acid (RA) which have been shown to induce neurite outgrowth and elevation of the activity of the cholinergic marker enzyme choline acetyltransferase (ChAT) (Mizuno et al. 1989; Matsuoka et al. 1989).

#### 3.3.2.2 Depression Aspects Studied in PC12 Cell Line

From many years, PC12 cells served and still serve as a platform for screening antidepressant activity of new compounds solely or in combination with primary neuronal cell cultures (hippocampal) or with behavioral animal tests. Many early studies on PC12 cell line in respect to depression were concentrated among the influences of glucocorticoids or antidepressants on monoamine metabolism and their receptor and transporter expression and neurotrophic factor levels (Divish et al. 1991; Foreman et al. 1992; McMahon and Sabban 1992; Dunigan and Shamoo 1995; Hindley et al. 1997; Zhu et al. 1997). The standard assays which have been used for many years for first screening of potential antidepressant activity in PC12 cells are (1) the CORT-induced cell death which assesses neuroprotective properties of the tested compounds as well as their mechanisms and (2) the NGF-stimulated neurite outgrowth where the putative antidepressants potentiate this effect. There are plenty of data demonstrating that clinically used antidepressants, natural compounds with already described antidepressant activity in in vivo models, and new compounds have shown positive results in both models (summarized in Tables 3.1 and 3.2). By using the neurite outgrowth assay in NGF-differentiated cells, it is also possible to test combination of different drugs/compounds in order to find synergistic effects in the promotion of neuritogenesis as has been shown, for example, for brexpiprazole + fluoxetine, but not for brexpiprazole + paroxetine (Ishima et al. 2015). CORT used to evoke cell damage in PC12 is used at different concentrations (10–500 μM) for different time exposures (2–48 h). Cell death induced by this agent is thought to be an apoptotic one (lower concentrations) or necrotic–apoptotic (higher concentrations) with increasing of $[Ca^{2+}]_i$, induction of mitochondrial pathway of apoptosis mediated by decrease in mitochondrial membrane potential (MMP), cytochrome c release, activation of caspases, and apoptotic DNA fragmentation (Wu et al. 2013a; Jiang et al. 2015). The neuroprotective effects of tested compounds are usually confirmed by biochemical cell viability/toxicity assays (MTT reduction and LDH release assays, respectively) and measurements of $[Ca^{2+}]_i$ and of apoptotic markers (e.g., caspase-3 activity, DNA fragmentation, Bcl-2/Bax ratio). In many studies, a more detailed mechanism has been studied in respect to detrimental action of GR agonists (CORT or DEX) and protective effects of antidepressants (e.g., an involvement of intracellular signaling pathways, induction of transcription factors, or effects on gene expression) (Foreman et al. 1992; Nakano et al. 2010; Heiser et al. 2013; Terada et al. 2014). Neuroprotective effects of putative antidepressant are also widely studied in other models of PC12 cell injury, e.g., mediated by oxidative stress (hydrogen peroxide ($H_2O_2$); 6-hydroxydopamine (6-OHDA); 1-methyl-4-phenylpyridinium (MPP+)) or glutamatergic overstimulation (glutamate, NMDA) (Table 3.3). Since PC12 cells are relatively easily transfected with liposomes/polymer reagents, thus, some reporter gene assays have been established in PC12 cells to screen particular mechanisms of psychoactive drugs or new compounds with suspected antidepressant activity. An example is the study by Schwaninger et al. (1995) where a rage of various antidepressant drugs were tested on the CREB/CRE-directed gene transcription by using transient transfections of reporter fusion genes in PC12 cells. In another study, promoters of neurofilaments, NF68 and NF200, linked with luciferase reporter gene (pNF68-Luc and pNF200-Luc) were applied in PC12 cells to optimize the compatibility of paired herbs in an ancient Chinese herbal decoction composed of four species, Kai-Xin-San (Yan et al. 2015). Other possible application of PC12 cells for studying depression is the exploration of the role of particular receptors on neuronal plasticity as has been demonstrated for α(2)-adrenergic receptors or Ndrg2 receptors (Takahashi et al. 2005; Karkoulias et al. 2006) or in cells where the hyperforin-mediated stimulation of neurite outgrowth was not observed in cells expressing a dominant-negative mutant of TRPC6 (Leuner et al. 2007). Also, physical interaction between various types of receptors, formation of heteroreceptors complexes, and their synergism in stimulating neuritogenesis can be investigated by using PC12 cells. Such interplay has been shown for adenosine A(2A) receptor and fibroblast growth factor receptor (FGFR) and for FGFR1 and

**Table 3.1** Neuroprotection mediated by natural compounds and antidepressants against corticosterone (CORT)-evoked cell death in PC12 cells

| | CORT | Methods and mechanisms | Other models | Reference |
|---|---|---|---|---|
| Extract of *Morinda officinalis* F.C. How (P6 oligosaccharides) (3; 10 μM)<br>DIM (1; 5 μM) | 10 μM for 5 days | DNA fragmentation<br>PI staining | – | Li et al. (2003b) |
| DIM (0.625–10 μM)<br>FLU (0.625–10 μM)<br>MOC (2.5–40 μM) | 200 μM for 48 h | MTT assay<br>$[Ca^{2+}]_i$ (DMI, FLU)<br>NGF mRNA (DMI, FLU) | – | Li et al. (2003c) |
| Extract of *Morinda officinalis* F.C. How inulin-type hexasaccharide (IHS) (0.625; 1.25 μM)<br>DIM (0.25; 1 μM) | 200 μM for 48 h | MTT assay<br>$[Ca^{2+}]_i$<br>NGF mRNA | – | Li et al. (2004) |
| Saponins extracted from Chaihu-jia-longgu-muli-tang (CLM: 25–100 μg/ml)<br>FLU (10 μM) | 10 μM for 48 h | MTT assay, LDH release<br>$[Ca^{2+}]_i$<br>NGF mRNA | Mice TST<br>Rat FST | Zhu et al. (2006) |
| Curcumin (0.01–1 μM)<br>DIM (0.1–10 μM)<br>pretreatment for 2 h | 200 μM for 2 h | MTT assay<br>Via ERK1/2 inhibition | – | Zhou et al. (2009) |
| Peony (*Paeonia* L.) glycosides (1–10 mg/l) | 200 μM for 48 h | caspase-3<br>bcl2/bax mRNA | – | Mao et al. (2009) |
| Extract of *Apocynum venetum* L. (25–100 μg/ml)<br>FLU (10 μM) | 10 μM for 48 h (serum-free medium) | MTT assay, LDH release<br>$[Ca^{2+}]_i$<br>BDNF mRNA<br>MAP4 mRNA | – | Zheng et al. (2011) |
| *Rosmarinus officinalis* L. polyphenols:<br>Luteolin (30 μM)<br>Carnosic acid (30 μM)<br>Rosmarinic acid (20 μM) | 200 μM for 48 h | MTT assay | Mice TST<br>MPK-1, TH, PH<br>Choline and Ach level | Sasaki et al. (2013) |
| Saikosaponins of *Bupleurum yinchowense* R.H. Shan & Yin Li (3.125–25 μg/ml) | 250 μM for 24 h | MTT assay, LDH release<br>$[Ca^{2+}]_i$, caspase-3<br>DNA fragmentation<br>Bcl-2/Bax<br>GRP78, GADD-153, XBP-1 | – | Li et al. (2013) |
| Extract of *Acanthopanax senticosus* Rupr. & Maxim (50–400 μg/ml) | 200 μM for 24 h | MTT assay, LDH release<br>$[Ca^{2+}]_i$<br>Annexin V-FITC PI labeling<br>BDNF mRNA, CREB | – | Wu et al. 2013a, b |

(continued)

**Table 3.1** (continued)

| | CORT | Methods and mechanisms | Other models | Reference |
|---|---|---|---|---|
| Venlafaxine (0.03–3 μM) | 500 μM for 24 h | MTT assay<br>DNA fragmentation via PI3-K/Akt/FoxO3a | Primary cortical neuronal cell cultures | Wang et al. (2013) |
| Agomelatine (1.25 μM)<br>Agomelatine analogs 4a and 5b (1.25 μM)<br>IMI (1.25 μM) | 200 μM for 48 h | MTT assay | C57 mice FST, active only IMI and analog 4a | Chang et al. (2014) |
| Naphthalene compounds (6 k, 6o, and 6p) (1.25 μM)<br>FLU (1.25 μM)<br>Agomelatine 1.25 μM<br>Mifepristone 1.25 μM | 200 μM for 24 h | MTT assay | $H_2O_2$ in PC12<br>Liver LO2 cells<br>Mice FST and OFT | Ang et al. (2014) |
| Extract of *Panax ginseng* C.A. Meyer | 250 μM for 24 h | MTT assay, LDH release<br>$[Ca^{2+}]_i$, MMP, ROS<br>DNA fragmentation<br>Caspase-3, -9, and -12<br>GR, Hsp 90, HDAC6, GRP78, GADD-153, XBP-1 | – | Jiang et al. (2015) |
| Extract of zhi-qiao (immature fruit of *Citrus aurantium* L.) (25–800 μg/ml) | 200 μM for 24 h | MTT assay, LDH release<br>Annexin V-FITC PI labeling<br>$[Ca^{2+}]_i$<br>BDNF mRNA, CREB | Mice TST and FST | Wu et al. (2015) |

*Ach* acetylcholine, *BDNF* brain-derived neurotrophic factor, *CREB* cAMP response element-binding protein, *CORT* corticosterone, *GADD-153* growth arrest and DNA damage-inducible protein 153, *GR* glucocorticoid receptor, *GRP78* glucose-regulated protein 78, *DIM* desipramine, *FLU* fluoxetine, *FST* forced-swimming test, *HDAC6* histone deacetylase 6, *Hsp* heat shock protein 90, *IMI* imipramine, *LDH* lactate dehydrogenase, *MMP* mitochondrial membrane potential, *MAP4* microtubule associated protein 4, *MTT* 3-(4,5-dimethy thiazol-2-yl)-2,5-diphenyl tetrazolium bromide, *MOC* moclobemid, *MPK-1* MAPK phosphatase 1, *NGF* nerve growth factor, *OFT* open field test, *PC* pyruvate carboxylase, *PI* propidium iodide, *ROS* reactive oxygen species, *TH* tyrosine hydroxylase, *TST* tail suspension test, *XBP-1* X-box DNA-binding protein-1

$5\text{-HT}_{1A}$ in the NGF-differentiated PC12 cells and in primary hippocampal neurons, with robust activation of MAPK/ERK1/2 pathway in both cases (Flajolet et al. 2008; Borroto-Escuela et al. 2012). Moreover, the synergism between FGFR1 and $5\text{-HT}_{1A}$ was confirmed in an animal study where acute and repeated intracerebroventricular treatment with FGF2 and 8-hydroxy-2-(dipropylamino)tetralin (8-OH-DPAT) produced highly significant antidepressant action in forced-swimming test (FST) (Borroto-Escuela et al. 2012), pointing to a good translation rate of results obtained in in vitro studies to in vivo one.

**Table 3.2** Stimulatory effect of various antidepressants and natural compounds on PC12 neurite outgrowth

| Compounds | NGF | Effect | Mechanisms | Other models | Reference |
|---|---|---|---|---|---|
| Sodium nitrite<br>Sodium nitroprusside | 2.5 ng/ml | + (0.1–10 mM)<br>+ (0.1–1 μM) | Via cGMP | Primary hippocampal neurons | Hindley et al. (1997) |
| Fluvoxamine<br>Sertraline<br>Paroxetine<br>SA4503<br>4-Phenyl-1-(4-phenylbutyl) piperidine (PPBP)<br>Dehydroepiandrosterone (DHEA) sulfate | 2.5 ng/ml | + (1–10 μM)<br>Inh. (10 μM)<br>Inh. (10 μM)<br>+ (1–10 μM<br>+ (0.1–10 μM)<br>+ (1–10 μM) | All via sigma-1 receptor (effect blocked by antagonists NE-100) effect SA4503 blocked by inhibitors of PLC-gamma, PI3-K, p38MAPK, JNK Ras/Raf/MAPK | – | Li et al. (2003a) |
| *Rosmarinus officinalis* L. extracts (1:20,000–1:5000)<br>Carnosic acid (2.4–6.8 μg/ml)<br>Rosmarinic acid (2.4–4.8 μg/ml)<br>NGF (50 μg/ml) | 0 ng/ml | +<br>+<br>+<br>+ | Via MAPK/ERK1/2<br>Ach and choline level<br>AChE activity | – | El Omri et al. (2010) |
| Ketamine<br>Imipramine | 0.625 ng/ml | + (10–10,000 nM)<br>+ (10–10,000 nM) | Via sigma 1 receptor (effect blocked by antagonists NE-100) | Mice FST | Robson et al. (2012) |
| Phenylbutenoid dimers from *Zingiber purpureum* Roscoe<br>NGF 10 ng/ml<br>bFGF 10 ng/ml | 0 ng/ml | + (10–30 μM)<br>+<br>+ | – | Primary cortical neurons<br>Hip neurogenesis in OBX mice | Matsui et al. (2012) |
| Isorhamnetin from *Ginkgo biloba* L.<br>NGF 50 ng/ml | 0.5 ng/ml | + (1–10 μM)<br>+ | Excluded: TrkA, PI3-K/Akt, ERK1/2<br>Via TrkA, PI3-K/Akt, ERK1/2 | – | Xu et al. (2012) |

(continued)

| | | | | | |
|---|---|---|---|---|---|
| Hyperforin | 0 ng/ml | + (0.3–10 μM) | Via TRPC6 and Ras/MEK/ERK, PI3K/Akt, and CAMKIV<br>CREB | Primary hippocampal neurons | Heiser et al. (2013) |
| Fluvoxamine<br>Sertraline<br>Fluoxetine<br>Escitalopram<br>Citalopram<br>Paroxetine<br>Duloxetine<br>Venlafaxine<br>Milnacipram<br>Mirtazapine | 2.5 ng/ml | + (1 and 10 μM)<br>Inh. (10 μM)<br>+ (1 and 10 μM)<br>+ (1 μM)<br>–<br>–<br>–<br>–<br>–<br>+ (10 μM) | Sigma 1 receptor<br>Sigma 1 receptor<br>Sigma 1 receptor<br>Via 5-$HT_{1A}$ | – | Ishima et al. (2014) |
| Brexpiprazole<br>Aripiprazole<br>Fluoxetine<br>Paroxetine | 2.5 ng/ml | + (0.1–1 μM)<br>+ (1 μM)<br>+ (1 μM)<br>Inh. (1 μM) | Via 5-$HT_{1A}$ and 5-$HT_{2A}$ receptors<br>Ca(2+)/IP3 and Hsp 90 | – | Ishima et al. (2015) |

*Ach* acetylcholine, *AChE* acetylcholine esterase, *bFGF* basic fibroblast growth factor, *CAMKIV* $Ca^{2+/}$ calmodulin-dependent protein kinase, *CREB* cAMP response element-binding protein, *FST* forced swimming test, *Hip* hippocampal, *Hsp 90* heat shock protein 90, *NGF* nerve growth factor, *OBX* olfactory bulbectomized, *TRPC6* transient receptor canonical 6

**Table 3.3** Neuroprotection mediated by antidepressants and natural compounds against cell damage induced by various factors in PC12 cells

| Tested compounds | Cell damage inducer | Methods and mechanisms | Other models | References |
|---|---|---|---|---|
| Agmatine (1–100 μM)<br>DIM (2.5 and 10 μM) | NMDA 200 μM for 24 h | MTT assay<br>LDH release<br>$[Ca^{2+}]_i$<br>Monoamine levels | Mice TST and FST<br>Rat FST | Li et al. (2003a) |
| Extract of *Hypericum perforatum* L. (20–100 μg/ml) | $H_2O_2$ 200 μM for 24 h | MTT assay<br>LDH release<br>ROS<br>DNA fragmentation | – | Lu et al. (2004) |
| Extract of:<br>*Hypericum perforatum* L. (10–100 μg/ml)<br>*Ginkgo biloba* L. (10–100 μg/ml)<br>*Apocynum venetum* L. (1–100 μg/ml) | 13-HPODE 10 μM for 1 h<br>diff. PC12 | ROS | – | Shirai et al. (2005) |
| AMI (50–100 μM)<br>FLU (50–100 μM)<br>Pretreatment for 24/48 h | $H_2O_2$ 200 μM for 24 h | MTT assay<br>SOD assay | – | Kolla et al. (2005) |
| DIM (1 and 10 μM)<br>FLU (0.625–10 μM)<br>MOC (2.5 and 10 μM) | NMDA 200 μM for 24 h | MTT assay<br>LDH release<br>$[Ca^{2+}]_i$ (DIM, FLU)<br>NOS activity | – | Li et al. (2006) |
| Ondamtanggamibang (ODG) (Oriental medicine)<br>Pretreatment for 12 h | $H_2O_2$ 400 μM for 1 h<br>6-OHDA 100 μM for 24 h | MTT assay<br>ROS<br>HO-1 | – | Lee et al. (2006) |
| extract of *Melissa officinalis* L. (60–80 μg/ml) | $H_2O_2$ 250 μM for 24 h | MTT assay<br>ROS assay<br>MAO-A assay | – | López et al. (2009) |
| AMI (5–25 μM)<br>Tranylcypromine (1–25 μM)<br>FLU (5–25 μM)—potentiation of cell damage | MPP(+) 500 μM for 24 h<br>diff. PC12 | MTT assay<br>ROS, Cyt.c<br>Caspase-3, Bax/Bcl-2<br>GSH | – | Han and Lee (2009) |
| Extract of *Hypericum perforatum* L. (12.5–50 μg/ml)<br>Pretreatment for 12 h | $H_2O_2$ 300 μM for 4 h | MTT assay<br>LDH release<br>DNA fragmentation | – | Zou et al. (2010) |
| Paeoniflorin (1–50 μM) | Glutamate 10 mM for 24 h | MTT assay<br>$[Ca^{2+}]_I$, ROS, SOD<br>Caspase-3<br>DNA fragmentation | – | Mao et al. (2010) |
| Paeoniflorin (1–50 μM)<br>DIM (10 μM) | NMDA 300 μM for 24 h | MTT assay<br>LDH release<br>$[Ca^{2+}]_i$<br>Calbindin-D28K mRNA | – | Mao et al. (2011) |

(continued)

**Table 3.3** (continued)

| Tested compounds | Cell damage inducer | Methods and mechanisms | Other models | References |
|---|---|---|---|---|
| Agomelatine (20 μM)<br>Duloxetine (10 μM) | $H_2O_2$ 250 μM for 4 h<br>diff. PC12 | ROS, MMP, $[Ca^{2+}]_i$<br>GSH<br>Via TRPM2 channel blocker of voltage-gated Ca(2+) channel (VGCC) | – | Akpinar et al. (2014) |

*AMI* amitriptyline, *DIM* desipramine, *FLU* fluoxetine, *FST* forced-swimming test, *GSH* glutathione, *HO-1* heme ooxygenase 1, *LDH* lactate dehydrogenase, *MAO-A* monoamine oxidase A, *MMP* mitochondrial membrane potential, *MTT* 3-(4,5-dimethy thiazol-2-yl)-2,5-diphenyl tetrazolium bromide, *MOC* moclobemid, *NMDA* *N*-methyl-D-aspartate, *NOS* nitric oxide synthase, *ROS* reactive oxygen species, *SOD* superoxide dismutase, *TPRM2* transient receptor potential melastatin 2, *TST* tail suspension test

### 3.3.3 Human Neuroblastoma SH-SY5Y Cells

The SH-SY5Y cell line is a thrice-cloned subline of the neuroblastoma cell line SK-N-SH which was established in 1970 from a bone marrow biopsy of a 4-year-old patient with neuroblastoma (Biedler et al. 1978). It exhibits a catecholaminergic phenotype including the ability to transport DA and NE, expresses enzymes for the synthesis and metabolism of DA and acetylcholine, and expresses receptors for DA and acetylcholine (Biedler et al. 1978; Willets et al. 1995). However, in respect to the presence of glutamatergic receptors in SH-SY5Y, there are contradictory data about the presence of active ionotropic glutamatergic receptors in these cells (Jantas et al. 2008, 2014b; Kritis et al. 2015). Neuronal-like differentiation of SH-SY5Y cells can be induced by various factors, for example, retinoic acid (RA), phorbol ester 12-*O*-tetradecanoylphorbol-13-acetate (TPA), BDNF, dibutyryl cAMP, or staurosporine (Tieu et al. 1999; Presgraves et al. 2004; Pregi et al. 2006; Sarkanen et al. 2007; Cheung et al. 2009). Retinoic acid is the most commonly used neuronal differentiation agent in SH-SY5Y cells, and it evokes many morphological (inhibition of proliferation, neurite elongation) and biochemical (changes in the expression of neuronal markers, various intracellular pathways, and bioenergetics parameters) changes (Lopez-Carballo et al. 2002; Miloso et al. 2004; Presgraves et al. 2004; Cheung et al. 2009; Agholme et al. 2010; Lopes et al. 2010; Schneider et al. 2011; Korecka et al. 2013). SH-SY5Y cells are the most frequently used cellular model in neurotoxicity and neuroprotection research especially in relation to Parkinson's disease; however, it is a matter of debate whether differentiated or undifferentiated SH-SY5Y cells are more appropriate (Presgraves et al. 2004; Jantas et al. 2008, 2013; Cheung et al. 2009; Lopes et al. 2010; Luchtman and Song 2010; Wenker et al. 2010; Yong-Kee et al. 2011). It has been reported that RA treatment of SH-SY5Y cells can activate pro-survival intrinsic molecular pathways, e.g., ERK1/2 and PI3-K/Akt, and can modulate the expression of pro- and antiapoptotic proteins, e.g., MnSOD and Bcl-2 protein family (Lopez-Carballo et al. 2002; Cheung et al. 2009; Miloso et al. 2004; Schneider et al. 2011; Wenker et al. 2010), which protect cells against neurotoxic agents and/or can camouflage neuroprotective effects of tested compounds.

#### 3.3.3.1 Methodology

SH-SY5Y cells are usually grown and maintained in DMEM with high glucose (4.5 g/ml) and L-glutamine (2 mM) which is supplemented with heat-inactivated 10 % FBS and antibiotics (50–100 U/ml penicillin and 10–100 μg/ml streptomycin) (Presgraves et al. 2004; Jantas et al. 2008; Yaniv et al. 2008). Other types of cell culture medium (e.g., EMEM, RPMI 1640) and additional supplements (e.g., supplement N2 or F12) can be used (Sarkanen et al. 2007; Tazik et al. 2009; Jaworska-Feil et al. 2010; Wenker et al. 2010). Cells are kept under standard cell culture conditions (at 37 °C in a humidified atmosphere of 5 % $CO_2$ in air). In order to obtain differentiated SH-SY5Y cells, cells are usually cultured with medium supplemented with RA (10 μM) for 7 days although shorter (2–4 days) or longer (14 days) time protocols can be used (Presgraves et al. 2004; Jantas et al. 2008; Cheung et al. 2009; Wenker et al. 2010).

#### 3.3.3.2 Depression Aspects Studied in SH-SY5Y Cells

SH-SY5Y cells can also be used as in vitro neuronal model for studying depression and antidepressant drug action mechanisms. Most research in this respect has been concentrated on testing the neuroprotective effects of various synthetic antidepressants or natural compounds and their mechanisms in models of cell damage induced by glucocorticoids (CORT or DEX), oxidative stress ($H_2O_2$, 6-OHDA, glutamate), trophic deprivation, pro-inflammatory compounds (e.g., IL-1α, IL-1β, TNFα), or proapoptotic factors (e.g., staurosporine, doxorubicin) (Johnson et al. 2010; Liu et al. 2010; Wenker et al. 2010; Seo et al. 2013; Hu et al. 2014b; Jantas et al. 2014a; Lucki et al. 2014). In some cases, SH-SY5Y cells were used together with animal studies to search for a detailed mechanism responsible for antidepressant effect of some endogenously expressed factors like peptide Pro-Leu-Gly-NH(2) (MIF-1) or 17β-estradiol (Khan et al. 2010; Adeosun et al. 2012). This cell line can be employed for the establishment of gene reporter assays; however, in this respect, other cell types are more convenient, since SH-SY5Y cells are rather difficult to transfect with high efficiency using liposome/polymer reagents. Nevertheless, the CRE-Luc/pGL4 plasmid has been transfected with lipofectamine 2000 to SH-SY5Y cells to measure the CRE-mediated gene expression by 6′-disinapoyl sucrose (DISS) (an oligosaccharide ester originated from the root of wild *Polygala tenuifolia* Willd.) which was mediated through MAPK and CaMKII pathways (Hu et al. 2014a). Another example is a study on the transcriptional activity of the dopamine D2 receptor gene promoter after various antidepressant drug treatments (Dziedzicka-Wasylewska and Solich 2004). An important aspect of depression, namely, neuronal plasticity, can also be explored in SH-SY5Y cells. The influence of various factors (e.g., antidepressants, calprotectin, noradrenaline, cholesterol) on cell proliferation rate (neurogenic potential) and RA-stimulated neurite outgrowth (neuritogenesis potential) was tested (Laifenfeld et al. 2002; Foster et al. 2006; Sarkanen et al. 2007; Jantas et al. 2014a). Also the efficiency of such compounds can be tested under disturbed conditions as has been shown, for example, for noradrenaline or various antidepressants, which prevented the DEX-evoked changes in proliferation rate and neuronal plasticity markers in SH-SY5Y cells (Yaniv et al. 2008; Leskiewicz et al. 2013). Interestingly, SH-SY5Y cells could also be used for the study of intracellular mechanisms evoked by non-pharmacological antidepressant approaches as has been demonstrated for magnetic stimulation of SH-SY5Y cells (Shaul et al. 2003). The last aspect which should be underlined in respect to the use of SH-SY5Y cell for studies on depression and antidepressants is its utility for testing cytotoxic effects of antidepressants and new compounds which could be an important factor in predicting possible side effects of new drugs or could be used for augmentation of chemotherapy efficiency (Levkovitz et al. 2005; Argov et al. 2009; Drzyzga et al. 2009; Bilir et al. 2010; Cavarec et al. 2013; Xue et al. 2013; Jantas et al. 2014a).

## *3.3.4 HT-22 Cell Line*

HT-22 cell line is a clonal mouse hippocampal cell line immortalized with a temperature-sensitive SV-40T antigen (Morimoto and Koshland 1990). The HT-22 clone was the most sensitive to glutamate of the 25 clones tested but it does not express ionotropic glutamate receptors (Maher and Davis 1996). Thus it has been used as a model to study mechanisms of oxidative neuronal cell death induced by high concentrations of glutamate (Maher and Davis 1996; Li et al. 1997; Tobaben et al. 2011; Doti et al. 2014; Oppermann et al. 2014). Also, other cell death-inducing factors can be administered to HT-22, and such models can be a platform for screening compounds with neuroprotective properties (e.g., $H_2O_2$, amyloid-beta protein, CORT, DEX, homocysteic acid) (Behl et al. 1995; Pfister et al. 2008; Tobaben et al. 2011; Xu et al. 2011; Grohm et al. 2012; Inkielewicz-Stepniak et al. 2012).

### 3.3.4.1 Methodology

HT-22 cells are cultured in DMEM supplemented with 10 % heat-inactivated FBS, 2 mM glutamine, antibiotics (100 U/ml penicillin, 100 mg/ml streptomycin), and optionally with 25 mM HEPES and are cultivated under standard mammalian cell culture conditions (at 37 °C in a humidified atmosphere of 5 % CO2 in air) (Tan et al. 1998; Tobaben et al. 2011). In some protocols, they are differentiated by culturing them at low density in Neurobasal/B27 medium, and then it is possible to study the impact of various factors on neurite elongation (Xu et al. 2011). Since this cell line is relatively good when transfected with traditional transfection reagents, it is often used in exploring mechanisms of cell death or neuroprotection (Grohm et al. 2012; Doti et al. 2014).

#### 3.3.4.2 Depression Aspects Studied in HT-22 Cells

Regarding depression research, HT-22 cells can be used for testing neuroprotective potency of various compounds with putative antidepressant activity against various detrimental agents, as, for example, has been shown for 17β-estradiol, melatonin, flavonoids, curcuminoids, as well as extracts of some plants against oxidative stress- or glutamate-induced cell damage (Behl et al. 1995; Wu et al. 2013a, b; Du et al. 2014; Jirásek et al. 2014; Kling et al. 2014). Since HT-22 cells are vulnerable for the DEX-induced inhibition of cell proliferation (concentrations 0.1–10 μM), various compounds could be tested in respect to their neurogenic potential or the mechanisms of sensitization by DEX for detrimental action of oxidative stress (Braun et al. 2000; Herr et al. 2003). Another aspect which is widely investigated in HT-22 cell line is mechanisms of regulation of GR-mediated gene expression by various factors (e.g., glucocorticoids, estradiol, retinoids) since, likewise in primary neuronal hippocampal cell cultures, HT-22 cell line expresses high-level glucocorticoid receptors (Wang et al. 2002, 2005; Brossaud et al. 2013; Malviya et al. 2013). Therefore, gene reporter assays have been established in HT-22 for testing the influence, for example, of some antidepressant drugs, estradiol, or melatonin on the GR-mediated gene expression (Herr et al. 2003; Quiros et al. 2008; Malviya et al. 2013). Interestingly, the mechanisms of mutual interactions between two important systems disturbed in depression, inflammatory system, and GR signaling have also been studied in HT-22 cells (Pace et al. 2011; Alboni et al. 2013, 2014). Since HT-22 cells have been shown to express essential cholinergic markers (the high-affinity choline transporter, choline acetyltransferase, vesicular acetylcholine transporter, and muscarinic acetylcholine receptors), thus, it can be also engaged as an in vitro model for defining the mechanisms in cognitive deficits which are common in depression and Alzheimer's disease (Liu et al. 2009).

### *3.3.5 AR-5 Cell Line*

The AR-5 cell line is an immortalized clonal rat amygdaloid cell line which was selected on the basis of CRF peptide release into the culture supernatant (Kasckow et al. 1999; Mulchahey et al. 1999). Thus, these cells have been studied concerning the influence of stress (CORT or DEX) or other factors (e.g., 17-β estradiol, nicotine) on CRF signaling (Kasckow et al. 1999; Mulchahey et al. 1999; Lalmansingh and Uht 2008). By using this cell line, it was possible to study a reciprocal regulation between CRF and neuropeptide Y which was shown to be mediated by the protein kinase A pathway (Sheriff et al. 2001).

#### 3.3.5.1 Methodology

The AR-5 cells are cultured in DMEM/F12 media supplemented with 10 % FBS, 1 % L-glutamine, 1 % sodium pyruvate, 1 % nonessential amino acids, and 1 % penicillin/streptomycin under standard mammalian cell culture conditions (at 37 °C in a humidified atmosphere of 5 % $CO_2$ in air) (Lalmansingh and Uht 2008).

#### 3.3.5.2 Depression Aspects Studied in AR-5 Cells

In respect to depression research, this cell line could be hypothetically used for testing the effects of various compounds with antidepressant activity on CRF signaling, since antagonism of CRF receptors is one of the drug targets for new antidepressants (Aubry 2013). Another possibility for the use of this cell line for depression research is exploring differential mechanisms of glucocorticoid-induced changes in AR-5 cell line in comparison to hippocampal HT-22 cell line which, to some extent, mimic the animal situation with impaired neuroplasticity in the hippocampal region and increased neuroplasticity in the amygdala (Thompson et al. 2015). In the study by Xu et al. (2011), it has been shown that CORT and 5-HT may differentially modulate neuronal morphology in the hippocampus and amygdala depending on the expression levels of the 5-HT7 receptor, which was increased by CORT in HT-22 after CORT administration and decreased in AR-5 cell line. The CORT evoked changes in morphology and 5-HT7 expression were normalized by 5-HT7 antagonist in HT-22 cells and 5-HT7 agonist in AR-5 cells (Xu et al. 2011). By usage of these two cell line models, it will be possible to screen agents with normalizing features of differential effects of stress (CORT) in the hippocampus and amygdala.

## 3.4 Glial Cell Cultures

There is increasing knowledge about the active role of glial cells in nutrient supply to neurons, maintenance of synaptic contact, and tissue homeostasis within the CNS as well as modulation of inflammatory processes which, when disturbed under various pathological situations, can lead to onset and/or progression of various neurological and neuropsychiatric diseases (Páv et al. 2008; Sofroniew and Vinters 2010). Since the hypothesis that abnormal functioning of glial cells could contribute to the pathophysiology of mood disorders, including depression, has been proposed a decade ago (Rajkowska and Miguel-Hidalgo 2007), an intense experimental research significantly improved our understanding on astrocytic functions in the nervous system, although many important questions remain still unanswered (Czéh and Di Benedetto 2013). In this respect, the use of primary glial cell cultures (mixed, astrocyte, oligodendrocyte, and microglia) and/or cell lines (rat glioma C6 cell line, human glioblastoma U87-MG, human astrocytoma U373-MG, murine BV2 microglial cell line) in combination with neuronal cell models and/or animal studies has enabled studies on the role of glial cells in the pathophysiology of depression and therapeutic action of antidepressant compounds (Czéh and Di Benedetto 2013).

### 3.4.1 *Primary Glial Cell Cultures*

Primary glial cell cultures are the most physiologically relevant cellular model to study the functions of glia and their influence on other cell types of the CNS under normal and pathological situations (Páv et al. 2008; Sofroniew and Vinters 2010). Since a wide range of receptors and transporters for all the major neurotransmitter systems (glutamatergic, GABAergic, adrenergic, serotonergic, dopaminergic, acetylcholinergic, and purinergic systems) as well as receptors for neurotrophic factors, cytokines, neuropeptides, and mineralocorticoid and glucocorticoid receptors are expressed in primary glial cell cultures (Páv et al. 2008; Sofroniew and Vinters 2010), this model is convenient to study not only the impact of various stressful conditions on these parameters but also the effect of used therapeutics.

#### 3.4.1.1 Methodology

Primary mixed glial cell cultures are usually prepared from cortices of 1/2-day-old rat/mice pubs. After dissection of cerebral cortices, meninges are removed and the isolated tissue is minced into small pieces and digested with trypsin (0.1–0.25 % for 25–30 min at 37 °C). After trypsinization, cells are mechanically triturated in the presence of 20 % fetal bovine serum and DNAse I (150 Kunitz units) and finally centrifuged for 5 min at 100×*g*. Next, cells are resuspended in DMEM containing 10–20 % FBS and 100 units/ml of penicillin and 100 μg/ml of streptomycin. Finally, cells are seeded into 25- or 75-cm$^2$ culture flasks coated with poly-ornithine or polylysine (Johann et al. 2008; Dhami et al. 2013; Jantas et al. 2014a). Mixed glia are incubated at 37 °C with 5 % $CO_2$ and are fed every 3–4 days with fresh DMEM with 10 % FBS. Mixed glial cell cultures could be used for experiments after 2 weeks of culture by trypsinization of cells and seeding them onto poly-ornithine-coated or polylysine-coated multi-well plates.

For the preparation of pure astrocyte cell cultures, after cells reaching about 80 % confluence, the flasks are shaken for 30 min. in order to remove loosely attached cells (neurons and microglia), and adherent cells are trypsinized and replated at a lower density into new T75 cm$^2$ flasks. This procedure is repeated three times and, after the last trypsinization, cells are seeded on poly-ornithine-coated or polylysine-coated multi-well plates. According to that procedure, 4-week-old astrocytes are used for experiments and obtained cell cultures characterized by about 90 % homogeneity of glial fibrillary acidic protein (GFAP)-positive cells with no neuronal (MAP-2-positive cells) contamination (Johann et al. 2008; Jantas et al. 2014a).

When comparative studies are performed in primary neuronal and glial cell cultures, it is plausible to culture both types of cells in neuronal medium (Neurobasal/B27) (Jantas et al. 2014a). For long-term experiments, astrocytes in cultures could be differentiated by treatment with dibutyryl cyclic AMP (Kong et al. 2002).

For the preparation of microglial cell cultures, mixed glial cell cultures after reaching confluency (usually after 10–12 days after seeding) are usually shaken overnight at 200 rpm, and floating cells are pelleted and subcultured at 400,000 cells/ml (100,000 cells/cm$^2$) on mixed glia-conditioned medium (Giulian and Baker 1986). The microglial cell cultures obtained by this procedure are of high purity but low yield. In order to obtain high-purity cultures with high yield, microglial cell culture procedure of mild trypsynization has been established (Saura et al. 2003). Confluent mixed glial cultures are subjected to mild trypsinization (0.05–0.12 %) in the presence of 0.2–0.5 mM EDTA and 0.5–0.8 mM $Ca^{2+}$. This procedure results in the detachment of an intact layer of cells containing virtually all the astrocytes, leaving undisturbed a population of firmly attached cells identified as 98 % microglia (Saura et al. 2003). These almost pure microglial preparations can be kept in culture for weeks and show proliferation and phagocytosis. This method allows for the preparation of highly enriched mouse or rat microglial cultures with ease and reproducibility and is especially convenient when high amounts of microglial protein/mRNA are required or in cases in which the starting material is limited, such as microglial cultures from transgenic animals (Saura et al. 2003).

#### 3.4.1.2 Depression Aspects Studied in Primary Mixed Glial Cell Cultures

In vitro studies have provided important evidences that astrocytes respond directly to antidepressant treatment (Czéh and Di Benedetto 2013). The early studies performed on glial cell cultures showed that after chronic exposure to antidepressant drugs, a downregulation in astrocyte beta receptors occurs with a time course that parallels to in vivo animal studies and clinical observations (Richardson and Hertz 1983). On the other hand, some observations made in in vitro system, e.g., influence of antidepressants on monoamine (NA and 5-HT) uptake, have not been confirmed in the animal brain (Hösli and Hösli 1995; Inazu et al. 2001). In primary cell cultures, it was possible to distinguish cell-specific effect for observations found in animal studies after antidepressant treatment as well as to demonstrate new mechanisms of their action which can be relevant for their therapeutic effects (Czéh and Di Benedetto 2013). For example, differential effect of various antidepressants on intracellular signaling pathways (e.g., MAPK/ERK1/2, PI3-K, NFκB), gene expression of astrocytes (GFAP, vimentin, aquaporin 4, or Ndrg2), the expression of receptors (VEGFR, TrkB), and the release of various trophic factors (BDNF, VEGF, GDNF) have been shown (Hughes et al. 2004; Mercier et al. 2004; Angelucci et al. 2011; Allaman et al. 2011; Kittel-Schneider et al. 2012; Czéh and Di Benedetto 2013). Based on these findings, it has been suggested that astrocyte-specific effects of antidepressants may contribute to normalize the trophic and metabolic support to neurons in major depression (Allaman et al. 2011; Czéh and Di Benedetto 2013).

The ability to induce astrocyte-specific increase in neurotrophic signaling has been shown for various natural compounds and extracts (e.g., scutellarin, caffeic acid ester, gastrodin, Kai-Xin-San) which can be an important part of their antidepressant and/or neuroprotective effect (Chai et al. 2013; Zhu et al. 2013). Also primary glial cell cultures have been used to test neuroprotective potential of various synthetic antidepressants or neuroprotective natural products, for example, in models of glutamate-, oxygen glucose deprivation-, or proapoptotic factors-induced glial cell damage (Gabryel et al. 2010; Lee et al. 2012; Jantas et al. 2014a; Lin et al. 2014; Zhang et al. 2014). A direct antagonistic interaction between stress (e.g., dexamethasone) and antidepressant (e.g., fluoxetine) on morphology and expression of neurotrophic factors and cytokines can be also detected in astrocyte cell cultures (Henkel et al. 2014). In pure one-type cells, it is possible to study in detail the mechanisms of additional targets for traditional antidepressants. For example, it has been shown that chronic treatment with various antidepressants could stimulate 5-$HT_{2B}$ receptor on astrocytes which consequently via transactivation of epidermal growth factor (EGF) receptor activate various intracellular pathways (e.g., ERK1/2 activation, cPLA2 expression) (Kong et al. 2002; Li et al. 2008, 2009; Zhang et al. 2010, 2014). Among other possible processes which could be investigated in primary glial cell culture system and are involved in depression etiopathogenesis is the influence of various classes of antidepressants on egigenetic regulation of gene expression (Perisic et al. 2010; Zimmermann et al. 2012; Yang et al. 2012). Some aspects of neuroplasticity could also be studied in pure glial cell cultures. In this respect, interesting observations were done in in vitro setting where some antidepressants were able to induce human mature astrocytes to differentiate into cells that express various neuron-specific markers (Cabras et al. 2010). However, this finding should be interpreted with caution because of the limitations of the in vitro experiments, but it suggests that antidepressant drugs may even guide the differentiation path of precursor cells. The advantage of the cell culture experiments is that they allow the clear distinction of cell-type-specific responses. In this respect, by using primary neuronal and astrocyte cell cultures, differential regulation of autophagy processes by various antidepressant drugs have been observed (Zschocke et al. 2011) pointing to the important role of astrocyte-mediated autophagy in maintenance of cellular energy homeostasis. Another aspect which can be explored by using neuronal, glial, and neuron-glial co-cultures is the influence of various factors on the astrocyte-mediated fostering of neurite outgrowth (e.g., Pizzurro et al. 2014).

Regarding the involvement of neuroinflammation in depression, the LPS-stimulated rodent microglial or mixed glial cells are used to demonstrate pro- or anti-inflammatory potential of various antidepressants or neuroprotective natural compounds with exploration of the detailed mechanisms responsible for these effects (e.g., MAPKs: ERK1/2, p38MAPK, JNK; NFµB) (Wang et al. 1999; Hashioka 2011; Chung et al. 2012; Zhang et al. 2012a, b; Czéh and Di Benedetto 2013; Obuchowicz et al. 2014). It should be underlined that the majority of the performed studies in respect to depression and antidepressant drug action have been done on cell cultures established from normal neonates where a pathological situation is usually modeled by exogenous application of various compounds implicated in depression pathogenesis (e.g., glucocorticoids, pro-inflammatory factors, glutamate) (Páv et al. 2008; Sofroniew and Vinters 2010). However, there are some attempts to study in in vitro models the cell-specific changes evoked by stressful conditions applied to animals and compare them with cells originated from non-stressed animals. In this respect, the study performed by Ślusarczyk et al. (2015) demonstrated that prenatal stress could lead to excessive microglial activation in offspring. It will be interesting to study how other CNS cells (e.g., neurons, astrocytes) are affected in such model as well as to extend this approach to other models of depression (e.g., maternal separation model).

## 3.4.2 *Rat C6 Glioma Cell Line*

### 3.4.2.1 Methodology

C6 glioma (astrocytoma) cells are grown in DMEM supplemented with 2 mM L-glutamine and 5–10 % FBS and antibiotics (100 U/ml penicillin and 100 mg/ml streptomycin) under standard cell culture conditions (at 37 °C in a humidified atmosphere of 5 % $CO_2$ in air) (Hisaoka et al. 2001; Golan et al. 2011; Di Benedetto et al. 2012).

### 3.4.2.2 Depression Aspects Studied in C6 Cell Line

Since C6 cells have been reported to possess high levels of glial cell line-derived neurotrophic factor (GDNF) synthesis and release, thus, it has been proposed as a good model for studying the effect of various classes of antidepressants on GDNF production and release (Hisaoka et al. 2001). Since amitriptyline relatively induced the highest effect on GDNF, in subsequent studies, the exact mechanisms of this process have been identified, e.g., induction of CREB phosphorylation and CRE-mediated gene expression, protein tyrosine kinase and ERK activation, and FGFR activation by the matrix metalloproteinase (MMP)-dependent shedding of FGFR ligands, such as FGF2 via PTX-sensitive Gαi/o (Hisaoka et al. 2007, 2008, 2011; Hisaoka-Nakashima et al. 2015). Other studies confirmed the results on increase of GDNF expression by antidepressants and demonstrated the crucial role of β-arrestin1 and CREB in this process (Golan et al. 2011). Also, some other agents with antidepressant activity (e.g., N-desalkylquetiapine) have been shown to induce GDNF in C6 cell via MAPK/ERK1/2 activation (Di Benedetto et al. 2012). Other aspects connected with depression which have been investigated in C6 cells are mechanisms of neuronal plasticity, for example, regulation of the excitatory amino acid carrier 1 (EAAC1) function by PKC or NMDA receptor-dependent regulation of EAAC1 which could provide a novel mechanism that may shape excitatory signaling during synaptic plasticity and/or excitotoxicity (González et al. 2002; Waxman et al. 2007). Moreover, in C6 cell, it has been shown that fluoxetine treatment regulates neuronal plasticity and neurite outgrowth by phosphorylating and activating CREB via the NCAM140 homophilic interaction-induced activation of the Ras-MAPK pathway (Choi et al. 2012). Another study showed that HDAC inhibitors promoted the differentiation of rat C6 glioma cells through the production of 5alpha-reduced neurosteroids, resulting in the enhancement of serotonin-stimulated BDNF gene expression as a consequence of promoting their differentiation (Morita et al. 2009). It is possible that differentiated glial cells may be implicated in preserving the integrity of neural networks as well as improving the function of neuronal cells in the brain. C6 cells transfected with β1 AR-green fluorescent protein (GFP) have been used to test the effect of compounds from St. John's wort extract (hyperforin and hyperoside) on β1-adrenergic receptor level (Jakobs et al. 2013). Since C6 cells differentiated by retinoic acid are vulnerable to glutamate toxicity, they have been used in parallel with neuroblastoma cells to test the neuroprotective effect of the extract from the leaves of *Withania somnifera* (L.) Dunal. In both models, this extract apart from inhibiting the glutamate-induced cell death, it reversed Glu-induced changes in Hsp70 and influenced neuroplasticity markers (neural cell adhesion molecule, NCAM, and its polysialylated form, PSA-NCAM) affected by glutamate (Kataria et al. 2012). Interestingly, prolonged treatment of primary astrocytes and C6 cells with TNFα has been shown to increase SERT gene expression and activity, an effect that is at least in part dependent on p38 MAPK activation (Malynn et al. 2013).

Apart from using rat C6 glioma cells, in some studies on the effect of various compounds on astrocytic neurotrophic production, human glioblastoma U87-MG, human astrocytoma U-373-MG, or human astroglial A172 cell lines have been used (Liu et al. 2013; Eissa et al. 2014; Yuhas et al. 2015).

### *3.4.3 BV2 Microglial Culture*

BV2 is a murine immortalized microglial cell line which has been generated by infecting primary microglial cell cultures with a v-raf/v-myc oncogene carrying retrovirus (J2) (Blasi et al. 1990). BV2 cells express nonspecific esterase activity and phagocytic ability, lack peroxidase activity, and secrete lysozyme and, following appropriate stimulation, also interleukin 1 and tumor necrosis factor (Blasi et al. 1990). Since BV2 cells retain most of the morphological, phenotypical, and functional properties described for freshly isolated microglial cells, thus, they are often used as a screening platform for new anti-inflammatory agents or, in combination with other cell types, to study mechanism of their neuroprotective action, e.g., in co-culture with primary hippocampal neurons, primary astrocyte cell cultures, HT-22 cells, SH-SY5Y cells, or PC12 cells (Kim et al. 2004; Lauro et al. 2008; Qu et al. 2012; Jin et al. 2013; Tao et al. 2014; Xu et al. 2015). BV2 cells together with primary neuronal cell cultures can be used to find new mechanisms as, for example, has been shown in the study by Marrs et al. (2010) where a new role of serine hydrolase alpha-beta-hydrolase domain 6 (ABHD6) as a rate-limiting step of 2-AG signaling in cannabinoid system has been discovered.

#### 3.4.3.1 Methodology

BV2 cells are usually maintained in DMEM supplemented with 10 % FBS, 2 mM L-glutamine, 100 U/ml penicillin, and 100 mg/ml streptomycin in a 5 % $CO_2$ incubator (Liu et al. 2011a, b). After reaching 80 % confluence, cells are seeded on multiwell plates in DMEM with 10 % FBS overnight. Cells are treated with LPS (0.1–1 μg/ml) and compounds are tested for various time exposure (18–96 h) in cell culture medium which is devoid of serum (Liu et al. 2011a, b, 2014). Among standard assays engaged in these studies are nitric oxide release assay, ROS production, cytokines mRNA and/or protein level (TNF-α, IL-1b, IL-6), activation of pro-inflammatory enzymes (e.g., cyclooxygenase (COX)-2, inducible nitric oxide synthase (iNOS)), and induction of intracellular pathways (MAPKs: ERK1/2, p38MAPK, JNK, NFκB).

#### 3.4.3.2 Depression Aspects Studied in BV2 Cell Line

In respect to depression studies in BV2 cell model, various antidepressant drugs (e.g., imipramine, fluoxetine, or paroxetine) were tested which, apart from demonstrating general anti-inflammatory effect against LPS-induced inflammation, they differentiated at the level of the signaling pathway responsible for the observed effects (NF-κB signaling vs. MAPK signaling). In another study using BV2 cells, it has been shown that a wide range of antidepressants inhibited proton currents induced by LPS, and by these mechanisms, they inhibited the microglial production of pro-inflammatory cytokines (Song et al. 2012). There are also contradictory studies demonstrating an exaggeration of pro-inflammatory response by some drugs, e.g., fluoxetine via NFκB-mediated pathway (Ha et al. 2006). BV2 cell could be also activated by other factors like, for example, amyloid-β, an effect attenuated, for example, by aromatic-turmerone (isolated from *Curcuma longa* L. with antidepressant activity) through blocking of NF-κB, JNK, and p38 MAPK signaling pathways (Park et al. 2012). Moreover, in a co-culture model, this compound protected hippocampal HT-22 cells from indirect neuronal toxicity induced by activated microglial cells. These results suggest that the therapeutic effects of antidepressants could be partially mediated by modulating microglial activation; however, by using only this model, it will be rather unlikely to find new compounds with antidepressant activity. An interesting approach is the co-culture of mouse bone marrow-derived mesenchymal stem cells with BV2 microglial cells where various reciprocal connections can be observed in respect to LPS-induced inflammatory response as well on neuronal differentiation process (Rahmat et al. 2013).

## 3.5 Conclusions and Future Directions

As has been described in this book chapter, various cell models can be employed for studies of depression and antidepressant drug mechanisms. Most often, they are combined with animal studies in order to search cell-specific mechanisms of antidepressant action of various compounds (Matsui et al. 2012). New mechanisms could also be detected first in cellular models; however, they need to be validated in vivo since cultured cells may express more receptors than cells in situ because of their altered gene expression profile. This aspect is one of the main limitations of cellular models. Also, some observations made in cellular models could be clinically irrelevant, as, for example, the influence of high concentrations (above 10 μM) of antidepressants on some processes (e.g., increase of intracellular calcium $[Ca^{2+}]_i$) (Shimizu et al. 1993). There is also doubt whether the results obtained from cell lines are relevant to physiological situation. Generally, parallel studies performed in cell lines and in primary cell culture systems demonstrate high relevance of the use of several cell lines as a first screening platform for searching for compounds with antidepressant or neuroprotective activity. Such system could, at a very early step, exclude compounds which potentially will be toxic or evoke side effects. Using reliable cell line models is also very important for first drug screen studies since the rules of using animals in research are becoming more restricted recently. In order to improve the searching potential of new antidepressant compounds, a set of standard assays performed in various cell lines should be proposed to cover most of the mechanisms which are important for depression medication (e.g., prevention of stress-induced changes in neuroplasticity, promotion of neurite outgrowth, neurogenic potential, neuroprotection, induction of expression of trophic factors, and prevention of neuroinflammation). The important area will be describing a role of human embryonic and somatic tissue-derived stem cell models for antidepressant drug discovery and toxicity testing (Hung et al. 2010; Hunsberger et al. 2015). Moreover patient-derived induced pluripotent stem cell (iPSC) models will offer an opportunity to identify novel disease mechanisms (Hunsberger et al. 2015).

**Acknowledgments** The author is funded by statutory funds of the Institute of Pharmacology Polish Academy of Sciences.

*Conflict of Interest*: The author confirms that this chapter contents have no conflict of interest.

# References

Abelaira HM, Réus GZ, Petronilho F, Barichello T, Quevedo J. Neuroimmunomodulation in depression: a review of inflammatory cytokines involved in this process. Neurochem Res. 2014;39:1634–9.

Adeosun SO, Albert PR, Austin MC, Iyo AH. 17β-estradiol-induced regulation of the novel 5-HT1A-related transcription factors NUDR and Freud-1 in SH SY5Y cells. Cell Mol Neurobiol. 2012;32:517–21.

Agholme L, Lindström T, Kågedal K, Marcusson J, Hallbeck M. An in vitro model for neuroscience: differentiation of SH-SY5Y cells into cells with morphological and biochemical characteristics of mature neurons. J Alzheimers Dis. 2010;20:1069–82.

Akpinar A, Uğuz AC, Nazıroğlu M. Agomelatine and duloxetine synergistically modulates apoptotic pathway by inhibiting oxidative stress triggered intracellular calcium entry in neuronal PC12 cells: role of TRPM2 and voltage-gated calcium channels. J Membr Biol. 2014; 247:451–9.

Alboni S, Gibellini L, Montanari C, Benatti C, Benatti S, Tascedda F, Brunello N, Cossarizza A, Pariante CM. N-acetyl-cysteine prevents toxic oxidative effects induced by IFN-α in human neurons. Int J Neuropsychopharmacol. 2013;16:1849–65.

Alboni S, Montanari C, Benatti C, Sanchez-Alavez M, Rigillo G, Blom JM, Brunello N, Conti B, Pariante MC, Tascedda F. Interleukin 18 activates MAPKs and STAT3 but not NF-κB in hippocampal HT-22 cells. Brain Behav Immun. 2014;40:85–94.

al-Damluji S, Krsmanovic LZ, Catt KJ. High-affinity uptake of noradrenaline in postsynaptic neurones. Br J Pharmacol. 1993;109:299–307.

Allaman I, Fiumelli H, Magistretti PJ, Martin JL. Fluoxetine regulates the expression of neurotrophic/growth factors and glucose metabolism in astrocytes. Psychopharmacology (Berl). 2011;216:75–84.

Alt A, Nisenbaum ES, Bleakman D, Witkin JM. A role for AMPA receptors in mood disorders. Biochem Pharmacol. 2006;71:1273–88.

Amaral MD, Pozzo-Miller L. TRPC3 channels are necessary for brain-derived neurotrophic factor to activate a nonselective cationic current and to induce dendritic spine formation. J Neurosci. 2007;27:5179–89.

Anacker C, Cattaneo A, Musaelyan K, Zunszain PA, Horowitz M, Molteni R, Luoni A, Calabrese F, Tansey K, Gennarelli M, Thuret S, Price J, Uher R, Riva MA, Pariante CM. Role for the kinase SGK1 in stress, depression, and glucocorticoid effects on hippocampal neurogenesis. Proc Natl Acad Sci U S A. 2013;110:8708–13.

Anderson G, Kubera M, Duda W, Lasoń W, Berk M, Maes M. Increased IL-6 trans-signaling in depression: focus on the tryptophan catabolite pathway, melatonin and neuroprogression. Pharmacol Rep. 2013;65:1647–54.

Andjus PR, Stevic-Marinkovic Z, Cherubini E. Immunoglobulins from motoneurone disease patients enhance glutamate release from rat hippocampal neurones in culture. J Physiol. 1997;504:103–12.

Ang W, Chen G, Xiong L, Chang Y, Pi W, Liu Y, Li C, Zheng J, Zhou L, Yang B, Deng Y, Yang S, Luo Y, Wei Y. Synthesis and biological evaluation of novel naphthalene compounds as potential antidepressant agents. Eur J Med Chem. 2014;82:263–73.

Angelucci F, Croce N, Spalletta G, Dinallo V, Gravina P, Bossù P, Federici G, Caltagirone C, Bernardini S. Paroxetine rapidly modulates the expression of brain-derived neurotrophic factor mRNA and protein in a human glioblastoma-astrocytoma cell line. Pharmacology. 2011; 87:5–10.

Argov M, Kashi R, Peer D, Margalit R. Treatment of resistant human colon cancer xenografts by a fluoxetine–doxorubicin combination enhances therapeutic responses comparable to an aggressive bevacizumab regimen. Cancer Lett. 2009;274:118–25.

Aubry JM. CRF system and mood disorders. J Chem Neuroanat. 2013;54:20–4.

Bakunina N, Pariante CM, Zunszain PA. Immune mechanisms linked to depression via oxidative stress and neuroprogression. Immunology. 2015. doi:10.1111/imm.12443. [Epub ahead of print].

Behl C, Widmann M, Trapp T, Holsboer F. 17-beta estradiol protects neurons from oxidative stress-induced cell death in vitro. Biochem Biophys Res Commun. 1995;216:473–82.

Berton O, Nestler EJ. New approaches to antidepressant drug discovery: beyond monoamines. Nat Rev Neurosci. 2006;7:137–51.

Biedler JL, Roffler-Tarlov S, Schachner M, Freedman LS. Multiple neurotransmitter synthesis by human neuroblastoma cell lines and clones. Cancer Res. 1978;38:3751–7.

Bilir A, Erguven M, Yazihan N, Aktas E, Oktem G, Sabanci A. Enhancement of vinorelbine-induced cytotoxicity and apoptosis by clomipramine and lithium chloride in human neuroblastoma cancer cell line SH-SY5Y. J Neurooncol. 2010;100:385–95.

Blasi E, Barluzzi R, Bocchini V, Mazzolla R, Bistoni F. Immortalization of murine microglial cells by a v-raf/v-myc carrying retrovirus. J Neuroimmunol. 1990;27:229–37.

Bleakman D, Alt A, Witkin JM. AMPA receptors in the therapeutic management of depression. CNS Neurol Disord Drug Targets. 2007;6:117–26.

Borroto-Escuela DO, Romero-Fernandez W, Mudó G, Pérez-Alea M, Ciruela F, Tarakanov AO, Narvaez M, Di Liberto V, Agnati LF, Belluardo N, Fuxe K. Fibroblast growth factor receptor 1–5-hydroxytryptamine 1A heteroreceptor complexes and their enhancement of hippocampal plasticity. Biol Psychiatry. 2012;71:84–91.

Brachman RA, McGowan JC, Perusini JN, Lim SC, Pham TH, Faye C, Gardier AM, Mendez-David I, David DJ, Hen R, Denny CA. Ketamine as a prophylactic against stress-induced depressive-like behavior. Biol Psychiatry. 2015;pii:S0006-3223(15)00360-1. doi:10.1016/j.biopsych.2015.04.022.

Braun S, Liebetrau W, Berning B, Behl C. Dexamethasone-enhanced sensitivity of mouse hippocampal HT22 cells for oxidative stress is associated with the suppression of nuclear factor-kappaB. Neurosci Lett. 2000;295:101–4.

Brewer GJ. Serum-free B27/neurobasal medium supports differentiated growth of neurons from the striatum, substantia nigra, septum, cerebral cortex, cerebellum, and dentate gyrus. J Neurosci Res. 1995;42:674–83.

Brewer GJ, Torricelli JR, Evege EK, Price PJ. Optimized survival of hippocampal neurons in B27-supplemented Neurobasal, a new serum-free medium combination. J Neurosci Res. 1993; 35:567–76.

Brossaud J, Roumes H, Moisan MP, Pallet V, Redonnet A, Corcuff JB. Retinoids and glucocorticoids target common genes in hippocampal HT22 cells. J Neurochem. 2013;125:518–31.

Brunn J, Wiroth V, Kowalski M, Runge U, Sabolek M. Valproic acid in normal therapeutic concentration has no neuroprotective or differentiation influencing effects on long term expanded murine neural stem cells. Epilepsy Res. 2014;108:623–33.

Cabras S, Saba F, Reali C, Scorciapino ML, Sirigu A, Talani G, Biggio G, Sogos V. Antidepressant imipramine induces human astrocytes to differentiate into cells with neuronal phenotype. Int J Neuropsychopharmacol. 2010;13:603–15.

Cattaneo A, Macchi F, Plazzotta G, Veronica B, Bocchio-Chiavetto L, Riva MA, Pariante CM. Inflammation and neuronal plasticity: a link between childhood trauma and depression pathogenesis. Front Cell Neurosci. 2015;31:9–40.

Cavarec L, Vincent L, Le Borgne C, Plusquellec C, Ollivier N, Normandie-Levi P, Allemand F, Salvetat N, Mathieu-Dupas E, Molina F, Weissmann D, Pujol JF. In vitro screening for drug-induced depression and/or suicidal adverse effects: a new toxicogenomic assay based on CE-SSCP analysis of HTR2C mRNA editing in SH-SY5Y cells. Neurotox Res. 2013;23:49–62.

Chai L, Guo H, Li H, Wang S, Wang YL, Shi F, Hu LM, Liu Y, Adah D. Scutellarin and caffeic acid ester fraction, active components of Dengzhanxixin injection, upregulate neurotrophins synthesis and release in hypoxia/reoxygenation rat astrocytes. J Ethnopharmacol. 2013; 150:100–7.

Chang Y, Pi W, Ang W, Liu Y, Li C, Zheng J, Xiong L, Yang T, Luo Y. Synthesis and evaluation of amide side-chain modified Agomelatine analogues as potential antidepressant-like agents. Bioorg Med Chem Lett. 2014;24:1672–6.

Chen Q, de Lecea L, Hu Z, Gao D. The hypocretin/orexin system: an increasingly important role in neuropsychiatry. Med Res Rev. 2015;35:152–97.

Cheung YT, Lau WK, Yu MS, Lai CS, Yeung SC, So KF, Chang RC. Effects of all-trans-retinoic acid on human SH-SY5Y neuroblastoma as in vitro model in neurotoxicity research. Neurotoxicology. 2009;30:127–35.

Choi MR, Oh DH, Kim SH, Jung KH, Das ND, Chai YG. Fluoxetine increases the expression of NCAM140 and pCREB in rat C6 glioma cells. Psychiatry Investig. 2012;9:180–6.

Chu CC, Wang JJ, Chen KT, Shieh JP, Wang LK, Shui HA, Ho ST. Neurotrophic effects of tianeptine on hippocampal neurons: a proteomic approach. J Proteome Res. 2010;9:936–44.

Chung HS, Kim H, Bae H. Phenelzine (monoamine oxidase inhibitor) increases production of nitric oxide and proinflammatory cytokines via the NF-κB pathway in lipopolysaccharide-activated microglia cells. Neurochem Res. 2012;37:2117–24.

Crochemore C, Lu J, Wu Y, Liposits Z, Sousa N, Holsboer F, Almeida OF. Direct targeting of hippocampal neurons for apoptosis by glucocorticoids is reversible by mineralocorticoid receptor activation. Mol Psychiatry. 2005;10:790–8.

Czéh B, Di Benedetto B. Antidepressants act directly on astrocytes: evidences and functional consequences. Eur Neuropsychopharmacol. 2013;23:171–85.

Dhami KS, Churchward MA, Baker GB, Todd KG. Fluoxetine and citalopram decrease microglial release of glutamate and D-serine to promote cortical neuronal viability following ischemic insult. Mol Cell Neurosci. 2013;56:365–74.

Di Benedetto B, Kühn R, Nothdurfter C, Rein T, Wurst W, Rupprecht R. N-desalkylquetiapine activates ERK1/2 to induce GDNF release in C6 glioma cells: a putative cellular mechanism for quetiapine as antidepressant. Neuropharmacology. 2012;62:209–16.

Divish MM, Sheftel G, Boyle A, Kalasapudi VD, Papolos DF, Lachman HM. Differential effect of lithium on fos protooncogene expression mediated by receptor and postreceptor activators of protein kinase C and cyclic adenosine monophosphate: model for its antimanic action. J Neurosci Res. 1991;28:40–8.

Doti N, Reuther C, Scognamiglio PL, Dolga AM, Plesnila N, Ruvo M, Culmsee C. Inhibition of the AIF/CypA complex protects against intrinsic death pathways induced by oxidative stress. Cell Death Dis. 2014;5, e993.

Drevets WC, Price JL, Furey ML. Brain structural and functional abnormalities in mood disorders: implications for neurocircuitry models of depression. Brain Struct Funct. 2008;213:93–118.

Drzyzga ŁR, Marcinowska A, Obuchowicz E. Antiapoptotic and neurotrophic effects of antidepressants: a review of clinical and experimental studies. Brain Res Bull. 2009;79:248–57.

Du G, Tu H, Li X, Pei A, Chen J, Miao Z, Li J, Wang C, Xie H, Xu X, Zhao H. Daphnetin, a natural coumarin derivative, provides the neuroprotection against glutamate-induced toxicity in HT22 cells and ischemic brain injury. Neurochem Res. 2014;39:269–75.

Duman RS. Role of neurotrophic factors in the etiology and treatment of mood disorders. Neuromolecular Med. 2004;5:11–25.

Duman RS, Heninger GR, Nestler EJ. A molecular and cellular hypothesis of depression. Arch Gen Psychiatry. 1997;54:597–606.

Dunigan CD, Shamoo AE. Li+stimulates ATP-regulated dopamine uptake in PC12 cells. Neuroscience. 1995;65:1–4.

Dziedzicka-Wasylewska M, Solich J. Neuronal cell lines transfected with the dopamine D2 receptor gene promoter as a model for studying the effects of antidepressant drugs. Brain Res Mol Brain Res. 2004;128:75–82.

Eissa TF, González-Burgos E, Carretero ME, Gómez-Serranillos MP. Phenolic content, antioxidant and astroprotective response to oxidative stress of ethanolic extracts of Mentha longifolia from Sinai. Nat Prod Commun. 2014;9:1479–82.

El Omri A, Han J, Yamada P, Kawada K, Abdrabbah MB, Isoda H. Rosmarinus officinalis polyphenols activate cholinergic activities in PC12 cells through phosphorylation of ERK1/2. J Ethnopharmacol. 2010;131:451–8.

Flajolet M, Wang Z, Futter M, Shen W, Nuangchamnong N, Bendor J, Wallach I, Nairn AC, Surmeier DJ, Greengard P. FGF acts as a co-transmitter through adenosine A(2A) receptor to regulate synaptic plasticity. Nat Neurosci. 2008;11:1402–9.

Foreman PJ, Taglialatela G, Jackson GR, Perez-Polo JR. Dexamethasone blocks nerve growth factor induction of nerve growth factor receptor mRNA in PC12 cells. J Neurosci Res. 1992;31:52–7.

Foster R, Kandanearatchi A, Beasley C, Williams B, Khan N, Fagerhol MK, Everall IP. Calprotectin in microglia from frontal cortex is up-regulated in schizophrenia: evidence for an inflammatory process? Eur J Neurosci. 2006;24:3561–6.

Freudenberg F, Celikel T, Reif A. The role of α-amino-3-hydroxy-5-methyl-4-isoxazolepropionic acid (AMPA) receptors in depression: central mediators of pathophysiology and antidepressant activity? Neurosci Biobehav Rev. 2015;52:193–206.

Gabryel B, Bielecka A, Stolecka A, Bernacki J, Langfort J. Cytosolic phospholipase $A_2$ inhibition is involved in the protective effect of nortriptyline in primary astrocyte cultures exposed to combined oxygen-glucose deprivation. Pharmacol Rep. 2010;62:814–26.

Gassen NC, Hartmann J, Zschocke J, Stepan J, Hafner K, Zellner A, Kirmeier T, Kollmannsberger L, Wagner KV, Dedic N, Balsevich G, Deussing JM, Kloiber S, Lucae S, Holsboer F, Eder M, Uhr M, Ising M, Schmidt MV, Rein T. Association of FKBP51 with priming of autophagy pathways and mediation of antidepressant treatment response: evidence in cells, mice, and humans. PLoS Med. 2014;11, e1001755.

Giulian D, Baker TJ. Characterization of ameboid microglia isolated from developing mammalian brain. J Neurosci. 1986;6:2163–78.

Golan M, Schreiber G, Avissar S. Antidepressants elevate GDNF expression and release from $C_6$ glioma cells in a β-arrestin1-dependent. CREB interactive pathway. Int J Neuropsychopharmacol. 2011;14:1289–300.

González MI, Kazanietz MG, Robinson MB. Regulation of the neuronal glutamate transporter excitatory amino acid carrier-1 (EAAC1) by different protein kinase C subtypes. Mol Pharmacol. 2002;62:901–10.

Gramowski A, Jügelt K, Stüwe S, Schulze R, McGregor GP, Wartenberg-Demand A, Loock J, Schröder O, Weiss DG. Functional screening of traditional antidepressants with primary cortical neuronal networks grown on multielectrode neurochips. Eur J Neurosci. 2006;24:455–65.

Greene LA, Tischler AS. Establishment of a noradrenergic clonal line of rat adrenalpheochromocytoma cells which respond to nerve growth factor. Proc Natl Acad Sci U S A. 1976; 73:2424–8.

Grohm J, Kim SW, Mamrak U, Tobaben S, Cassidy-Stone A, Nunnari J, Plesnila N, Culmsee C. Inhibition of Drp1 provides neuroprotection in vitro and in vivo. Cell Death Differ. 2012; 19:1446–58.

Ha E, Jung KH, Choe BK, Bae JH, Shin DH, Yim SV, Baik HH. Fluoxetine increases the nitric oxide production via nuclear factor kappa B-mediated pathway in BV2 murine microglial cells. Neurosci Lett. 2006;397:185–9.

Hamby ME, Sofroniew MV. Reactive astrocytes as therapeutic targets for CNS disorders. Neurotherapeutics. 2010;7:494–506.

Han YS, Lee CS. Antidepressants reveal differential effect against 1-methyl-4-phenylpyridinium toxicity in differentiated PC12 cells. Eur J Pharmacol. 2009;604:36–44.

Hannan MA, Kang JY, Mohibbullah M, Hong YK, Lee H, Choi JS, Choi IS, Moon IS. Moringa oleifera with promising neuronal survival and neurite outgrowth promoting potentials. J Ethnopharmacol. 2014a;152:142–50.

Hannan MA, Mohibbullah M, Hong YK, Nam JH, Moon IS. Gelidium amansii promotes dendritic spine morphology and synaptogenesis, and modulates NMDA receptor-mediated postsynaptic current. In Vitro Cell Dev Biol Anim. 2014b;50:445–52.

Hashimoto K. Inflammatory biomarkers as differential predictors of antidepressant response. Int J Mol Sci. 2015;16:7796–801.

Hashioka S. Antidepressants and neuroinflammation: can antidepressants calm glial rage down? Mini Rev Med Chem. 2011;11:555–64.

Heiser JH, Schuwald AM, Sillani G, Ye L, Müller WE, Leuner K. TRPC6 channel-mediated neurite outgrowth in PC12 cells and hippocampal neurons involves activation of RAS/MEK/ERK, PI3K, and CAMKIV signaling. J Neurochem. 2013;127:303–13.

Henkel AW, Alali H, Devassy A, Alawadi MM, Redzic ZB. Antagonistic interactions between dexamethasone and fluoxetine modulate morphodynamics and expression of cytokines in astrocytes. Neuroscience. 2014;280:318–27.

Herr AS, Tsolakidou AF, Yassouridis A, Holsboer F, Rein T. Antidepressants differentially influence the transcriptional activity of the glucocorticoid receptor in vitro. Neuroendocrinology. 2003;78:12–22.

Hindley S, Juurlink BH, Gysbers JW, Middlemiss PJ, Herman MA, Rathbone MP. Nitric oxide donors enhance neurotrophin-induced neurite outgrowth through a cGMP-dependent mechanism. J Neurosci Res. 1997;47:427–39.

Hisaoka K, Nishida A, Koda T, Miyata M, Zensho H, Morinobu S, Ohta M, Yamawaki S. Antidepressant drug treatments induce glial cell line-derived neurotrophic factor (GDNF) synthesis and release in rat C6 glioblastoma cells. J Neurochem. 2001;79:25–34.

Hisaoka K, Takebayashi M, Tsuchioka M, Maeda N, Nakata Y, Yamawaki S. Antidepressants increase glial cell line-derived neurotrophic factor production through monoamine-independent activation of protein tyrosine kinase and extracellular signal-regulated kinase in glial cells. J Pharmacol Exp Ther. 2007;321:148–57.

Hisaoka K, Maeda N, Tsuchioka M, Takebayashi M. Antidepressants induce acute CREB phosphorylation and CRE-mediated gene expression in glial cells: a possible contribution to GDNF production. Brain Res. 2008;1196:53–8.

Hisaoka K, Tsuchioka M, Yano R, Maeda N, Kajitani N, Morioka N, Nakata Y, Takebayashi M. Tricyclic antidepressant amitriptyline activates fibroblast growth factor receptor signaling in glial cells: involvement in glial cell line-derived neurotrophic factor production. J Biol Chem. 2011;286:21118–28.

Hisaoka-Nakashima K, Miyano K, Matsumoto C, Kajitani N, Abe H, Okada-Tsuchioka M, Yokoyama A, Uezono Y, Morioka N, Nakata Y, Takebayashi M. Tricyclic antidepressant amitriptyline-induced glial cell line-derived neurotrophic factor production involves pertussis toxin-sensitive Gαi/o activation in astroglial cells. J Biol Chem. 2015;pii:jbc.M114.622415.

Hösli E, Hösli L. Autoradiographic studies on the uptake of 3H-noradrenaline and 3H-serotonin by neurones and astrocytes in explant and primary cultures of rat CNS: effects of antidepressants. Int J Dev Neurosci. 1995;13:897–908.

Hu Y, Liu MY, Liu P, Dong X, Boran AD. Neuroprotective effects of 3,6'-disinapoyl sucrose through increased BDNF levels and CREB phosphorylation via the CaMKII and ERK1/2 pathway. J Mol Neurosci. 2014a;53:600–7.

Hu Y, Zhou XJ, Liu P, Dong XZ, Mu LH, Chen YB, Liu MY, Yu BY. Antidepressant and neuroprotective effect of the Chinese herb kaixinsan against lentiviral shRNA knockdown brain-derived neurotrophic factor-induced injury in vitro and in vivo. Neuropsychobiology. 2014b;69(3):129–1239.

Hughes EG, Maguire JL, McMinn MT, Scholz RE, Sutherland ML. Loss of glial fibrillary acidic protein results in decreased glutamate transport and inhibition of PKA-induced EAAT2 cell surface trafficking. Brain Res Mol Brain Res. 2004;124:114–23.

Hung CW, Liou YJ, Lu SW, Tseng LM, Kao CL, Chen SJ, Chiou SH, Chang CJ. Stem cell-based neuroprotective and neurorestorative strategies. Int J Mol Sci. 2010;11:2039–55.

Hunsberger JG, Efthymiou AG, Malik N, Behl M, Mead IL, Zeng X, Simeonov A, Rao M. Induced pluripotent stem cell models to enable in vitro models for screening in the central nervous system. Stem Cells Dev. 2015;24(16):1852–64.

Iadarola ND, Niciu MJ, Richards EM, Vande Voort JL, Ballard ED, Lundin NB, Nugent AC, Machado-Vieira R, Zarate Jr CA. Ketamine and other N-methyl-D-aspartate receptor antagonists in the treatment of depression: a perspective review. Ther Adv Chronic Dis. 2015; 6:97–114.

Inazu M, Takeda H, Ikoshi H, Sugisawa M, Uchida Y, Matsumiya T. Pharmacological characterization and visualization of the glial serotonin transporter. Neurochem Int. 2001;39:39–49.

Inkielewicz-Stepniak I, Radomski MW, Wozniak M. Fisetin prevents fluoride- and dexamethasone-induced oxidative damage in osteoblast and hippocampal cells. Food Chem Toxicol. 2012; 50:583–9.

Ip FC, Ng YP, An HJ, Dai Y, Pang HH, Hu YQ, Chin AC, Harley CB, Wong YH. Ip NYCycloastragenol is a potent telomerase activator in neuronal cells: implications for depression management. Neurosignals. 2014;22:52–63.

Ishima T, Fujita Y, Hashimoto K. Interaction of new antidepressants with sigma-1 receptor chaperones and their potentiation of neurite outgrowth in PC12 cells. Eur J Pharmacol. 2014; 727:167–73.

Ishima T, Futamura T, Ohgi Y, Yoshimi N, Kikuchi T, Hashimoto K. Potentiation of neurite outgrowth by brexpiprazole, a novel serotonin-dopamine activity modulator: a role for serotonin 5-HT1A and 5-HT2A receptors. Eur Neuropsychopharmacol. 2015;25:505–11.

Jakobs D, Hage-Hülsmann A, Prenner L, Kolb C, Weiser D, Häberlein H. Downregulation of β1 -adrenergic receptors in rat C6 glioblastoma cells by hyperforin and hyperoside from St John's wort. J Pharm Pharmacol. 2013;65:907–15.

Jantas D, Pytel M, Mozrzymas JW, Leskiewicz M, Regulska M, Antkiewicz-Michaluk L, Lason W. The attenuating effect of memantine on staurosporine-, salsolinol- and doxorubicin-induced apoptosis in human neuroblastoma SH-SY5Y cells. Neurochem Int. 2008;52:864–77.

Jantas D, Roman A, Kusmierczyk J, Lorenc-Koci E, Konieczny J, Lenda T, Lason W. The extent of neurodegeneration and neuroprotection in two chemical in vitro models related to

Parkinson's disease is critically dependent on cell culture conditions. Neurotox Res. 2013; 24:41–54.

Jantas D, Krawczyk S, Lason W. The predominant protective effect of tianeptine over other antidepressants in models of neuronal apoptosis: the effect blocked by inhibitors of MAPK/ERK1/2 and PI3-K/Akt pathways. Neurotox Res. 2014a;25:208–25.

Jantas D, Greda A, Golda S, Korostynski M, Grygier B, Roman A, Pilc A, Lason W. Neuroprotective effects of metabotropic glutamate receptor group II and III activators against MPP(+)-induced cell death in human neuroblastoma SH-SY5Y cells: the impact of cell differentiation state. Neuropharmacology. 2014b;83:36–53.

Jaworska-Feil L, Jantas D, Leskiewicz M, Budziszewska B, Kubera M, Basta-Kaim A, Lipkowski AW, Lason W. Protective effects of TRH and its analogues against various cytotoxic agents in retinoic acid (RA)-differentiated human neuroblastoma SH-SY5Y cells. Neuropeptides. 2010;44:495–508.

Jett JD, Boley AM, Girotti M, Shah A, Lodge DJ, Morilak DA. Antidepressant-like cognitive and behavioral effects of acute ketamine administration associated with plasticity in the ventral hippocampus to medial prefrontal cortex pathway. Psychopharmacology (Berl). 2015;232(17):3123–33.

Jiang Y, Li Z, Liu Y, Liu X, Chang Q, Liao Y, Pan R. Neuroprotective effect of water extract of Panax ginseng on corticosterone-induced apoptosis in PC12 cells and its underlying molecule mechanisms. J Ethnopharmacol. 2015;159:102–12.

Jin ML, Park SY, Kim YH, Park G, Lee SJ. Acanthopanax senticosus exerts neuroprotective effects through HO-1 signaling in hippocampal and microglial cells. Environ Toxicol Pharmacol. 2013;35:335–46.

Jirásek P, Amslinger S, Heilmann J. Synthesis of natural and non-natural curcuminoids and their neuroprotective activity against glutamate-induced oxidative stress in HT-22 cells. J Nat Prod. 2014;77:2206–17.

Johann S, Kampmann E, Denecke B, Arnold S, Kipp M, Mey J, Beyer C. Expression of enzymes involved in the prostanoid metabolism by cortical astrocytes after LPS-induced inflammation. J Mol Neurosci. 2008;34:177–85.

Johnson S, Tazik S, Lu D, Johnson C, Youdim MB, Wang J, Rajkowska G, Ou XM. The New Inhibitor of Monoamine Oxidase, M30, has a Neuroprotective Effect Against Dexamethasone-Induced Brain Cell Apoptosis. Front Neurosci. 2010;4:180.

Karkoulias G, Mastrogianni O, Ilias I, Lymperopoulos A, Taraviras S, Tsopanoglou N, Sitaras N, Flordellis CS. Alpha 2-adrenergic receptors decrease DNA replication and cell proliferation and induce neurite outgrowth in transfected rat pheochromocytoma cells. Ann N Y Acad Sci. 2006;1088:335–45.

Kasckow JW, Regmi A, Sheriff S, Mulchahey J, Geracioti Jr TD. Regulation of corticotropin-releasing factor messenger RNA by nicotine in an immortalized amygdalar cell line. Life Sci. 1999;65:2709–14.

Kataria H, Wadhwa R, Kaul SC, Kaur G. Water extract from the leaves of Withania somnifera protect RA differentiated C6 and IMR-32 cells against glutamate-induced excitotoxicity. PLoS One. 2012;7, e37080.

Khan RS, Yu C, Kastin AJ, He Y, Ehrensing RH, Hsuchou H, Stone KP, Pan W. Brain activation by peptide Pro-Leu-Gly-NH(2) (MIF-1). Int J Pept. 2010;2010:537639.

Kim WK, Jang PG, Woo MS, Han IO, Piao HZ, Lee K, Lee H, Joh TH, Kim HS. A new anti-inflammatory agent KL-1037 represses proinflammatory cytokine and inducible nitric oxide synthase (iNOS) gene expression in activated microglia. Neuropharmacology. 2004;47:243–52.

Kittel-Schneider S, Kenis G, Schek J, van den Hove D, Prickaerts J, Lesch KP, Steinbusch H, Reif A. Expression of monoamine transporters, nitric oxide synthase 3, and neurotrophin genes in antidepressant-stimulated astrocytes. Front Psychiatry. 2012;3:33.

Kling B, Bücherl D, Palatzky P, Matysik FM, Decker M, Wegener J, Heilmann J. Flavonoids, flavonoid metabolites, and phenolic acids inhibit oxidative stress in the neuronal cell line

HT-22 monitored by ECIS and MTT assay: a comparative study. J Nat Prod. 2014; 77:446–54.

Kolla N, Wei Z, Richardson JS, Li XM. Amitriptyline and fluoxetine protect PC12 cells from cell death induced by hydrogen peroxide. J Psychiatry Neurosci. 2005;30:196–201.

Kong EK, Peng L, Chen Y, Yu AC, Hertz L. Up-regulation of 5-HT2B receptor density and receptor-mediated glycogenolysis in mouse astrocytes by long-term fluoxetine administration. Neurochem Res. 2002;27:113–20.

Korecka JA, van Kesteren RE, Blaas E, Spitzer SO, Kamstra JH, Smit AB, Swaab DF, Verhaagen J, Bossers K. Phenotypic characterization of retinoic acid differentiated SH-SY5Y cells by transcriptional profiling. PLoS One. 2013;8, e63862.

Kritis AA, Stamoula EG, Paniskaki KA, Vavilis TD. Researching glutamate - induced cytotoxicity in different cell lines: a comparative/collective analysis/study. Front Cell Neurosci. 2015;9:91.

Krystal JH, Mathew SJ, D'Souza DC, Garakani A, Gunduz-Bruce H, Charney DS. Potential psychiatric applications of metabotropic glutamate receptor agonists and antagonists. CNS Drugs. 2010;24:669–93.

Laifenfeld D, Klein E, Ben-Shachar D. Norepinephrine alters the expression of genes involved in neuronal sprouting and differentiation: relevance for major depression and antidepressant mechanisms. J Neurochem. 2002;83:1054–64.

Lalmansingh AS, Uht RM. Estradiol regulates corticotropin-releasing hormone gene (crh) expression in a rapid and phasic manner that parallels estrogen receptor-alpha and -beta recruitment to a 3',5'-cyclic adenosine 5'-monophosphate regulatory region of the proximal crh promoter. Endocrinology. 2008;149:346–57.

Lauro C, Di Angelantonio S, Cipriani R, Sobrero F, Antonilli L, Brusadin V, Ragozzino D, Limatola C. Activity of adenosine receptors type 1 Is required for CX3CL1-mediated neuroprotection and neuromodulation in hippocampal neurons. J Immunol. 2008;180:7590–6.

Lee MS, Lee J, Kwon DY, Kim MS. Ondamtanggamibang protects neurons from oxidative stress with induction of heme oxygenase-1. J Ethnopharmacol. 2006;108:294–8.

Lee YH, Kim SH, Kim Y, Lim Y, Ha K, Shin SY. Inhibitory effect of the antidepressant imipramine on NF-κB-dependent CXCL1 expression in TNFα-exposed astrocytes. Int Immunopharmacol. 2012;12:547–55.

Leskiewicz M, Jantas D, Regulska M, Kaczanowska J, Basta-Kaim A, Budziszewska B, Kubera M, Lason W. Antidepressants attenuate the dexamethasone-induced decrease in viability and proliferation of human neuroblastoma SH-SY5Y cells: a involvement of extracellular regulated kinase (ERK1/2). Neurochem Int. 2013;63:354–62.

Leuner K, Kazanski V, Müller M, Essin K, Henke B, Gollasch M, Harteneck C, Müller WE. Hyperforin—a key constituent of St. John's wort specifically activates TRPC6 channels. FASEB J. 2007;21:4101–11.

Leuner K, Li W, Amaral MD, Rudolph S, Calfa G, Schuwald AM, Harteneck C, Inoue T, Pozzo-Miller L. Hyperforin modulates dendritic spine morphology in hippocampal pyramidal neurons by activating Ca(2+) -permeable TRPC6 channels. Hippocampus. 2013;23:40–52.

Levkovitz Y, Gil-Ad I, Zeldich E, Dayag M, Weizman A. Differential induction of apoptosis by antidepressants in glioma and neuroblastoma cell lines: evidence for p-c-Jun, cytochrome c, and caspase-3 involvement. J Mol Neurosci. 2005;27:29–42.

Li Y, Maher P, Schubert D. A role of 12-lipoxygenase in nerve cell death caused by glutathione depletion. Neuron. 1997;19:453–63.

Li YF, Gong ZH, Cao JB, Wang HL, Luo ZP, Li J. Antidepressant-like effect of agmatine and its possible mechanism. Eur J Pharmacol. 2003a;469:81–8.

Li YF, Gong ZH, Yang M, Zhao YM, Luo ZP. Inhibition of the oligosaccharides extracted from Morinda officinalis, a Chinese traditional herbal medicine, on the corticosterone induced apoptosis in PC12 cells. Life Sci. 2003b;72:933–42.

Li YF, Liu YQ, Huang WC, Luo ZP. Cytoprotective effect is one of common action pathways for antidepressants. Acta Pharmacol Sin. 2003c;24:996–1000.

Li YF, Liu YQ, Yang M, Wang HL, Huang WC, Zhao YM, Luo ZP. The cytoprotective effect of inulin-type hexasaccharide extracted from Morinda officinalis on PC12 cells against the lesion induced by corticosterone. Life Sci. 2004;75:1531–8.

Li YF, Zhang YZ, Liu YQ, Wang HL, Cao JB, Guan TT, Luo ZP. Inhibition of N-methyl-D-aspartate receptor function appears to be one of the common actions for antidepressants. J Psychopharmacol. 2006;20:629–35.

Li B, Zhang S, Zhang H, Nu W, Cai L, Hertz L, Peng L. Fluoxetine-mediated 5-HT2B receptor stimulation in astrocytes causes EGF receptor transactivation and ERK phosphorylation. Psychopharmacology (Berl). 2008;201:443–58.

Li B, Zhang S, Li M, Hertz L, Peng L. Chronic treatment of astrocytes with therapeutically relevant fluoxetine concentrations enhances cPLA2 expression secondary to 5-HT2B-induced, transactivation-mediated ERK1/2 phosphorylation. Psychopharmacology (Berl). 2009; 207:1–12.

Li ZY, Guo Z, Liu YM, Liu XM, Chang Q, Liao YH, Pan RL. Neuroprotective effects of total saikosaponins of Bupleurum yinchowense on corticosterone-induced apoptosis in PC12 cells. J Ethnopharmacol. 2013;148:794–803.

Li H, Mao S, Wang H, Zen K, Zhang C, Li L. MicroRNA-29a modulates axon branching by targeting doublecortin in primary neurons. Protein Cell. 2014;5:160–9.

Lin CJ, Chen TH, Yang LY, Shih CM. Resveratrol protects astrocytes against traumatic brain injury through inhibiting apoptotic and autophagic cell death. Cell Death Dis. 2014;5, e1147.

Liu J, Li L, Suo WZ. HT-22 hippocampal neuronal cell line possesses functional cholinergic properties. Life Sci. 2009;84:267–71.

Liu P, Hu Y, Guo DH, Wang DX, Tu HH, Ma L, Xie TT, Kong LY. Potential antidepressant properties of Radix Polygalae (Yuan Zhi). Phytomedicine. 2010;17:794–9.

Liu B, Zhang H, Xu C, Yang G, Tao J, Huang J, Wu J, Duan X, Cao Y, Dong J. Neuroprotective effects of icariin on corticosterone-induced apoptosis in primary cultured rat hippocampal neurons. Brain Res. 2011a;1375:59–67.

Liu D, Wang Z, Liu S, Wang F, Zhao S, Hao A. Anti-inflammatory effects of fluoxetine in lipopolysaccharide(LPS)-stimulated microglial cells. Neuropharmacology. 2011b;61:592–9.

Liu AJ, Wang SH, Hou SY, Lin CJ, Chiu WT, Hsiao SH, Chen TH, Shih CM. Evodiamine induces transient receptor potential vanilloid-1-mediated protective autophagy in U87-MG astrocytes. Evid Based Complementary Alternat Med. 2013;13:354840.

Liu RP, Zou M, Wang JY, Zhu JJ, Lai JM, Zhou LL, Chen SF, Zhang X, Zhu JH. Paroxetine ameliorates lipopolysaccharide-induced microglia activation via differential regulation of MAPK signaling. J Neuroinflammation. 2014;11:47.

Lodge D, Mercier MS. Ketamine and phencyclidine: the good, the bad and the unexpected. Br J Pharmacol. 2015;172(17):4254–76. doi:10.1111/bph.13222.

Lopes FM, Schröder R, da Frota Jr ML, Zanotto-Filho A, Müller CB, Pires AS, Meurer RT, Colpo GD, Gelain DP, Kapczinski F, Moreira JC, Fernandes Mda C, Klamt F. Comparison between proliferative and neuron-like SH-SY5Y cells as an in vitro model for Parkinson disease studies. Brain Res. 2010;1337:85–94.

López V, Martín S, Gómez-Serranillos MP, Carretero ME, Jäger AK, Calvo MI. Neuroprotective and neurological properties of Melissa officinalis. Neurochem Res. 2009;34:1955–61.

Lopez-Carballo G, Moreno L, Masia S, Perez P, Barettino D. Activation of the phosphatidylinositol 3-kinase/Akt signaling pathway by retinoic acid is required for neural differentiation of SH-SY5Y human neuroblastoma cells. J Biol Chem. 2002;277:297–304.

Lopizzo N, Bocchio Chiavetto L, Cattane N, Plazzotta G, Tarazi FI, Pariante CM, Riva MA, Cattaneo A. Gene-environment interaction in major depression: focus on experience-dependent biological systems Front. Psychiatry. 2015;6:68.

Lu YH, Du CB, Liu JW, Hong W, Wei DZ. Neuroprotective effects of Hypericum perforatum on trauma induced by hydrogen peroxide in PC12 cells. Am J Chin Med. 2004;32:397–405.

Luchtman DW, Song C. Why SH-SY5Y cells should be differentiated. Neurotoxicology. 2010; 31:164–5.

Lucki A, Klein E, Karry R, Ben-Shachar D. Dexamethasone in the presence of desipramine enhances MAPK/ERK1/2 signaling possibly via its interference with β-arrestin. J Neural Transm. 2014;121:289–98.

Maher P, Davis B. The role of monoamine metabolismin oxidative glutamate toxicity. Neuroscience. 1996;16:6394–401.

Malviya SA, Kelly SD, Greenlee MM, Eaton DC, Duke BJ, Bourke CH, Neigh GN. Estradiol stimulates an anti-translocation expression pattern of glucocorticoid co-regulators in a hippocampal cell model. Physiol Behav. 2013;122:187–92.

Malynn S, Campos-Torres A, Moynagh P, Haase J. The pro-inflammatory cytokine TNF-α regulates the activity and expression of the serotonin transporter (SERT) in astrocytes. Neurochem Res. 2013;38:694–704.

Mao QQ, Ip SP, Ko KM, Tsai SH, Zhao M, Che CT. Peony glycosides protect against corticosterone-induced neurotoxicity in PC12 cells. Cell Mol Neurobiol. 2009;29:643–7.

Mao QQ, Zhong XM, Feng CR, Pan AJ, Li ZY, Huang Z. Protective effects of paeoniflorin against glutamate-induced neurotoxicity in PC12 cells via antioxidant mechanisms and Ca(2+) antagonism. Cell Mol Neurobiol. 2010;30:1059–66.

Mao QQ, Zhong XM, Li ZY, Huang Z. Paeoniflorin protects against NMDA-induced neurotoxicity in PC12 cells via Ca2+ antagonism. Phytother Res. 2011;25:681–5.

Marrs WR, Blankman JL, Horne EA, Thomazeau A, Lin YH, Coy J, Bodor AL, Muccioli GG, Hu SS, Woodruff G, Fung S, Lafourcade M, Alexander JP, Long JZ, Li W, Xu C, Möller T, Mackie K, Manzoni OJ, Cravatt BF, Stella N. The serine hydrolase ABHD6 controls the accumulation and efficacy of 2-AG at cannabinoid receptors. Nat Neurosci. 2010;13:951–7.

Matsui N, Kido Y, Okada H, Kubo M, Nakai M, Fukuishi N, Fukuyama Y, Akagi M. Phenylbutenoid dimers isolated from Zingiber purpureum exert neurotrophic effects on cultured neurons and enhance hippocampal neurogenesis in olfactory bulbectomized mice. Neurosci Lett. 2012;513:72–7.

Matsuoka I, Mizuno N, Kurihara K. Cholinergic differentiation of clonal rat pheochromocytoma cells (PC12) induced by retinoic acid: increase of choline acetyltransferase activity and decrease of tyrosine hydroxylase activity. Brain Res. 1989;502:53–60.

Mattson MP. Energy intake and exercise as determinants of brain health and vulnerability to injury and disease. Cell Metab. 2012;16:706–22.

McEwen BS, Chattarji S. Molecular mechanisms of neuroplasticity and pharmacological implications: the example of tianeptine. Eur Neuropsychopharmacol. 2004;14 Suppl 5:S497–502.

McEwen BS, Chattarji S, Diamond DM, Jay TM, Reagan LP, Svenningsson P, Fuchs E. The neurobiological properties of tianeptine (Stablon): from monoamine hypothesis to glutamatergic modulation. Mol Psychiatry. 2010;15:237–49.

McKernan DP, Dinan TG, Cryan JF. "Killing the Blues": a role for cellular suicide (apoptosis) in depression and the antidepressant response? Prog Neurobiol. 2009;88:246–63.

McMahon A, Sabban EL. Regulation of expression of dopamine beta-hydroxylase in PC12 cells by glucocorticoids and cyclic AMP analogues. J Neurochem. 1992;59:2040–7.

Mercier G, Lennon AM, Renouf B, Dessouroux A, Ramauge M, Courtin F, Pierre M. MAP kinase activation by fluoxetine and its relation to gene expression in cultured rat astrocytes. J Mol Neurosci. 2004;24:207–16.

Miller AH, Maletic V, Raison CL. Inflammation and its discontents: the role of cytokines in the pathophysiology of major depression. Biol Psychiatry. 2009;65:732–41.

Miloso M, Villa D, Crimi M, Galbiati S, Donzelli E, Nicolini G, Tredici G. Retinoic acid-induced neuritogenesis of human neuroblastoma SH-SY5Y cells is ERK independent and PKC dependent. J Neurosci Res. 2004;75:241e252.

Mizuno N, Matsuoka I, Kurihara K. Possible involvements of intracellular Ca2+ and Ca2+-dependent protein phosphorylation in cholinergic differentiation of clonal rat pheochromocytoma cells (PC12) induced by glioma-conditioned medium and retinoic acid. Brain Res Dev Brain Res. 1989;50:1–10.

Moosavi M, Maghsoudi N, Zahedi-Asl S, Naghdi N, Yousefpour M, Trounce IA. The role of PI3/Akt pathway in the protective effect of insulin against corticosterone cell death induction in hippocampal cell culture. Neuroendocrinology. 2008;88:293–8.

Morimoto BH, Koshland DE. Induction and expression of long and short term neurosecretory potentiation in a neural cell line. Neuron. 1990;5:875–8.
Morita K, Gotohda T, Arimochi H, Lee MS, Her S. Histone deacetylase inhibitors promote neurosteroid-mediated cell differentiation and enhance serotonin-stimulated brain-derived neurotrophic factor gene expression in rat C6 glioma cells. J Neurosci Res. 2009;87:2608–14.
Mulchahey JJ, Regmi A, Sheriff S, Balasubramaniam A, Kasckow JW. Coordinate and divergent regulation of corticotropin-releasing factor (CRF) and CRF-binding protein expression in an immortalized amygdalar neuronal cell line. Endocrinology. 1999;140:251–9.
Musazzi L, Rimland JM, Ieraci A, Racagni G, Domenici E, Popoli M. Pharmacological characterization of BDNF promoters I, II and IV reveals that serotonin and norepinephrine input is sufficient for transcription activation. Int J Neuropsychopharmacol. 2014;17:779–91.
Nakano M, Osada K, Misonoo A, Fujiwara K, Takahashi M, Ogawa Y, Haga T, Kanai S, Tanaka D, Sasuga Y, Yanagida T, Asakura M, Yamaguchi N. Fluvoxamine and sigma-1 receptor agonists dehydroepiandrosterone (DHEA)-sulfate induces the Ser473-phosphorylation of Akt-1 in PC12 cells. Life Sci. 2010;86:309–14.
Nestler EJ. Antidepressant treatments in the 21st century. Biol Psychiatry. 1998;4:526–33.
Nestler EJ, Barrot M, DiLeone RJ, Eisch AJ, Gold SJ, Monteggia LM. Neurobiology of depression. Neuron. 2002;34:13–25.
Newton SS, Duman RS. Regulation of neurogenesis and angiogenesis in depression. Curr Neurovasc Res. 2004;1:261–7.
Niciu MJ, Ionescu DF, Richards EM, Zarate Jr CA. Glutamate and its receptors in the pathophysiology and treatment of major depressive disorder. J Neural Transm. 2014;121:907–24.
Nishimura T, Ishima T, Iyo M, Hashimoto K. Potentiation of nerve growth factor-induced neurite outgrowth by fluvoxamine: role of sigma-1 receptors, IP3 receptors and cellular signaling pathways. PLoS One. 2008;3:2558.
Niswender CM, Conn PJ. Metabotropic glutamate receptors: physiology, pharmacology, and disease. Annu Rev Pharmacol Toxicol. 2010;50:295–322.
Obuchowicz E, Bielecka AM, Paul-Samojedny M, Pudełko A, Kowalski J. Imipramine and fluoxetine inhibit LPS-induced activation and affect morphology of microglial cells in the rat glial culture. Pharmacol Rep. 2014;66:34–43.
Okugawa G, Omori K, Suzukawa J, Fujiseki Y, Kinoshita T, Inagaki C. Long-term treatment with antidepressants increases glucocorticoid receptor binding and gene expression in cultured rat hippocampal neurones. J Neuroendocrinol. 1999;11:887–95.
Oppermann S, Schrader FC, Elsässer K, Dolga AM, Kraus AL, Doti N, Wegscheid-Gerlach C, Schlitzer M, Culmsee C. Novel N-phenyl-substituted thiazolidinediones protect neural cells against glutamate- and tBid-induced toxicity. J Pharmacol Exp Ther. 2014;350:273–89.
Ortega F, Berninger B, Costa MR. Primary culture and live imaging of adult neural stem cells and their progeny. Methods Mol Biol. 2013;1052:1–11.
Pace TW, Hu F, Miller AH. Activation of cAMP-protein kinase A abrogates STAT5-mediated inhibition of glucocorticoid receptor signaling by interferon-alpha. Brain Behav Immun. 2011;25:1716–24.
Paez-Pereda M, Hausch F, Holsboer F. Corticotropin releasing factor receptor antagonists for major depressive disorder. Expert Opin Investig Drugs. 2011;20:519–35.
Pałucha-Poniewiera A, Wierońska JM, Brański P, Stachowicz K, Chaki S, Pilc A. On the mechanism of the antidepressant-like action of group II mGlu receptor antagonist, MGS0039. Psychopharmacology (Berl). 2010;212:523–35.
Pałucha-Poniewiera A, Wierońska JM, Brański P, Burnat G, Chruścicka B, Pilc A. Is the mGlu5 receptor a possible target for new antidepressant drugs? Pharmacol Rep. 2013;65:1506–11.
Pałucha-Poniewiera A, Szewczyk B, Pilc A. Activation of the mTOR signaling pathway in the antidepressant-like activity of the mGlu5 antagonist MTEP and the mGlu7 agonist AMN082 in the FST in rats. Neuropharmacology. 2014;82:59–68.
Pani G, De Vos WH, Samari N, de Saint-Georges L, Baatout S, Van Oostveldt P, Benotmane MA. MorphoNeuroNet: an automated method for dense neurite network analysis. Cytometry A. 2014;85:188–99.

Pao LH, Lu SW, Sun GG, Chiou SH, Ma KH. Three Chinese herbal medicines promote neuro-proliferation in vitro, and reverse the effects of chronic mild stress on behavior, the HPA axis, and proliferation of hippocampal precursor cell in vivo. J Ethnopharmacol. 2012;144:261–9.

Pariante CM, Hye A, Williamson R, Makoff A, Lovestone S, Kerwin RW. The antidepressant clomipramine regulates cortisol intracellular concentrations and glucocorticoid receptor expression in fibroblasts and rat primary neurones. Neuropsychopharmacology. 2003;28: 1553–61.

Park SY, Jin ML, Kim YH, Kim Y, Lee SJ. Anti-inflammatory effects of aromatic-turmerone through blocking of NF-κB, JNK, and p38 MAPK signaling pathways in amyloid β-stimulated microglia. Int Immunopharmacol. 2012;14:13–20.

Partin KM. AMPA receptor potentiators: from drug design to cognitive enhancement. Curr Opin Pharmacol. 2015;20:46–53.

Paul IA, Skolnick P. Glutamate and depression: clinical and preclinical studies. Ann N Y Acad Sci. 2003;1003:250–72.

Paul RK, Singh NS, Khadeer M, Moaddel R, Sanghvi M, Green CE, O'Loughlin K, Torjman MC, Bernier M, Wainer IW. (R, S)-Ketamine metabolites (R, S)-norketamine and (2S,6S)-hydroxynorketamine increase the mammalian target of rapamycin function. Anesthesiology. 2014;121(1):149–59.

Páv M, Kovárů H, Fiserová A, Havrdová E, Lisá V. Neurobiological aspects of depressive disorder and antidepressant treatment: role of glia. Physiol Res. 2008;57:151–64.

Pepin MC, Beaulieu S, Barden N. Antidepressants regulate glucocorticoid receptor messenger RNA concentrations in primary neuronal cultures. Brain Res Mol Brain Res. 1989;6:77–83.

Perisic T, Zimmermann N, Kirmeier T, Asmus M, Tuorto F, Uhr M, Holsboer F, Rein T, Zschocke J. Valproate and amitriptyline exert common and divergent influences on global and gene promoter-specific chromatin modifications in rat primary astrocytes. Neuropsychopharmacology. 2010;35:792–805.

Pfister JA, Ma C, Morrison BE, D'Mello SR. Opposing effects of sirtuins on neuronal survival: SIRT1-mediated neuroprotection is independent of its deacetylase activity. PLoS One. 2008;3, e4090.

Pizzurro DM, Dao K, Costa LG. Diazinon and diazoxon impair the ability of astrocytes to foster neurite outgrowth in primary hippocampal neurons. Toxicol Appl Pharmacol. 2014;274: 372–82.

Pochwat B, Pałucha-Poniewiera A, Szewczyk B, Pilc A, Nowak G. NMDA antagonists under investigation for the treatment of major depressive disorder. Expert Opin Investig Drugs. 2014;23:1181–92.

Polajnar M, Zerovnik E. Impaired autophagy: a link between neurodegenerative and neuropsychiatric diseases. J Cell Mol Med. 2014;18:1705–11.

Pregi N, Vittori D, Perez G, Pe´rez Leiro's C, Nesse A. Effect of erythropoietin on staurosporine-induced apoptosis and differentiation of SH-SY5Y neuroblastoma cells. Biochim Biophys Acta. 2006;1763:238–46.

Presgraves SP, Tariq A, Borwege S, Joyce JN. Terminally differentiated SHSY5Y cells provide a model system for studying neuroprotective effects of dopamine agonists. Neurotox Res. 2004;5:579–98.

Qu WS, Tian DS, Guo ZB, Fang J, Zhang Q, Yu ZY, Xie MJ, Zhang HQ, Lü JG, Wang W. Inhibition of EGFR/MAPK signaling reduces microglial inflammatory response and the associated secondary damage in rats after spinal cord injury. J Neuroinflammation. 2012;9:178.

Quiros I, Mayo JC, Garcia-Suarez O, Hevia D, Martin V, Rodríguez C, Sainz RM. Melatonin prevents glucocorticoid inhibition of cell proliferation and toxicity in hippocampal cells by reducing glucocorticoid receptor nuclear translocation. J Steroid Biochem Mol Biol. 2008;110:116–24.

Raedler TJ. Inflammatory mechanisms in major depressive disorder. Curr Opin Psychiatry. 2011; 24:519–25.

Rahmat Z, Jose S, Ramasamy R, Vidyadaran S. Reciprocal interactions of mouse bone marrow-derived mesenchymal stem cells and BV2 microglia after lipopolysaccharide stimulation. Stem Cell Res Ther. 2013;4:12.

Rajkowska G, Miguel-Hidalgo JJ. Gliogenesis and glial pathology in depression. CNS Neurol Disord Drug Targets. 2007;6:219–33.

Rantamäki T, Yalcin I. Antidepressant drug action – from rapid changes on network function to network rewiring. Prog Neuropsychopharmacol Biol Psychiatry. 2015;pii:S0278-5846(15)00128-1. doi:10.1016/j.pnpbp.2015.06.001.

Richardson JS, Hertz L. The effects of antidepressant drugs on adenylyl cyclase linked beta adrenergic binding sites on mouse astrocytes in primary cultures. Prog Neuropsychopharmacol Biol Psychiatry. 1983;7:675–80.

Robson MJ, Elliott M, Seminerio MJ, Matsumoto RR. Evaluation of sigma (σ) receptors in the antidepressant-like effects of ketamine in vitro and in vivo. Eur Neuropsychopharmacol. 2012;22:308–17.

Sanacora G, Rothman DL, Mason G, Krystal JH. Clinical studies implementing glutamate neurotransmission in mood disorders. Ann N Y Acad Sci. 2003;1003:292–308.

Sanacora G, Treccani G, Popoli M. Towards a glutamate hypothesis of depression: an emerging frontier of neuropsychopharmacology for mood disorders. Neuropharmacology. 2012; 62:63–77.

Sarkanen JR, Nykky J, Siikanen J, Selinummi J, Ylikomi T, Jalonen TO. Cholesterol supports the retinoic acid-induced synaptic vesicle formation in differentiating human SH-SY5Y neuroblastoma cells. J Neurochem. 2007;102:1941–52.

Sasaki K, El Omri A, Kondo S, Han J, Isoda H. Rosmarinus officinalis polyphenols produce antidepressant like effect through monoaminergic and cholinergic functions modulation. Behav Brain Res. 2013;238:86–94.

Saura J, Tusell JM, Serratosa J. High-yield isolation of murine microglia by mild trypsinization. Glia. 2003;44:183–9.

Schildkraut JJ. The catecholamine hypothesis of affective disorders: a review of supporting evidence. 1965. J Neuropsychiatry Clin Neurosci. 1995;7:524–33.

Schneider L, Giordano S, Zelickson BR, Johnson MS, Benavides GA, Ouyang X, Fineberg N, Darley-Usmar VM, Zhang J. Differentiation of SH-SY5Y cells to a neuronal phenotype changes cellular bioenergetics and the response to oxidative stress. Free Radic Biol Med. 2011;51:2007–17.

Schramm J, Schulte D. A fast and simple differentiation protocol to study the pro-neurogenic activity of soluble factors in neurospheres. Neurosci Lett. 2014;562:69–74.

Schwaninger M, Schöfl C, Blume R, Rössig L, Knepel W. Inhibition by antidepressant drugs of cyclic AMP response element-binding protein/cyclic AMP response element-directed gene transcription. Mol Pharmacol. 1995;47:1112–8.

Seo MK, Cho HY, Lee CH, Koo KA, Park YK, Lee JG, Lee BJ, Park SW, Kim YH. Antioxidant and proliferative activities of bupleuri radix extract against serum deprivation in SH-SY5Y cells. Psychiatry Investig. 2013;10:81–8.

Shaul U, Ben-Shachar D, Karry R, Klein E. Modulation of frequency and duration of repetitive magnetic stimulation affects catecholamine levels and tyrosine hydroxylase activity in human neuroblastoma cells: implication for the antidepressant effect of rTMS. Int J Neuropsycho pharmacol. 2003;6:233–41.

Sheriff S, Dautzenberg FM, Mulchahey JJ, Pisarska M, Hauger RL, Chance WT, Balasubramaniam A, Kasckow JW. Interaction of neuropeptide Y and corticotropin-releasing factor signaling pathways in AR-5 amygdalar cells. Peptides. 2001;22:2083–9.

Shimizu M, Nishida A, Hayakawa H, Yamawaki S. Ca2+ release from inositol 1,4,5-trisphosphate-sensitive Ca2+ store by antidepressant drugs in cultured neurons of rat frontal cortex. J Neurochem. 1993;60:595–601.

Shirai M, Kawai Y, Yamanishi R, Terao J. Approach to novel functional foods for stress control 5. Antioxidant activity profiles of antidepressant herbs and their active components. J Med Invest. 2005;52(Suppl):249–51.

Ślusarczyk J, Trojan E, Głombik K, Budziszewska B, Kubera M, Lasoń W, Popiołek-Barczyk K, Mika J, Wędzony K, Basta-Kaim A. Prenatal stress is a vulnerability factor for altered morphology and biological activity of microglia cells. Front Cell Neurosci. 2015;9:82.

Sofroniew MV. Reactive astrocytes in neural repair and protection. Neuroscientist. 2005; 11:400–7.

Sofroniew MV, Vinters HV. Astrocytes: biology and pathology. Acta Neuropathol. 2010; 119:7–35.

Song JH, Marszalec W, Kai L, Yeh JZ. Narahashi T (2012) Antidepressants inhibit proton currents and tumor necrosis factor-α production in BV2 microglial cells. Brain Res. 2012;1435:15–23.

Storch A, Lester HA, Boehm BO, Schwarz J. Functional characterization of dopaminergic neurons derived from rodent mesencephalic progenitor cells. J Chem Neuroanat. 2003;26:133–42.

Takahashi K, Yamada M, Ohata H, Honda K, Yamada M. Ndrg2 promotes neurite outgrowth of NGF-differentiated PC12 cells. Neurosci Lett. 2005;388:157–62.

Tan S, Wood M, Maher P. Oxidative stress induces a form of programmed cell death with characteristics of both apoptosis and necrosis in neuronal cells. J Neurochem. 1998;71:95–105.

Tao L, Zhang F, Hao L, Wu J, Jia J, Liu JY, Zheng LT, Zhen X. 1-O-tigloyl-1-O-deacetyl-nimbolinin B inhibits LPS-stimulated inflammatory responses by suppressing NF-κB and JNK activation in microglia cells. J Pharmacol Sci. 2014;125:364–74.

Tazik S, Johnson S, Lu D, Johnson C, Youdim MB, Stockmeier CA, Ou XM. Comparative neuroprotective effects of rasagiline and aminoindan with selegiline on dexamethasone-induced brain cell apoptosis. Neurotox Res. 2009;15:284–90.

Terada K, Kojima Y, Watanabe T, Izumo N, Chiba K, Karube Y. Inhibition of nerve growth factor-induced neurite outgrowth from PC12 cells by dexamethasone: signaling pathways through the glucocorticoid receptor and phosphorylated Akt and ERK1/2. PLoS One. 2014;9, e93223.

Thompson SM, Kallarackal AJ, Kvarta MD, Van Dyke AM, LeGates TA, Cai X. An excitatory synapse hypothesis of depression. Trends Neurosci. 2015;38:279–94.

Tieu K, Zuo DM, Yu PH. Differential effects of staurosporine and retinoic acid on the vulnerability of the SH-SY5Y neuroblastoma cells: involvement of Bcl-2 and p53 proteins. J Neurosci Res. 1999;58:426–35.

Tobaben S, Grohm J, Seiler A, Conrad M, Plesnila N, Culmsee C. Bid-mediated mitochondrial damage is a key mechanism in glutamate-induced oxidative stress and AIF-dependent cell death in immortalized HT-22 hippocampal neurons. Cell Death Differ. 2011;18:282–92.

Vialou V, Feng J, Robison AJ, Nestler EJ. Epigenetic mechanisms of depression and antidepressant action. Annu Rev Pharmacol Toxicol. 2013;53:59–87.

Wahane SD, Hellbach N, Prentzell MT, Weise SC, Vezzali R, Kreutz C, Timmer J, Krieglstein K, Thedieck K, Vogel T. PI3K-p110-alpha-subtype signalling mediates survival, proliferation and neurogenesis of cortical progenitor cells via activation of mTORC2. J Neurochem. 2014;130:255–67.

Wang CN, Shiao YJ, Lin YL, Chen CF. Nepalolide A inhibits the expression of inducible nitric oxide synthase by modulating the degradation of IkappaB-alpha and IkappaB-beta in C6 glioma cells and rat primary astrocytes. Br J Pharmacol. 1999;128:345–56.

Wang X, Pongrac JL, DeFranco DB. Glucocorticoid receptors in hippocampal neurons that do not engage proteasomes escape from hormone-dependent down-regulation but maintain transactivation activity. Mol Endocrinol. 2002;16:1987–98.

Wang X, Wu H, Lakdawala VS, Hu F, Hanson ND, Miller AH. Inhibition of Jun N-terminal kinase (JNK) enhances glucocorticoid receptor-mediated function in mouse hippocampal HT22 cells. Neuropsychopharmacology. 2005;30(2):242–9.

Wang Y, Shi X, Qi Z. Hypericin prolongs action potential duration in hippocampal neurons by acting on K+ channels. Br J Pharmacol. 2010;159:1402–7.

Wang H, Zhou X, Huang J, Mu N, Guo Z, Wen Q, Wang R, Chen S, Feng ZP, Zheng W. The role of Akt/FoxO3a in the protective effect of venlafaxine against corticosterone-induced cell death in PC12 cells. Psychopharmacology (Berl). 2013;228:129–41.

Wang H, Zhang R, Qiao Y, Xue F, Nie H, Zhang Z, Wang Y, Peng Z, Tan Q. Gastrodin ameliorates depression-like behaviors and up-regulates proliferation of hippocampal-derived neural stem cells in rats: involvement of its anti-inflammatory action. Behav Brain Res. 2014; 266:153–60.

Watson S, Gallagher P, Porter RJ, Smith MS, Herron LJ, Bulmer S; North-East Mood Disorders Clinical Research Group, Young AH, Ferrier IN. A randomized trial to examine the effect of mifepristone on neuropsychological performance and mood in patients with bipolar depression. Biol Psychiatry. 2012;72:943–9.

Waxman EA, Baconguis I, Lynch DR, Robinson MB. N-methyl-D-aspartate receptor-dependent regulation of the glutamate transporter excitatory amino acid carrier 1. J Biol Chem. 2007;282:17594–607.

Wenker SD, Chamorro ME, Vota DM, Callero MA, Vittori DC, Nesse AB. Differential antiapoptotic effect of erythropoietin on undifferentiated and retinoic acid-differentiated SH-SY5Y cells. J Cell Biochem. 2010;110:151–61.

Willets JM, Lambert DG, Lunec J, Griffiths HR. Studies on the neurotoxicity of 6,7-dihydroxy-1-methyl-1,2,3,4-tetrahydroisoquinoline (salsolinol) in SH-SY5Y cells. Eur J Pharmacol. 1995;293:319–26.

Wu F, Li H, Zhao L, Li X, You J, Jiang Q, Li S, Jin L, Xu Y. Protective effects of aqueous extract from Acanthopanax senticosus against corticosterone-induced neurotoxicity in PC12 cells. J Ethnopharmacol. 2013a;148:861–8.

Wu Q, Cho JG, Lee DS, Lee DY, Song NY, Kim YC, Lee KT, Chung HG, Choi MS, Jeong TS, Ahn EM, Kim GS, Baek NI. Carbohydrate derivatives from the roots of Brassica rapa ssp. campestris and their effects on ROS production and glutamate-induced cell death in HT-22 cells. Carbohydr Res. 2013b;372:9–14.

Wu M, Zhang H, Zhou C, Jia H, Ma Z, Zou Z. Identification of the chemical constituents in aqueous extract of zhi-qiao and evaluation of its antidepressant effect. Molecules. 2015; 20:6925–40.

Wuwongse S, Cheng SS, Wong GT, Hung CH, Zhang NQ, Ho YS, Law AC, Chang RC. Effects of corticosterone and amyloid-beta on proteins essential for synaptic function: implications for depression and Alzheimer's disease. Biochim Biophys Acta. 2013;1832:2245–56.

Xu Y, Zhang C, Wang R, Govindarajan SS, Barish PA, Vernon MM, Fu C, Acharya AP, Chen L, Boykin E, Yu J, Pan J, O'Donnell JM, Ogle WO. Corticosterone induced morphological changes of hippocampal and amygdaloid cell lines are dependent on 5-HT7 receptor related signal pathway. Neuroscience. 2011;182:71–81.

Xu SL, Choi RC, Zhu KY, Leung KW, Guo AJ, Bi D, Xu H, Lau DT, Dong TT, Tsim KW. Isorhamnetin, a flavonol aglycone from Ginkgo biloba L, induces neuronal differentiation of cultured PC12 cells: potentiating the effect of nerve growth factor. Evid Based Complementary Alternat Med. 2012;2012:278273.

Xu Z, Wu J, Zheng J, Ma H, Zhang H, Zhen X, Zheng LT, Zhang X. Design, synthesis and evaluation of a series of non-steroidal anti-inflammatory drug conjugates as novel neuroinflammatory inhibitors. Int Immunopharmacol. 2015;25:528–37.

Xue R, Jin ZL, Chen HX, Yuan L, He XH, Zhang YP, Meng YG, Xu JP, Zheng JQ, Zhong BH, Li YF, Zhang YZ. Antidepressant-like effects of 071031B, a novel serotonin and norepinephrine reuptake inhibitor. Eur Neuropsychopharmacol. 2013;23:728–41.

Yan L, Xu SL, Zhu KY, Lam KY, Xin G, Maiwulanjiang M, Li N, Dong TT, Lin H, Tsim KW. Optimizing the compatibility of paired-herb in an ancient Chinese herbal decoction Kai-Xin-San in activating neurofilament expression in cultured PC12 cells. J Ethnopharmacol. 2015;162:155–62.

Yang X, Ewald ER, Huo Y, Tamashiro KL, Salvatori R, Sawa A, Wand GS, Lee RS. Glucocorticoid-induced loss of DNA methylation in non-neuronal cells and potential involvement of DNMT1 in epigenetic regulation of Fkbp5. Biochem Biophys Res Commun. 2012;420: 570–5.

Yaniv SP, Ben-Shachar D, Klein E. Norepinephrine-glucocorticoids interaction does not annul the opposite effects of the individual treatments on cellular plasticity in neuroblastoma cells. Eur J Pharmacol. 2008;596:14–24.

Yeoh JW, Campbell EJ, James MH, Graham BA, Dayas CV. Orexin antagonists for neuropsychiatric disease: progress and potential pitfalls. Front Neurosci. 2014;8:36.
Yong-Kee CJ, Salomonczyk D, Nash JE. Development and validation of a screening assay for the evaluation of putative neuroprotective agents in the treatment of Parkinson's disease. Neurotox Res. 2011;19:519–26.
Yuhas Y, Ashkenazi S, Berent E, Weizman A. Immunomodulatory activity of ketamine in human astroglial A172 cells: possible relevance to its rapid antidepressant activity. J Neuroimmunol. 2015;282:33–8.
Zhang S, Li B, Lovatt D, Xu J, Song D, Goldman SA, Nedergaard M, Hertz L, Peng L. 5-HT2B receptors are expressed on astrocytes from brain and in culture and are a chronic target for all five conventional 'serotonin-specific reuptake inhibitors'. Neuron Glia Biol. 2010;6:113–25.
Zhang F, Zhou H, Wilson BC, Shi JS, Hong JS, Gao HM. Fluoxetine protects neurons against microglial activation-mediated neurotoxicity. Parkinsonism Relat Disord. 2012a;18 Suppl 1:S213–7.
Zhang H, Liu B, Wu J, Xu C, Tao J, Duan X, Cao Y, Dong J. Icariin inhibits corticosterone-induced apoptosis in hypothalamic neurons via the PI3-K/Akt signaling pathway. Mol Med Rep. 2012b;6:967–72.
Zhang R, Peng Z, Wang H, Xue F, Chen Y, Wang Y, Wang H, Tan Q. Gastrodin ameliorates depressive-like behaviors and up-regulates the expression of BDNF in the hippocampus and hippocampal-derived astrocyte of rats. Neurochem Res. 2014;39:172–9.
Zhang GF, Liu WX, Qiu LL, Guo J, Wang XM, Sun HL, Yang JJ, Zhou ZQ. Repeated ketamine administration redeems the time lag for citalopram's antidepressant-like effects. Eur Psychiatry. 2015;30:504–10.
Zheng M, Liu C, Pan F, Shi D, Ma F, Zhang Y, Zhang Y. Protective effects of flavonoid extract from Apocynum venetum leaves against corticosterone-induced neurotoxicity in PC12 cells. Cell Mol Neurobiol. 2011;31:421–8.
Zhou H, Li X, Gao M. Curcumin protects PC12 cells from corticosterone-induced cytotoxicity: possible involvement of the ERK1/2 pathway. Basic Clin Pharmacol Toxicol. 2009;104: 236–40.
Zhu WH, Conforti L, Millhorn DE. Expression of dopamine D2 receptor in PC-12 cells and regulation of membrane conductances by dopamine. Am J Physiol. 1997;273:C1143–50.
Zhu W, Ma S, Qu R, Kang D. Antidepressant-like effect of saponins extracted from Chaihu-jia-longgu-muli-tang and its possible mechanism. Life Sci. 2006;79(8):749–56.
Zhu KY, Xu SL, Choi RC, Yan AL, Dong TT, Tsim KW. Kai-xin-san, a chinese herbal decoction containing ginseng radix et rhizoma, polygalae radix, acori tatarinowii rhizoma, and poria, stimulates the expression and secretion of neurotrophic factors in cultured astrocytes. Evid Based Complementary Alternat Med. 2013;2013:731385.
Zimmermann N, Zschocke J, Perisic T, Yu S, Holsboer F, Rein T. Antidepressants inhibit DNA methyltransferase 1 through reducing G9a levels. Biochem J. 2012;448:93–102.
Zou YP, Lu YH, Wei DZ. Protective effects of a flavonoid-rich extract of Hypericum perforatum L. against hydrogen peroxide-induced apoptosis in PC12 cells. Phytother Res. 2010;24 Suppl 1:S6–10.
Zschocke J, Rein T. Antidepressants encounter autophagy in neural cells. Autophagy. 2011;7: 1247–8.
Zschocke J, Zimmermann N, Berning B, Ganal V, Holsboer F, Rein T. Antidepressant drugs diversely affect autophagy pathways in astrocytes and neurons-dissociation from cholesterol homeostasis. Neuropsychopharmacology. 2011;36:1754–68.
Zusso M, Debetto P, Guidolin D, Giusti P. Cerebellar granular cell cultures as an in vitro model for antidepressant drug-induced neurogenesis. Crit Rev Neurobiol. 2004;16:59–65.

# Chapter 4
# Animal Models of Depression

**Eva González-Trujano, Fabiola Domínguez, Juan Manuel Gallardo, Christian Guerra-Araiza, Julia J. Segura-Uribe, and Sandra Orozco-Suárez**

**Abstract** Stress is a concept related to biological responses produced in all living organisms when facing environmental changes. These biological responses allow them to adapt. Chronic stress is a major risk factor in the development of several chronic diseases that has a high social and economic impact in many countries. Within these diseases is depression. Through a better understanding of the neurobiology of stress and depressive episodes, better preventative therapies and antidepressants with fewer side effects may be developed. Animal models of chronic stress have provided very important information through the pathophysiology of depression-associated behaviors that are similar to the behavior of depressed patients. This chapter reviews several depression assays involving acute stress (e.g., forced swim test), models consisting of prolonged physical or social stress (e.g., social defeat), models of secondary depression, genetic models, and experiments designed to elucidate the mechanisms of antidepressant action. These paradigms are critically evaluated in relation to the disease, validity and replicability, the molecular insights that they, and their capacity to offer the next generation of therapeutics for depression.

---

E. González-Trujano
Dirección de Investigaciones en Neurociencias, Instituto Nacional de Psiquiatría Ramón de la Fuente Muñiz, Mexico DF, Mexico

F. Domínguez
Centro de Investigación Biomédica de Oriente, Instituto Mexicano del Seguro Social, Metepec, Mexico

J.M. Gallardo
Unidad de Investigación Médica en Enfermedades Nefrológicas, Hospital de Especialidades, Centro Médico Nacional, Siglo XXI, Instituto Mexicano del Seguro Social, 06720 Ciudad de México, Mexico
e-mail: jmgallardom@gmail.com

J.J. Segura-Uribe • S. Orozco-Suárez (✉)
Unidad de Investigación Médica en Enfermedades, Hospital de Especialidades, Centro Médico Nacional Siglo XXI, Instituto Mexicano del Seguro Social, Av. Cuauhtémoc 330 Col. Doctores, 06720 Ciudad de México, Mexico
e-mail: sorozco5@hotmail.com

C. Guerra-Araiza
Unidad de Investigación Médica en Farmacología, Centro Médico Nacional Siglo XXI, Instituto Mexicano del Seguro Social, Mexico DF, Mexico



C. Grosso (ed.), *Herbal Medicine in Depression*,
DOI 10.1007/978-3-319-14021-6_4

**Keywords** Animal models • Behavioral testing • Depression • Stress

## 4.1 Introduction

Depression is a porque se repite disorder that has been present throughout the history of humanity and has been approached from different views according to time, thinking, and scientific and technological advances (Guimarães et al. 2006). In the present day, depression is classified in the Diagnostic and Statistical Manual of Mental Disorders (DSM-IV) as a mood disorder (American Psychiatric Association 1994). Within its major symptoms are anhedonia; episodes of irritability, anxiety, and fear; and changes in appetite, body weight, sleep, and psychomotor efficiency. All of these may be accompanied by decreased self-esteem and increased feelings of guilt and ideas of death which risk suicide (American Psychiatric Association 1994). Depression occurs with high frequency in the general population. It is estimated that 15–20 % of the world population has suffered some form of depression at some stage of their lives. It occurs most often in women (10–25 %) compared to men (5–12 %), in a ratio of 2 to 1 (Breslau et al. 1995). Multiple studies in humans and in experimental animals have tried to establish the neurobiological causes of depression and the effect of antidepressant therapies (Cryan et al. 2002).

## 4.2 Study of Depression in Rodents

Animal models offer the best approach to understanding the biopsychosocial processes of stress in organisms. However, no animal model can reflect the wide range of symptoms of depressive disorders, a limitation that is important to remember in order not to confuse their reach (Abelaira et al. 2013). Nevertheless, it can be said that animal models are currently the best alternative to experimentally approach these disorders. It is difficult to develop an animal model reproducing the symptoms of depression in patients. Animals have no consciousness of themselves; self-reflection and consideration are characteristics of depression, low self-esteem, or suicidality, but hardly accessible in animals. However, depression and other mental disorders consist of endophenotypes that can be played independently and assessed in animals (Touma et al. 2009). An ideal animal model provides an opportunity to understand the molecular, genetic, and epigenetic factors that can lead to depression. Using animal models, the underlying molecular alterations and the causal relationship between genetic or environmental disturbances and depression can be examined, which allows a better understanding of the pathology of depression. In addition, animal models are essential to identify new therapies for depression (Sloman 2008).

## 4.3 Endophenotypes in Animal Models of Depression

The following endophenotypes have been described in animal models.

### *4.3.1 Anhedonia*

Loss of interest is a core symptom of depression. Anhedonia can be evaluated in rodents by the sucrose preference test or intracranial self-stimulation (Knapman et al. 2010a).

### *4.3.2 Behavioral Distress*

Behavioral despair can be evaluated with tests such as forced swimming test (FST) or tail suspension test (TST) (Knapman et al 2010b).

### *4.3.3 Changes in Appetite or Weight Gain*

Depression is often associated with changes in appetite or weight gain, which are easily measured in rodents (Knapman et al. 2010a).

### *4.3.4 Neuroendocrine Disorders*

Disorders of the hypothalamus-pituitary-adrenal (HPA) axis are one of the most consistent symptoms of major depression. The functionality of the HPA axis can be evaluated by the dexamethasone suppression test (Samsom and Wong 2015).

### *4.3.5 Alterations in Sleep Architecture*

Alterations in circadian rhythm and especially in sleep architecture are often observed in depression. In rodents, it is accessible through electroencephalography (Samsom and Wong 2015).

### *4.3.6 Behavior Related to Anxiety*

Anxiety is a symptom of high prevalence in depression. Therefore, animal models of depression are often associated with anxiety (Samsom and Wong 2015).

## 4.4 Experimental Models of Depression

Animal models play a central role in all areas of biomedical research (van der Staay et al. 2009). Biomedical research is based on at least three sources of knowledge: first, the sick man who is the natural source of knowledge in the clinic; second, the body is the source of knowledge of pathology; and finally, experimental animals are the source of knowledge of pathophysiology. An experimental model is defined as any system, logical, physical, or biological, able to simulate the whole or a part of the process we intend to study. The purposes of using animal models for the study of neurobehavioral disorders are usually:

(a) To translate the insights from the preclinical animal study to the clinical (and vice versa), through identifying new targets, pathways, and mechanisms of drug action (Gould and Gottesman 2006).
(b) To improve our understanding of the substrates and the mechanisms that control normal behavior and abnormal underlying, i.e., the brain-behavior relation is made experimentally by induction of dissociations between processes, subprocesses, and influences of modulation, either pharmacologically, or through damage to neural tissue, or by using animals with genetic deficiency. Researching the natural origin and consequences should help to meet the pathophysiological mechanisms of disease.
(c) To assess the effects of putative neuroprotective, anti-degenerative, revalidation-supporting, mental health promoting, and/or cognition-enhancing compounds or treatments, and the risks (safety, teratology, toxicology) associated with these treatments (Matthews et al. 2005).

Animals, such as mice, rats, rabbits, dogs, and monkeys, are widely used as surrogates for humans in fundamental medical research (Langley 2009). In fact, many animal models have proven useful in elucidating the basic mechanisms of human disease processes; others have assisted in the development of novel and more effective treatments (Matthews et al. 2005). However, model development requires a multidisciplinary approach, where preclinical and clinical experts should establish a set of scientific criteria. McKinney and Bunney (1969) have described that an experimental model should possess at least the following minimal requirements:

(a) To be reasonably analogous to the human disorder in its symptomatology (face validity)
(b) To cause behavioral changes that can be monitored objectively

(c) To produce behavioral changes that are reversed by the same treatment modalities that are effective in humans (predictive validity)
(d) Should be reproducible between investigators

The DSM-IV (American Psychiatric Association 1994) defines a depressive episode patient through the presence of at least one of the two main symptoms of psychiatric pathology: anhedonia, a depressive behavior characterized by a significant decrease in the ability to experience pleasure, and depressed mood for at least 2 weeks (Mesulam 2000). In this context, there are no animal models of depression because this disease is defined to humans. However, in experimental animals depressive behavior modeling in various paradigms of chronic stress can develop preclinical studies for antidepressant drugs.

The aim of models for chronic stress is to generate anhedonia and depressive behaviors for a period of time, as in the patient who is experiencing a depressive episode. In parallel, in these models of depressive behaviors, other disturbances in behavior are produced, such as decreased motivation, changes in the and sleep disturbances, increased aggression and anxiety, reduced sexual behavior, and increased submission and decreased social interaction (Cryan and Holmes 2005; Insel 2007; Krishnan and Nestler 2011). Based on that, many variants in chronic stress models were developed to produce anhedonia and then describe such depressive behaviors (Duman 2010) (Table 4.1).

### *4.4.1 Learned Helplessness (LH)*

Essential features of major depressive episodes are either depressed mood or a marked loss of interest and pleasure in nearly all activities. Communication of changes in mood and thoughts during depressive episodes relies on language and is not testable in animal models. Nevertheless, LH is a well-established stress model of major depression with good construct and face validity (Willner 1997; Vollmayr and Henn 2001), where uncontrollable and unpredictable stress induces LH in rats (Seligman and Beagley 1975; Vollmayr and Henn 2001). Its duration for more than 1 week is comparable to the duration of human depressive episodes if the rats' much shorter life span is considered (Henn et al. 2002). In a paradigm of moderate stress, it has been reported that LH is observed only in a portion of susceptible animals, thus modeling the human susceptibility to depression (Vollmayr and Henn 2001). Helplessness is a failure to avoid punishment or obtain rewards even though they are under the agent's control. The etiology and consequences of helplessness have been studied extensively in the animal learning literature using the learned helplessness paradigm. The phenomenon was first demonstrated in dogs in the context of testing the two-factor learning theory using the shuttle-box escape task (Lieder et al. 2013). In the current classical version of the task, three animals are compared. A master rat experiences electrical shocks. These come on unpredictably, but can be terminated by some action, for instance, turning a wheel. A yoked rat is exposed to the exact

**Table 4.1** Summary of animal models used for the study of the main types of stress and depression

| Mouse behavioral model | Ethological correlate (Morris 1984; Wall and Messier 2000; Prut and Belzung 2003; Ruiz et al. 2007; Samsom and Wong 2015) |
|---|---|
| *Distress based* | *In unconditioned response (motor and/or exploratory behavior in unfamiliar situations)* |
| Forced swimming test (FST) | FST is based on the observation that animals develop an immobile posture in an inescapable water-filled cylinder<br>In this test, the immobility is interpreted as a passive coping strategy stress or depressive-like behavior |
| Tail suspension test (TST) | Increased immobility possibly related to behavioral despair or coping with stress |
| *Reward based* | *Can measure the emotional and state motivation of rodents* |
| Sucrose preference test | Mouse preference for sugar vs. normal water. Reduced preference for sweet solution represents anhedonia |
| Morris water maze (MWM) | Latency to find a hidden platform in a pool of murky water and memory for platform position<br>It tests spatial learning and memory |
| Intracranial self-stimulation | Intracranial self-stimulation can be used in rodents to understand how drugs affect the function of the brain's reward system<br>It can measure the affective state and motivation of rodent subjects |
| *Anxiety based* | *The test reflects the anxiolytic effects of antidepressants* |
| Three-chamber social interaction test | Time spent with mouse vs. object tests social motivation; time spent with familiar mouse vs. new mouse tests social novelty<br>It tests social memory |
| Open-field test (OFT) | OFT tests exploration and motor activity, willingness to enter the center of the field<br>It measures anxiety |
| Dark box | It is based on the innate aversion of rodents to the light and brightness and on their spontaneous exploratory behavior |
| Elevated plus maze (EPM) | In EPM model, a decreased time in open arms of a maze with open and enclosed arms states anxiety |
| *Other models* | |
| Olfactory bulbectomy | Chronic stress caused by chronic sensory deprivation |
| Psychostimulant withdrawal | Following the withdrawal of drugs such as amphetamine or cocaine, rodents show behavioral changes that are very similar to some aspects of depression in humans, such as anhedonia |
| Genetic-modified mice | Knockout mice are modified in some target molecules such as A2a adrenergic receptor, glucocorticoid receptor, cAMP response element binding to express some behaviors associated with depression disorder |

same sequence of shocks that are delivered to the master rat, but has no action available to terminate the shocks. A third rat, the control, is not exposed to shocks. Compared to the controls, the yoked rats are impaired at acquiring new instrumental responses, but the master rats are either unaffected or may even show an improvement (Maier and Watkins 2005). This model can be used to measure the overall performance of exhaust mice with different mutations in target genes that depression can affect and

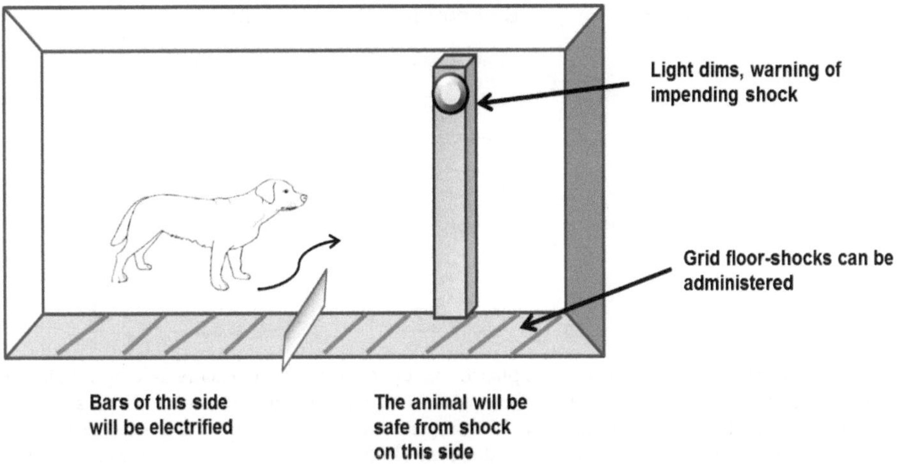


**Fig. 4.1** Learned helplessness (LH). In this paradigm, helplessness is induced in healthy animals by exposure to inescapable electric shocks. Helplessness is then measured by the subsequent failure to escape escapable shocks in a novel environment

develop vulnerability to a depressed type. For this excellent face and predictive validation, LH makes an interesting model to study the pathophysiology of depression. The disadvantage is that it requires LH strong factors to induce behavioral phenotypes, which pose no ethical stress problems. In addition, most of the symptoms do not persist long enough after the cessation of uncontrollable disturbance (Fig. 4.1) (Vollmayr et al. 2004).

### *4.4.2 Forced Swim Test (FST)*

FST is one of the most commonly applied tests for assessing antidepressant-like behavior, as well as the most widely used paradigm to assess "depression" and antidepressant-related phenotypes in genetically altered mice. Drug sensitivity depends on the strain and test used (Cryan et al. 2005a). This protocol takes place in over 48 h and is followed by the video analysis of the behavior. The swim test involves the scoring of active (swimming and climbing) or passive (immobility) behavior when rodents are forced to swim in a cylinder from which there is no escape (Lucki 1997). There are two versions that are used, namely, the traditional and modified FSTs, which differ in their experimental setup. The traditional FST measures swimming time and immobile time through the subtraction of one video frame from its neighboring frame to find the differences between these two frames. The modified FST measures swimming time, climbing time, escaping, immobile, and diving (divided into passive and active diving) by automatically identifying animal's body parts and detecting the movement of those body parts to determine animal's behaviors (Armario et al. 1988). For both versions, a pretest of 15 min

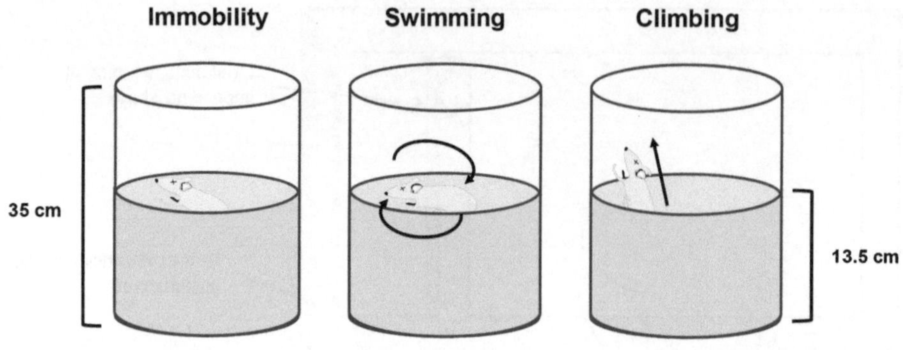


**Fig. 4.2** Forced swimming test. A rat is placed in a cylinder of 35 cm usually 24 cm in diameter with approximately 13.5 cm of water; the process is usually divided into two sessions: before exposure test (pretest) with a duration of 15 min, and, after 24 h, the direct exposure or (test) is performed with a duration of 5 min. During this time allotted for the behavioral observation of rat, latency time which corresponds to the swimming time of the animal from the time it is placed on the water surface until it remains immobile is recorded, as well as the immobility time. Immobility is considered when the animal performs only the minimum movements necessary to keep its head above water. Pretest session in rats induces prolonged float during the test session time; this prolonged float time has been interpreted as a state of hopelessness reflected by the animal

(although a number of laboratories have used a 10-min pretest with success) is included, as this accentuates the different behaviors in the 5-min swim test following drug treatment. Reduction in passive behavior is interpreted as an antidepressant-like effect of the manipulation, provided that it does not increase general locomotor activity, which could be responsible for a false positive result in the FST (Fig. 4.2).

### 4.4.3 *Tail Suspension Test (TST)*

TST is a derivative of the FST and is based on the fact that a mouse suspended by the tail alternates periods of agitation and immobility similar (but not identical) to that observed in the FST (Cryan et al. 2005b). Mice are suspended by the tail, using an adhesive scotch tape, to a hook connected to a strain gauge that picks up all movements of the mouse and transmits them to a central unit which calculates the total duration of immobility during a 6-min test (Steru et al. 1985). In both tests, TST and FST, animals are placed in an inescapable situation and the antidepressant-like activity is expressed by the decrease of immobility duration. Both test procedures are equi-sensitive to antidepressant drugs of different classes in the strain of animals used (Fig. 4.3) (Ripoll et al. 2003; Kulkarni and Dhir 2007).

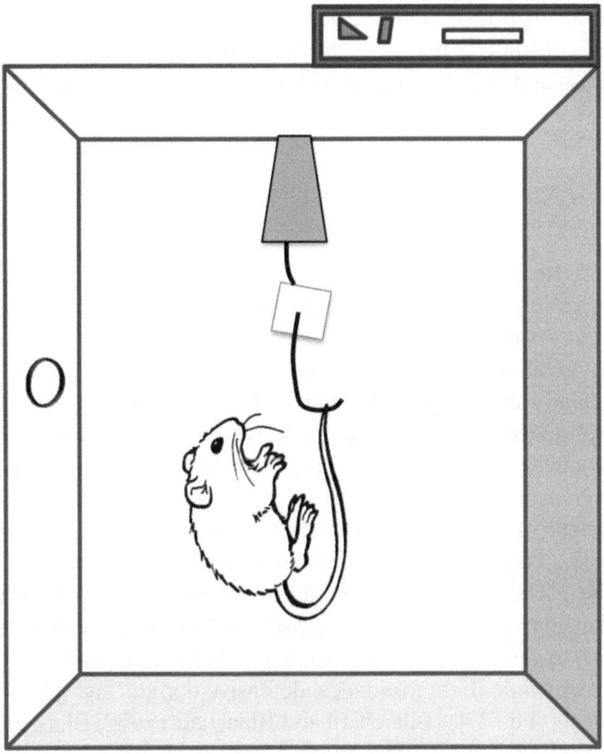

**Fig. 4.3** Tail suspension test. Rodents, suspended by their tail, using an adhesive cap to a horizontal bar, innately attempt to escape from this aversive situation. The test (usually 5–6 min) the struggling movements are recorded. However, following failed attempts to escape, they experience despair and become immobile. The magnitude of immobility is considered to be correlated with the depressive-like state of the subjects

### *4.4.4 The Chronic Mild Stress*

CMS paradigm has been established as a model that is sensitive to chronic antidepressant treatment and emphasizes the predominant role of stress in the etiological cause of depression (Willner 2005). The CMS paradigm involves the exposure of animals to a series of mild and unpredictable stressors (isolation or crowded housing, food or water deprivation, disruption of dark-light cycle, tiling of home cages, dampened bedding, etc.), during at least 2 weeks (Figs. 4.2 and 4.3). CMS has been reported to result in long-lasting changes in behavioral, neurochemical, neuroimmune, and neuroendocrinological parameters resembling dysfunctions observed in depressed patients. The CMS paradigm interferes with brain reward functions, including decreased sucrose preference and intracranial self-stimulation, reflecting anhedonia that is reversed by chronic but not acute antidepressant treatment (Monleon et al. 1995). CMS is a paradigm developed to study neural changes that result from stress of a more chronic nature (Willner 1997; 2005). This paradigm aims to model a chronic depressive-like state that develops gradually over time in response to stress and is thus considered more naturalistic in the induction.

### 4.4.5 Social Stress

Social stress models are suggested to represent an induction of a depressive-like state that may be more relevant to human depression compared with models that employ acute or severe stressors. In these models, social conflict is created between male animals (Martinez et al. 1998). In animal models, this leads to a variety of behavior changes, particularly involving emotionality-linked behaviors such as anxiety, defensiveness, and substance self-administration, as well as social and sexual behaviors producing many changes in the brain, affecting neuronal structure and survival as well as chemical neurotransmission, as happens in human depression (Blanchard et al. 2001). The most commonly used subjects of social stress laboratory research are mice and rats, in which agonistic behavior is a very obvious and salient component of most social grouping studies. The behaviors seen during agonistic interactions comprise two groups: one that may be used to infer victory and the other, defeat (Blanchard and Blanchard 1990). Aggressive or offensive components include lateral attack, chase, and standing on top of, while flight/avoidance, defensive upright, and lying on the back are defensive elements (Fig. 4.4) (Blanchard and Blanchard; 1977; Blanchard et al. 2001).

### 4.4.6 Olfactory Bulbectomy

The bilateral removal of the olfactory bulbs rats, mice, or even hamsters has demonstrated behavioral, neurochemical, neuroendocrine, and neuroimmune alterations, many of which are correlated with changes observed in major depression (Del Cañizo et al. 2008; Kelly et al. 1997). The most consistent behavioral change caused by this model is a hyperactive response related to an increase in defensive behaviors (Stock et al. 2001) or alterations in aversively motivated behavior (Primeaux and Holmes 1999), reversed almost exclusively by chronic, but not acute, antidepressant treatment. This model possesses the quality of prediction of known antidepressant compounds following repeated administration and is reliable between laboratories (Kelly et al. 1997).

## 4.5 Advantages of Experimental Models of Depression

Modeling mood disorders as depression in animals a major challenge considering the complex nature of the disease. There are models and paradigms that have proven extremely useful, not only with respect to the identification and improvement of

**Fig. 4.4** Social stress. In rodents, chronic social stress paradigm used to reproduce more hippocampal atrophy and memory impairment is the movement restriction. The animal is introduced into well-ventilated acrylic or mesh restrictor daily for 21 days, and this model of chronic stress produces, besides hippocampal atrophy and memory impairment, decreased weight gain does not affect locomotor activity of animals

therapeutic substances but also regarding the validation of neurobiological underpinnings (Dedic et al. 2011). In this disorder, as well as other mood disorders, certain endophenotypes can be reproduced independently and evaluated in animals; these include physiological, endocrinological, neuroanatomical, and behavioral alterations, many of which can be measured in mice (Deussing 2006). These paradigms are capable of consistently generating behavioral changes reminiscent of symptoms of depression, which can be reversed by antidepressant treatment (Deussing 2006). An animal model of depression which more closely incorporates disease etiology and predisposition is the learned helplessness paradigm (Seligman and Beagley 1975; Vollmayr and Henn 2001). Humans experiencing social defeat show increased symptoms of depression, loneliness, anxiety, social withdrawal, and a loss of self-esteem (Björkqvist 2001). Therefore, social conflict models in animals have gained increasing attention, as they might prove useful to study certain endophenotypes of depression (Blanchard et al. 2001; Deussing 2006). A major advantage of the most commonly used experimental models of depression, such as FST and TST, is that they are simple, are inexpensive, and may allow for automation (Deussing 2006).

## 4.6 Limitations of Animal Models

According to Langley (2009), there are two major uncertainties when animal models are used to understand human diseases. Firstly, there are significant species differences in anatomy, metabolism, physiology, or pharmacology caused by underlying genetic variations, including regulatory genes. This means that even minor molecular differences may be amplified when extrapolated to the cell, organ, or species levels. These variations between species can, and do, regularly confound the translation of laboratory animal results to humans. A second major cause of uncertainty with animal models is that human illnesses are researched in animals precisely because there is a lack of knowledge about them. An animal model is usually developed on the basis of a narrow range of human symptoms, selected at a time when researchers often do not know which disease characteristics are the most important or even which are the causes rather than the outcomes of illness.

It is exceedingly difficult to envision an animal model that perfectly recapitulates the symptoms of depression in human patients. Animals not only lack consciousness of self, self-reflection, and consideration of others, but also hallmarks of the disorder such as depressed mood, low self-esteem, or suicidality are hardly accessible in nonhumans. Likewise, antidepressant screening paradigms, such as the FST, also do not represent a model of depression but should rather be regarded as drug-screening assays. It is a fact that a major drawback of the TST is that its application is restricted to mice and limited to strains that do not tend to climb their tail what otherwise confuses interpretation of behavioral measures (Mayorga and Lucki 2001). Moreover, neither the TST nor the FST the slow onset of antidepressant action as it is observed in depressed patients.

## 4.7 Conclusion

A major problem in the establishment of a suitable animal model of depression is that the development of such a model requires a better understanding of the etiology of the disorders (Blanchard et al. 2001). It is important to point out the importance of discriminating between an animal model, a behavioral test, and an antidepressant screening paradigm. Animal models of human disease have proven of considerable value in elucidating basic pathophysiological mechanisms and in developing novel treatments, but they are expected to show sufficient construct, face, and predictive validity (Garner 2005). Although progress has been made to identify the underlying molecular mechanism of depression, the lack of specific markers and biomarkers still represents a major drawback. In contrast to many other illnesses such as cancer, infectious diseases, cardiovascular arrest, diabetes, stroke, etc., depression cannot simply be diagnosed by making use of the usual methods, such as assessing blood pressure, glucose levels, inflammatory agents, heart rate, and others. Clinicians still fully rely on observations and verbal communication. Therefore, behavioral tests that are used to assess phenotypic alterations relevant to depression in animals are of extreme importance (Dedic et al. 2011).

**Acknowledgments** We would like to thank the CYTED program (Action 112RT0460. "CORNUCOPIA" thematic network) for the support to the accomplishment of this work.

## References

Abelaira HM, Réus GZ, Quevedo J. Animal models as tools to study the patho-physiology of depression. Rev Bras Psiquiatr. 2013;35(Suppl2):S112–20.

American Psychiatric Association. DSM-IV, diagnostic and statistical manual of mental disorders. 4th ed. Washington, DC: APA; 1994. p. 345–59.

Armario A, Gavaldà A, Martí O. Forced swimming test in rats: effect of desipramine administration and the period of exposure to the test on struggling behavior, swimming, immobility, and defecation rate. Eur J Pharmacol. 1988;158:207–12.

Björkqvist K. Social defeat as a stressor in humans. Physiol Behav. 2001;73:435–42.

Blanchard DC, Blanchard RJ. Aggressive behavior in the rat. Behav Biol. 1977;21:197–224.

Blanchard DC, Blanchard RJ. Behavioral correlates of chronic dominance-subordination relationships of male rats in a seminatural situation. Neurosci Biobehav Rev. 1990;14:455–62.

Blanchard RJ, McKittrick CR, Blanchard DC. Animal models of social stress: effects on behavior and brain neurochemical systems. Physiol Behav. 2001;73:261–71.

Breslau N, Schultz L, Peterson E. Sex differences in depression: a role for preexisting anxiety. Psychiatr Res. 1995;58:1–12.

Cryan J, Holmes A. The ascent of mouse: advances in modeling human depression and anxiety. Nat Rev Drug Discov. 2005;4:775–90.

Cryan JF, Markou A, Lucki I. Assessing antidepressant activity in rodents: recent developments and future needs. Trends Pharmacol Sci. 2002;23:238–45.

Cryan JF, Valentino RJ, Lucki I. Assessing substrates underlying the behavioral effects of antidepressants using the modified rat forced swimming test. Neurosci Biobehav Rev. 2005a; 29:547–69.

Cryan JF, Mombereau C, Vassout A. The tail suspension test as a model for assessing antidepressant activity: review of pharmacological and genetic studies in mice. Neurosci Biobehav Rev. 2005b;29:571–625.

Dedic N, Walser SM, Deussing JM. Mouse models of depression. In: Uehara T, editor. Psychiatric disorders—trends and developments. Rijeka: InTech; 2011. p. 185–222.

Del Cañizo LJF, López MD, Lledó GE, et al. Design of experimental models in surgical investigation. Actas Urol Esp. 2008;32:27–40.

Deussing JM. Animal models of depression. Drug Discov Today Dis Models. 2006;3:375–83.

Duman CH. Models of depression. Vitam Horm. 2010;82:1–22. doi:10.1016/S0083-6729(10)82001-1.

Garner J. Stereotypies and other abnormal repetitive behaviors: potential impact on validity, reliability, and replicability of scientific outcomes. ILAR. 2005;46:106–17.

Gould T, Gottesman I. Psychiatric endophenotypes and the development of valid animal models. Genes Brain Behav. 2006;5:113–9.

Guimarães F, Joca S, Padovan C, Molina V. Mood disorders. In: Brandão M, Graeff F, editors. Neurobiology of mental disorders. New York, NY: Nova Science; 2006. p. 95–124.
Henn FA, Edwards E, Anderson D, Vollmayr B. Psychotherapy and antidepressant treatment of depression: evidence for similar neurobiological mechanisms. World Psychiatry. 2002;1: 115–7.
Insel TR. From animal models to model animals. Biol Psychiatry. 2007;62:1337–9.
Kelly JP, Wrynn AC, Leonard BE. The olfactory bulbectomized rat as a model of depression: an update. Pharmacol Ther. 1997;74:299–316.
Knapman A, Heinzmann JM, Holsboer F, Landgraf R, Touma C. Modeling psychotic and cognitive symptoms of affective disorders: disrupted latent inhibition and reversal learning deficits in highly stress reactive mice. Neurobiol Learn Mem. 2010a;94:145–52.
Knapman A, Heinzmann JM, Hellweg R, Holsboer F, Landgraf R, Touma C. Increased stress reactivity is associated with cognitive deficits and decreased hippocampal brain-derived neurotrophic factor in a mouse model of affective disorders. J Psychiatr Res. 2010b;44:566–75.
Krishnan V, Nestler EJ. Animal models of depression: molecular perspectives. Curr Top Behav Neurosci. 2011;7:121–47.
Kulkarni SK, Dhir A. Effect of various classes of antidepressants in behavioral paradigms of despair. Prog Neuropsychopharmacol Biol Psychiatry. 2007;31:1248–54.
Langley G. The validity of animal experiments in medical research. Rev Semestr Droit Anim. 2009;1:161–8.
Lieder F, Goodman ND, Huys QJM. Learned helplessness and generalization. Berlin: Paper presented at the cognitive science conference; 2013.
Lucki I. The forced swimming test as a model for core and component behavioral effects of antidepressant drugs. Behav Pharmacol. 1997;8:523–32.
Maier SF, Watkins LR. Stressor controllability and learned helplessness: the roles of the dorsal raphe nucleus, serotonin, and corticotropin-releasing factor. Neurosci Biobehav Rev. 2005;29:829–41.
Martinez MM, Calvo-Torrent MA, Pico-Alfonso MA. Social defeat and subordination as models of social stress in laboratory rodents: a review. Aggress Behav. 1998;24:241–56.
Matthews K, Christmas D, Swan J, Sorrell E. Animal models of depression: navigating through the clinical fog. Neurosci Biobehav Rev. 2005;29:503–13.
Mayorga AJ, Lucki I. Limitations on the use of the C57BL/6 mouse in the tail suspension test. Psychopharmacology (Berl). 2001;155:110–2.
McKinney Jr WT, Bunney Jr WE. Animal model of depression. I. Review of evidence: implications for research. Arch Gen Psychiatry. 1969;21:240–8.
Mesulam MM. Principles of behavioral and cognitive neurology. 2nd ed. New York, NY: Oxford University Press; 2000.
Monleon S, D'Aquila P, Parra A, Simon VM, Brain PF, Willner P. Attenuation of sucrose consumption in mice by chronic mild stress and its restoration by imipramine. Psychopharmacology (Berl). 1995;117:453–7.
Morris R. Developments of a water-maze procedure for studying spatial learning in the rat. J Neurosci Methods. 1984;11:47–60.
Primeaux SD, Holmes PV. Role of aversively motivated behavior in the olfactory bulbectomy syndrome. Physiol Behav. 1999;67:41–7.
Prut L, Belzung C. The open field as a paradigm to measure the effects of drugs on anxiety-like behaviors: a review. Eur J Pharmacol. 2003;463:3–33.
Ripoll N, Paul David DJ, Dailly E, Hascoët M, Bourin M. Antidepressant-like effects in various mice strains in the tail suspension test. Behav Brain Res. 2003;143:193–200.
Ruiz JC, Soler MJ, Fuentes I, Tomás P. Intellectual functioning and memory deficits in schizophrenia. Compr Psychiatry. 2007;48:276–82.
Samsom JN, Wong AH. Schizophrenia and depression co-morbidity: what we have learned from animal models. Front Psychiatry. 2015;6:13.
Seligman ME, Beagley G. Learned helplessness in the rat. J Comp Physiol Psychol. 1975; 88:534–41.

Sloman L. A new comprehensive evolutionary model of depression and anxiety. J Affect Disord. 2008;106:219–28.

Steru L, Chermat R, Thierry B, Simon P. The tail suspension test: a new method for screening antidepressants in mice. Psychopharmacology (Berl). 1985;85:367–70.

Stock HS, Hand GA, Ford K, Wilson MA. Changes in defensive behaviors following olfactory bulbectomy in male and female rats. Brain Res. 2001;903:242–6.

Touma C, Fenzl T, Ruschel J, Palme R, Holsboer F, Kimura M, et al. Rhythmicity in mice selected for extremes in stress reactivity: behavioural, endocrine and sleep changes resembling endophenotypes of major depression. PLoS One. 2009;4, e4325.

van der Staay FJ, Arndt SS, Nordquist RE. Evaluation of animal models of neurobehavioral disorders. Behav Brain Funct. 2009;5:11.

Vollmayr B, Henn FA. Learned helplessness in the rat: improvements in validity and reliability. Brain Res Protoc. 2001;8:1–7.

Vollmayr B, Bachteler D, Vengeliene V, Gass P, Spanagel R, Henn F. Rats with congenital learned helplessness respond less to sucrose but show no deficits in activity or learning. Behav Brain Res. 2004;150:217–21.

Wall PM, Messier C. Ethological confirmatory factor analysis of anxiety-like behaviour in the murine elevated plus-maze. Behav Brain Res. 2000;114:199–212.

Willner P. Validity, reliability and utility of the chronic mild stress model of depression: a 10-year review and evaluation. Psychopharmacology (Berl). 1997;134:319–29.

Willner P. Chronic mild stress (CMS) revisited: consistency and behavioral-neurobiological concordance in the effects of CMS. Neuropsychobiology. 2005;52:90–110.

# Chapter 5
# Prevalence of Use of Herbal Medicines and Complementary and Alternative Medicine in Europe

**Lucian Hritcu and Oana Cioanca**

**Abstract** In recent years, anxiety and depression became two major health problems for younger active population (25–45 years old) in Europe. Consecutively, the demand for herbal drugs has increased considerably as a safer alternative or complementary therapy to conventional medicine. After carefully examining the available literature with particular attention to herbal medicines used in Europe for anxiety and depression, we tried to summarize the bioactive compounds, in vitro and in vivo proven pharmacological activities, indications, and side effects of *Hypericum perforatum*, *Verbena officinalis*, *Tilia* sp., *Ocimum basilicum*, *Melissa officinalis*, and *Valeriana officinalis*. This selection comprises most common used plants both in modern and folk medicines by Europeans, to treat nervousness, disturbance of sleep, and similar psychological problems. Therefore, this chapter provides sufficient baseline information about some of the common herbal medicines used in European traditional and modern therapy for depression and anxiety that could be used by specialists in the medical and pharmaceutical field.

**Keywords** Herbal medicines • Pharmacology • Anxiety • Depression • Therapy • Europe

## 5.1 Introduction

Depression as well as anxiety has many causes and variable symptoms (Barlow and Campbell 2000, Kessler and Wittchen 2002). Most of the patients suffering from minor to mild problems tend to use complementary and alternative medicine (CAM)

L. Hritcu (✉)
Department of Biology, Alexandru Ioan Cuza University,
Bd. Carol I, No. 11, Iasi 700506, Romania
e-mail: hritcu@uaic.ro

O. Cioanca (✉)
Department of Pharmacognosy-Phytotherapy, Faculty of Pharmacy, University of Medicine and Pharmacy "Gr. T. Popa", 16 University Str., Iasi 700117, Romania
e-mail: oana.cioanca@gmail.com



C. Grosso (ed.), *Herbal Medicine in Depression*,
DOI 10.1007/978-3-319-14021-6_5

with or without a recommendation from a specialist. In emerging economies and developing countries from East and Southeast Europe, people depend a lot on relatives and media as resources of knowledge (Ballenger et al. 2001). For example, in Romania many patients read newspapers and magazines and look in the Internet to find a proper "natural treatment." At the same time, immigrants from India and African countries in the European Union (EU) use the same traditional remedies as those in their country of origin. Nevertheless, medicinal plants are not totally safe, especially when used for treatment of serious conditions on a long term without proper advice.

Although there are many reviews published in the last decade concerning the use of herbal remedies in Europe, there are still many flaws and problems regarding the methodology. There is a huge variability of data that provides a range of prevalence rates depending mostly on assumptions (Menniti-Ippolito et al. 2002, Nissen et al. 2012). Moreover, it is reasonable to assume that low response rates included in the surveys might generate a falsely high value rate (Fischer et al. 2012, 2014). Furthermore, there is heterogeneity in regard to the definition, providers, and legislation of CAM and its use across the whole EU (Fischer et al. 2014). Most of the data are mainly from EU members. Also European research in this field is limited and the funding is scarce; for instance, in the UK, less than 0.009 % of national medical research funds are spent on CAM research, while over 10 % of the UK population uses CAM (Bodeker et al. 2005, Posadzki et al. 2012, 2013). Still, all available data show that herbal medicines are the most popular among the CAM, followed by homeopathy and aromatherapy (Buono et al. 2001, Eardley et al. 2012). Homeopathy is particularly popular in France, where over 70 % of the specialists (physicians and pharmacists) prescribe it to their patients (Damase-Michel et al. 2004).

Therefore, it is of high importance to provide scientifically proven data not only to those who can afford it but to whoever needs it, specialist or patient.

Countries like Germany and Austria have a long history of herbal medicine use and production (Bücker et al. 2008, Härtel and Volger 2004); nevertheless, there are developed countries such as the UK where expenditure on CAM rises up to £1.6 billion/year (Lewith 2007), even though researchers from Complementary Medicine, Peninsula Medical School, Exeter, Devon, consider that there is little solid evidence that the use of CAM is high in the UK (Thomas and Coleman 2004).

Of all herbal remedies indicated for depression, *Hypericum perforatum* L. is the one that has been researched more extensively. This is why the prescriptions for such preparations tripled between 1993 and 2003 in Germany, where the first-line treatment for depression was and still is St. John's wort (Linde et al. 2008). Lately, the same plant is used in observational studies on children and adolescences. Doses of 300–900 mg/day were effective and well tolerated in the 2-week treatments. Additionally, bitter herbs were prescribed as a tonic; a few drops of tinctures are thought to stimulate especially the liver and stomach functions, thus helping the body to cope better in a stressful environment.

For combined cases of depression with anxiety, most of the xenobiotics need special attention because some prescribed drugs in anxiety may aggravate depressive behavior and the other way around. The European traditional medicine indicates the association of general herbal treatment with soothing teas that promote sleep. Such calming and relaxing tea formulas comprise borage leaves (*Borago officinalis* L.), oatstraw (*Avena sativa* L.), vervain (*Verbena officinalis* L.), lime blossom (*Tilia* sp.), lemon balm (*Melissa officinalis* L.), skullcap (*Scutellaria lateriflora* L.), chamomile (*Chamomilla recutita* L.), and/or hop cones (*Humulus lupulus* L.). Most of these medicinal and aromatic plants are stress and anxiety relievers and are administered morning and night infused in boiling water for 3–6 months. For shorter periods (a week), these extracts may be given up to three times per day.

However, for some of the mentioned vegetal products, there are insufficient data to support their use in therapy for certain indications. Others, like the ethanol extract of *Borago officinalis* L. aerial parts, were recently identified as specific serotonin transporter (SERT) binding agent, whereas *Trigonella foenum-graecum* L. seeds ($IC_{50}$=4 μg/ml), leaf ethanolic extract of *Apium graveolens* L. ($IC_{50}$=5 μg/ml), and the water extract of aerial parts of *Calluna vulgaris* (L.) Hull ($IC_{50}$=8 μg/ml) showed a concentration-dependent activity in monoamine oxidase-A (MAO-A) assay (Jäger et al. 2013). Also, due to its high hepatotoxic profile, borage might be excluded from further investigations, unless a specific extraction method is established to isolate the pyrrolizidine alkaloids from the extracts/preparations used in testing and therapy.

Thus, our study comes to reinforce the recommendations of some of the most used medicinal plants in European traditional and modern therapy for depression and anxiety.

## 5.2 Search Methods

Europe PubMed Central, PubMed, Science Direct, Wiley, SpringerLink, Google Scholar.

## 5.3 References


- Barlow D, Campbell L. Mixed anxiety-depression and its implications for models of mood and anxiety disorders. Comprehens Psychiatry 2000;41(2 Suppl 1): 55–60.
- Kessler R, Wittchen H. Patterns and correlates of generalized anxiety disorder in community samples. J Clin Psychiatry. 2002,63(Suppl 8):4–10.
- Ballenger J, Davidson J, Lecrubier Y, Nutt D, Borkovec T, Rickels K, et al. Consensus statement on generalized anxiety disorder from the International Consensus Group on Depression and Anxiety. J Clin Psychiatry. 2001;62 (Suppl 11):53–8.
- Menniti-Ippolito F, Gargiulo L, Bologna E, Forcella E, Raschetti R. Use of unconventional medicine in Italy: a nation-wide survey. Eur J Clin Pharmacol. 2002;58:61–4.

- Nissen N, Schunder-Tatzber S, Weidenhammer W, Johannessen H. What attitudes and needs do citizens in Europe have in relation to complementary and alternative medicine? Res Complement Med. 2012;19(Suppl 2):9–17.
- Fischer F, Lewith G, Witt C, Linde K, von Ammon K, Cardini F, et al. High prevalence but limited evidence in complementary and alternative medicine: guidelines for future research. BMC Complement Altern Med. 2014;14:46.
- Fischer HF, Junne F, Witt C, von Ammon K, Cardini F, Fønnebø V, et al. Key issues in clinical and epidemiological research in complementary and alternative medicine – a systematic literature review. Res in Complement Med. 2012;19(Suppl 2):51–60.
- Fischer F, Lewith G, Witt CM, Linde K, Ammon K, Cardini F, et al. A research roadmap for complementary and alternative medicine – what we need to know by 2020. Res Complement Med. 2014;21:e1–e16.
- Bodeker G, Ong CK, Grundy C, Burford G, Shein K. WHO Global Atlas of traditional, complementary and alternative medicine: WHO Kobe Centre; 2005.
- Posadzki P, Alotaibi A, Ernst E. Prevalence of use of complementary and alternative medicine (CAM) by physicians in the UK: a systematic review of surveys. Clin Med. 2012;12:505–12.
- Posadzki P, Watson L, Alotaibi A, Ernst E. Prevalence of complementary and alternative medicine (CAM)-use in UK paediatric patients: a systematic review of surveys. Complement Ther Med. 2013;21:224–31.
- Buono MD, Urciuoli O, Marietta P, Padoani W, De Leo D. Alternative medicine in a sample of 655 community-dwelling elderly. J Psychosom Res. 2001;50:147–54.
- Eardley S, Bishop FL, Prescott P, Cardini F, Brinkhaus B, Santos-Rey K, et al. A systematic literature review of complementary and alternative medicine prevalence in EU. Res Complement Med. 2012;19(Suppl 2):18–28.
- Damase-Michel C, Vié C, Lacroix I, Lapeyre-Mestre M, Montastruc JL. Drug counselling in pregnancy: an opinion survey of French community pharmacists. Pharmacoepidemiol Drug Safety. 2004;13:711–5.
- Bücker B, Groenewold M, Schoefer Y, Schäfer T. The use of complementary alternative medicine (CAM) in 1001 German adults: results of a population-based telephone survey. Gesundheitswesen. 2008;70:e29–e36.
- Härtel U, Volger E. Inanspruchnahme und Akzeptanz klassischer Naturheilverfahren und alternativer Heilmethoden in Deutschland – Ergebnisse einer repräsentativen Bevölkerungsstudie. Res Complement Med. 2004;11: 327–34.
- Lewith GT. Funding for CAM. Br Med J. 2007;335:951.
- Thomas K, Coleman P. Use of complementary or alternative medicine in a general population in Great Britain. Results from the National Omnibus Survey. J Public Health. 2004;26:152–7.
- Linde K, Berner M, Kriston L. St John's wort for major depression. Cochrane Database Syst Rev. 2008;8:CD000448.
- Jäger AK, Gauguin B, Andersen J, Adsersen A, Gudiksen L. Screening of plants used in Danish folk medicine to treat depression and anxiety for affinity to the serotonin transporter and inhibition of MAO-A. J Ethnopharmacol. 2013;145: 822–5.

## 5.4 *Hypericum perforatum* L.

### *5.4.1 (Hypericaceae)*

1) Common names:
   St. John's wort (UK), Johanniskraut (Germany), Le millepertuis (France), Sint-janskruid (Netherlands), Iperico (Italy), Sunătoarea (Romania).
2) Photograph of the species (Fig. 5.1).
3) Description of the species:

   A) Habitat and world distribution:
      Native to Europe and the eastern hemisphere, *Hypericum* was naturalized to North America. Many species of *Hypericum* (over 400) are known throughout the world, starting from Europe and continuing to North and South America, Australia, New Zealand, and Eastern Asia. It grows wild near forests and sunny areas, roadsides, and meadows on dry, calcareous, or siliceous soils. The flowering period starts at the end of June and finishes late July with the seed ripening.
   B) Morphology of the plant:
      The plant is perennial with a regular height of 20–100 cm, sometimes growing up to 1 m. The rhizomes are woody and with stolon, and the stems are erect, round, with two edges, hard, and woody in the lower part while spreading many branches in the upper part. The leaves are small, opposite, oval, narrow, and sessile with a deep green color; seen in the transparency, they give the impression of having many small holes due to transparent oil glands, hence the name *perforatum*. The star-shaped yellow flowers are arranged in corymb, pentamerous; petals have tiny black spots on the edges, and stamens are numerous, long, and prominent. The fruit is a capsule with three lodges containing brown-blackish, oval and cylindrical seeds.

**Fig. 5.1** Flowers of *Hypericum perforatum* L. Photograph taken in Presov, Slovakia, in 2014 by Oana Cioanca

C) Medicinal parts:

The drug, *Hyperici herba*, consists of dried upper part of the plant with flowers, avoiding as much as possible the fructifications and the lignified parts of the stalk. It has a faint, aromatic smell and a bitter taste.

D) Chemical constituents:

The plant has a complex chemical composition; the main compounds responsible for the biological effects are:

| Chemical class | Individual components | Plant part/ organ |
|---|---|---|
| Anthraquinone derivatives | 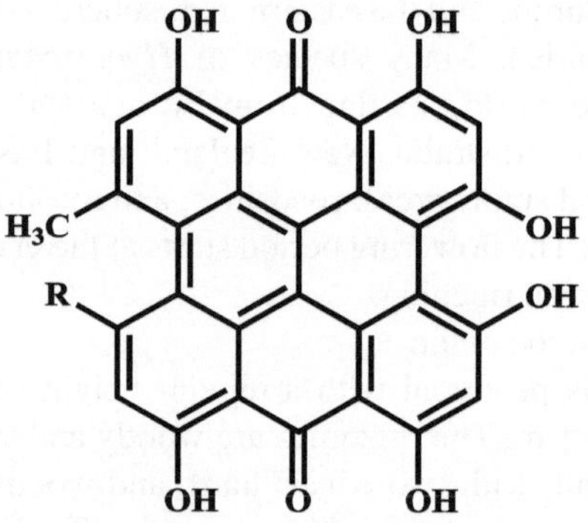 **Hypericin R = $CH_3$** **Pseudohypericin R = $CH_2OH$** | Flowers, upper part of the stem with leaves |
| Phloroglucinol derivatives | 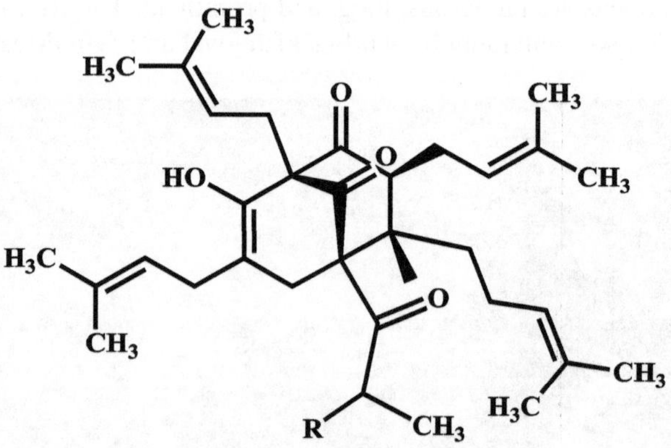 **Hyperforin R = $CH_3$** **Adhyperforin R = $CH_2CH_3$** | Flowers, upper part of the stem with leaves |

(continued)

| | | |
|---|---|---|
| Flavonoids | 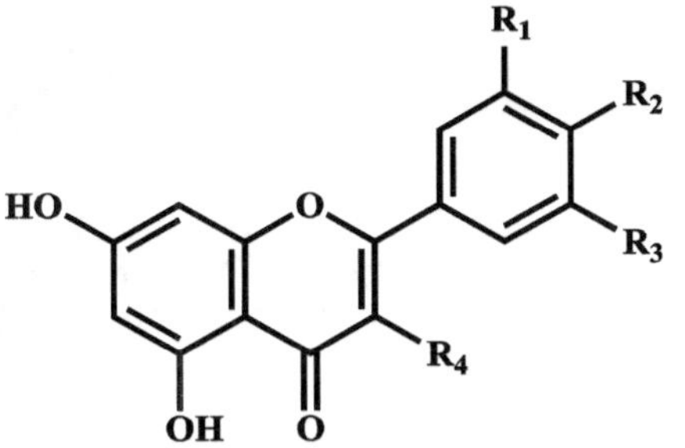<br>**Quercetin: $R_1$, $R_2$, $R_4$= OH; $R_3$= H**<br>**Kaempferol: $R_1$, $R_3$ = H; $R_2$, $R_4$ = OH**<br>**Luteolin: $R_1$, $R_2$ = OH; $R_3$, $R_4$ = H**<br>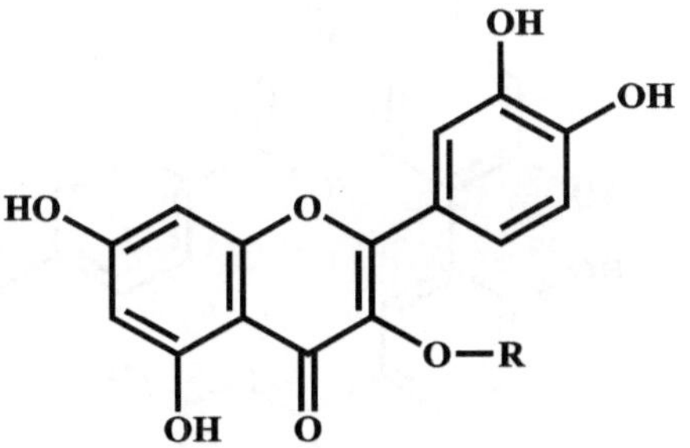<br>**Rutin: R = rutinose**<br>**Hyperoside: R = galactose** | Flowers, upper part of the stem with leaves |
| Phenolic acids | 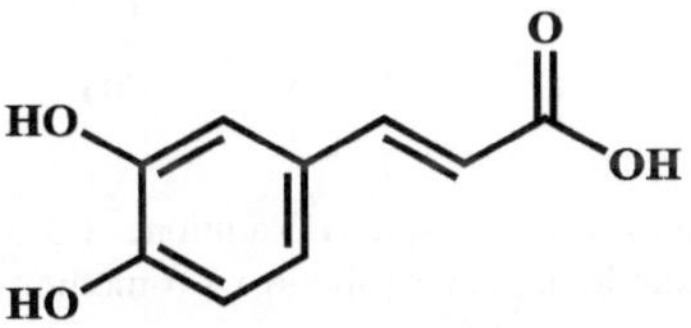<br>Caffeic acid<br>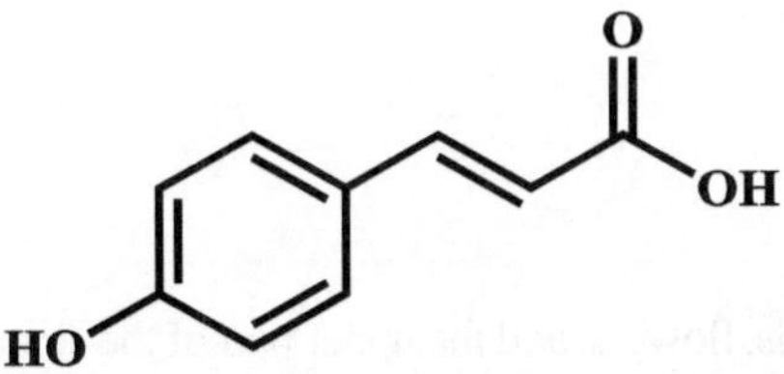<br>*p*-Coumaric acid | Flowers, upper part of the stem with leaves |

(continued)

| | | |
|---|---|---|
| | 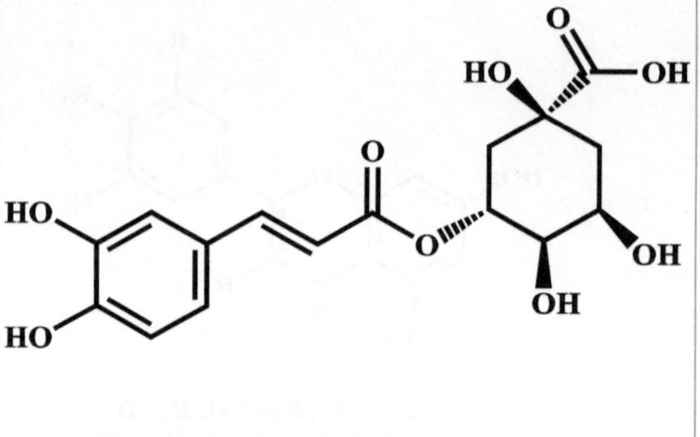 Chlorogenic acid | |
| Biflavonoids | 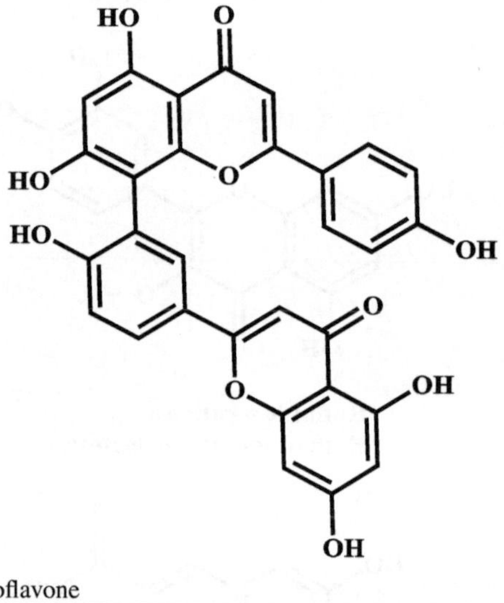 Amentoflavone | Flowers, upper part of the stem with leaves |
| Xanthones | 1,3,6,7-Tetrahydroxy-8-prenyl xanthone, 1,3,6,7-tetrahydroxy-2-prenyl xanthone, 1,3,7-trihydroxy-6-methoxy-8-prenyl xanthone, paxanthone | |
| Essential oils/ terpenes | α-Pinene, β-pinene, myrcene, limonene, caryophyllene, humulene | Flowers, upper part of the stem with leaves |
| Tannins | | Aerial parts |

Only the flower buds, flowers, and the upper part of the stem with branches and leaves have a high content of anthrones. Therefore, when harvesting, the lack of fruits, seeds, and woody stems (lower part) should be taken into account to ensure the pharmaceutical quality (min. 0.04 % hypericin) of the raw material used for extraction.

4) Pharmacology and bioactivities:
   A) In vitro:
      A mechanism investigated in the last two years was the influence of ethanolic extracts of St. John's wort (rich in hyperforin, 3.6 %) on transient receptor

potential cation channel 2 (TRPM2) channels and the influx of $Ca^{2+}$. The results indicated that the *Hypericum* extract decreased the TRPM2 channels and voltage-gated calcium channel (VGCC)-dependent $Ca^{2+}$ influx in rat dorsal root ganglion neurons. In addition, *Hypericum perforatum* exhibited protection against $H_2O_2$-induced oxidative stress, increasing at the same time the total glutathione levels in the dorsal root ganglion neurons. Therefore, the authors concluded that the *Hypericum* extract acts as a modulator against oxidative stress-induced by $Ca^{2+}$ mobilization through TRPM2 and VCGG channels in the dorsal root ganglions. This finding may provide an explanation for the use of *Hypericum perforatum* in neuropathic and peripheral pain (Nazıroğlu et al. 2014).

B) In vivo:
Preclinical animal studies suggest that *Hypericum* is effective in three major biochemical systems relevant for **antidepressant activity** (Ernst 1995), namely, the inhibition of the reuptake system for serotonin (5-HT), noradrenaline (NA), and dopamine (DA) into the presynaptic neuron. It is the only antidepressant capable of inhibiting the reuptake of 5-HT, NA, and DA with similar potencies (Calapai et al. 2001, Bonetto et al. 2007). For example, a team from Yugoslavia has tested three types of extracts in a mouse model of intestinal motility and pentobarbital-induced sleep. Their results indicated that the highest antidepressant and stimulant effects on the central nervous system (CNS) were induced by the ethanol and ethyl acetate extracts. Moreover, the ethanol extract exhibited a spasmolytic activity on the smooth muscles of the gastrointestinal (GI) tract, whereas the ethyl acetate fraction showed a significant analgesic activity. These data confirm that hypericin- and hyperforin-derived compounds are responsible for antidepressant and analgesic effects, while the spasmolytic activity is mainly given by the flavonoids (Jakovljević et al. 2000).

A great number of double-blind studies were done in the last two decades. There is a great variability in terms of methods, extracts/preparations, and standards that were used. The most known are the German investigations related to standardized extracts.

A review of 18 controlled clinical trials in which ethanol or methanol extracts of *Hyperici herba* were used as a treatment for depression indicated that only nine placebo-controlled and three comparison trials met the methodological inclusion criteria. The results show that *Hypericum* extracts were superior to placebo for the symptomatic treatment of depression as measured by the Hamilton Depression Rating Scale. Maprotiline (75 mg daily), imipramine (50–75 mg daily), and other antidepressants were used as standards. Therefore, by comparison to these classical drugs, the extracts have a similar therapeutic profile. Some flaws that were reported included no intention to treat analysis, lack of control over compliance, and insufficient description of the extract or placebo used (Linde et al. 2008).

Another review of 12 double-blind, placebo-controlled, and three comparison clinical trials assessed the efficacy of the herb in depression and the methodology used in the studies. The conclusions were that the therapeutic dose of 300 mg standardized extract in 0.9 mg hypericin administered three times daily for 4–8 weeks is efficient for mild to moderate depression, whereas no dose-finding studies had been conducted. Also, there is a lack of data in regard to patients with severe depression. However, the comparison studies concluded that the daily dose of 75 mg maprotiline or 30 mg amitriptyline was too low. Still, further trials of longer duration in comparison with higher doses of standard antidepressants are warranted (Volz 1997).

5) Indications and usage:

In dyspeptic and hepatobiliary disorders, the vegetal product is recommended as infusion due to its choleretic, cholagogue, and antidiarrheal properties.

The lipophilic compounds from *Hyperici herba* extracted as a macerate in vegetable oil (known as *Oleum Hyperici coctum* or *Hypericum* butter) have healing and anti-inflammatory properties, thus supporting the folkloric topical uses as treatment of muscle pain, burns, wounds, and bruises. Additionally, St. John's wort extracts are among the most widely prescribed antidepressants (especially in Germany) (Linde et al. 2005). Yet, many preparations of *Hypericum* are sold on the free market. For such preparations, the approval procedures are not so strict; therefore, the pharmaceutical industry is not forced to conduct clinical trials. Still, numerous studies on *Hypericum* extracts, especially standardized extracts in hypericin and hyperforin, were published in the last 20 years (Cervo et al. 2002).

Extracts standardized in hypericin are components of antidepressant remedies such as Hyperforat® (tablets, drops, ampoules), Jarsin® (tablets) or Psychotonin® (tincture), and Laif®600 (STW 3, one of the few extracts standardized both in hypericin, hyperforin, and flavone derivatives). However, sometimes the plant product is also used in the composition of preparations indicated in micturition disorders encountered in urology, for example, Inconturina® (drops) and Rhoival® (tablets, drops).

6) Precautions and side effects:

Concurrent administration of cheese, salted porbeagle, beer, and wine with St. John's wort preparations should be avoided. Food such as those described may be ingested at a distance of 2 h after administration.

Co-administration of many classes of drugs with *H. perforatum* extracts imposes precaution due to possible interactions. *H. perforatum* has been confirmed as a CYP450 isozyme 3A4 inductor, therefore increasing the metabolic transformation of all the medicines with a similar pathway. CYP3A4 is the most abundant hepatic and intestinal phase I enzyme that metabolizes immunosuppressive (cyclosporine) and antiviral drugs, non-sedating antihistamines, calcium channel blockers, estrogen, benzodiazepines, macrolide antibiotics, warfarin, etc.

St. John's wort is contraindicated in pregnancy (uterotonic activity in experimental animals was observed) and lactation (it inhibits pituitary prolactin secretion). Hypericin showed in vitro mutagenic and carcinogenic potential, but further in vivo studies did not confirm these activities. There is insufficient data in regard to children and adolescent use and dosage.

Photoactive hypericin derivatives may induce photosensitivity reactions to prolonged exposure to sunlight or UV radiation. Normally, the therapeutically effective doses are 30–50 times smaller than the photosensitizing doses; however, blonde individuals should be cautious when using *Hypericum* preparations for a long time. Due to this activity, it should not be administered in combination with other photosensitizing drugs such as tetracycline or chlorpromazine.

The co-administration of St. John's wort extracts with other classes of antidepressants should be avoided (Izzo 2004).

Alcohol consumption should be discouraged during the treatment with *Hypericum*.

There is data suggesting that *Hypericum* preparation used should be discarded at least 10 days prior to surgery.

7) Dosage:
   A) Internal use:

      The recommended dosage as antidepressant treatment is 300–900 mg extract (LI160/WS5572)/day or equivalent to 0.2–1 mg hypericin/day, respectively, or 4 g dry plant/day. This equivalency calculation is based on the manufacturing procedures of standardized extracts, where from 4 g of plant (dry weight), one can obtain 300–900 mg of extract containing up to 1 mg hypericin. Underdosing is equivalent to failure in the installation of the pharmacological activity.

      The antidepressant activity begins to manifest after about 8–10 days since starting the treatment and becomes maximal after 4–6 weeks of use. Extracts or preparations with *Hypericum* should be administered 6–18 months, like any other antidepressant.

      For dyspeptic and hepatobiliary disorders, the infusion with St. John's wort is prepared using 2–4 g of ground plant product to which 150 ml of boiling water are added, allowed to stand covered for 10 min, and then filtered. Usually, the dose is 1–2 cups of freshly brewed infusion, morning and evening.

   B) External use:

      Oily *Hypericum* preparations are recommended topically as needed to treat burns and inflammation.

## 5.5 References


- Ernst E. St. John's Wort, an anti-depressant? A systematic, criteria-based review. Phytomed. 1995;2:67–71.
- Calapai G, Crupi A, Firenzuoli F, Inferrera G, Squadrito F, Parisi A, et al. Serotonin, norepinephrine and dopamine involvement in the antidepressant action of *Hypericum perforatum*. Pharmacopsychiatry. 2001;34:45–9.
- Bonetto N, Santelli L, Battistin L, Cagnin A. Serotonin syndrome and rhabdomyolysis induced by concomitant use of triptans, fluoxetine and hypericum. Cephalalgia. 2007;27:1421–3.
- Jakovljević V, Popović M, Mimica-Dukić N, Sabo A, Gvozdenović L. Pharmacodynamic study of *Hypericum perforatum* L. Phytomed. 2000;7:449–53.
- Nazıroğlu M, Çiğ B, Özgül C. Modulation of oxidative stress and $Ca^{2+}$ mobilization through TRPM2 channels in rat dorsal root ganglion neuron by *Hypericum perforatum*. Neuroscience. 2014;263:27–35.
- Linde K, Berner M, Kriston L. St John's wort for major depression. Cochrane Database Syst Rev. 2008;8:CD000448.

- Volz HP. Controlled clinical trials of *Hypericum* extracts in depressed patients – an overview. Pharmacopsychiatry. 1997,30:72–6.
- Linde K, Berner M, Egger M, Mulrow C. St John's Wort for depression. 2005.
- Cervo L, Rozio M, Ekalle-Soppo C, Guiso G, Morazzoni P, Caccia S. Role of hyperforin in the antidepressant-like activity of *Hypericum perforatum* extracts. Psychopharmacology. 2002;164:423–8.
- Izzo A. Drug interactions with St. John's Wort (*Hypericum perforatum*): a review of the clinical evidence. Int J Clin Pharmacol Ther. 2004;42:139–48.


## 5.6 *Valeriana officinalis* L..

### *5.6.1 (Valerianaceae)*

1) Common names:

   Valerian (UK), cat's love (UK), Baldrian (Germany), Baldrianwurzel (Germany), fragrant valerian (UK), garden heliotrope (USA), great wild valerian (UK), Valeriană (Romania).
2) Photograph of the species (Fig. 5.2).
3) Description of the species:
   A) Habitat and world distribution:

      Approximately 200 *Valeriana* species are available, but only a few are or were used medicinally, such as *V. fauriei* Briquet (Japanese valerian),

**Fig. 5.2** Flowers of *Valeriana officinalis* L. Photograph taken in Neamt County, Romania, in 2010 by Oana Cioanca

*V. wallichii* DC (Indian valerian), and *V. edulis* Nutt ex. Torr. & Gray. In commerce, *V. edulis* Nutt. ex Torr. & Gray is known as *Valeriana mexicana*. Plants bearing this common name should not be confused with *V. mexicana* DC., which is in fact *V. sorbifolia* H.B.K. var. *mexicana* (DC) F.G. Mey.

*Valeriana officinalis* is native to Europe and Western Asia (throughout temperate and subpolar Eurasian area). It grows well either in full sun or part shade damp locations (ditches and stream banks), with moist, well-drained, and humus-rich soils.

B) Morphology of the plant:

Valerian is a tall perennial herb whose underground portion consists of a vertical yellowish brown or dark brown rhizome bearing numerous roots and one or more stolons. The aerial portion consists of a rosette of leaves at the soil surface (first year), then a cylindrical hollow, and channeled stem up to 2 m high, branched in the terminal region, bearing opposite exstipulate, pinnatisect, cauline leaves with clasping petioles. Pentamerous pink or white small flowers are grouped into inflorescences (racemes of cymes). The fruits are oblong-ovate, four-ridged, single-seeded achenes. When fresh, the roots and rhizomes have a fragrant smell, but by drying, it changes to a disagreeable tint.

C) Medicinal parts:

*Valerianae radix* consists of the underground parts of *Valeriana officinalis* L. including the rhizomes, roots, and stolons, dried at temperatures below 40 °C. It has a strong unpleasant smell noticeable especially after processing (cutting, grounding, etc.).

D) Chemical constituents:

The chemical composition of valerian roots varies greatly depending on the subspecies, variety, age of the plant, and pedoclimatic conditions. The main groups of components are presented below:

| Chemical class | Individual components | Plant part/ organ |
|---|---|---|
| Essential oil with more than 150 components (0.3–1 %) | 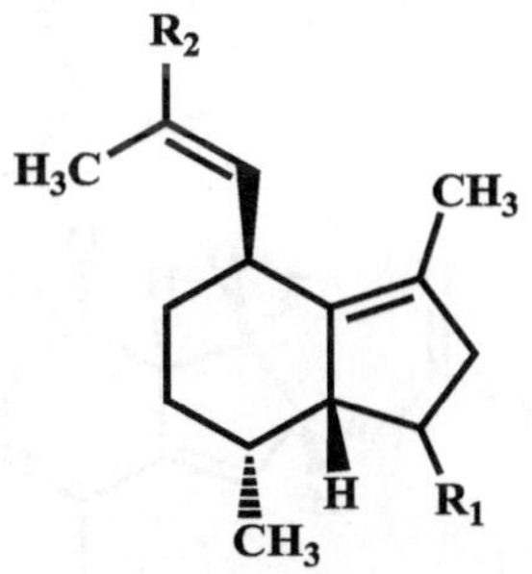<br>**Valerenal: $R_1 = H, R_2 = CHO$**<br>**Valerenic acid: $R_1 = H, R_2 = COOH$**<br>**Acetoxyvalerenic acid: $R_1 = OAc, R_2 = COOH$** | Rhizomes with roots |

(continued)

| | | |
|---|---|---|
| | 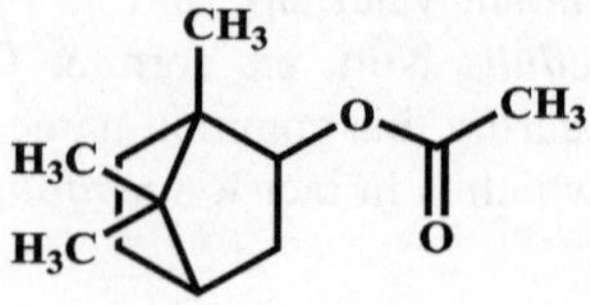<br>Bornyl acetate<br>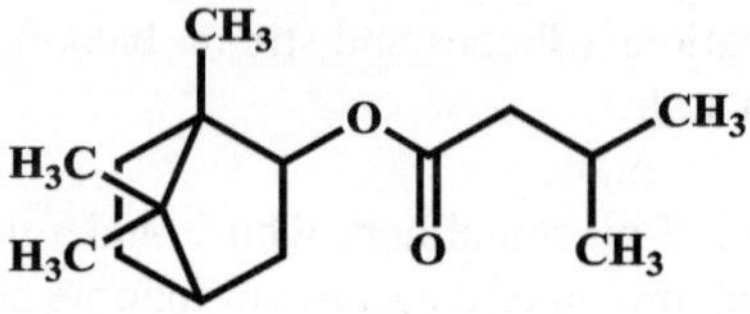<br>Bornyl isovalerate | |
| Valepotriates (non-glycosidic bicyclic iridoid monoterpene epoxy esters) (0.5–2 %) | 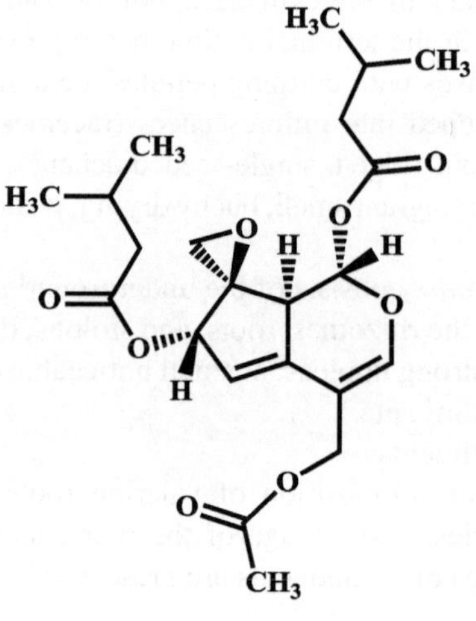<br>Valtrate<br>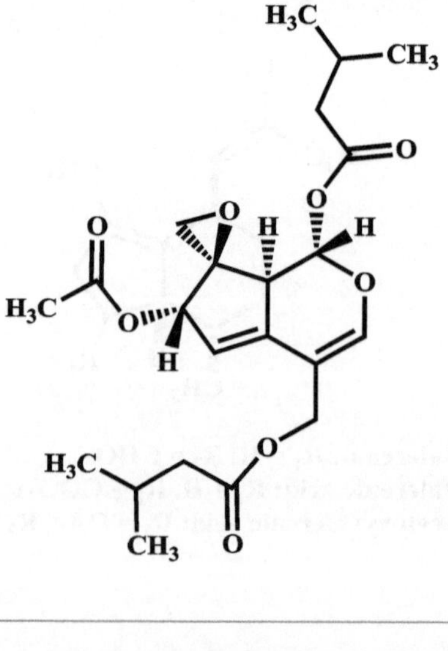<br>Isovaltrate | Especially in the fresh rhizomes, but also in the dry rhizomes Metabolites found in the dry and processed rhizomes |

(continued)

| | | |
|---|---|---|
| | 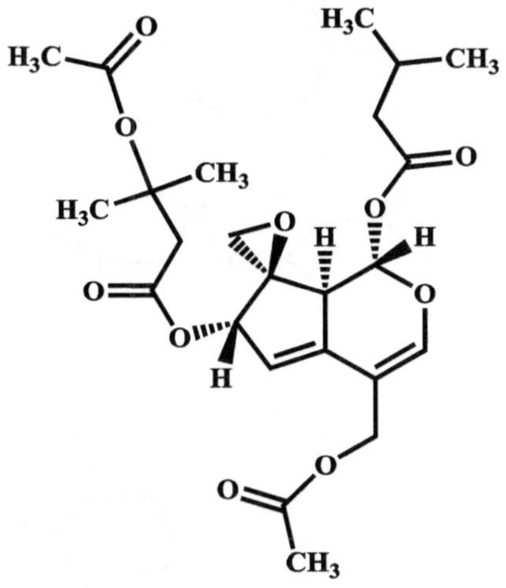<br>Acevaltrate<br>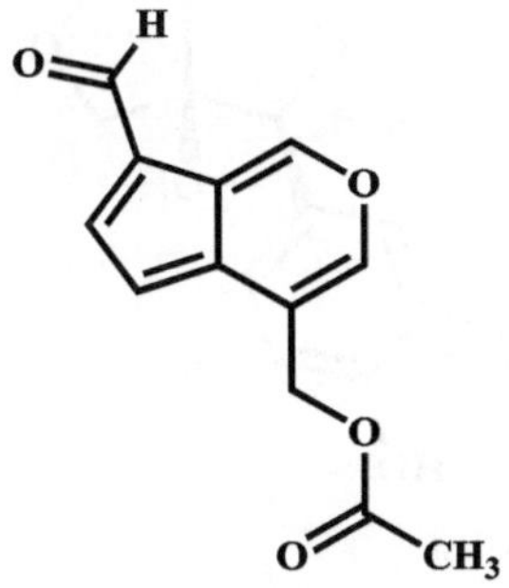<br>Baldrinal<br>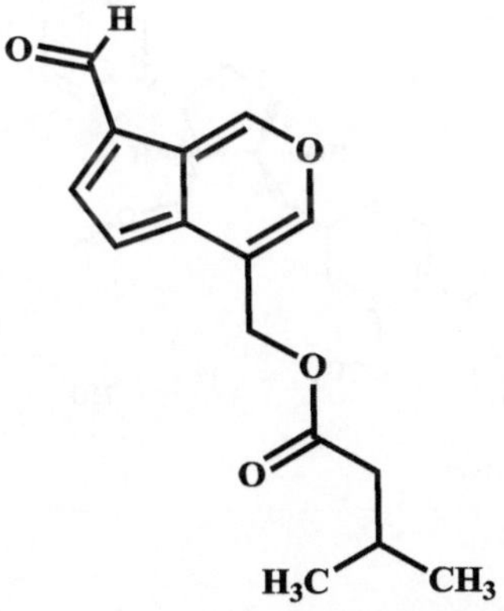<br>Homobaldrinal | |

(continued)

| | | |
|---|---|---|
| | 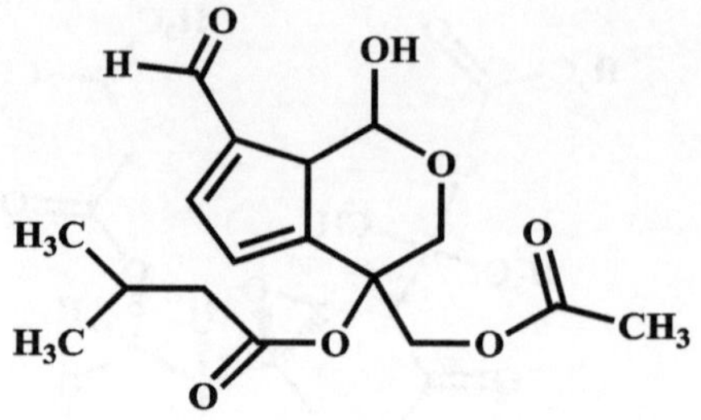<br>Valtroxal | |
| Lignans | 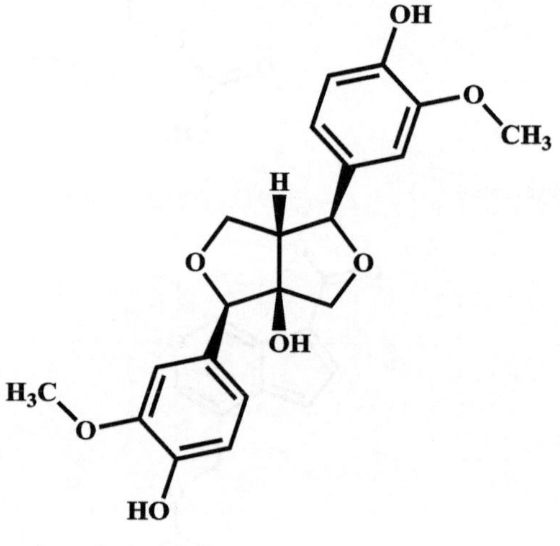<br>Hydroxypinoresinol<br>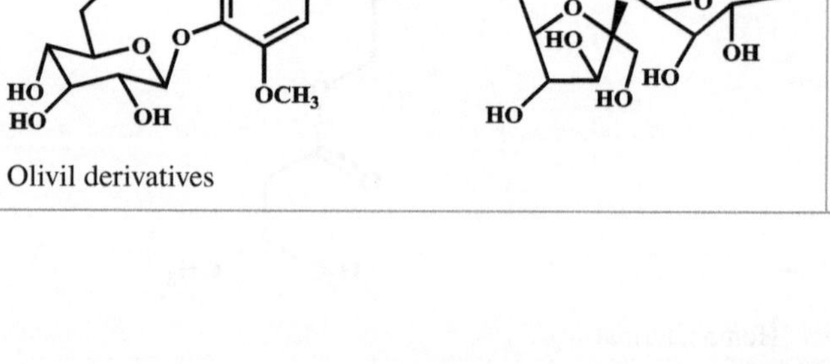<br>Olivil derivatives | Rhizomes with roots |

(continued)

| | | |
|---|---|---|
| Flavonoids | 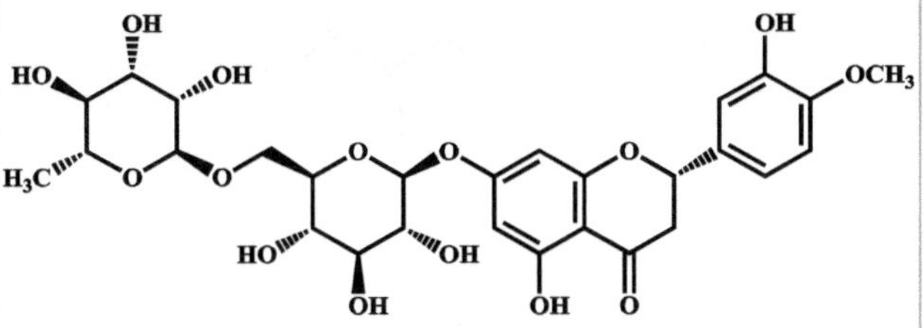 Hesperidin 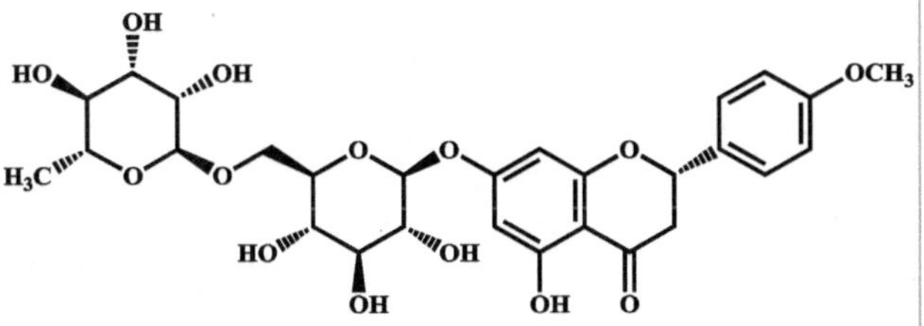 Linarin 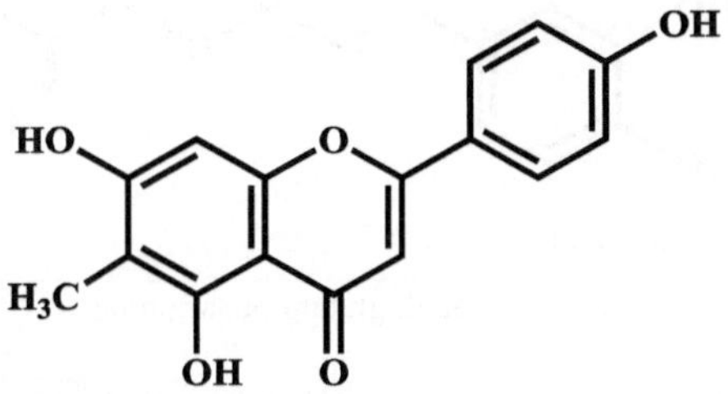 6-Methyl apigenin | Rhizomes with roots |
| Phenolic acids | 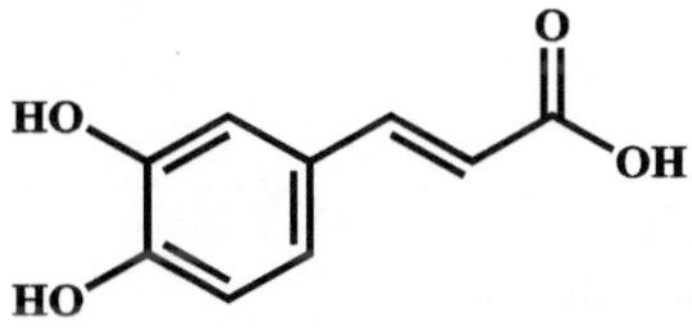 Caffeic acid 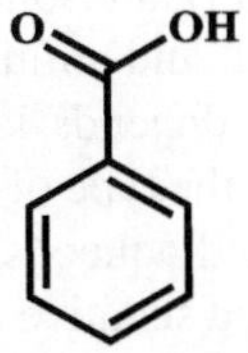 Benzoic acid | Rhizomes with roots |

(continued)

| | | |
|---|---|---|
| | 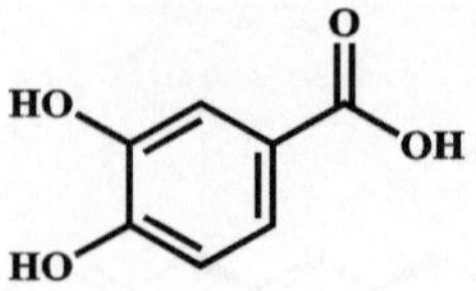 Protocatechuic acid | |
| | 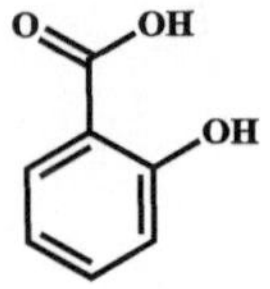 Salicylic acid 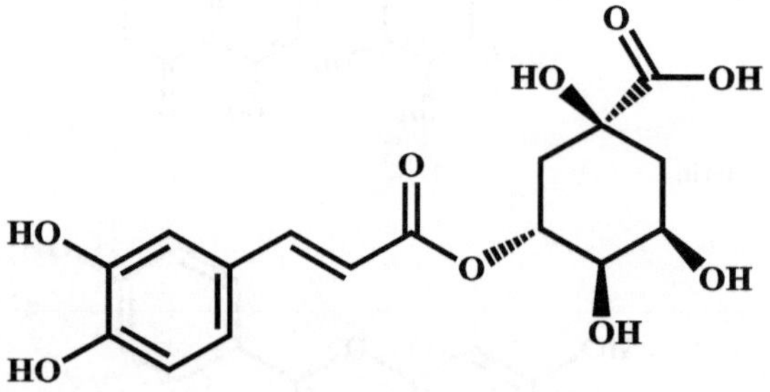 Chlorogenic acid | |
| Amino acids | Gamma-aminobutyric acid, glutamine, arginine | Rhizomes with roots |
| Alkaloids (0.01–0.05 %) | Actinidin (the most important), chatinin, valerianin | Rhizomes with roots |
| Resin | No examples are given in literature | Rhizomes with roots |
| Tannins | | Rhizomes with roots |

4) Pharmacology and bioactivities:

Various studies have demonstrated the sedatives, anxiolytics and hypno-inducing properties of valerian (Hattesohl et al. 2008, Jung et al. 2014). Also, antispasmodic, muscle relaxant, and antiulcer effects have been proven as well. However, the clinical efficacy depends largely on the processing of the plant material (rhizomes with roots), the type of extract, and the dose. Some investigations of both hydroalcoholic and aqueous extracts obtained from valerian root, and isolated constituents showed sedative and anxiolytic properties, which helps promote sleep and improving the CNS activity (Marder et al. 2003).

A) In vitro:

In vitro tests demonstrated the binding of valerian extracts to GABA receptors, adenosine receptors, barbiturate, and benzodiazepine receptors (Cavadas et al. 1995). Although, both hydroalcoholic and aqueous extracts showed affinity for the $GABA_A$ receptors, a clear correlation between any of the known isolated components from valerian roots and $GABA_A$ binding activity was impossible to prove. Still, 6-methylapigenin and hesperidin were considered to be ligands for the benzodiazepine binding site of the $GABA_A$ receptor according to Wasowski et al. (Wasowski et al. 2002). Moreover, aqueous extracts of *Valerianae radix* inhibit the reuptake and stimulate the release of radiolabeled GABA in synaptosomes (on rat isolated brain cortex), thus inducing an increase of GABA in the synaptic cleft with possible enhanced biochemical effects of this compound, not to mention that GABA as a constituent of valerian could be responsible also for this activity (Santos et al. 1994a, b, c). However, the high content of GABA in some valerian root extracts cannot explain in vivo pharmacologic effects since it does not cross the blood-brain barrier.

A high affinity for $5\text{-}HT_{1A}$ receptor was evidenced for lignans, especially for hydroxypinoresinol, which plays a role in anxiolytic and sleep induction activity.

In vitro agonistic activity on A1 adenosine receptors was shown both for some olivil derivate and a methanolic (45 %) valerian root extract, thus inducing a sedative effect (Schumacher et al. 2002).

Although many studies on the isolated compounds were published, the results could not fully explain the complex pharmacological mechanism of the total extracts from valerian roots. Therefore, the phytocomplex (considered as the total of the chemical substances that could be isolated in an extract) should be considered as the active ingredient for *Valerianae radix*.

B) In vivo:

In the elevated plus-maze test at a dose of 1 mg/kg i.p., both 6-methylapigenin and hesperidin (isolated from a valerian root extract) exhibited anxiolytic-like effects without sedative properties. Additionally, 2 mg/kg i.p. of hesperidin increased the thiopental sleeping time in mice, significantly enhanced when hesperidin and 6-methylapigenin were combined (Marder et al. 2003).

Dose-dependent sedative and sleep-enhancing effects were demonstrated for linarin, markedly potentiated for its combination with valerenic acid (Fernández et al. 2004).

Several pharmacological animal models were used to investigate valerian extracts (aqueous, hydroalcoholic), and the results indicated that several classes of substances (sesquiterpenes, derived lignans, iridoids, flavonoids) with concerted action and synergisms play a significant role on the sedative and anxiolytic effects. Data from animal experiments are consistent and in accordance with clinical observations.

The clinical data showed that valerian preparations improve the sleep quality with a gradual onset of efficacy rather than a general sedative effect exerted especially in pathological conditions and less healthy subjects (Hattesohl et al. 2008). However, after a single dose of valerian, subjective perception of sleep changes and symptoms are improved gradually in 2–4 weeks. Additionally, valerian preparations do not induce characteristic side effects of the conventional medication, often characterized by the rebound effect (benzodiazepines).

Evidence-based (EEG studies) clinical trials on healthy volunteers and females with sleep disorders were published. Although not all results are consistent, there is evidence of mild to moderate sedative or tranquilizing effect of valerian root extracts (Wagner et al. 1980). In a placebo-controlled, double-blind study of crossover design, the effects of a single dose of 1200 mg valerian root extract (drug-extract ratio (DER) 5–7:1, solvent not specified; corresponding to 7.2 g of plant) on 12 healthy subjects (average age: $53.7 \pm 5.6$ years) were compared to that of $1 \times 10$ mg diazepam. The results showed an increase for valerian root extract in the delta and theta bands and a decreased power in the beta band (European Medicines Agency 2007).

A meta-analysis of randomized placebo-controlled trials published on MEDLINE, the Cochrane Library, EMBASE, and BIOSIS databases in 2010 established the effectiveness of *Valeriana* extracts on insomnia compared with a placebo. The conclusion was that valerian is effective for a subjective improvement of insomnia (Taibi et al. 2007), although its effectiveness has not been demonstrated with quantitative or objective measurements; therefore further clinical studies are required.

5) Indications and usage:

Valerian is recommended as a mild sedative and sleep-promoting agent to treat anxiety, restlessness, and insomnia. Also, extracts are used as adjuvant in spasms of smooth muscle and gastrointestinal pains of nervous origin.

Medicines containing valepotriates have other indications than valerian extract preparations; thus, those containing didrovaltrate and/or homovaltrate (monoenes) are mainly used as tranquilizers during the day, while pharmaceuticals with valtrate, isovaltrate, and acevaltrate (dienes) are thymoleptics and anxiety relievers.

The prescription of drugs containing valerian extracts should be carefully assessed in regard to the processed species. Therefore, preparations obtained from *Valeriana officinalis* (European) are recommended mainly as sedatives and sleep inducers, while *Valeriana wallichii* and *Valeriana edulis* extracts should be indicated as thymoleptics. A dose of 500–600 mg dry extract (DER 5:1, equivalent to a min. 50 mg valepotriates) is necessary for a therapeutic effect.

Combinations of valerian extract with other plant products with similar or complementary effects are often found on the market. Among them, a preparation consisting of 32 mg *Passiflora* extract, 60 mg *Hypericum* extract, and 28 mg valerian extract is administered as an antidepressant, anxiolytic, and sleep inducer. Also, this preparation was reported to reduce social isolation, anxiousness, depression, and sleep disorders in a group of children aged 6–12 years with emotional and behavioral disorders (Jorm et al. 2006).

6) Precautions and side effects:

There are no significant side effects for valerian extracts, but due to insufficient data *Valerianae radix* preparations should not be used during pregnancy or lactation. For the same reasons, such preparations should be avoided for children less than 12 years of age without medical supervision.

Minor side effects (headaches, excitability, uneasiness, and insomnia) have been associated with chronic use of valerian extracts. Very large doses (up to 20 times the recommended dose) may cause bradycardia and arrhythmias and decrease intestinal motility.

7) Dosage:
   A) Internal use.

      Adults and children over 12 years old should use dried hydroalcoholic extracts (ethanol 40–70 % v/v) equivalent to 2–3 g of vegetable (single dose) three times daily for anxiety and a single dose of 400–900 mg extract ½–1 h before bedtime for sleep disorders.

      Maximum daily dosage is 4 divided doses. The optimal therapeutic effect is obtained within 2–4 weeks.

      As tea 3–5 g (2 teaspoon) chopped valerian root is brewed with 150 ml boiling water, allow to stand in a covered container for 4–8 h (valepotriates are poorly soluble in water), stirring occasionally. Patients should know that although at first it is difficult to drink the tea due to the unpleasant smell, after 4–5 days, they will become used to it. Administration in the evening is at least 20 min before bedtime, during which any exciting activity should be avoided, preferring to sleep through relaxation training that will ensure sleep onset.

      Such preparations act gentle and do not normally interfere with daily activities that require attention. Nevertheless, association with other CNS depressants should be avoided.

## 5.7 References


- Jung H, Yoo D, Kim W, Nam S, Kim J, Choi J, et al. *Valeriana officinalis* root extract suppresses physical stress by electric shock and psychological stress by nociceptive stimulation-evoked responses by decreasing the ratio of monoamine neurotransmitters to their metabolites. BMC Complement Alternative Med 2014;14:476.
- Hattesohl M, Feistel B, Sievers H, Lehnfeld R, Hegger M, Winterhoff H. Extracts of *Valeriana officinalis* L. s.l. show anxiolytic and antidepressant effects but neither sedative nor myorelaxant properties. Phytomedicine. 2008;15:2–15.
- Marder M, Viola H, Wasowski C, Fernández S, Medina JH, Paladini AC. 6-Methylapigenin and hesperidin: new valeriana flavonoids with activity on the CNS. Pharmacol Biochem Behav. 2003;75:537–45.
- Cavadas C, Araujo I, Cotrim MD, Amaral T, Cunha AP. In vitro study on the interaction of *Valeriana officinalis* L. extracts and their amino acids on GABAA receptor in rat brain. Arzneimittelforschung. 1995;45:753–5.
- Wasowski C, Marder M, Viola H, Medina JH, Paladini AC. Isolation and Identification of 6-Methylapigenin, a competitive ligand for the brain GABAA receptors, from *Valeriana wallichii*. Plant Med. 2002;68:934–6.
- Santos MS, Ferreira F, Cunha AP, Carvalho AP, Macedo T. An aqueous extract of valerian influences the transport of GABA in synaptosomes. Plant Med. 1994;60:278–9.

- Santos MS, Ferreira F, Cunha AP, Carvalho AP, Ribeiro C. Synaptosomal GABA release as influenced by valerian root extract--involvement of the GABA carrier. Arch Int Pharmacodyn Thér. 1994;327:220–31.
- Santos MS, Ferreira F, Faro C, Pires E, Carvalho AP, Cunha AP, et al. The amount of GABA present in aqueous extracts of valerian is sufficient to account for [3H]GABA release in synaptosomes. Plant Med. 1994;60:475–6.
- Schumacher B, Scholle S, Hölzl J, Khudeir N, Hess S, Müller C. E. Lignans isolated from valerian: identification and characterization of a new olivil derivative with partial agonistic activity at A1 adenosine receptors. J Nat Product. 2002;65:1479–85.
- Fernández S, Wasowski C, Paladini AC, Marder M. Sedative and sleep-enhancing properties of linarin, a flavonoid-isolated from *Valeriana officinalis*. Pharmacol Biochem Behav. 2004;77:399–404.
- Wagner H, Jurcic K, Schaette R. Comparative studies on the sedative action of *Valeriana* extracts, valepotriates and their degradation products Plant Med. 1980;37:358–62.
- European Medicines Agency, Assessment report on *Valeriana officinalis* L., Radix. 2007;8.
- Taibi DM, Landis CA, Petry H, Vitiello MV. A systematic review of valerian as a sleep aid: safe but not effective. Sleep Med Rev. 2007;11:209–30.
- Jorm A, Allen N, O'Donnell C, Parslow R, Purcell R, Morgan A. Effectiveness of complementary and self-help treatments for depression in children and adolescents. Med J Aust. 2006;185:368–72.


## 5.8 *Ocimum basilicum* L.

### *5.8.1 (Lamiaceae)*

1) Common names:

   Basil (UK), common basil (UK), sweet basil (UK), Basilic (France), Herbe royale (France), Saint Joseph's wort (UK), Basilikum (Germany), Busuioc (Romania), Vasilikos (Greece), Bosilek (Bulgaria), Alfavaca (Portugal), Albahaca (Spain).
2) Photograph of the species (Fig. 5.3).
3) Description of the species:

   It is estimated that there are many varieties and subspecies of the genus *Ocimum*, but, although the same, species may present both morphological (sometimes quite important) and chemical composition differences:

   A) Habitat and world distribution:

      *Ocimum basilicum* L. is annual or short-lived perennial herb originally native to India and other regions of Asia. It is spread in several Mediterranean countries, including Turkey, and was known to the Greeks and Romans since

**Fig. 5.3** Flowers of *Ocimum basilicum* L. Photograph taken in Iasi, Romania, in 2014 by Oana Cioanca

ancient times. Today, the plant is extensively cultivated both for food and therapeutic purposes. It grows well on moist, well-drained, fertile soil in the sun or part shade in hot climates. It is sensitive to frost.

B) Morphology of the plant:

The plant grows 20–60 cm high with a straight four-edged (square) stem freely branched at the top. It has opposite, ovate-acuminate, glossy, leaves, glabrous, slightly serrated to base, with a deep green color. Some varieties have partly red or entirely purple leaves. Flowers are white or purple-pink borne in simple terminal racemes; the corolla is bilabial. The fruits are small, black capsules containing four very small ovoid seeds. Freshly dried vegetable product shows a characteristic aroma, pleasant with a slightly spicy and salty flavor.

C) Medicinal parts:

The plant parts above the ground are used to obtain the dry drug *Basilici herba* used for therapeutic purposes.

D) Chemical constituents:

Although tannins and essential oil are considered both as main groups, the volatile fraction is definitively most studied. Still, the amounts of the primary constituents depend on the variety, chemo-type, and geographical distribution.

| Chemical class | Individual components | Plant part/ organ |
|---|---|---|
| Essential oil (0.04–0.70 %) | 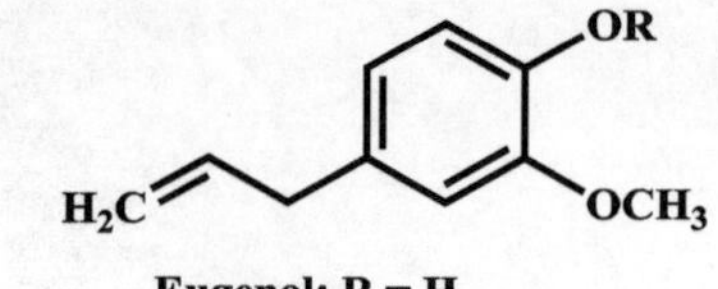<br>**Eugenol: R = H**<br>**Methyleugenol: R = $CH_3$**<br>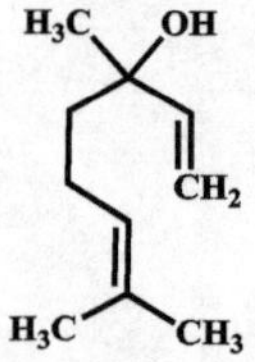<br>Linalool<br>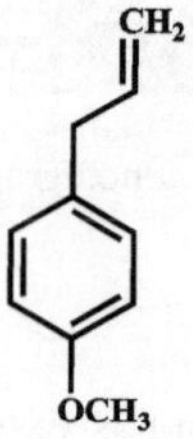<br>Estragole<br>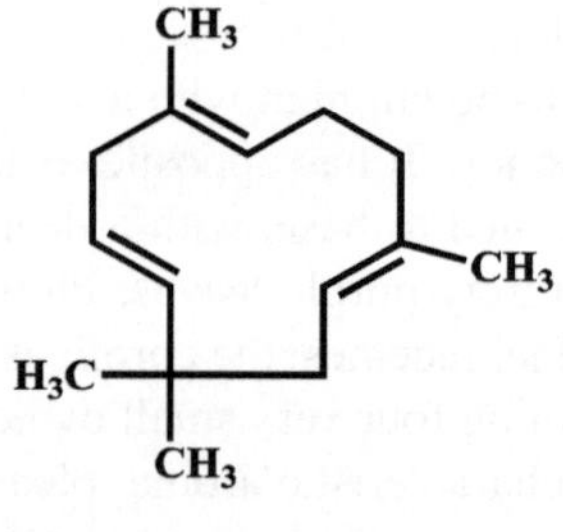<br>α-Caryophyllene<br>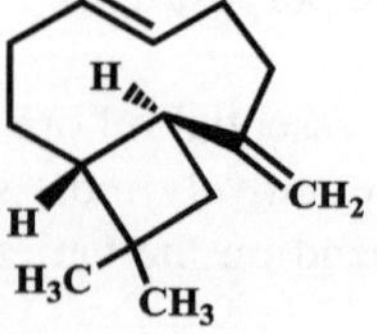<br>β-Caryophyllene | Leaves, stem, and flowers |

(continued)

| | | |
|---|---|---|
| Tannins | | Leaves, stem, and flowers |
| Flavonoids | 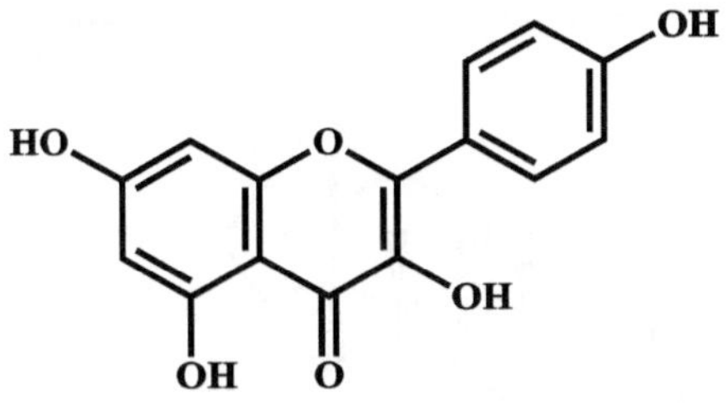<br>Kaempferol derivatives<br>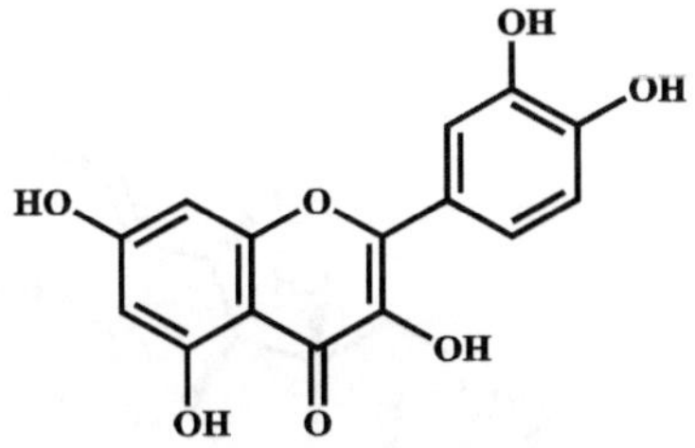<br>Quercetin derivatives | Leaves, stem, and flowers |
| Phenolic acids | 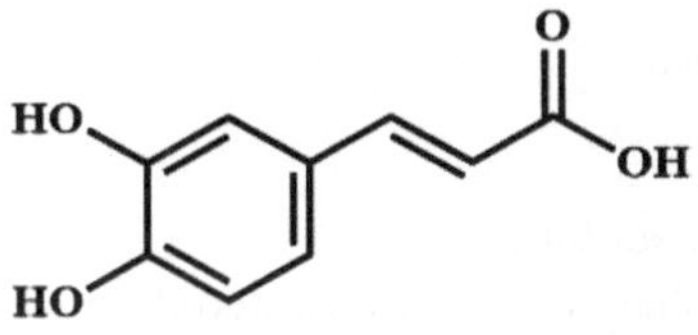<br>Caffeic acid<br>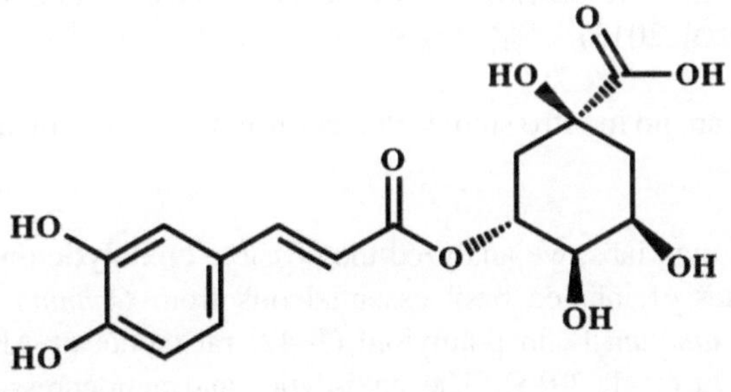<br>Chlorogenic acid | Leaves, stem, and flowers |

(continued)

| | | |
|---|---|---|
| Phytosterols | 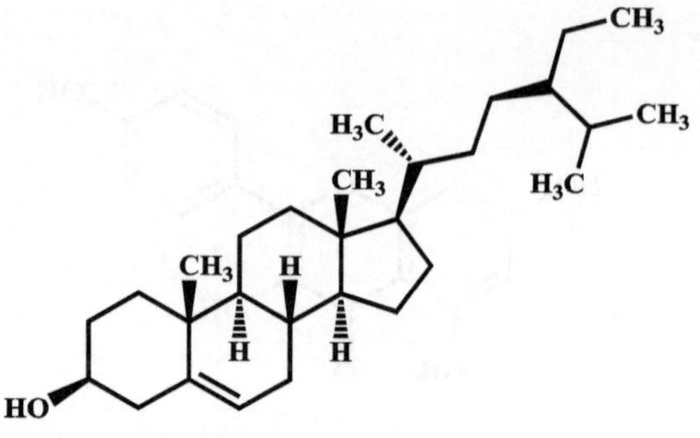<br>Beta-sitosterol<br>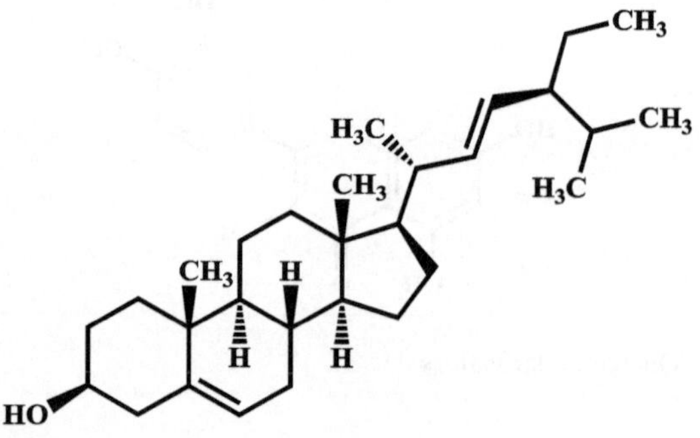<br>Stigmasterol | Leaves, stem, and flowers |

4) Pharmacology and bioactivities:

Most of the published studies are centered on *Ocimum sanctum* (the Indian basil, Tulsi) and less on *Ocimum basilicum*. There are some published studies in Europe regarding the chemical composition of the essential oil and flavonoid content of *Basilici herba* extracts (Chalchat and Özcan 2008, Chang et al. 2008, Gradinariu et al. 2013).

A) In vitro:

There are no in vitro studies that confirm the anxiolytic and antidepressant effects.

B) In vivo:

In our own labs, we analyzed the possible **anxiolytic and antidepressant proprieties** of inhaled basil essential oils from *Ocimum basilicum* L. and *Ocimum sanctum* L. in β-amyloid (1–42) rat model of Alzheimer's disease (Gradinariu et al. 2015). The anxiolytic- and antidepressant-like effects of inhaled basil essential oils were studied by elevated plus-maze and forced swimming tests. The results showed that exposure to basil essential oils significantly improved the behavior of the animals, suggesting anxiolytic- and antidepressant-like effects. Therefore, multiple exposures to basil essential oils (1 %, accepted in aromatherapy) can be a useful mean to counteract anxiety and depression (Gradinariu et al. 2015). Abdoly et al. (2012) used FST to evaluate the antidepressant effect of a *O. basilicum* extract in albino male Wistar rats that had been exposed to electromagnetic field (50 Hz) for 8 weeks. At the end of 8th week, rats which have been fed with *O. basilicum* extract (1.5 g/kg body weight), showed a decrease in immobility score and an increase in swimming score in comparison with control group, but climbing was similar for both groups.

The **toxicity** of two extracts obtained from *Ocimum basilicum* and *O. tenuiflorum* (by cold hydroalcoholic extraction for 7 days, under mechanical stirring) was evaluated on young rats that received the 1/10 diluted extract in the drinking water, for 60 days. The results showed that the rats treated with basil extracts had increased values of red blood cells, hemoglobin, and packed cell volume, whereas the values for mean corpuscular volume, mean hemoglobin concentration, and mean corpuscular hemoglobin concentration were slightly higher than normal. These parameters indicate a low toxicity for chronic use of basil ethanolic extracts.

There are no clinical studies up to date, but this remains a field open for further research.

5) Indications and usage:

The main uses of basil (all aerial part or just the leaves) are mostly due to the presence of essential oil and flavonoid derivatives. The plant product is commonly used as eupeptic, stomachic, antiflatulent, antispasmodic, and antitussive. It is also used as a diuretic, mild sedative, and sleep inducer.

Externally, it is used as gargles in laryngeal disorders, but also as mouth antiseptic. Alcoholic extracts embedded in ointments may be used in the treatment of difficult healing wounds.

Some anthelmintic effect has been described for the pure essential oil, but its main use is in perfumery.

6) Precautions and side effects:

Although there are no known side effects, basil is not recommended for therapeutic use during pregnancy and lactation and should not be given to children younger than 6 years old.

7) Dosage:

A) Internal use:

Basil tea is prepared from 2.4 g plant in a cup (150–200 ml) of hot water; allow standing for 10–15 min, and then strain. For flatulence and bloating, 3–4 cups a day between meals should be recommended for a maximum of 8 days. If necessary, the treatment may be resumed after two weeks of interruption.

## 5.9 References


- Abdoly M, Farnam A, Fathiazad F, Khaki A, Khaki AA, Ibrahimi A, Afshari F, Rastgar H. Antidepressant-like activities of *Ocimum basilicum* (sweet Basil) in the forced swimming test of rats exposed to electromagnetic field (EMF). Afr J Pharmacy Pharmacol. 2012;6:211–15.

- Chalchat JC, Özcan MM. Comparative essential oil composition of flowers, leaves and stems of basil (*Ocimum basilicum* L.) used as herb. Food Chem. 2008;110:501–3.
- Chang X, Alderson PG, Wright CJ. Solar irradiance level alters the growth of basil (*Ocimum basilicum* L.) and its content of volatile oils. Environ Exp Bot. 2008;63:216–23.
- Gradinariu V, Cioanca O, Gille E, Aprotosoaie AC, Hrițcu L, Hăncianu M. The chemical profile of basil biovarieties and its implication on the biological activity. Farmacia. 2013;61:532–639.
- Gradinariu V, Cioanca O, Hritcu L, Trifan A, Gille E, Hancianu M. Comparative efficacy of *Ocimum sanctum* L. and *Ocimum basilicum* L. essential oils against amyloid beta (1–42)-induced anxiety and depression in laboratory rats. Phytochem Rev. 2015;14:567–575.


## 5.10 *Melissa officinalis* L.

### *5.10.1 (Lamiaceae)*

1) Common names:

Balm (UK), lemon balm (UK), cure all (UK), Herzkraut (Germany), Melissenblätter (Germany), Melissenkraut (Germany), Melissa (Russia), sweet balm (UK), Roiniță/Lamâiță (Romania).

**Fig. 5.4** Flowers of *Melissa officinalis* L. Photograph taken in Botanical Garden, Iasi, Romania, in 2006 by Oana Cioanca

2) Photograph of the species (Fig. 5.4).
3) Description of the species:
   A) Habitat and world distribution:
      *Melissa* is a perennial herb native to the eastern Mediterranean and Western Asia region and is cultivated in Central, Eastern, and Western Europe and the USA. Lemon balm grows in clumps and spreads vegetatively, as well as by seed. In temperate zones, the stems of the plant die off at the start of the winter, but shoot up again in spring. In the wild it prefers deserted areas, near fences, or ditches.
   B) Morphology of the plant:
      As an herb, lemon balm grows up to 60–120 cm, the stem is glabrous on the lower part and hairy at the top. Its leaves are opposite, petiolate, ovate to obovate-oval, base cuneate or cordate at the base, and densely pilose on both surfaces. The flowers are grouped in axillary pseudoverticilate with five to six flowers showing bilabial corolla, white to yellowish white, or lilac. The fruits are ovoid-oblong elongated with chestnut color.
   C) Medicinal parts:
      For therapeutic purposes, one may use balm leaves, *Melissae folium*, and the essential oil, *Melissae aetheroleum*, isolated by steam distillation.
   D) Chemical constituents:
      The main constituents of lemon balm leaves are the hydroxycinnamic acids (up to 6 % rosmarinic acid), and the essential oil composed mostly of monoterpenes (over 60 %) and sesquiterpenes (35 %).

| Chemical class | Individual components | Plant part/organ |
|---|---|---|
| Essential oil (0.06–0.80 %) | 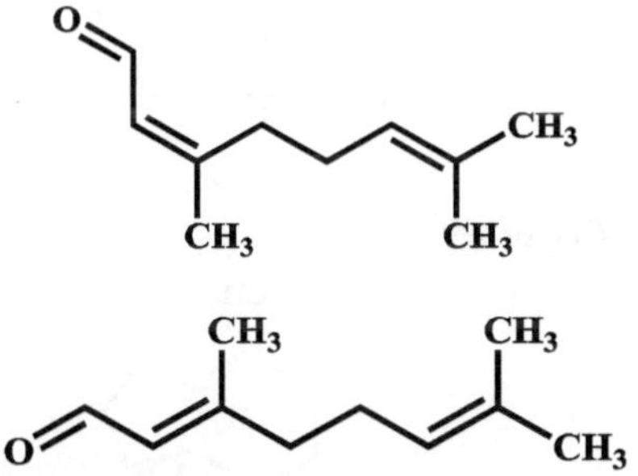<br>Citral (a mixture of the isomers neral and geranial)<br>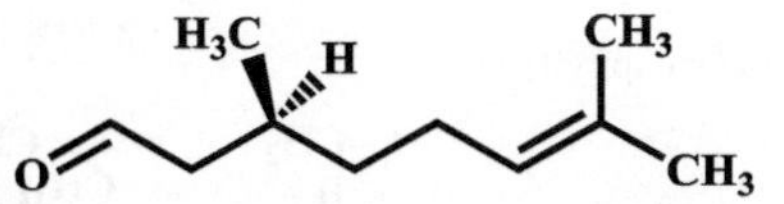<br>Citronellal | Leaves |

(continued)

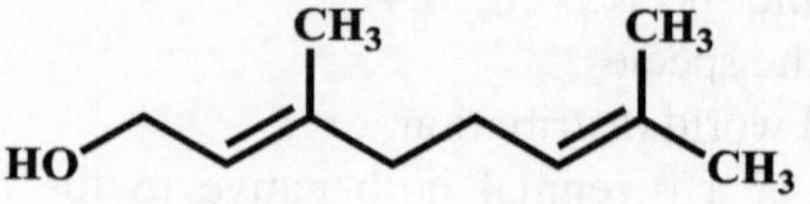
Geraniol

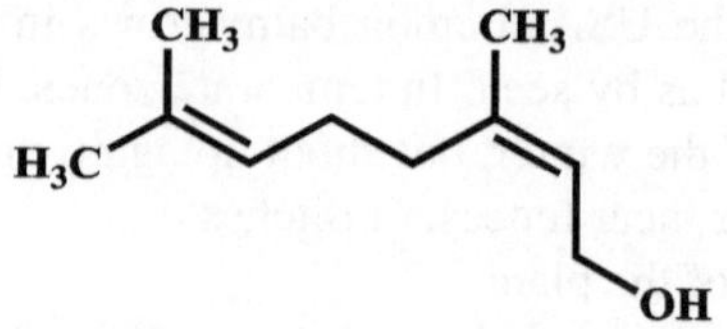
Nerol

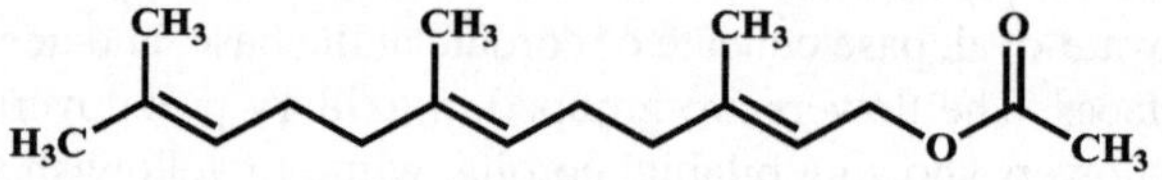
Farnesyl acetate

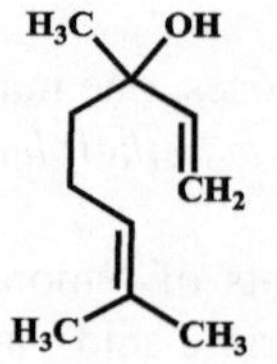
Linalool

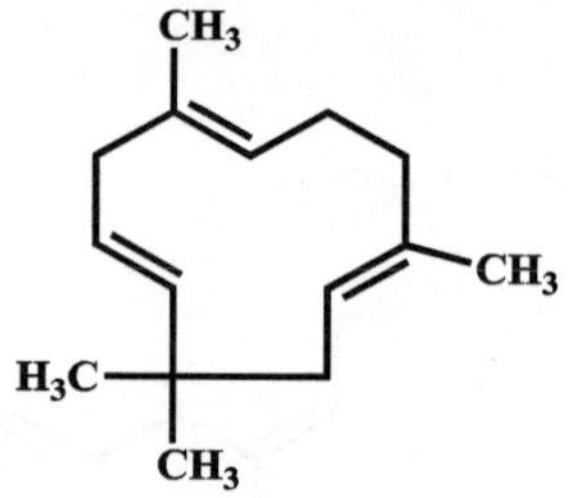
α-Caryophyllene

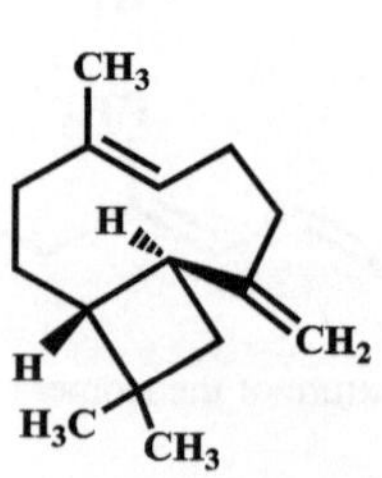
β-Caryophyllene

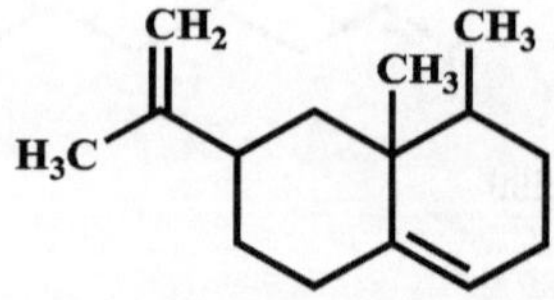
Eremophilene

(continued)

| | | |
|---|---|---|
| Hydroxycinnamic acids | 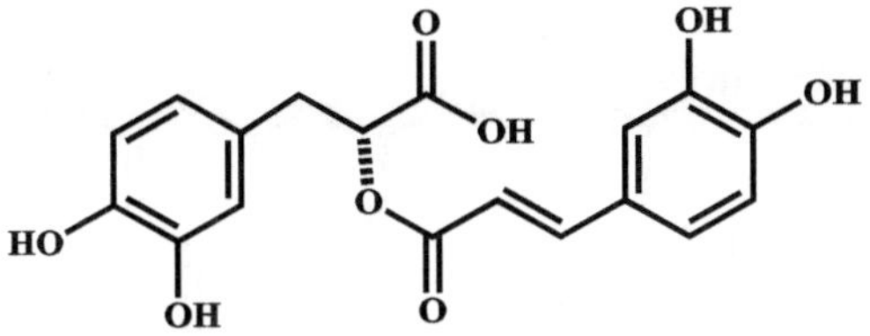 Rosmarinic acid and *p*-coumaric, caffeic, and chlorogenic acids | Leaves |
| Flavonoids | 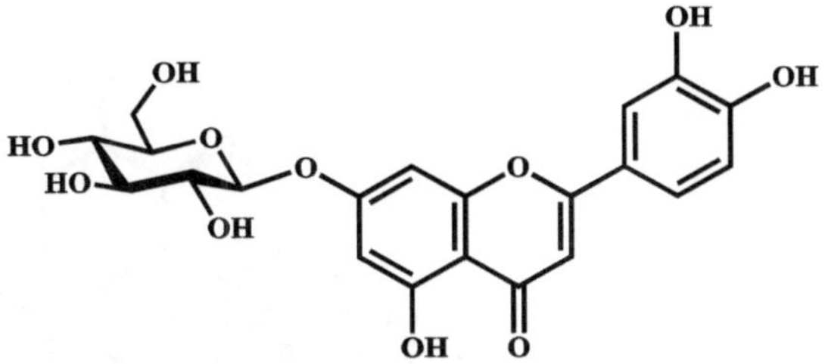 Cynaroside 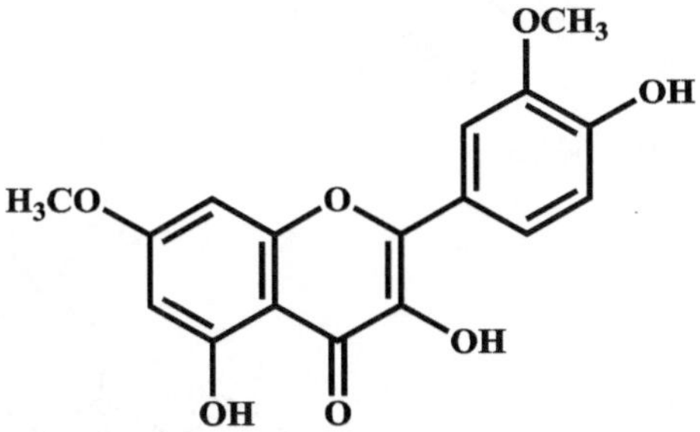 Rhamnazin 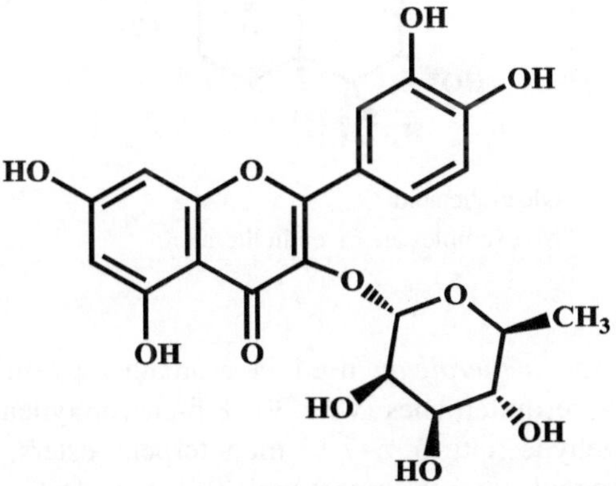 Quercitrin | Leaves |

(continued)

| | | |
|---|---|---|
| | 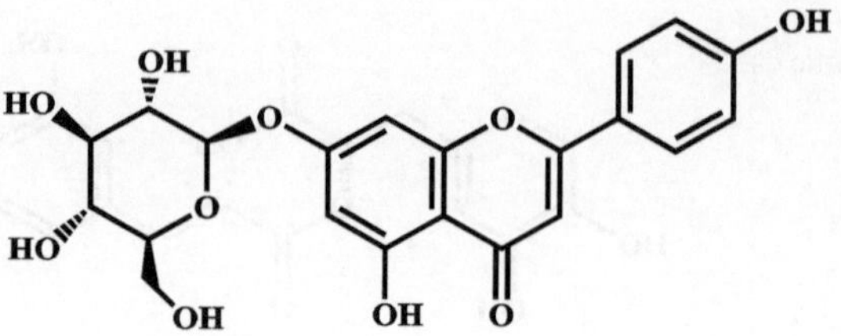<br>Apigenin-7-*O*-glucoside | |
| Tannins | | Leaves |
| Triterpenes | 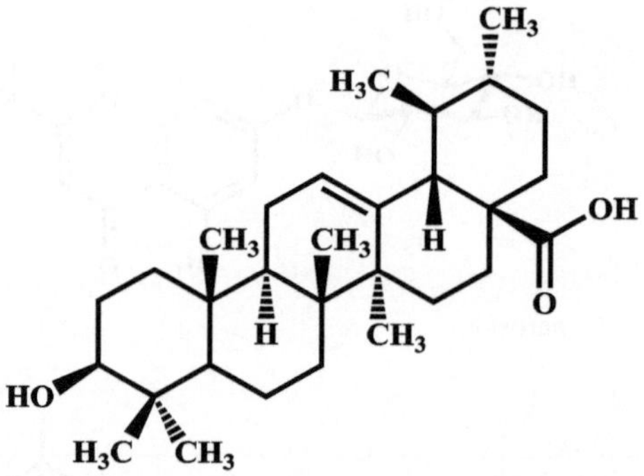<br>Ursolic acid<br>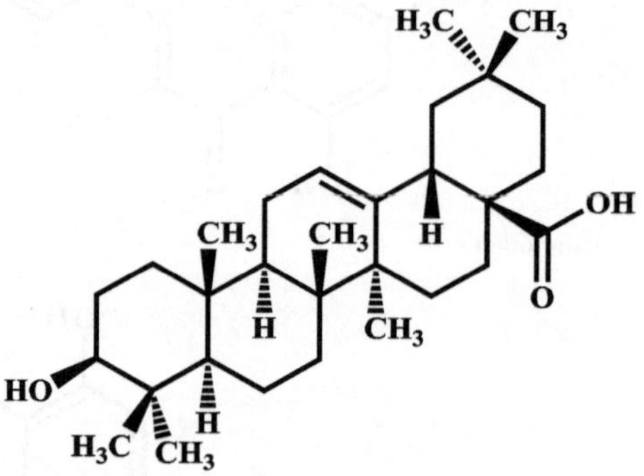<br>Oleanolic acid | Leaves |
| Resins | No examples are given in literature | Leaves |
| Pectin | | Leaves |

*Melissae aetheroleum* used in aromatherapy must have a content of 40–60 % sesquiterpenes (with 30 % β-caryophyllene), 25–55 % monoterpene aldehyde (citral), 5–7 % monoterpene esters, monoterpene ketones, monoterpenols, and sesquiterpenols (less than 15 %).

4) Pharmacology and bioactivities:
   A) In vitro:
      There are no in vitro studies that confirm the anxiolytic and antidepressant effects.
   B) In vivo:
      *Melissa* sedative activity is determined primarily by monoterpene components of the essential oil with impact on the limbic system (Yoo et al. 2011). There is a high probability that the occurrence of sedative and psychodynamic mechanisms induced by olfactory characteristics of the product plays an important role as well. For *Melissae folium* alcoholic extracts, anxiolytic effects mediated by the action of GABA transaminase inhibition have been demonstrated. These effects are dose dependent and are attributed to rosmarinic acid and oleanolic and ursolic acids (Awad et al. 2009). A clinical study with a standardized methanolic extract (7.95 rosmarinic acid, 0.86 triterpenes, and 0.69 % luteolin and hesperidin derivatives) showed significant improvement of associated anxiety symptoms and improvement of mood and induction of sleep. Anxiolytic effects do not affect circadian activity and are not accompanied by specific reactions of benzodiazepines. However, higher doses than 600 mg extract/day may increase anxiety and negatively impact the mood.

      Some pharmacological research data showed that *Melissa officinalis* extracts could improve cognitive performance, especially by reducing anxiety (Ibarra et al. 2010). The mechanism of action is based on enhancing cholinergic transmission.

      Altogether, this species anxiolytic activity is more intense than its antidepressant effect.
5) Indications and usage:
   *Melissa officinalis* is considered a good eupeptic and carminative, with spasmolytic properties; therefore, it should be used in the treatment of stomach pain and gastrointestinal disorders of psychosomatic origin (spastic colitis, irritable bowel syndrome, and dyspepsia).

   Preparations with lemon balm extracts should be administered orally as a carminative for gastrointestinal disorders and as a sedative for treatment of nervous disturbances of sleep.

   Ointments are given externally for the symptomatic treatment of mouth ulcers (herpes labialis).
6) Precautions and side effects:
   Although there are no human studies to confirm antithyroid effects, people with thyroid dysfunction should use lemon balm with caution. Due to lack of clinical data, lemon balm preparations should not be used in pregnant women and children below 12 years. Minor allergic reactions have been reported in some cases, leading to contact dermatitis. *Melissa* should not be associated with alcohol, barbiturates, or sedatives due to its own sedative effects that could induce a more intense sedation.

7) Dosage:
   A) Internal use:
      The dried leaves can be prepared as an infusion, a tincture (1:5, ethanol 45–53 % v/v), and also as dried aqueous and alcoholic extracts (45–53 %, v/v) to be included in different preparations.
      The recommended dosage in adults is 1.5–4.6 g leaves up to three times a day (infusion); alcoholic extracts (tincture) should be indicated in amounts of 2–6 ml up to three times daily. The extracts are used in doses equivalent to those for infusion dosage.
   B) External use:
      The aqueous extract (70:1) of *Melissae folium* is nowadays prepared as a cream 1 % or as a lip balm (LomaHerpan®) indicated for symptomatic treatment of herpes labialis.

## 5.11 References


- Yoo D, Choi J, Kim W, Yoo KY, Lee C, Yoon Y, et al. Effects of *Melissa officinalis* L. (lemon balm) extract on neurogenesis associated with serum corticosterone and GABA in the mouse dentate gyrus. Neurochem Res. 2011;36:250–7.
- Awad R, Muhammad A, Durst T, Trudeau VL, Arnason JT. Bioassay-guided fractionation of lemon balm (*Melissa officinalis* L.) using an in vitro measure of GABA transaminase activity. Phytotherapy Res. 2009;23:1075–81.
- Ibarra A, Feuillere N, Roller M, Lesburgere E, Beracochea D. Effects of chronic administration of *Melissa officinalis* L. extract on anxiety-like reactivity and on circadian and exploratory activities in mice. Phytomedicine. 2010;17:397–403.


## 5.12 *Tilia* sp..

### *5.12.1 (Malvaceae)*

1) Common names:
   Lime (UK), lime tree (UK), linden (UK), lind (UK), Linde (Germany), Tilleul (France), Tiglio (Italy), Tei (Romania).
2) Photograph of the species (Fig. 5.5).
3) Description of the species:
   A) Habitat and world distribution:
      *Tilia genus* comprises around 45 species, all deciduous trees, native to the Northern temperate regions. It is spread from Europe (*T. cordata* Mill., *T. tomentosa* Moench, *T. platyphyllos* Scop., *Tilia x europaea* L.) to North America and China. Currently, the largest suppliers of the vegetal product are China, Bulgaria, Poland, Russia, Romania, and Turkey.
      The trees grow well in full sun to part shade moist, fertile, and well-drained loams, but adapt easily to different soil conditions. The maintenance is low and is preferred in urban conditions due its high tolerance and adaptability.

**Fig. 5.5** Flowers of *Tilia* sp. Photograph taken in Prague, Czech Republic, in 2014 by Oana Cioanca

B) Morphology of the plant:

The trees are high up to 40 m, a broad-columnar crown; the branches divide into numerous ramifications covered with large, heart-shaped and asymmetrical leaves. On the inferior epidermis, there are groups of surface trichrome that are grouped near the veins: reddish-brown for *T. cordata* Mill. and smooth silvery-white for *T. platyphyllos* Scop. Fragrant pale yellow pentamerous are borne in drooping cymes, and the peduncles are partly united to a leaflike elliptical shape bract.

C) Medicinal parts:

The vegetal products used in therapy, *Tiliae flos* and/or *Tiliae flos cum bracteis*, are represented by a mixture of the flowers without or with bracts of *T. cordata* Mill., *T. tomentosa* Moench, and *T. platyphyllos* Scop. When fresh, the odor is pleasant; when dry, faint. The taste is sweet-mucilaginous.

D) Chemical constituents:

Linden flowers are rich in flavones and essential oil, whereas bracts main compounds are represented by mucilage.

| Chemical class | Individual components | Plant part/ organ |
|---|---|---|
| Essential oil (0.02–0.1 %) | 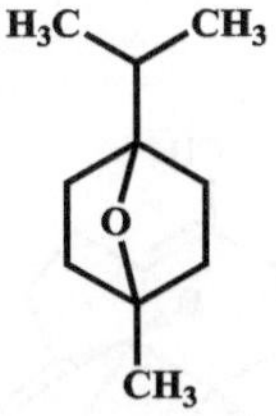  1,8-Cineole | Flowers |

(continued)

| | | |
|---|---|---|
| | 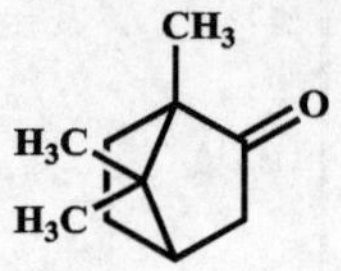<br>Camphor | |
| | 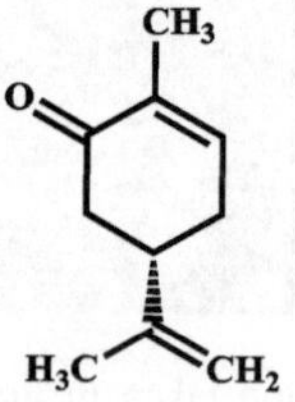<br>Carvone | |
| | 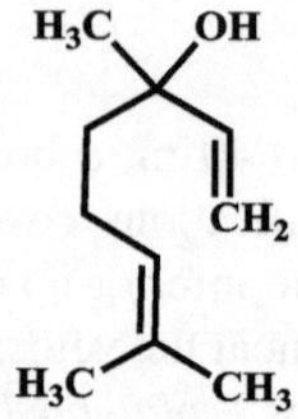<br>Linalool | |
| | 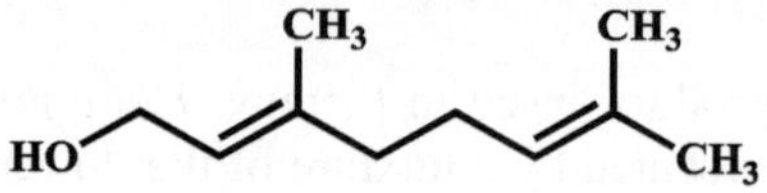<br>Geraniol | |
| | 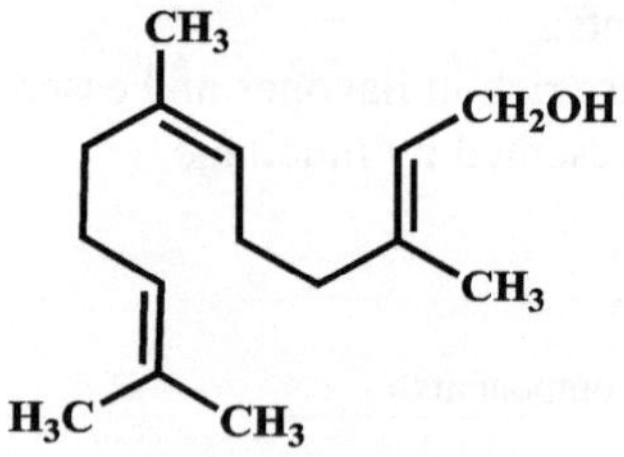<br>Farnesol | |
| | 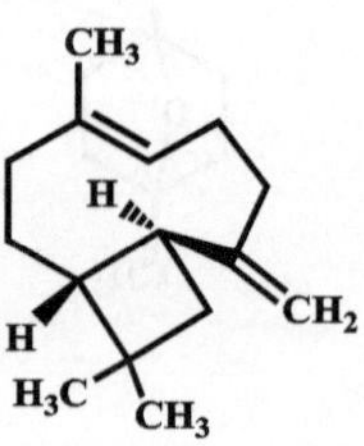<br>β-Caryophyllene | |

(continued)

| | | |
|---|---|---|
| Flavonoids (1 %) | 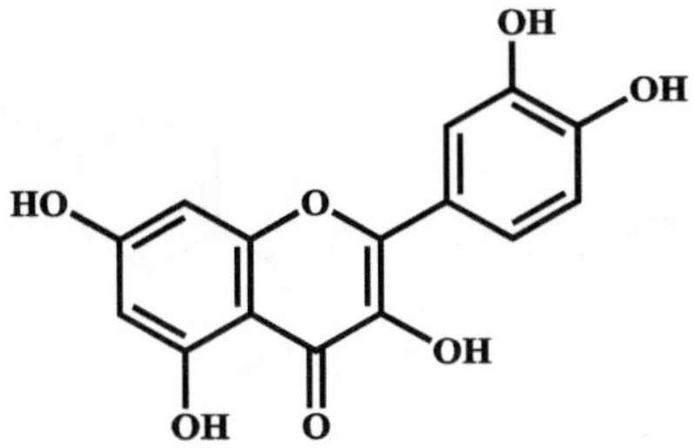<br>Quercetin<br>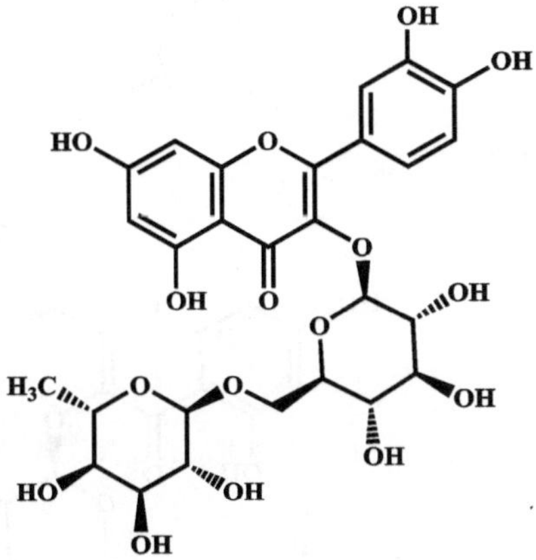<br>Rutin<br>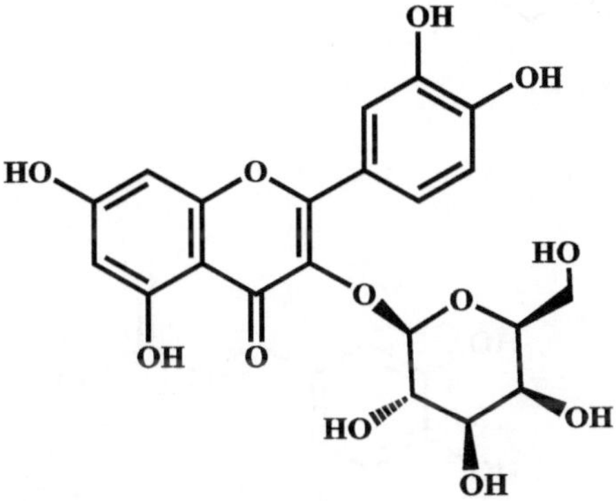<br>Hyperoside<br>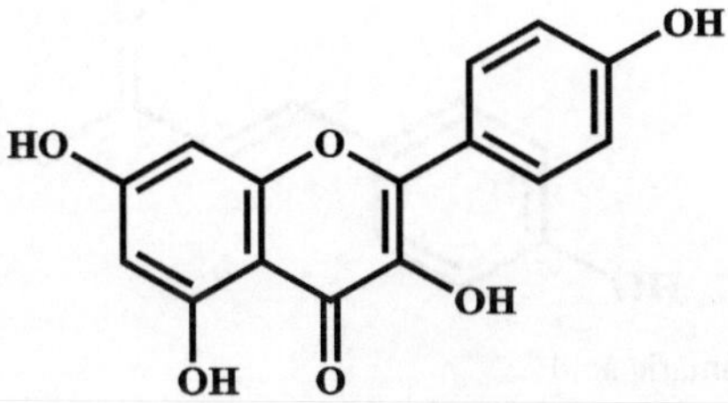<br>Kaempferol | Flowers |

(continued)

| | | |
|---|---|---|
| | 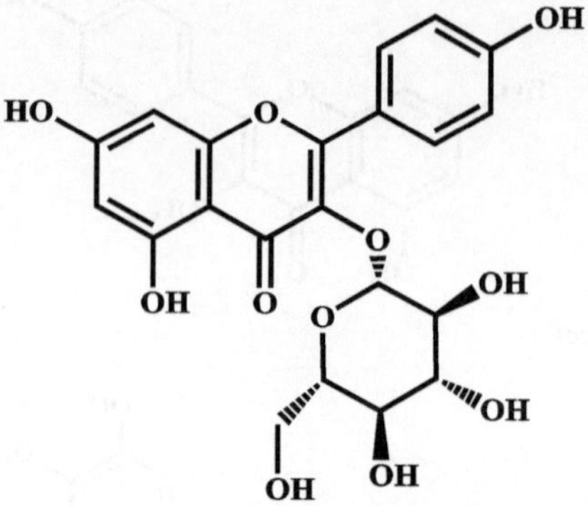<br>Astragalin<br>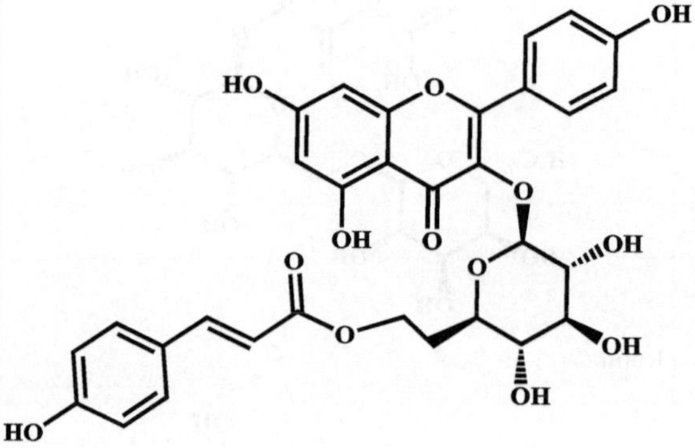<br>Tiliroside | |
| Mucilaginous polysaccharides (3 %) | | Bracts |
| Phenolic acids | 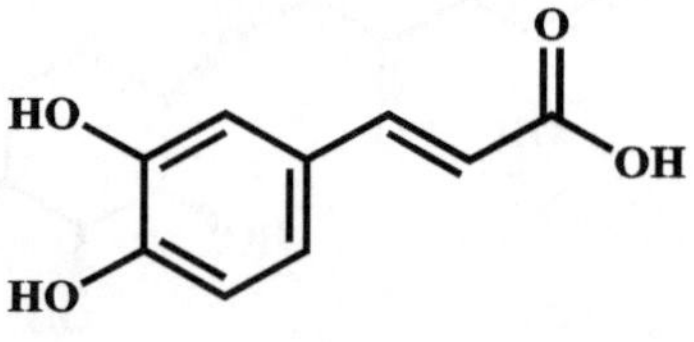<br>Caffeic acid<br>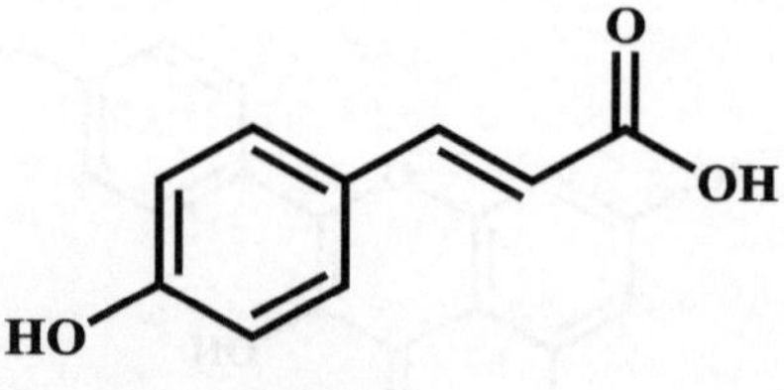<br>*p*-Coumaric acid | Flowers and bracts |

(continued)

| | | |
|---|---|---|
| | 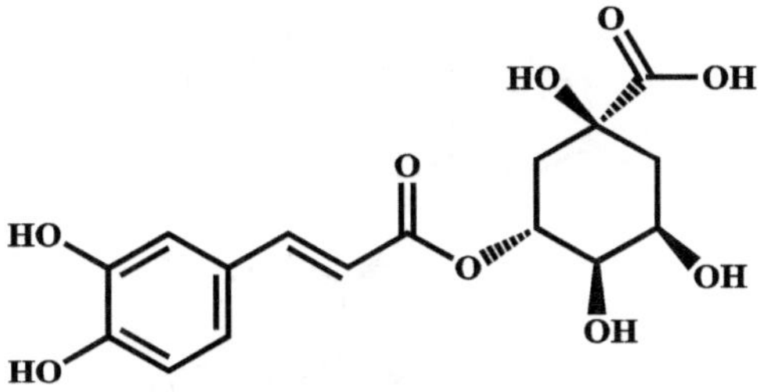<br>Chlorogenic acid<br>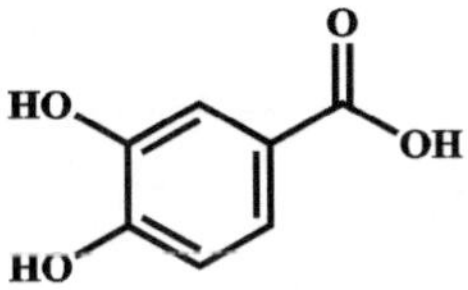<br>Protocatechuic acid | |
| Mixt tannins (2 %) | 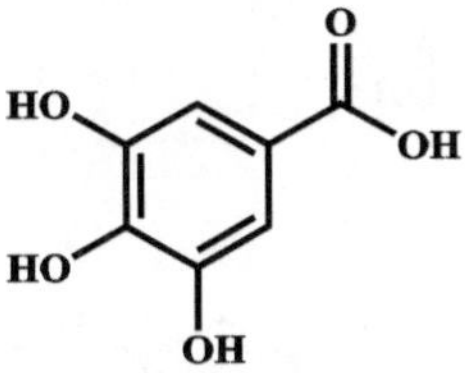<br>Tannins based on Gallic acid<br>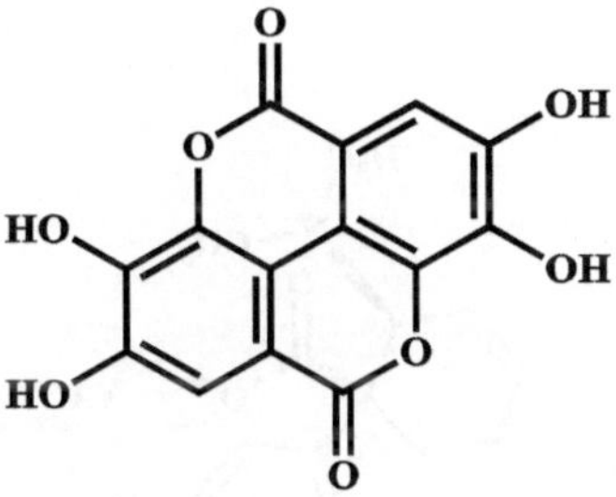<br>Tannins based on Ellagic acid | Flowers and bracts |

(continued)

| | | |
|---|---|---|
| Coumarins | 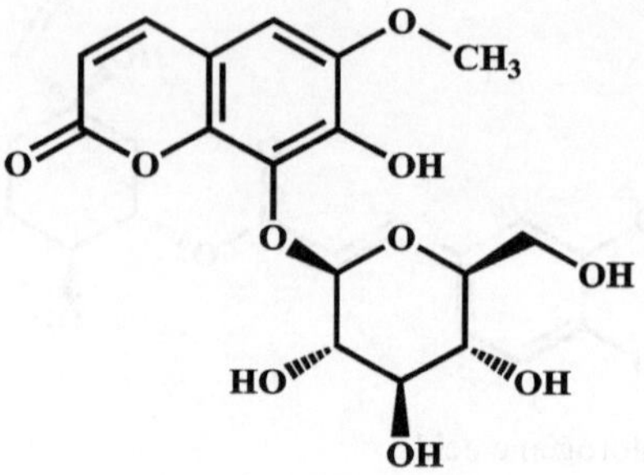<br>Fraxin<br>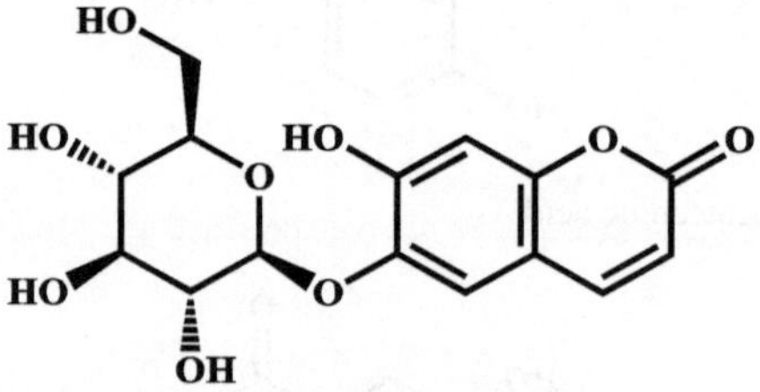<br>Esculin<br>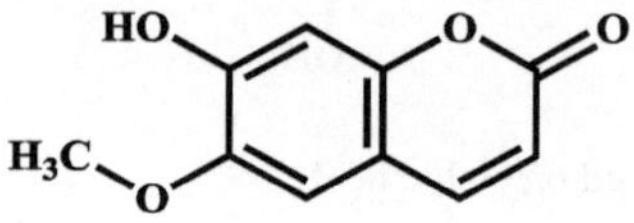<br>Scopoletin | Flowers and bracts |
| Triterpene | 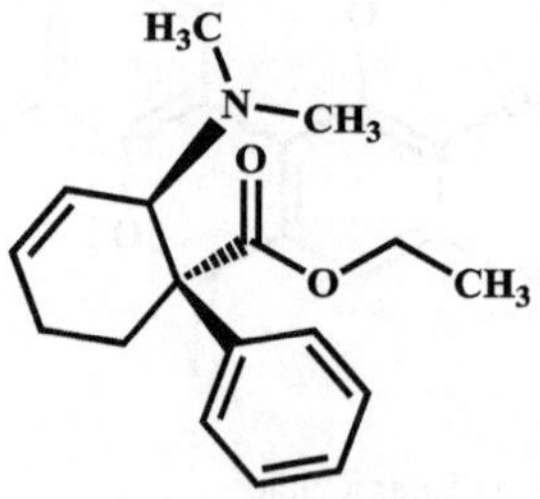<br>Tilidine | Flowers and bracts |

(continued)

| | | |
|---|---|---|
| Sterols | 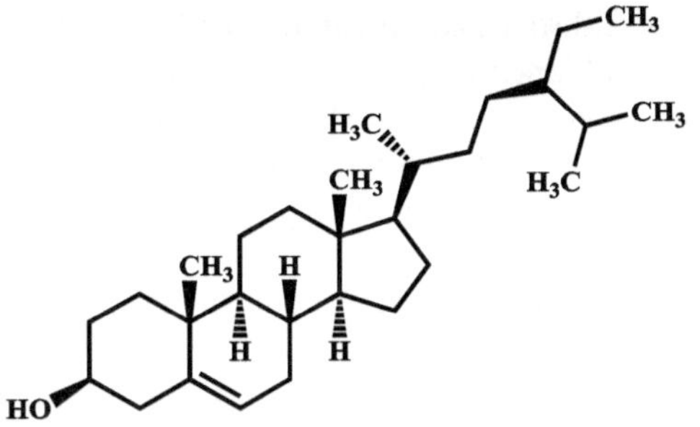<br>Beta-sitosterol<br>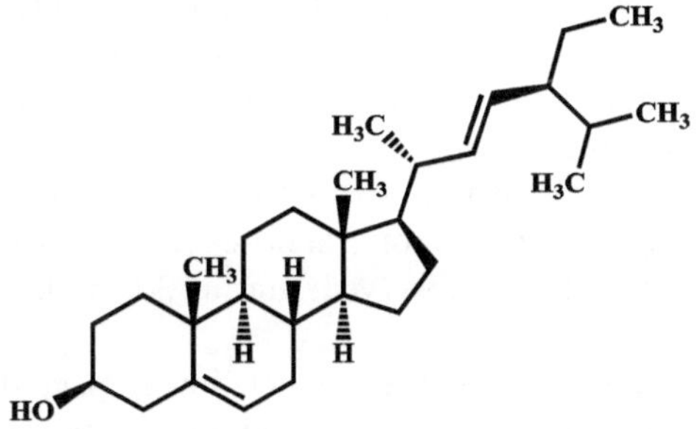<br>Stigmasterol | Flowers and bracts |

4) Pharmacology and bioactivities:

*Tilia* species are recognized worldwide for their soothing and calming effects from traditional medicine uses. But at the moment, the mechanism of action of linden flowers and its active principles are still unknown and scientifically unexplained. To date farnesol, an alcohol sesquiterpenes found in the essential oil, is indicated as the responsible compound for the anxiolytic effect, along with the mild sedative properties.

By calming psychomotor excitement, the phyto-tranquilizer action is to reduce anxiety and the state of tension by balancing the emotional behavior and emotional reactions. Moreover, lime has a slight hypno-inducing effect in psychogenic neurosis and sleeplessness. Additionally, lime-based phytopreparations act gentle and exhibit an antispasmodic effect relaxing tight muscles by inducing a good quality sleep without developing dependency subsequently.

The topical use of lime extracts (compresses) has an emollient, adsorbent, and refreshing activity, attributed to mucilaginous substances. There is no scientific data to support this traditional use, but the mechanism is somewhat explained by the effect of mucilage on nerve endings reducing local inflammation. In pharyngitis (dry and sore throat), administration of linden tea increases the stimulation of cough threshold, resulting an antitussive effect.

Although lime is used in large scale in Europe, most of the studies on these species are published by South American researchers. Therefore, only the few data available in Europe are going to be presented below.

A) In vitro:

There are no in vitro studies that confirm the anxiolytic and antidepressant effects.

B) In vivo:

Most of the studies that are available describe the **anxiolytic** and **sedative** activity of different hexane and methanol extracts in animal models (Noguerón-Merino et al. 2015). The results confirmed that the highest potential corresponds to methanolic extracts and is directly proportional to the dose. The maximum activity recorded 60 min after administration was similar to diazepam. By chromatographic methods, three phytosterols were isolated from the methanolic extract of which β-sitosterol was the main compound. Administered to mice at doses ranging from 1 to 10 mg/kg i.p., an anxiolytic effect was observed, and at 30 mg/kg a sedative response appeared.

In 2004, a team of Turkish researchers investigated the anti-inflammatory and analgesic action of kempferol-3,7-*O*-α-dirhamnoside and quercetol-3,7-*O*-α-dirhamnoside isolated from the leaves of *Tilia argentea* Desf. ex DC. The results indicated that both compounds possessed powerful anti-inflammatory and antinociceptive activity at a dose of 50 mg/kg, per os, without inducing any acute toxicity and without inducing gastric problems (Toker et al. 2004).

5) Indications and usage:

The approved uses of linden flower are for colds and cold-related coughs, upper respiratory catarrh, irritable coughs, hypertension, and restlessness.

6) Precautions and side effects:

No side effects have been reported; however, the sensitivity to each preparation should be verified prior to administration.

7) Dosage:

A) Internal use:

As a remedy to relieve cough and upper respiratory problems, or to lower the fever in influenza, the infusion is prepared from 2 to 4 g of flowers in 150 ml boiling water (at least 30 min to brew). The recommended dose is at least two infusions freshly prepared to be drank as hot as possible in the second half of the day. Such remedy is especially recommended for children.

Below, we present a couple of mixtures with other medicinal plants indicated for:

– Colds

| |
|---|
| Rp. *Sambuci flos* |
| *Tiliae flos* |
| *Chamomillae flos* ad 75 g |
| M.f. *species* |

D.s. internally, 2–3 teaspoons of the mixture to 250 ml hot boiling water; three cups per day

– Depression and anxiety, gastrointestinal spasms

| | |
|---|---|
| Rp. *Tilia flos* | 5p |
| *Droserae herba* | 1p |
| *Serpylli herba* | 2p |
| *Pini strobuli* | 2p |
| M.f. infus. | |

D.s. internally, 1 spoon of grounded plants to 250 ml hot boiling water; to be administered as needed

B) External use:

Warm or cold poultices are indicated for small wound, scrapes, and minor inflammations. Infusion or hydroalcoholic extracts are indicated as mouth washes and gargles.

## 5.13 References


- Noguerón-Merino MC, Jiménez-Ferrer E, Román-Ramos R, Zamilpa A, Tortoriello J, Herrera-Ruiz M. Interactions of a standardized flavonoid fraction from *Tilia americana* with serotoninergic drugs in elevated plus maze. J Ethnopharmacol. 2015;164:319–27.
- Toker G, Küpeli E, Memisoğlu M, Yesilada E. Flavonoids with antinociceptive and anti-inflammatory activities from the leaves of *Tilia argentea* (silver linden). J Ethnopharmacol. 2004;95:393–7.


## 5.14 *Verbena officinalis* L.

### *5.14.1* (Verbenaceae)

1) Common names:

Vervain (UK), Echtes Eisenkraut (Germany), IJzerhard (Netherlands), Læge-Jernurt (Denmark), Železník lekársky (Slovakia), Rohtorautayrtti (Finland), Verbină/Sporiş (Romania).

**Fig. 5.6** Flowers of *Verbena officinalis* L. Photograph taken in Iasi County, Romania, in 2006 by Oana Cioanca

2) Photograph of the species (Fig. 5.6).
3) Description of the species:

   A) Habitat and world distribution:
      *Verbena officinalis* is a perennial herb native to Europe, but it is naturalized even in North America, Egypt, Palestine, Sudan, Ethiopia, tropical east Africa, and south Africa. It grows in ruderal limey soils; propagation is by root cuttings or seeds.
   B) Morphology of the plant:
      It grows up to a meter high, with an upright habitus. Stems are square, red-green, with very short fine hairs. It grows a few branches at the top, and only two opposite leaves underneath the branches. The leaves are lobed with rounded ends, sessile. Flowers are pink-white to pale lilac with five petals, grouped in spikes with opposite pairs near the top of each branch. The fruits are dark brown small nuts.
   C) Medicinal parts:
      *Verbenae herba* represents the dried parts (stem, leaves and flowered tops) of the herb used in therapy.
   D) Chemical constituents:
      There are some recent studies that confirm the presence of certain compounds in the aerial parts of the plant.

| Chemical class | Individual components | Plant part/organ |
|---|---|---|
| Iridoid glycosides | 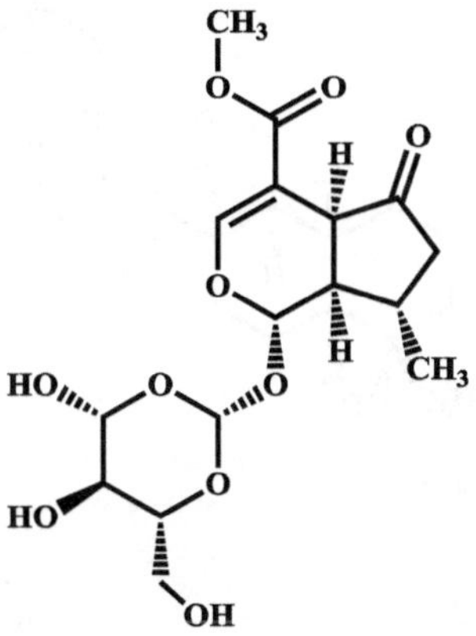 Verbenalin 1.49–2.73 % 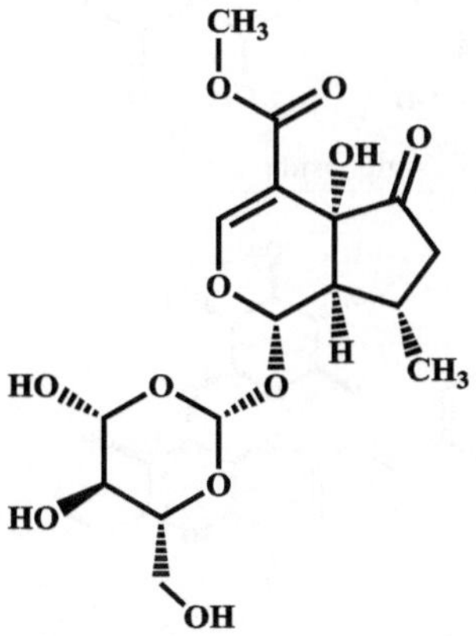 Hastatoside 0.5 % | Aerial parts |
| Triterpenes | 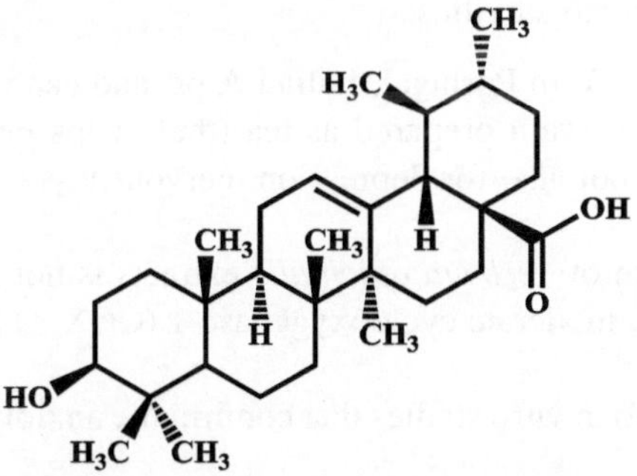 Ursolic acid | Aerial parts |

(continued)

| | | |
|---|---|---|
| | 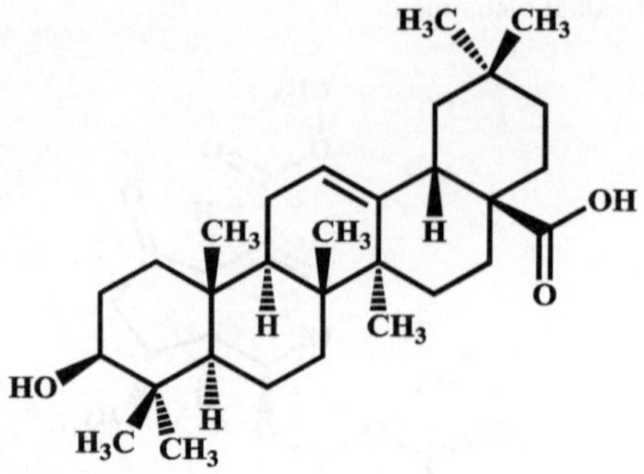 Oleanolic acid | |
| Flavonoids | 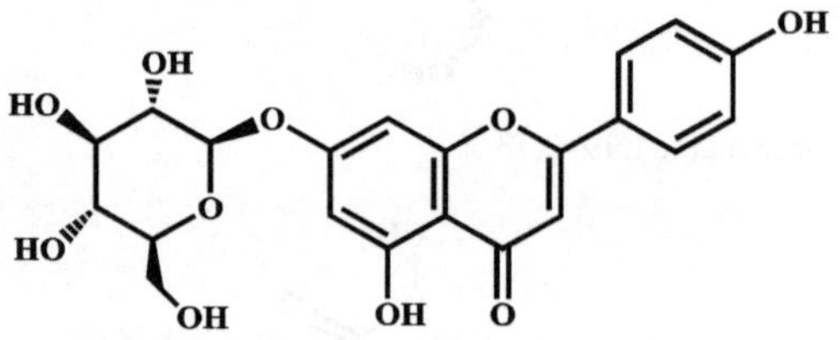 Apigenin-7-*O*-glucoside | Aerial parts |
| Phenylpropanoids | 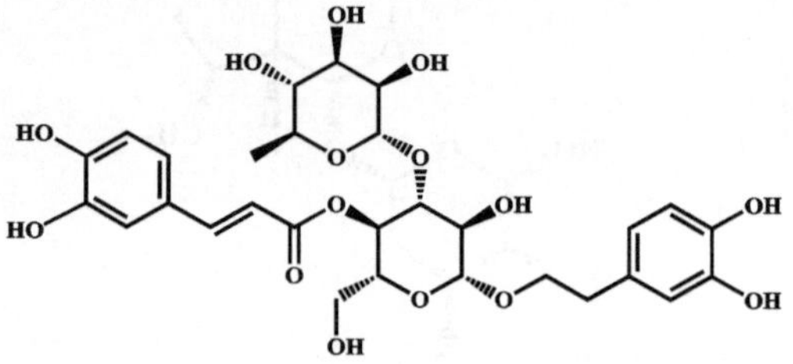 Verbascoside and isoverbascoside, eukovoside | Aerial parts |

4) Pharmacology and bioactivities:

Several surveys from Portugal, Italian Alps, and central Italy sustain the use of aerial parts of vervain prepared as tea (three cups per day) for coughs and asthma and gout, but also for depression, nervousness (Liu et al. 2012), stress, and insomnia.

The mechanism of *Verbena officinalis* extracts is not known, still, but it has been reported as a moderate cyclooxygenase-1 (COX-1) inhibitor.

A) In vitro:

There are no in vitro studies that confirm the anxiolytic and antidepressant effects.

B) In vivo:

Jawaid et al. (2015) evaluated the antidepressant activity of a methanolic extract from the leaves of *V. officinalis* in mice using the tail suspension test

(TST) and forced swimming test (FST). The results obtained suggested that the extract possessed potential antidepressant effects which could be of therapeutic interest.

5) Indications and usage:

Today, *Verbena officinalis* is used as a nervine agent (Lai et al. 2006) with sedative and anxiolytic properties; as a thymoleptic specifically for the depression and convalescence after influenza, in postnatal or postoperative depression; and as a nerve tonic in chronic diseases.

Topically, ointments and poultices could be used for wounds, throat inflammation, insect bites, and rheumatic conditions.

6) Precautions and side effects:

Due to insufficient data (only one study), vervain should not be used during pregnancy.

In vitro studies showed that vervain inhibits nonheme iron absorption; therefore, it should not be administered with food or drinks.

7) Dosage:

A) Internal use:

2–4 g three times a day of dried herb should be prepared as an infusion and drunk.

The plant can be used as a tonic for the nerves in combination with *Matricaria chamomilla*, *Tilia* sp., *Mentha piperita*, and *Lavandula angustifolia*.

B) External use:

Ointments are to be applied for 3 days and continue after 3 remission days.

## 5.15 References


- Liu Z, Xu Z, Zhou H, Cao G, Cong XD, ZhangY, et al. Simultaneous determination of four bioactive compounds in *Verbena officinalis* L. by using high-performance liquid chromatography. Pharmacognosy Mag. 2012;8:162–5.
- Lai SW, Yu MS, Yuen WH, Chang RCC. Novel neuroprotective effects of the aqueous extracts from *Verbena officinalis* Linn. Neuropharmacol. 2006;50: 641–50.
- Jawaid T, Imam SA, Kamal M. Antidepressant activity of methanolic extract of *Verbena officinalis* Linn. plant in mice. Asian J Pharm Clin Res. 2015;8:308–10.

# Chapter 6
# Medicinal Plants from Countries of Former USSR Used for Treatment of Depression

**Nazim Mamedov and Nilufar Mamadalieva**

**Abstract** Flora of ex-Soviet Union countries contains more than 17,600 higher plants species, and 260 of them are used in conventional medicine. Twenty five plant species are used conventionally and traditionally for the treatment of depression. Here, we summarize the information on the antidepressant effects of *Hypericum perforatum* L., *Eleutherococcus senticosus* Maxim, *Leonurus cardiaca* L., *Leonurus turkestanicus* V.I. Kreczetovicz & Kuprianova, *Nepeta cataria* L., *Oplopanax elatus* Nakai, *Patrinia intermedia* (Hornem.) Roem et Schult., and *Rhaponticum carthamoides* (Willd.) Iljin.

**Keywords** Depression • Antidepressant drugs • Medicinal plants • Distribution • Chemical constituents • Pharmacological activities

## 6.1 Introduction

Flora of ex-USSR contains 17,600 plus species of higher plants; some 2500 of them have been considered as medicinal plants and used in traditional medicines. The official medical system of the former Soviet Union classified 260 species as medicinal plants for use as galenicals and for use as supplements in combination with synthetic compounds (Mamedov and Craker, 1997).

There are a number of medicinal plants and compositions that possess antidepressant activity clinically comparable to the synthetic antidepressants. About 25 medicinal plants are used in traditional and conventional medicine for treatment of depression (Komarov 1934–1960, Lovkova et al. 1989, Sokolov and Zamotayev 1984, ). Most popular among them are *Hypericum perforatum* L., *Eleutherococcus*

N. Mamedov (✉)
Stockbridge School of Agriculture, University of Massachusetts at Amherst, Amherst, MA, USA
e-mail: mamedov@cas.umass.edu

N. Mamadalieva
Institute of the Chemistry of Plant Substances of Academy of Sciences of Uzbekistan, Tashkent, Uzbekistan
e-mail: nmamadalieva@yahoo.com



C. Grosso (ed.), *Herbal Medicine in Depression*,
DOI 10.1007/978-3-319-14021-6_6

*senticosus* Maxim, *Leonurus cardiaca* L., *Leonurus turkestanicus* V.I. Kreczetovicz and Kuprianova, *Nepeta cataria* L., *Oplopanax elatus* Nakai, *Passiflora incarnata* L., *Patrinia intermedia* (Hornem.) Roem et Schult., *Rhaponticum carthamoides* (Willd.) Iljin, and *Valeriana officinalis* L. All of them are widely distributed in the wild flora except for *Passiflora incarnata*, which is native to South America, but cultivated in the humid subtropics of Caucasus (Georgia and Azerbaijan) and in Crimea (Ukraine). Some of these plants are found effective in experimental as well as in clinical cases of depression. However, except *Hypericum perforatum*, more detailed clinical studies are required for the plants showing in vivo antidepressant activity, so that depression can be treated effectively by use of plant-based formulations.

## 6.2 Search Methods

SciFinder, PubMed, Toxline, Handbooks, textbooks and Pharmacopoeias, E-library.ru, Scopus, CNKI (www.cnki.net), Web of Science, www.fips.ru, and the Espacenet patent databases.

## 6.3 References


- Komarov VL (ed). Flora of the USSR, Vol. 1–30 (in Russian), Moscow: Academy of Sciences of USSR; 1934–1960.
- Lovkova M, Rabinovich A, Ponomaryova S, Buzuk G, Sokolova S, Why do plants cure? (in Russian), Moscow: Nauka; 1989, p 255.
- Mamedov N, Craker L. Medicinal and aromatic plants in the former USSR. J Herbs Spices Med Plant. 1997;4: 1–2.
- Sokolov S, Zamotayev I. Directory of medicinal plants (in Russian). Moscow: Medicina. 1984; p 463.


## 6.4 *Hypericum perforatum* L.

### *6.4.1* (Hypericaceae)

1) Common names:

   English name: St. John's wort, Saint John's wort, Klamath weed, goat weed
   Russian name: Zveroboy prodyryavlennyy
   Uzbek name: Dalachoy, choycho'p, choyo't

**Fig. 6.1** *Hypericum perforatum*. Photo taken by Paulo Ventura Araújo and used with permission

Azerbaijani name: Dazi otu
Kyrgyz name: Kozonokcholuu sary chay chop
Persian name: Alaf-e Chai or Gol-e-Raee

2) Synonyms:
*Hypericum komorovii* Gorschk., *Hypericum nachitschevanicum* Grossh
3) Photograph of the plant (Fig. 6.1)
4) Description of the species:

A) Habitat and world distribution:
*Hypericum perforatum* L. (Hypericaceae, Clusiaceae, Guttiferae), commonly known as St. John's wort, is a perennial herbaceous of Eurasian origin. The plant is abundant in the wild throughout Europe, Russia, Caucasus, Iran, Central Asia, India, Mongolia, China, and Japan, growing on grasslands, meadows, and edges of forests. The plant spreads rapidly and can invade pastures, disturbed sites, roads, and sparse woods and is considered as an aggressive weed on pastures of New Zealand, Australia, and North America (Damirov et al., 1983; Hammermann and Grom, 1976).
B) Morphology of the plant:
Stems are erect, profusely branched, two-edged, and up to 1 m high. Root is slender and woody, with abundant fibrous secondary roots. The rhizomes often present giving rise to new shoots. Dark green leaves are linear-elliptic, entire, often black dotted, particularly at the margins. Rubbed between the

fingers, the small black dots will reveal a red pigment. Number of bright, transparent dots can be noted if a leaf is held against the light, which contains colourless secretions consisting of volatile oil and resin.

Two special features, the transparent dots on leaves and two raised lines down the stem, make identification of this plant easier. Inflorescence is a cymose panicle with numerous delicate golden-yellow flowers. Flowers are 3 cm in diameter; have five green sepals, five golden-yellow petals, and many stamens; and are associated to three tufts and one pistil. Flowering occurs from June to August. Fruit is a three-celled capsule with many seeds. Seeds are cylindrical with rounded to pointed ends and dark brown; the surface is pitted and 1 mm long. The seeds ripe in September to October. The golden-yellow star-like flowers make St. John's wort one of the most beautiful wild plants (Hammermann and Grom 1976,Hutchens 1991).

C) Medicinal part:
Herb

D) Chemical constituents:
From a phytochemical point of view, St. John's wort is one of the best-investigated medicinal plants. The herb contains **naphthodianthrone derivatives** (hypericin, pseudohypericin), **phloroglucinol derivatives** (hyperforin, adhyperforin, furohyperforin, oxepahyperforin) (Richard 2014, Li and Cao 2012), **flavonoid derivatives** (quercetin-3-*O*-galactoside, quercetin-3-O-rutinoside, isorhamnetin, rhamnetin, quercetin-3-*O*-glucoside, quercetin, kaempferol, myricetin, luteolin, astilbin, 6-*O*-acetyl-quercetin-3-*O*-alloside, catechin, epicatechin, cyanidin) (Butterweck et al. 2000, Naeem et al. 2010, Aybastier et al. 2013), proanthocyanidins (procyanidin B2), **biflavonoid derivatives** (biapigenin, amentoflavone) (Nahrstedt and Butterweck 1997), **xanthone derivatives** (kielcorin, 1,3,5,6-tetrahydroxyxanthone, 1,5,6-trihydroxy-3-methoxyxanthone, ferrxanthone, brasilixanthone B, neolancerin, 1,6-dihydroxy-5-methoxy-4′,5′-dihydro-4′,4′,5′-trimethylfurano-(2′,3′:3,4)-xanthone, and 4,6-dihydroxy-2,3-dimethoxyxanthone) (Crockett et al. 2011, Li et al. 2013), **phenolic acids** (protocatechuic acid, caffeic acid, chlorogenic acid, ferulic acid, *p*-hydroxybenzoic acid and vanillic acid) (Maslennikov et al. 2014), **carotenoids** and **essential oil** (2-methyl-octane, α-pinene, β-caryophyllene, caryophyllene oxide, *n*-tetradecanol, α-ylangene, germacrene-D, *n*-octane, α-selinene, *n*-dodecanol, bicyclogermacrene, spathulenol, etc.) (Hajdari et al. 2011, 2014, Sharopov et al. 2010), ***n*-alkenes/*n*-alkanols** (1-tetracosanol, 1-hexacosanol), **amino acids** (GABA), **melatonin**, **sterols** (β-sitosterol and ergosterol) (Pravdivtseva et al. 2011), **vitamins** (vitamin A, vitamin C, choline, nicotinamide, nicotinic acid) (Dostalek and Stark 2012), **pectin polysaccharides** (Zlobin et al. 2012), **heavy metals and microelements**, **fatty acids**, **and resins** (Mihaljev et al. 2014, Barnes et al. 2001, Dostalek and Stark 2012, Hostettmann and Wolfender 2005).

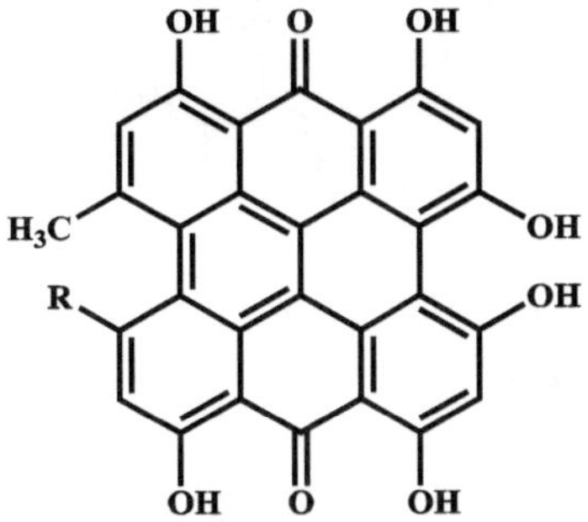

**Hypericin** R = $CH_3$

**Pseudohypericin** R = $CH_2OH$

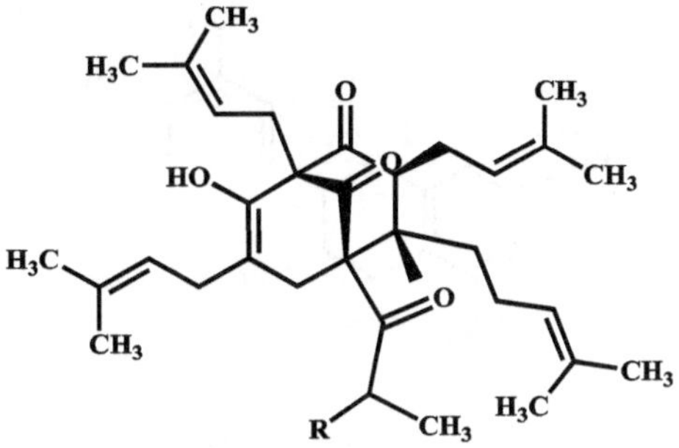

**Hyperforin** R = $CH_3$
**Adhyperforin** R = $CH_2CH_3$

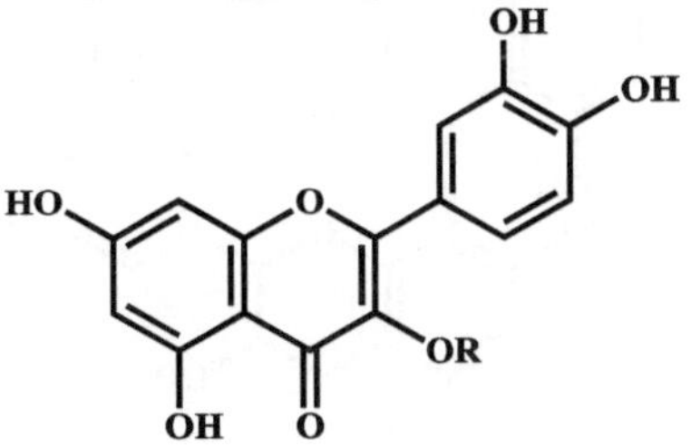

R = galactose: Quercetin-3-*O*-galactoside
R = rhamnose: quercetin-3-*O*-rhamnoside
R = glucose: Quercetin-3-*O*-glucoside
R = rutinose: Quercetin-3-*O*-rutinoside

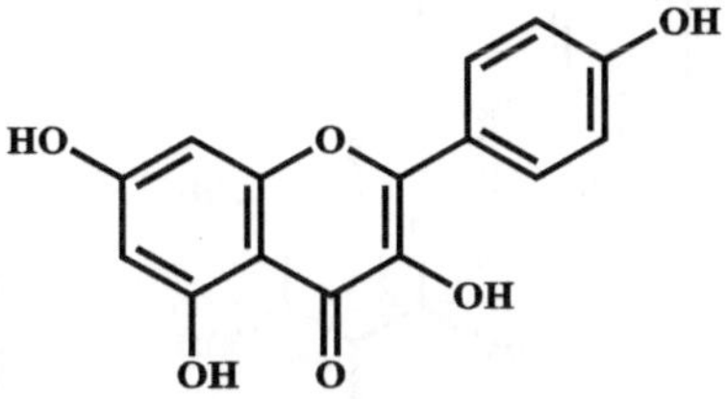

Kaempferol

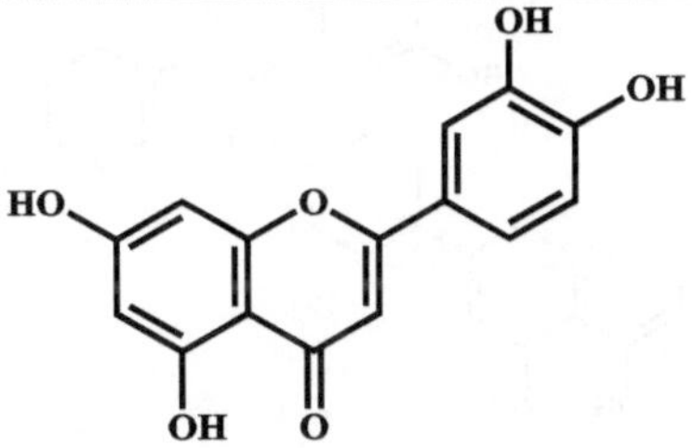

Luteolin

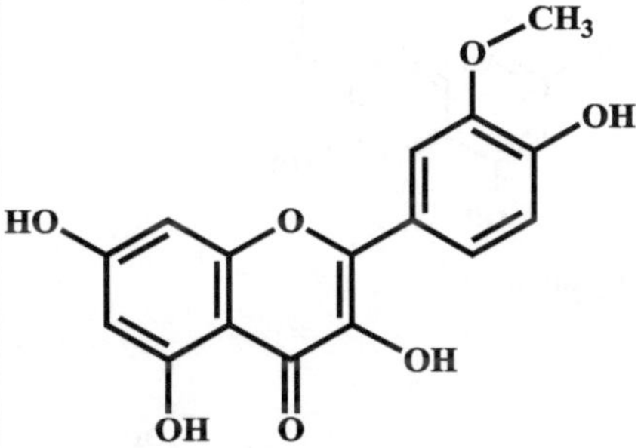

Isorhamnetin

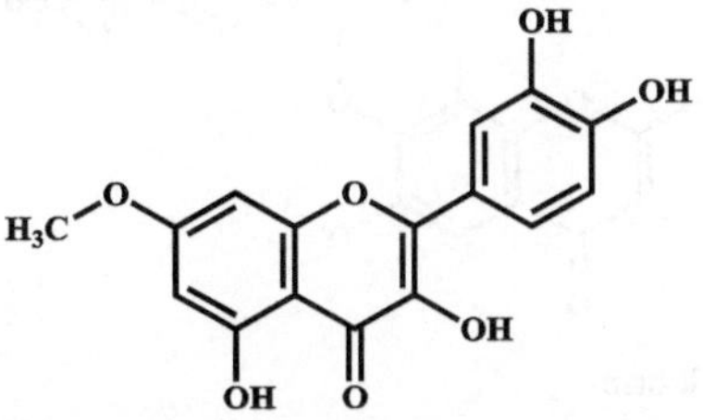

Rhamnetin

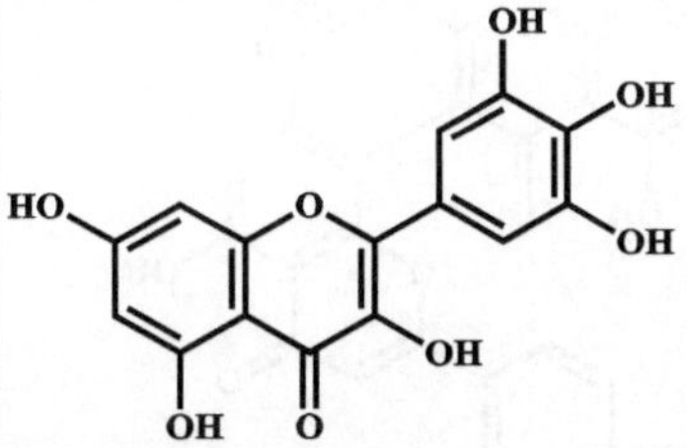

Myricetin

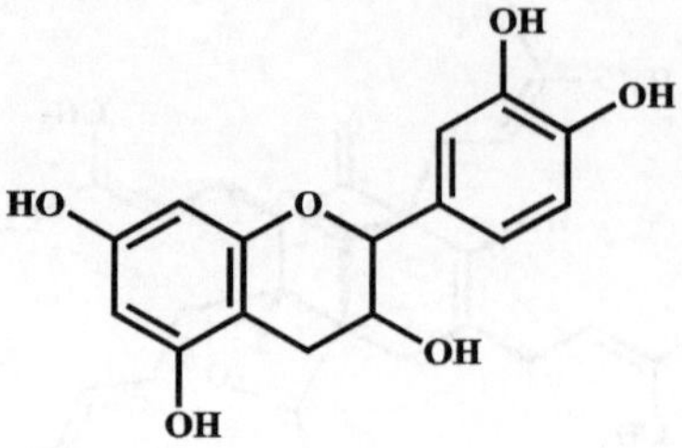

Catechin

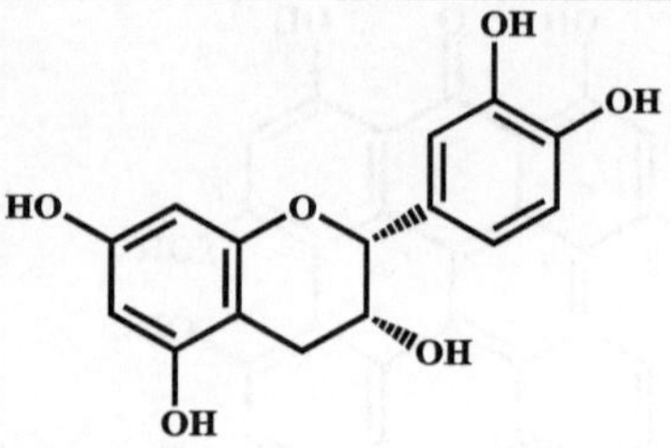

Epicatechin

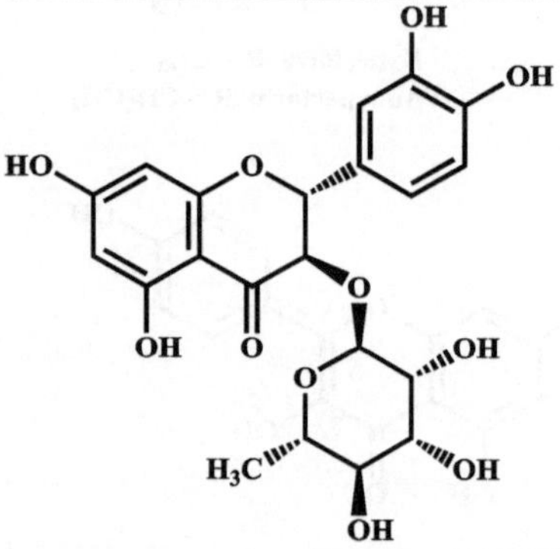

Astilbin

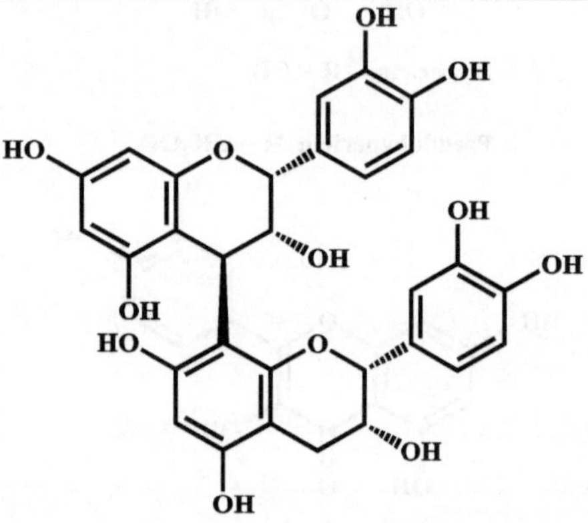

Procyanidin B2

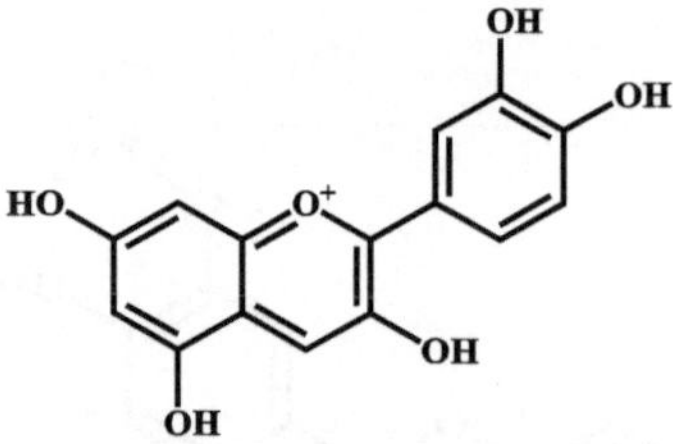

Cyanidin

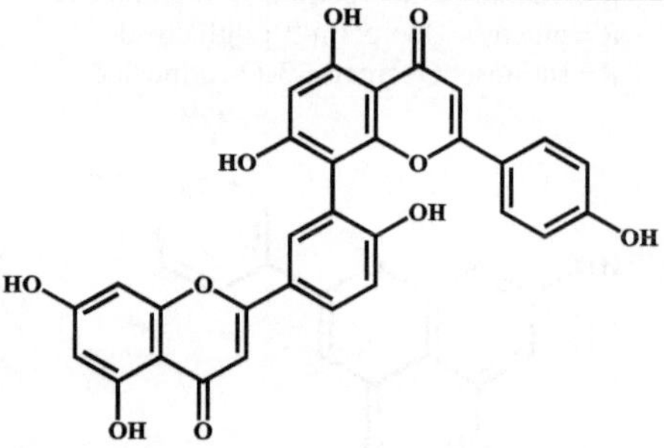

Amentoflavone

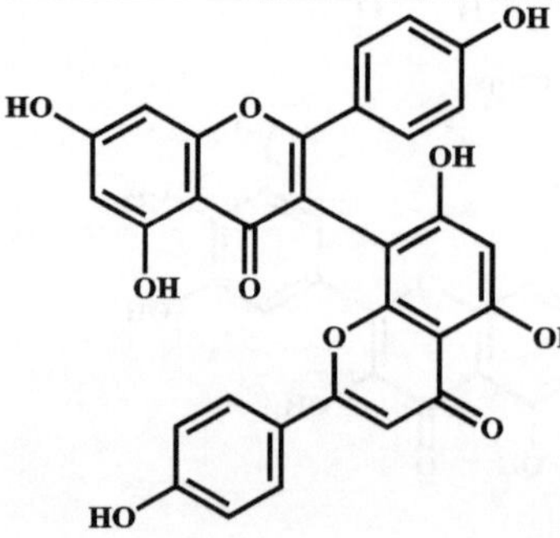

Biapigenin

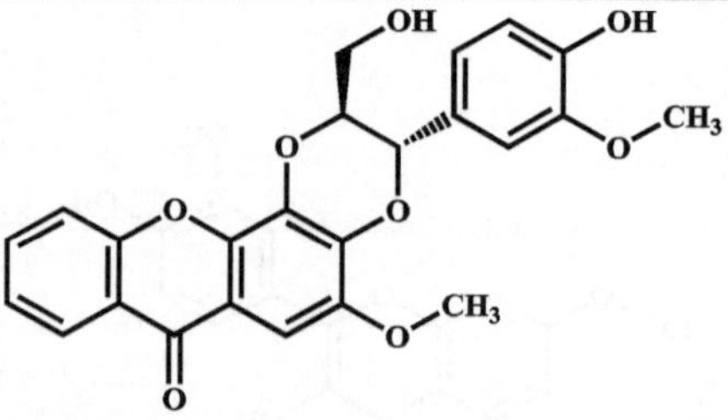

Kielcorin

| | |
|---|---|
| 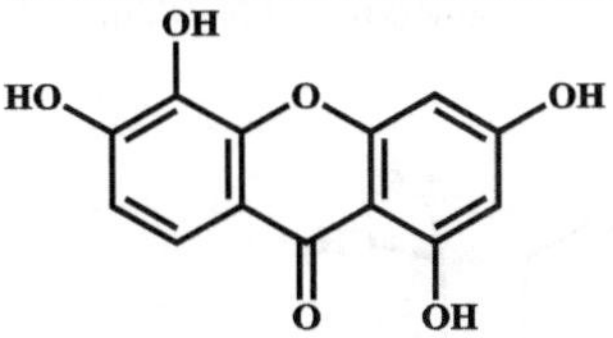 1,3,5,6-Tetrahydroxyxanthone | 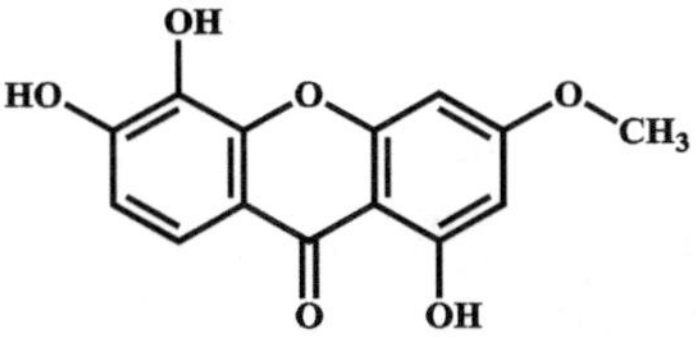 1,5,6-Trihydroxy-3-methoxyxanthone |
| 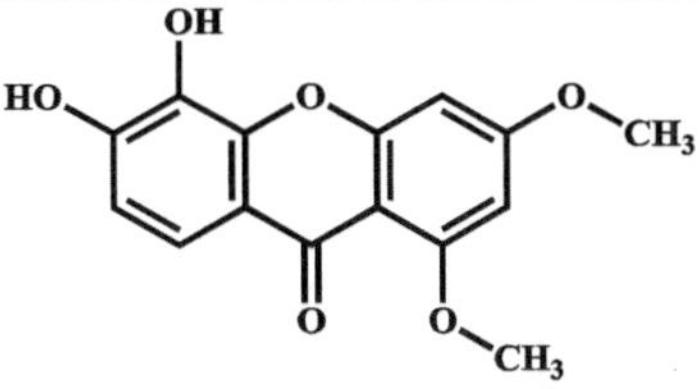 Ferrxanthone | 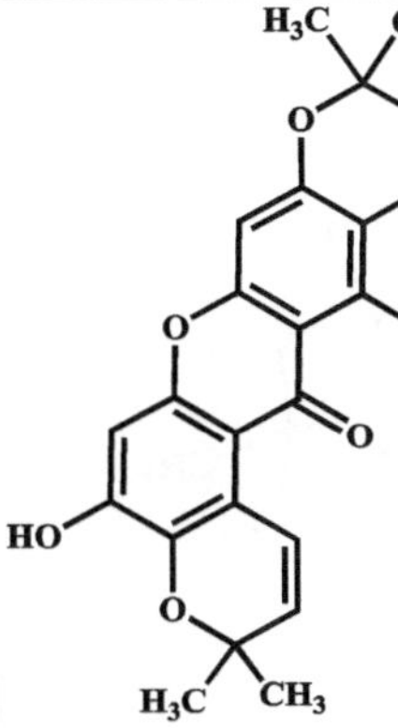 Brasilixanthone B |
| 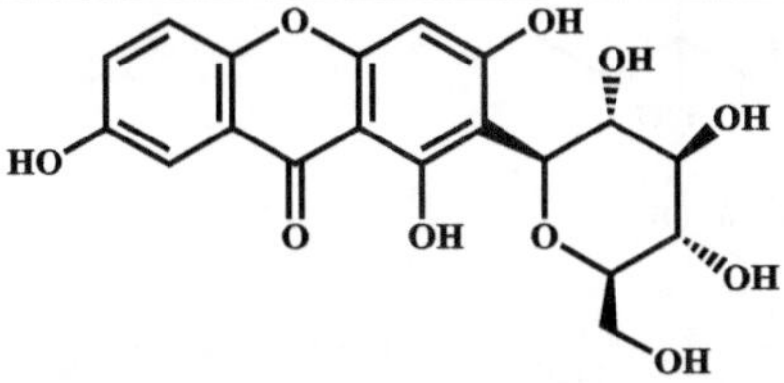 Neolancerin | 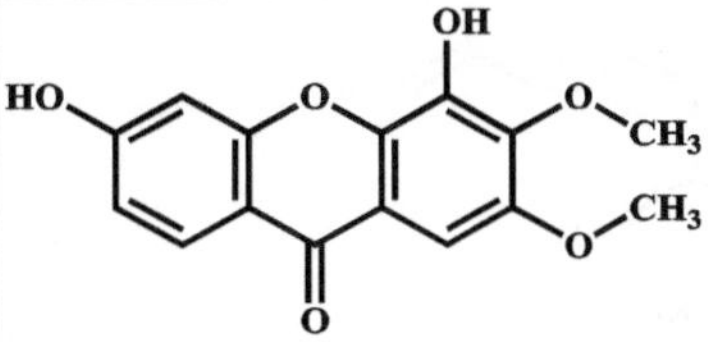 4,6-Dihydroxy-2,3-dimethoxyxanthone |
| 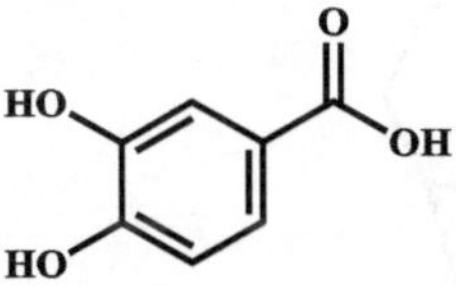 Protocatechuic acid | 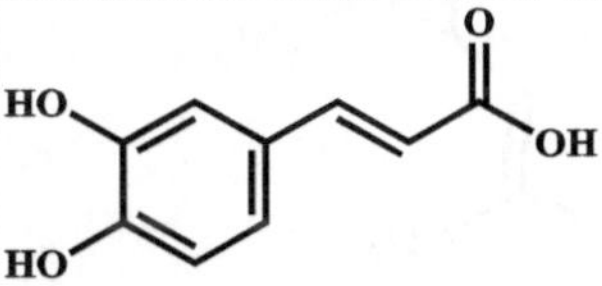 Caffeic acid |
| 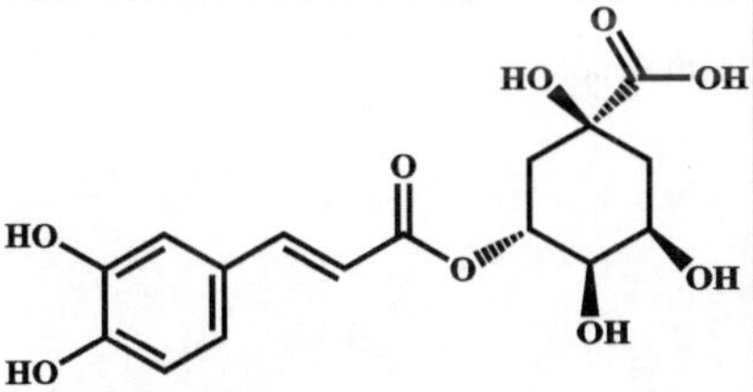 Chlorogenic acid | 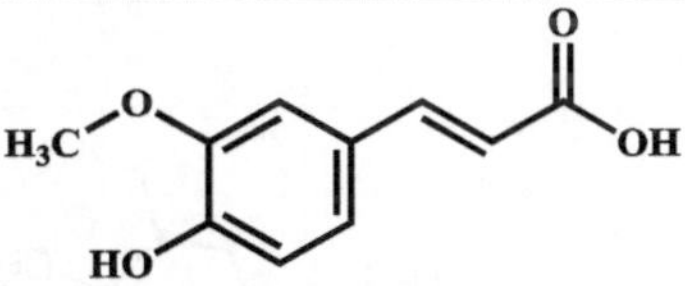 Ferulic acid |
| 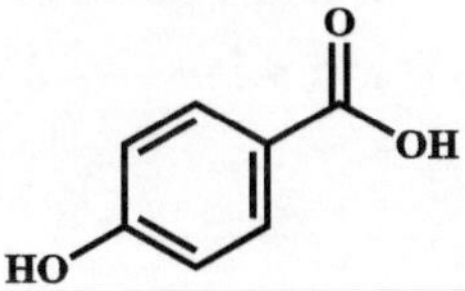 4-Hydroxybenzoic acid | 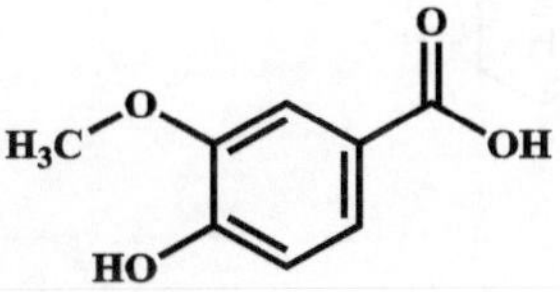 Vanillic acid |

| | |
|---|---|
| 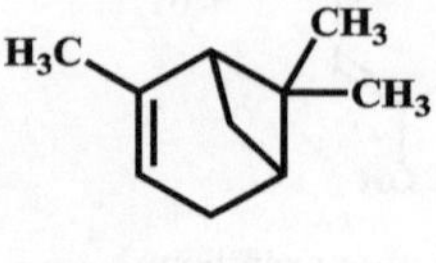 α-Pinene | 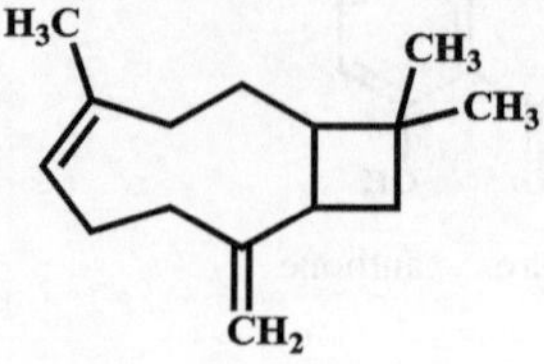 β-Caryophyllene |
| 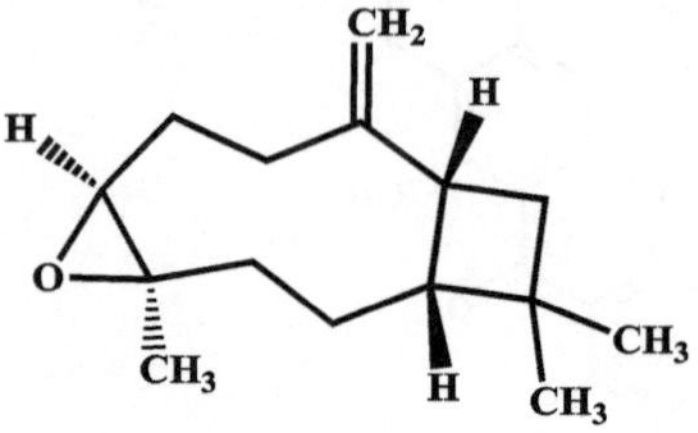 β-Caryophyllene oxide | 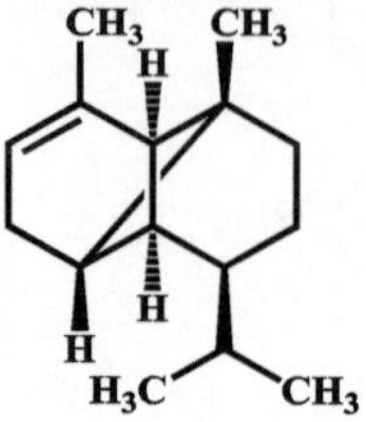 α-ylangene (=α-copaene) |
| 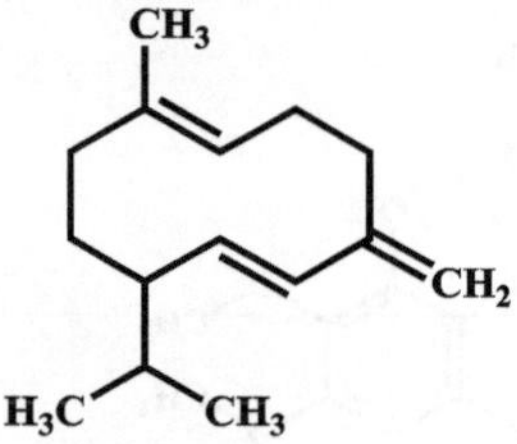 Germacrene-D | 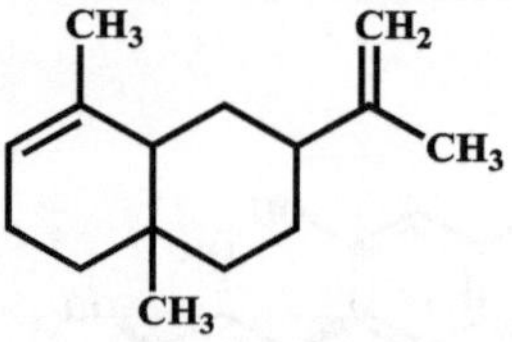 α-Selinene |
| 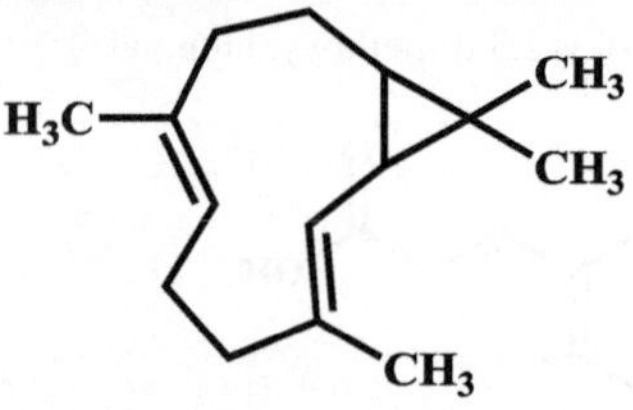 Bicyclogermacrene | 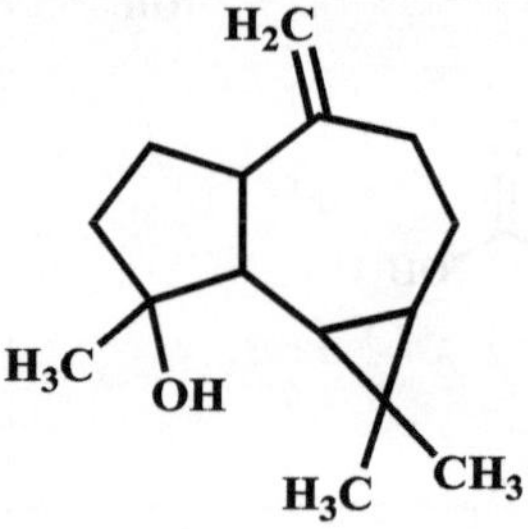 Spathulenol |
| 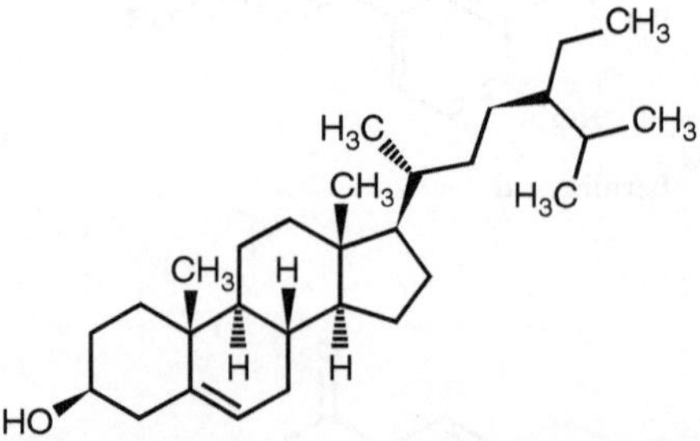 β-Sitosterol | |

5) Pharmacology and bioactivities:

A) In vitro and in vivo antidepressant effect:
The precise mechanism of action for the antidepressant effect of St. John's wort is unclear. Hypericin and pseudohypericin, compounds unique in select species of the *Hypericum* genus, were originally thought to be responsible for the antidepressant activities of *H. perforatum* extracts (Bilia et al. 2002). Recent experimental and clinical evidence showed that the phloroglucinol hyperforin, which is present in only select species of *Hypericum*, is at least partially responsible for the antidepressant activity (Bilia et al. 2002). Additionally, interactions of constituents in *H. perforatum* preparations may enhance the antidepressant bioactivities of *H. perforatum. H. perforatum* is a well-established antidepressant medicinal plant (Linde et al. 2008) in part due to the inhibition of the serotonin reuptake to the presynaptic neuron as also confirmed by the data shown by Muller (2003). *H. perforatum* extracts can be effective in the treatment of generalized anxiety disorder, somatoform disorders, sleep disorders, obsessive-compulsive disorder, psychoautonomic disturbances, depressed mood, nervousness, and seasonal affective disorder (Russo et al. 2014; Wurglics and Schubert-Zsilavecz 2006).
There are many review articles published about antidepressant effect of *H. perforatum* in the peer-reviewed journals (Barnes et al. 2001; Butterweck et al. 2000; 2003; Muller et al. 2003; Linde et al. 2008; Bilia et al. 2002, 2010; Jat 2013, Crupi et al. 2013, Gharge et al. 2009, Pilkingtona et al. 2006, Wurglics and Schubert-Zsilavecz 2006).

B) Other CNS-related activities:
St. John's wort species have shown prominent in vitro inhibition of acetylcholinesterase activity; they could be further investigated as potential substances in preventing Alzheimer's disease (Bozin et al. 2013). Hypericin may interfere with the processes of polymerization of beta-amyloid peptide responsible for the onset of Alzheimer's disease (Griffith et al. 2010).
*H. perforatum* is used in traditional medicine for its anticonvulsant property. Anticonvulsant activity of the aqueous and ethanolic extracts of *H. perforatum* aerial parts in mice has been shown (Hosseinzadeh et al. 2005).
*H. perforatum* has been functionally used as analgesic agent in traditional medicine. The analgesic effect of *H. perforatum* preparation was evaluated using chemically induced pain models in adult albino mice. The intraperitoneal (i.p.) administration of 30–100 mg/kg of the preparation produced significant analgesic effect (75 %) in acetic acid-induced writhing and formalin licking tests (Bukahri et al. 2004).
In other test, *H. perforatum*, administered orally, displayed antinociceptive activity in the tail electric stimulation and hot plate tests. The antinociceptive activity was observed with 25 mg/kg and a maximal increase in hot plate latency by 50 % (compared to 73.2 and 77.8 % increases by 5 or 10 mg/kg fluoxetine, respectively) (Abdel-Salam 2005).

The pretreatment with *H. perforatum* significantly potentiated an antinociceptive effect of morphine (0.3 mg/kg, s.c.), although concentrations of morphine in plasma and brain were not significantly changed by these herbal extracts. *H. perforatum* exert significant antinociceptive effects in the formalin test of mice (Uchida et al. 2008). The extract from an Indian variety of *H. perforatum* potentiated the analgesic activities of pentazocine and aspirin (Kumar et al. 2001).

*H. perforatum* counteracted the NO donor-induced pain hypersensitivity and meningeal activation by blocking PKC-mediated pathways involving NF-B, CREB, and STAT1 (Galeotti and Ghelardini 2013). Galeotti et al. (2014) suggested that St. John's wort seed extract may become new therapeutic perspectives for painful diabetic peripheral neuropathy. These results validate the traditional use of the plant as analgesic and other conditions associated with pain such as trauma, burns, rheumatism, and neuralgia.

Vandenbogaerde et al. (2000) reported that the total extract of *H. perforatum* increases the locomotor activity in the open field and exerts anxiolytic activity in the light–dark test, whereas the single components (protohypericin, hypericin, and pseudohypericin) did not show any effect. Interestingly, the anxiolytic activity of the total extract of *H. perforatum* was blocked by pretreatment of rats with the benzodiazepine antagonist flumazenil, hence suggesting an implication of benzodiazepine receptor activation in the anxiolytic effect of *H. perforatum* extract. Electrophysiological studies, performed to gain more information on the mechanism of action, showed that hypericin reduced the GABA-activated chloride currents, while pseudohypericin did an opposite effect. Furthermore, both hypericin and pseudohypericin inhibited the activation of NMDA receptors.

Oral administration of a standardized 50 % ethanolic extract of Indian *H. perforatum* 100 and 200 mg kg−1 once daily for 3 days in rat models of learning and memory resulted in significant attenuation of scopolamine- and sodium nitrite-induced impaired retention of active avoidance. This was comparable to that seen with intraperitoneal piracetam 500 mg $kg^{-1}$ (a nootropic agent) (Kumar et al. 2000).

6) Indications and usage:
For the treatment of anxiety, depression, cuts, and burns, as well as cancer, inflammation-related disorders, bacterial and viral diseases, and as an antioxidant and neuroprotective agent (Klemow et al. 2011).
7) Precautions and side effects:
Hypersensitivity to the active substance. Hypericum dry extract induces the activity of CYP3A4, CYP2C9, CYP2C19, and P-glycoprotein. The concomitant use of cyclosporine, tacrolimus, amprenavir, indinavir, and other protease inhibitors, irinotecan, and warfarin is contraindicated. Special care should be taken in case of concomitant use of all drug substances, the metabolism of which is influenced by CYP3A4, CYP2C9, CYP2C19, or P-glycoprotein (e.g., amitriptyline, fexofenadine, benzodiazepines, methadone, simvastatin, digoxin, finasteride), because a reduction of plasma concentrations is possible. Prior to a surgery,

possible interactions with products used during general and regional anaesthesia should be identified, and if necessary, the herbal medicinal product should be discontinued. *Hypericum* dry extract may contribute to serotonergic effects when combined with other antidepressants such as serotonin reuptake inhibitors (e.g., sertraline, paroxetine, nefazodone), buspirone (anxiolytic) or with triptans (anti-migraine drugs). During the treatment, intense UV exposure should be avoided (EMEA, 2008).

8) Dosage:

   A) Internal use:
      Oral use. According to EMEA (2008):

      – Dry extract (DER 3-7:1), extraction solvent methanol (80 % v/v); single dose, 300–600 mg; dosage frequency, 1–3 times daily; daily dose, 600–1800 mg
      – Dry extract (DER 3-6:1), extraction solvent ethanol (80 % v/v); single dose, 900 mg; dosage frequency, one single daily dose; daily dose, 900 mg
      – Dry extract (DER 2.5–8:1), extraction solvent ethanol (50–68 % v/v); 612 mg once daily or single dose, 250–650 mg; dosage frequency, 2–3 times daily; daily dose, 500–1200 mg

   B) External use:
      Topical treatment of superficial wounds, burns, and dermatitis (Sosa et al. 2007; Ute et al. 2014)

## 6.5 References


- Abdel-Salam OME. Anti-inflammatory, anti-nociceptive, and gastric effects of *Hypericum perforatum* in rats. ScientificWorldJournal. 2005;5: 585–95.
- Aybastier Onder, Sahin Saliha, Demir Cevdet. Response surface optimized ultrasonic-assisted extraction of quercetin and isolation of phenolic compounds. From *Hypericum perforatum* L. by column chromatography. Separ Sci Tech. 2013; 48(11): 1665–74.
- Barnes J, Anderson LA, Phillipson D (2001). St John's Wort (*Hypericum perforatum* L.): a review of its chemistry, pharmacology and clinical properties. J Pharm Pharmacol, 53: 583–600.
- Bilia AR, Gallori S, Vincieri FF. St. John's wort and depression: Efficacy, safety and tolerability-an update. Life Sci. 2002;70: 3077–96.
- Bilia, Anna Rita, Bergonzi, Maria Calmilla, Karioti, Anastasia, Isacchi, Benedetta. In: Awaad, Amani S, Singh VK, Govil JN, editors. St. John's wort (*Hypericum perforatum* L.). Recent Progress in Medicinal Plants. 2010;28, 235–70.
- Bozin Biljana, Kladar Nebojsa, Grujic Nevena, Anackov Goran, Samojlik Isidora, Gavaric Neda, Conic Branislava Srdenovic. Impact of origin and biological source on chemical composition, anticholinesterase and antioxidant

properties of some St. John's wort species (*Hypericum* spp., Hypericaceae) from the Central Balkans. Molecules 2013;18(10): 11733–50.

- Bukahri IA, Dar A, Alam Khan R. Antinociceptive activity of methanolic extracts of St. John's Wort (*Hypericum perforatum*) preparation. Pak J Pharmaceut Sci. 2004;17(2): 13–9.
- Butterweck V, Jurgenliemk G, Nahrstedt A, Winterhoff H. Flavonoids from *Hypericum perforatum* show antidepressant activity in the forced swimming test. Planta Med. 2000; 66:3–6.
- Butterweck Veronika. Mechanism of action of St John's Wort in depression what is known? CNS Drugs. 2003; 17 (8): 539–62.
- Crockett SL, Poller B, Tabanca N, Pferschy-Wenzig EM, Kunert O, Wedge DE, Bucar F. Bioactive xanthones from the roots of *Hypericum perforatum* (common St John's wort). J Sci Food Agric. 2011;91(3):428–34.
- Crupi Rosalia, Abdel Kareem Abusamra Yousef, Spina Edoardo, Calapai Gioacchino. Preclinical data supporting/refuting the use of *Hypericum perforatum* in the treatment of depression. CNS Neurol Disord. Drug Targets. 2013; 12(4): 474–86
- Damirov I, Prilipko L, Shukurov J, Kerimov Y. Medicinal Plants of Azerbaijan. Baku: Maarif; 1983.
- Dostalek M, Stark AK. St. John's Wort (*Hypericum perforatum* L.). In: Anzenbacher Pavel, Zanger Ulrich M, editors. Metabolism of drugs and other xenobiotics. 2012; 583–610.
- EMEA. Community herbal monograph on *Hypericum perforatum* L., herba (well-established medicinal use). 2008.
- Galeotti N, Ghelardini C. St. John's wort reversal of meningeal nociception: a natural therapeutic perspective for migraine pain. Phytomedicine. 2013;20(10): 930–8.
- Galeotti N, Maidecchi A, Mattoli L, Burico M, Ghelardini C. St. John's Wort seed and feverfew flower extracts relieve painful diabetic neuropathy in a rat model of diabetes. Fitoterapia. 2014;92:23–33.
- Gharge Deepali, Pavan Todkar, Sunil Bavaskar, Dhabale Pandurang. Hyperforin as a natural antidepressant: an overview. J Pharm Res. 2009; 2(9): 1373–5.
- Griffith TN, Varela-Nallar L, Dinamarca MC, Inestrosa NC. Neurobiological effects of Hyperforin and its potential in Alzheimer's disease therapy. Curr Med Chem 2010;17: 391–406.
- Hajdari Avni, Mustafa Behxhet, Nebija Dashnor, Kashtanjeva Albana, Widelski Jaroslaw, Glowniak Kazimierz, Novak Johannes. Essential oil composition and variability of *Hypericum perforatum* L. from wild population in Kosovo. Curr Issues Pharm Med Sci 2014; 27(1): 51–4.
- Hammermann AF, Grom II. Wild growing medicinal plants of the USSR. Moscow: Medicina; 1976.
- Hosseinzadeh H, Karimi GR, Rakhshanizadeh M. Anticonvulsant effect of *Hypericum perforatum*: role of nitric oxide. J Ethnopharmacol. 2005;98(1-2): 207–8.

- Hostettmann K, Wolfender JL. Phytochemistry. In: Müller WE editor. St. John's Wort and its active principles in depression and anxiety. Basel: Birkhäuser. 2005; 5–20.
- Hosni K, Msaada K, Ben Taarit M, Chahed T, Marzouk B. Volatile constituents of two *Hypericum* species from Tunisia. Nat Prod Commun. 2011; 6(11): 1731–4.
- Hutchens AR. Indian herbalogy of North America. Boston: Shambhala; 1991.
- Jat Lala Ram. Hyperforin: a potent anti-depressant natural drug. Int J Pharm Pharmaceut Sci. 2013; 5(Suppl 3): 9–13.
- Klemow KM, Bartlow A, Crawford J, Kocher N, Shah J, Ritsick M. Chapter 11: medical attributes of St. John's Wort (*Hypericum perforatum*). In: Benzie IFF, Wachtel-Galor S, editors. Herbal medicine: biomolecular and clinical aspects. 2nd edn. Boca Raton, FL: CRC Press; 2011.
- Kumar V, Jaiswal AK, Singh PN, Bhattacharya SK. Anxiolytic activity of Indian *Hypericum perforatum* Linn: an experimental study. Indian J Exp Biol. 2000; 38(1):36–41.
- Kumar V, Singh PN, Bhattacharya SK. Anti-inflammatory and analgesic activity of Indian *Hypericum perforatum* L. Indian J Exp Biol. 2001;39(4): 339–43.
- Linde K, Berner MM, Kriston L. St John's wort for major depression, Cochrane Database Syst Rev 2008: CD000448
- Li Yan, Cao Xueli. Isolation and characterization of furohyperforin from *Hypericum perforatum* L. by high-speed countercurrent chromatography. J Liq Chromatogr Relat Technol. 2012; 35(18): 2558–66.
- Li Wei, Sun Ya Nan, Yan Xi Tao, Yang Seo Young, Choi Chun Whan, Hyun Jin Won, Kang Hee Kyoung, Paek Kee Yoeup, Kim Young Ho. Isolation of xanthones from adventitious roots of St. John's Wort (*Hypericum perforatum* L.) and their antioxidant and cytotoxic activities. Food Sci Biotechnol 2013; 22(4): 945–9.
- Maslennikov PV, Chupakhina GN, Skrypnik LN. The content of phenolic compounds in medicinal plants of a botanical garden (Kaliningrad oblast). Biol Bull. 2014; 41(2): 133–8.
- Mihaljev Zeljko, Zivkov-Balos Milica, Cupic Zeljko, Jaksic Sandra. Levels of some microelements and essential heavy metals in herbal teas in Serbia. Acta Poloniae Pharmaceutica. 2014;71(3): 385–91.
- Muller WE. Current St John's wort research from mode of action to clinical efficacy. Pharmacol Res 2003; 47: 101–9.
- Naeem Ismat, Saddiqe Zeb, Patel Asmita, Hellio Claire. Analysis of flavonoid and antimicrobial activity of extracts of *Hypericum perforatum*. Asian J Chem. 2010; 22(5): 3596–600.
- Pilkingtona Karen, Boshnakova Anelia, Richardson Janet. St John's wort for depression: time for a different perspective? Complement Ther Med. 2006;14: 268—81.
- Pravdivtseva OE, Kurkin VA. Khimiya. Sterols of *Hypericum perforatum* L. Herbs. Rastitel'nogo Syr'ya. 2011; (4, Pt. 2): 333–4.

- Richard Jean-Alexandre. Chemistry and biology of the polycyclic polyprenylated acylphloroglucinol hyperforin. Eur J Org Chem. 2014; 2014(2): 273–99.
- Russo Emilio, Scicchitano Francesca, Whalley Benjamin J, Mazzitello Carmela, Ciriaco Miriam, Esposito Stefania, Patane Marinella, Upton Roy, Pugliese Michela, Chimirri Serafina, et al. *Hypericum perforatum*: pharmacokinetic, mechanism of action, tolerability, and clinical drug-drug interactions. Phytother Res. 2014; 28(5): 643–55.
- Sharopov Farukh S, Gulmurodov Isomiddin S, Setzer William N. Essential oil composition of *Hypericum perforatum* L. and *Hypericum scabrum* L. growing wild in Tajikistan. J Chem Pharmaceut Res 2010;2(6): 284–90.
- Sosa S, Pace R, Bornancin A, Morazzoni P, Riva A, Tubaro A, Della Loggia R. Topical anti-inflammatory activity of extracts and compounds from *Hypericum perforatum* L. J Pharm Pharmacol. 2007;59:703–9
- Uchida S, Hirai K, Hatanaka J, Hanato J, Umegaki K, Yamada S. Antinociceptive effects of St. John's wort, *Harpagophytum procumbens* extract and Grape seed proanthocyanidins extract in mice. Biol Pharm Bull. 2008; 31(2):240–5.
- Ute Woelfle, Guenter Seelinger, Christoph M Schempp. Topical application of St. John's Wort (*Hypericum perforatum*). Planta Medica. 2014;80(2/3): 109–20.
- Vandenbogaerde A, Zanoli P, Puia G, Truzzi C, Kamuhabwa A, De Witte P, Merlevede W, Baraldi M. Evidence that total extract of *Hypericum perforatum* affects exploratory behavior and exerts anxiolytic effects in rats. Pharmacol Biochem Behav. 2000;65(4):627–33.
- Wurglics Mario, Schubert-Zsilavecz Manfred. *Hypericum perforatum*: a 'modern' herbal antidepressant pharmacokinetics of active ingredients. Clin Pharmacokinet 2006; 45(5): 449–68.
- Zlobin AA, Martinson EA, Ovechkina IA, Durnev EA, Ovodova RG, Litvinets SG. Composition and properties of pectin polysaccharides of St. John's wort *Hypericum perforatum* L. Russ J Bioorg Chem. 2012; 38(7): 697–701.


## 6.6 *Eleutherococcus senticosus* (Rupr et Maxim.) Maxim

### *6.6.1* (Araliaceae)

1) Common names:
   English name: Siberian ginseng, Eleuthero, Russian root, Russian ginseng root, Devil's shrub, wild pepper
   Russian name: Svobodnoyagodnik, eleuterokok kolyuchiy, sibirskiy jenshen
   Chinese name: Ci-wu-jia
   Korean name: Gasiogalpi
   Japanese name: Ezo-ukogi

**Fig. 6.2** *Eleutherococcus senticosus*. Photo taken by Denis Kochetkov and used with permission

2) Synonyms:
   *Acanthopanax asperatus* Franch. & Sav., *Acanthopanax senticosus* (Rupr. & Maxim.) Harms, *Eleutherococcus asperatus* (Franch. & Sav.) Koidz., and *Hedera senticosa* Rupr. & Maxim
3) Photograph of the species (Fig. 6.2)
4) Description of the species:

   A) Habitat and world distribution:
      *Eleutherococcus senticosus* (Rupr et Maxim.) Maxim grows at elevations of up to 800 m or more above sea level. This plant is distributed in Amur, Khabarovsk, Primorye, Sakhalin of Russian Far East, Hebei, Heilongjiang, Henan, Jilin, Liaoning, Shaanxi, Shanxi, and Sichuan of China and also in Japan and Korea. *E. senticosus* is broadly tolerant to soils, growing in sandy, loamy, and heavy clay soils with acid, neutral, or alkaline chemistry and including soils of low nutritional value. It can tolerate sun or dappled shade and some degree of pollution (Huang et al. 2011).
   B) Morphology of the plant:
      This herb grows in mixed and coniferous mountain forests, forming low undergrowth, or is found in groups in thickets and edges. Shrubs of *E. senticosus* are up to 6 m tall. Branches are with dense to scattered, slender, terete, bristle-like prickles. Petiole is 3–12 cm, slender, and sometimes with fine prickles; petiolule of central leaflet is 0.6–1.22 cm and typically has brownish pubescent; leaflets (3–5) are elliptic obovate or oblong, with 5–13 × 3–7 cm, papery, abaxially pubescent on veins, adaxially with scattered hairs, secondary veins six or seven pairs, conspicuous on both surfaces, base broadly cuneate, margin sharply biserrate, and apex shortly acuminate or acuminate. Inflorescence terminal, solitary or compound umbel, borne on

leafy shoots, frequently with 2–6 umbels together; peduncles with 5–7 cm, glabrous; pedicels with 1–2 cm, glabrous or slightly pubescent at base. Calyx subentire or with five inconspicuous teeth, glabrous. Corolla is purple yellow. Ovary 5-carpellate; styles united into a column. Fruit is ovoid to globose, ca. 8 mm; styles are persistent, ca. 1.5 mm. Flowering occurs from June to July; fruits occur from August to October (Brekhman and Fulder 1980, Huang et al. 2011).

C) Medicinal parts:
Roots and aerial parts

D) Chemical constituents:
These species contain **triterpenoid saponins** (3,4-seco-lupane-20(29)-ene-3,28-dioic acid; ciwujianosides A1-A4, B, C1-C4, D1-D3, and E; hederasaponin B; 3-*O*-β-rhamnopyranosyl-(1→2)-α-arabin; opyranosyl oleanolic acid; saponin PE; 30-norolean-12,20(29)-dien-28-oic acid; 29-hydroxyoleanolic acid, 3β-{*O*-β-D- glucopyranosyl-(1→3)-*O*-β-D-galactopyranosyl-(1→4)-{*O*-α-L-rhamnopyranosyl-(1→2)}-*O*-β-D-glucuronopyranosyl}-16α-hydroxy-13β,28-epoxyoleanan; 3β-{*O*-α-L-20 rhamnopyranosyl-(1→4)-*O*-α-Lrhamnopyr-anosyl-(1→4)-[*O*-α-L-rhamnopyranosyl-(1→2)]-*O*-β-D glucopyranosyl-(1→x)-*O*-β-D-glucuronopyranosyl}-16α-hydroxy-13β,28-epoxyoleanane; acanthopanaxosides A, B, C, and E; sessiloside; taurosideH1;chiisanoside;3-*O*-α-rhamnopyranosyl-(1→2)-α-arabinopyranosyl mesembryanthemoidigenic acid; silphioside F; copteroside B; hederagenin 3-*O*-β-D-glucuronopyranoside-6′-*O*-methyl ester; gypsogenin 3-*O*-β-D-glucuronide; silphioside G; ilexoside XLVIII; hederagenin 3-*O*-β-D-glucuronopyranosyl methyl ester-28-*O*-β-D- glucopyranoside; 3-*O*-β-D-glucuronopyranosylgypsogenin; hederagenin 28-*O*-β-D-glucopyranoside; hemsgiganoside B; anhuienside C; eleutheroside L; hederagenin-3-*O*-β-glucuronopyranoside; oleanolic acid-3-*O*-β-glucuronopyranoside; and friedelin), **lignans** (neociwujiaphenol, ciwujiatone, sesamin, eleutheroside E, E2, (+)-pinoresinol-*O*-β-D-glucoside, (+)-syringaresinol-*O*-β-D-glucoside, (+)-pinoresinol-di-*O*-β-D-glucoside, (+)-medioresinol-di-*O*-β-D-glucoside, syringin (eleutheroside B)), **coumarins** (isofraxidin, isofraxidin-7-*O*-β-D--glucoside, scopoletin, eleutheroside B2), **flavonoids** (quercitrin, hyperoside, quercetin, rutin), **essential oils** (isocaryophyllene (=β-caryophyllene), caryophyllene oxide, *n*-octadecanol, manoyl oxide, 9,17-octadecadienal, tetradecanal, humulene oxide, humulene (= α-caryophyllene), β-farnesene, (E,E)-2.4-decadienal, linalool, *p*-cymene, 2-*n*-pentylfuran, α-pinene, *n*-heptaldehyde, etc.), **phenolics** (chlorogenic acid, 1,5-*O*-dicaffeoylquinic acid, 3,5-*O*-dicaffeoylquinic acid, 4,5-*O*-dicaffeoylquinic acid, caffeic acid, coniferyl aldehyde, caffeic acid ethyl ester, vanillic acid, syringic acid, 5-hydroxymethylfurfural protocatechuic acid), and **other compounds** (isovanillin, tyrosol, octadecanoic acid, daucosterol, amygdaloside, methyl oleate, palmitic acid, 2,6-dimethoxy-pbenzoquinone, feruloyl sucrose, geniposide, oplopanone B, eleutherazine A, B) (Huang et al. 2011).

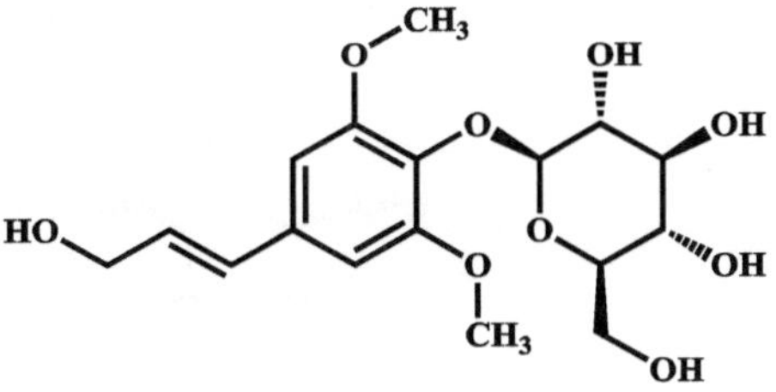

Eleutheroside B

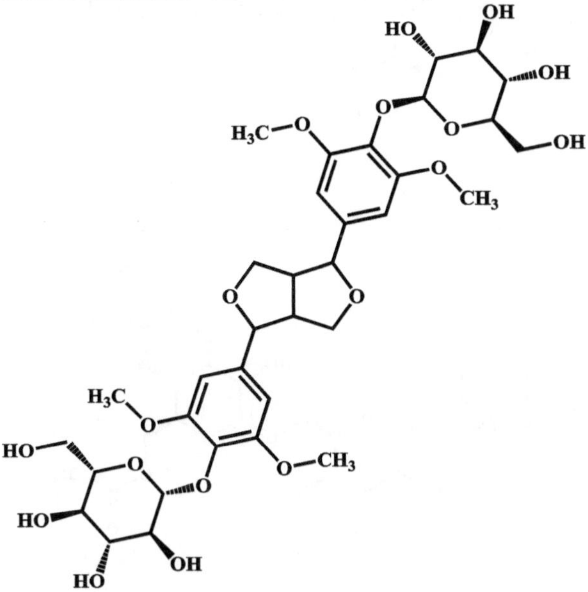

Eleutheroside E

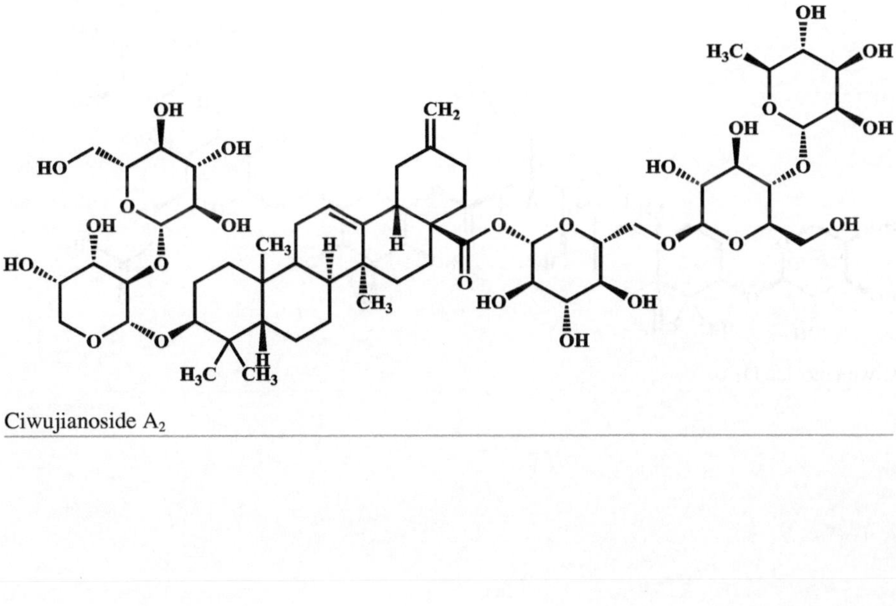

Ciwujianoside $A_2$

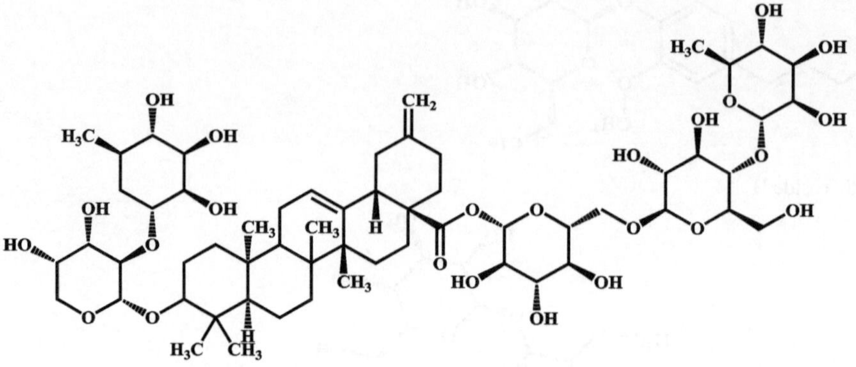

Ciwujianoside B

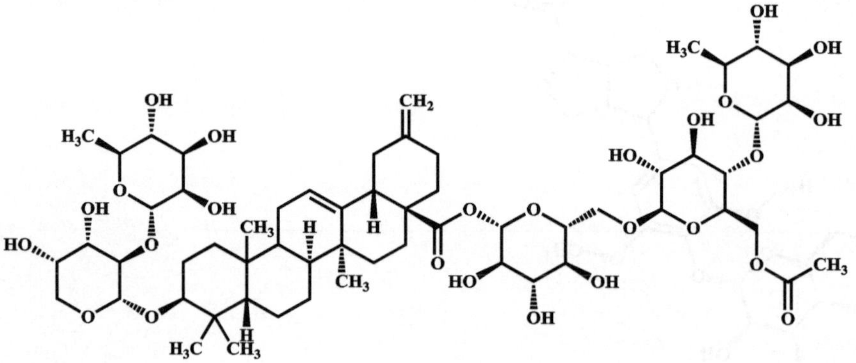

Ciwujianoside C2

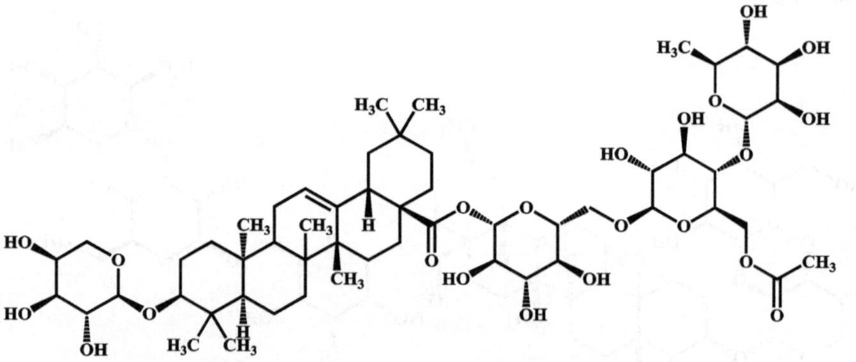

Ciwujianoside D1

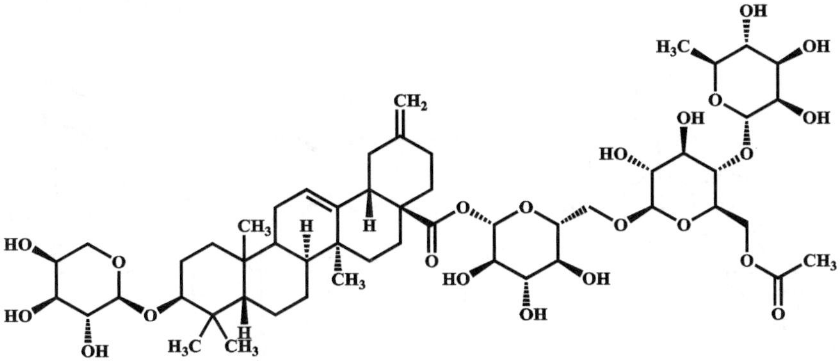

Ciwujianoside D2

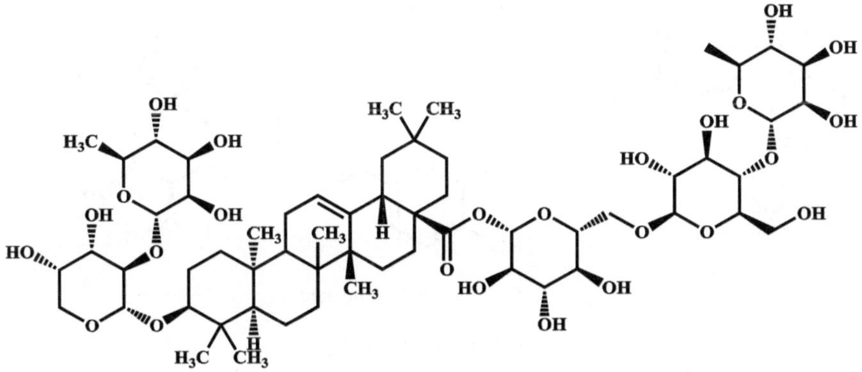

Hederasaponin B

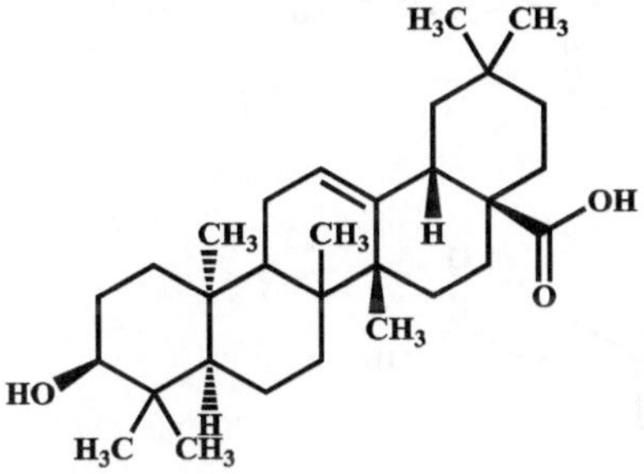

Oleanolic acid

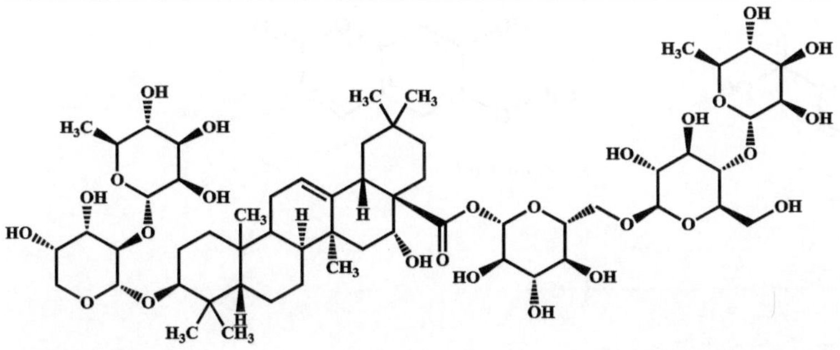

Tauroside H1

| | |
|---|---|
| 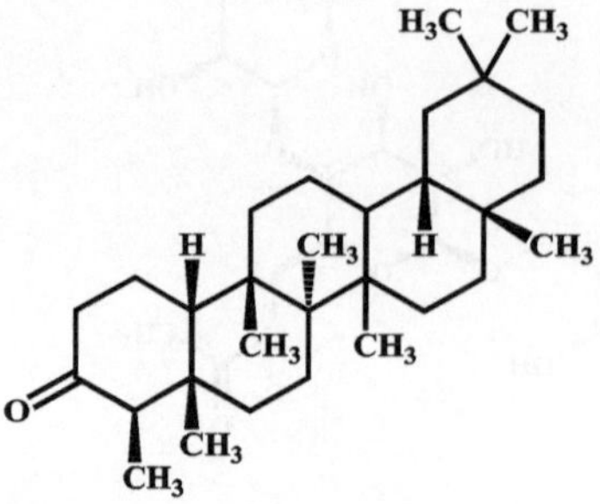 Friedelin | 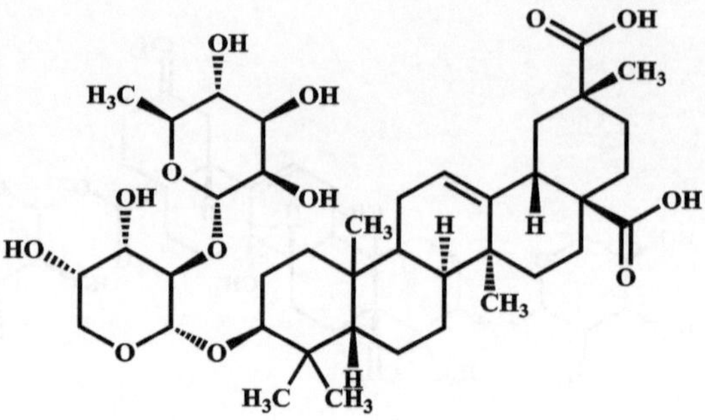 Acanthopanaxoside C |
| 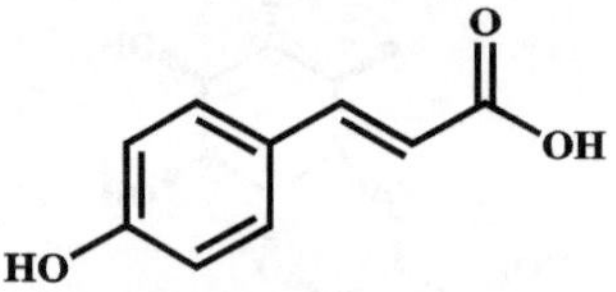 *p*-Coumaric acid | 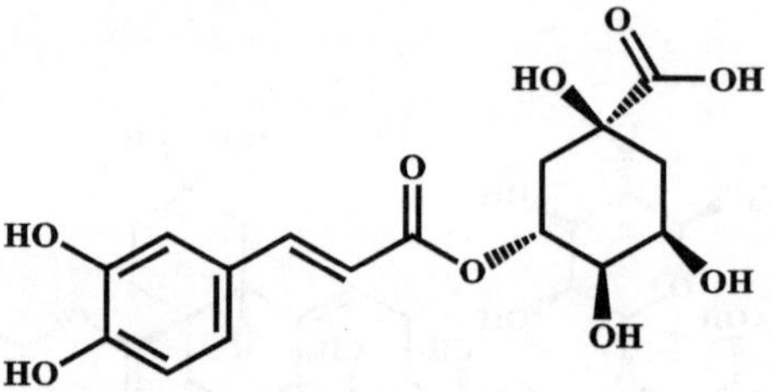 5-*O*-Caffeoylquinic acid (=Chlorogenic acid) |
| 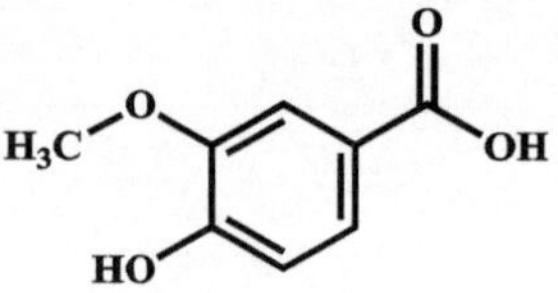 Vanillic acid | 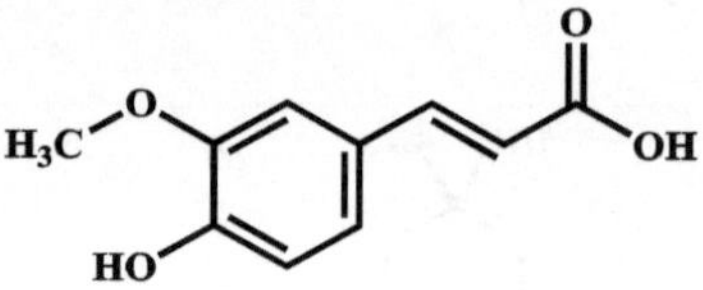 Ferulic acid |
| 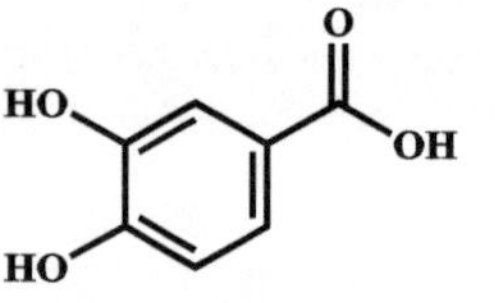 Protocatechuic acid | 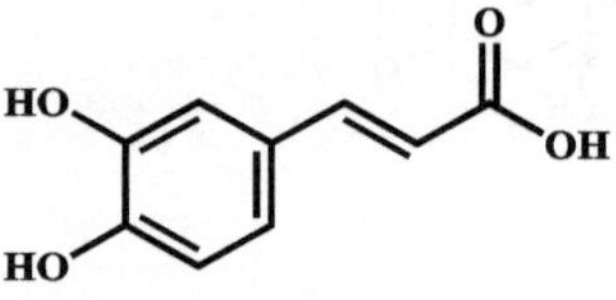 Caffeic acid |
| 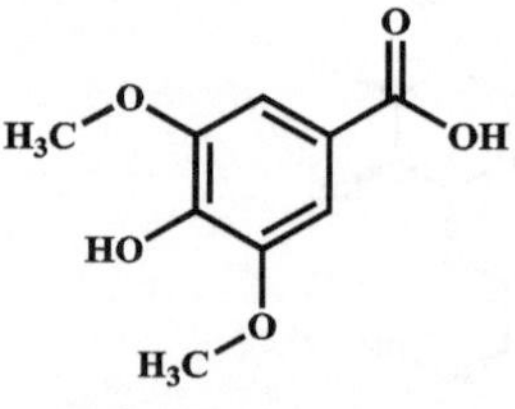 Syringic acid | 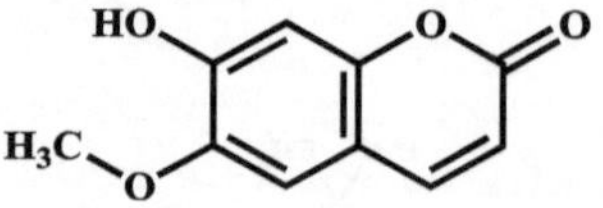 Scopoletin |

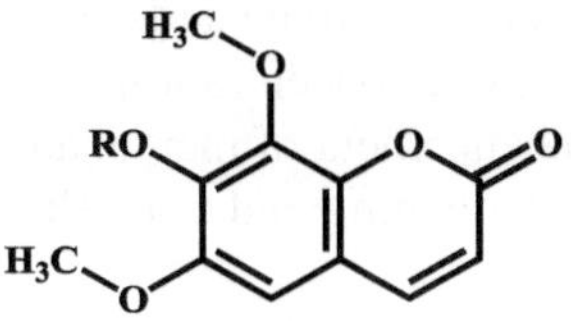

Isofraxidin: R = H
Isofraxidin-7-*O*-glucoside: R = glucose

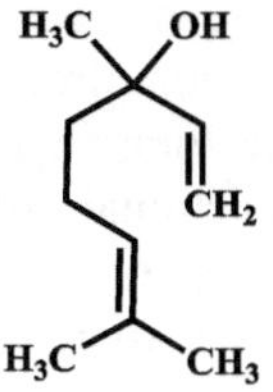

Linalool

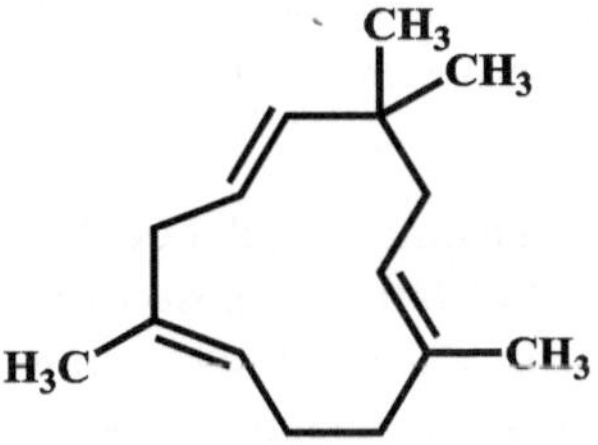

α-Caryophyllene

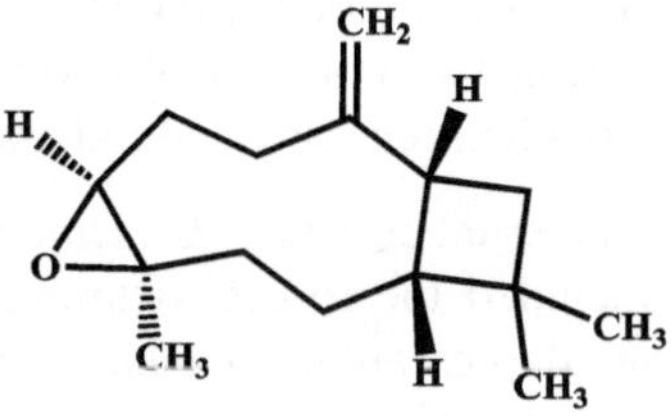

β-Caryophyllene oxide

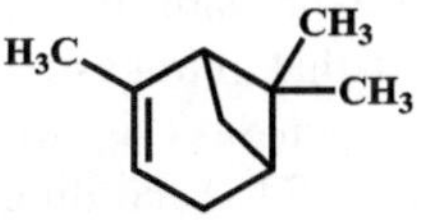

α-Pinene

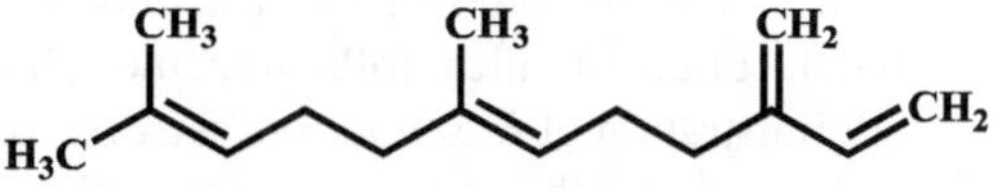

β-Farnesene

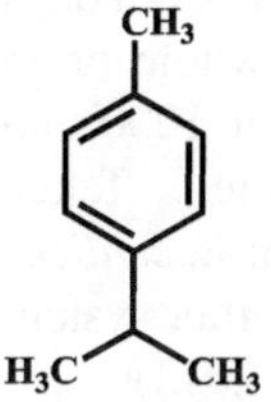

*p*-Cymene

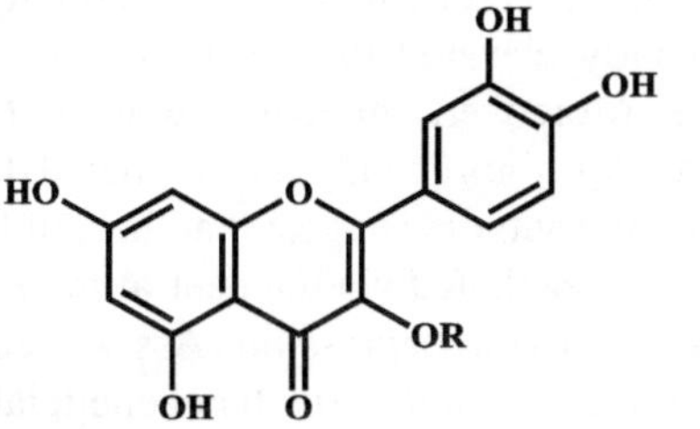

Quercetin: R = H
Quercetin-3-*O*-galactoside: R = galactose
Quercetin-3-*O*-rhamnoside: R = rhamnose
Quercetin-3-*O*-rutinoside: R = rutinose

5) Pharmacology and bioactivities:

A) In vitro and in vivo studies:

*E. senticosus* is a widely used plant in folk and traditional medicines for the cardiovascular, immune, nervous, and endocrine systems, and according to the researchers Brekhman and Dardymov (1969), it has the ability to increase the nonspecific resistance state, which characterizes it as an adaptogenic substance.

More than hundreds of other articles related to research work on *E. senticosus* are summarized in various monographs and reviews. The following indications have been reported in literature for *E. senticosus* root (Assessment report 2013):

- As a tonic in case of decreased performance such as fatigue and sensation of weakness, exhaustion, tiredness, and loss of concentration.
- As a prophylactic and restorative tonic for enhancement of mental and physical position for functional asthenia, strengthening and normalizing, as well as convalescence.
- As adaptogen, to increase body resistance to such stressful exposures as heat, cold, physical exhaustion, viruses, bacteria, chemicals, extreme working conditions, noise, and pollution.
- *Eleutherococcus* root was commonly used in Russia in oncology hospital departments to increase the tolerance of the patients to the adverse effects of chemotherapy and radiation therapy.

The antidepressant effects of *E. senticosus* studied using animal models of depression included the forced swimming (FST) and tail suspension tests (TST). The antidepressant mechanism of *E. senticosus* was explored by monitoring the levels of monoamine neurotransmitters including serotonin (5-hydroxytrylamine, 5-HT), norepinephrine (NE), and dopamine (DA), as well as cAMP response element-binding (CREB) protein expression in the whole brain of mice following the TST. Results showed that intragastric administration of *E. senticosus* at a dose of 2000 mg/kg for seven days significantly reduced the duration of immobility in both the FST and TST. According to the authors, these results indicate that *E. senticosus* possesses antidepressant-like properties. Pretreatment with 2000 mg/kg of *E. senticosus* for seven days significantly elevated the levels of 5-HT, NE, and DA in the whole brain of mice. Moreover, aqueous extract of *E. senticosus* at doses of 1000 and 2000 mg/kg significantly up-regulated the level of CREB protein. Taken together, the authors suggest that the antidepressant mechanism of *E. senticosus* may be mediated via the central monoaminergic neurotransmitter system and CREB protein expression (Assessment report 2013; Jin et al.2013).

Moreover, *E. senticosus* has beneficial effects against other CNS-related disorders. Tohda et al. (2008) investigated the effects of *E. senticosus* methanol and water extracts on the regeneration of neurites and the reconstruction of synapses in rat-cultured cortical neurons damaged by amyloid beta Aβ-(25–35). Results showed that extracts of *E. senticosus* protect against neuritic atrophy and cell death under Aβ-(25–35) treatment, and one of active constituents may be eleutheroside B. Later the same research group (Bai et al. 2011) published a similar study where a comprehensive evaluation of constituents was conducted to explore active components from *E. senticosus* which can protect against neuritic atrophy induced by amyloid β Aβ-(25–35) in cultured rat cortical neurons. The ethyl acetate, *n*-butanol, and water fractions from the methanol extract of *E. senticosus* showed protective effects against Aβ-induced neuritic atrophy. Twelve compounds were isolated from the active fractions and identified. Among them, eleutheroside B, eleutheroside E, and isofraxidin showed obvious protective effects against Aβ-induced atrophies of axons and dendrites at 1 and 10 μM (Bai et al. 2011).

Jin et al. (2013a) examined the mechanism of *E. senticosus* (ethyl acetate fraction from aqueous extract of fruits) in antineuroinflammatory and neuroprotective processes, and authors concluded that *E. senticosus* serves as a potential therapeutic agent for neuronal disorders.

Bu et al. (2005) suggest that *E. senticosus* has a neuroprotective effect by inhibiting inflammation and microglial activation in brain ischaemia. Chen et al. (2011) observed that *Eleutherococcus* saponins could enhance the viability of spinal motor neurons and have protective effects on hypoxic neurons.

Bocharov et al. (2008) conclude that *E. senticosus* should be used as therapeutic modulators in neurodegenerative disorders including Parkinson's disease. Liu et al. (2012) studied the neuroprotective effect of an extract of *E. senticosus* against 1-methyl-4-phenyl-1,2,3,6-tetrahydropyridine (MPTP)-induced mice model of Parkinson's disease, and its mechanism was studied. Li et al. (2013), who had summarized herbal medicines with anti-Parkinsonian activities, has included *E. senticosus* among compounds reported to be effective on Parkinson's disease experimental models in vivo.

6) Indications and usage:
   Traditional herbal medicinal product for symptoms of asthenia such as fatigue and weakness (EMEA 2014)
7) Precautions and side effects:
   Few adverse reactions have been reported. *E. senticosus* is well tolerated and free of side effects and can be taken for long periods of time. High doses of *E. senticosus* are associated with irritability, insomnia, and anxiety. In human trials, few adverse reactions are reported (Bohn et al. 1987, Cicero et al. 2004). Adverse reactions have included skin eruptions, headache, diarrhoea, hypertension, and pericardial pain in rheumatic heart patients. *E. senticosus* was associated with increased aggressive behaviour in mice (Lewis et al. 1983).
8) Dosage:
   A) Internal use:
      Doses of powdered root 1–4 g per day have been used in trials. Doses of *E. senticosus* extracts are recommended at less than 1 g/day (Brekhman et al. 1991, Gaffney et al. 2001). More information about dosage is given in the Assessment Report (2013).
   B) External use:
      Not found.

## 6.7 References


- Assessment report on *Eleutherococcus senticosus* (Rupr. et Maxim.) Maxim. radix. EMA/HMPC/680615/2013. P.1-51.
- Bai Y, Tohda C, Zhu S, Hattori M, Komatsu K. Active components from Siberian ginseng (*Eleutherococcus senticosus*) for protection of amyloid beta (25-35) induced neuritic atrophy in cultured rat cortical neurons. J Nat Med. 2011; 65(3-4):417–23.

- Bocharov EV, Kucherianu VG, Bocharova OA, Karpova RV. Neuroprotective features of phytoadaptogens. Vestn Ross Akad Med Nauk. 2008;(4):47–50.
- Bohn B, Nebe CT, Birr C. Flow-cytometric studies with *Eleutherococcus senticosus* extract as an immunomodulatory agent. Arzneimittelforschung. 1987; 37:1193–6.
- Brekhman II, Dardymov IV. New substances of plant origin which increase non-specific resistance. Ann Rev Pharmacol. 1969; 9:419–30.
- Brekhman I, Fulder S. Man and biologically active substances: the effect of drugs, diet, and pollution on health. Oxford: Pergamon Press; 1980.
- Brekhman I, Favorov VV, Abramov AK, et al. Comparative study of *Eleutherococcus* preparations made from raw material of various origin. Farmatsiia. 1991; 40:39–41.
- Bu Y, Jin ZH, Park SY, Baek S, Rho S, Ha N, Park SK, Kim H. Siberian ginseng reduces infarct volume in transient focal cerebral ischaemia in Sprague-Dawley rats. Phytother Res. 2005;19(2):167–9.
- Chen JF, Fan J, Tian XW, Tang TS. Protective effects of two constituents of chinese herbs on spinal motor neurons from embryonic rats with hypoxia injury. Afr J Tradit Complement Altern Med. 2011;9(2):234–41.
- Cicero AF, Derosa G, Brillante R, Bernardi R, Nascetti S, Gaddi A. Effects of Siberian ginseng (*Eleutherococcus senticosus* Maxim.) on elderly quality of life: a randomized clinical trial. Arch Gerontol Geriatr Suppl. 2004; 9:69–73.
- EMEA. Community herbal monograph on *Eleutherococcus senticosus* (Rupr. et Maxim.) Maxim. radix. EMA/HMPC/680618/2013; 2014.
- Gaffney BT, Hugel HM, Rich PA. The effects of *Eleutherococcus senticosus* and *Panax ginseng* on steroidal hormone indices of stress and lymphocyte subset numbers in endurance athletes. Life Sci. 2001;70:431–42.
- Huang L, Zhao H, Huang B, Zheng C, Peng W, Qin L. *Acanthopanax senticosus*: review of botany, chemistry and pharmacology. Pharmazie. 2011; 66(2): 83–97.
- Jin L, Wu F, Li X, Li H, Du C, Jiang Q, You J, Li S, Xu Y. Anti-depressant effects of aqueous extract from *Acanthopanax senticosus* in Mice. Phytother Res. 2013;27(12):1829–33.
- Jin ML, Park SY, Kim YH, Park G, Lee SJ. *Acanthopanax senticosus* exerts neuroprotective effects through HO-1 signaling in hippocampal and microglial cells. Environ Toxicol Pharmacol. 2013a;35(2):335–46.
- Lewis WH, Zenger VE, Lynch RG. No adaptogen response of mice to ginseng and *Eleutherococcus* infusions. J Ethnopharmacol. 1983; 8:209–14.
- Li XZ, Zhang SN, Liu SM, Lu F. Recent advances in herbal medicines treating Parkinson's disease. Fitoterapia. 2013;84:273–85.
- Liu SM, Li XZ, Huo Y, Lu F. Protective effect of extract of *Acanthopanax senticosus* Harms on dopaminergic neurons in Parkinson's disease mice. Phytomedicine. 2012;19(7):631–8.
- Tohda C, Ichimura M, Bai Y, Tanaka K, Zhu S, Komatsu K. Inhibitory effects of *Eleutherococcus senticosus* extracts on amyloid beta(25-35) induced neuritic atrophy and synaptic loss. J Pharmacol Sci. 2008;107(3):329–39

## 6.8 *Leonurus cardiaca* L

### *6.8.1* (Lamiaceae)

1) Common names:

   English name: Motherwort, lion's tail, lion's ear, heart wort, heart gold, heart heal, heart herb, lion's ear, lion's tail, throw wort, and mother weed

   Russian: Pustyrnyk serdechniy, Pustyrnyl volosistyy, Pustyrnyk mohnatyy, Serdechnaya trava

   Uzbek name: Arslonquyruq

   Azerbaijani name: Dem otu

2) Synonyms:
   Synonyms: *Leonurus villosus* Desf. et Spreng., *Leonurus campestris* Andrz., *Leonurus canescens* Dumort, *Leonurus trilobatus* (lam.) Dulac, *Cardiaca vulgaris* Moench, *Cardiaca trilobata* Lam., and *Cardiaca vulgaris* Moench (Assessment report 2010)
3) Photograph of the plant (Fig. 6.3).
4) Description of the species:

   A) Habitat and world distribution:
   *Leonurus cardiaca* L. is a perennial herb widespread in Europe and Scandinavia, commonly found in rural areas throughout the lowlands, foothills, gardens, ruins, roadsides, waste ground, rubbish dumps, as well as in Asia to the Himalayas and eastern Siberia and has been introduced to North America and northern Africa.
   B) Morphology of the plant:
   It has a height of 30–120 cm. Stem is upper part branching, four edged, and short haired–almost glabrous, usually partly with purple shades. Flowers are dull purple pink, pale purple, or white, small, and clustered in axils of upper leaves. Corolla is irregular (zygomorphic), light red, occasionally white, 8–12 mm (0.32–0.48 in.) long, fused, bilabiate, and long tubed. Upper lip is convex, top white haired, approx. 4 mm long; lower lip is three lobed, downward curved, with red spots. Calyx is campanulate (bell-shaped), five veined, and five lobed; lobes are sharp pointed. Stamens are four, of which two short and two long. Gynoecium is composed of two fused carpels. Inflorescence is a spike-like group composed of axillary whorls, terminating stem, and branches. Leaves are opposite and stalked. Blade is ovate–cordate and usually five lobed, margin is large toothed, underside is light grey, and top is sparsely haired. Inflorescence's subtending bracts are usually shallowly three lobed. Its fruit is four-parted schizocarp. Mericarps are three-edged brown, upper surface hairy. Flowering occurs from June to September.
   C) Medicinal part:
   Herb

**Fig. 6.3** *Leonurus cardiaca*. Photo taken by Peter M. Dziuk and used with permission

D) Chemical constituents:

In the aerial parts of *L. cardiaca*, there are compounds belonging to the group of terpenes: **monoterpenes** (iridoids), **diterpenes** (of clerodane, furanolabdane, and labdane types), **triterpenes** (ursolic and oleanolic acids), **nitrogen-containing compounds** (leonurine, stachydrine), **phenylpropanoids** (lavandulifolioside), as well as **flavonoids**, **phenolic acids**, **sterols**, and **tannins** (Tasdemir et al. 1998, Wojtyniak et al. 2013, Zhogova et al. 2014).

Terpenes are represented by monoterpenoids (iridoids) (leonuride, ajugoside, galiridoside, and reptoside) (Weinges et al. 1973, Wichtl 2004), bitter diterpenoids of clerodane (Brieskorn and Hofmann 1979) and of the labdane type (15-*O*-ethylleopersin C, 15-*O*-methylleopersin C, 15-epi-*O*-methylleopersin C) (Malakov et al. 1985, Agnihotri et al. 2008), furanolabdane derivatives (leosibiricin, leocardin, 19-acetoxypregaleopsine) (Knöss and Zapp 1998, Papanov et al. 1998), and triterpenes (ursolic acid, oleanolic acid, corosolic acid, euscaphic acid and ilelatifol D) (Janicsák et al. 2006, Ali et al. 2007).

The nitrogen-containing compounds are pyrrolidine-type alkaloids (stachydrine) and stereoisomers of 4-hydroxystachydrine (betonicine and turicin) (Kuchta et al. 2014, Romanowski 1960), imines guanidine derivative (leonurine and 4-guanidino-1-butanol) (Gulubov and Chervenkova 1970, Kuchta et al. 2012), and amine choline (Reuter and Diehl 1970).

The phenylpropanoid lavandulifolioside or arabinoside of verbascoside was isolated from L. *cardiaca.*

Phenolic acids: chlorogenic, rosmarinic, caffeic, *p*-coumaric, *p*-hydroxybenzoic, vanillic, ferulic acid, as well as phenolic glycoside: caffeic acid 4-rutinoside (Tschesche et al. 1980, Bernatoniene et al. 2009, Wojtyniak 2011, Zhogova et al. 2014).

Flavonoids: quercetin-*O*-glucosides and kaempferol-*O*-glycosides (rutin, hyperoside, quercitrin, isoquercitrin, astragalin), as well as apigenin-7- or 5- or 4′-*O*-glycosides; quinqueloside; flavonoid *C*-glycosides: vitexin and isovitexin; free aglycones (quercetin, kaempferol, genkwanin); genkwanin is proposed as a chemotaxonomic marker of the *Leonurus* genus (Schultz and Alhyane 1973, Kartnig et al. 1985, Bernatoniene et al. 2009).

The volatile oil mainly contains sesquiterpenes (germacrene-D, epi-cedrol, β-caryophyllene, α-humulene, spatulenol, α-guaiol acetate), monoterpenes (α-pinene, dehydro-1,8 cineole) (Mockutë et al. 2005; Morteza-Semnani et al. 2008), and thymol, borneol, and α-phellandrene (Ali et al. 2012). The seeds mainly contain fatty acids: laballenic acid, 20:1delta9c acid, phlomic acid, etc. (Aizetmüller 1998).

Sterols are represented by β-sitosterol and stigmasterol (Senatore et al. 1991).

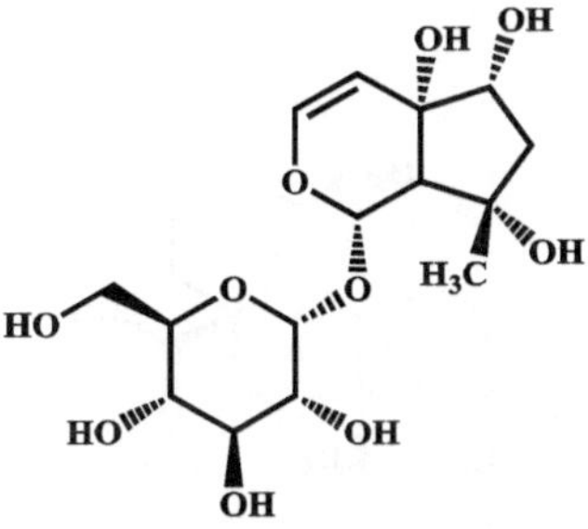

Leonuride

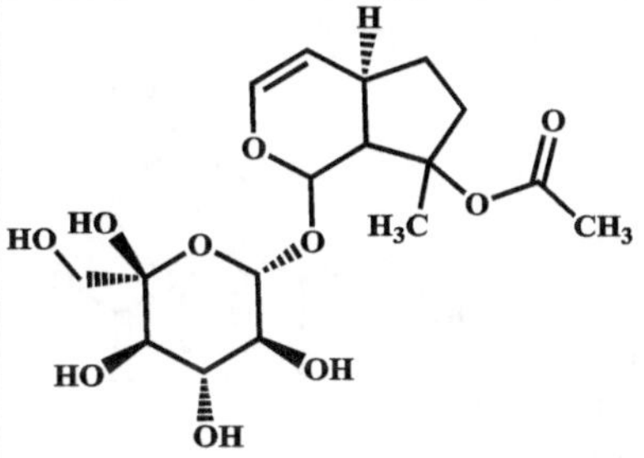

Ajugoside

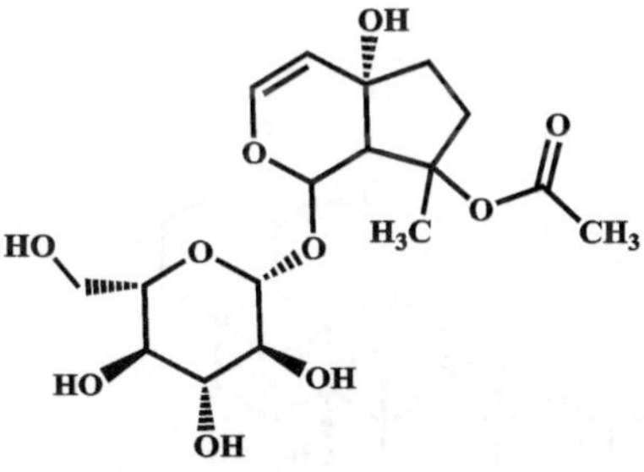

Reptoside

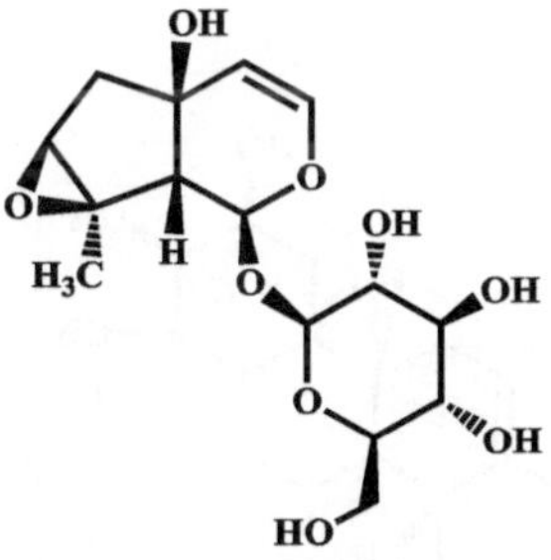

Galiridoside

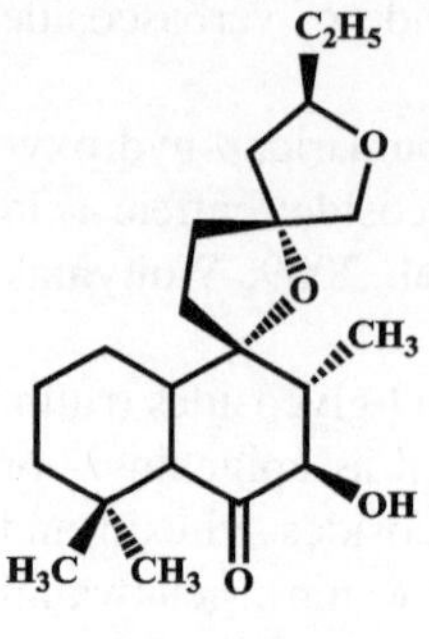

15-*O*-Ethylleopersin C

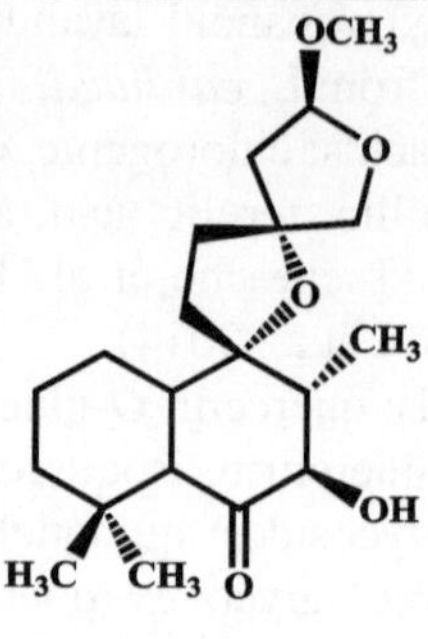

15-*O*-methylleopersin C

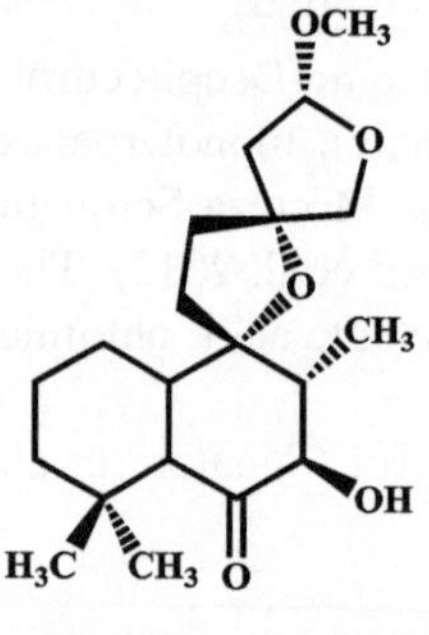

15-epi-*O*-methylleopersin C

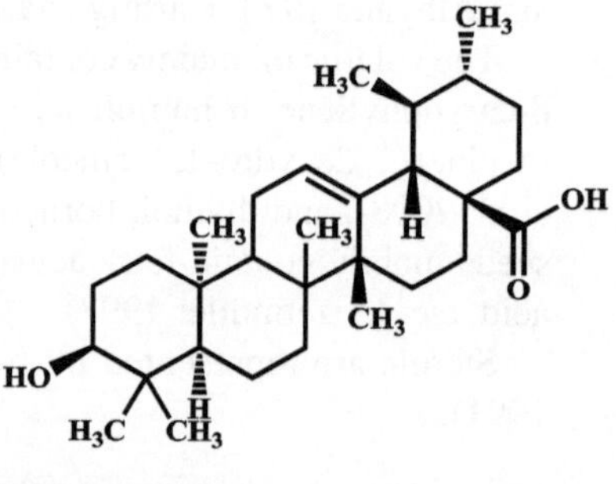

Ursolic acid

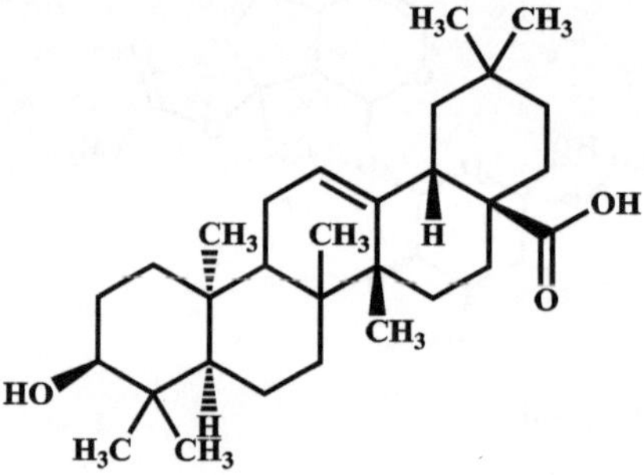

Oleanolic acid

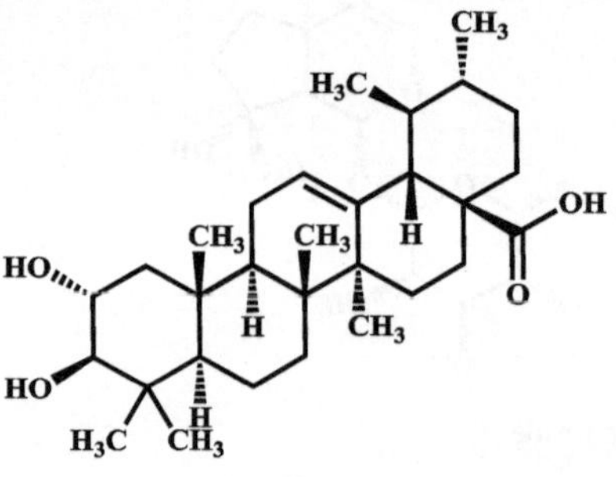

Corosolic acid

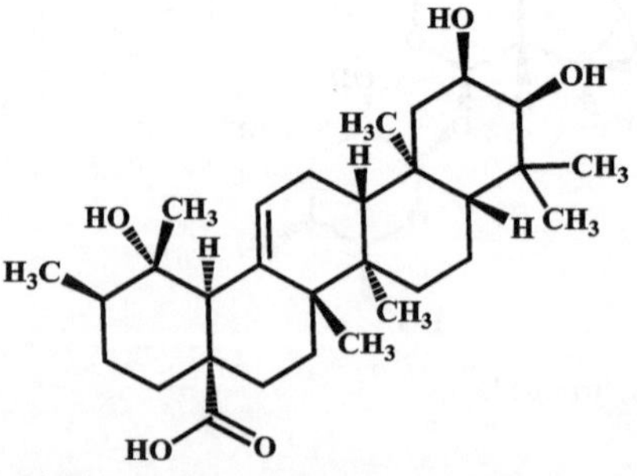

Euscaphic acid

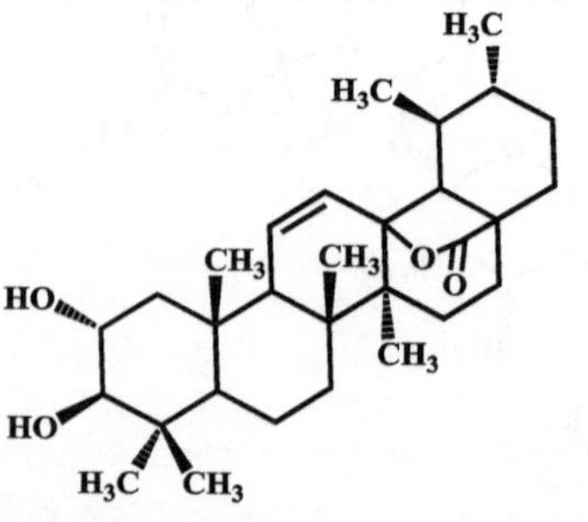

Ilelatifol D

| | |
|---|---|
| 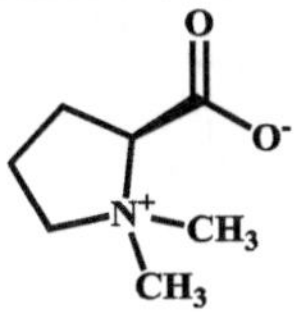 Stachydrine | 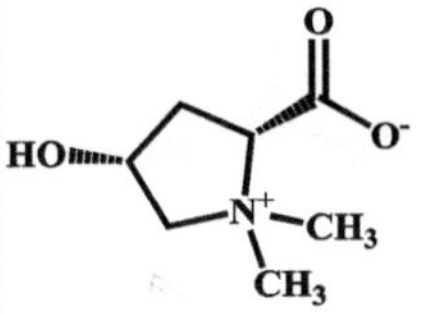 Turicin |
| 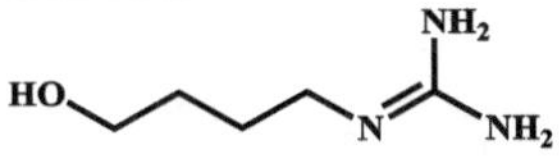 4-Guanidino-1-butanol | 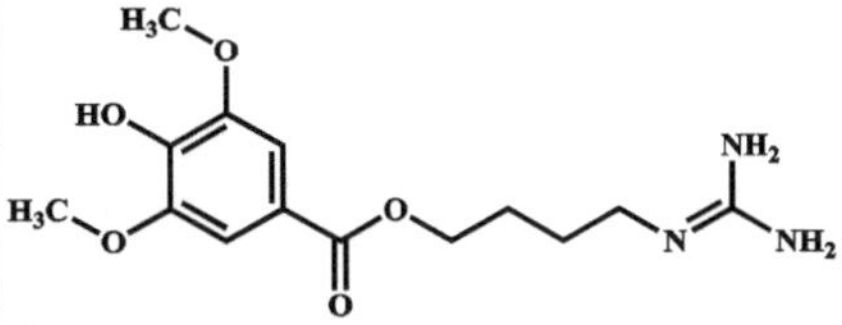 Leonurine |
| 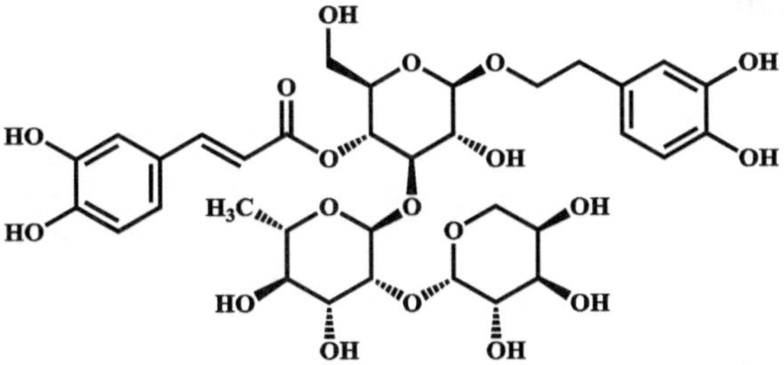 Lavandulifolioside | |
| 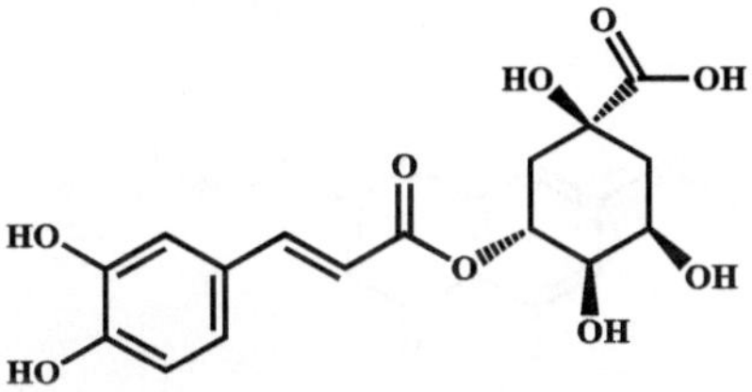 Chlorogenic acid | 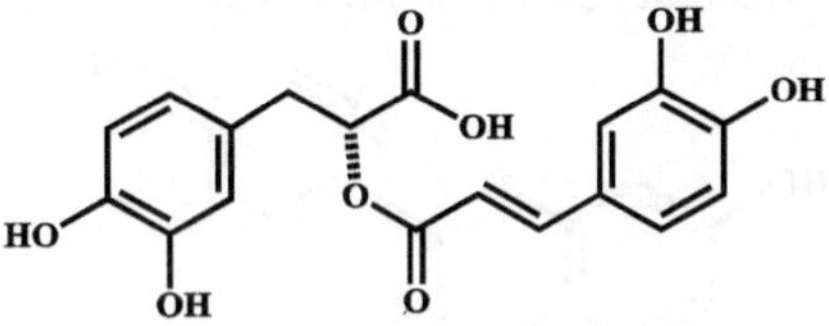 Rosmarinic acid |
| 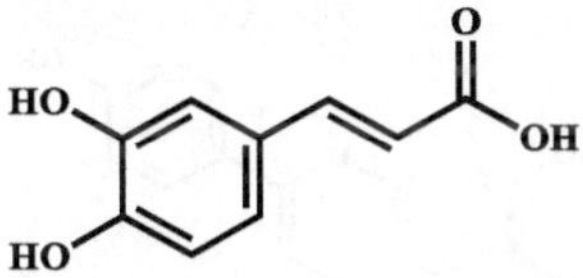 Caffeic acid | 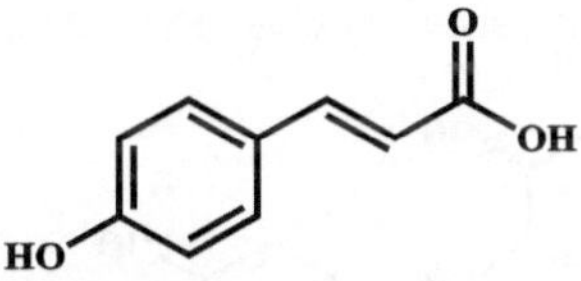 *p*-Coumaric acid |
| 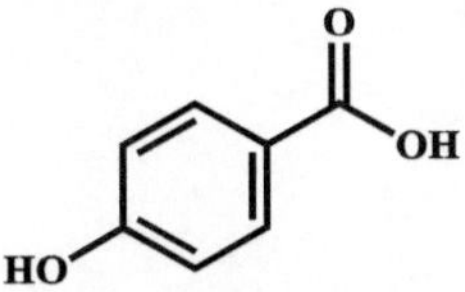 4-Hydroxybenzoic acid | 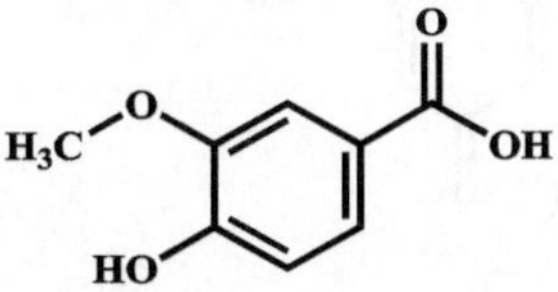 Vanillic acid |

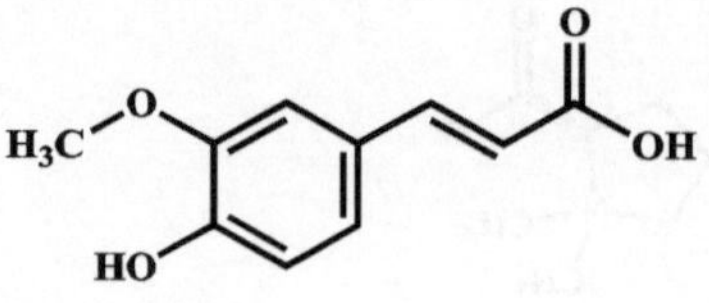

Ferulic acid

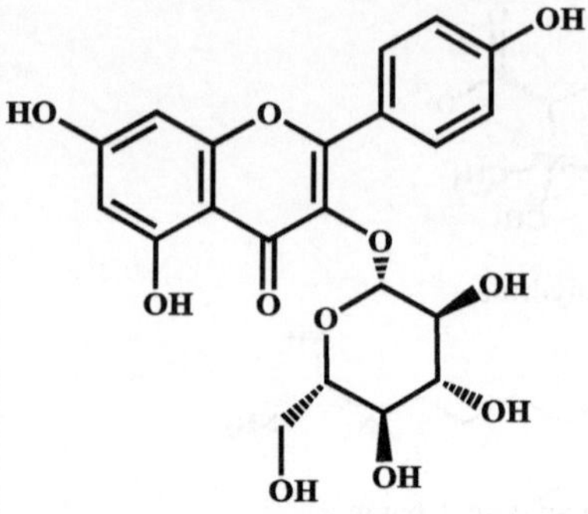

Astragalin

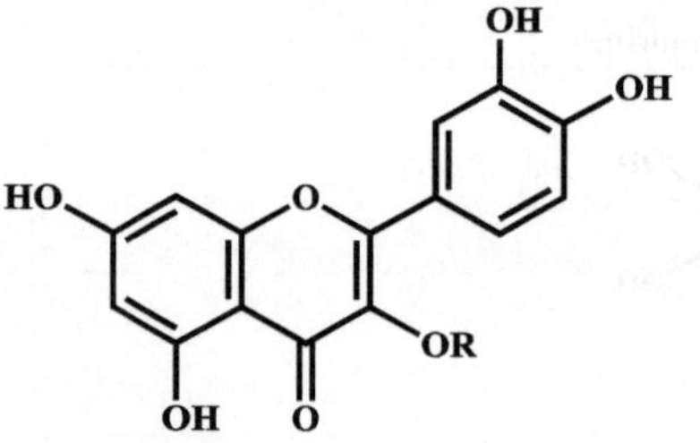

R = H: Quercetin
R = galactose: Quercetin-3-*O*-galactoside
R = rhamnose: quercetin-3-*O*-rhamnoside
R = rutinose: Quercetin-3-*O*-rutinoside

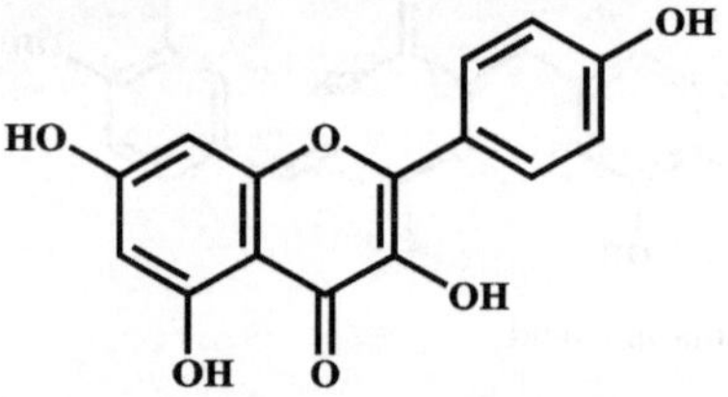

Kaempferol

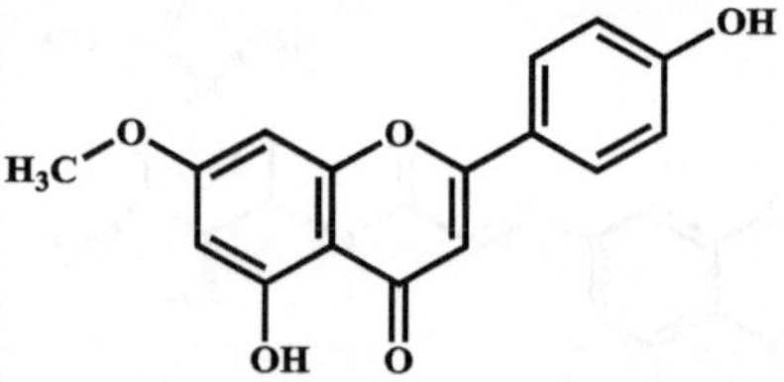

Genkwanin

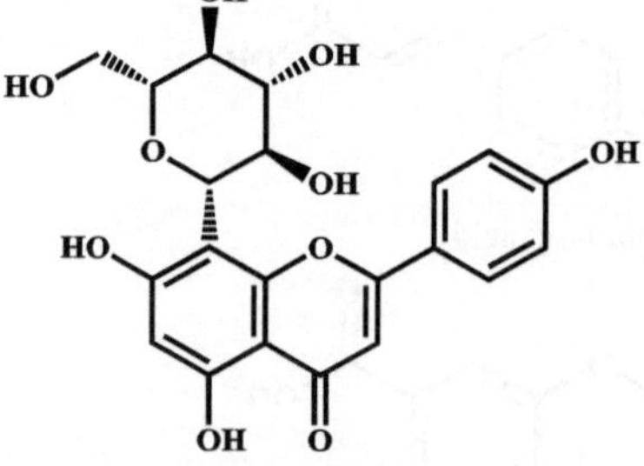

Vitexin

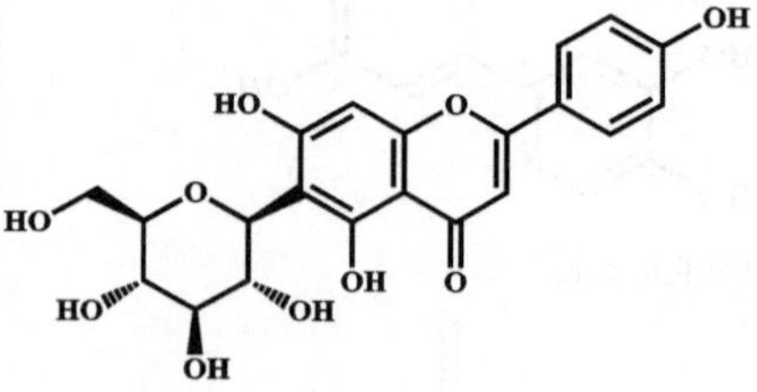

Isovitexin

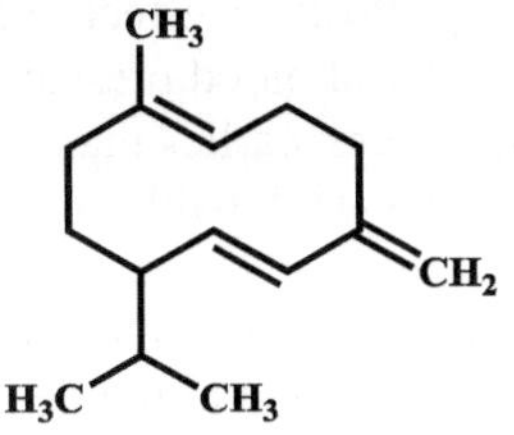


Germacrene-D

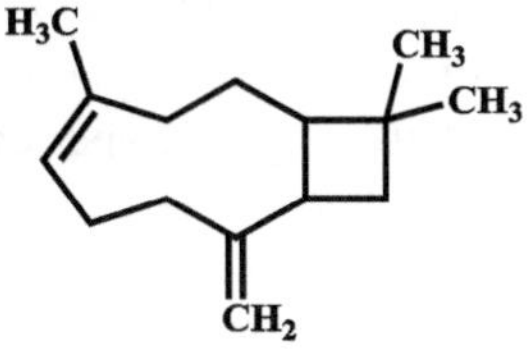


beta-Caryophyllene

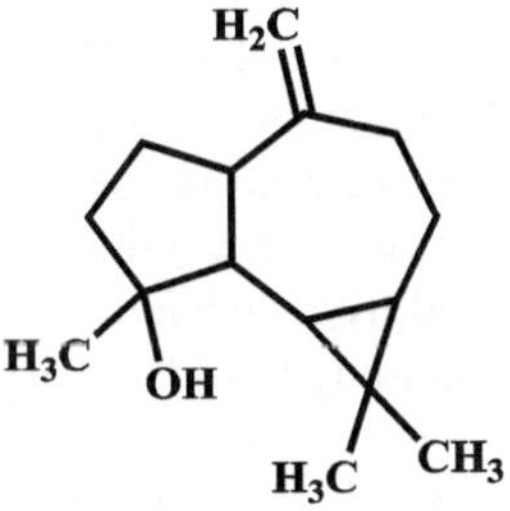


Spathulenol

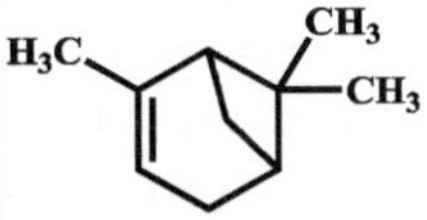


α-Pinene

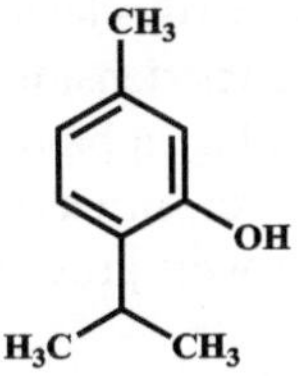


Thymol

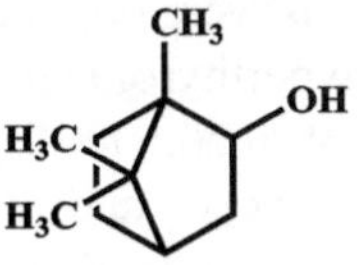


Borneol

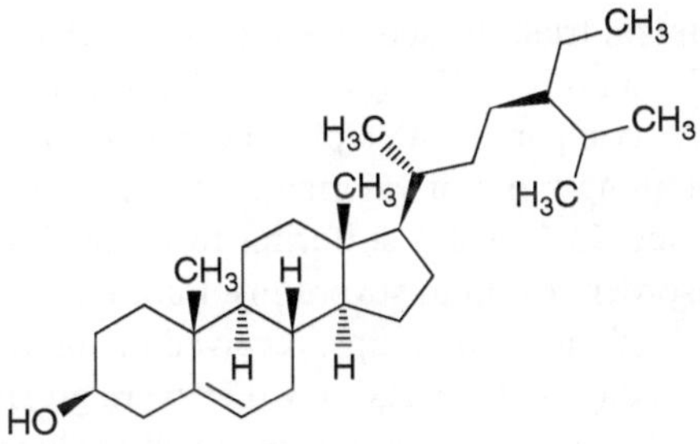


β-Sitosterol

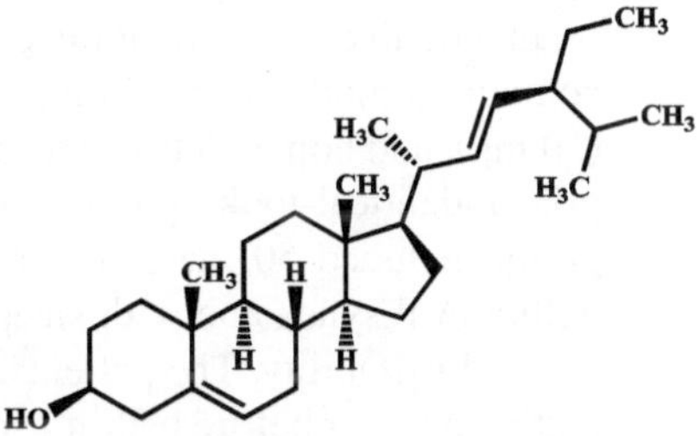


Stigmasterol

5) Pharmacology and bioactivities:

A) In vivo studies:

Sedative and hypotensive activity of *L. cardiaca* has been demonstrated in clinical trials. A study was carried out to assess the clinical efficacy of *Leonurus* oil extract in patients with arterial hypertension stages 1 and 2, accompanied by anxiety and sleep disorders. Fifty patients were treated for

28 days with 1200 mg *Leonurus* oil extract per day. Positive effects of *Leonurus* oil extract on psycho-emotional status and arterial blood pressure in patients with stage 1 hypertension were observed 1 week earlier than in patients with stage 2 hypertension. According to the Clinical Global Impression (CGI) scale, a significant improvement in the symptoms of anxiety and depression was observed in 32 % of patients, a moderate improvement in 48 % and a weak effect in 8 %; 12 % of patients did not respond to therapy. Side effects were minimal in all groups. *Leonurus* oil extract may therefore be a potentially effective therapeutic agent for patients with arterial hypertension and concurrent psycho-neurological disorders (Shikov et al. 2011).

Results of old studies on various preparations of motherwort (tincture, alcoholic liquid extract, dry extract, and infusion) used in a group of 100 patients (aged 18–65) suffering from nervous cardiovascular conditions, hyperthyroidism, atherosclerosis and heart diseases, and hypertension were published. After 3–4 days of treatment with tablets containing liquid and dry extracts, reduced palpitations and sensations of tightness in the heart area and weakened tachycardia and nervous excitability in people suffering from hyperthyroidism were observed. The best results of the treatment with motherwort preparations were obtained in the group suffering from nervous heart conditions and hyperthyroidism. In the patients with essential hypertension, after 8–10 days of treatment, there was a decrease of 8–20 mmHg in blood pressure, and headaches and dizziness were alleviated. However, in cases with heavy heart ailments, therapy with preparations of motherwort proved to be ineffective (Rossyjskij, 1949).

Sedative and hypotensive activity of *L. cardiaca* has been demonstrated in clinical trials (Wojtyniak et al. 2013):

In a double-blind randomized clinical trial, the sedative effects of tablets containing motherwort (50 mg), valerian root (170 mg), lemon balm leaves (50 mg), and hop pellets (50 mg) were compared with a placebo which comprised identical-looking tablets containing 5 mg of valerian root. The study group included 50 males (average age 45.6 years) suffering from alcohol withdrawal syndrome with sleep disorders (mild to severe insomnia), anxiety, and irritability. The patients, divided into two groups, received the preparation an hour before bedtime, once a day, and the next day they were given the placebo. Compared with the placebo, a significant improvement in sleep quality and a decrease in the frequency of awakenings and nightmares proved a mild sedative effect of the product used; however, it caused drowsiness the following day. Motherwort could therefore, to some extent, be helpful in disorders associated with alcohol abstinence (Widy-Tyszkiewicz and Schmida 1997).

A study to determine if the administration of sedatives, including a tincture of motherwort, decreases the limitation of the ability of the retina to distinguish colours caused by a state of anxiety was also carried out. The experiment involved 26 healthy volunteers with a diagnosed state of nervous anxiety, divided into three groups, and control group consisting of 12 patients without emotional disorders. A decrease in anxiety and an improved ability to distinguish colours, both after the application of tofizopam (benzodiazepine derivative) (10 days) and the tincture of motherwort, were noted, yet the anxiolytic effect of tofizopam persisted longer (up to 1 month after cessation of use) in comparison with the tincture. The positive effect of the treatment on vision may have resulted from an impact on the GABAergic system in the retina and the brain structures connected with it (Ovanesov 2005).

In another experiment, 21 young patients (divided into three groups) with mild symptoms of anxiety and depression were administered melatonin, a tincture of motherwort or a placebo for 10 days (ten healthy volunteers constituted the control group). The quality of sleep and emotional state of the patients as well as their retinal function, i.e., the threshold of excitability to light stimuli and the sensomotoric response time of the process of vision, were evaluated before and after the application of the preparations. The administration of melatonin resulted in increased retinal sensitivity to light and an accelerated sensomotoric reaction. The impact on the process of vision after the administration of the tincture of motherwort was statistically insignificant, and sleep quality improved in some patients only. The anxiolytic activity of the tinctures of motherwort was confirmed, but it turned out to be weaker in comparison with that of melatonin (Ovanesov et al. 2006).

An oil extract of motherwort ($4\times300$ mg capsules daily for 28 days) was administered to 50 patients with a first (22 patients)- and second (28 patients)-degree hypertension and symptoms such as anxiety and sleep disorders. In patients with a first degree of hypertension, a reduction in the symptoms of anxiety, emotional instability, headaches, and sleep disorders was achieved. After 21 days, there was a significant decrease in and normalization of blood pressure (from 145/96 to 130/87), and the patients reported they experienced less fatigue and improved mood and activity. However, their reduction in heart rate (from 81.7 to 75.4) was not statistically significant. A considerable drop in blood pressure (from 153/103 to 142/92) in patients with a second degree of hypertension occurred a week later than the first group. The psycho-emotional condition of the patients (anxiety, emotional liability, headache, and sleep disturbances) was improved seven days before their blood pressure lowered. The hypotensive, anxiolytic, and calming effect may have been caused by iridoids, while the lack of antiarrhythmic action may have resulted from the absence of alkaloids in the prepared extract (Shikov et al. 2010).

6) Indications and usage:
Traditional herbal medicinal product used to relieve symptoms of nervous cardiac complaints such as palpitations (EMEA, 2010).

7) Precautions and side effects:
*Leonurus cardiaca* contains the alkaloid stachydrine. Because of its ability to stimulate uterine contractions, it is not recommended during pregnancy. Side effects include diarrhoea, stomach irritation, uterine bleeding, sleepiness, and allergic reactions. Contact with the skin can cause rashes and increased sensitivity

to the sun. Taking motherwort by mouth during pregnancy should be avoided. Motherwort can stimulate the uterus and might cause a miscarriage. Motherwort can interfere with the treatment of heart conditions. If you have a heart condition, get your healthcare provider's advice before starting motherwort. Motherwort can slow down (sedate) the central nervous system (CNS). There is a concern that when combined with anaesthesia and other medications used during and after surgery, motherwort might cause too much sedation. Stop using motherwort at least 2 weeks before a scheduled surgery. *L. cardiaca* may be able to stimulate uterine blood flow, which is why some women try it for starting their periods. If you have a uterine bleeding condition, use motherwort with caution, since it might make this condition worse.

Although no studies have been performed with *L. cardiaca* on toxicity in reproduction, the data on leonurine (present in *L. cardiaca*) and the studies in Asian *Leonurus* species point to a risk and confirm the contraindication in pregnancy that is widely and consistently found in several handbooks.

No studies on genotoxicity and carcinogenicity have been found.

8) Dosage:

A) Internal use:

Commission E recommends daily use (1–3 times) of infusions prepared from 4.5 g of dry motherwort herb in nervous cardiac conditions and as an adjuvant for thyroid hyperfunction. No health hazards or side effects are known if the prescribed therapeutic dosages are administered properly. Excessive use may interfere with the existing therapy of cardiac disorders, such as cardiac glycosides. Sensitive individuals may experience an allergic reaction. Motherwort is reputed to affect the menstrual cycle. In view of the documented in vitro uterotonic activity, the use of motherwort during pregnancy and lactation should be avoided. It is also not recommended to individuals under 18 years of age (Blumenthal et al. 2000, Wojtyniak et al. 2013).

B) External use:

Not found.

## 6.9 References


- Agnihotri VK, ElSohly HN, Smillie TJ, Khan IA, Walker LA. New labdane diterpenes from Leonurus cardiac. Planta Med. 2008;74:1288–90.
- Aizetmüller K, Tsevegüren N. Phlomic acid in Lamioidae seed oils. Lamiales Newslett. 1998;6:13–6
- Ali M, Salimpour F, Danaei Darzikolaei M, Akhoondi S, Kianoosh S. The quantitative comparison of essential oil composition of an Iranian endemic plant [*Leonurus cardiaca* subsp. persicus (Boiss.) Rech. F.] vs. previous studied population. Ann Biol Res. 2012;3(2): 1117–24.
- Assessment report on *Leonurus cardiaca*, herba. Committee on Herbal Medicinal Products (HMPC). EMA/HMPC/127430/2010. P. 1-35.

- Bernatoniene J, Kucinskaite A, Masteiková R, Kalveniene Z, Kasparaviciene G, Savickas A. The comparison of antioxidative kinetics in vitro of the fluid extract from maiden hair tree, motherwort and hawthorn. Acta Pol Pharm 2009;66(4): 415–21.
- Bernatoniene Jurga, Kopustinskiene Dalia M, Jakstas Valdas, Majiene Daiva, Baniene Rasa, Kursvietiene Lolita, Masteikova Ruta, Savickas Arunas, Toleikis Adolfas, Trumbeckaite Sonata. The effect of *Leonurus cardiaca* herb extract and some of its flavonoids on mitochondrial oxidative phosphorylation in the heart. Planta Medica. 2014; 80(7): 525–32.
- Blumenthal M, Goldberg A, Brinckmann J. Herbal Medicine Expanded Commission E Monographs. Integrative Medicine Communication. Austin, Texas: American Botanical Council. 2000;267–269.
- Brieskorn CH, Hofmann R. Labiatenbitterstoffe: ein Clerodanderivat aus *Leonurus cardiaca* L. Tetrahedron Lett. 1979;27: 2511–2.
- EMEA. Community herbal monograph on *Leonurus cardiaca* L., herba. EMA/HMPC/127428/2010.
- Gulubov AZ, Chervenkova VB. Structure of alkaloids from *Leonurus cardiaca*. Mat Fiz Khim Biol. 1970;8(1): 129–32.
- Janicsák G, Veres K, Kakasy AZ, Máthé I. Study of the oleanolic and ursolic acid contents of some species of the Lamiaceae. Bioch Syst Ecol. 2006;34: 392–6.
- Kartnig T, Gruber A, Menzinger S. Flavonoid-O-glycosides from the herbs of *Leonurus cardiaca*. J Nat Prod. 1985;48(3):494–507.
- Knöss W, Zapp J. Accumulation of furanic labdane diterpenes in *Marrubium vulgare* and *Leonurus cardiaca*. Plant Med. 1998;64: 357–61.
- Kuchta K, Melichar K, Ritter M, Dhein S, Rauwald HW. The antiarrhythmic effects of *Leonurus cardiaca* on cardiac electrophysiology, 7th Joint Meeting of GA, AFERP, ASP, PSE & SIF abstracts. Plant Med. 2008;74:326.
- Kuchta K, Ortwein J, Rauwald HW. *Leonurus japonicus*, *Leonurus cardiaca*, *Leonotis leonurus*: a novel HPLC study on the occurrence and content of the pharmacologically active guanidino derivative leonurine. Pharmazie. 2012; 67(12): 973–9.
- Kuchta Kenny, Ortwein Jutta, Hennig Lothar, Rauwald Hans Wilhelm. 1H-qNMR for direct quantification of stachydrine in *Leonurus japonicus* and *L. cardiac*. Fitoterapia. 2014; 96: 8–17.
- Malakov PY, Papanov GY, Jakupovic J, Grenz M, Bohlmann F. The structure of leocardin, two epimers of a diterpenoids from *Leonurus cardiaca*. Phytochemistry 1985;24: 2341–3.
- Masteiková R, Muselík J, Bernatoniene J, Majiene D, Savickas A, Malinauskas F, Bernatoniene R, Peciura R, Chalupová Z, Dvorácková K. Antioxidacní aktivita tinktur pripravených z hlohových plodu a nati srdecníku. [Antioxidant activity of tinctures prepared from hawthorn fruits and motherwort herb.] Ceská a Slovenská farmacie: casopis Ceské farmaceutické spolecnosti a Slovenské farmaceutické spolecnosti. 2008;57:35–8.
- Miłkowska-Leyck K, Filipek B, Strzelecka H. Pharmacological effects of lavandulifolioside from *Leonurus cardiaca*. J Ethnopharmacol. 2002;80:85–90.

- Mockutë D, Bernotienë G, Judzentienë A. Storage-induced changes in essential oil composition of *Leonurus cardiaca* L. plants growing in Vilnius and of commercial herbs. Chimija. 2005;16(2): 29–32.
- Morteza-Semnani K, Saeedi M, Akbarzadeh M. The essential oil composition of *Leonurus cardiaca*. J Essent Oil Res. 2008;20:107–9.
- Ovanesov KB. The effect of tofisopam and tinctura leonuri on the color-discrimination function in young humans. Eksp Klin Farmakol. 2005;68(3): 56–9.
- Ovanesov KB, Ovanesova IM, Arushanian EB. Effects of melatonin and motherwort tincture on the emotional state and visual functions in anxious subjects. Eksp Klin Farmakol. 2006;69(6):17–19.
- Papanov GY, Malakov PY, Rodriguez B, de la Torre MC. A prefuranic labdane diterpene from *Leonurus cardiaca*. Phytochemistry. 1998;47(6):1149–1151.
- Qi J, Hong ZY, Xin H, Zhu YZ. Neuroprotective effects of leonurine on ischemia/reperfusion-induced mitochondrial dysfunctions in rats' cerebral cortex. Biol Pharm Bull. 2010;33(12):1958–64.
- Reuter G, Diehl HJ. Medicinal plants of species *Leonurus* and their active substances. Pharmazie. 1970;25: 586–8.
- Romanowski H. Isolation and analysis of alkaloids from (*Leonurus cardiaca* L.) by chromatographic method. Acta Pol Pharm. 1960;17(1): 13–22.
- Rossyjskij DM. New cardiac drugs – preparations from the Motherwort. Klin Med. 1949;27: 83.
- Schultz OE, Alhyane M. Components of *Leonurus cardiaca* L. Sci Pharm. 1973;41: 149–155.
- Senatore F, De Feo V, De Simone F, Mscisz A, Mrugasiewicz K, Gorecki P. Sterols from *Leonurus cardiaca* L. in different geographical areas. Herba Pol. 1991;37(1): 3–7.
- Shikov AN, Pozharitskaya ON, Makarov VG, Demchenko DV, Shikh EV. Effect of *Leonurus cardiaca* oil extract in patients with arterial hypertension accompanied by anxiety and sleep disorders. Phytother Res. 2010;25(4): 540–543.
- Shikov, Alexander N.; Pozharitskaya, Olga N.; Makarov, Valery G.; Demchenko, Dmitry V.; Shikh, Evgenia V. Effect of *Leonurus cardiaca* oil extract in patients with arterial hypertension accompanied by anxiety and sleep disorders. Phytother Res. 2011; 25(4), 540–3.
- Tasdemir D, Calis I, Sticher O. Labdane diterpenes from *Leonurus persicus*. Phytochemistry. 1998;49:137–43
- Tschesche R, Diederich A, Jha HC. Caffeic acid 4-rutinoside from *Leonurus cardiaca*. Phytochemistry. 1980;19: 2783.
- Weinges K, Kloss P, Henkels WD. Naturstoffe aus Arzneipflanzen, XVIII Isolierung und Konstitutionsaufklärung eines neuen C15-Iridoidglucosids aus *Leonurus cardiaca* L. Liebigs Ann Chem. 1973;4: 566–72.
- Wichtl M. Herbal drug and phytopharmaceuticals: a handbook for practice on scientific basis. Stuttgart: Scientific Publishers; 2004.
- Widy-Tyszkiewicz E, Schmida RA. Randomized double blind study of sedative effects of phytotherapeutic containing valerian, hops, balm and motherwort versus placebo. Herba Pol. 1997;43: 154–9.

- Wojtyniak K. Master thesis: phenolic compounds in motherwort herb *Leonurus cardiaca*. Poznań: Poznań University of Medical Science; 2011.
- Wojtyniak K, Szymański M, Matławska I. *Leonurus cardiaca* L. (Motherwort): a review of its phytochemistry and pharmacology. 2013; 27(8):1115–20.
- Zhogova AA, Perova IB, Samylina IA, Eller KI, Ramenskaya GV. Identification and quantitative determination of the main biologically active substances in motherwort herb by HPLC-mass spectrometry. Pharmaceut Chem J. 2014; 48(7), 461–6.


## 6.10 *Leonurus turkestanicus* V.I. Kreczetovicz and Kuprianova

### *6.10.1* (Lamiaceae)

1) Common names:

English name: Turkestan motherwort
Russian name: Pustyrnik turkestanskiy
Uzbek name: Turkiston arslon quyrug'i
Kyrgyz name: Turkstan duloy chalkany
Kazakh name: Sasyqshep Turkistan (Eisenman et al. 2013)
Azerbaijani name: Turkestan shirguyrugu or Turkestan aslanguyrugu (Damirov et al. 1983)

2) Synonyms:
*Leonurus cardiaca* ssp. *turkestanicus* (V.Krecz. & Kuprian.) Rech. fil. and *Leonurus cardiaca* L. subsp. *turkestanicus* (V.I.Krecz. & Kuprian) Rech.f. Fl. Iranica [Rechinger]
3) Photograph of the species (Fig. 6.4)
4) Description of the species:

A) Habitat and world distribution:
It is common in mountain regions of Central Asia (Kyrgyzstan, Kazakhstan, Tajikistan, Turkmenistan, Uzbekistan) and Mongolia, Western Himalayas, Tibet, Xinjiang, and Iran. The tau zone is 1000–2000 m. Found on stony, floodplains, streamsides shallow-soiled slopes, among trees and shrubs (Eisenman et al. 2013, Vvedenskiy 1961).
B) Morphology of the plant:
*L. turkestanicus* grows in isolation, sometimes forming thickets. It grows as a weed, along fallows, on waste grounds, along the roads, in road cracks, in residential areas, and along the fences. It is found growing in small bunches among dumetums, on wood glades, and wood skirts, on pastures.

It is a perennial herbaceous plant with a ligneous rootstock. Stems are 50–150 cm tall, purple red, branched, and pubescent or glabrous. Leaves are opposite, petiolate, wide ovate to nearly circular in outline, and palmatipartite; lobes are pinnately divided into broadly lanceolate lobules. Flowers are sessile,

**Fig. 6.4** *Leonurus turkestanicus*. Photo taken by Eugene Davkaev and used with permission

15–20 per vertillaster, forming spiciform inflorescences. Bracts are awl- like and pubescent. Calyx is 8–9 mm long, funnelform, and short pubescent, with triangular spinescent lobes. Corolla is pink–lilac, ca. 1 cm long, two lipped, and villous outside. Fruits are triquetrous nutlets and light brown. Upper corolla lip is obovate, while lower lip is with three lobes, middle lobe larger than lateral lobes. Nutlets are grey brown, triquetrous, oblong, ca. 2 mm, base cuneate, apex truncate, and puberulent. Flowering occurs in Jul–Aug, fruits occur in Aug–Sep (Eisenman et al. 2013).

C) Medicinal parts:
Aerial parts of the plants collected during flowering

D) Chemical constituents:
The plant contains **flavonoids** (genkwanin, rutin, hyperoside, quercitrin), **iridoids** (8-acetylharpagide, harpagide, 6-deoxy-8-acetylharpagide), **phenolics** (chlorogenic acid, caffeic acid, caffeic acid derivatives, rosmarinic acid, etc.), **alkaloids** (leonurine, stachydrine), **triterpenoid** (oleanolic acid), **essential oils** (alpha-pinene, benzaldehyde, caryophyllene, limonene), and diterpene, tannins, saponins, resins, bitter substances, and other compounds (betaine, β-sitosterin) (Eisenman et al. 2013, Gusakova and Umarov 1972, Isaev et al. 2011, Kholmatov et al. 1998, Khalmatov 1964, Pulatova 1969, Khalmatov and Kosimov 1994).

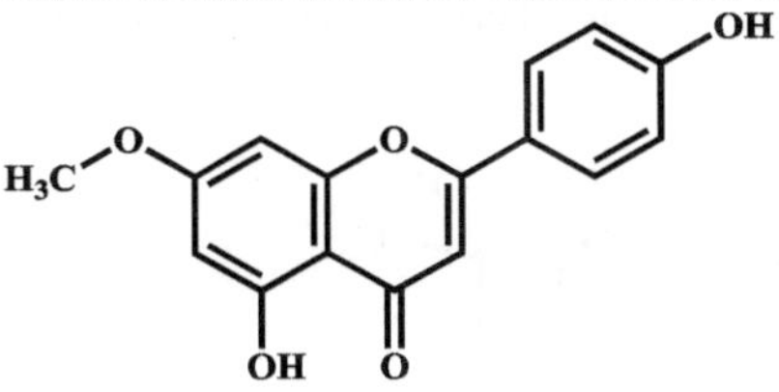

Genkwanin

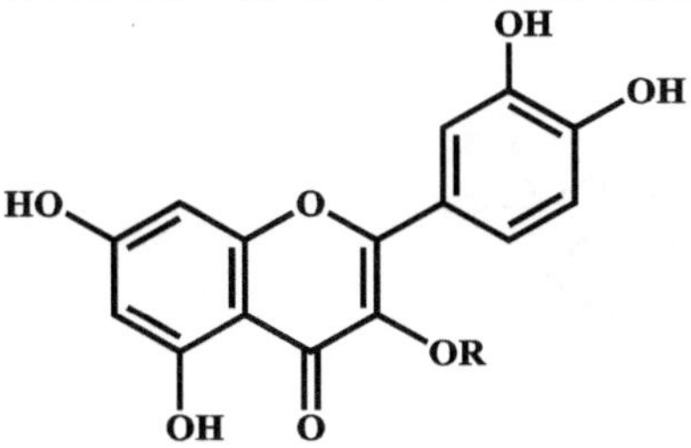

R = galactose: Quercetin-3-*O*-galactoside
R = rhamnose: quercetin-3-*O*-rhamnoside
R = rutinose: Quercetin-3-*O*-rutinoside

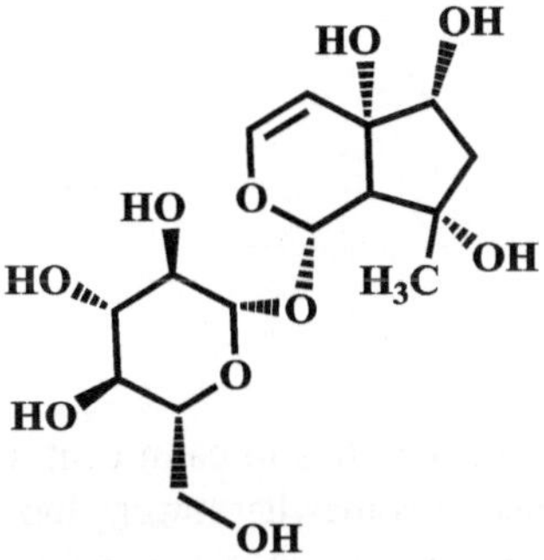

Harpagide

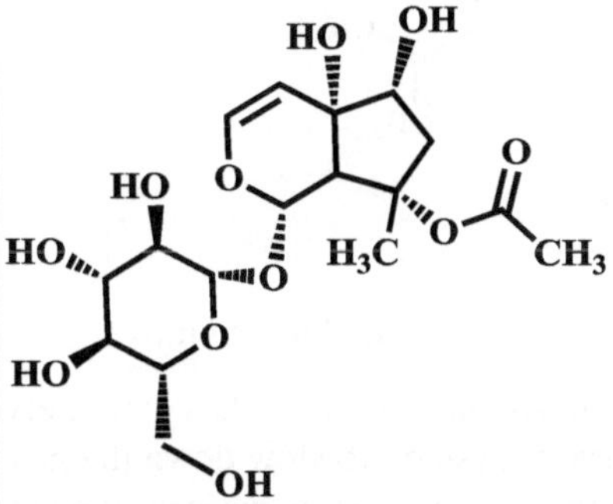

8-Acetylharpagide

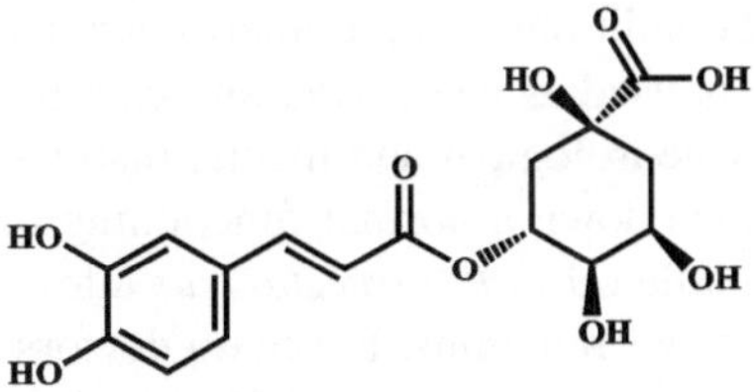

Chlorogenic acid

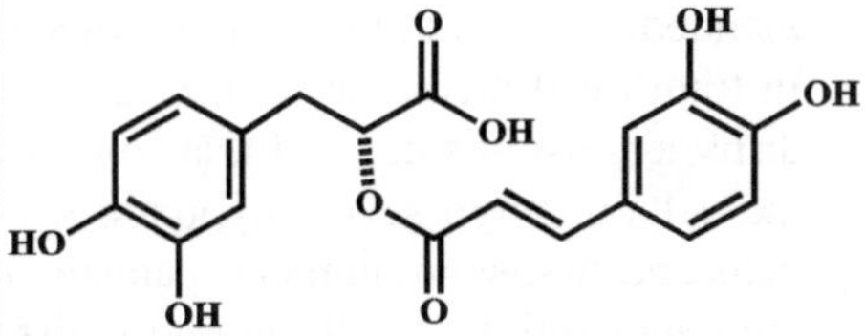

Rosmarinic acid

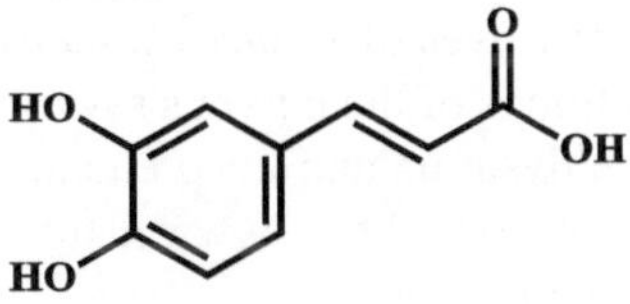

Caffeic acid

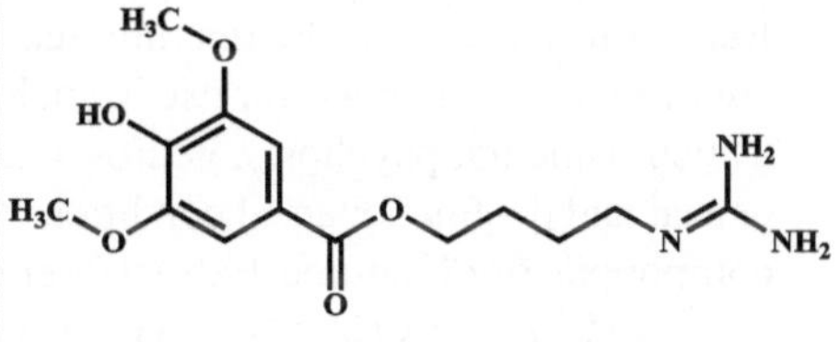

Leonurine

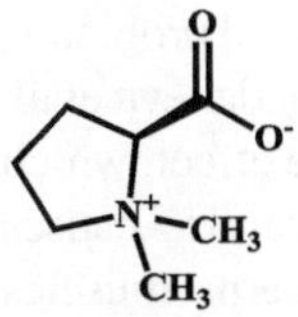

Stachydrine

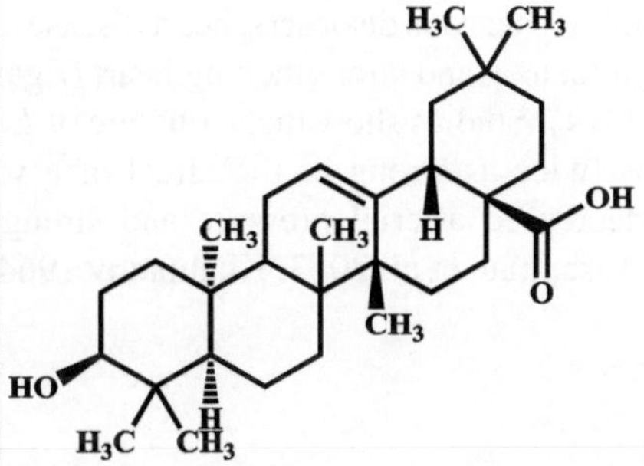

Oleanolic acid

| | |
|---|---|
| 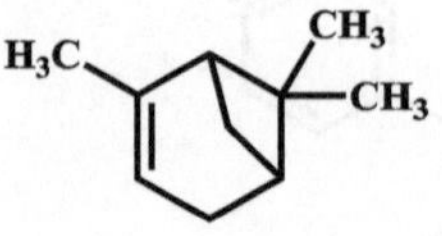 α-Pinene | 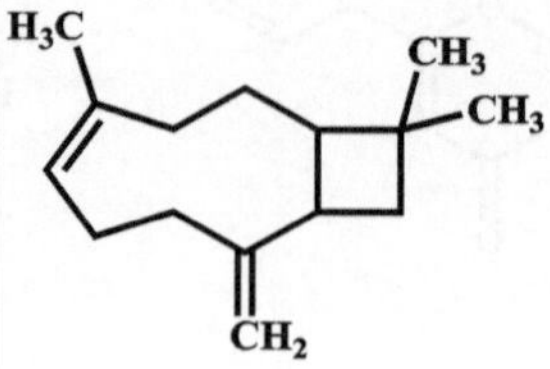 beta-Caryophyllene |
| 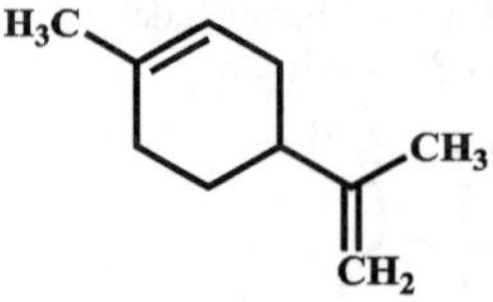 Limonene | |

5) Pharmacology and bioactivities:

*Leonurus turkestanicus* has extremely powerful characteristics to calm central nervous system, to slow down the heartbeat, and also to somewhat lower blood pressure. *L. turkestanicus* is used mainly for postmenopausal symptoms, including irregular heartbeat and anxiety. The tea from *L. turkestanicus* is recommended to use in multiple neurological heart conditions and for atherosclerosis. In traditional medicine, *L. turkestanicus* tea is used as energizing and calming drink against anxiety and fear. As a result of neurological and mental distress, the cells undergo severe hypoxia, which brings down a normal functioning of cells and tissues in human organism. The volatile oil of *L. turkestanicus* allows the regeneration of cells and normalization of cell functions. Based on doctors' observations, the preparations of *L. turkestanicus* inhibit some of the functions of central nervous system many times more than a tincture of valerian root. Based on that ability, the tincture, tea, and extract of leaves of *L. turkestanicus* are often recommended for use in such serious conditions of the nervous system as neurasthenia, psychosis, neurosis, and vegetative dystonia and for premenopausal and postmenopausal conditions (Eisenman et al. 2013, http://www.naturalcompounds.org/Featured-Extracts/Leonurus-Turkestanicus.html].

*L. turkestanicus* showed a wide spectrum of cytotoxic, diaphoretic, antispasmodic, tonic, sedative, and hypotensive activities. The plant is used to treat neuralgia, nervous disorders, heart diseases, stomach aches, gynaecological problems, influenza, and strengthening heart (Egamberdieva et al. 2013, Mamadalieva et al. 2014). Studies show that a tincture of *L. turkestanicus* has a sedative effect, which is twice as strong as the effect of a valerian tincture. The tincture also causes decreased arterial pressure and strengthens the contraction of uterus muscles (Eisenman et al. 2013, Khalmatov 1964).

6) Indications and usage:
   A decoction of the aboveground parts is used to treat heart, stomach, and nervous system diseases (Eisenman et al. 2013; Khalmatov 1964). A tea and an infusion of the aboveground parts are used to treat nervous disorders, hypertension, hysteria, epilepsy, tachycardia, gastrointestinal, and female diseases and are used as soporific, anti-inflammatory, diaphoretic, and laxative remedies (Eisenman et al. 2013; Khodzhimatov 1989). It is stomachic, laxative, antispasmodic, diaphoretic, emmenagogue. The plant is used in absent or painful menstruation, premenstrual tension, and menopausal flushes. It is hypnotic and sedative and is used as a cardiac tonic. Studies in China have shown that motherwort extracts show antiplatelet aggregation actions and decrease the levels of blood lipids. It is used on the treatment of nervous cardiac disorders and as adjuvant for thyroid hyperfunction. It is antispasmodic. The British Herbal Compendium indicated its use for patients who have neuropathic cardiac disorders and cardiac complaints of neurological origin. The herb is a slow-acting adjuvant in functional and neurogenic heart diseases. Its sedative and spasmolytic properties combine well with *Valeriana officinalis* or other cardioactive substances.
7) Precautions and side effects:
   Possible side effects could be allergy and indigestion.

## 6.11 References


- Damirov IA, Prilipko LI, Shukurov JZ, Karimov YB. Medicinal plants of Azerbaijan (in Russian), Baku: Maarif. 1983; 319.
- Egamberdieva D, Mamadalieva N, Khodjimatov O, Tiezzi A. Medicinal plants from Chatkal Biosphere Reserve used for folk medicine in Uzbekistan. Med Aromat Plant Sci Biotechnol. 2013;7(1):56–64.
- Eisenman SW, Zaurov DE, Struwe L. Medicinal plants of Central Asia: Uzbekistan and Kyrgyzstan. New York: Springer. 2013;340.
- Gusakova SD, Umarov AU. The oils of some plants of the family Labiatae. Chem Nat Compd. 1972;8(1): 23–8.
- Isaev IM, Agzamova MA, Isaev MI. Genkwanin and iridoid glycosides from *Leonurus turkestanicus*. Chem Nat Compd. 2011;47(1), 132–4.
- Khalmatov KK. Dikorastushchiye Lekarstvenniye Rasteniya Uzbekistana (Wild-growing medicinal plants of Uzbekistan) (in Russian). Tashkent: Meditsina; 1964.
- Khalmatov KK, Kosimov AI. Dorivor Usimliklar (Malummotnoma) [Medicinal plants (references)] (in Uzbek). Tashkent: Ibn Sino; 1994.
- Khodzhimatov M. Dikorastushchiye Lekarstvennuiye Rasteniya Tadzhikistana (Wild-growing medicinal plants of Tadjikistan) (in Russian). Dushanbe: Glavnaya nauchnaya redaktsiya Tadzhikskoi Sovetckoi entsiklopedii; 1989.

- Kholmatov Kh, Kharlamov IA, Mavlankulova ZI. Lekarstvennye rasteniya Tsentralnoy Azii (Medicinal plants of Central Asia). Tashkent: Med Lit Publishing Ibn Sino;1998;296.
- Mamadalieva NZ, Mamedov NA, Craker LE, Tiezzi A. Ethnobotanical uses and cytotoxic activities of native plants from the Lamiaceae family in Uzbekistan. Acta Horticulturae. USA: International Society for Horticultural Science. 2014;1030:61–70.
- Pulatova TP. Alkaloid content of some plants of the family Labiatae. Chem Nat Compd. 1969;5(1):55.
- Vvedenskiy A. Flora of Uzbekistan. Tashkent: Fan AS RUz. 1961;5: 363–4.


## 6.12 *Nepeta cataria* L

### *6.12.1* (Lamiaceae)

1) Common names:
   English name: Catnip, catnep, catmint, catswort, catrup, cataria
   Russian name: Koshachaya myata, koshachaya trava, kotovnik limonniy, kotovnik myatniy, koshechnik, koshachaya trava, shandra dushistaya, bolyachaya trava, polevoy balzam
   Uzbek name: Zufo, mushuk zufo
   Azerbaijani name: Pishik nanesi
   Persian name: Pudana
2) Synonyms:
   *Nepeta cataria* var. *citriodora* Beck
3) Photograph of the plant (Fig. 6.5)

**Fig. 6.5** *Nepeta cataria*. Photo taken by D. Gordon E. Robertson and used with permission

4) Description of the species:

A) Habitat and world distribution:
*Nepeta cataria* L. (family Labiatae), commonly known as catnip, is a herbaceous plant and is a native of southeast Europe, Orient, Eurasia, and Western temperate Himalayas and widely naturalized in North America, south-eastern Australia, and possibly elsewhere. It is often found growing in disturbed areas. The foliage has a minty odour that is often considered mildly unpleasant by humans, but is very attractive to many cats.

B) Morphology of the plant:
*Nepeta cataria* is a perennial herb, 30–100 cm in height, and branched above. The leafy stems are grey pubescent (downy). Leaves are heart shaped and toothed with covered hairs. The leaves (3–8 cm) are simple (lobed or unlobed but not separated into leaflets). Leaf arrangement is opposite: there are two leaves per node along the stem. The edge of the leaf blade has teeth. There are five petals, sepals, or tepals in the flower. Flowers are whitish or pale lilac, dotted with pink or purple; the lower lip is slightly toothed. Flowering occurs from July to September. The fruit is dry but does not split open when ripe. Fruit length is 1.3–1.7 mm.

C) Medicinal parts:
Stem, leaves, and flowers

D) Chemical constituents: **Essential oils**.
4aα,7α,7aα-nepetalactone, 4aα,7α,7aβ-nepetalactone, 4aα,7β,7aα-nepetalactone, 4aα,7α,7aβ-3-nepetalactone, 3,4β-dihydro-4aα,7α,7aα-nepetalactone, 5,9-dehydronepetalactone, dihydronepetalactone, epinepetalactone, Z,E-nepetalactone and E,Z-nepetalactone, ascaridole, β-bourbonene, camphor, δ-3-carene, β-caryophyllene, β-caryophyllene oxide, citronellal, citronellyl acetate, β-citronellol, citronellol, α-citral, 1,8-cineol, α-copaene, 3,7-dimethyl-6-octen-1-ol formate, 3,7-dimethyl-1-oxabicyclo[3,3,0]oct-2-ene, β-farnesene, geranyl acetate, geranial, geranic acid, germacrene-D, hexenyl benzoate, α-humulene (=α-caryophyllene), humulene oxide, linalool, linolenic acid, 3-octanone, neral, nerol, nerol oxide, α-pinene, β-pinene, piperitone, α-terpineol, α-terpinolene, α-thujene, thymol, thymol methyl ether, trans-rose oxide, rose-oxide, etc. were found in the oil of catnip (Adiguzel et al. 2009; Baranauskiene et al. 2003; Baser et al. 2001; Bourrel et al. 1993; Chalchat and Lamy 1997; Domokos et al. 1994; Gilani et al. 2009; Kim et al. 2006; Klimek and Modnicki 2005; Malizia et al. 1996; Mohammadi and Mohammad Jamal 2011; Morteza-Semnani and Saedii 2004; Regnier et al. 1967a, b; Safaei-Ghomi et al. 2009; Saeidnia et al. 2014; Sastry et al. 1972; Wang et al. 2007; Wesolowska et al. 2011). **Iridoids and iridoid glucosides** (nepetolglucosylester, 5-epideoxyloganic acid, 1,5,9-epideoxyloganic acid, nepetaside, (1R,5R,8S,9S)-deoxyloganic acid, nepetariaside) (Handjieva et al. 1996; Murai et al. 1984, 1987; Tagawa and Murai 1980, 1983; Xie et al. 1988) and **triterpenes** (ursolic acid, oleanolic acid, daucosterol (β-sitosterol 3-*O*-glucoside), β-sitosterol, campesterol, α-amyrin, and β-amyrin) were isolated from *Nepeta cataria* var. *citriodora* (Janicsa´k et al. 2006; Klimek and Modnicki 2005). **Phenolic compounds**

**and flavonoids** (luteolin-7-*O*-glucuronide, luteolin-7-*O*-glucurono-(1 → 6)-glucoside, apigenin-7-*O*-glucuronide, luteolin, apigenin, apigenin-7-*O*-glucoside, luteolin-7-*O*-glucoside, 8-hydroxycirsiliol, 8-hydroxycirsimaritin or isothymusin, cirsimaritin, genkwanin, caffeoyl phenylethanoid glycosides (teucrioside, verbascoside, and lamiuside A) and **other compounds** have been isolated from catnip herb (Jamzad et al. 2003; Modnicki et al. 2007; Prescott et al. 2011). In *N. cataria*, amino acids were detected, including aspartic acid, threonine, arginine, histidine, isoleucine, glutamic acid, glutamine, alanine, phenylalanine, valine, glycine, leucine, serine, proline, methionine, asparagine, ornithine, and lysine (Grytsyk et al. 2013); 14 **micro- and macrominerals** (Al, Ca, Co, Cr, Cu, Fe, K, Li, Mg, Mn, Na, P, Se, and Zn) were detected in *Nepeta cataria* (Petenatti et al. 2011).

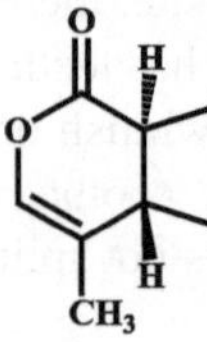

*Trans*, *cis*-nepetalactone

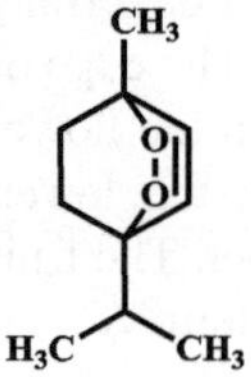

Ascaridole

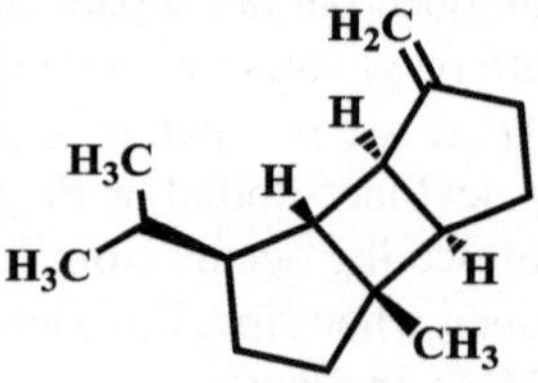

β-Bourbonene

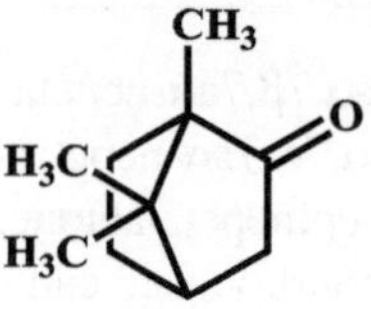

Camphor

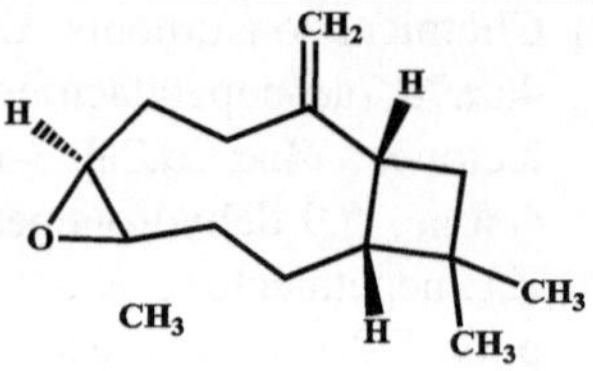

β-Caryophyllene oxide

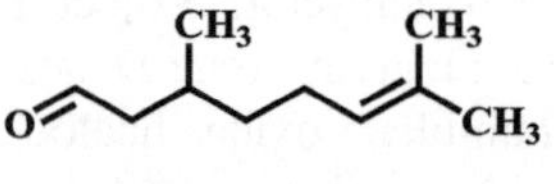

Citronellal

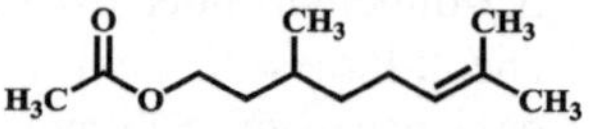

Citronellyl acetate

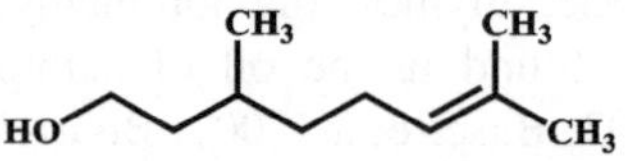

β-Citronellol

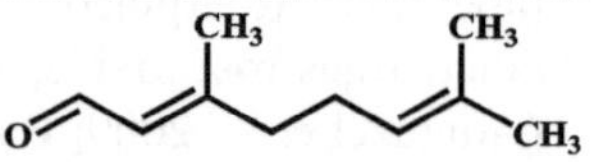

Citral (a mixture of the isomers neral and geranial)

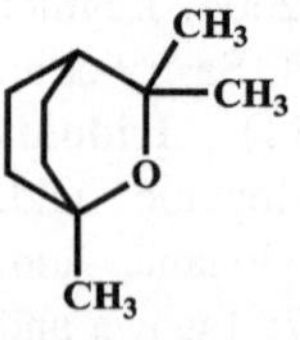

1,8-Cineole

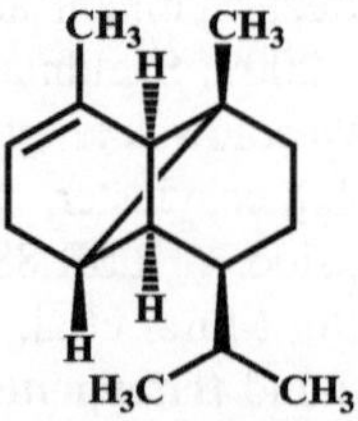

α-Ylangene (= α-Copaene)

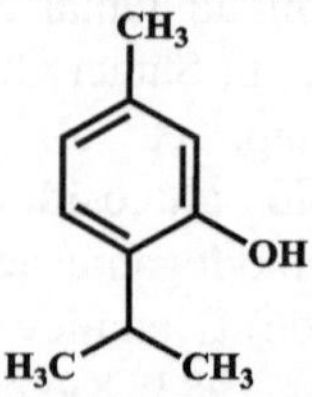

Thymol

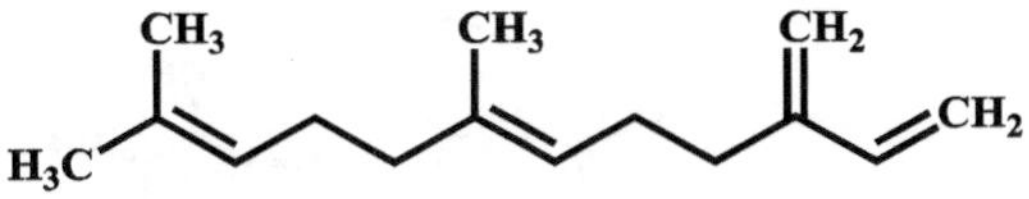

β-Farnesene

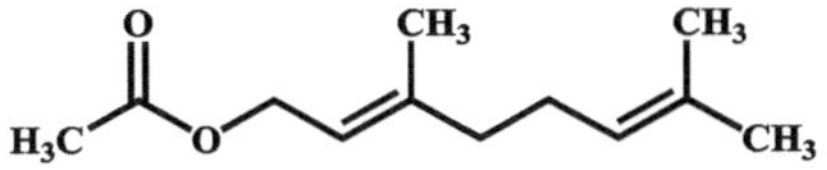

Geranyl acetate

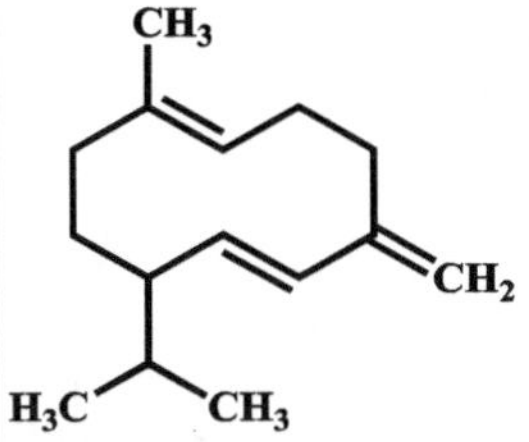

Germacrene-D

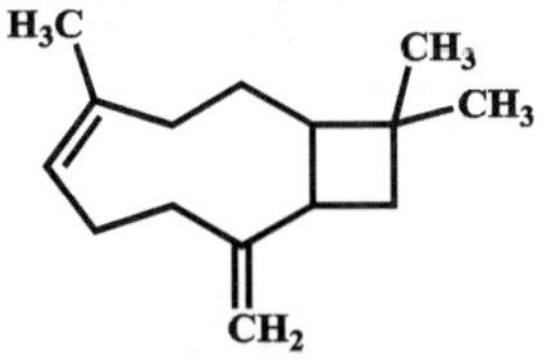

β-Caryophyllene

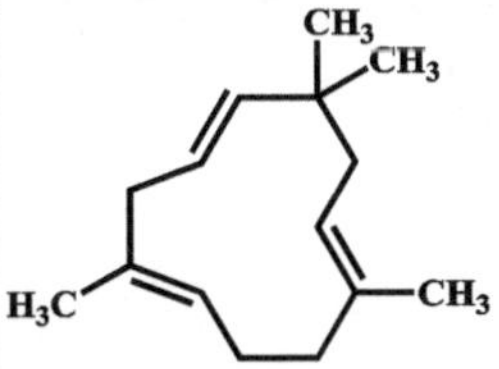

α-Caryophyllene (= α-humulene)

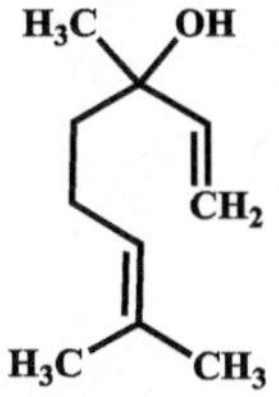

Linalool

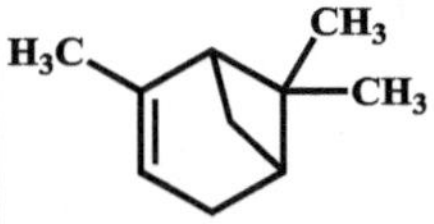

α-Pinene

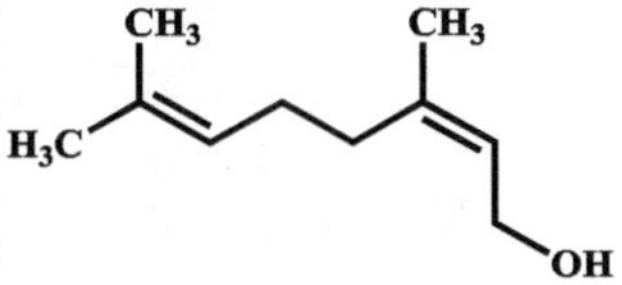

Nerol

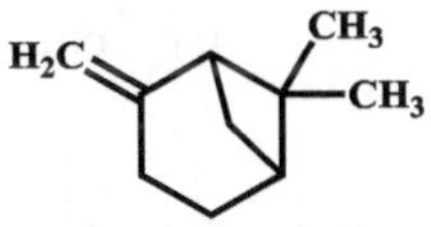

β-Pinene

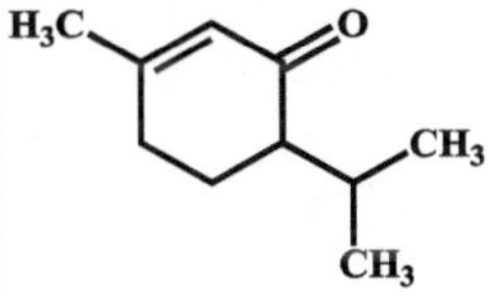

Piperitone

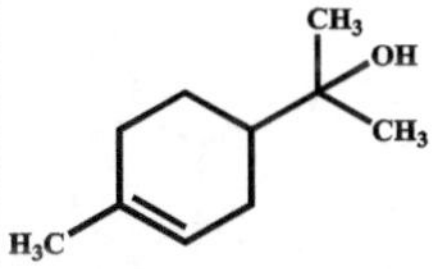

α-Terpineol

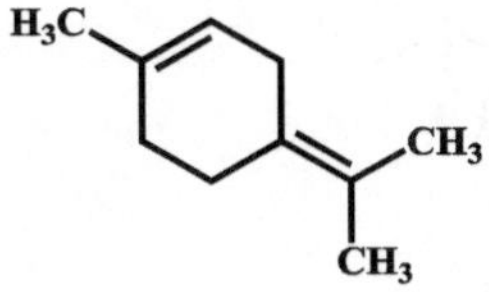

α-Terpinolene

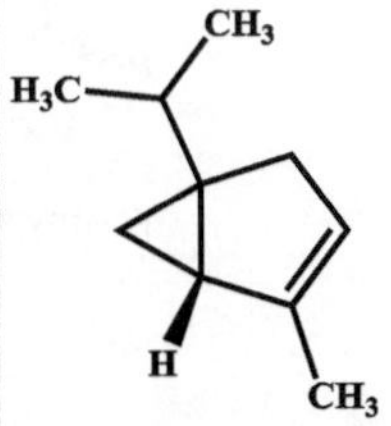

α-Thujene

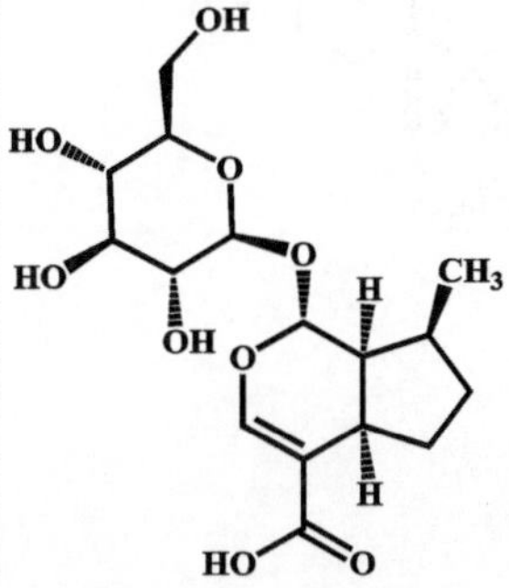

5-Epideoxyloganic acid

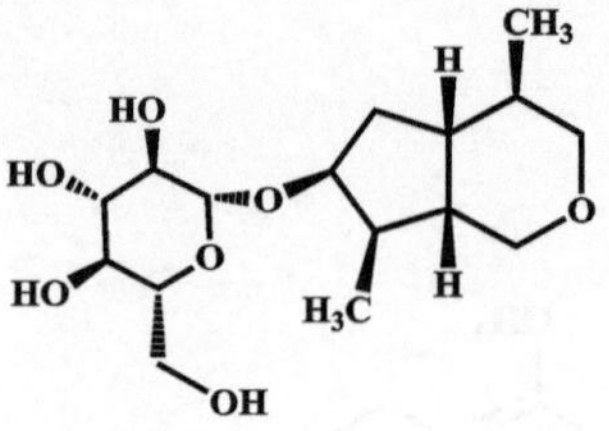

Nepetaside

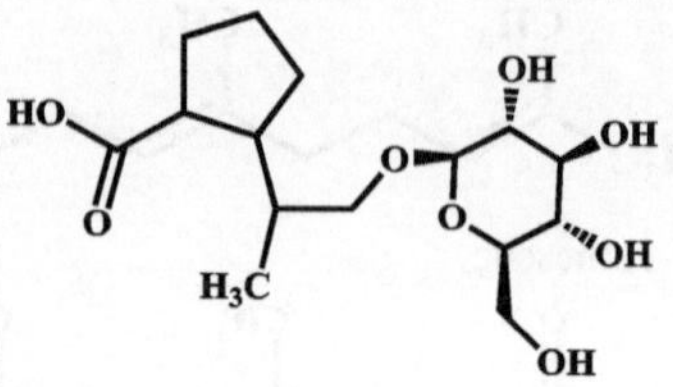

Nepetariaside

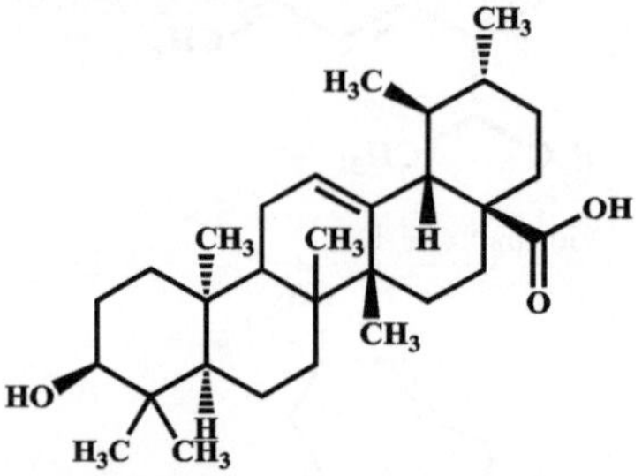

Ursolic acid

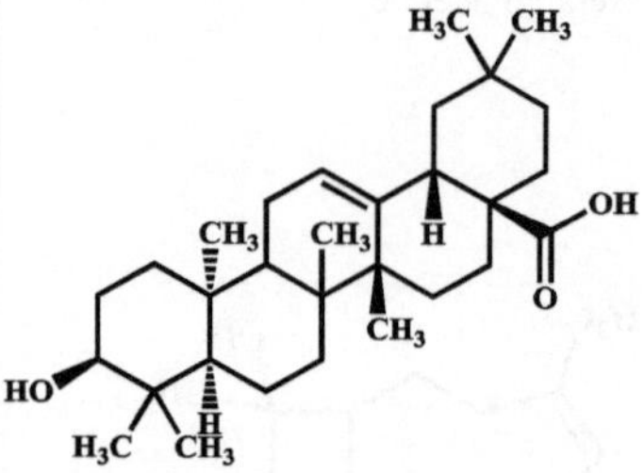

Oleanolic acid

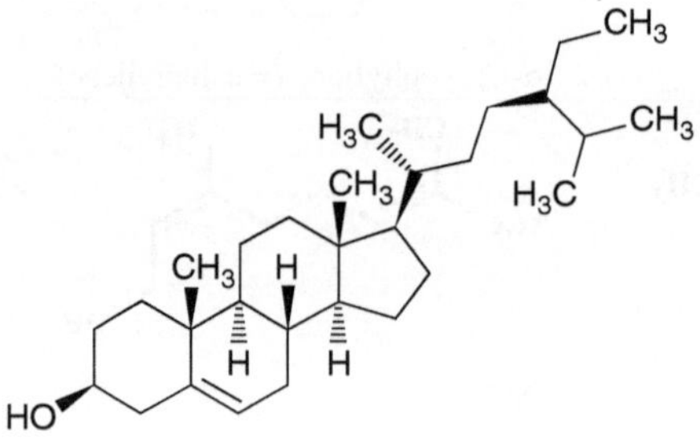

β-Sitosterol

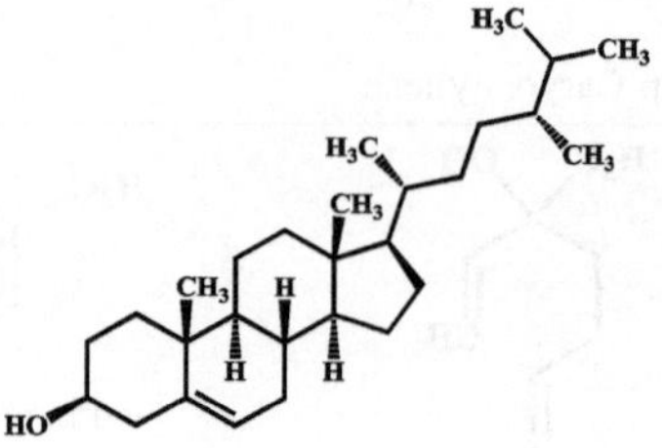

Campesterol

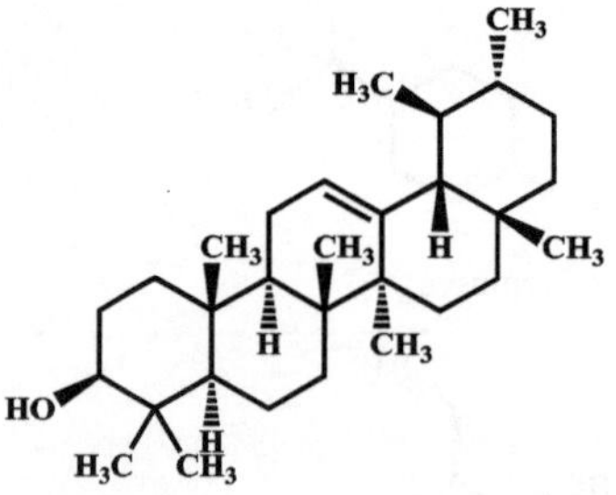

α-Amyrin

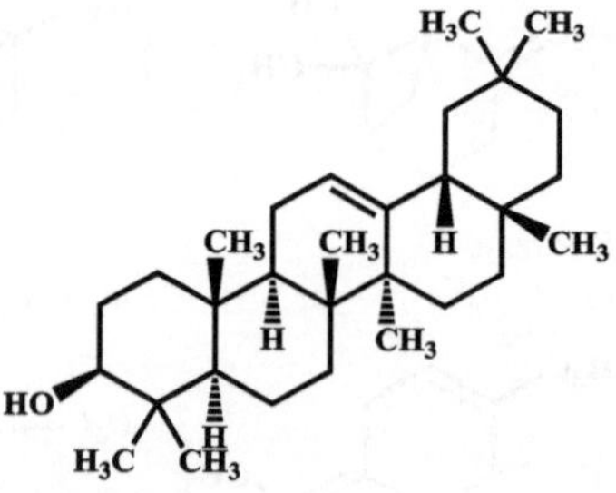

β-Amyrin

| | |
|---|---|
| 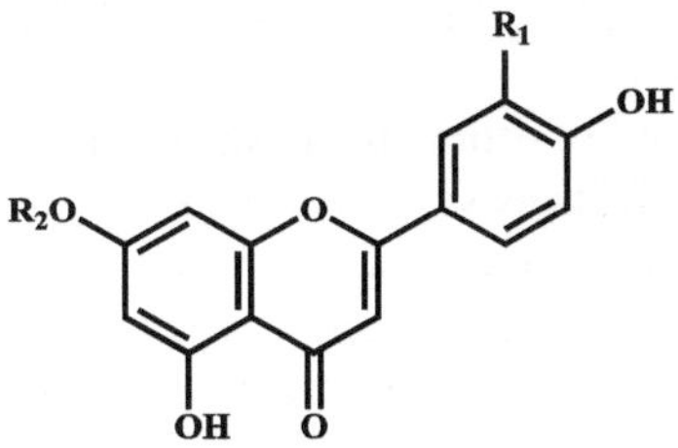<br>Luteolin: $R_1$=OH, $R_2$=H<br>Apigenin: $R_1$=$R_2$=H<br>Luteolin-7-*O*-glucuronide: $R_1$=OH, $R_2$=glucuronic acid<br>Apigenin-7-*O*-glucuronide: $R_1$=H, $R_2$=glucuronic acid<br>Apigenin-7-*O*-glucoside: $R_1$=H, $R_2$=glucose<br>Luteolin-7-*O*-glucoside: $R_1$=OH, $R_2$=glucose | 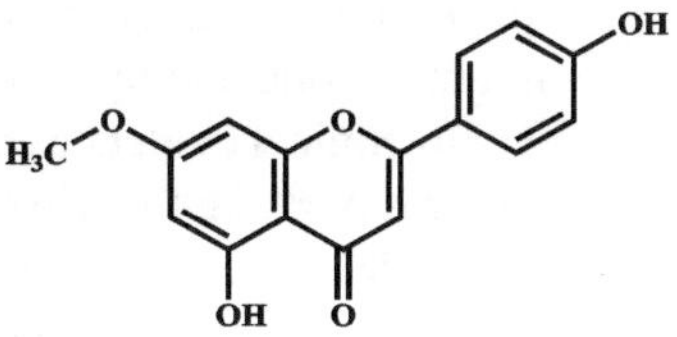<br>Genkwanin |
| 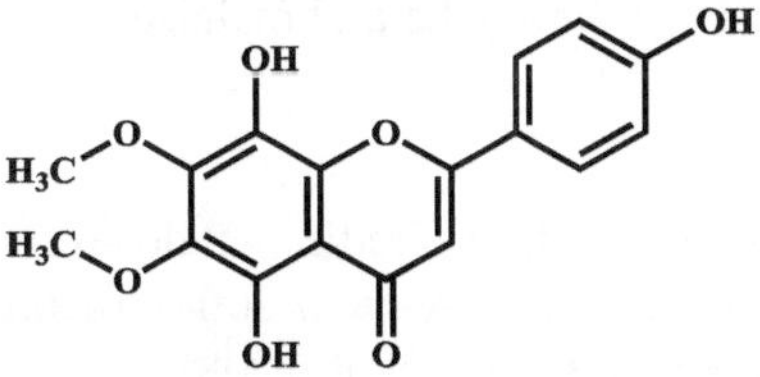<br>Isothymusin | |

5) Pharmacology and bioactivities:

A) In vivo studies:

Several central nervous system effects related to antidepressants have been described for *N. cataria*. Cauffield and Forbes (1999) reported that the *N. cataria* leaf is used as a dietary supplement for the treatment of depression, anxiety, and sleep disorders. Additionally, Bhat and Moskovitz (2009) showed that *N. cataria* is effective in the treatment of stress, tension, and insomnia. Comparing the *N. cataria* and *Melissa officinalis* effects, these plants have similar pharmacological profiles as central nervous system antidepressants (Zeichen et al. 2004).

The studies of Bernardi et al. (2010) indicated that chow enriched with 10 % *N. cataria* and the apolar extract have antidepressant properties, especially after repeated administration. Additionally, this study indicated that the active principles responsible for the antidepressant-like effects of *N. cataria* are present in the apolar extract. These findings corroborate reports of *N. cataria* for the treatment of depressive disorders as a complementary and alternative medicine. Moreover, the results suggest that an effective treatment should be with the apolar extract using repeated administration.

In traditional medicine, this plant was used as a sedative, spasmolytic, and tonic remedy (Klimek and Modnicki 2005). Catnip is also said to soothe the nervous system and calm the stomach, aiding both diarrhoea and colic. *Nepeta cataria* contain tannin and volatile oil which are aromatic, carminative, tonic, diaphoretic, refrigerant, emmenagogue, antiseptic, stimulant, and useful in infantile colic and hysteria (Sarkar et al. 1995).

No clinical evidence exists to support the use of catnip as a CNS stimulant. Older reports on the use of catnip for CNS effects exist, with euphoric states being described; however, the identity of the plants has been questioned (Jackson and Reed 1969; Petersik et al. 1969). There is one case report of an infant with CNS depression after consuming a large amount of catnip (Osterhoudt et al. 1997).

6) Indications and usage:
Used for nervousness and as sedative, tonic and stimulant, digestive, gases, wounds, menstruation induction, cough and whooping cough, anti-haemorrhoidal, migraine, diuretic (Neves et al. 2009)
7) Precautions and side effects:
Data are limited. Excessive ingestion may result in headache and malaise.
8) Dosage:

A) Internal use:
There is no clinical evidence to guide dosage of catnip. Traditional doses for sedation require 4 g of herb, usually given as a tea. A 15 % lotion of the essential oil has been used as an insect repellent (Spero et al. 2008).
B) External use:
Not found.

## 6.13 References


- Adiguzel A, Ozer H, Sokmen M, Gulluce M, Sokmen A, Kilic H, Sahin F, Baris O. Antimicrobial and antioxidant activity of the essential oil and methanol extract of *Nepeta cataria*. Pol J Microbiol. 2009;58: 1: 69–76.
- Baranauskiene R, Venskutonis RP, Demyttenaere JC. Sensory and instrumental evaluation of catnip (*Nepeta Cataria* L.) aroma. J Agric Food Chem 2003;51: 13: 3840-8.
- Baser KHC, Kirimer N, Kurkcuoglu M, Demirci B. Essential oils of *Nepeta* species growing in Turkey. Chem Nat Compd 2001;36: 4: 356–9.
- Bernardi MM, Kirsten TB, Salzgeber SA, Ricci EL, Romoff P, Lago JHG, Lourenço LM. Antidepressant-like effects of an apolar extract and chow enriched with *Nepeta cataria* (catnip) in mice. Psychol Neurosci. 2010; 3: 251–8.
- Bhat RD, Moskovitz G. Herbal medicinal teas from South Africa. Int J Exp Bot. 2009;78: 67–73.
- Bourrel C, Perineau F, Michel G, Bessiere JM. Catnip (*Nepeta cataria* L.) essential oil: Analysis of chemical constituents, bacteriostatic and fungistatic properties. J Essent Oil Res. 1993;5(2): 159–67.
- Cauffield JS, Forbes HJ. Dietary supplements used in the treatment of depression, anxiety, and sleep disorders. Lippincotts Primary Care Practice. 1999;3: 290–304

- Chalchat Jean-Claude, Jacques Lamy. Chemical composition of the essential oil isolated from wild catnip *Nepeta cataria* L. cv. *citriodora* from the drome region of France. J Essent Oil Res. 1997;9(5): 527–32.
- Domokos J, Peredi J, Halasz-Zelnik K. Characterization of seed oils of dragonhead (*Dracocephalum moldavica* L.) and catnip (*Nepeta cataria* var. *citriodora* Balb.). Ind Crops Prod. 1994;3(1,2): 91–4.
- Gilani AH, Shah AJ, Zubair A, Khalid S, Kiani J, Ahmed A, Rasheed M, Ahmad VU. Chemical composition and mechanisms underlying the spasmolytic and bronchodilatory properties of the essential oil of *Nepeta cataria* L. J Ethnopharmacol 2009;121(3): 405–11.
- Grytsyk LM, Tuchak NI, Stasiv TG, Grytsyk AR. Amino acid composition of the alchemilla L. genus plants and *Nepeta cataria* L. Pharma Innovat 2013;2(4): 50–3.
- Handjieva Nedjalka V, Simeon S Popov, Ljubka N. Evstatieva. Constituents of essential oils from *Nepeta cataria* L., *N. grandiflora* M.B. and *N. nuda* L. J. Essent. Oil Res. 1996;8: 6 639–43.
- Jackson B, Reed A. Catnip and the alteration of consciousness. JAMA. 1969; 207(7):1349–50.
- Jamzad Ziba, Renee J Grayer, Geoffrey C Kite, Monique SJ Simmonds, Martin Ingrouille, Adel Jalili. Leaf surface flavonoids in iranian species of *Nepeta* (Lamiaceae) and some related genera. Biochem Syst Ecol. 2003;31:6:587–600.
- Janicsak Gabor, Katalin Veres, Andras Zoltan Kakasy, Imre Mathe. Study of the oleanolic and ursolic acid contents of some species of the Lamiaceae. Biochem Syst Ecol. 2006;34: 5: 392–6.
- Kim JH, Jung DH, Park HK. The Composition of essential oil from *Nepeta cataria* and its effect on microorganism. J Ecol Field Biol. 2006;29 (4): 381–7.
- Klimek B, Modnicki D. Terpenoids and sterols from *Nepeta cataria* L. var. *citriodora* (Lamiaceae). Acta Pol Pharm. 2005;62(3): 231–5.
- Malizia Ruben A, Jose S Molli, Daniel A Cardell, Juan A. Retamar. Volatile constituents of the essential oil of *Nepeta cataria* L. Grown in Cordoba Province (Argentina). J Essent Oil Res. 1996;8(5): 565–7.
- Modnicki D, Tokar M, Klimek B. Flavonoids and phenolic acids of *Nepeta cataria* L. var. *citriodora* (Becker) Balb. (Lamiaceae). Acta Pol Pharm. 2007; 64(3): 247–52.
- Mohammadi Shirin, Mohammad Jamal Saharkhiz. Changes in essential oil content and composition of catnip (*Nepeta cataria* L.) during different developmental stages. J Essent Oil Bear Plants 2011;14(4): 396–400.
- Morteza-Semnani Katayoun, Madjid Saeedi. Essential oil composition of *Nepeta cataria* L. and *Nepeta crassifolia* Boiss. and Buhse from Iran. J Essent Oil Bear Plants 2004;7(2): 120–4.
- Murai F, Tagawa M, Damtoft S, Jensen SR, Nielsen BJ. (1r,5r,8 s,9 s)-Deoxyloganic acid from *Nepeta cataria*. Chem Pharm Bull (Tokyo) 1984;32(7): 2809–14.
- Murai Fujio, Motoko Tagawa, Hiroyuki Inouye, Toshimasa Ishida, Masatoshi Inoue. A new iridoid glucoside, nepetariaside, from *Nepeta cataria*. Chem Pharm Bull. 1987;35(6):2533–7.

- Neves JM, Matos C, Moutinho C, Queiroz G, Gomes LR. Ethnopharmacological notes about ancient uses of medicinal plants in Trás-os-Montes (northern of Portugal). J Ethnopharmacol. 2009; 124: 270–83.
- Osterhoudt KC, Lee SK, Callahan JM, Henretig FM. Catnip and the alteration of human consciousness. Vet Hum Toxicol. 1997; 39(6):373–5.
- Petenatti Marta E, Elisa m Petenatti, Luis A Del Vitto, Mauricio R. Teves, Nestor O. Caffini, Eduardo J. Marchevsky, Roberto G. Pellerano. Evaluation of macro and microminerals in crude drugs and infusions of five herbs widely used as sedatives. Rev Bras Farmacogn. 2011;21(6): 1144–9.
- Petersik JT, Poundstone J, Estes JW, et al. Of cats, catnip, and Cannabis. JAMA. 1969;208:360.
- Prescott Thomas AK, Nigel C Veitch, Monique SJ Simmonds. Direct inhibition of calcineurin by caffeoyl phenylethanoid glycosides from *Teucrium chamaedrys* and *Nepeta cataria*. J. Ethnopharmacol. 2011;137(3): 1306–10.
- Regnier Frederick E, Edmund J Eisenbraun, George R Waller. Nepetalactone and epinepetalactone from *Nepeta cataria*. Phytochemistry 1967a;6(9): 1271–81.
- Regnier Frederick E, George R Waller, Edmund J Eisenbraun. Studies on the composition of the essential oils of three *Nepeta* species. Phytochemistry 1967b;6(9): 1281–9.
- Saeidnia Soodabeh, Ahmad Reza Gohari, Azita Haddadi, Gholamreza Amin, Marjan Nikan, Abbass Hadjiakhoondi. Presence of monoterpene synthase in four labiatae species and solid-phase microextraction- gas chromatography-mass spectroscopy analysis of their aroma profiles. Pharmacogn Res. 2014;6(2): 138–142, 3.
- Safaei-Ghomi Javad, Zahra Djafari-Bidgoli, Hossein Batooli. Volatile constituents analysis of *Nepeta Cataria* from central Iran. Chem Nat Compd. 2009;45(6): 913–5.
- Sarkar M, Rashmi R Vikramaditya, Varma PN. Pharmacognosy of *Nepeta cataria*. Anc Sci Life 1995;14(4): 225–34.
- Sastry SD, Springstube WR, Waller GR. Identification of 5,9-dehydronepetalactone, a new monoterpene from *Nepeta cataria*. Phytochemistry. 1972;11(1): 453–5.
- Spero NC, Gonzalez YI, Scialdone MA, Hallahan DL. Repellency of hydrogenated catmint oil formulations to black flies and mosquitoes in the field. J Med Entomol. 2008; 45(6):1080–6.
- Tagawa Motoko, Fujio Murai. A new iridoid glucoside, nepetolglucosylester from *Nepeta cataria*. Planta Med. 1980;39(2): 144–7.
- Tagawa M, Murai F. 5-Epideoxyloganic acid from *Nepeta cataria*. Planta Med 1983;47(2): 109–11.
- Wang M, Cheng KW, Wu Q, Simon JE. Quantification of nepetalactones in catnip (*Nepeta cataria* L.) by HPLC coupled with ultraviolet and mass spectrometric detection. Phytochem Anal 2007;18(2): 157–60.
- Wesolowska A, Jadczak D, Grzeszczuk M. GC-MS Analysis of lemon catnip (*Nepeta cataria* L. Var. *citriodora* Balbis) Essential Oil Acta Chromatogr. 2011;23(1): 169–80.

- Xie Shan, Shinichi Uesato, Hiroyuki Inouye, Tetsuro Fujita, Fujio Murai, Motoko Tagawa, Tetsuro Shingu. Absolute structure of nepetaside, a new iridoid glucoside from *Nepeta cataria*. Phytochemistry 1988;27(2): 469–72.
- Zeichen R, Gargiulo S, Carena M, Bindstein E. Estudió farmacológico comparativo de dos espécies argentinas: *Nepeta cataria* L.(Labiatae) y Melissa officinalis L.(Labiatae). Boletín Latinoamericano Y del Caribe de Plantas Medicinales y Aromaticas. 2004;3: 103–6.


## 6.14 *Oplopanax elatus* (Nakai) Nakai

### *6.14.1* (Araliaceae)

1) Common names:

   Russian name: Zamanikha
   Chinese name: Ci shen

2) Synonym: *Echinopanax elatus* Nakai
3) Photograph of the plant (Fig. 6.6)

**Fig. 6.6** *Oplopanax elatus*. Photo taken by Denis Kochetkov and used with permission

4) Description of the species:

A) Habitat and world distribution:
It is mainly distributed in Russia, in the south of Primorye, while outside of Russia it grows in the northern part of the Korean Peninsula and in the temperate regions of Northern China (Shikov et al. 2014).

B) Morphology of the plant:
It is a deciduous shrub with a spiny stem of about 1 m (rarely to 2–3 m) height, with large palmately compound leaves and a long creeping rhizome. Inflorescence is axillary and 7–18 cm long; peduncles are 2.5–6 cm and densely covered with long rust-coloured hairs and prickles; umbels are small, 9–13 mm across, in oblong raceme or a slightly branching panicle at tips of main and lateral axes; fruit is yellow red and 7–12 mm long. The plant is endangered in Russia due to intense uncontrolled harvesting and is included in the Red Book of the Russian Federation. The roots are collected in September–October after ripening of the fruits (Shikov et al. 2014).

C) Medicinal parts:
Roots and rhizomes

D) Chemical constituents:
**Saponins** (cirenshenosides or cirensenosides I, J, K, L, M, N, O, P, Q, R, S, T, U, and V and their glycosides I, II, III, and A–D) (Wang et al. 1993, 1994a, b, 1996a–c, 1997a–c, 2004), **anthraquinones** (chrysophanol, physcion, emodin, rhein, and aloe-emodin) (Xu and Liang 1998; Wang et al. 2009), **lignans** (3,3′-dimethoxy-4,9,9′-trihydroxy-4′,7-epoxy-5′8-lignan-4,9-bis-O-β-D-glucopyranoside, 5-methoxylariciresinol-4-O-β-D-glucopyranoside, (+) 5′-methoxy isolariciresinol-3α-O-β-D-glucopyranoside, (+) isolariciresinol-3α-O-β-D-glucopyranoside) (Dou et al. 2009), **phenylpropanoids** (3-*O*-biosides of kaempferol and quercetin, 4-(3-hydroxy-1-propenyl)-2,6-dimethoxyphenyl-β-D-glucopyranoside (syringin), 4-(3-hydroxy-1-propenyl)-2-methoxyphenyl-β-D-glucopyranoside (coniferin), 4-(3-hydroxypropyl)-2-methoxyphenyl-β-D-glucopyranoside, naringenin, hesperidin, quercetin, *trans*-cinnamic acid, 3-*O*-caffeoylquinic acid, 1-*O*-caffeoylquinic acid, *p*-hydroxybenzoic acid, chlorogenic acid, caffeic acid, syringic acid, p-coumaric acid, ferulic acid) (Dou et al. 2009; Murav'ev et al. 1973; Kim et al. 2007; Wang and Xu 1993), and **essential oils**. A total of nineteen volatile ingredients were characterized in the roots and rhizomes of *O. elatus*. The stem contained 3.1 % of the oil, and 32 compounds were identified. Of these, α-pinene, octanal, 6, 6-dimethyl- 2-methylene-norpinane, and 5-methyl hexanal were the four most abundant volatile ingredients (Wu et al. 2007; Zhang et al. 1999).

**Other constituents**, such as sterols and their glycosides (β-sitosterol, stigmasterol, stigmasterol-3-*O*-β-D-glucopyranoside VI, sitosterol-3-*O*-β-Dglucopyranoside), polyacetylenes (oploxynes A and B, oplopandiol and falcarindiol), sugars (sucrose and glucose), fatty acids (6,9-octadecadienoic acid, palmitoleic acid and heneicosanoic acid), organic acids, aliphatic acids, oligosaccharides, coumarins, alkaloids, resins, and trace elements, have also been reported from *O. elatus* (Dou et al. 2009; Murav'ev and Dzhumaev 1973; Shikov et al. 2014).

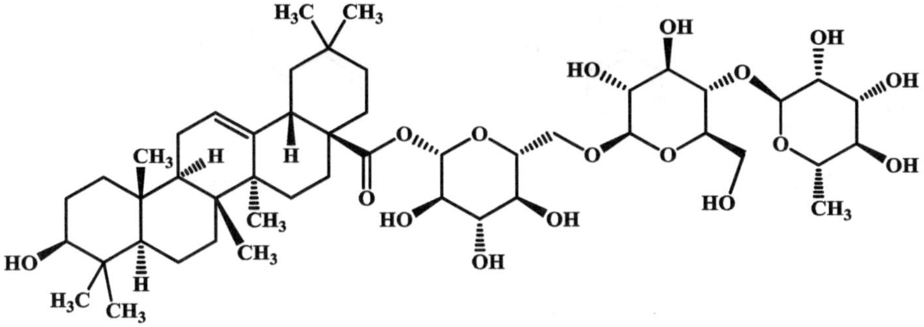

Cirensenoside O

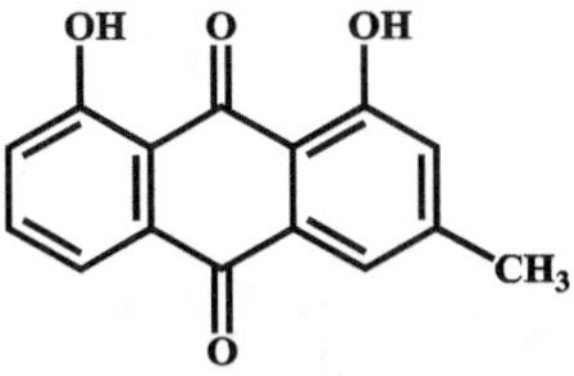

Chrysophanol

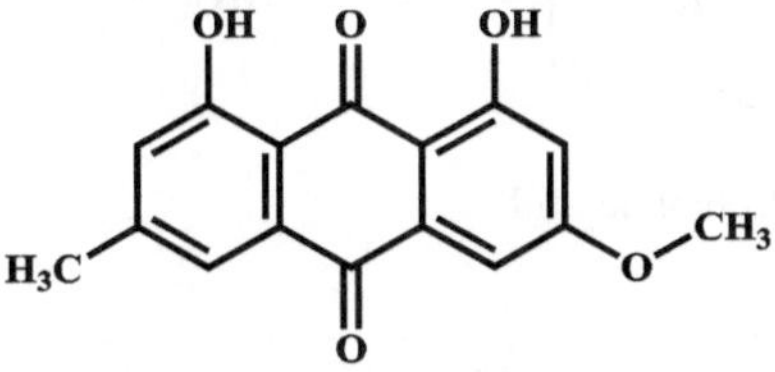

Physcion

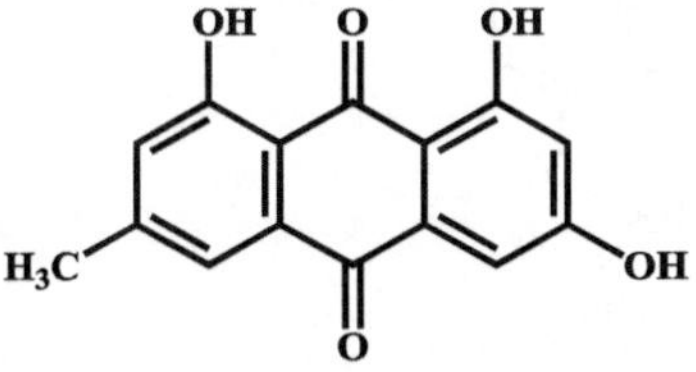

Emodin and its derivatives

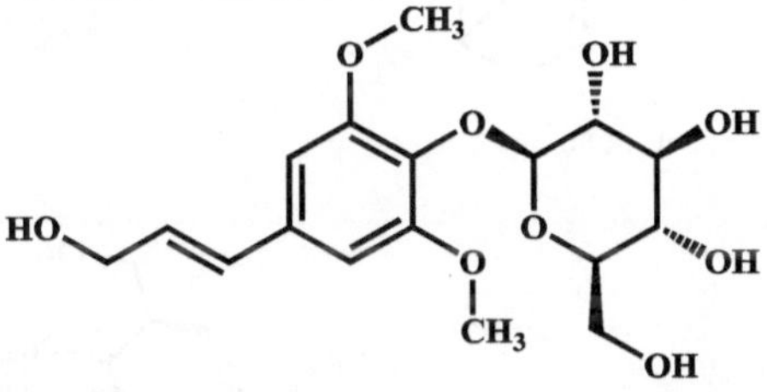

Syringin

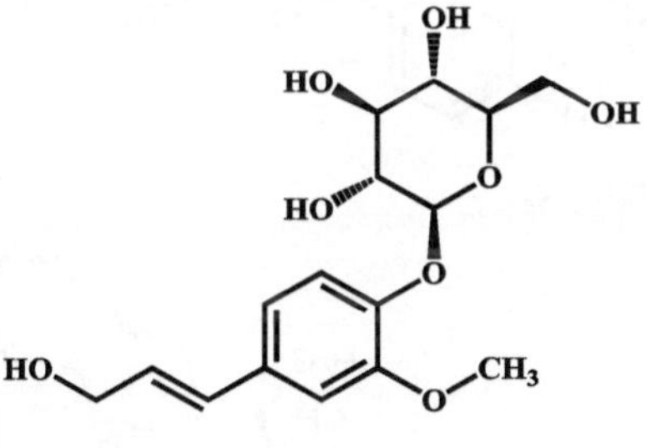

Coniferin

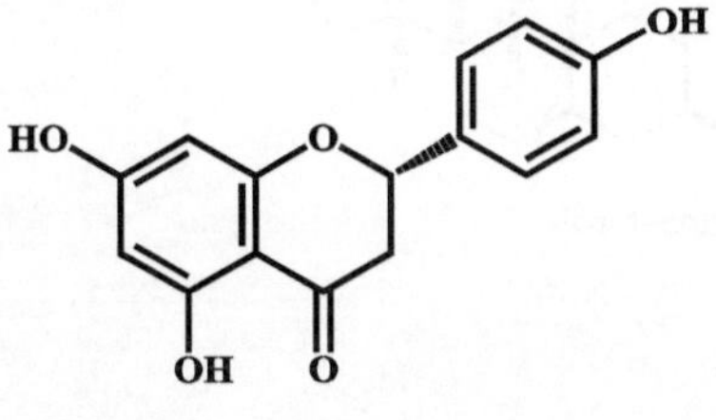

Naringenin

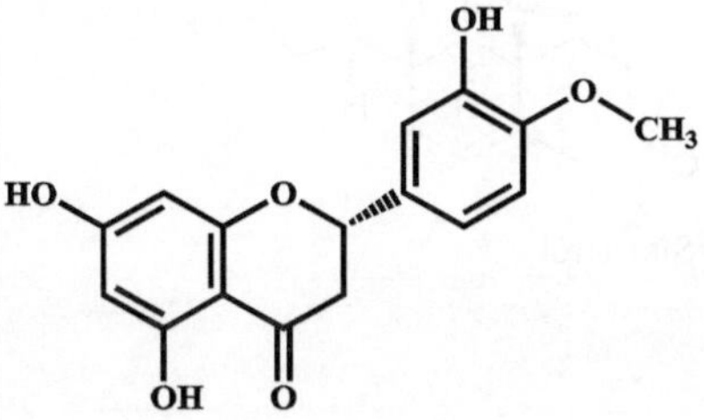

Hesperetin

| | |
|---|---|
| 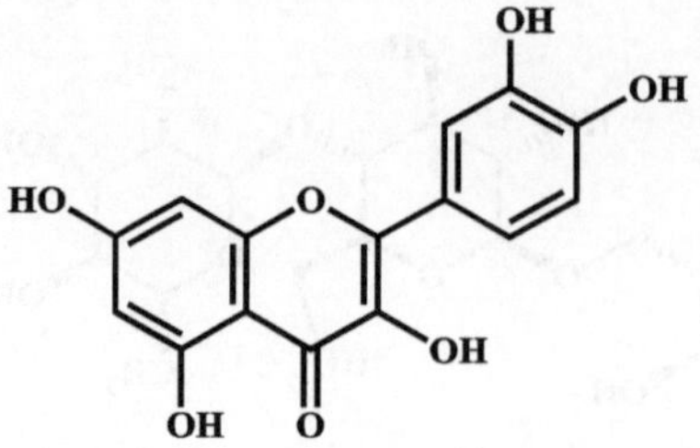 Quercetin | 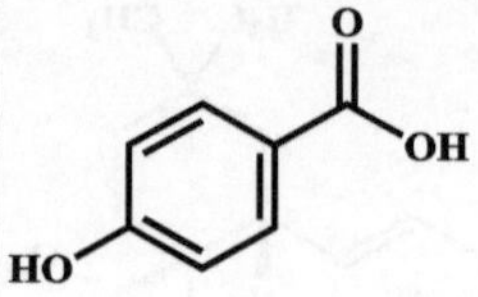 4-Hydroxybenzoic acid |
| 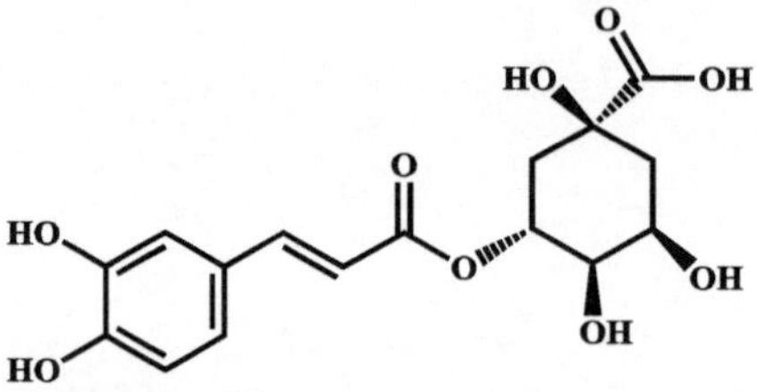 Chlorogenic acid | 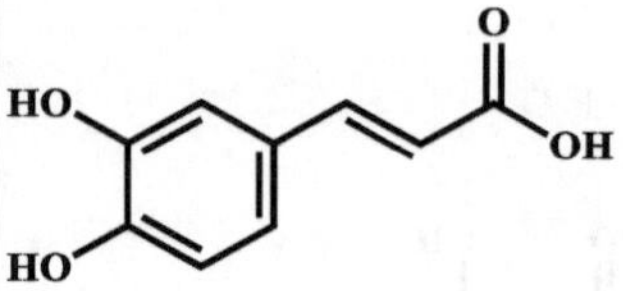 Caffeic acid |
| 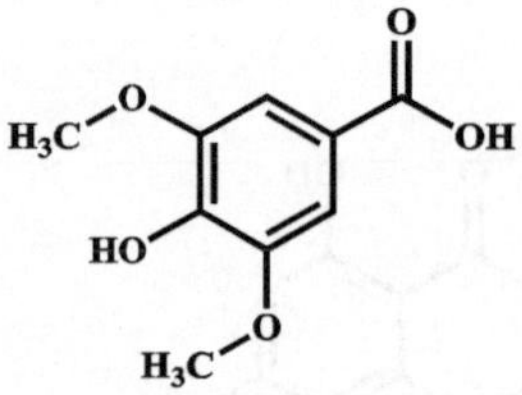 Syringic acid | 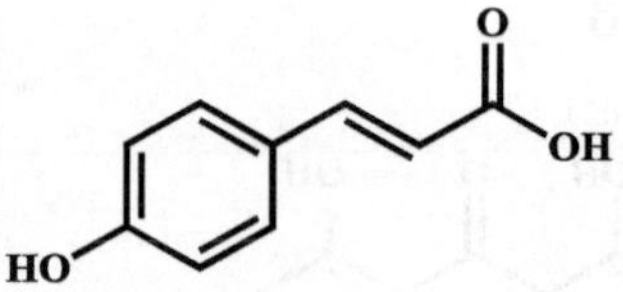 *p*-Coumaric acid |
| 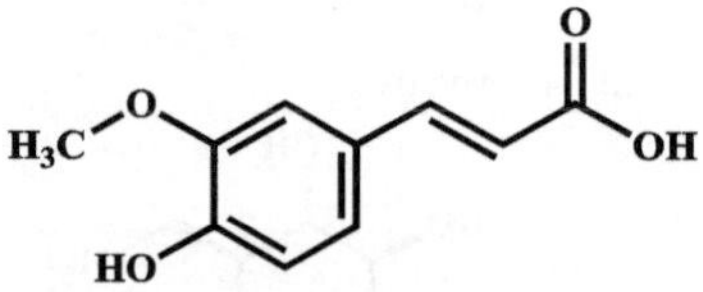 Ferulic acid | 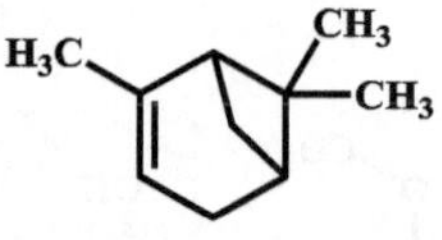 α-Pinene |
| 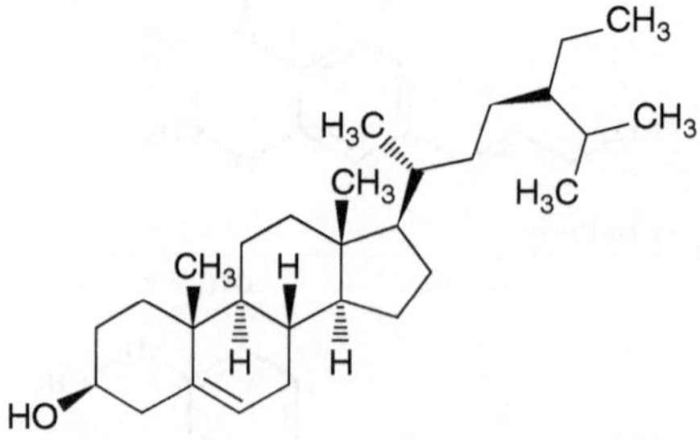 β-Sitosterol | 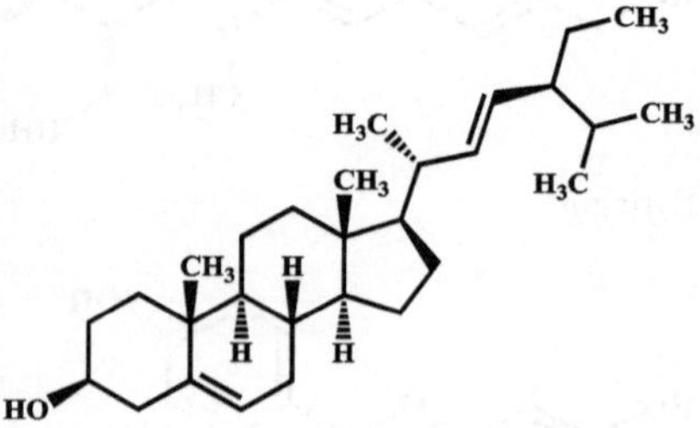 Stigmasterol |

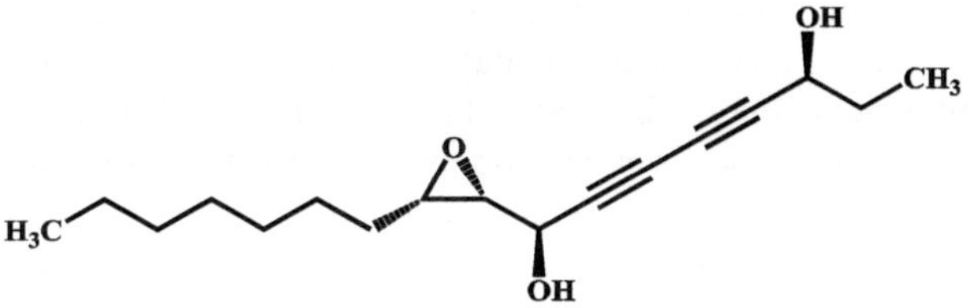


Oploxyne A

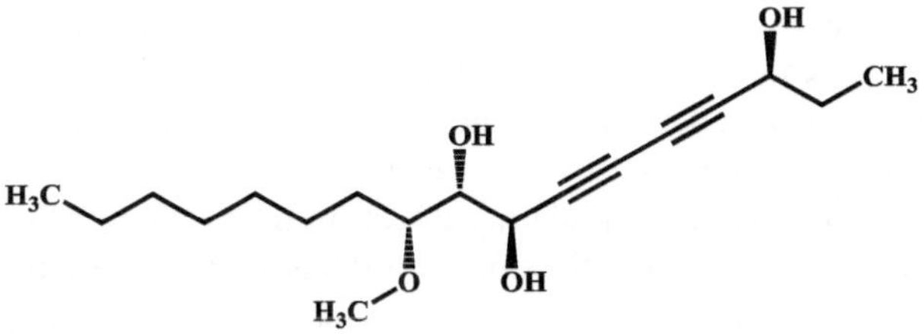


Oploxyne B

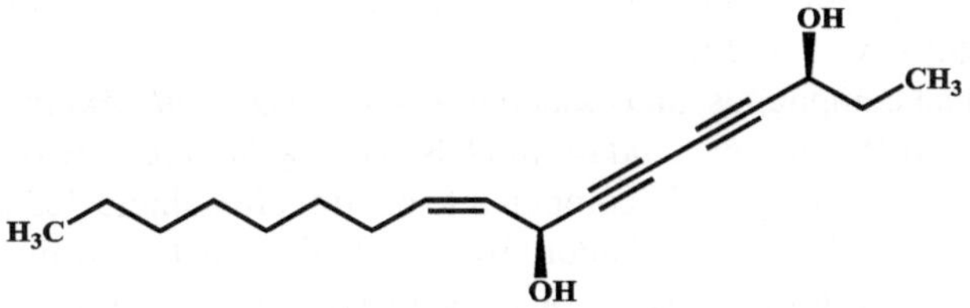


Oplopandiol

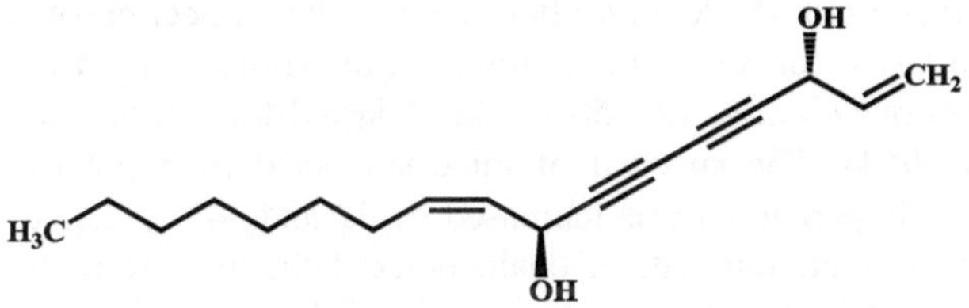


Falcarindiol

5) Pharmacology and bioactivities:

This plant (herb, roots, and stems) is used as medicine in Russia for invigorating vital energy, strengthening yang, and stimulating central nervous system. Clinically, it is used to treat cardiovascular disease, diabetes, rheumatism, neurasthenia, mental depression, schizophrenia, impotence, etc.

An intensive study of *O. elatus* was initiated in the All-Union Institute of Medicinal Plants (Moscow, USSR) in the early 1950s. Since 1955, the EtOH (70 %) tincture from the roots and rhizome of *O. elatus* (1:5) has been officially approved for therapeutic use in the USSR as tonic and antidiabetic in mild cases of diabetes. The effect of a tincture of the roots is, in general, similar to that of the tincture of ginseng. It is employed primarily to stimulate the central nervous system in asthenia conditions and depression. It is also useful to treat impotence, emaciation, hypotension, and physical and mental fatigue. *O. elatus* is not included in the State Pharmacopoeia of the USSR 11th edition, which follows the Russian Federation nowadays, but separate monographs for the roots and rhizome and tincture are supplemented to the pharmacopoeia.

In Chinese medicine, it has been used for treating neurasthenic, hypopiesis, schizophrenia, cardiovascular diseases, diabetes mellitus, and rheumatism. It also possesses antifungal, fever-relieving, pain-relieving, and anti-ageing activities. The stem of *O. elatus* has been traditionally used in China as an analgesic medicine to treat arthritis. In Korea, it has been traditionally used for treating a variety of ailments, such as asthenia, depressive states, diabetes, cough, rheumatism, gastrointestinal disorders, and wounds (Shikov et al. 2014).

The ethanol extract of *O. elatus* has stimulating effect on maturation of the reproductive tract of female rats. Positive therapeutic effects of *O. elatus* on asthenic and astheno-depressive states (particularly in exogenous depression) in psychiatric diseases have been reported. Administration of the tincture (1.5–2.0 mL, 2–3 times per day) for 7 days produced stimulation of the CNS in patients with asthenia (Shikov et al. 2014).

The extracts, individual compounds, and essential oils of the *O. elatus* exhibited adaptogen activity to different mice and rat models. The swimming time of mice to exhaustion was increased by 36 % compared to control after single dose per oral administration of 5 $mL.kg^{-1}$ of ethanol (10 %) liquid extract from the roots of *O. elatus* (Brekhman 1963). An excitation effect of a tincture (10–200 μg) was observed in frogs (Turova and Aleshkina 1964). The locomotor activity of mice was significantly increased 15–20 min after subcutaneous injection of a tincture (5 mL $kg^{-1}$) and lasted for 30–40 min. However, the stimulating effect of the purified glycosides of *O. elatus* (25–50 mg $kg^{-1}$) lasted for 2 h in mice (Turova and Aleshkina 1964). The survival of mice subjected to hypobaric hypoxia and hypoxia with hypercapnia was increased by 42 and 34 %, respectively, after intragastric pretreatment with a dealcoholized tincture from the roots of *O. elatus* (2.5 g $kg^{-1}$) over 7 days. After 7 days of consecutive oral administration, *O. elatus* root dry extract (2.5 g $kg^{-1}$) significantly increased the vertical activity of mice by 46 % and ambulation time. Oral administration of *O. elatus* EtOH (40 %) extract (10 g $kg^{-1}$) to rats caused a marked depletion of adrenal ascorbic acid in intact rats, while no effect was observed in hypophysectomized rats. The authors suggested that *O. elatus* has a stimulatory action on the hypophyseal-adrenal system. The psycho-stimulating effect of an ethanolic tincture (40 %) was observed in animals after injection of sodium barbiturate. The sleeping time of animals was decreased by 35 %. Immediately after injection of 0.15 mg $g^{-1}$ of sodium 5, 5-diethylbarbiturate, the EtOH extract of *O. elatus* roots was intraperitoneally injected to albino mice at a dose of 0.5 g $kg^{-1}$, and the sleeping time of the animals was decreased by 20 % (Shikov et al. 2014).

The vertical activity of mice with pentylenetetrazole-induced seizures was increased by 88 %, and the ambulation time was increased by 81 % after 7 days of treatment with *O. elatus* root dry extract at 2.5 g $kg^{-1}$. The essential oil from *O. elatus*, as a 0.1 mL volatile oil/mL emulsion, injected intraperitoneally at a dose of 0.67 mL $kg^{-1}$ could keep the animal quiet, tame, and with an absence of locomotor activity. When the dose was increased to 2.76 mL $kg^{-1}$, the rats were prostrate with drooping eyelids, but they could still respond to outside stimuli. The *O. elatus* essential oil emulsion had a good synergistic effect on antagonism with pentobarbital, chloral hydrate, and chlorpromazine to pentylenetetrazole-induced and electroconvulsive convulsions (Qu et al. 1984; Shikov et al. 2014).

6) Indications and usage:
-treatment of chronic fatigue syndrome, neurasthenia, hypopiesis, schizophrenia, cardiovascular diseases, diabetes mellitus and rheumatism (Shikov et al. 2014).
7) Precautions and side effects:
The Sprague-Dawley rats were treated with the 50 % EtOH extract of *O. elatus* by gavage at 0, 500, 1000, and 2000 mg $kg^{-1}$. There were no significant changes in body and organ weights during the experimental period. The haematological analysis and blood chemistry data revealed no toxic effects from the extracts. Pathologically, neither gross abnormalities nor histopathological changes were observed comparing between the control and treated rats of both sexes (Shikov et al. 2014).
8) Dosage:
   A) Internal use:
   The tincture of *O. elatus* is recommended in Russia for internal administration at a dose of 30–40 drops, 2–3 times a day for 25–30 days as a CNS stimulant and adaptogen.
   B) External use:
   Not found.

## 6.15 References


- Brekhman II. Comparative data about pharmacological effects of ginseng, *Eleutherococcus*, *Echinopanax* and *Aralia mandshurica*. In: Materialy k izucheniju zhen-shenja i drugih lekarstvennih rasteniy Dalnego Vostoka, Primorskoe knizhnoe izdatelstvo, Vladivostok, 1963;5: 219–27.
- Dou DQ, Hu XY, Zhao YR, Kang TG, Liu FY, Kuang HX, Smith DC. Studies on the anti-psoriof *Oplopanax elatus* Nakai. Nat Prod Res. 2009; 23(4): 334–42.
- Kim JH, Eom SH, Lee HS, Kim JK, Yu CY, Kwon YS, Lee JK, Kim MJ. Assessment on antioxidant properties of *Oplopanax elatus* Nakai in vitro. Korean J. Med Crop Sci. 2007;15(2): 112–9.
- Murav'ev IA, Dzumaev MA. Phytochemical study of roots and rhizomes of *Echinopanax elatum*. Rastit Resur. 1973; 9(4): 535–9.
- Qu S, Wu Y, Wang Yi, et al. Inhibitory effect of tall oplopanax (*Oplopanax elatus*) oil on the central nervous system. Chin Tradit Herb Drugs. 1984;15(6): 19–20.
- Shikov AN, Pozharitskaya ON, Makarov VG, Yang WZ, Guo DA. *Oplopanax elatus* (Nakai) Nakai: chemistry, traditional se and pharmacology. Chin J Nat Med. 2014; 12(10): 721–9.

- Turova AD, Aleshkina Y.A. Experimental study of *Echinopanax elatum*. In: Medicinal preparations from plants. Moscow; 1964.
- Wang GS, Xu JD. Chemical study of glycosides of *Oplopanax elatus* Nakai. Chin Pharm J, 1993; 28 (10): 593–4.
- Wang GS, Zhao CF, Xu JD, et al. Studies on the glycosides in the leaves of *Oplopanax elatus* Nakai (I). Chem Res Chin Univ. 1994a; 10 (3): 185–192.
- Wang GS, Xu JD, Murayama T, et al. Isolation and structure elucidation of new glycosides from the leaves of *Oplopanax elatus* Nakai. Chin Sci Bull. 1994b, 39 (20): 1970–2.
- Wang GS, Chen YP, Xu JD, et al. Determination of the content of saponins in the leaves of *Oplopanax elatus* Nakai. Ginseng Res. 1996a, 2: 34–5.
- Wang GS, Chen YP, Xu JD, et al. Isolation and structure elucidation of cirensenosides O and P from the leaves of *Oplopanax elatus* Nakai. Acta Pharm Sin. 1996b; 31 (12): 940–4.
- Wang GS, Meng Q, Xu JD, et al. Glycosides in the leaves of *Oplopanax elatus* Nakai. Chin Pharm J. 1996c, 31 (9): 522–4.
- Wang GS, Xu JD, Zhang L. Chemical studies on the glycosides in the leaves of tall Oplopanax (*Oplopanax elatus*) (VI). Chin Tradit Herb Drugs. 1997a; 28 (7): 390–2.
- Wang GS, Xu JD. Chemical studies on the glycosides in the leaves of *Oplopanax elatus* Nakai (IV). Chem Res Chin Univ. 1997b;13(1): 34–8.
- Wang GS, Xu JD, Tetsuya M, et al. Isolation and structure elucidation of cirensenosides Q and R. Chin Pharm J. 1997c;22(2): 101–3.
- Wang GS, Yang XH, Xu JD. Structures of four new triterpenoid saponins from the leaves of *Oplopanax elatus* Nakai. Acta Pharm Sin. 2004;39(5): 354–8.
- Wang HJ, Yan MM, Wu X, et al. Study on the chemical constituents of the stocks of *Oplopanax elatus* Nakai. Lishizhen Med Mater Med. 2009;20(3): 678–9.
- Wu X, Yan MM, Liu WH, et al. Super-critical fluid extraction and analysis of the volatile constituents in *Oplopanax elatus* Nakai. Changchunzhongyiyaodaxue Xuebao. 2007; 23(2): 28–9.
- Xu S, Liang HQ. Anthraquinones from *Oplopanax elatus* Nakai. Chin Tradit Herb Drugs. 1998;29(4): 222–3.


## 6.16 *Patrinia intermedia* (Hornem.) Roem et Schult

### *6.16.1* (Valerianaceae*)*

1) Common names:

English name: Unknown
Russian name: Patriniya srednyaya, Kamennaya valeriana
Kyrgyz name: Orto patriniya

**Fig. 6.7** *Patrinia intermedia*. Photo taken by Viktor Kolesnikov and used with permission

2) Synonyms:
   *Fedia intermedia* Hornem., *Fedia rupestris* var. *intermedia* (Hornem.) Vahl, *Patrinia nudiuscula* Fisch
3) Photograph of the plant (Fig. 6.7)
4) Description of the species:

   A) Habitat and world distribution:
      *Patrinia intermedia* (Hornem.) Roem et Schult. (Valerianaceae) is distributed throughout mainly Russia (Siberia) (Altai and Pamir) and Central Asia (Tian Shan, Zhetysu, or Semirechye region, Kazakhstan, Kyrgyzstan). Plant grows on stony floodplains of mountain rivers, on stony slopes, and in steppe and forest-meadow belts of mountains (1000–3000 m). Population status is commonly found in loosely arranged groups.
   B) Morphology of the plant:
      It is a herbaceous perennial, with taproot that branches towards the top. Stems are single or few, 20–75 cm tall, up to 3 cm in diameter, and short haired. Leaves are opposite, 4–18 cm long, 2–5 cm wide, grey green, and glabrous; basal leaves are long petiolate, elongate oblong, strongly dentate, and pinnatilobate or pinnatisect; lower stem leaves are sessile, in 2–5 pairs, and pinnatisect; upper stem leaves are lanceolate ovate and 3 nerved. Inflorescence is corymbiform to paniculate. Corolla is bright yellow, campanulate, and 5 lobed. Fruits have slightly hairy achenes. Flowers have

four stamens. Flowering occurs in June–July; fruits occur from July to September (Eisenman et al. 2013).

C) Medicinal parts:
    Roots and rhizomes

D) Chemical constituents.
    In the roots of *P. intermedia*, there are compounds belonging to the group of **triterpene saponins** (oleanolic acid, patrizid A, patrinoside B, C, D, interoside B) (Babash and Perel'son, 1982; Bukharov et al. 1967,1968, 1969a– c; Khorlin and Ivanova 1965; Matveeva 1970; Matveeva and Abubakirov 1964; Serova and Utkin 1959; Sokol'skaya 1951a–c, 1956), **fatty acids** (palmitic, stearic, oleic, and linolenic acids) (Dolya 1987), **phenolics acids** and **flavonoids** (caffeic acid, chlorogenic acids, kaempferol, quercetin and rutin) (Larchenko et al. 1979), and **nitrogenous bases**, **inulin**, **organic acids**, **tannins**, **essential oils**, and **alkaloids** (Ivanova 1964; Khodzhimatov 1989, Sokolov 1990).

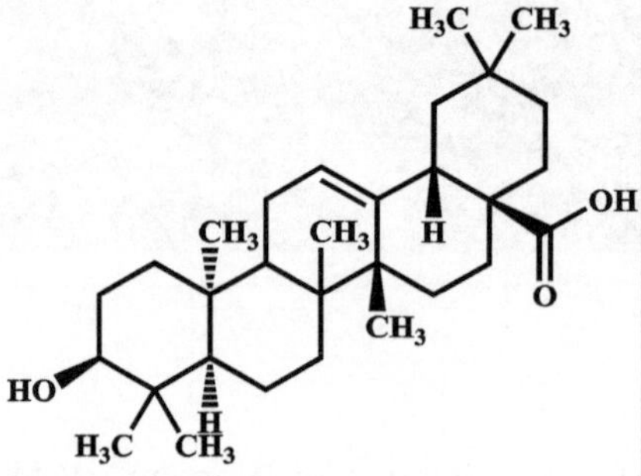

Oleanolic acid

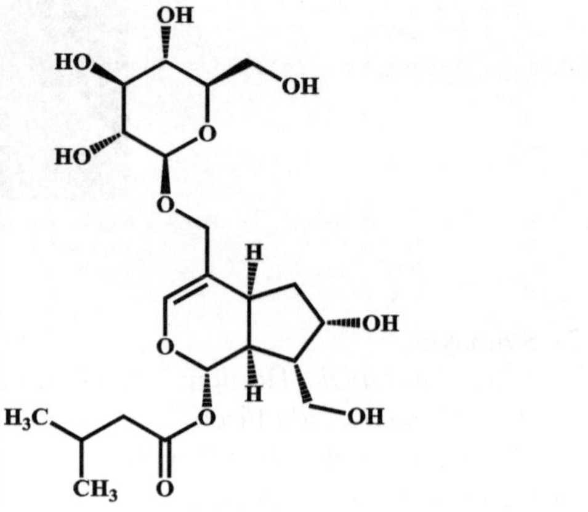

Patrizid A

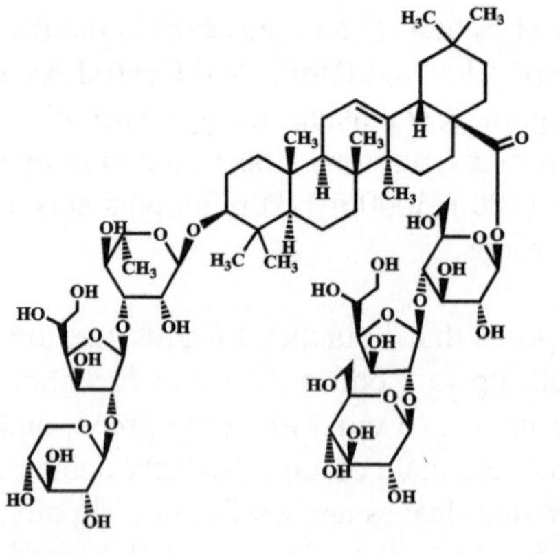

Patrinoside C

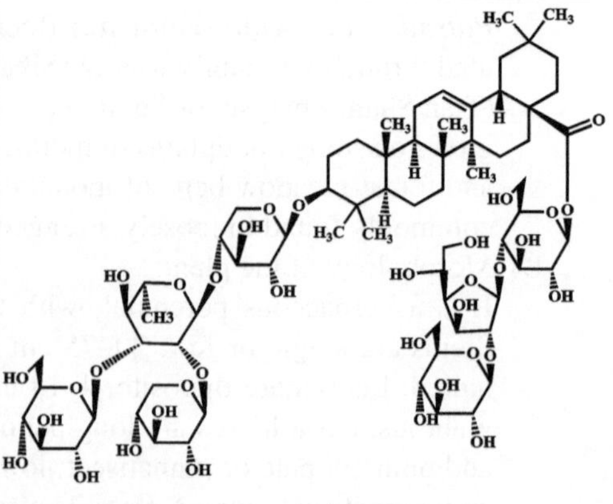

Patrinoside D

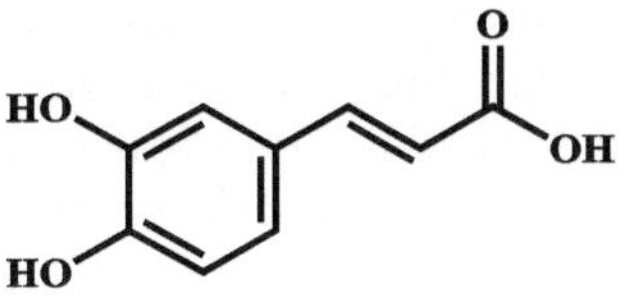

Caffeic acid

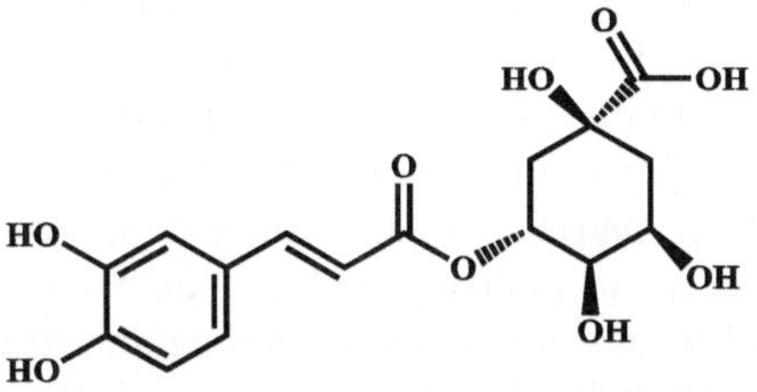

Chlorogenic acid

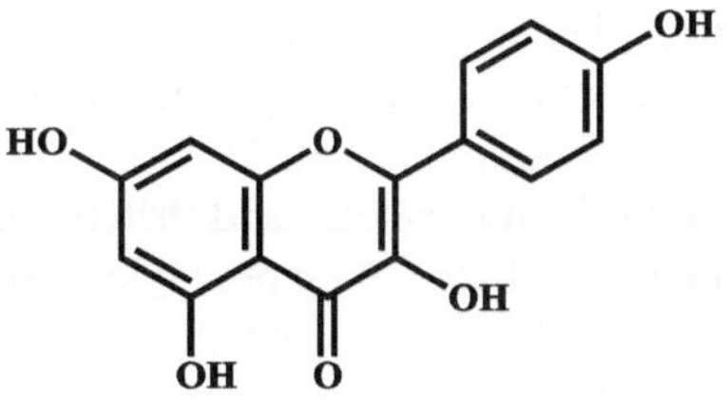

Kaempferol

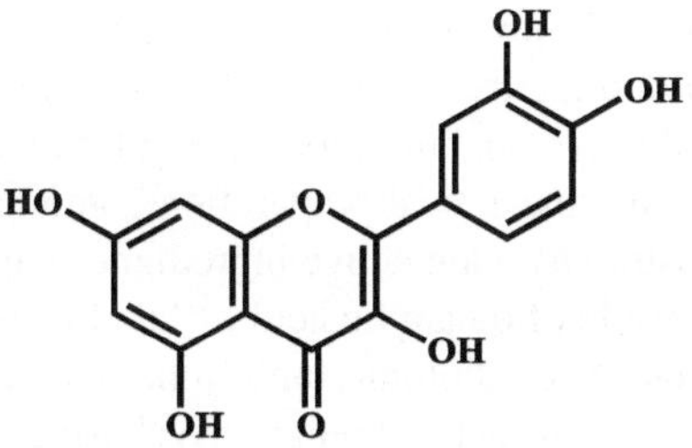

Quercetin

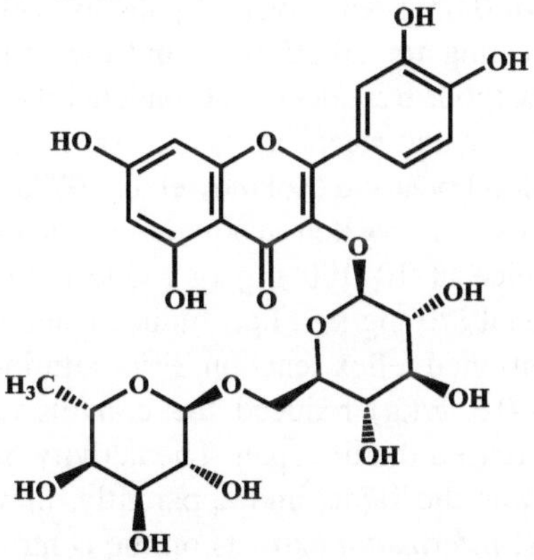

Rutin

5) Pharmacology and bioactivities:

Rhizomes and roots of *P. intermedia* have 50 % more of calming properties of valerian root. An infusion or decoction of the roots is used similarly to valerian root to treat nervous excitement and cardiac neurosis (Turova and Sapozhnikova 1984). In traditional medicine, *P. intermedia* is used for treatment of tuberculosis of lungs, jaundice, fever, scrofula, and gastroenteritis.

The sedative effect of this species is nearly twice as strong as that of Valeriana (Tolmachev 1976). The roots of this species reduce excitability of the nervous system. Clinical tests showed that application of an alcohol infusion stopped or noticeably reduced chest pain as well as nervous and cardiovascular excitation caused by hypodermic introduction of caffeine (Akopov 1990). The biological activity is due to the presence of saponins, and removal of the saponins from the tincture leads to complete loss of the pharmacological properties (Ivanova 1963).

An Herba Patriniae preparation, PATRIMIN, has tranquilizing, blood pressure-lowering, blood coagulation-inhibiting, blood platelet coagulation-preventing, and blood fat-reducing effects. The preparation is a dried extract of root of *P. intermedia* and contains patrinoside D, patrinoside B, patrinoside C, and 10–19 % of oleanolic acid. It is taken after meals in dosage of 0.1 g/tablet each time, 2–3 times daily for continuous 2–3 weeks, and administered again after being suspended for 10–14 days. The preparation has low toxicity and wide biological activity range and is effective in tranquilizing mind, lowering blood pressure, inhibiting blood coagulation, resisting platelet aggregation, and reducing blood fat and can be used for treating diseases of the nervous system, cardiovascular system, and blood coagulation system (Kolkhir et al. 2002).

A pharmaceutical preparation contains heptaoside of oleanolic acid (triterpene glycoside) as main active ingredient in an amount of 0.025–0.05 g per tablet. The heptaoside of oleanolic acid isolated from *P. intermedia* has tranquilizing, antihypertensive, cardiotonic, and antiatherosclerotic effects and can be used for the treatment of neurasthenia, psychasthenia, autonomic nerve disorder, nervous superexcitation, cardiovascular system neurosis, hypertension, and insomnia (Sokolov et al. 1977). Preparation in doses of 0.05 g was given to patients with vegetative vascular dysfunction daily after eating for 20–25 days and then for 10–14 days more after a 10–14-day lapse without treatment. The patients then showed decreased irritability, emotional liability and sleeplessness, cessation of headaches, and eliminated variations in the blood pressure (Sokolov et al. 1977a).

The acidic and neutral triterpene glycosides isolated from the roots of *P. intermedia* when administered s.c. to mice at 10–100 mg/kg reduced the spontaneous motor activity, prolonged hexanal (70 mg/kg., i.p.)-induced sleep, and, in rats at the same dose, inhibited conditioned reflex reaction. Prior administration of the triterpene glycosides (0.2–0.3 g/kg) reduced the convulsive action of strychnine (1 mg/kg., s.c.) and increased the life span. The activity of the neutral and acid triterpene glycosides was the same, and apparently, they are responsible for the observed effects of *P. intermedia* extracts on the central nervous system (Larik et al. 1968).

Decoction, alcoholic extract, and galenical preparations induce vomiting experimentally. Additionally, they showed positive activity on schizophrenia, antihypertensive properties in early stage of hypertension, and antibacterial and parasitocytic properties. The essential oil is active against dermatomes. The formulation "Patrin" (containing 65 % of patrinoside D) showed sedative, cardiotonic, and hypotensive properties and is recommended for treating hypertension, aggravated by atherosclerosis. Patrinoside reduced conditional reflector apparatus in animals (Sokolov 1990).

6) Indications and usage:
   The roots are used, like valerian, to treat nervous excitement and cardiac neurosis (Zaurov et al. 2013).
7) Precautions and side effects:
   Not found
8) Dosage:

   A) Internal use:

      1. 10 g of 20 % tincture of *Patrinia* (Tinctura *Patriniae intermediate*) mixed with 10 g of 20 % tincture of *Convallaria majalis*, 15–20 drops taken in one cup of water 2–3 times a day as an aid for insomnia and cardiac neurosis
      2. 10 g of 20 % tincture of *Patrinia* mixed with 10 g of 20 % tincture of valerian and 10 g of 20 % tincture of lily of the valley, 15–20 drops taken in one cup of water 2–3 times a day as calming aid
      3. 10 g of chopped rhizome in one cup of boiled water infused for 2 h and drained. One table spoon taken 3–4 times a day during anxiety
      4. 20 drops of 20 % alcohol tincture of *Patrinia* (Tinctura *Patriniae intermediate*) taken three times a day 30 min before meals for neurosis treatment
      5. 10 g of chopped rhizomes of *Patrinia* mixed in one cup of boiled water, infused, and drained. One table spoon taken 3–4 times a day as calming

   B) External use:
      Not found

## 6.17 References


- Akopov EI, The major domestic herbs and their application. Tashkent: Meditsina, 1990;442.
- Babash TA, Perel'son ME. Quantitative determination of the aglycone of the total patrinosides in the roots of *Patrinia intermedia*. Chem Nat Compound. 1982;18(5):587–90.
- Bukharov VG, Talan VA, Karlin V Glycosides of *Patrinia intermedia*. Chem Nat Compound. 1967;3(3):180.
- Bukharov VG, Talan A, Karlin V. Nontriterpenoid glycosides of *Patrinia intermedia*. Chem Nat Compound. 1968;4(4): 184–5.
- Bukharov VG, Karlin V, Talan VA. Triterpene glycosides of *Patrinia intermedia*. Chem Nat Compound. 1969a;5(1):17–19.
- Bukharov VG, Karlin V, Talan VA. The triterpene glycosides of *Patrinia intermedia* Roem et Schult. Chem Nat Compound.1969b; 5(2): 72–5.
- Bukharov VG, Karlin V, Talan VA. The triterpene glycosides of *Patrinia intermedia* Roem et Schult. Chem Nat Compound. 1969c; 5(2): 76–8.
- Dolya VS. Characterization of glyceridic oils in three genera of the Valerianaceae. Rastitel'nye Resursy. 1987; 23(3):440–9.

- Eisenman SW, Zaurov DE, Struwe L. Medicinal plants of Central Asia: Uzbekistan and Kyrgyzstan. New York: Springer. 2013;340.
- Ivanova VM. Study of the roots and rhizomes of *Patrinia intermedia* as a source for new medicinal preparations. Synopsis of dissertation for the degree of Ph.D. in Pharmacology, Moscow (in Russian). 1963.
- Ivanova VM. The chemical composition of the aboveground portions of *Patrinia intermedia* and a chromatographic method of isolating its saponins. Izuch. i Ispol's Lekarstv Rastit. Resursov SSSR (Leningrad: Med.) Sb. 1964; 328–31.
- Khodzhimatov M. Dikorastushchiye Lekarstvennuiye Rasteniya Tadzhikistana (Wild-growing medicinal plants of Tadjikistan). Dushanbe: Glavnaya nauchnaya redaktsiya Tadzhikskoi Sovetckoi entsiklopedii (in Russian). 1989.
- Khorlin A Ya, Ivanova VM. Triterpene saponins.XIV. Saponins of *Patrinia intermedia* Izvestiya. Akademii Nauk SSSR, Seriya Khimicheskaya.1965;2: 307–13.
- Kolkhir VK, Voskoboinikova IV, Tuzhilkin VV, Bykov VA. A herba patriniae preparation with tranquilizing, blood pressure lowering, blood coagulation inhibiting, blood platelet coagulation preventing, and blood fat reducing effects. Russ. 2002; RU 2189241 C1 20020920.
- Larchenko VA, Fursa NS, Drozd GA. Phenolic compounds of Siberian species of the genus *Patrinia*. Chem Nat Compound. 1979;15(6): 757.
- Lapik AS, Bukharov VG, Talan VA, Karlin VV. The action of triterpenic glycosides of *Patrinia intermedia* Roem. et Schult on the central nervous system. Farmakologiya i Toksikologiya, 1968;31(6):650–2.
- Matveeva AV, Abubakirov NK. Saponins of *Patrinia intermedia*. Uzbekskii Khimicheskii Zhurnal. 1964; 8(5): 43–6.
- Matveeva AV. Simplified method for isolating patriside A from *Patrinia intermidia*. Fiziol Aktiv Soedin Rast Kirg. 1970; 75–80.
- Serova NA, Utkin LM. Sapogenin of *Patrinia* roots. Zhurnal Obshchei Khimii. 1959; 29: 336–8.
- Sokol'skaya AM.Sapogenin of the roots of *Patrinia intermedia*. Zhurnal Obshchei Khimii. 1951a; 21: 962–7.
- Sokolskaya AM. The saponin of the roots of *Patrinia intermedia*. Zhurnal Obshchei Khimii. 1951b;21: 1049–51.
- Sokolskaya AM. The sapogenin of the roots of *Patrinia intermedia*. Zhurnal Obshchei Khimii. 1951c;21: 1053–7.
- Sokol'skaya AM. Saponins of the Kazakhstan flora. Aptechnoe Delo. 1956; 5(6): 22–4.
- Sokolov PD. Plant resources of USSR: Flowering plants, their chemical composition, uses. Family Caprifoliaceae – Plantaginaceae. Leningrad: Nauka. 1990; 5:21.
- Sokolov SY, Turova AD, Ivanova VM, Novak-Ivanova RM, Zinkevich EP, Selavri TV, Fokina AV, Belova F, Kogan VI. Pharmaceutical preparation containing heptaoside of oleanolic acid as main active ingredient and having tranquilizing, antihypertensive, cardiotonic, and antiatherosclerotic effects. Fr. Demande. 1977a; FR 2326929 A1 19770506.

- Sokolov SYa, Turova AD, Ivanova VM, Zinkevich EP, Selavri TV, Fokina AV, Belova LF, Kogan VI. Pharmaceutical composition, containing the triterpene glycoside oleanolic acid heptaoside, with tranquillizing, hypotensive, cardiotonic, and antisclerotic action. Ger Offen. (1977b; DE 2543053 A1 19770331.
- Tolmachev AI. Atlas Arealov I Resursov Lekarstvennuikh Rastenii SSSR (Atlas of distributions and resources of medicinal plants of the USSR), Moscow (in Russian). 1976.
- Turova AD, Sapozhnikova EN. Lekarstvennuiye Rasteniya SSSR i ikh Premeneniye (Medicinal plants of the USSR and their uses), 4th edn. Meditisina, Moscow (in Russian). 1984.
- Zaurov DE, Belolipov IV, Kurmukov AG, Sodombekov IS, Akimaliev AA, Eisenman SW. The medicinal plants of Uzbequistan and Kyrgyzstan. In: Sasha W. Eisenman, David E. ZaurovSasha W. Eisenman, David E. Zaurov, editors. Medicinal plants of Central Asia: Uzbekistan and Kyrgyzstan. Lena Struwe. Springer. 2013.


## 6.18 *Rhaponticum carthamoides* (Willd.) Iljin

### *6.18.1* (Asteraceae)

1) Common names:
   English name: Maral root, Russian Leuzea, Maralu
   Russian name: Levzeya saflorovidnaya, Rapontikum saflorovidnyy, Bolshegolovik saflorovidnaya, Stemakanta saflorovidnaya, Maraly koren
   Uzbek name: Mahsarsimon levzeya
2) Synonyms:
   *Leuzea carthamoides* (Willd.)., *Cnicus carthamoides* Willd., *Stemmacantha carthamoides* (Willd.) Dittrich, *Centaurea carthamoides*, *Cirsium carthamoides*, *Leuzea rhapontica*
3) Photograph of the species (Fig. 6.8)
4) Description of the species:

   A) Habitat and world distribution:
      *Rhaponticum carthamoides* (Willd.) Iljin is a perennial herbaceous species from the *Asteraceae* (Compositae) family, which is distributed throughout Central Asia, mainly Southern Siberia (Altai and Western Sayani Mountains) and Kazakhstan (Kholmatov et al. 1998, Shreter 1975). This plant is endemic to Southern Siberia, where it is found abundantly growing wild on the slopes of the Altai and Sayani Mountains. It is commonly found in high alpine and subalpine meadows located 1200–1900 m above sea level. Originally an endemic plant of southern Siberia, now it grows widely in Central and Eastern Europe. During the last few decades, the plant has been introduced to various regions of Central and Eastern Europe, where it is now widely grown for its marked medicinal properties (Opletal et al. 1997).

**Fig. 6.8** *Rhaponticum carthamoides*. Photo taken *by* Petr Kocna and used with permission

B) Morphology of the plant:
*Rhaponticum carthamoides* arises from a woody rhizome, which terminates in a mass of wiry roots spanning 20–40 cm. The leaves are alternately situated and usually oval or elliptical in shape. The inflorescences appear in the second year and have a height which range between 80 and 170 cm. Situated on the inflorescence is the capitulum (compact cluster of sessile flowers) which has a diameter of 3–6 cm and magenta florets; the length and width of the seeds are ~5–8 mm and ~2–4 mm, respectively (usually found in a mass of 1000). This erect growth plant can reach heights of approximately 2 m.

C) Medicinal parts:
Often used aerial parts (stem, leaves, flowers, seeds) and roots of this plant

D) Chemical constituents:
**Phytoecdysteroids** (20–hydroxyecdysone, polypodine B, makisterone A, 2-deoxyecdysterone, integristerone A, integristerone B, taxisterone, ajugasterone C, α-ecdysone, lesterone, rapisterone D, inokosterone, rapisterone, 20-hydroxyecdysone 2,3;20,22-diacetonide, 20-hydroxyecdysone 2,3-monoacetonide, 20-hydroxyecdysone 20,22-monoacetonide, 22-oxo-20-hydroxyecdysone, 24(28)-dehydromakisterone A, (24Z)-29-hydroxy-24(28)-dehydromakisterone C, carthamosterone, rubrosterone, dihydro-rubrosterone, posterone, isovitexirone, leuzeasterone, makisterone C, polypodine B 20,22-acetonide, rapisterone B, rapisterone C, rapisterone D 20-acetate, 24(24-)[Z]-dehydroamarasterone B, polypodine B-22-benzoate, carthamosterone A, carthamosterone B, amarasterone A, carthamoleusterone, 24(28)-dehydroamarasterone B, 22-deoxy-28-hydroxymakisterone C, 3-epi-20-hydroxyecdysone, 24-epi-makisterone A, 14-epi-ponasterone A 22-glucoside, 5-α-20-hydroxyecdysone, 20-hydroxyecdysone 2-acetate,

20-hydroxyecdysone 3-acetate, 1β-hydroxymakisterone C, 26-hydroxymakisterone C, 15-hydroxyponasterone A, inokosterone 20,22-acetonide, integristerone A 20,22-acetonide, turkesterone) (Kokoska and Janovska 2009), **flavonoids and related compounds** (6-hydroxykaempferol-7-*O*-(6′-*O*-acetyl-β-D-glucopyranoside), patuletin, 6-hydroxykaempferol-7-*O*-glucoside, quercetagitrin, 6-methoxykempferol, quercetin-5-*O*-glucoside, isorhamnetin-5-*O*-glucoside, quercetin-3,3′-dimethyl ether, quercetin, quercetagetin, luteolin, kaempferol, isorhamnetin, quercetin-3-methyl ether, quercetin-5-*O*-β-D-galactoside, isorhamnetin-5-*O*-α-L-rhamnoside, quercetagetin-7-*O*-β-glucopyranoside, 6-hydroxykaempferol-7-*O*-β-glucopyranoside, quercetagetin-7-*O*-(6”-*O*-acetyl-β-glucopyranoside), 6-methoxykaempferol-3-*O*-β-glucopyranoside,6-hydroxykaempferol-7-*O*-(6”-*O*-acetyl-β-D-glucopyranoside), quercimeritrin, apigenin, eriodictyol, eriodictyol-7-β-glucopyranoside, hesperetin, chrysanthemin, cyanin) (Kokoska and Janovska, 2009), and **phenolic acids** (benzoic acid, *m*-hydroxybenzoic acid, *p*-hydroxybenzoic acid, salicylic acid, protocatechuic acid, gentisic acid, vanillic acid, gallic acid, syringic acid, *o*-coumaric acid, *p*-coumaric acid, caffeic acid, ferulic acid, sinapic acid, *o*-hydroxyphenylacetic acid, *p*-hydroxyphenylacetic acid, chlorogenic, neochlorogenic, isochlorogenic acid) (Kokoska and Janovska, 2009). In regard to **other groups** of compounds found in *R. carthamoides*, various polyacetylenes and sesquiterpene lactones (cynaropicrin, repdiolide, chlorojanerin), repensolide (cebellin E), and janerin have been isolated from roots of the plant. Vereskovskii et al. (1978) have found triterpenoid glycosides, namely, rhaponticosides A, B, C, D, E, F, G, and H, in the roots and aerial parts of *R. carthamoides*. Triterpenoid components reported from the literature for underground parts of the plant are parkeol and parkeyl acetate known also as carthamenol and carthamenyl acetate, respectively (Kokoska and Janovska 2009). **Phytosterols** (campesterol, stigmasterol, cholesterol, D7-avenasterol, β-sitostanol, β-sitosterol), triterpenes (**β-amyrin**, **α-amyrin**), **fatty alcohols**, and **acids** (Biscup et al. 2013; Stransky et al.1998; Khomova et al. 1995), and **essential oils** (Yankulov 1962; Geszprych and Weglarz 2002; Havlik et al. 2009).

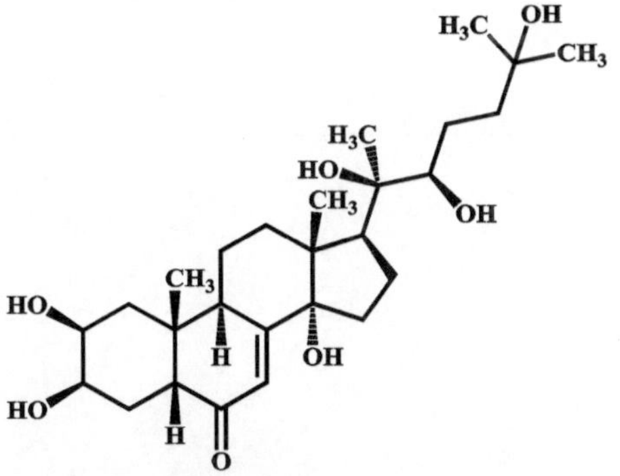


20-hydroxyecdysone

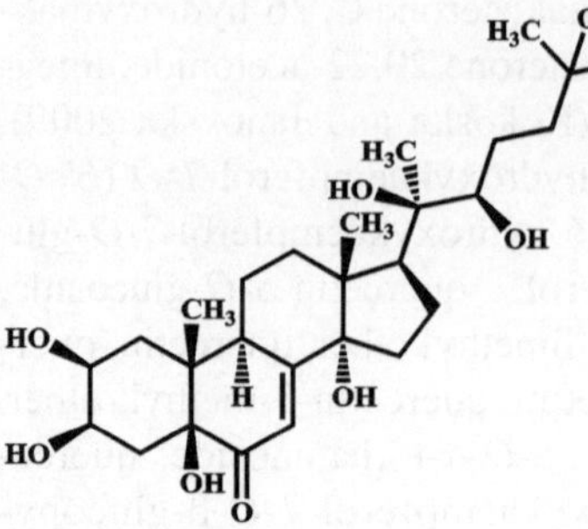


Polypodine B

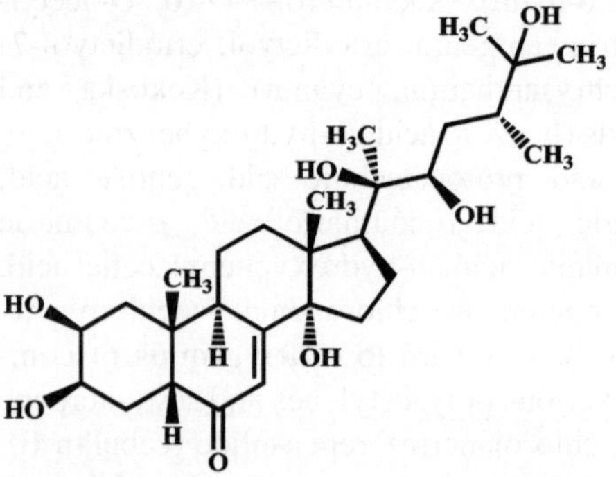


Makisterone A

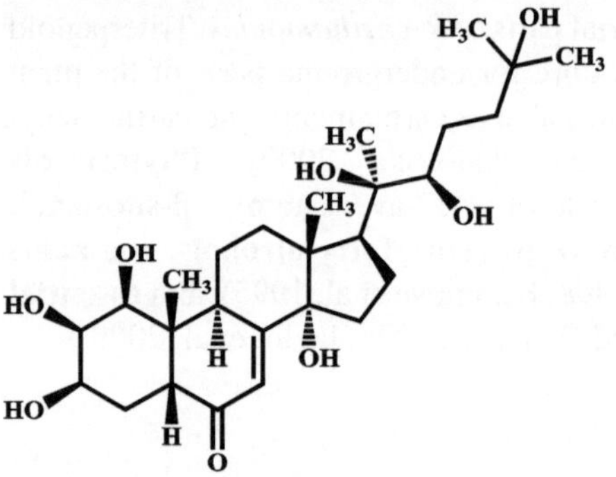


Integristerone A

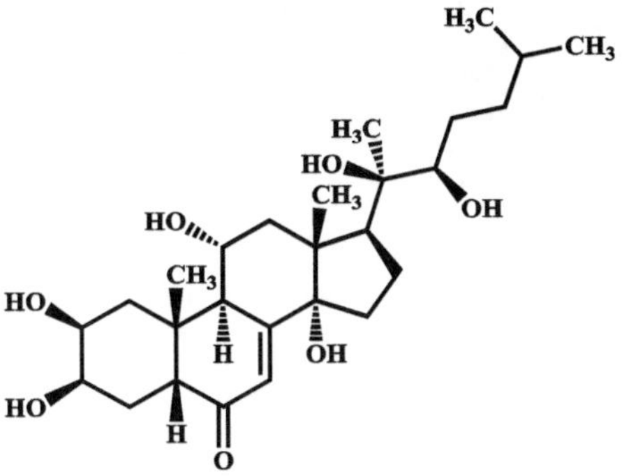


Ajugasterone C

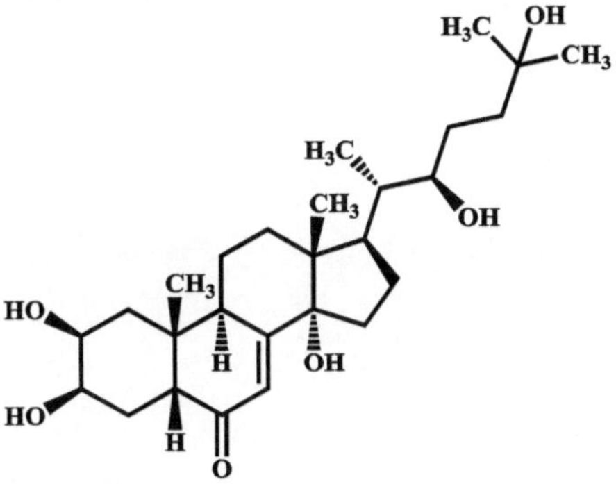


α-Ecdysone

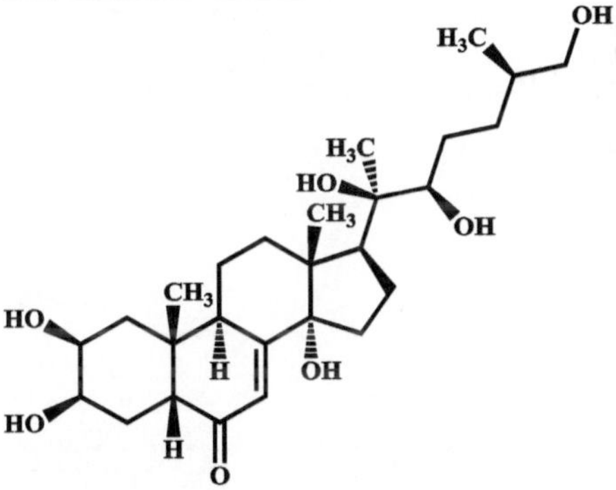


Inokosterone

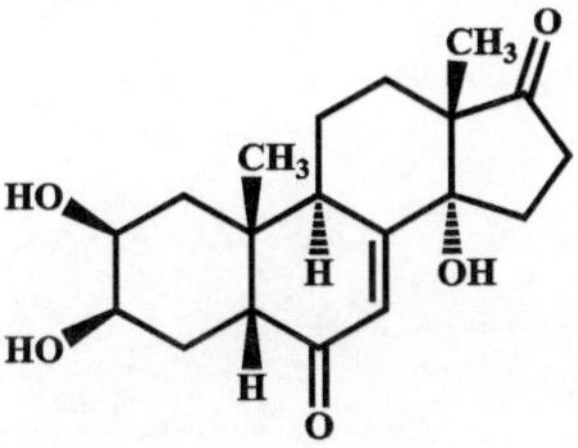


Rubrosterone

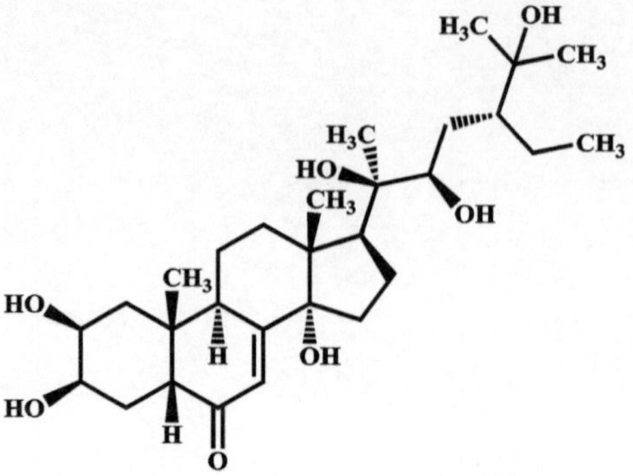

Makisterone C

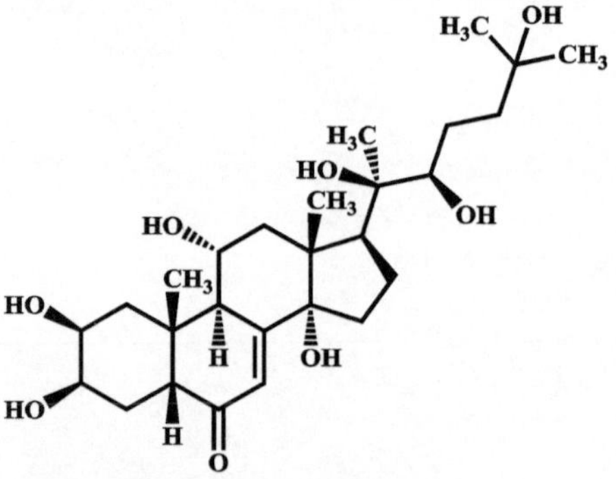

Turkesterone

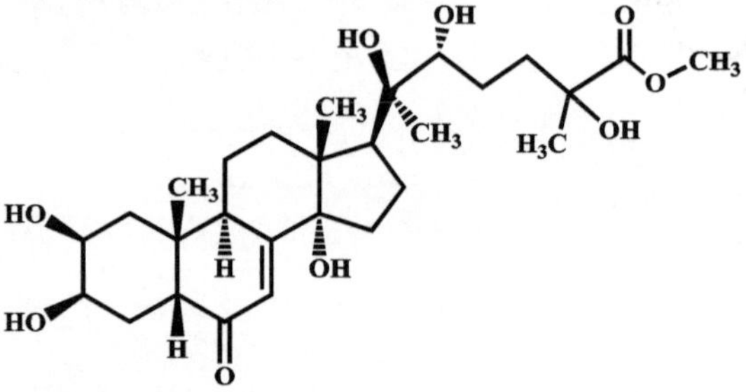

Carthamosterone B

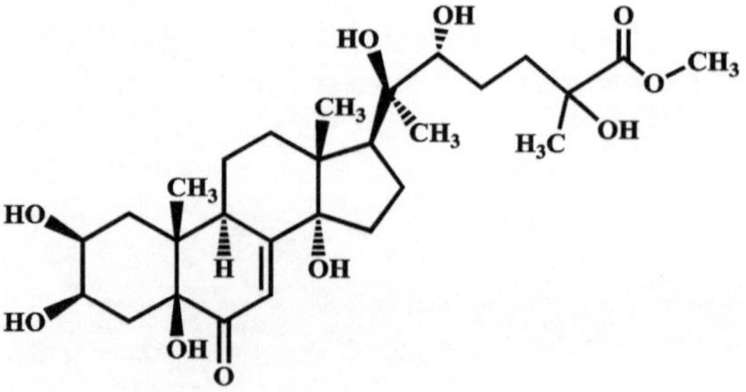

Carthamosterone A

| | |
|---|---|
| 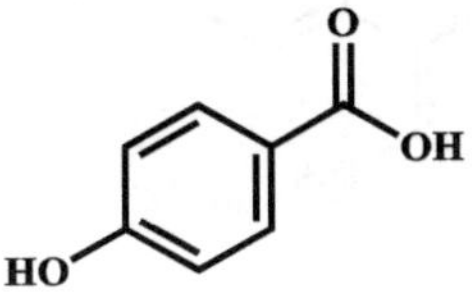<br>4-Hydroxybenzoic acid | |
| 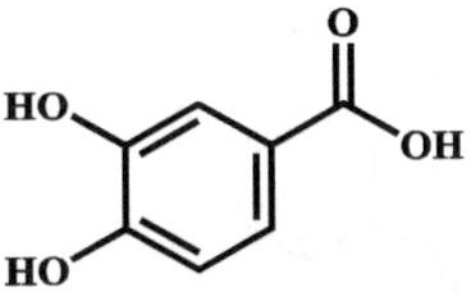<br>Protocatechuic acid | |
| 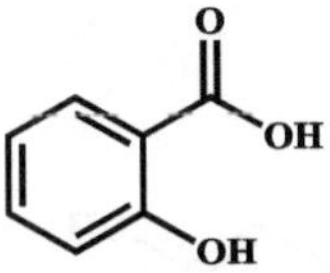<br>Salicylic acid | 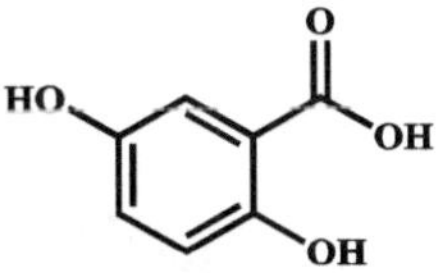<br>Gentisic acid |
| 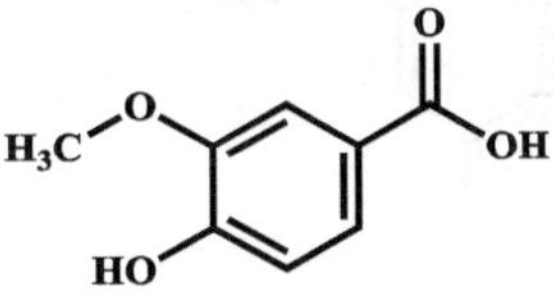<br>Vanillic acid | 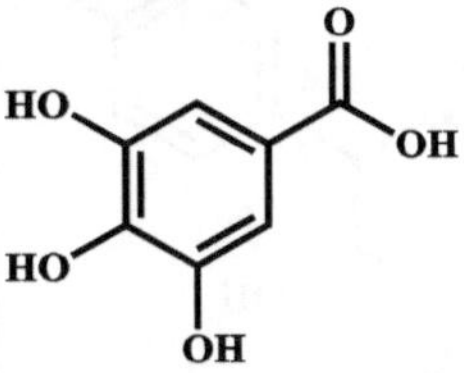<br>Gallic acid |
| 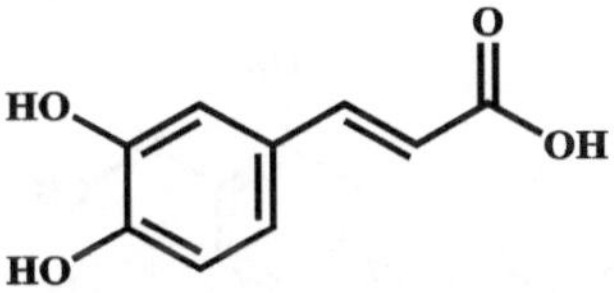<br>Caffeic acid | 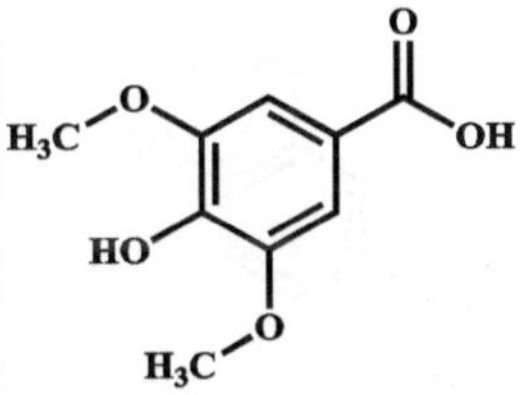<br>Syringic acid |
| 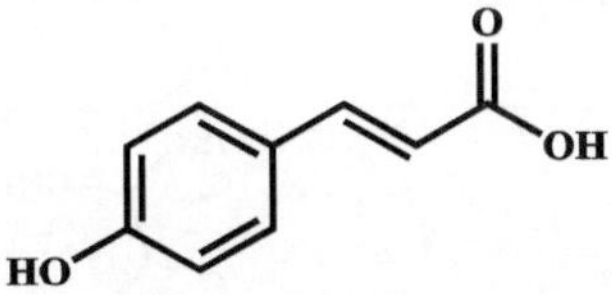<br>*p*-Coumaric acid | 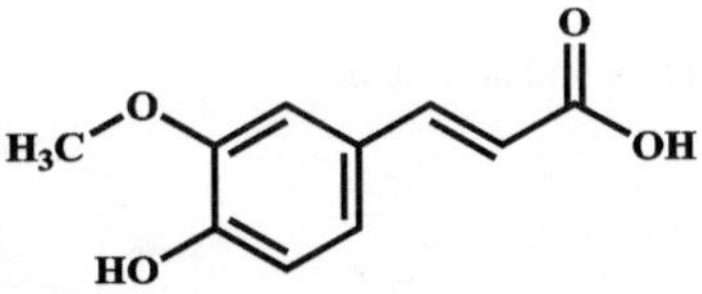<br>Ferulic acid |

| | |
|---|---|
| 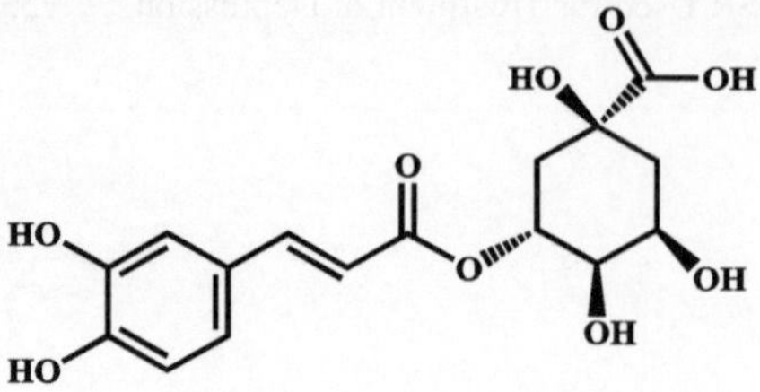 Chlorogenic acid | 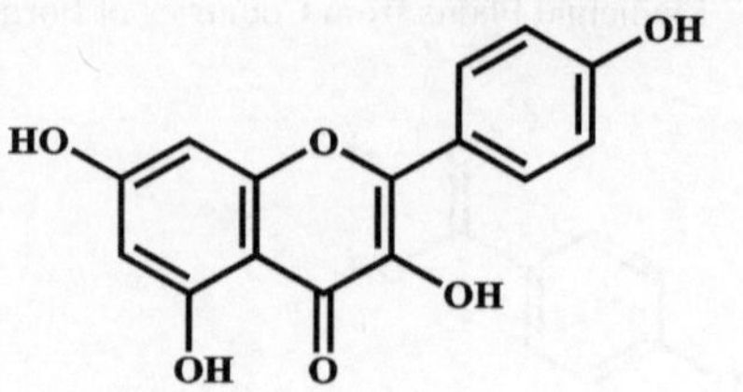 Kaempferol derivatives |
| 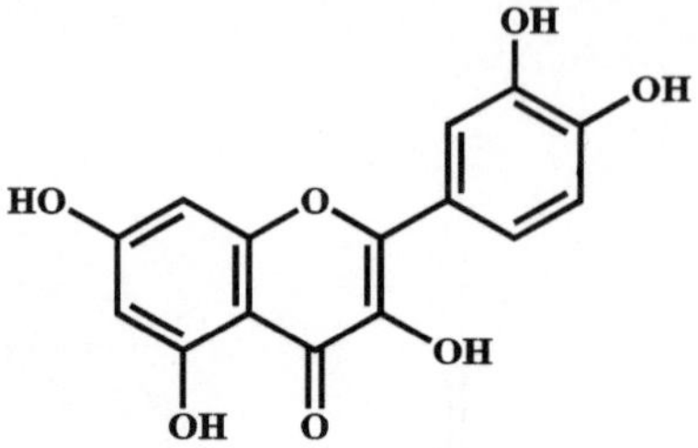 Quercetin derivatives | 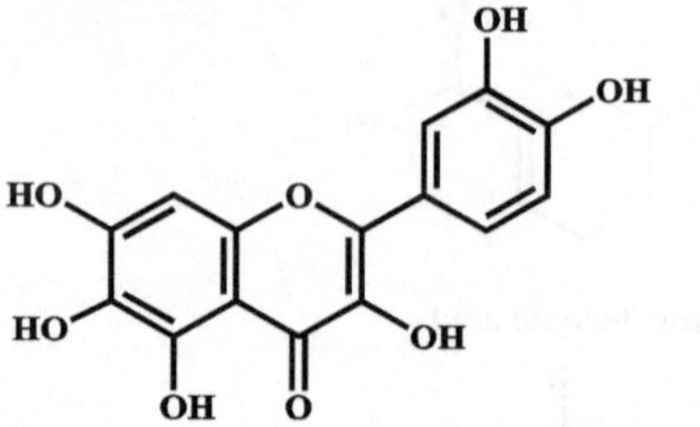 Quercetagetin derivatives |
| 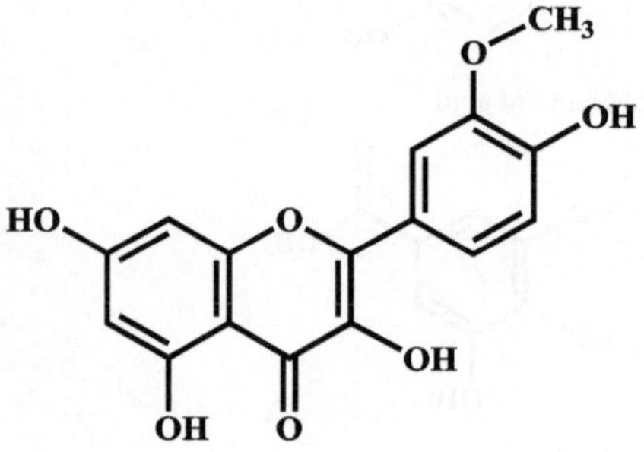 Isorhamnetin derivatives | 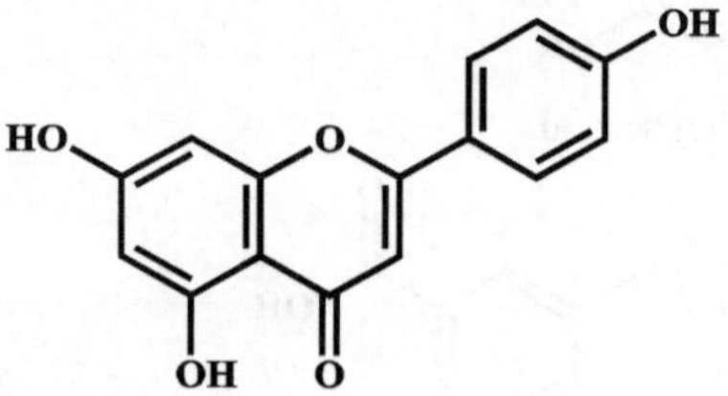 Apigenin |
| 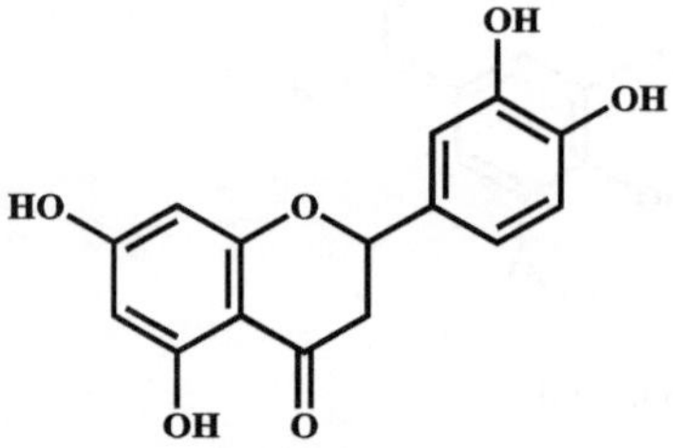 Eriodictyol and derivatives | 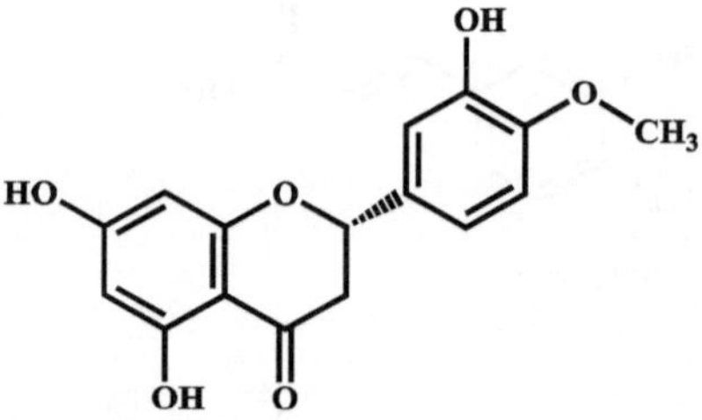 Hesperetin |
| 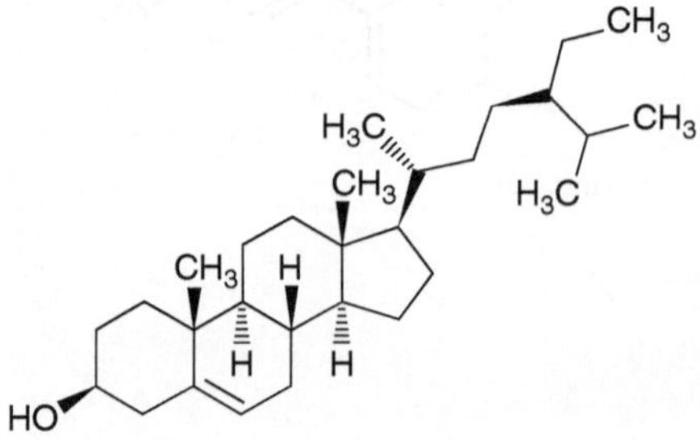 β-Sitosterol | 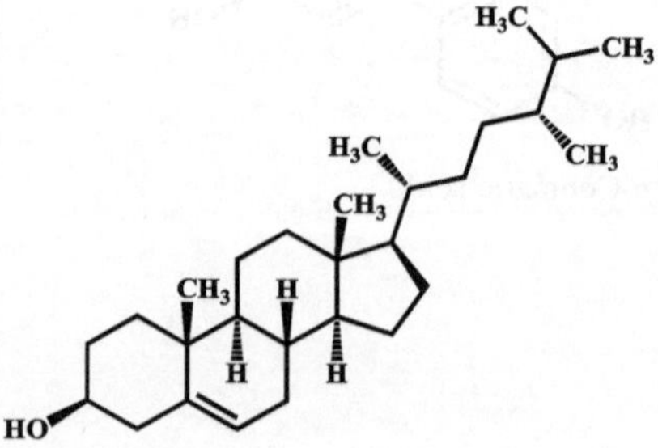 Campesterol |

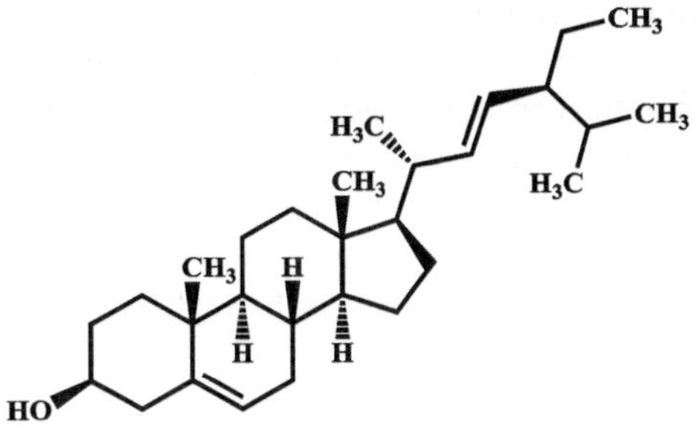

Stigmasterol

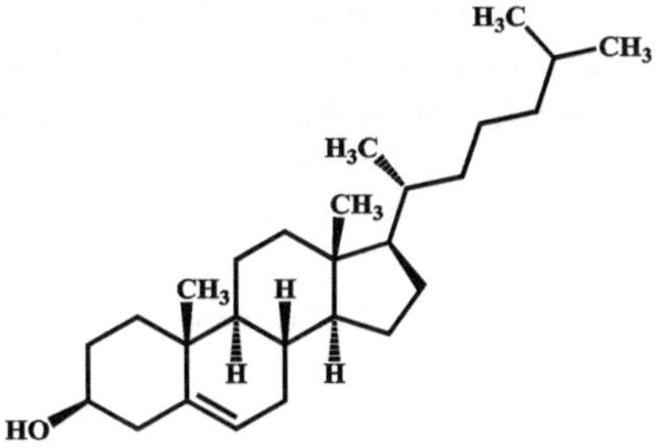

Cholesterol

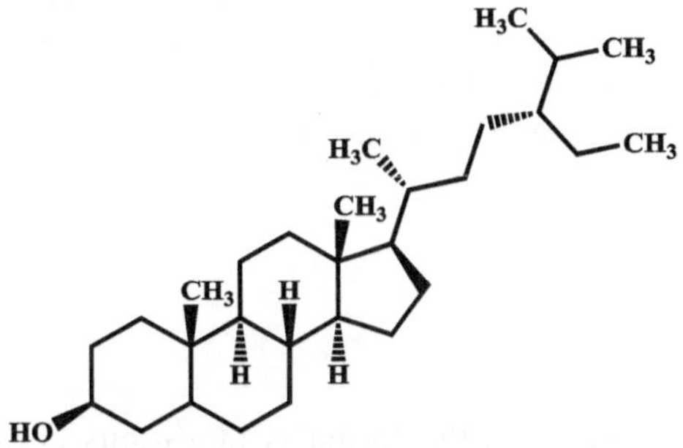

β-Sitostanol

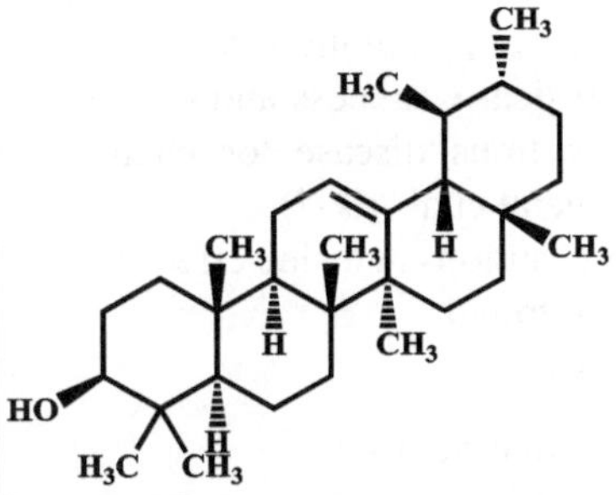

α-Amyrin

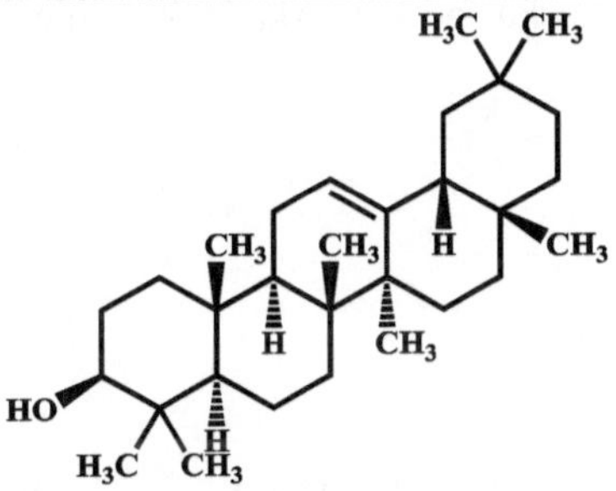

β-Amyrin

5) Pharmacology and bioactivities:

*Rhaponticum carthamoides* (Willd) Iljin has a long story of use in Siberian folk medicine, mostly to treat overstrain and common weakness after illnesses and as a stimulant and a remedy against male sex dysfunction. The principal bioactive constituents of this plant are ecdysteroids, flavonoids, and phenolic acids (Dushkin et al. 2014).

The water–ethanol extract from roots of *R. carthamoides* at 500 mg/kg doses increased the locomotor activity, showed a clear-cut tendency towards antagonizing the narcotic effects of chloral hydrate and towards increasing CNS excitability, and improved the learning and memory indices in rats (Petkov et al. 1984).

The root ethanol extract demonstrated a significant improvement in learning and memory ability in rats (Mosharrof, 1987), prevented destructive changes and decreased the density of synapses in the cerebral cortex in rats with cerebral ischaemia (Logvinov et al. 2001), and possessed selective stress-reducing effects in rats with pre-existing tendencies towards anxiety (Yamamotova et al. 2007).

Decoction of *R. carthamoides* root promoted a correction of depressive states in human and also improved health conditions of patients with gastrointestinal pain of somatic origin (Ibatov 1995).

6) Indications and usage:
   Currently, the extracts or some compounds from roots and rhizomes are used for their adaptogenic and tonic properties in various dietary supplements or nutraceutical preparations to promote muscle growth, treat impotency, and eliminate physical weakness and mental weariness, as well as for recovery after surgery, infectious disease, or chemical intoxication (Kokoska and Janovska 2009, Opletal et al. 1997).
7) Precautions and side effects:
   Not found
8) Dosage:
   A) Internal use:
      *R. carthamoides* roots are also included in the formulas of various non-alcoholic beverages, cosmetic, and bath products. Dried underground or aerial parts are included in herbal teas (Kokoska and Janovska 2009; Opletal et al. 1997).
   B) External use:
      Not found.

## 6.19 References


- Bespalov VG, Aleksandrov VA, Yaremenko KV, Davydov VV, Lazereva NL, Limarenko AYu, Slepyan LI, Petrov AS, Troyan DN. The inhibitory effect of phytoadaptogenic drugs bioginseng, *Eleutherococcus senticosus* and *Rhaponticum carthamoides* on the development of N-nitrosomethylurea induced tumours of the nervous system in rats. Voprosy Onkologii. 1992;38: 1073–1080.
- Biskup E, Szynklarz B, Golebiowski M, Borsuk K, Stepnowski P, Lojkowska E. Composition and biological activity of *Rhaponticum carthamoides* extracts obtained from plants collected in Poland and Russia. J Med Plant Res. 2013;7(11): 687–95.
- Dushkin M, Khrapova M, Kovshik G, Chasovskikh M, Menshchikova E, Trufakin V, Shurlygina A, Vereschagin E. Effects of *Rhaponticum carthamoides* versus *Glycyrrhiza glabra* and *Punica granatum* extracts on metabolic syndrome signs in rats. BMC Complement Alternat Med. 2014;14:33.

- Geszprych A, Weglarz Z. Composition of essential oil from underground and aboveground organs of *Rhaponticum carthamoides* (Willd.) Iljin. Herba Pol. 2002;48:188–92.
- Hamburger M, Gaube F, Wolfl S, Pusch L, Kroll P, Riese U, Schrenk D. Effects of *Leuzea carthamoides* DC on human breast cancer MCF-7 cells detected by gene expression profiling. Planta Med. 2006;72: 992.
- Havlik J, Budesinsky M, Kloucek P, Kokoska L, Valterova I, Vasickova S, Zeleny V. Norsesquiterpene hydrocarbon, chemical composition and antimicrobial activity of *Rhaponticum carthamoides* root essential oil. Phytochemistry. 2009; 70: 414–8.
- Ibatov AN. Application of *Leuzea carthamoides* rhizome decoction in therapy of alcoholic patients with depressive state. Zh. Nevropatol. Psikhiatr. Im. S. S. Korsakova 1995; 95: 78–9.
- Kholmatov Kh, Kharlamov IA, Mavlankulova ZI. Lekarstvennye rasteniya Tsentralnoy Azii (Medicinal Plants of Central Asia). Tashkent: Med Lit Publishing Ibn Sino. 1998;296.
- Khomova TV, Gusakova SD, Glushenkova AI. Lipids from ekdisten production wastes. Chem Nat Compd. 1995;31: 172–4.
- Kokoska L, Janovska D. Chemistry and pharmacology of *Rhaponticum carthamoides*: a review. Phytochemistry. 2009;70:842–55.
- Lamer-Zarawska E, Serafinowicz W, Gasiorowski K, Brokos B. Immunomodulary activity of polysaccharide-rich fraction from *Rhaponticum carthamoides* leaves. Fitoterapia. 1996: 67: 371–2.
- Logvinov SV, Pugachenko NV, Potapov AV, Krasnov EA, Plotnikov MB, Maslov MYu, Aliev OI, Tyukavkina NA. Ischemia-induced changes in synaptoarchitectonics of brain cortex and their correction with ascovertin and leuzea extract. Bull Exp Biol Med. 2001;132:1017–20.
- Mosharrof AH. Effects of extract from *Rhaponticum carthamoides* (Willd) Iljin (Leuzea) on learning and memory in rats. Acta Physiol Pharmacol Bulg. 1987; 13:37–42.
- Opletal L, Sovova M, Dittrich M, Solich P, Dvorak J, Kratky F, Cerovsky J, Hofbauer J. Phytotherapeutic aspects of diseases of the circulatory system. 6. *Leuzea carthamoides* (WILLD.) DC: the present state of research and possible use of the taxon. Ceska Slov Farm. 1997;46:247–55.
- Petkov V, Roussinov K, Todorov S, Lazarova M, Yonkov D, Draganova S. Pharmacological investigations on *Rhaponticum carthamoides*. Planta Med. 1984;50: 205–9.
- Shreter AI. Medicinal plants of Soviet Far East (in Russian). Moscow: Nauka. 1975.
- Stransky K, Nemec V, Slama K. Lipid composition of the seeds of an ecdysteroid-containing plant, *Leuzea carthamoides* (Willd.) DC (Asteraceae). Russ J Plant Physiol. 1998;45: 333–8.
- Syrov VN, Kurmukov AG. Anabolic activity of phytoecdysone–ecdysterone, isolated from *Rhaponticum carthamoides* (Willd.) Iljin. Farmakol. Toksikol. 1976;39:690–3.

- Vereskovskii VV, Kintya PK, Shapiro DK, Chekalinskaya II. Triterpene glycosides of *Rhaponticum carthamoides* cultivated in Belorussia. Chem Nat Compd. 1978;13:484–5.
- Vershinina SF. Concerning the effect of the *Leuzea carthamoides* extract and sarcolysin on the course of lympholeucosis in mice. Voprosy Onkologii. 1967; 13: 99–101.
- Yamamotova A, Pometlova M, Harmatha J, Raskova H, Rokyta R. The selective effect of N-feruloylserotonins isolated from *Leuzea carthamoides* on nociception and anxiety in rats. J Ethnopharmacol. 2007;112:368–74.
- Yankulov DB Kepova Vrkhu razvitieto dobiva i khimichniya cctav na korenite na levzeiyata [*Rhaponticum carthamoides* (Willd) Iljin] otglezhdana u nas. Farmaciya. 1962;12:36–9.

# Chapter 7
# Chinese Herbal Medicine Used Against Depression in China

**Dandan Feng, Xiangping Lin, Shu Yang, Zhaoyu Yang, Zian Xia, Zhiqing Li, Tao Tang, Rong Fan, Chunhu Zhang, and Yang Wang**

**Abstract** In China, depression remains a disabling condition with high prevalence and a large clinical burden.

Taken together, this chapter gathered the most common herbs for depressive therapy in China through introducing each Chinese herbal medicine (CHM) description, chemical constituents, pharmacology and bioactivities, indications and usage, precautions and side effects, and dosage. The aim is to provide CHM information to treat depression for everyone who needs. Thus, information about *Acanthopanax senticosus* (Rupr. et Maxim.) Harms, *Acorus tatarinowii* Schott, *Allium macrostemon* Bunge, *Aloe barbadensis* Miller, *Centella asiatica* (L.) Urban, *Corydalis yanhusuo* W. T. Wang, *Crocus sativus* L., *Cyperus rotundus* L., *Citrus aurantium* L., *Ginkgo biloba* L., *Panax ginseng* C. A. Mey., *Hypericum perforatum* L., *Nelumbo nucifera* Gaertner, *Paeonia lactiflora* Pall., *Polygala tenuifolia* Willd., *Polygonum multiflorum* (Thunberg) Haraldson, *Bupleurum chinense* de Candolle or *Bupleurum scorzonerifolium* Willdenow, *Ligusticum chuanxiong* Hort., *Curcuma longa* L., *Rhodiola rosea* L., *Valeriana officinalis* L., and *Ziziphus jujuba* Mill. was gathered in this chapter.

**Keywords** Depression • Antidepressant effect • Herb • Traditional Chinese medicine • China

D. Feng • X. Lin • S. Yang • Z. Yang • Z. Xia • Z. Li
T. Tang • R. Fan • C. Zhang • Y. Wang (✉)
Institute of Integrated Traditional Chinese and Western Medicine, Xiangya Hospital, Central South University, Changsha 410008, China
e-mail: wangyang_xy87@163.com



C. Grosso (ed.), *Herbal Medicine in Depression*,
DOI 10.1007/978-3-319-14021-6_7

## 7.1 Introduction

Depression is the most prevalent psychiatric disorder (Flint and Kendler 2014). In addition to mortality associated with suicide, depression increases the risk for cardiovascular disorders, stroke, Alzheimer's disease, epilepsy, diabetes, and cancer (Lang nd Borgwardt 2013; Koponen et al. 2015; Yang et al. 2015). The World Health Organization (WHO) has ranked depression as the fourth leading cause of disability worldwide and estimates that it will be the second leading cause by 2020 (Kessler and Bromet 2013). In China, depression remains a disabling condition with high prevalence and a large clinical burden (Van Orden et al. 2015). It is one of the highest burdens of diseases with a cost of 51,370 million Renminbi (RMB) at 2002 prices, and 8090 million RMB were direct costs, accounting for about 16 % of the total cost of depression (Hu et al. 2007).

Antidepressant drugs are the first-line treatment for depression (Anthes 2014). Unfortunately there still remains a large portion of depressed patients who do not make a full response and experience relapse, often resulting in the worsening and hospitalization after antidepressant drug administration (Mulder 2015). The reason is that newer antidepressants differ from the older drugs in decreasing the incidence of unwanted side effects and/or narrowing the neurochemical target, rather than by introducing novel mechanisms of action (Willner et al. 2013). Consequently above reasons lead to the urgent demand of more effective antidepressant therapies with less adverse effects.

Chinese herbal medicine (CHM), as the main therapy of complementary and alternative medicine, has a long history of more than 8,000 years (Yeung et al. 2014; Drasar and Moravcova 2004). It has been increasingly introduced into depressive treatment over recent decades (Liu et al. 2015; Fava 2010). For example, in Taiwan, more than 40 % of individuals prescribed with antidepressant treatments for depression used CHM services (Pan et al. 2013). There are growing preclinical and clinical trials, which show beneficial efficacy for CHM to treat depression (Saki et al. 2014). A recent meta-analysis showed that CHM monotherapy was better than placebo and as effective as antidepressants in reducing the Hamilton Depression Rating Scale score; CHM was associated with less adverse events than synthetic antidepressants, and adding Chinese herbal medicine to synthetic antidepressants reduced adverse events (Yeung et al. 2014). It was revealed that CHM-treated patients experienced remarkably fewer incidences of physical tiredness, headache, palpitation, dry mouth, and constipation; all these adverse events are often reported in psychotropic treatment (Liu et al. 2015).

CHMs are complex multicomponent mixtures from herbs that contain several pharmacologically active ingredients and act on multiple targets. Thus, they possess a broader therapeutic profile than most synthetic drugs and cause fewer side effects related to their mechanisms of action (Liu et al. 2015). In addition, CHM can offer major opportunities for finding novel antidepressant targets and mechanisms, which are potential candidates in the development of new antidepressants (Sun et al. 2015).

In China, traditional Chinese medicine (TCM) doctors always judge which syndromes a patient belongs to through inspection, auscultation and olfaction, inquiry, and palpation (four diagnostic methods of TCM); then they give medical care to the patient using CHM according to the previous diagnosis (Song et al. 2015). Consequently TCM treatments are often prescribed based on individuals' pattern diagnoses (Yeung et al. 2015). It is evident that TCM doctors may only grasp TCM theory of each herb rather than its pharmacological actions. Thus it is requisite to understand the multiple effective actions of each herb to help modern TCM doctors prescribe CHM for depressive treatment better than before.

Search methods: *Pharmacopoeia of the People's Republic of China* 2010, *Flora of China*, Tropicos, PubMed, China National Knowledge Infrastructure (CNKI), Web of Science, TCM Wiki, Baidu, Wanfang Database.

## 7.2 References

Flint J, Kendler KS. The genetics of major depression. Neuron. 2014; 81: 484–503.

Lang UE, Borgwardt S. Molecular mechanisms of depression: perspectives on new treatment strategies. Cellular physiology and biochemistry. Int J Exp Cell Physiol Biochem Pharmacol. 2013;31: 761–77.

Koponen H, Kautiainen H, Leppanen E, Mantyselka P, Vanhala M. Association between suicidal behavior and impaired glucose metabolism in depressive disorders. BMC Psychiatry, 2015;15: 163.

Yang L, Jia CX, Qin P. Reliability and validity of the Center for Epidemiologic Studies Depression Scale (CES-D) among suicide attempters and comparison residents in rural China. BMC Psychiatry. 2015;15: 76.

Kessler RC, Bromet EJ. The epidemiology of depression across cultures. Annu Rev Public Health. 2013;34, 119–38.

Van Orden KA, Chen S, O'Riley A, Conwell Y. Course of late-life depression in China is chronic and unremitting. Int J Geriatr Psychiatry. 2015;30: 409–15.

Hu TW, He Y, Zhang M, Chen N. Economic costs of depression in China. Social Psychiatry Psychiatr Epidemiol. 2007; 42: 110–6.

Anthes E. Depression: a change of mind. Nature. 2014;515: 185–7.

Mulder R. Depression relapse: importance of a long-term perspective. Lancet. 2015; 386:10–12.

Willner P, Scheel-Kruger J, Belzung C. The neurobiology of depression and antidepressant action. Neurosci Biobehav Rev. 2013;37:2331–71.

Yeung WF, Chung KF, Ng KY, Yu YM, Ziea ET, Ng BF. A systematic review on the efficacy, safety and types of Chinese herbal medicine for depression. J Psychiatr Res. 2014;57:165–75.

Drasar P, Moravcova J. Recent advances in analysis of Chinese medical plants and traditional medicines. J Chromatograph B. 2004;812:3–21.

Liu LY, Feng B, Chen J, Tan QR, Chen ZX, Chen WS, Wang PR, Zhang ZJ. Herbal medicine for hospitalized patients with severe depressive episode: a retrospective controlled study. J Affect Disord. 2015;170: 71–7.

Fava M. Using complementary and alternative medicines for depression. J Clin Psychiatr. 2010;71: e24.

Pan YJ, Cheng IC, Yeh LL, Cho YM, Feng J. Utilization of traditional Chinese medicine in patients treated for depression: a population-based study in Taiwan. *Complement Ther Med*, 2013; 21:215–23.

Saki K, Bahmani M, Rafieian-Kopaei M. The effect of most important medicinal plants on two importnt psychiatric disorders (anxiety and depression)-a review. Asian Pac J Trop Med. 2014;7S1:S34–42.

Sun CL, Geng CA, Yin XJ, Huang XY, Chen JJ. Natural products as antidepressants documented in Chinese patents from 1992 to 2013. J Asian Nat Prod Res. 2015;17, 188–98.

Song J, Liu X, Deng Q, Dai W, Gao Y, Chen L, Zhang Y, Wang J, Yu M, Lu P, Guo R. A network-based approach to investigate the pattern of syndrome in depression. Evid Based Complement Alternat Med. 2015;2015:768249.

Yeung WF, Chung KF, Ng KY, Yu YM, Zhang SP, Ng BF, Ziea ET. Prescription of Chinese herbal medicine in pattern-based traditional Chinese medicine treatment for depression: a systematic review. Evid Based Complement Alternat Med. 2015;2015: 160189.

## 7.3 *Acanthopanax senticosus* (Rupr. et Maxim.) Harms

### *7.3.1* Acanthopanacis Senticosi Radix et Rhizoma seu Caulis

#### 7.3.1.1 (Araliaceae)

1) Common names:
   Ci-wu-jia, Ci-mu-bang, Kan-guai-bang-zi (Pinyin in Chinese), Siberian ginseng, Eleuthero root.
2) Photograph of the species (Fig. 7.1)
3) Description of the species:
   A) Habitat and world distribution:
      Scrub fields, forests, roadsides, valleys; below 2000 m. Hebei, Heilongjiang, Henan, Jilin, Liaoning, Shaanxi, Shanxi, Sichuan of China, Amur,

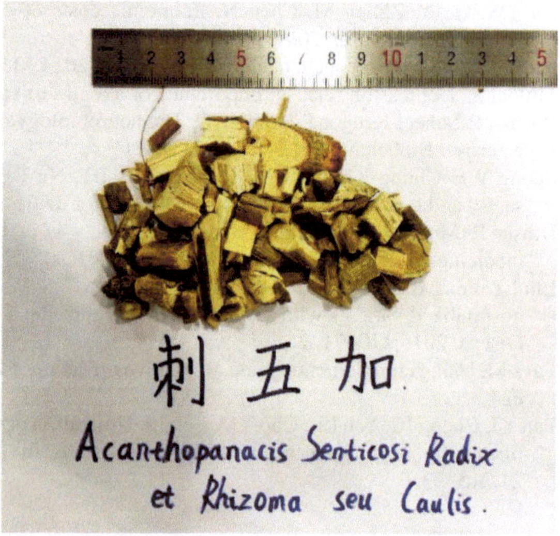


**Fig. 7.1** *Acanthopanacis Senticosi Radix et Rhizoma seu Caulis*

Khabarovsk, Primorye, Sakhalin of Russian Far East, Japan, and Korea (Flora of China Editorial Committee 2007; Deyama et al. 2001).

B) Morphology of the plant:
Shrubs, to 6 m tall. Branches with dense to scattered, slender, terete, bristle-like prickles. Petiole with 3–12 cm, slender, sometimes with fine prickles; petiolule of central leaflet with (0.6-) 1.2–2 cm, frequently brownish pubescent; leaflets (3-)5, elliptic-obovate or oblong, with dimensions of 5–13 × 3–7 cm, papery, abaxially pubescent on veins, adaxially with scattered hairs, secondary veins 6 or 7 pairs, conspicuous on both surfaces, base broadly cuneate, margin sharply biserrate, apex shortly acuminate or acuminate. Inflorescence terminal, solitary or compound umbel type, borne on leafy shoots, normally with 2–6 umbels together; peduncles with 5–7 cm, glabrous; pedicels with 1–2 cm, glabrous or slightly pubescent at base. Calyx subentire or with 5 inconspicuous teeth, glabrous. Corolla purple-yellow color. Ovary pentacarpellate, styles united into a column. Fruit ovoid-globose, with ca. 8 mm; styles persistent, ca. 1.5 mm. Flowering: June–July, fruiting: August–October (Flora of China Editorial Committee 2007).

C) Medicinal parts:
The root and the rhizome (or the stem) of *Acanthopanax senticosus* are being used.

D) Chemical constituents:
**Lignans** (neociwujiapenol, ciwujiatone, sesamin, eleutheroside E, eleutheroside E2, syringin (=eleutheroside B), syringaresinol, feruloyl sucrose, (+)-pinoresinol-*O*-β-D-glucoside, (+)-syringaresinol-*O*-β-D-glucoside, (+)-pinoresinol-di-*O*-β-D-glucoside, (+)-medioresinol-di-*O*-β-D-glucoside, etc.); **coumarins** (isofraxidin, isofraxidin-7-*O*-glucoside, scopoletin, eleutheroside B2); **flavonoids** (quercetin, quercetin-3-*O*-galactoside, quercetin-3-*O*-rhamnoside, quercetin-3-*O*-rutinoside); **phenolic acids** (5-caffeoylquinic acid, 1,5-di-*O*-caffeoylquinic acid, 3,5-di-*O*-caffeoylquinic acid, 4,5-di-*O*-caffeoylquinic acid, caffeic acid, syringic acid, *p*-coumaric acid, ferulic acid, protocatechuic acid); **triterpenoid saponins** (ciwujianosides $A_1$, $A_2$, $A_3$, $A_4$, B, $C_1$, $C_2$, $C_3$, $C_4$, $D_1$, $D_2$, $D_3$, and E; acanthopanaxosides A, B, C, and E; oleanolic acid derivatives, friedelin, eleutheroside L, sessiloside, copteroside B, silphioside G, etc.); **volatile compounds** (isocaryophyllene, linalool, α-pinene, caryophyllene oxide, *n*-octadecanol, manoyl oxide, 9, 17-octadecadienal, tetradecanal, humulene oxide, humulene, β-farnesene, (*E*, *E*)-2,4-decadienal, *p*-cymene, 2-*n*-pentylfuran, *n*-heptaldehyde, etc.); **other compounds** (octadecanoic acid, daucosterol, methyl oleate, etc.) (Huang et al. 2011).

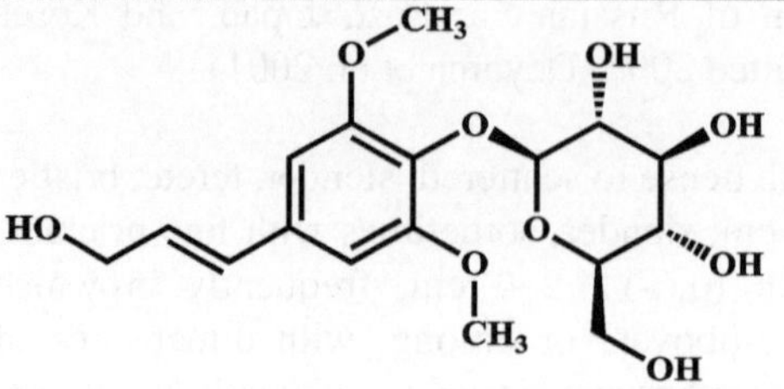


Eleutheroside B

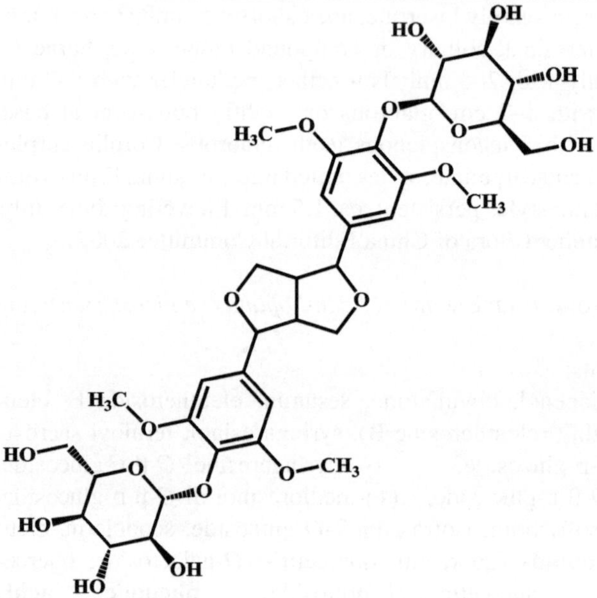


Eleutheroside E

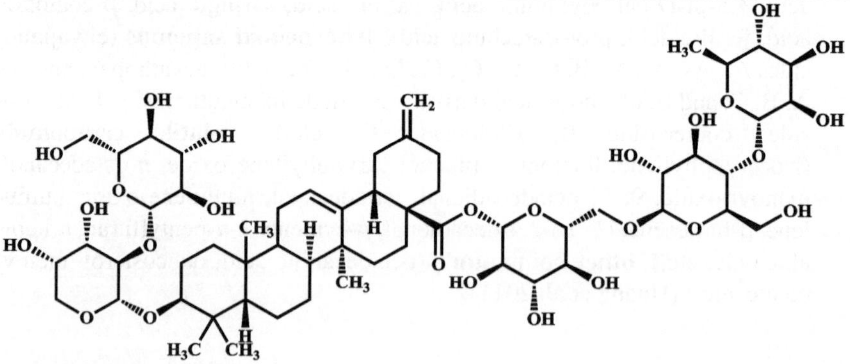


Ciwujianoside $A_2$

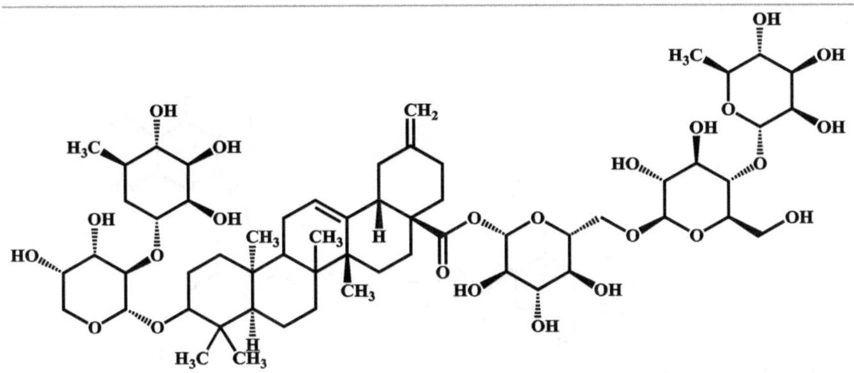

Ciwujianoside B

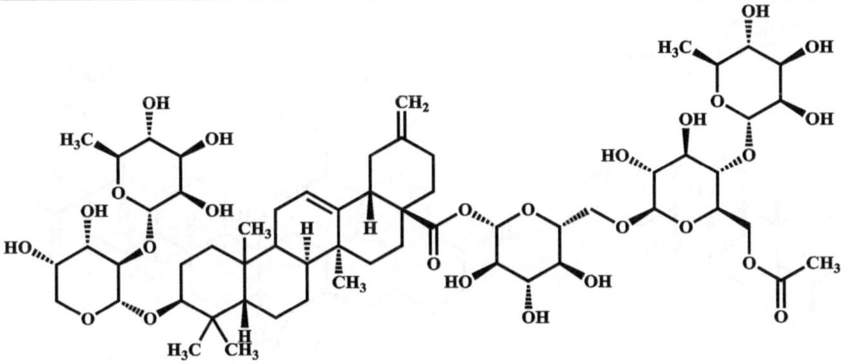

Ciwujianoside C2

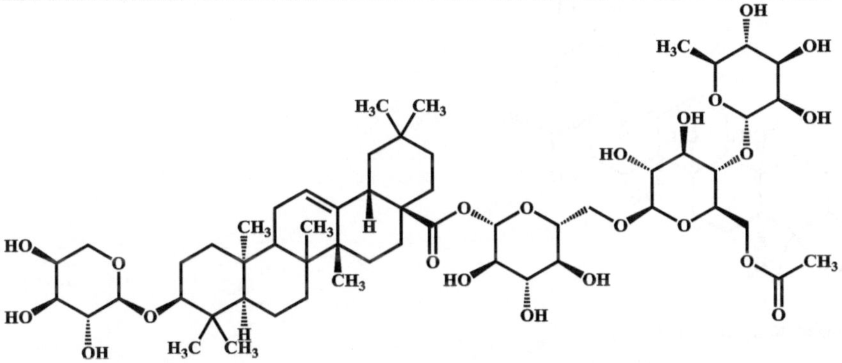

Ciwujianoside D1

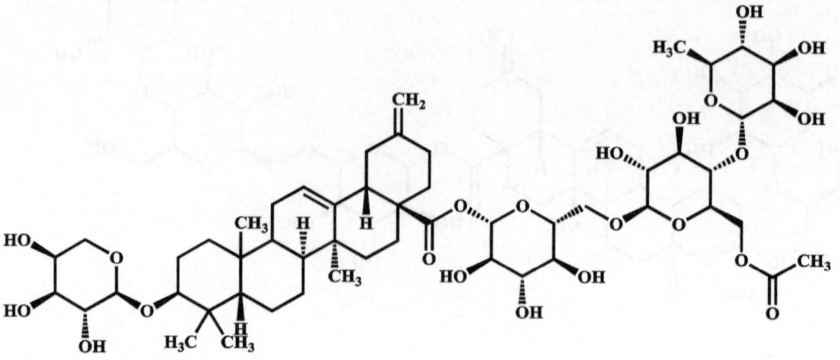

Ciwujianoside D2

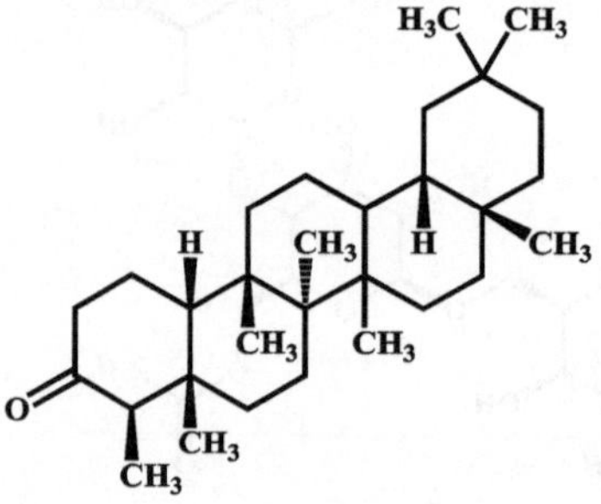

Friedelin

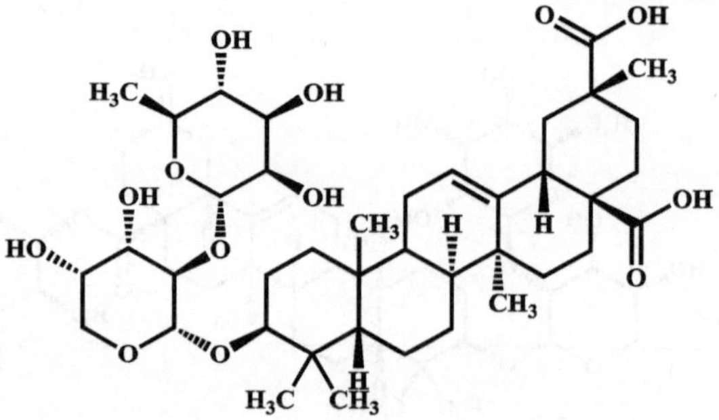

Acanthopanaxoside C

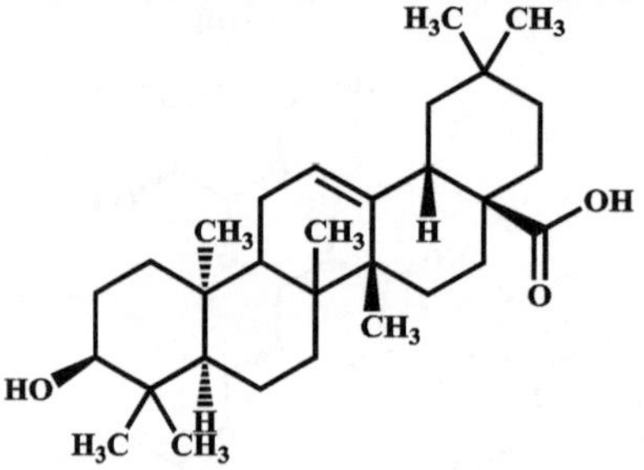

Oleanolic acid derivatives

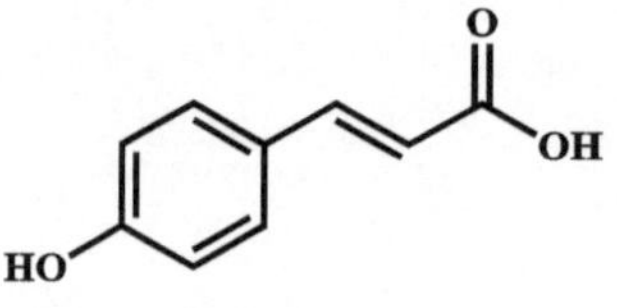

*p*-Coumaric acid

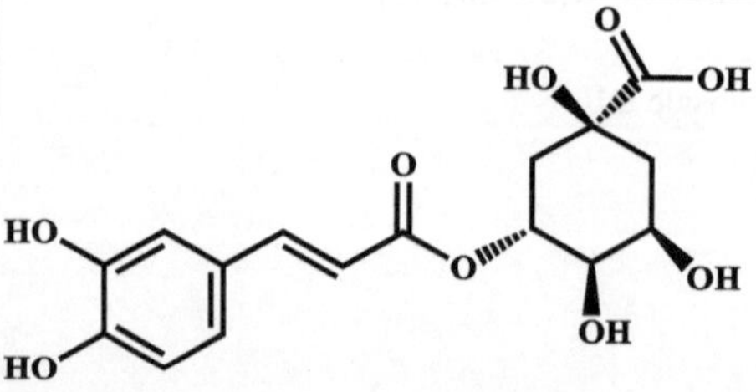

5-*O*-Caffeoylquinic acid (=Chlorogenic acid)

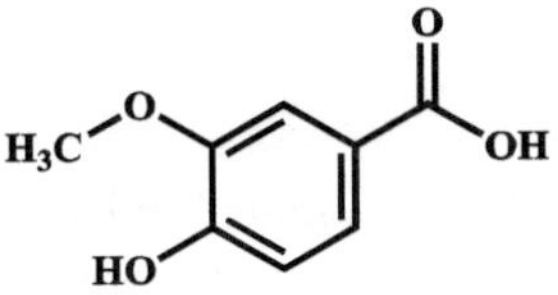


Vanillic acid

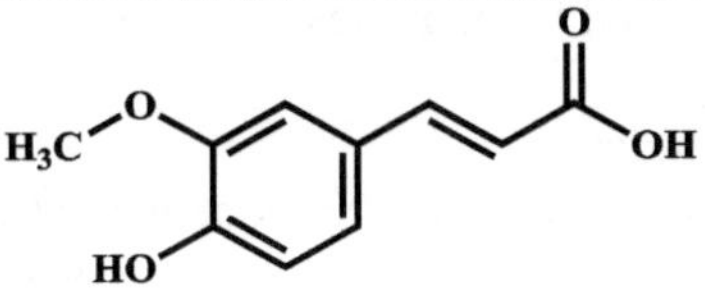


Ferulic acid

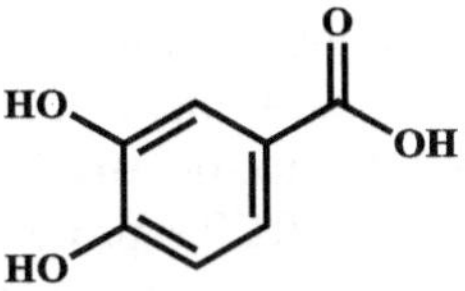


Protocatechuic acid

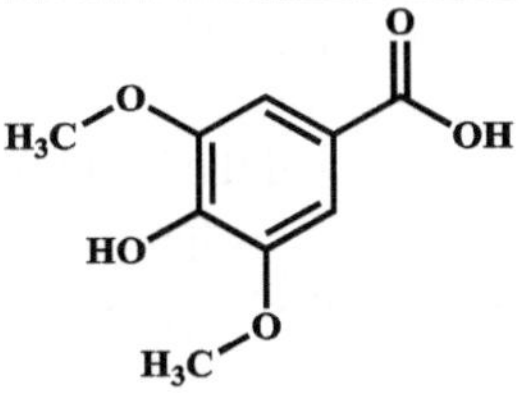


Syringic acid

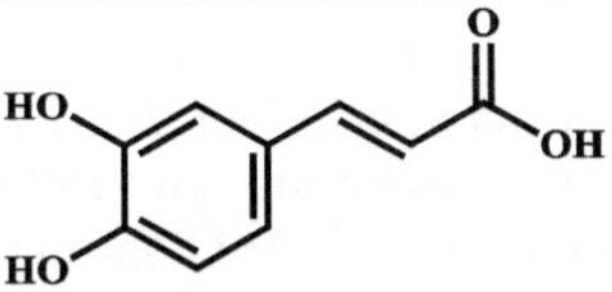


Caffeic acid

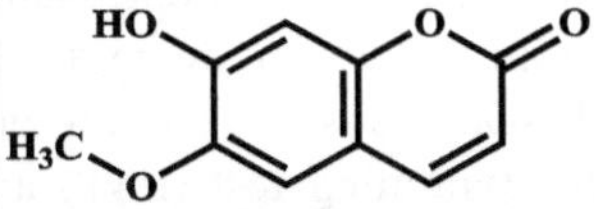


Scopoletin

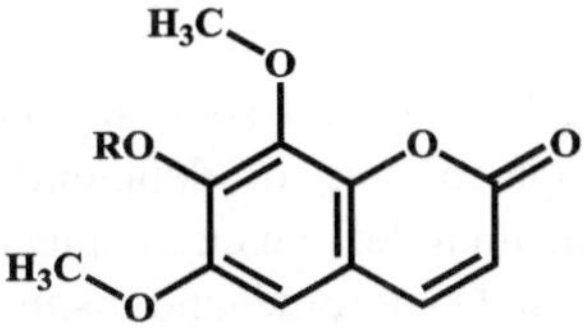


Isofraxidin: R = H
Isofraxidin-7-*O*-glucoside: R = glucose

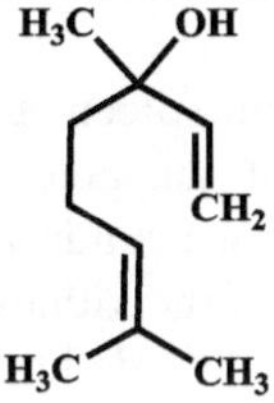


Linalool

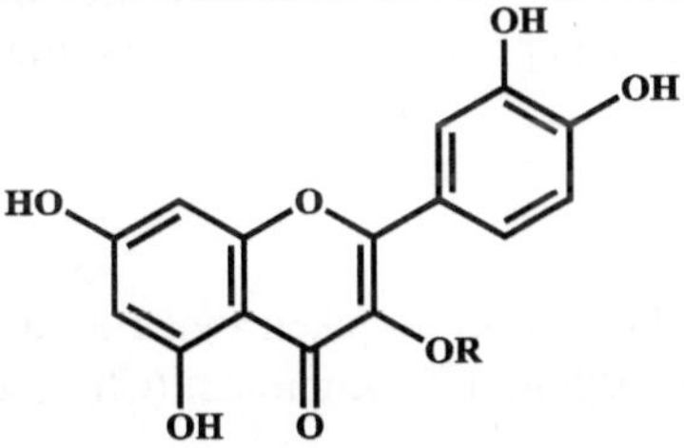


Quercetin: R = H
Quercetin-3-*O*-galactoside: R = galactose
Quercetin-3-*O*-rhamnoside: R = rhamnose
Quercetin-3-*O*-rutinoside: R = rutinose

4) Pharmacology and bioactivities:

   A) In vitro studies:
      A study suggested that the aqueous extract from *Acanthopanax senticosus* exerts a neuroprotective effect on corticosterone-induced neurotoxicity in PC12 cells, which may be one of the acting mechanisms that account for its in vivo antidepressant activity. In the study, different concentrations of the extract significantly increased the cell viability, decreased the LDH release, suppressed the apoptosis of PC12 cells, attenuated the intracellular $Ca^{2+}$ overloading, and upregulated the brain-derived neurotrophic factor (BDNF) mRNA level and cAMP response element-binding (CREB) protein expression compared with the corresponding corticosterone-treated group (Wu et al. 2013).

   B) In vivo studies:
      A study indicated that the antidepressant effects may be mediated via the central monoaminergic neurotransmitter system and CREB protein expression in mice. *A. senticosus* extract can reduce the duration of immobility in both the forced swimming test (FST) and the tail suspension test (TST); elevate the levels of serotonin (5-HT), norepinephrine (NE), and dopamine (DA) in the whole brain of mice; and upregulate the level of CREB protein (Jin et al. 2013).

5) Indications and usage:
   Invigorating qi, strengthening the spleen, tonifying kidney, and soothing the nerves. Treat hypodynamia and inappetence of spleen–lung qi deficiency, chronic cough and deficient dyspnea syndrome of the dual deficiency of spleen and kidney, waist and knee pain of the kidney deficiency syndrome, and insomnia and dreaminess syndrome of the deficiency of heart and spleen (Chinese Pharmacopoeia Commission 2010).
6) Precautions and side effects:
   It is contraindicated in case of yin deficiency. It can increase human digoxin serum level by suppressing P-glycoprotein and peptide transporter activities (Huang et al. 2011).
7) Dosage:

   A) Internal use:
      9–27 g/day, decocted in water for oral administration, taken twice a day, generally for 5–7 days, modificated according to symptoms (Chinese Pharmacopoeia Commission 2010).
   B) External use:
      Not found.

## 7.4 References

Flora of China Editorial Committee. Flora of China. Beijing and St. Louis: Science Press and Missouri Botanical Garden Press. 2007;13: 468.
Deyama T, Nishibe S, Nakazawa Y. Constituents and pharmacological effects of Eucommia and Siberian ginseng. Acta Pharmacol Sinica. 2001;22(12): 1057–70.
Huang L, Zhao H, Huang B, Zheng C, Peng W, Qin L. *Acanthopanax senticosus*: review of botany, chemistry and pharmacology. Pharmazie. 2011; 66(2), 83–97.
Wu F, Li H, Zhao L, Li X, You J, Jiang Q, Li S, Jin L, Xu Y. Protective effects of aqueous extract from *Acanthopanax senticosus* against corticosterone-induced neurotoxicity in PC12 cells. J Ethnopharmacol. 2013;148(3): 861–8.
Jin L, Wu F, Li X, Li H, Du C, Jiang Q, You J, Li S, Xu Y. Anti-depressant effects of aqueous extract from *Acanthopanax senticosus* in mice. Phytother Res. 2013;27(12): 1829–33.
Chinese Pharmacopoeia Commission. Pharmacopoeia of People's Republic of China. Beijing: China Medical Science Press. 2010.

## 7.5 *Acorus tatarinowii* Schott

### *7.5.1* Acorus Tatarinowii Rhizoma

#### 7.5.1.1 (Araceae)

1) Common names:
   Rhizoma acori graminei, Grassleaf sweetflag rhizome, Shi-chang-pu, Chang-pu-ye, Yao-chang-pu, Shui-jian-cao (Pinyin in Chinese).
2) Photograph of the species (Fig. 7.2)
3) Description of the species:

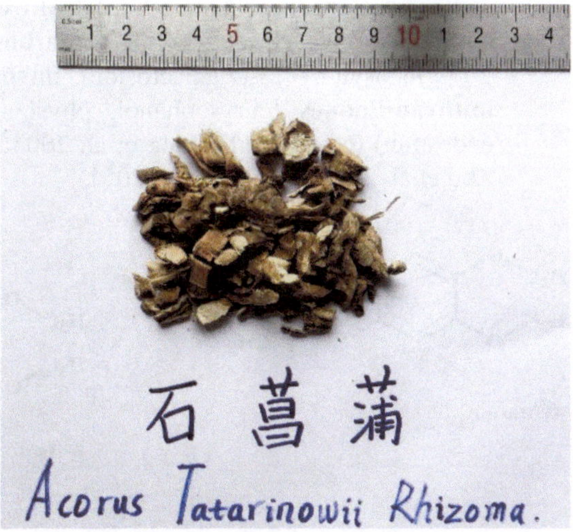


**Fig. 7.2** *Acorus Tatarinowii Rhizoma*

A) Habitat and world distribution:
   Dense forests, moist rocky stream banks, meadows; at an altitude of 20–2600 m. Located in the provinces south of the Yellow River, northeast of India, and northern Thailand (Flora of China Editorial Committee 2010).
B) Morphology of the plant:
   Perennial herb; rhizome slender, 5-10 × 0.4-0.6(-0.8) cm, aromatic. Leaves several, dark green, ensiform, (15-) 20–45(-55) × (0.3-) 0.5–1(-1.4) cm, midrib lacking, apex acuminate. Peduncle compressed triangular, (4-) 9–20(-24) cm. Spathe green, leaflike, (8-) 10–24(-25) × 0.2–0.4(-0.5) cm. Spadix straight or slightly curved, narrowly cylindric to subcylindric, (3-) 4-10(-14) × (0.3-) 0.4–0.6(-0.7) cm, densely flowered. Flowers yellowish or yellow-green to somewhat whitish, 1.8–2 mm in diam. seen from above; tepals ± oblong, 1.5–2 × 0.7–1 mm, keeled, membranous, apex rounded or acute; filaments oblong, flat, ca. 1.5 mm, anthers yellow, 0.4–0.5 mm in diam.; pollen grains ca. 15 μm, exine more densely foveolate; gynoecium obconic-cylindric, (2-)2.5–3 × (1.2-)1.8–2 mm, with conic, spongy apex and stigma very small. Infructescence yellowish to yellow-white at maturity, 1–1.5 cm in diam., berries densely arranged. Berry obovoid-globose, few to several seeded, 3–3.5 × 2.2–2.5 mm. Seed ellipsoid, (2-)2.5–3 × (0.8-)1–1.2 mm, with many long (3–4 mm) bristles (longer than seed itself); testa light brown, smooth. Flowering and fruiting in February–July (Flora of China Editorial Committee 2010).
C) Medicinal parts:
   The dry rhizome of perennial herbaceous plant *Acorus tatarinowii* Schott (Flora of China Editorial Committee 2010).
D) Chemical constituents:
   **Volatile compounds** (β-asarone, α-asarone, 1,2-dimethoxy-4-(1-propenyl)-benzene, 1,2-dimethoxy-4-(2-propenyl)-benzene, camphene, α-pinene, β-pinene, *p*-cymene, γ-terpinene, linalool, camphor, α-terpineol, γ-gurjunene, cedrene, α-copaene, calamenene, 1-hydroxy-7(11), 9-guaiadien-8-one, etc.), **spiroalkaloids** (acortatarins A, B), **xanthones** (4, 5, 8-trimethoxy-xanthone-2-*O*-glucosyl(1 → 2)-*O*-galactoside), **flavones** (luteolin-6,8-*C*-diglucoside), **anthraquinones** (chrysophanol, physcion, emodin), **lignans** (galbacin, eudesmin) (Chen 2011; Deng et al. 2004; Ganjewala and Srivastava 2011; Zhu et al. 2010; Wagner et al. 2011).

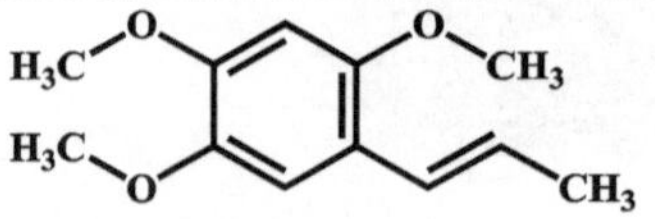


α-Asarone

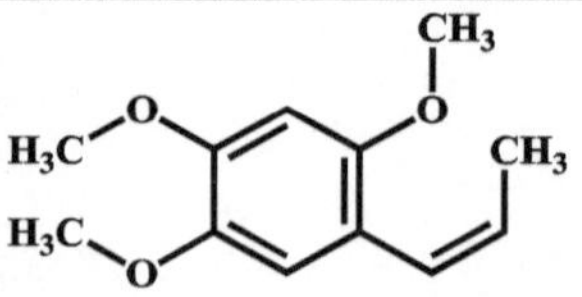


β-Asarone

| | |
|---|---|
| 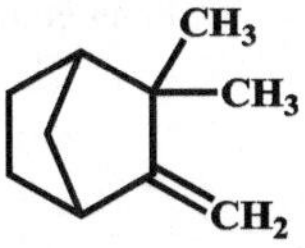<br>Camphene | 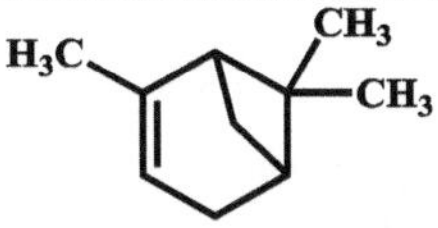<br>α-Pinene |
| 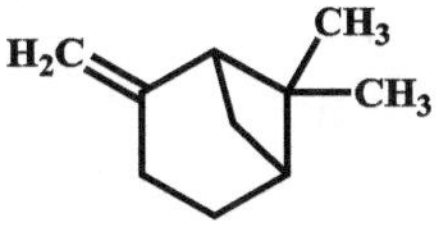<br>β-Pinene | 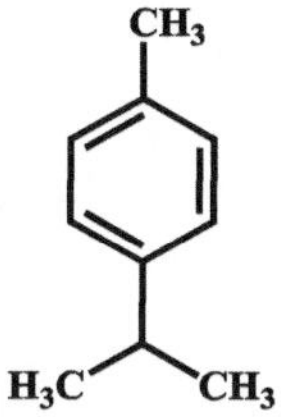<br>*p*-Cymene |
| 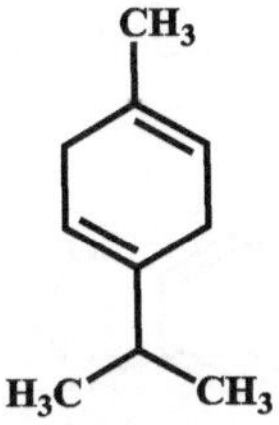<br>γ-Terpinene | 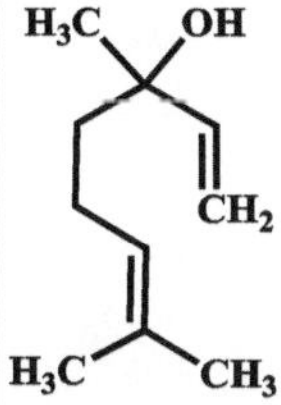<br>Linalool |
| 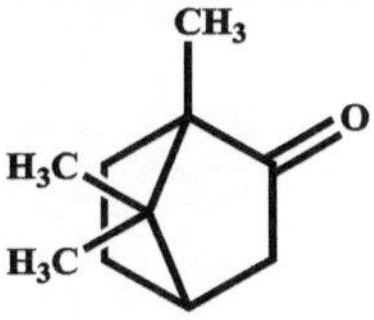<br>Camphor | 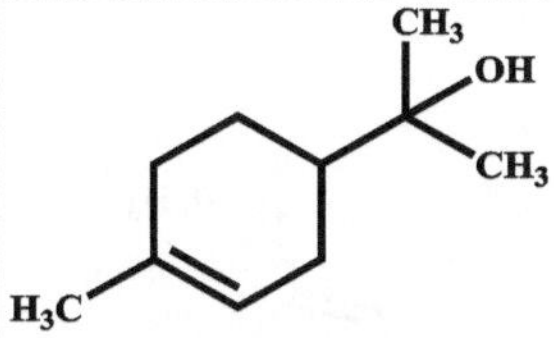<br>α-Terpineol |
| 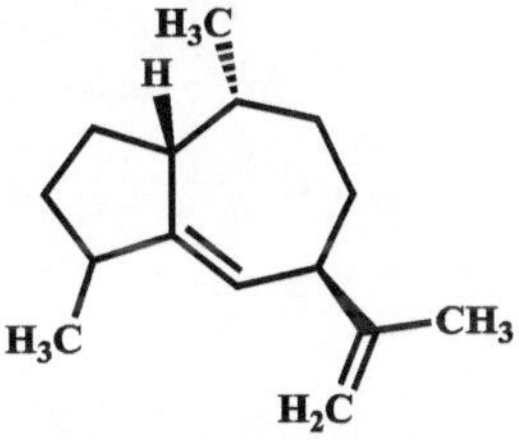<br>γ-Gurjunene | 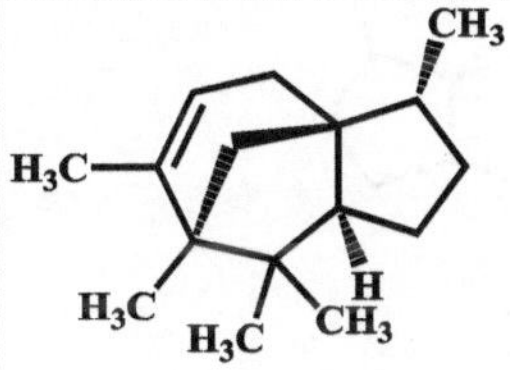<br>Cedrene |
| 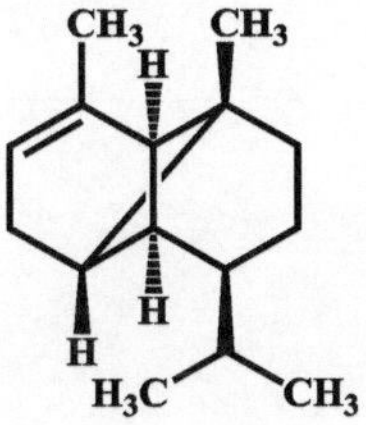<br>α-Copaene | 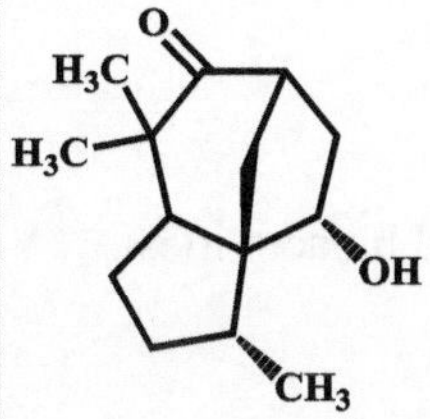<br>Calamenene |

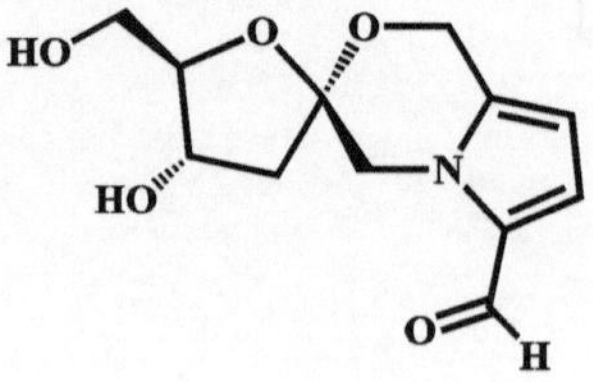

Acortatarin A

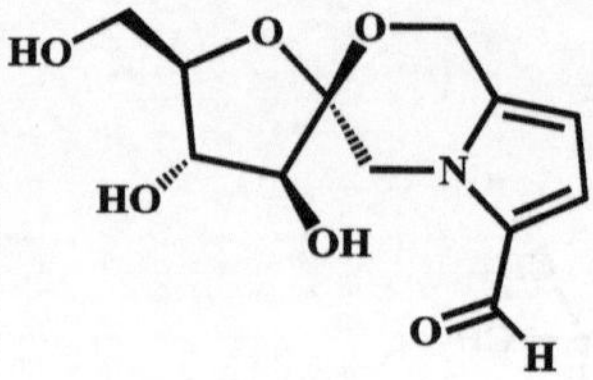

Acortatarin B

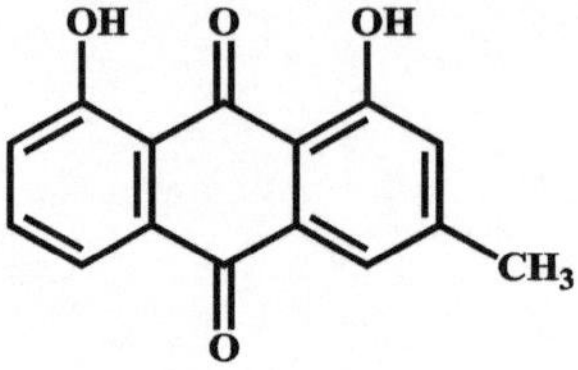

Chrysophanol

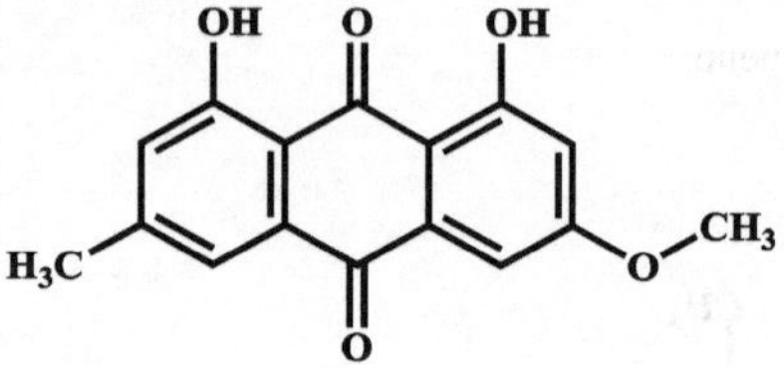

Physcion

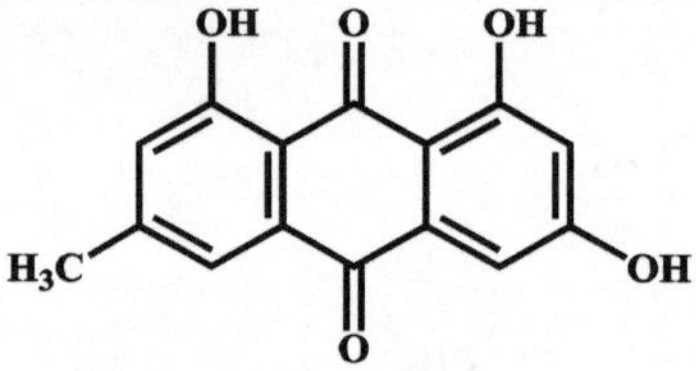

Emodin

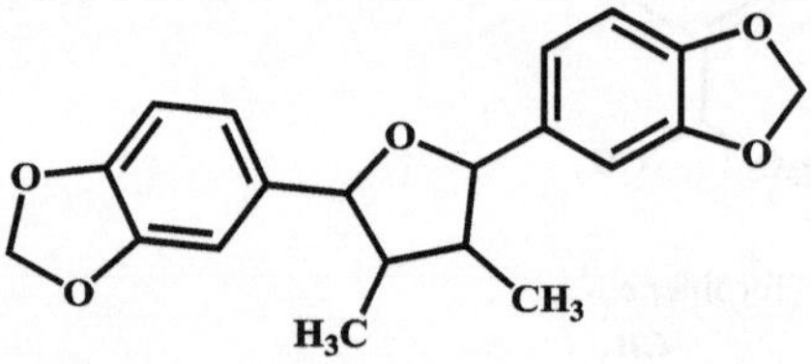

Galbacin

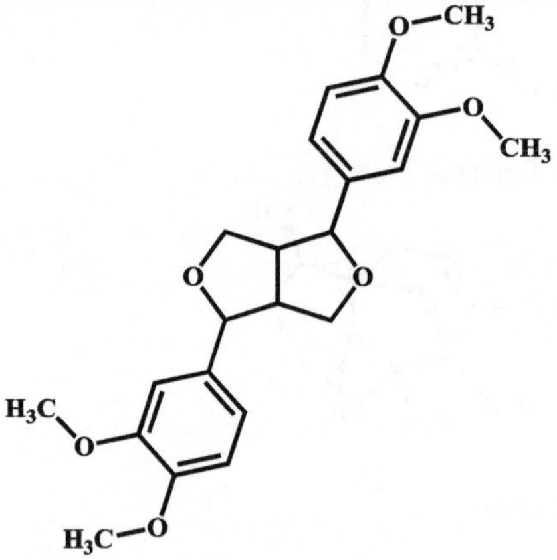

Eudesmin

4) Pharmacology and bioactivities:

A) In vitro studies:

Not found.

B) In vivo studies:

*Acorus tatarinowii* Schott (Shi-chang pu) is an herb widely used in traditional Chinese medicines. It is not only an anti-dementia and effective in improving learning and memory but also an antidepressant, a sedative, an anticonvulsant, and an antiepileptic (Zhu et al. 2014).

The essential oil and asarones from the rhizomes of *A. tatarinowii* can cure depression, since they reduced the immobility time in FST and TST (Han et al. 2013). However, no change in ambulation was observed in the OFT.

Physical exercise is often terminated not due to muscle fatigue but to the increase in serotonin (5-HT) concentrations in the brain during prolonged exercise. The antifatigue action of *A. tatarinowii* Schott can be through inhibition of the exercise-induced synthesis of 5-HT and tryptophan hydroxylase 2 (TPH2) expression and prevention of the exercise-induced decrease of 5-$HT_1B$ expression in the dorsal raphe (Zhu et al. 2014).

Eudesmin has significant anticonvulsant and sedative effects, and the mechanism of eudesmin may be related to upregulation of $GABA_A$ and glutamate decarboxylase (GAD65) expressions, and anti-apoptosis of neurons in the brain of chronic epileptic rats (Liu et al. 2015).

One of the most common models of Parkinson's disease is unilateral 6-hydroxydopamine (6-OHDA) injection into the medial forebrain bundle (MFB). β-Asarone treatment recovered the decrease in the OFT and rotarod tests observed in the 6-OHDA group. Rats from 6-OHDA model group exhibited a significant increase in initiation time and stepping time test, which were reversed by β-asarone treatment. β-Asarone also increased the homovanillic acid (HVA), 3,4-dihydroxyphenylacetic acid (DOPAC), and 5-hydroxyindole acetic acid (5-HIAA) levels in striatum but not the DA and 5-HT levels; increased tyrosine hydroxylase (TH) positive cells; and declined α-synuclein. β-Asarone could also interfere with autophagy activation by affecting the expression of Beclin-1, Bcl-2, JNK, and p-JNK (Zhang et al. 2014a).

In other study, it was shown that aspartic acid, glutamic acid, glycine, and GABA levels were increased in cerebral ischemia rats. *A. tatarinowii* Schott can enter the cerebral parenchyma through blood–brain barrier and cut down glutamic acid and aspartic acid increased during cerebral ischemia. As a result, *A. tatarinowii* can reduce the neurotoxicity of the excessive excitatory amino acid to avoid the secondary impairment of neurons (Zhang et al. 2014b).

5) Indications and usage:

Induce resuscitation, eliminate phlegm and promote intelligence, resolve dampness, and harmonize stomach. Treat block pattern of coma, retention of dampness in middle energizer, insomnia and amnesia, tinnitus, and deafness (Chinese Pharmacopoeia Commission 2010).

6) Precautions and side effects:

Hyperactivity of yang due to yin deficiency, sweating, and spermatorrhea (Chinese Pharmacopoeia Commission 2010).

7) Dosage:

A) Internal use:

5–10 g/day, decocted in water for oral administration, taken twice a day, generally for 4–5 days, modificated according to symptoms (Chinese Pharmacopoeia Commission 2010).

B) External use:

Use fresh herb in appropriate amount for washing with decoction or make into powder and use as paste (Chinese Pharmacopoeia Commission 2010).

## 7.6 References


Flora of China Editorial Committee. Flora of China. Beijing and St. Louis: Science Press and Missouri Botanical Garden Press. 2010;23: 1–2.

Chen F. The chemical constituents and pharmacological effects of plants of Acorus. The World Science and Technology Modernization of Traditional Chinese Medicine, 2011;13(6), 1013–7.

Deng C, Li N, Zhang X. Rapid determination of essential oil in *Acorus tatarinowii* Schott. By pressurized hot water extraction followed by solid-phase microextraction and gas chromatography-mass spectrometry. J Chromatograph A. 2004;1059:149–55.

Ganjewala D, Srivastava AK. An update on chemical composition and bioactivities of *Acorus* species. Asian J Plant Sci. 2011;10(3): 182–9.

Zhu M, Tan N, Ji C, Xu J, He W, Zhang Y. Chemical constituents from petroleum ether fraction of ethanol extract of *Acorus tatarinowii*. Zhongguo Zhong Yao Za Zhi. 2010;35(2): 173–6.

Wagner H, Bauer R, Melchart D, Xiao PG, Staudinger A. Chromatographic fingerprint analysis of herbal medicines. Vienna, Austria: Springer. 2011:903–21.

Zhu M, Zhu H, Tan N. The effects of *Acorus tatarinowii* Schott on 5-HT concentrations, TPH2 and 5-HT1B expression in the dorsal raphe of exercised rats. J Ethnopharmacol. 2014;158:431–6.

Han P, Han T, Peng W, Wang XR. Antidepressant-like effects of essential oil and asarone, a major essential oil component from the rhizome of *Acorus tatarinowii*. Pharmaceut Biol. 2013;51(5): 589–94.

Liu H, Song Z, Liao DG. Anticonvulsant and Sedative Effects of Eudesmin isolated from *Acorus tatarinowii* on mice and rats. Phytother Res. 2015;29(7):996–1003.

Zhang S, Gui X, Huang L, Deng M, Fang R, Ke X, He Y, Li L Fang Y. Neuroprotective effects of β-asarone against 6-hydroxy dopamine-induced Parkinsonism via JNK/Bcl-2/Beclin-1 pathway. *Mol Neurobiol*. 2014a;1–12.

Zhang WT, Chai D, Xu S, Liu P, Liu K. Effect of *Acorus tatarinowii* Schott on the amino acid neurotransmitters in the striatum focal cerebral ischemia in rat. *Zhongguo Ying Yong Sheng Li Xue Za Zhi*, 2014b;30(4): 352–6.

Chinese Pharmacopoeia Commission. Pharmacopoeia of People's Republic of China. Beijing: China Medical Science Press. 2010.

## 7.7 *Allium macrostemon* Bunge

### *7.7.1* Allii Macrostemonis Bulbus

#### 7.7.1.1 (Liliaceae)

1) Common names:
   Long stamen onion bulb, Chinese wild onion, Wild garlic, Scallions hoary head, Xie-bai, Xiao-gen-suan (Pinyin in Chinese).
2) Photograph of the species (Fig. 7.3)
3) Description of the species:

   A) Habitat and world distribution:
      Hills, slopes, valleys, plains; near sea level up to 1600 m (to 3000 m in Xizang and Yunnan). Distribution in Japan, Korea, Mongolia, Russia (Far East), and throughout China except Hainan, Qinghai, and Xinjiang (Flora of China Editorial Committee 2000).
   B) Morphology of the plant:
      Bulb solitary, subglobose, diameter of 0.7–1.5(-2) cm, usually with bulbils at base; tunic blackish, papery or membranous, entire. Leaves with 2–5 mm wide, semiterete or 3-angled-semiterete, abaxially strongly 1-angled, fistulose, adaxially channeled. Scape with 3–70 cm, terete, covered with leaf sheaths for 1/4–1/3 its length. Spathe bivalved, persistent. Umbel hemispheric to globose, densely flowered, bearing bulblets and flowers or only bulblets. Pedicels subequal, 3–5× as long as perianth, bracteolate. Perianth pale purple to pale red color; segments oblong ovate to oblong lanceolate, equal, 4-5.5 × 1.2–2 mm; inner ones usually shorter than outer ones.

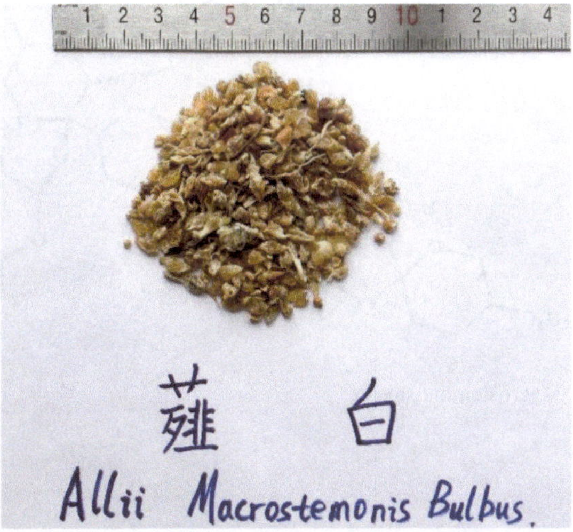


**Fig. 7.3** *Allii Macrostemonis Bulbus*

Filaments equal, 1/2 as long as to slightly shorter than perianth segments, connate and adnate to perianth segments, base triangular; inner ones ca. 1/2 as wide as outer at base. Ovary subglobose, with concave nectaries covered by hoodlike projections at base. Style exserted. Flowering and fruiting: May–July (Flora of China Editorial Committee 2000).

C) Medicinal parts:
The dry bulb of *Allium macrostemon* Bunge or *Allium chinensis* G. Don has been used.

D) Chemical constituents:
Until now, 50 saponins, 35 volatile compounds, 5 nitrogenous compounds, and 9 other compounds have been identified in *A. macrostemon*. **Steroid saponins** (macrostemonosides A, B, C, D, E, F, G, I, L, S, O, P; elephanoside E; (22S)-cholest-5-ene-1β,3β,16β,22-tetraol-1-*O*-α-L-rhamnopyranosyl-16-*O*-β-D-glucopyranoside; (25R)-26-*O*-β-D-glucopyranosyl-22-hydroxy-furost-5(6)-ene-3β,26-diol-3-*O*-β-D-glucopyranosyl(1 → 2)[β-D-glucopyranosyl (1 → 3)]-β-D-glucopyranosyl(1 → 4)-β-D-galactopyranoside,(25R)-26-*O*-β-D-glucopyranosyl-22-hydroxy-5β-furost-3β,26-diol-3-*O*-β-D-glucopyranosyl(1 → 2)-β-D-galactopyranoside; etc.) (Cheng et al. 2013; Chen et al. 2005; Qin et al. 2016); **nitrogen compounds** (adenosine, thymidine, etc.); **volatile oil** (methyl propyl disulfide, dimethyl trisulfide, methyl allyl trisulfide, 1,3-dithiane, diallyl disulfide, diallyl thiosulfinate, etc.) (Liu et al. 2014); **polysaccharides** (Zhang et al. 2015); **other compounds** (amino acids, minerals, and long-chain fatty acids such as palmitic acid, linoleic acid, etc.) (Chen et al. 2005; Zhang and Gao 2003).

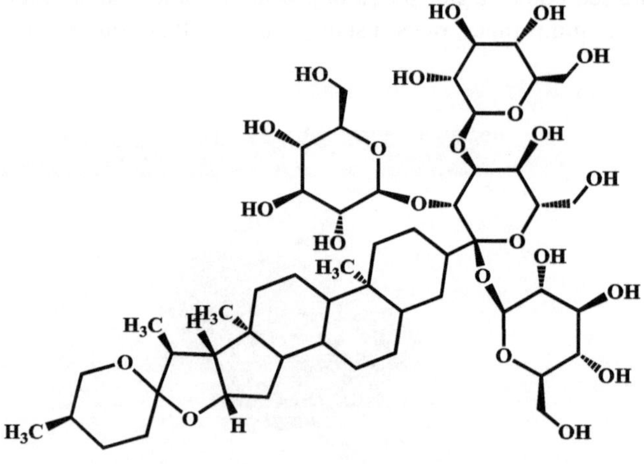

Macrostemonoside A

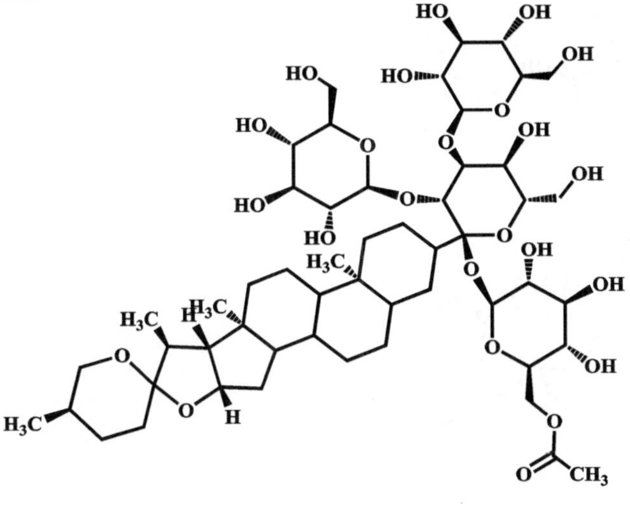

Macrostemonoside D

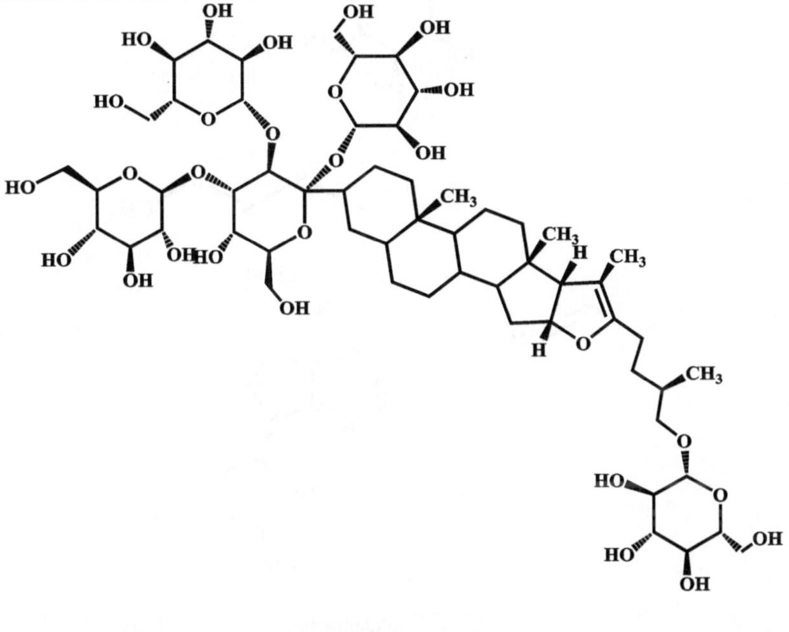

Macrostemonoside E

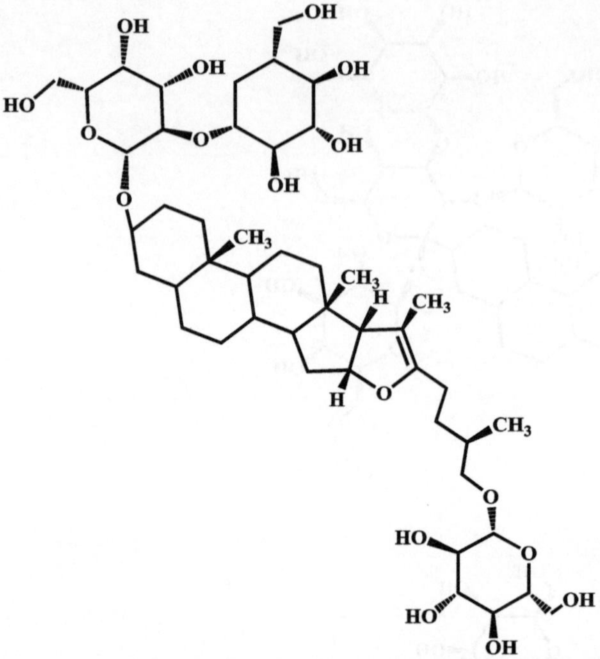

Macrostemonoside F

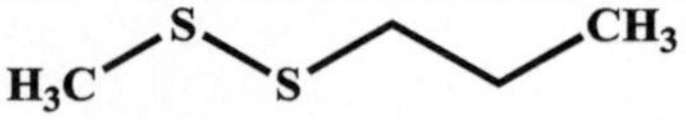

Methyl propyl disulfide

Dimethyl trisulfide

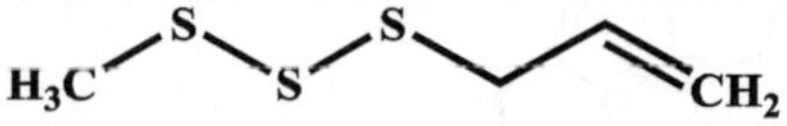

Methyl allyl trisulfide

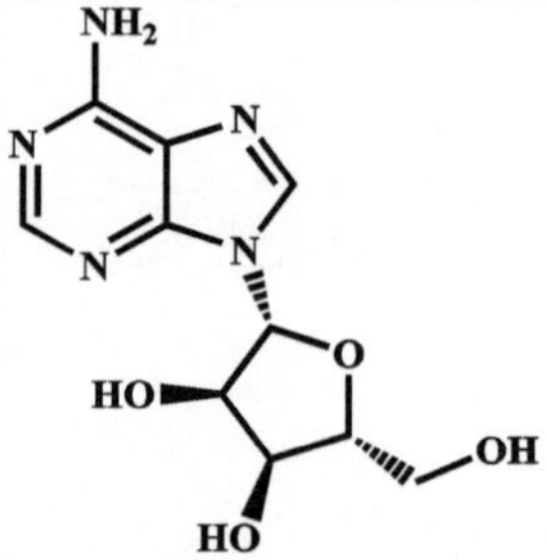

Adenosine

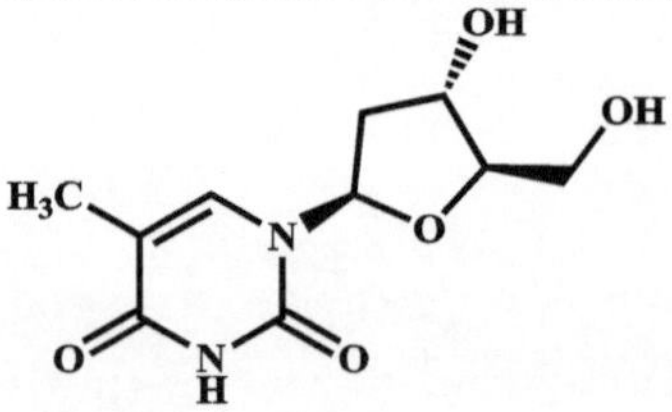

Thymidine

4) Pharmacology and bioactivities:

   A) In vitro studies:
      Not found.
   B) In vivo studies:
      A study has showed that the extract of *A. macrostemon* can alleviate the symptoms of depression in a rat model of depression induced by exposure to chronic immobilization stress. The extract statistically increased struggle time and decreased immobile time than rats in the depression group during the TST. It also returned dysregulated metabolites to their normal levels, such as plasma lysophosphatidylcholines, acylcarnitines, and the phospholipase A2 activity, indicating that an inflammatory condition occurred in depressed rats (Chen et al. 2014).

5) Indications and usage:
   Activate yang and dissipate nodulation, move qi and relieve stagnation. Treat chest impediment and heart pain; abdominal fullness and distention; and pain, diarrhea, and dysentery with tenesmus (Chinese Pharmacopoeia Commission 2010).
6) Precautions and side effects:
   It should be applied with caution in qi deficiency syndrome (Chinese Pharmacopoeia Commission 2010).
7) Dosage:

   A) Internal use:
      5–10 g/day, fresh 30–60 g/day, decocted in water for oral administration, taken twice a day, generally for 5–7 days, modificated according to symptoms, or into pills and powder (Chinese Pharmacopoeia Commission 2010).
   B) External use:
      Use fresh herb in appropriate amount. Smash and apply to affected areas (Chinese Pharmacopoeia Commission 2010).

## 7.8 References

Flora of China Editorial Committee. Flora of China. Beijing and St. Louis: Science Press and Missouri Botanical Garden Press. 2000;1–431.

Cheng SB, Wang Y, Zhang YF, Wang Y. Steroidal saponins from Allii Macrostemonis Bulbus. Chinese Trad Herb Drugs. 2013;44(9): 1078–81.

Chen HF, Wang NL, Yao XS. Study on bioactive steroidal saponins of Allium macrostemon Bunge. *Chinese* J Med Chem. 2005;15(3), 142–7.

Qin ZF, Dai Y, Yao ZH, He LL, Wang QY, Geng JL, Chen HF, Yao XS. Study on chemical profiles and metabolites of *Allii Macrostemonis Bulbus* as well as its representative steroidal saponins in rats by ultra-performance liquid chromatography coupled with quadrupole time-of-flight tandem mass spectrometry. Food Chem. 2016;192: 499–515.

Liu XC, Liu Q, Zhou L, Liu ZL. Evaluation of larvicidal activity of the essential oil of *Allium macrostemon* Bunge and its selected major constituent compounds against *Aedes albopictus* (Diptera: Culicidae). Parasites Vectors. 2014;7: 184.

Zhang Z, Wang F, Wang M, Ma L, Ye H, Zeng X. A comparative study of the neutral and acidic polysaccharides from *Allium macrostemon* Bunge. Carbohydr Polym. 2015;117: 980–7.

Zhang Q, Gao E. Advances in the study of *Allium macrostemon* Bunge. Zhongguo Zhong Yao Za Zhi. 2003;28(2): 105–7.
Chen S, Wei C, Gao P, Kong H, Jia Z, Hu C, Dai W, Wu Y, Xu G. Effect of *Allium macrostemon* on a rat model of depression studied by using plasma lipid and acylcarnitine profiles from liquid chromatography/mass spectrometry. J Pharmaceut Biomed Analysis. 2014;89: 122–9.
Chinese Pharmacopoeia Commission. Pharmacopoeia of People's Republic of China. Beijing: China Medical Science Press. 2010

## 7.9 *Aloe barbadensis* Miller

### *7.9.1* Aloe

#### 7.9.1.1 (Liliaceae)

1) Common names:

   Aloe vera, Lu-hui, Nu-hui, Xiang-dan, Lao-wei (Pinyin in Chinese).

2) Photograph of the species (Fig. 7.4)
3) Description of the species:

   A) Habitat and world distribution:
   It grows mainly in the dry areas of Africa, Asia, Europe, and America (Surjushe et al. 2008). Ideal environment to grow this plant is a tropical climate and low rainfall areas (Mangaiyarkarasi et al. 2015).

   B) Morphology of the plant:
   Succulent herbs. Stems short, suckering freely to form dense clumps. Leaves with 15–35(-50)×4–5(-7) cm, subbasal, slightly distichous in seedlings and new shoots, erect, pale green, sometimes with pale spots in very young plants, linear-lanceolate, margin sporadically spiny-dentate, apex bi- or tridentate, pointed. Inflorescence erect, with 60–90 cm; peduncle to 2 cm thick; raceme with

**Fig. 7.4** *Aloe*

30–40 × 5–6 cm, sometimes with 1 or 2 ascending branches, numerous flowered; bracts whitish, broadly lanceolate, ca. 10 × 5–6 mm, veins 5–7, apex acute. Flowers reflexed; pedicel ca. 1/2 as long as bract. Perianth pale yellow mottled with red, slightly ventricose, 2.5(-3) cm, outer lobes free for ca. 1.8 cm, slightly recurved at apex. Stamens exserted by 4–5 mm. Style conspicuously exserted (Flora of China Editorial Committee 2000).

C) Medicinal parts:
The leaves of *Aloe barbadensis* Miller, *Aloe ferox* Miller, or other relative congeneric species are being used (Chinese Pharmacopoeia Commission 2010).

D) Chemical constituents:
*Aloe vera* (*Aloe barbadensis* Miller) contains 75 potentially active constituents: vitamins, enzymes, minerals, sugars, lignin, saponins, salicylic acids, and amino acids (Surjushe et al. 2008). **Carbohydrates** (mannose-6-phosphate, acemannan, etc.); **anthraquinones** (aloin, emodin, etc.); **sterols** (cholesterol, campesterol, β-sitosterol, and lupeol); **vitamins** (vitamins A, C, E, B1, B2, B3, B6, B9, and B12 and choline); **enzymes** (alliinase, alkaline phosphatase, amylase, bradykinase, carboxypeptidase, catalase, cellulase, lipase, and peroxidase); **minerals** (calcium, chromium, copper, selenium, magnesium, manganese, potassium, sodium, and zinc); **others** (20 human-required amino acid, 7 essential amino acids, salicylic acid, lignin, saponins) (Surjushe et al. 2008).

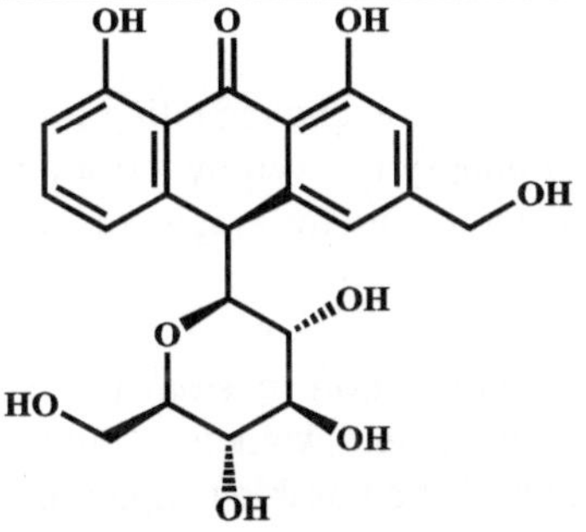

Aloin

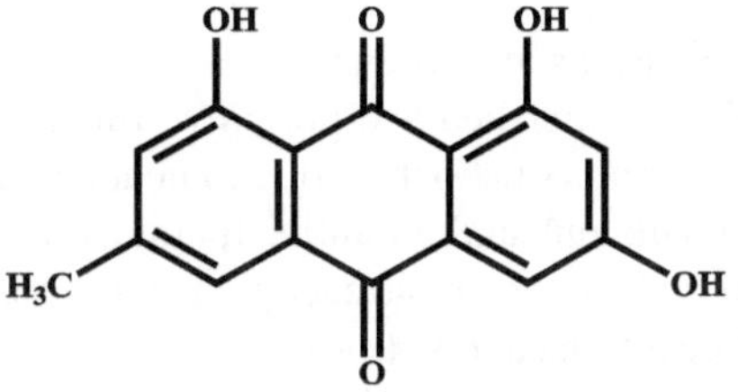

Emodin

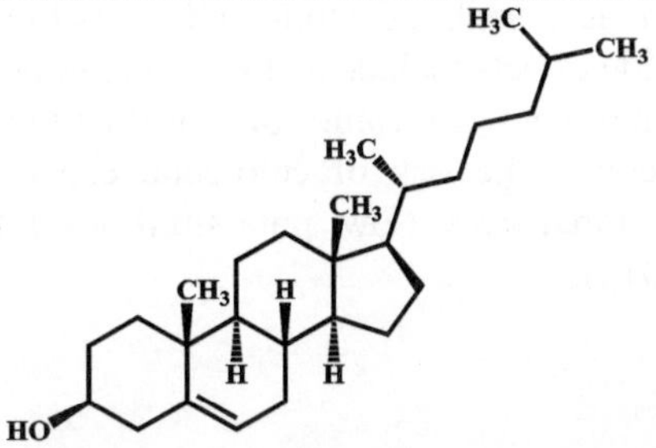

Cholesterol

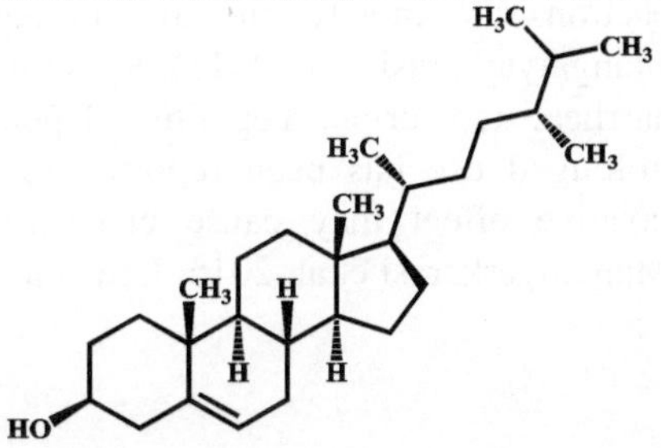

Campesterol

| | |
|---|---|
| 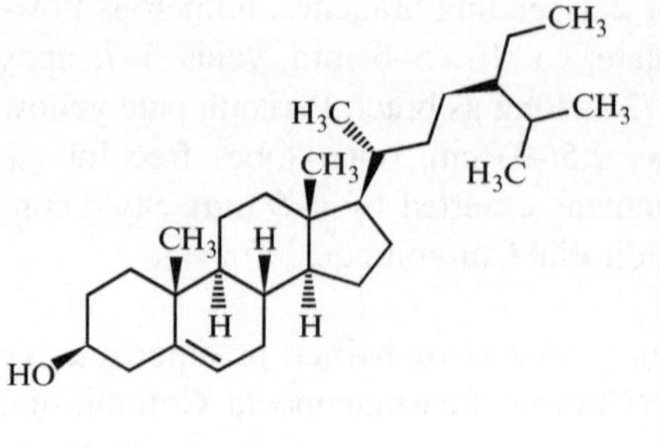 β-Sitosterol | 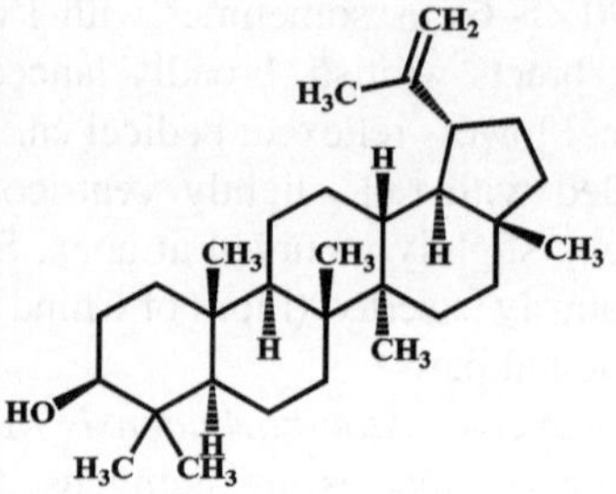 Lupeol |

4) Pharmacology and bioactivities:

   A) In vitro studies:
      Not found.
   B) In vivo studies:
      *Aloe barbadensis* enhanced learning and memory, and also alleviated depression in mice. As compared to distilled water, *Aloe barbadensis*, at 200 and 400 mg/kg, p.o., increased the acquisition and retention step-down latency in the passive avoidance task; at 400 mg/kg, reduced the transfer latency in the elevated plus maze; and at 100, 200, and 400 mg/kg, decreased the period of immobility significantly in the FST as well as in TST (Halder et al. 2013).

5) Indications and usage:
   Making diarrhea and purgative function, clearing liver fire, and expelling ascarid. Treat constipation due to heat accumulation, epilepsy induced by terror and convulsion, and infantile malnutrition. External treatment for ringworm and sores (Chinese Pharmacopoeia Commission 2010).
6) Precautions and side effects:
   It is contraindicated in cases of patients with hypofunction of stomach and spleen, poor appetite, diarrhea, as well as pregnant women (Chinese Pharmacopoeia Commission 2010). For external use, it's advisable to apply in a small area of skin first to test for possible allergic reaction.

   Topical side effects include redness, burning, and stinging sensation. Allergic reactions are mostly due to anthraquinones, such as aloin and barbaloin (Mangaiyarkarasi et al. 2015). Systemic side effects include abdominal cramps, diarrhea, red urine, hepatitis, dependency, or worsening of constipation. Prolonged use has been reported to increase the risk of colorectal cancer. Laxative effect may cause electrolyte imbalances (low potassium levels) (Mangaiyarkarasi et al. 2015; Lee et al. 2014).

7) Dosage:

   A) Internal use:
   2–5 g/day, in pill or powder form, taken 2–3 times a day, generally for 5–7 days, modificated according to symptoms (Chinese Pharmacopoeia Commission 2010).
   B) External use:
   Use fresh herb in appropriate amount. Smash and apply to affected areas (Chinese Pharmacopoeia Commission 2010).

## 7.10 References

Surjushe A, Vasani R, Saple DG. Aloe vera: a short review. Ind J Dermatol Venereol Leprol. 2008;53(4): 163–6.

Mangaiyarkarasi SP, Manigandan T, Elumalai M, Cholan PK, Kaur RP. Benefits of Aloe vera in dentistry. *J Pharm Bioallied Sci*. 2015;7(Suppl 1): S255–9.

Flora of China Editorial Committee. Flora of China. Beijing and St. Louis: Science Press and Missouri Botanical Garden Press. 2000;24:160.

Chinese Pharmacopoeia Commission. Pharmacopoeia of People's Republic of China. Beijing: China Medical Science Press. 2010.

Halder S, Mehta AK, Mediratta PK. Aloe vera improves memory and reduces depression in mice. Nutr Neurosci. 2013;16(6): 250–4.

Lee J, Lee MS, Nam KW. Acute toxic hepatitis caused by an aloe vera preparation in a young patient: a case report with a literature review. Korean J Gastroenterol. 2014;64(1): 54–8.

## 7.11 *Centella asiatica* (L.) Urban

### *7.11.1* Centellae herba

#### 7.11.1.1 (Apiaceae)

1) Common names:
   Herba centellae asiaticae, Asiatic pennywort herb, *Centella asiatica*, Ma-ti-cao, Ji-xue-cao, Lei-gong-gen, Lao-wan-jiang, Tong-qian-cao, Da-jin-qian-cao, Tie-deng-zhan (Pinyin in Chinese).
2) Photograph of the species (Fig. 7.5)
3) Description of the species:

   A) Habitat and world distribution
   Shady, wet, grassy places, river margins; 200–1900 m. Anhui, Fujian, Guangdong, Guangxi, Hubei, Hunan, S Jiangsu, Jiangxi, S Shaanxi, Sichuan, Taiwan, Yunnan, and Zhejiang of China. Widespread throughout tropical and subtropical countries worldwide, Bhutan, India, Indonesia, Japan, Korea, Laos, Malaysia, Myanmar, Nepal, Pakistan, Thailand, Vietnam, Sri Lanka,

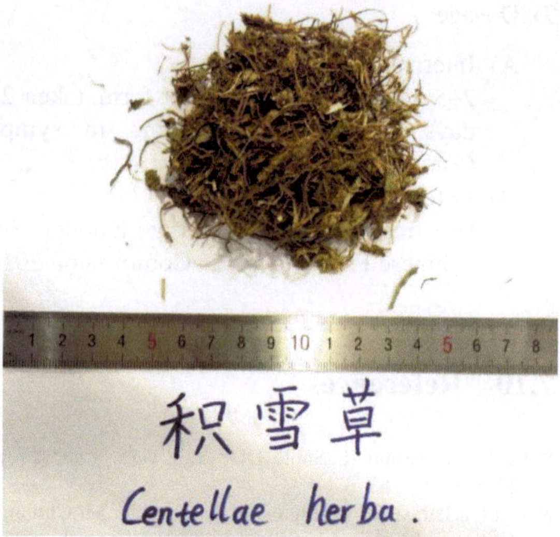


**Fig. 7.5** *Centellae herba*

Madagascar, South Africa, and South Pacific and Eastern Europe (Gohil et al. 2010; Flora of China Editorial Committee 2005).

B) Morphology of the plant:

Perennial herb. Petiole with 0.5–10(–30) cm; leaf blade orbicular or reniform, with 1–4.5 × 1.5–5 cm, palmate veins 5–7, prominent, both surfaces glabrous or abaxially sparsely pubescent on the veins, base broadly cordate, coarsely toothed. Peduncles 2–4, clustered axillary, 0.2–1.5 cm; bracts two (rarely three), ovate, with 3–4 × 2.1–3 mm, persistent in fruit; umbels 3–4-flowered, capitate. Flowers sessile or subsessile with white or rose-tinged petals. Fruit with 2.1–3 × 2.2–3.6 mm. Flowering and fruiting in April–October (Flora of China Editorial Committee 2005).

C) Medicinal parts:

The whole plant or the whole grass roots of *Centella asiatica* (L.) Urban.

D) Chemical constituents:

**Saponins** (asiaticoside, asiatic acid, madecassoside, madecassic acid, isothankuniside, centelloside, thankuniside, brahmoside, and brahminoside); **essential oil** (β-caryophyllene, *trans*-β-farnesene, germacrene D); **phytosterols** (campesterol, sitosterol, stigmasterol); **free amino acids** (alanine, serine, aminobutyrate, aspartate, glutamate, lysine, and threonine); **flavonoids** (quercetin and kaempferol derivatives); **fatty acids** (linoleic acids, linolenic, oleic, palmitic, and stearic acids); **other constituents** (abundant tannins, mucilages, resins, hydrochotine, vallarine) (Gohil et al. 2010).

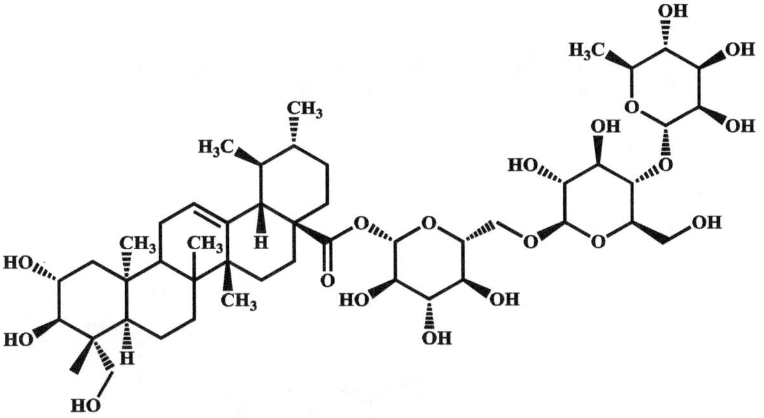

Asiaticoside

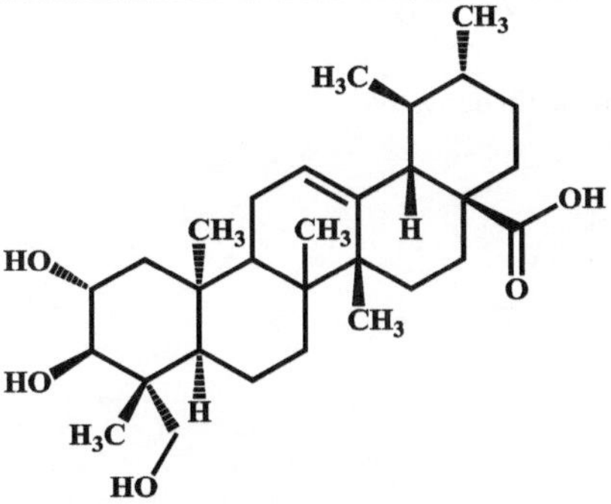

Asiatic acid

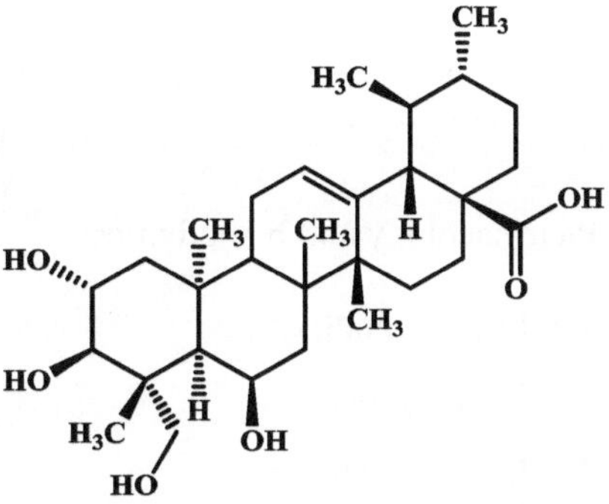

Madecassic acid

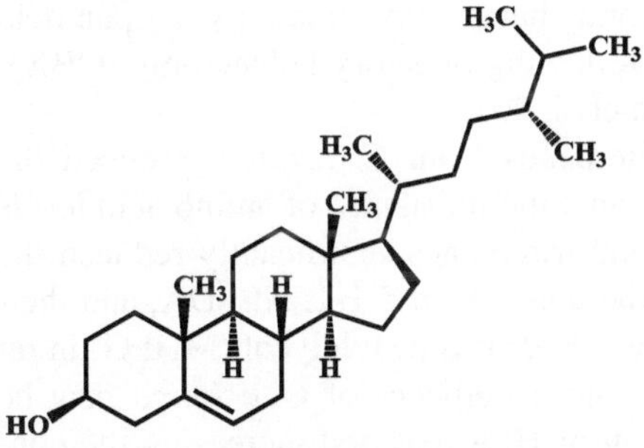

Campesterol

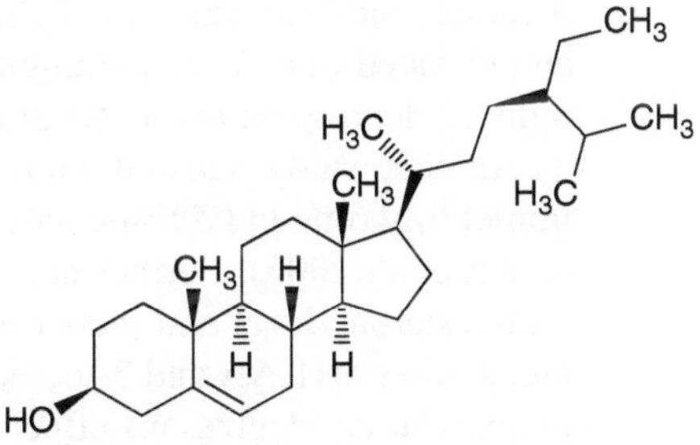

β-Sitosterol

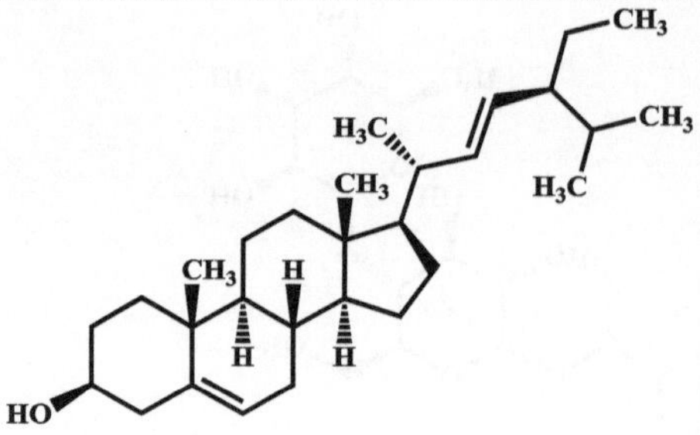


Stigmasterol

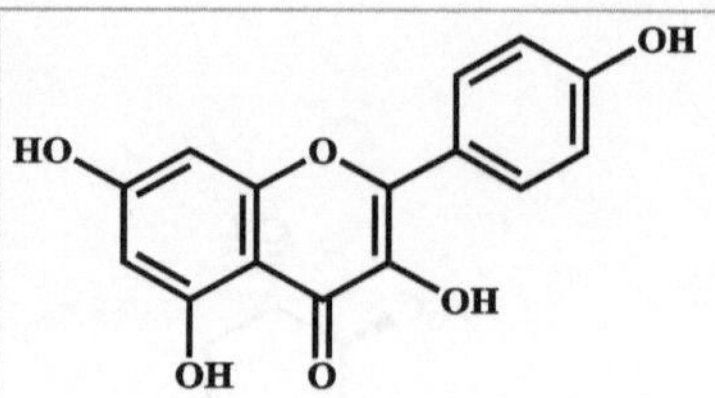


Kaempferol derivatives

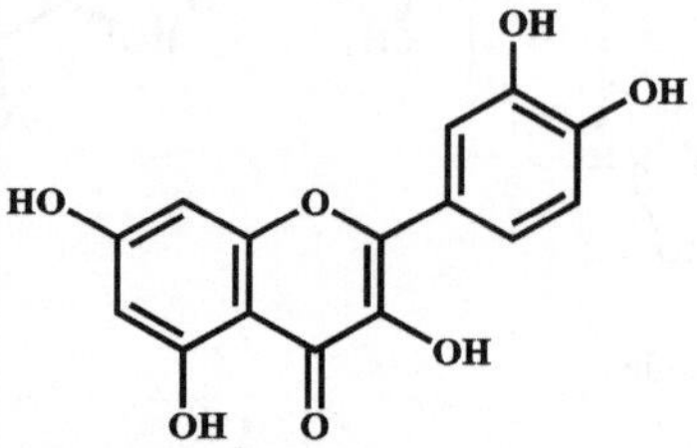


Quercetin derivatives

4) Pharmacology and bioactivities:

   A) In vitro studies:

      Not found.

   B) In vivo studies:

A standardized extract of *Centella asiatica* L. leaves (INDCA) showed dose-dependent reversal of physiological changes (i.e., body weight, body temperature, and heart rate) and behavioral changes (hyperactivity in open field and elevated plus maze paradigm) following olfactory bulbectomy (OBX)-induced depression in rats (Padmaja et al. 2012).

Another study showed total triterpenes from *C. asiatica* reduced the immobility time in FST and ameliorated the imbalance of amino acid levels (Chen et al. 2003). Furthermore, total triterpenes significantly reduced the corticosterone level and increased the contents of 5-HT, NE, DA, and their metabolites 5-HIAA and 3-methoxy-4-hydroxyphenylglycol (MHPG) in rat brain. The antidepressant effect of total triterpenes of *C. asiatica* may be involved in ameliorating the function of HPA axis and increasing the contents of monoamine neurotransmitters (Chen et al. 2005).

The administration of different doses (100 mg and 200 mg/kg) of the ethanolic extract of *C. asiatica* in rats and mice was able to induce antidepressant effects. In FST, the extract can decrease the immobility time in rats with mild sedative effect. The decrease in the immobility time is accompanied with the increase in swimming time. The ethanolic extract significantly reduced the number of head dippings and number of line crossings by mice in hole-board test, which were the indicator of exploratory behavior (Selvi et al. 2012).

5) Indications and usage:
Clear heat and promote diuresis, resolve swelling, and remove toxicity. Treat jaundice due to damp-heat, heatstroke, and diarrhea, urolithic stranguria, stranguria due to hematuria, sores and carbuncle, and traumatic injury (Chinese Pharmacopoeia Commission 2010).
6) Precautions and side effects:
It is contraindicated for deficiency-cold syndrome. Side effects are rare but may include skin allergy and burning sensations (with external use), headache, stomach upset, nausea, dizziness, and extreme drowsiness which tend to occur with high doses of the herb. The fresh plant may have a low potential for skin irritation. A contact dermatitis has been reported on a few occasions using topical preparations. Consuming in excess can cause headaches and transient unconsciousness (Gohil et al. 2010).
7) Dosage:
   A) Internal use:
   9–15 g/day (15–30 g/day for fresh product), decocted in water for oral administration, taken twice a day, generally for 5–7 days, modificated according to symptoms (Chinese Pharmacopoeia Commission 2010).
   B) External use:
   Use fresh herb in appropriate amount. Pound or crush to extract the juice and apply to affected areas (State Administration of Traditional Chinese Medicine 1999).

## 7.12 References


Gohil KJ, Patel JA, Gajjar AK. Pharmacological review on *Centella asiatica*: a potential herbal cure-all. Ind J Pharmaceut Sci. 2010;72(5): 546–56.

Flora of China Editorial Committee. Flora of China. Beijing and St. Louis: Science Press and Missouri Botanical Garden Press: 2005;14: 18.

Padmaja K, Urmila A, Subhash B, Arulmozhi S, Mohan V, Prasad K. Antidepressant effects of standardized extract of *Centella asiatica* L in olfactory bulbectomy model, Biomed Aging Pathol. 2012;2(2): 48–53.

Chen Y, Han T, Qin L, Rui Y, Zheng H. Effect of total triterpenes from *Centella asiatica* on the depression behavior and concentration of amino acid in forced swimming mice. *Zhong Yao Cai*, 2003;26(12): 870–3.

Chen Y, Han T, Rui Y, Yin M, Qin L, Zheng H. Effects of total triterpenes of *Centella asiatica* on the corticosterone levels in serum and contents of monoamine in depression rat brain. Zhong Yao Cai. 2005;28(6):492–6.

Selvi TP, Senthil Kumar M, Rajesh R, Kathiravan T. Antidepressant activity of ethanolic extract of leaves of *Centella asiatica*. Linn by in vivo methods. Asian J Res Pharmaceut Sci. 2012;2(2): 76–9.

Chinese Pharmacopoeia Commission. Pharmacopoeia of People's Republic of China. Beijing: China Medical Science Press. 2010.

Chinese Materia Medica Commission. State administration of traditional Chinese medicine. Shanghai: Shanghai Scientific and Technical Publishers. 1999.

# 7.13 *Corydalis yanhusuo* W.T. Wang

## *7.13.1* Corydalis Rhizoma

### 7.13.1.1 (Papaveraceae)

1) Common names:

   Yan-hu-suo, Xuan-hu, Yuan-hu (Pinyin in Chinese).

2) Photograph of the species (Fig. 7.6)
3) Description of the species:

   A) Habitat and world distribution:
      Grassland of hill areas. Anhui, Henan (Tanghe, Xinyang), Hubei, Hunan, Jiangsu, and Zhejiang; cultivated in Beijing, Gansu, Shaanxi, Sichuan, and Yunnan (Flora of China Editorial Committee 2008; Su and Wu 1985).
   B) Morphology of the plant:
      Perennial herb. Tuber yellow color, rounded, with (0.5-)1–2.5 cm in diameter. Stems erect, 10–30 cm high, with 1 (sometimes 2) scale leaves, with 2 or often 3–4 cauline leaves, scales and lower cauline leaves often with axillary tuber. Leaves biternate or nearly triternate; leaflets with 2–2.5 cm × 5–8 mm, entire to deeply trifid with lanceolate entire acute segments; lowermost cauline leaf frequently long petiolate and vaginate at base. Raceme lax, with 5–15 flowers; bracts lanceolate or narrowly ovate, with 5–12 × 2-5 mm, margin entire, apex acute, sometimes lower bracts slightly divided. Pedicel ca. 10 mm at flowering stage, larger in fruit (to 20 mm). Flowers amaranth. Sepals small and caducous. Outer petals with broad dentate limbs, emarginate with a mucro in notch; spur of upper petal upcurved, cylindric, 11–13 mm;

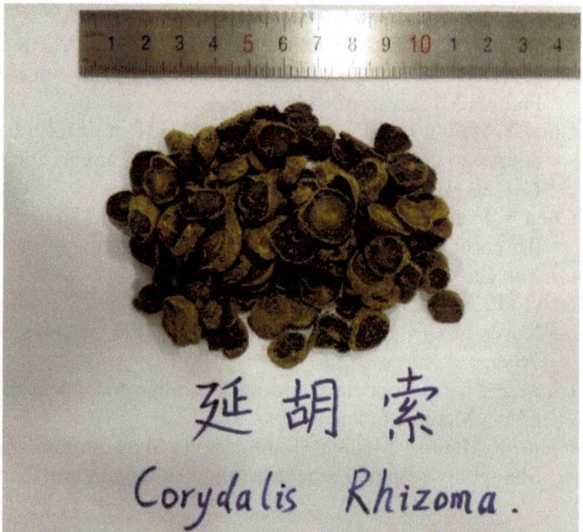


**Fig. 7.6** *Corydalis Rhizoma*

nectary extended through ca. 1/2 of spur, apex obtuse; lower petal with short claw; inner petals with 8–9 mm, claw longer than petal lobes. Stigma nearly orbicular, papillae. Capsule linear, with 20–28 mm. Seeds in 1 row. Flowering and fruiting: April–June (Flora of China Editorial Committee 2008; Su and Wu 1985).

C) Medicinal parts:
The dry tuber of *Corydalis yanhusuo* W. T. Wang has been used.

D) Chemical constituents:
**Alkaloids** (protopine, tetrahydrocolumbamine, corydaline, dehydrocorydaline, α-allocryptopine, coptisine, palmatine, berberine, D,L-tetrahydropalmatine, tetrahydroberberine, tetrahydrocoptisine) (Wu et al. 2014; Gao et al. 2014).

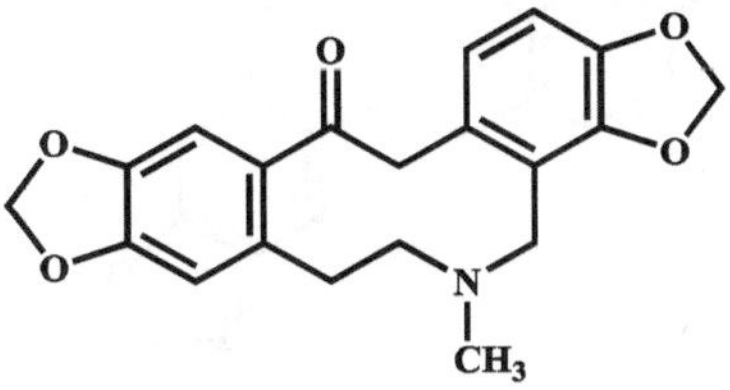

Protopine

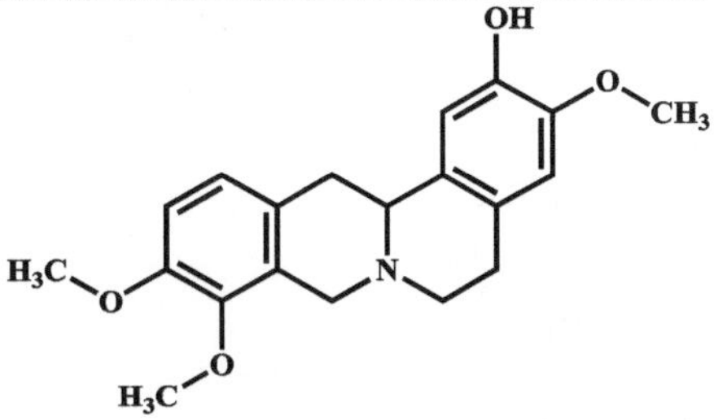

Tetrahydrocolumbamine

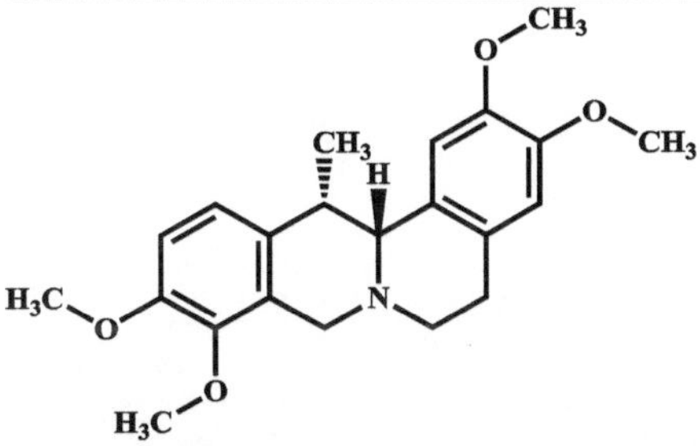

Corydaline

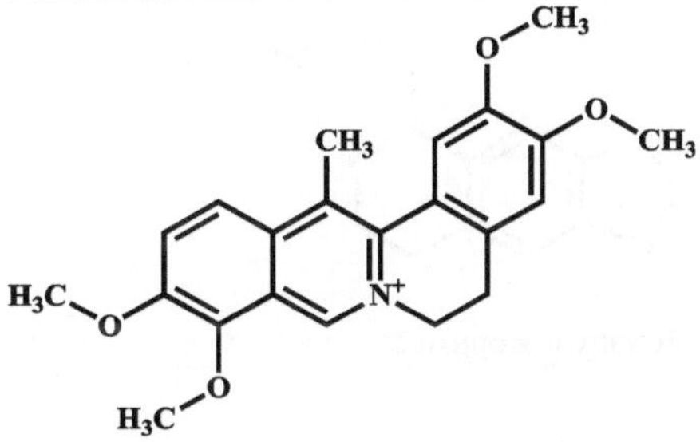

Dehydrocorydaline

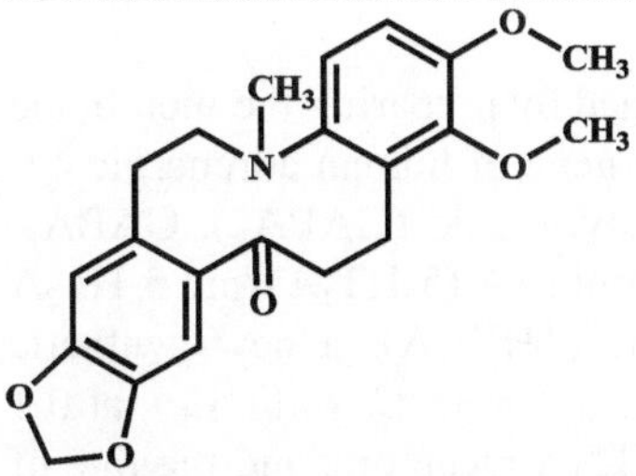

α-Allocryptopine

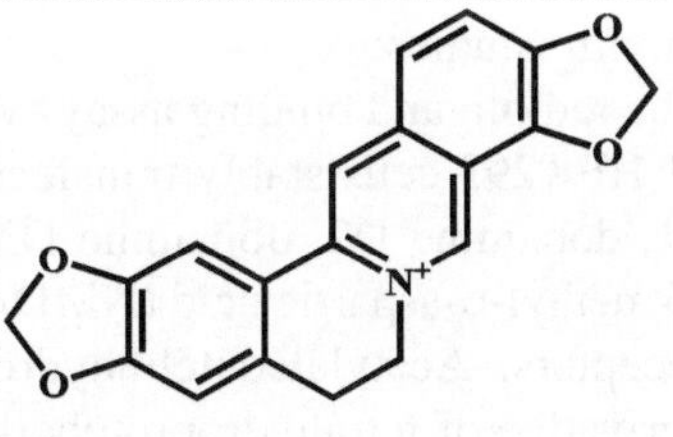

Coptisine

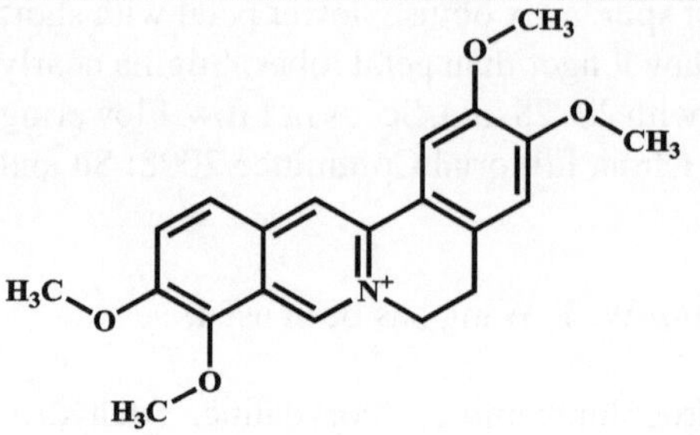

Palmatine

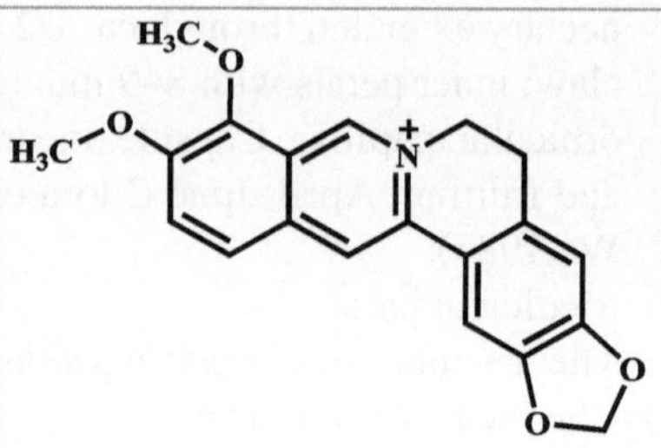

Berberine

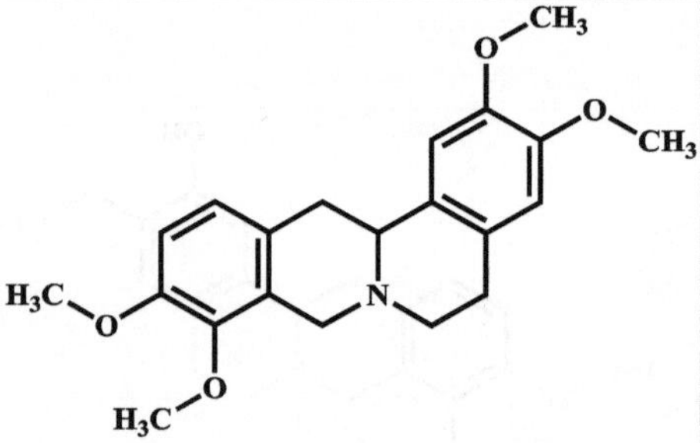

Tetrahydropalmatine

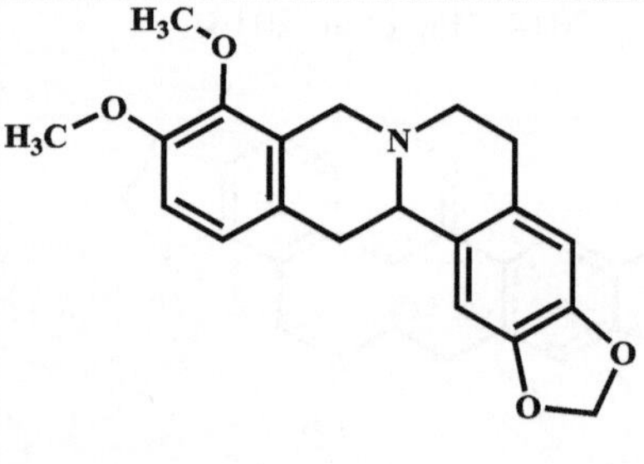

Tetrahydroberberine

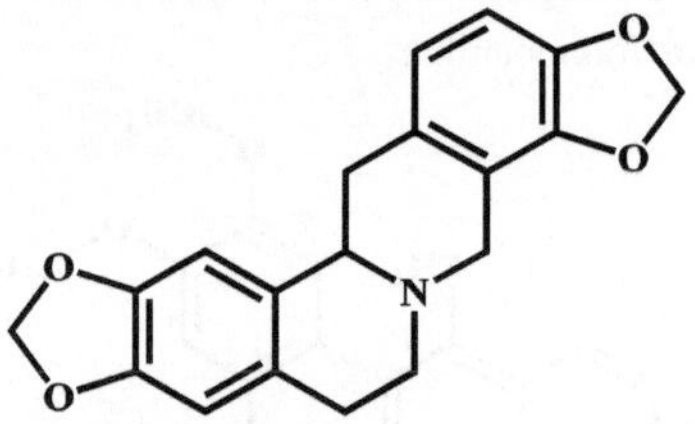

Tetrahydrocoptisine

4) Pharmacology and bioactivities:

A) In vitro studies:

The radioligand binding assays were performed by preparing the membrane of HEK293 cells stably transfected with the gene of human adrenergic α1, β1, dopamine D2, dopamine D3, γ-aminobutyric acid ($GABA_A$), $GABA_B$, *N*-methyl-D-aspartic acid (NMDA) and serotonin 1A ($5\text{-}HT_1A$) and $5\text{-}HT_2A$ receptors. Acetylated tetrahydroberberrabine (THBr-A), a new synthetic derivative of tetrahydroprotoberberines extracted from *C. yanhusuo*, at 0.1 nM, mainly interacted with D2, D3, and $5\text{-}HT_1A$ receptors, and the rate of inhibition was 96.43, 77.61, and 78.31 %, respectively (Liang Hui-chun et al. 2013).

B) In vivo studies:

Spontaneous activity, hole-board, elevated plus maze, and light-dark transition tests were used to assess the locomotor activity and anxiolytic effect of the synthetic acetylated tetrahydroberberrabine. This compound showed its dual dopamine $D_2$ antagonistic and serotonin 5-$HT_1$A partial agonistic actions (Liang Hui-chun et al. 2013).

Daily L-tetrahydropalmatine (50 mg/kg) administration significantly increased the number and duration of open arm visits in the elevated plus maze test, reduced the anxiety index, increased the risk assessment, and increased the number of head dips over the borders of the open arms after single prolonged stress. Antidepressant activity of L-tetrahydropalmatine also blocked the decrease in neuropeptide Y (NPY) and the increase in corticotrophin-releasing factor (CRF) expression in the hypothalamus. It exerts pronounced anxiolytic-like and antidepressant effects on the development of the behavioral and biochemical symptoms associated with posttraumatic stress disorder (PTSD), indicating its prophylactic potential (Bombi et al. 2014).

5) Indications and usage:

Activate blood, promote flow of Qi, and alleviate pain. Treat the pain of heart, chest, hypochondria and abdomen, dysmenorrhea, amenorrhea, postpartum tormina due to blood stasis and qi stagnation, and traumatic injuries and swelling pain (Chinese Pharmacopoeia Commission 2010).

6) Precautions and side effects:

It has strong power to injure the function of blood, so the pregnant woman must be cautious when using it (Chinese Pharmacopoeia Commission 2010).

7) Dosage:

A) Internal use:

3–10 g/day, decocted in water for oral administration. Make powder and take 1–3 g/day. Taken twice a day, generally for 5–7 days, modificated according to symptoms. Processed with vinegar can enhance the action of alleviating pain (Chinese Pharmacopoeia Commission 2010).

B) External use:

Not found.

## 7.14 References

Flora of China Editorial Committee. Flora of China. Beijing and St. Louis: Science Press and Missouri Botanical Garden Press. 2008;7: 315.

Su Z, Wu C. A study on the classification, distribution, evolutionary trends and uses of Chinese *Corydalis* subgen. Capnites DC. Acta Botanica Yunnanica. 1985;7(3): 270–1.

Wu HW, Waldbauer K, Tang LY, Xie LW, McKinnon R, Zehl M, Yang HJ, Xu HY, Kopp B. Influence of vinegar and wine processing on the alkaloid content and composition of the traditional Chinese medicine *Corydalis* Rhizoma (Yanhusuo). Molecules. 2014;19(8): 11487–504.

Gao Y, Hu S, Zhang M, Li L, Lin YN. Simultaneous determination of four alkaloids in mice plasma and brain by LC–MS/MS for pharmacokinetic studies after administration of *Corydalis* Rhizoma and Yuanhu Zhitong extracts. J Pharmaceut Biomed Analysis. 2014;92: 6–12.

Liang Hui-chun, Li Nuo-min, Xu Yu-hui, Ye En-mao, Zhang Jian, Yu Bo-yang,Yang Zheng. Anxiolytic effect and receptor binding properties of acetylated tetrahydroberberrabine. *Chinese* J Pharmacol Toxicol. 2013;27(6): 931–40.

Bombi L, Bongjun S, Mijung Y, Insop S, Hyejung L, Dae-Hyun H. L-Tetrahydropalmatine ameliorates development of anxiety and depression-related symptoms induced by single prolonged stress in rats. Biomol Therapeut. 2014;22(3): 213–22.

Chinese Pharmacopoeia Commission. Pharmacopoeia of People's Republic of China. Beijing: China Medical Science Press. 2010.

## 7.15 *Crocus sativus* L.

### *7.15.1* Croci Stigma

#### 7.15.1.1 (Iridaceae)

1) Common names:
   Saffron, Xi-hong-hua, Fan-hong-hua, Zang-hong-hua (Pinyin in Chinese).
2) Photograph of the species (Fig. 7.7)
3) Description of the species:

   A) Habitat and world distribution:
      Adapted to cool and wet climate. Cultivated throughout China [widely cultivated elsewhere, probably of hybrid origin in East Mediterranean region] (Flora of China Editorial Committee 2000).

**Fig. 7.7** *Croci Stigma*

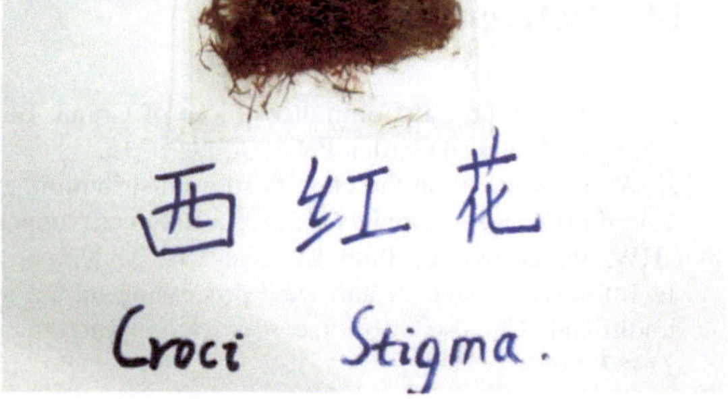

B) Morphology of the plant:
Small, perennial, and cormose herb. Corms ca. 3 cm in diameter; tunic light yellow or yellowish brown, reticulate fibrous. Leaves 9–15, with 15–20 cm × 2–3 mm. Flowers one or two, fragrant. Perianth purple; segments oblanceolate, 4–5 cm, apex obtuse. Stamens ca. 2.5 cm; anthers yellow, apex slightly curved, pointed. Style tri-branched at apex; branches recurved, deep red, 2.5 cm or more. Fertile seeds are not produced (Flora of China Editorial Committee 2000).

C) Medicinal parts:
The dry stigmas of *Crocus sativus* L. are being used.

D) Chemical constituents:
**Carotenoids** (phytoene, phytofluene, tetrahydrolycopene, zeaxanthin, β-carotene, crocetin, dimethylcrocetin, all-*trans*-crocetin-mono-(β-D-gentiobiosyl) ester, all-*trans*-crocetin-di-(β-D-gentiobiosyl) ester (=crocin), all-*trans*-crocetin-β-D-gentiobiosyl-β-D-glucosyl ester, all-*trans*-crocetin-di-(β-D-glucosyl) ester, *trans*-crocetin-1-al 1-*O*-β-gentiobiosyl ester; (4*R*)-4-hydroxy-2,6,6-trimethylcyclohex-1-ene-carbaldehyde-4-*O*-[β-D-glucopyranosyl(1 → 3)-β-D-glucopyranoside]; 13-*cis*-crocetin-di-(β-D-gentiobiosyl) ester, crocetin (β-gentiobiosyl) (β-neapolitanosyl) ester, 13-*cis*-crocetin-β-D-gentiobiosyl-β-D-glucosyl ester; crocetin monomethyl ester, etc.); **triterpenoid saponins** (azafrine 1, azafrine 2); **monoterpenoids** (crocusatins A-I, safranal, etc.) (Pfander and Schurtenberger 1982; Rubio-Moraga et al. 2011; Li and Wu 2002; Tarantilis et al. 1998; Van Calsteren et al. 1997; Tung and Shoyama 2013; Straubinger et al. 1997; Pfister et al. 1996; Zarghami and Heinz 1971); **flavonoids** (isorhamnetin-4'-*O*-neohesperidoside, kaempferol-3-*O*-glucoside, quercetin-3-*O*-glucoside, kaempferol, kaempferol-3-*O*-sophoroside, etc.) (Li and Wu 2002; Straubinger et al. 1998); **furans** ((4*R*)-4-hydroxy-dihydrofuran-2-one-*O*-β-D-glucopyranoside, (4S)-4-hydroxy-dihydrofuran-2-one-*O*-β-D-glucopyranoside, 2-formyl-5-methoxyfuran) (Li and Wu 2002; Straubinger et al. 1998; Gao et al. 1999); **nucleosides and alkaloids** (5-methyluracil, harman, adenosine, nicotinamide, 1-(9H-β-carbolin-1-yl)-3,4,5-trihydroxypentan-1-one, pyridin-3-yl-methanol, tribulusterine, uracil) (Li and Wu 2002; Gao et al. 1999; Li and Wu 2003; Li et al. 2004); **cyclohexane and cyclohexene derivatives** ((1*R*)-3,5,5-trimethylcyclohex-3-enol-O-β-D-glucopyranoside, (4*R*)-4-(hydroxyl-methyl)-3,5,5-trimethylcyclohex-2-enone-4-β-D-glucoside, (4*S*)-4-(hydroxyl-methyl)-3,5,5-trimethylcyclohex-2-enone-4-β-D-glucoside, etc.) (Li and Wu 2002; Straubinger et al. 1997; Zarghami and Heinz 1971; Straubinger et al. 1998; Li and Wu 2003; Li et al. 2004); **other compounds** (nonacosane, (2E, 4E)-2-methyl-6-oxohepta-2, 4-dienoic acid β-gentiobiosyl ester, (3*S*), 4-dihydroxybutyric acid, etc.) (Li and Wu 2002; Straubinger et al. 1997; Li et al. 2004).

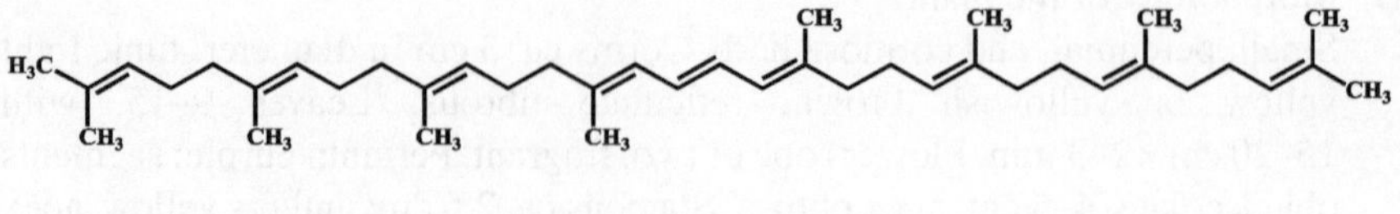

Phytoene

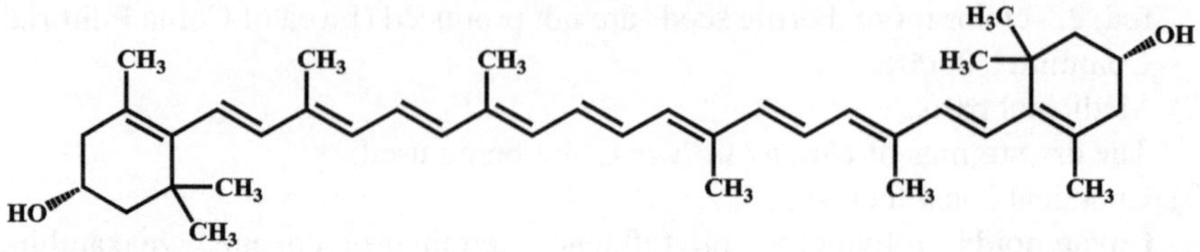

Zeaxanthin

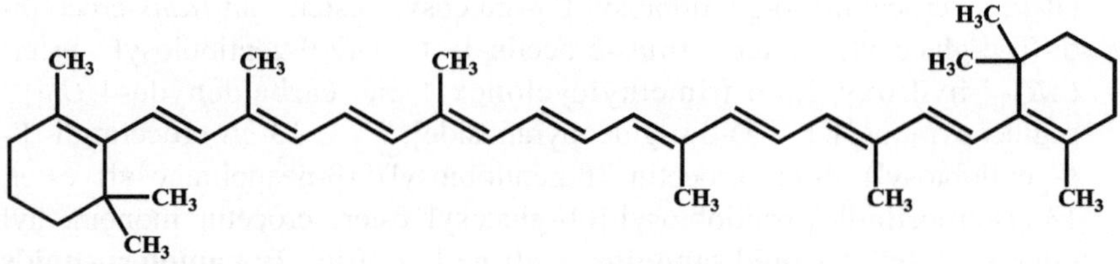

β-Carotene

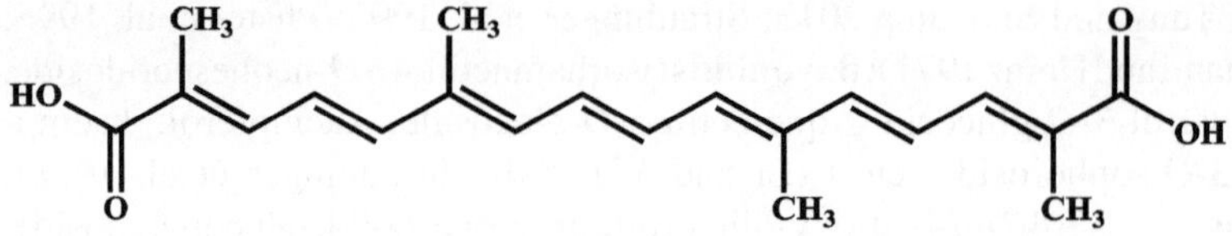

Crocetin

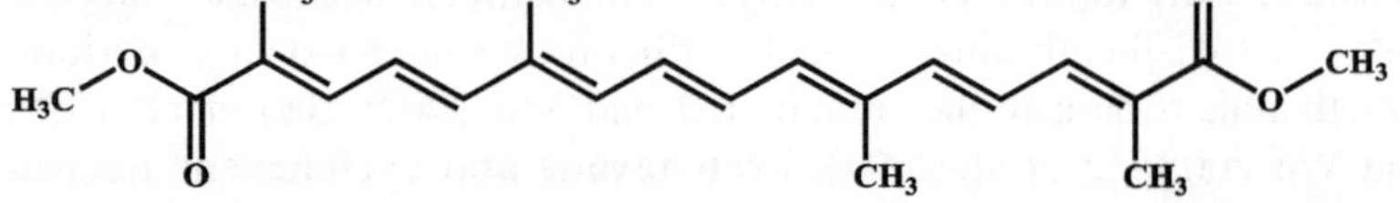

Dimethylcrocetin

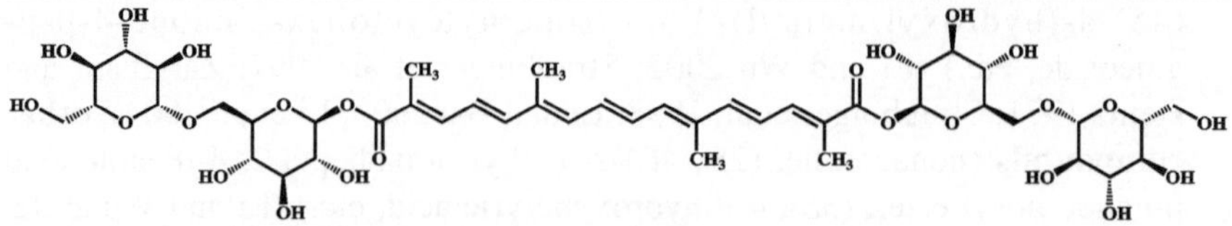

Crocin

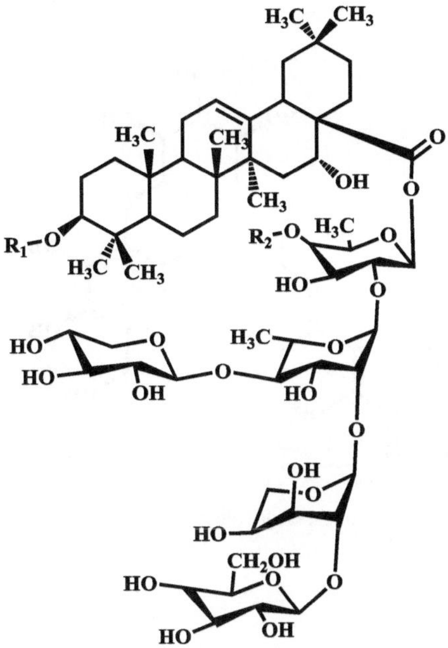

$R_1$ =

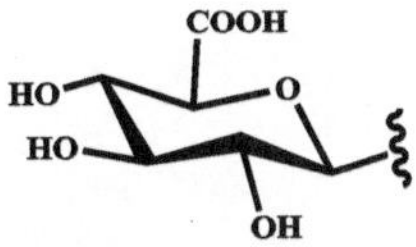

(1)

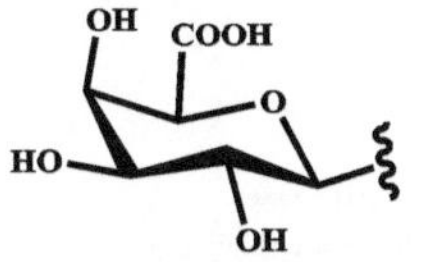

(2)

$R_2$=

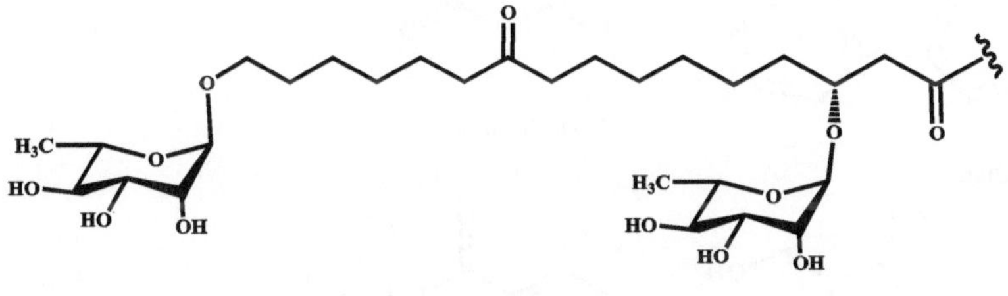

Azafrines 1 and 2

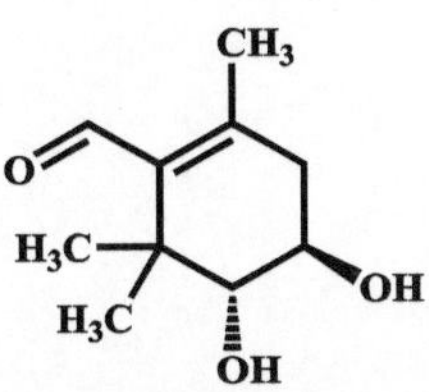

Crocusatin K

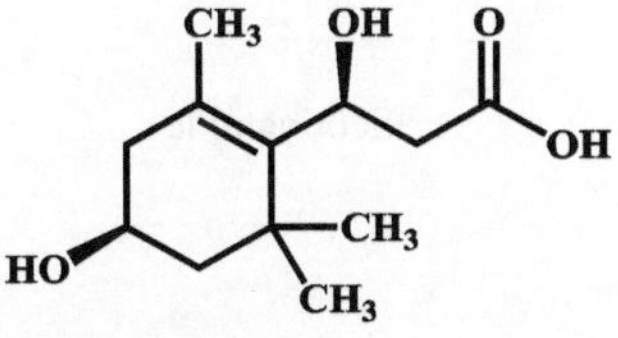

Crocusatin H

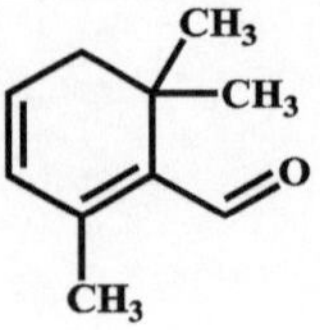

Safranal

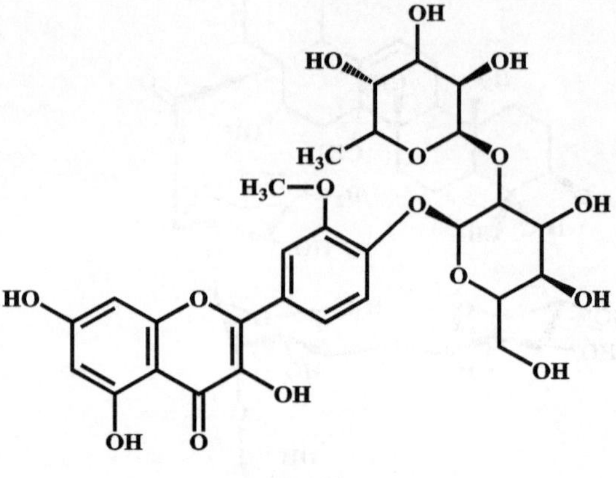

Isorhamnetin-4′-*O*-neohesperidoside

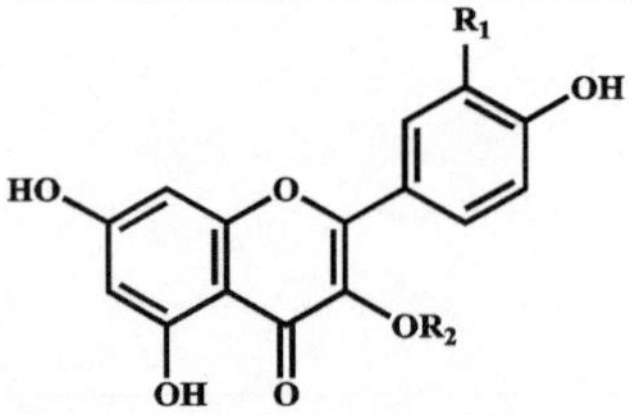

Kaempferol: $R_1$ = H; $R_2$ = H
Kaempferol-3-*O*-glucoside: $R_1$ = H; $R_2$ = glucose
Kaempferol-3-*O*-sophoroside: $R_1$ = H; $R_2$ = sophorose
Quercetin-3-*O*-glucoside: $R_1$ = OH; $R_2$ = glucose

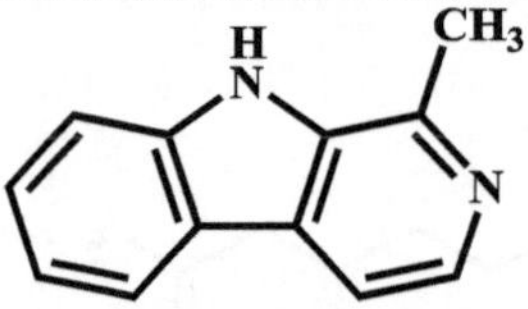

Harman

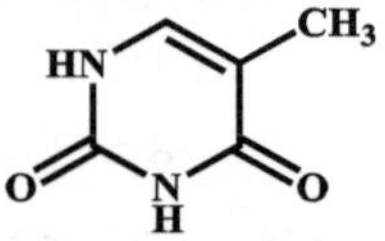

5-Methyluracil

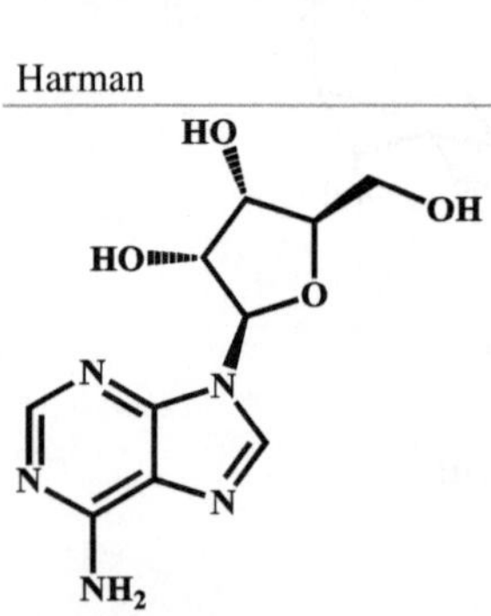

Adenosine

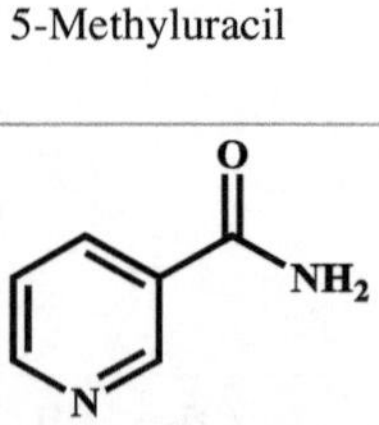

Nicotinamide

4) Pharmacology and bioactivities:

   A) In vitro studies:
      Not found.
   B) In vivo studies:
      A study has assessed the efficacy of crocin, the main active constituent of saffron, as an adjunctive treatment in major depressive disorder (MDD) patients. The result of the study demonstrated the effect of crocin in depression and could be administered in treatment of MDD patients (Talaei et al. 2015). In addition, petal of *C. sativus* was found to be effective, similar to fluoxetine in the treatment of mild to moderate depression (Akhondzadeh Basti et al. 2007). A research investigated the molecular mechanism of antidepressant effect of aqueous extract of saffron and its effect on the levels of BDNF, VGF, CREB, and p-CREB in rat hippocampus. The study found that aqueous extract of saffron has antidepressant effects and the mechanism of its antidepressant effect may be due to increasing the levels of BDNF, VGF, CREB, and p-CREB in rat hippocampus (Ghasemi et al. 2015).

5) Indications and usage:
   Activate blood and resolve blood stasis, relieve melancholy and alleviate pain, and remove pathogenic heat from the blood and toxic material from the body. Treat amenorrhea, traumatic injury, hematemesis and abdominal pain due to blood stasis after delivery, palpitation, absent-mindedness, typhoid, melancholy, and chest distension (Chinese Pharmacopoeia Commission 2010).
6) Precautions and side effects:
   It is contraindicated to pregnant women (Chinese Pharmacopoeia Commission 2010).
7) Dosage:

   A) Internal use:
      1–3 g/day, decocted in water for oral administration, taken twice a day, generally for 5–7 days, modificated according to symptoms or soaked in wine (Chinese Pharmacopoeia Commission 2010).
   B) External use:
      Not found.

## 7.16 References

Flora of China Editorial Committee. Flora of China. Beijing and St. Louis: Science Press and Missouri Botanical Garden Press. 2000; 24: 313.

Pfander H, Schurtenberger H. Biosynthesis of C20-carotenoids in *Crocus sativus*. Phytochemistry. 1982;21(5): 1039–42.

Rubio-Moraga A, Gerwig GJ, Castro-Diaz N, Jimeno ML, Escribano J, Fernandez JA, Kamerling JP. Triterpenoid saponins from corms of Crocus sativus: Localization, extraction and characterization. Industrial Crops Products. 2011;34(3):1401–9.

Li CY, Wu TS. Constituents of the stigmas of *Crocus sati*vus and their tyrosinase inhibitory activity. J Nat Product. 2002;65(10): 1452–6.

Tarantilis PA, Beljebbar A, Manfait M, Polissiou M. FT-IR, FT-Raman spectroscopic study of carotenoids from saffron (*Crocus sativus* L.) and some derivatives. Spectrochim Acta A. 1998;54(4): 651–7.

Van Calsteren MR, Bissonnette MC, Cormier F, Dufresne C, Ichi T, Yves LeBlanc JC, Perreault D, Roewer I. Spectroscopic characterization of crocetin derivatives from *Crocus sativus* and *Gardenia jasminoides*. J Agric Food Chem. 1997;45(4): 1055–61.

Tung NH, Shoyama Y. New minor glycoside components from saffron. J Nat Product. 2013;67(3): 672–6.

Straubinger M, Jezussek M, Waibel R, Winterhalter P. Novel glycosidic constituents from saffron. J Agric Food Chem. 1997;45(5): 1678–81.

Pfister S, Meyer P, Steck A, et al. (1996). Isolation and structure elucidation of carotenoid-glycosyl esters in gardenia fruits (*Gardenia jasminoides* Ellis) and saffron (*Crocus sativus* Linn.). J Agric Food Chem. 44(9): 2612–5.

Zarghami NS, Heinz DE. Monoterpene aldehydes and isophorone-related compounds of saffron. Phytochemistry. 1971;10(11): 2755–61.

Straubinger M, Bau B, Eckstein S, Fink M, Winterhalter P. Identification of novel glycosidic aroma precursors in saffron (*Crocus sativus* L.). J Agric Food Chem. 1998;46(8): 3238–43.

Gao WY, Li YM, Zhu DY. Phenolic glucosides and a γ-lactone glucoside from the sprouts of *Crocus sativus*. Plant Med. 1999;65(5): 425–7.

Li CY, Wu TS. Constituents of the pollen of Crocus sativus L. and their tyrosinase inhibitory activity. Chem Pharmaceut Bull. 2003;50(10): 1305–9.

Li CY, Lee EJ, Wu TS. Antityrosinase principles and constituents of the petals of *Crocus sativus*. J Nat Product. 2004;67(3): 437–40.

Talaei A, Hassanpour Moghadam M, Sajadi Tabassi SA, Mohajeri SA. Crocin, the main active *saffron* constituent, as an adjunctive treatment in major depressive disorder: A randomized, double-blind, placebo-controlled, pilot clinical trial. J Affect Disord. 2015;174: 51–6.

Akhondzadeh Basti A, Moshiri E, Noorbala AA, Jamshidi AH, Abbasi SH, Akhondzadeh S. Comparison of petal of *Crocus sativus* L. and fluoxetine in the treatment of depressed outpatients: a pilot double-blind randomized trial. Progress Neuropsychopharmacol Biol Psychiatry. 2007;31(2): 439–42.

Ghasemi T, Abnous K, Vahdati F, Mehri S, Razavi BM, Hosseinzadeh H. Antidepressant effect of *Crocus sativus* aqueous extract and its effect on CREB, BDNF, and VGF transcript and protein levels in rat hippocampus. Drug Res. 2015;65(7): 337–43.

Chinese Pharmacopoeia Commission. Pharmacopoeia of People's Republic of China. Beijing: China Medical Science Press. 2010.

## 7.17 *Cyperus rotundus* L.

### *7.17.1* Cyperi Rhizoma

#### 7.17.1.1 (Cyperaceae)

1) Common names:
   Nutgrass galingale rhizome, Xiang-fu, Suo-cao, Xiang-Fu-zi, Lei-Gong-tou (Pinyin in Chinese).
2) Photograph of the species (Fig. 7.8)

**Fig. 7.8** *Cyperi Rhizoma*

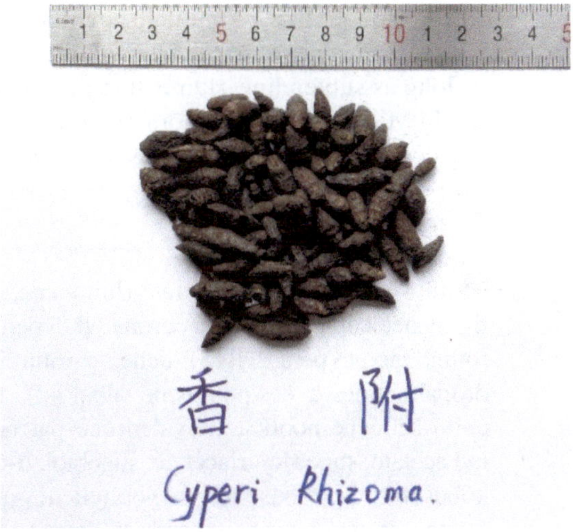


3) Description of the species:

A) Habitat and world distribution:
Grasslands, wet or dry areas on mountain slopes, stream margins, along trails, sandbanks, ditch margins, water margins in valleys, paddy field margins; near sea level up to 2100 m. Anhui, Chongqing, Dongsha Qundao, Fujian, South of Gansu, Guangdong, Guangxi, Guizhou, Hainan, Hebei, Henan, Hubei, Hunan, Jiangsu, Jiangxi, South of Liaoning, Nansha Qundao, South of Shaanxi, Shandong, Shanxi, Sichuan, Taiwan, Xisha Qundao, Southeast of Xizang, Yunnan, and Zhejiang [other countries: Afghanistan; Bhutan; India; Indonesia; Japan; Kazakhstan; Korea; Kyrgyzstan; Malaysia; Myanmar; Nepal; Pakistan; Papua New Guinea; Philippines; Sri Lanka; Tajikistan; Thailand; Uzbekistan; Vietnam; Africa; Southwest of Asia; Australia; Central, North, and South of America; Europe; Indian Ocean islands; Madagascar; Pacific islands] (Flora of China Editorial Committee 2008).

B) Morphology of the plant:
Perennial herb. Stolons slender, with ellipsoidal tubers. Culms solitary, rarely two laxly tufted, 15–90 cm tall, slightly slender, triquetrous, smooth, base swollen into a tuber, leaved at basal part. Leaves equaling or shorter than culm; sheath brown, usually disintegrating into fibers; leaf blade bluish-green, with 2–5 mm wide, more or less flat. Involucral bracts, 2 or 3(-5), longer to sometimes shorter than inflorescence. Inflorescence, a simple or compound anthelia; rays (2 or) 3–10, mostly to 12 cm, unequal in length, spreading. Spikes obdeltoid, with 3–10 slightly laxly arranged spikelets. Spikelets obliquely spreading, linear, with 1–3 cm × 1.5–2 mm, 8–28-flowered; rachilla wings white, slightly broad, hyaline. Glumes blood-red to purplish brown on both surfaces but middle green, subdensely imbricate, ovate to oblong ovate, ca. 3 mm, 5–7-veined (fading some distance before margin), apex acute to obtuse and muticous. Stamens three; anthers

linear; connective prominent beyond anthers. Style long; stigmas three, longer than style, exserted from glume. Fruit, nutlet obovoid oblong, 1/3-2/5 as long as subtending glume, three-sided, puncticulate. Flowering and fruiting in May–November (Flora of China Editorial Committee 2010).

C) Medicinal parts:
The dry rhizome of *Cyperus rotundus* L. has been used (Flora of China Editorial Committee 2010).

D) Chemical constituents:
**Mono- and sesquiterpenes** (limonene, cyperol, isocyperol, 1,8-cineole, β-pinene, camphene, α-cyperone, β-cyperone, cyperene, cyproterone, cyperotundone, cypera-2,4(15)-diene, α-rotunol, β-rotunol, rotundine (=tetrahydropalmatine, a sesquiterpene alkaloid), rotundone, β-selinene, α-selinene, epi-α-selinene, nootkatone, valencene, patchoulenone, isopatchoulenone, sugeonyl acetate, sugetriol triacetate, sugebiol, α-copaene, copadiene, caryophyllene, kobusone, isokobusone, caryophyllene oxide, isocurcumenol, carophylla-6-one, epoxy-quaine, rotundone, ylanga-2,4(15)-diene, γ-gurjunene epoxide, *trans*-calamenene, δ-cadinene, γ-calacorene, α-muurolene, γ-muurolene, cadalene, mustakone, (E)-pinocarveol, myrtenal, dihydrocarvone, verbenone, (E)-carveol, epi-guaidiol A, guaidiol A, cyperenoic acid, etc.) (Xu et al. 2010; Sonwa and König 2001; Tam et al. 2007; Kilani et al. 2008; Nima et al. 2008; Zhu et al. 1997; Kilani et al. 2005; Tang and Eisenbrand 2010; Xu et al. 2009); **flavonoids** (kaempferol, luteolin, quercetin, pinoquercetin, ginkgetin, sciadopitysin) (Xu et al. 2010); **iridoid glycosides** (ipolamiide, 6β-hydroxyipolamiide, rotundusides A-F) (Mohamed 2015; Zhou and Zhang 2013; Lin et al. 2015).

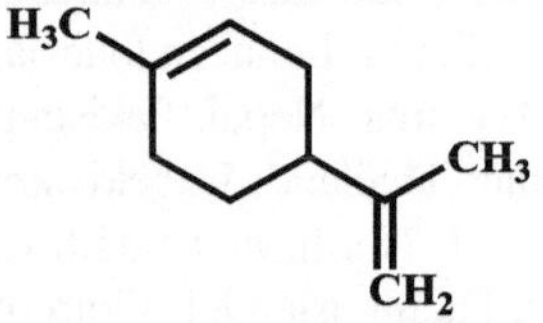


Limonene

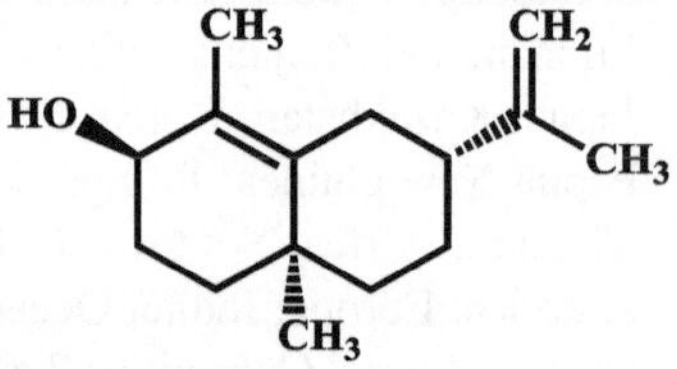


Cyperol

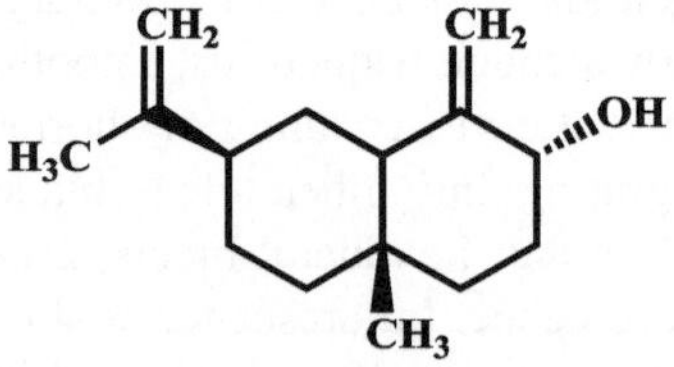


Isocyperol

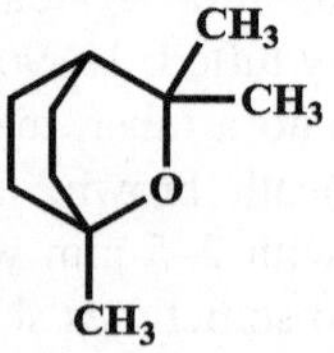


1,8-Cineole

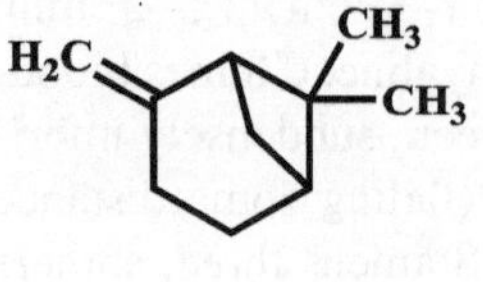


β-Pinene

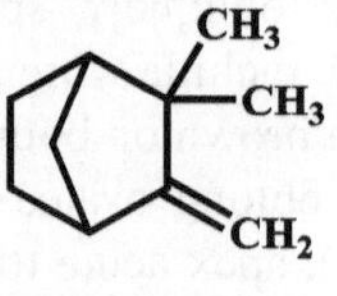


Camphene

| | |
|---|---|
| 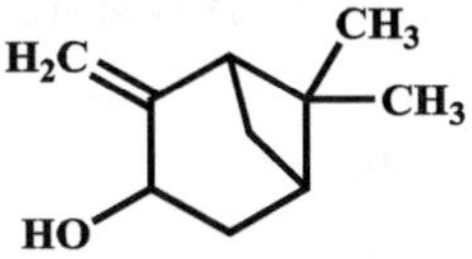<br>Pinocarveol | 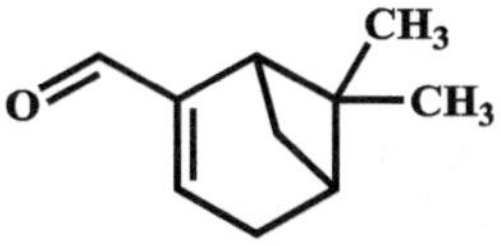<br>Myrtenal |
| 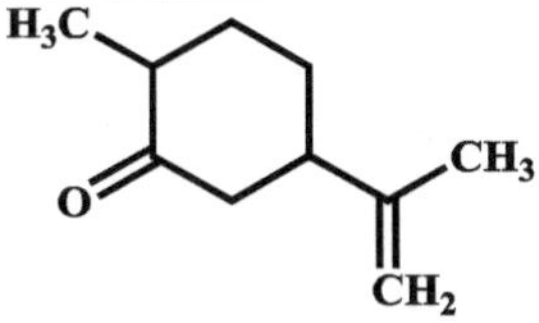<br>Dihydrocarvone | 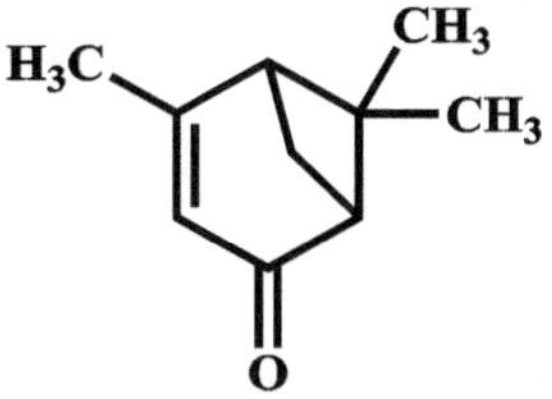<br>Verbenone |
| 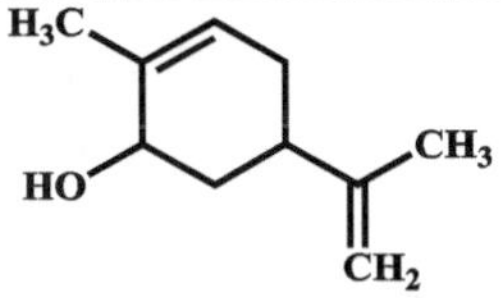<br>*E*-Carveol | 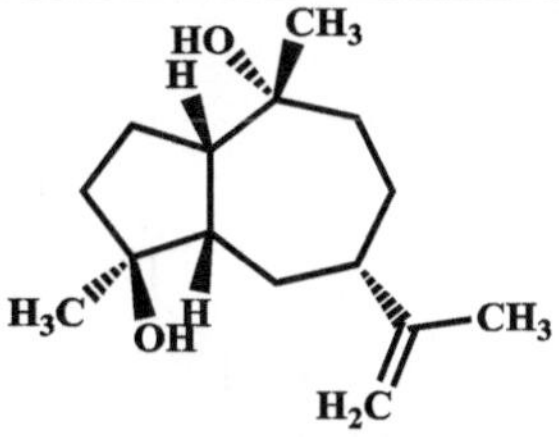<br>epi-Guaidiol A |
| 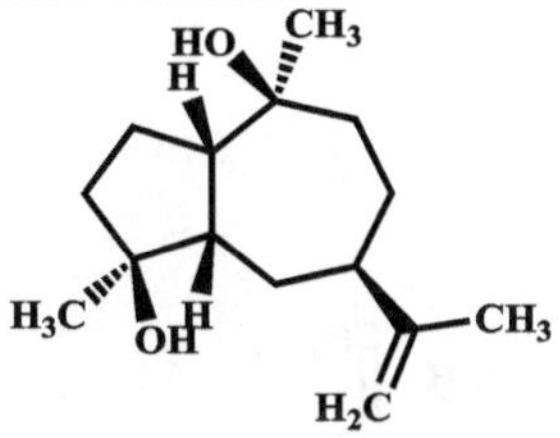<br>Guaidiol A | 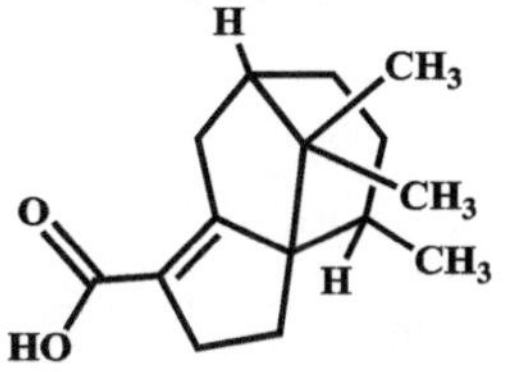<br>Cyperenoic acid |
| 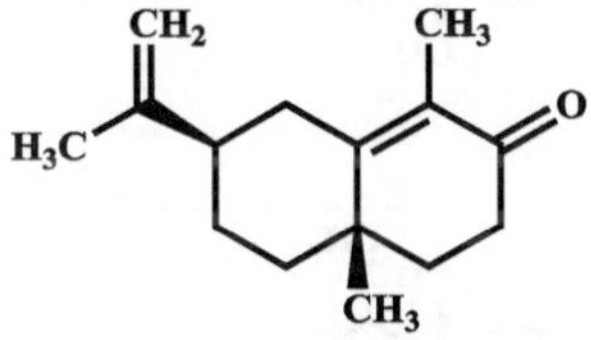<br>α-Cyperone | 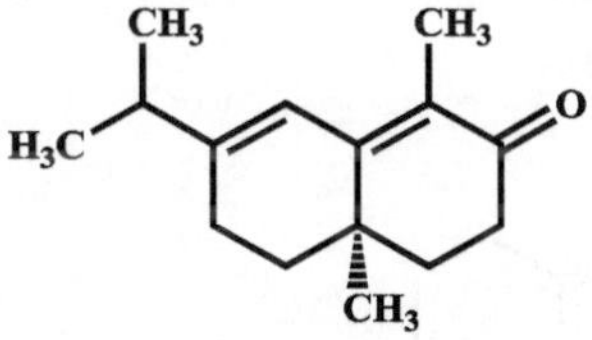<br>β-Cyperone |
| 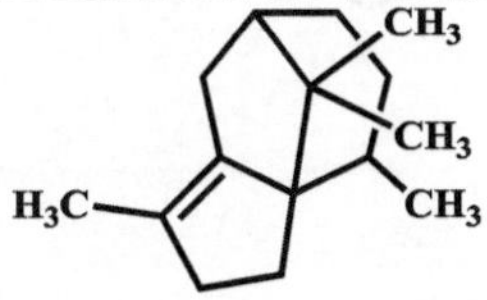<br>Cyperene | 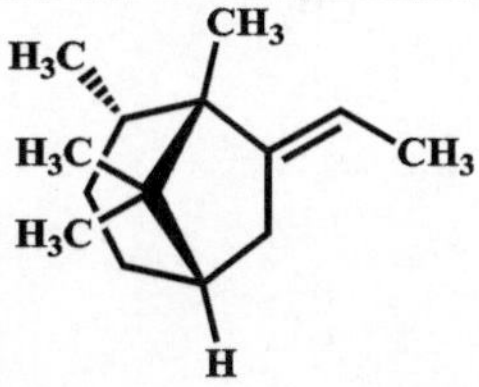<br>Cyproterone |

| | |
|---|---|
| 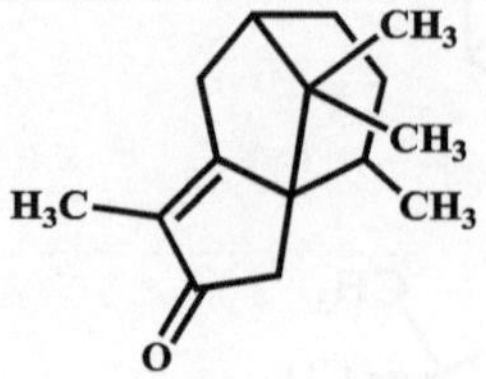<br>Cyperotundone | 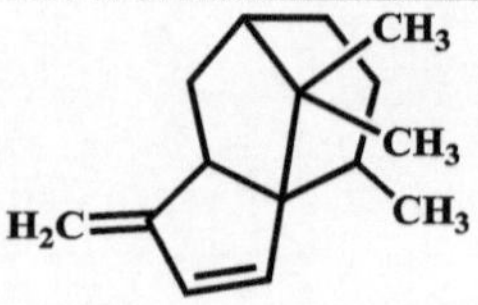<br>Cypera-2,4(15)-diene |
| 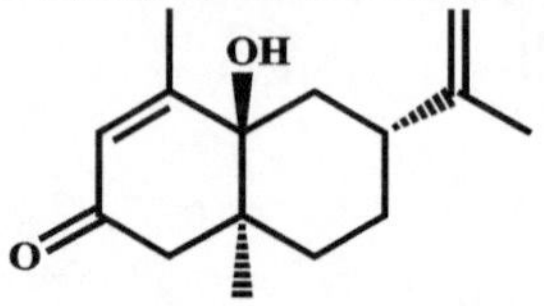<br>α-Rotunol | 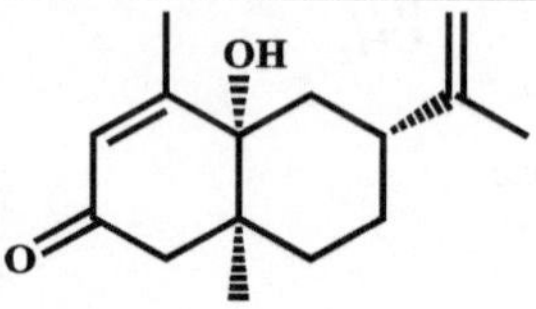<br>β-Rotunol |
| 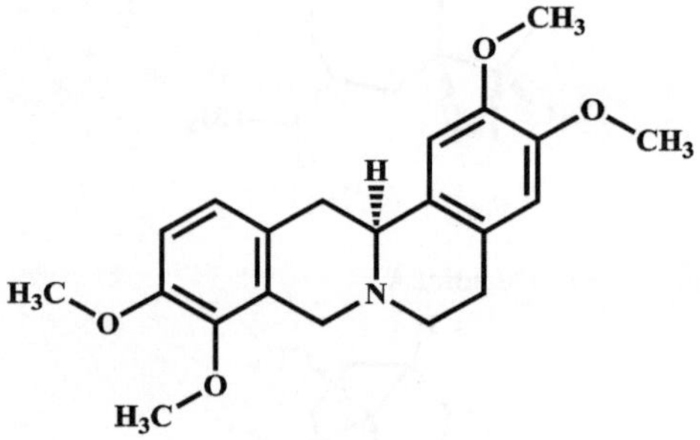<br>Tetrahydropalmatine | 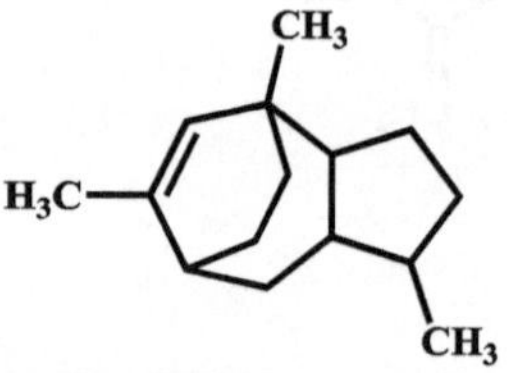<br>Rotundone |
| 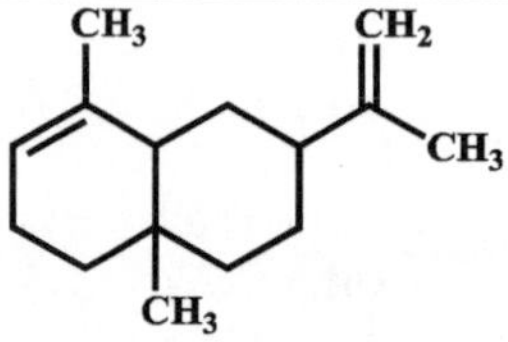<br>α-Selinene | 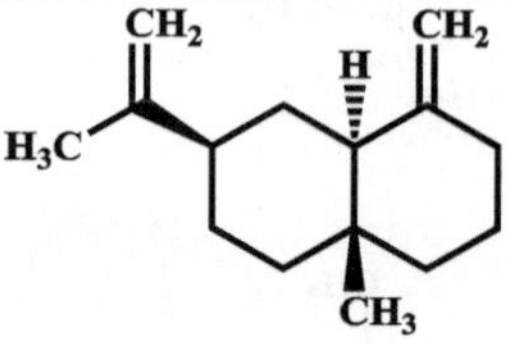<br>β-Selinene |
| 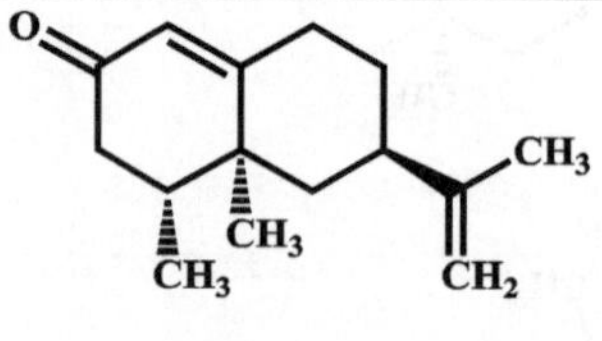<br>(+)-Nootkatone | 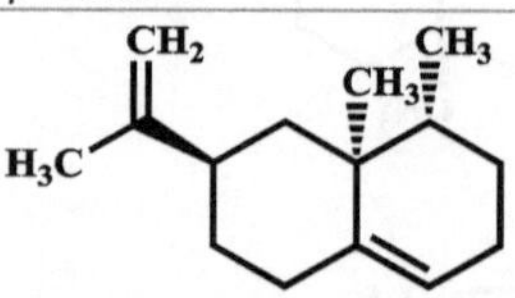<br>Valencene |

| | |
|---|---|
| 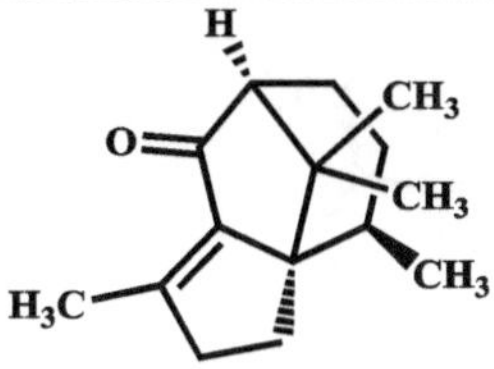 Patchoulenone | 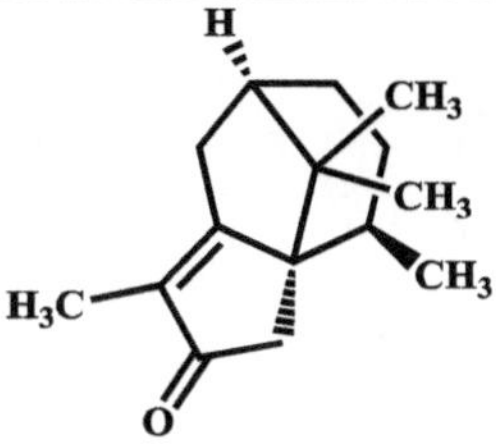 Isopatchoulenone |
| 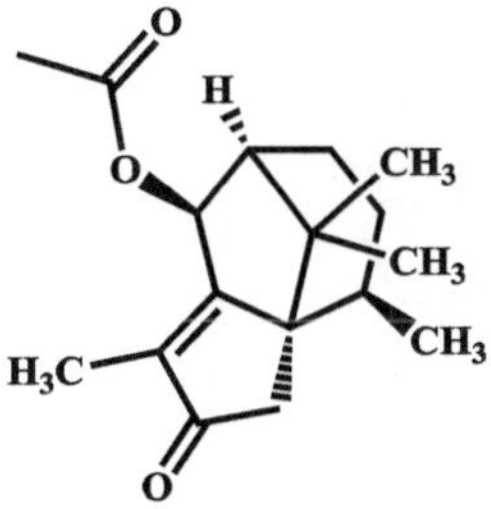 Sugeonyl acetate | 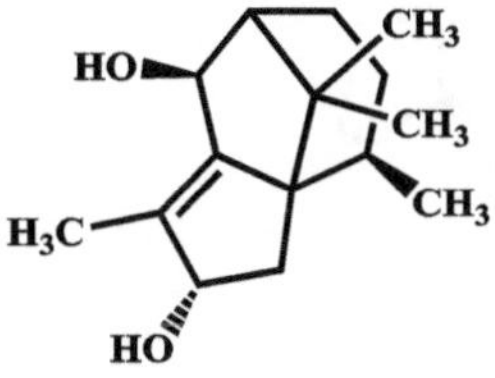 Sugebiol |
| 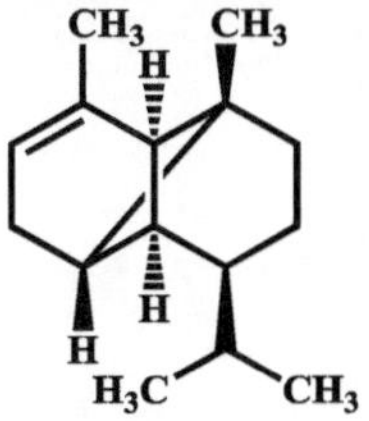 α-Copaene | 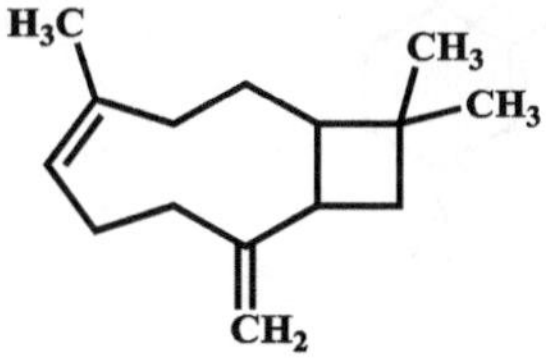 beta-Caryophyllene |
| 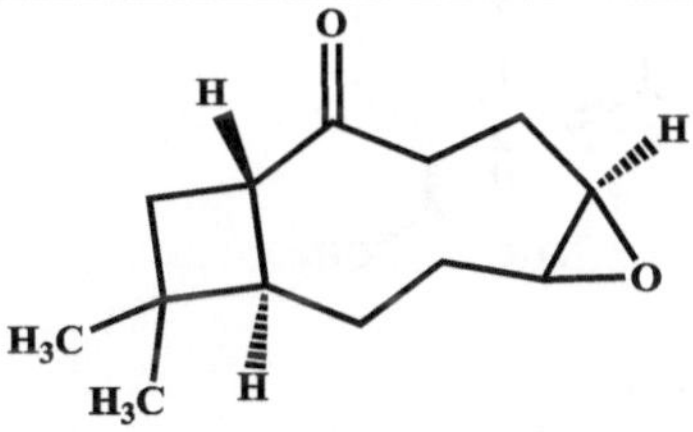 Kobusone | 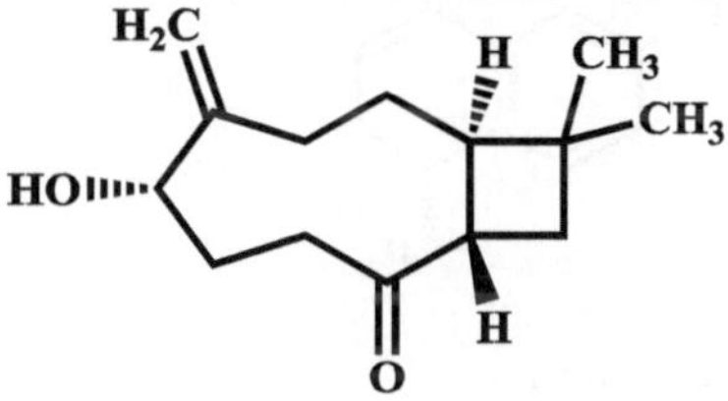 Isokobusone |
| 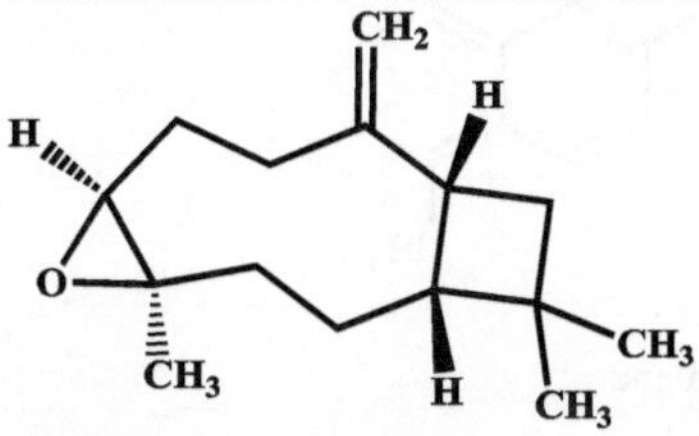 beta-Caryophyllene oxide | 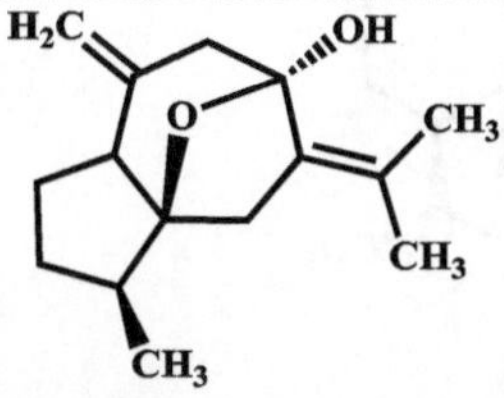 Isocurcumenol |

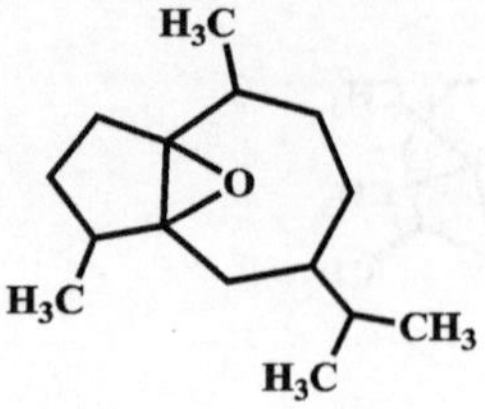

Epoxy-quaine

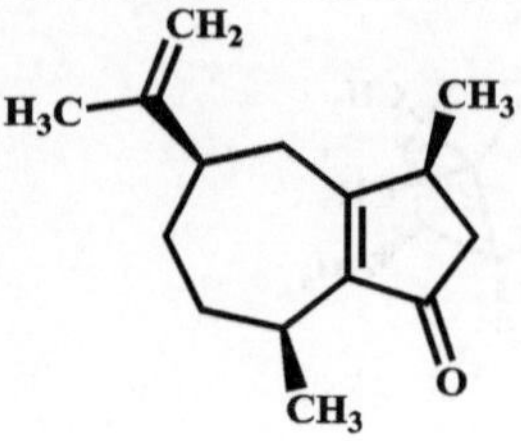

Rotundone

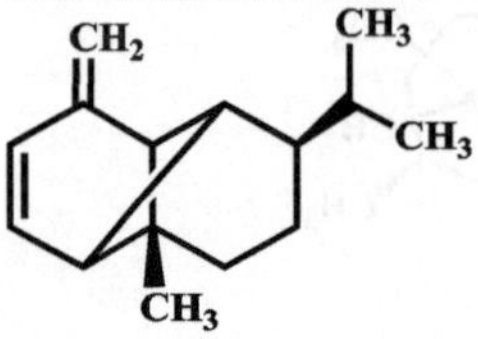

Ylanga-2,4(15)-diene

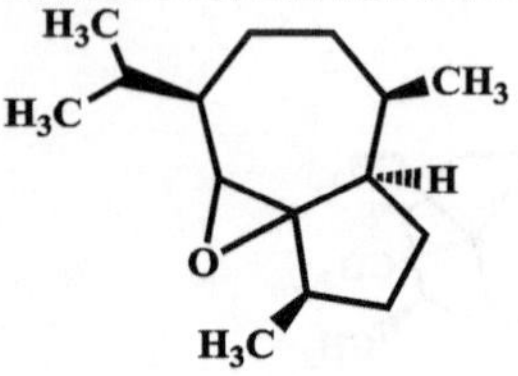

γ-Gurjunene epoxide

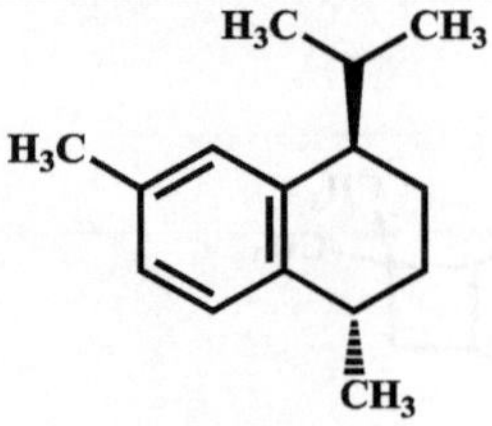

*trans*-Calamenene

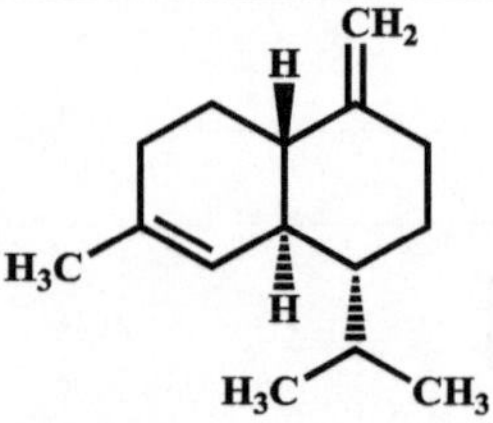

γ-Cadinene

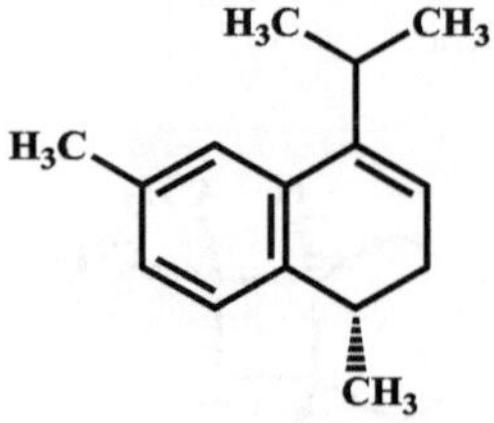

γ-Calacorene

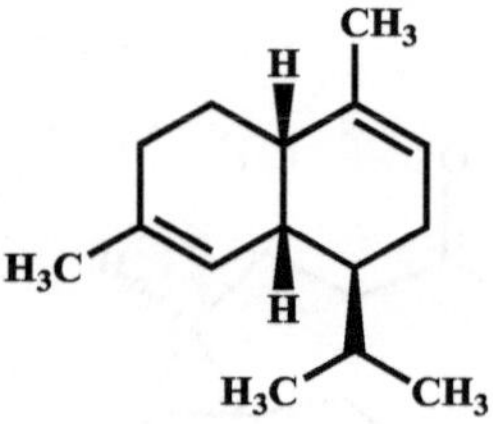

α-Muurolene

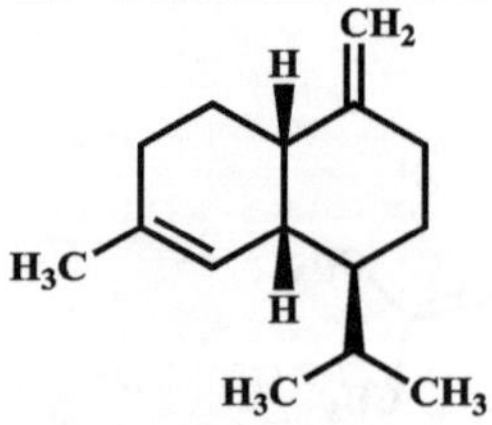

γ-Muurolene

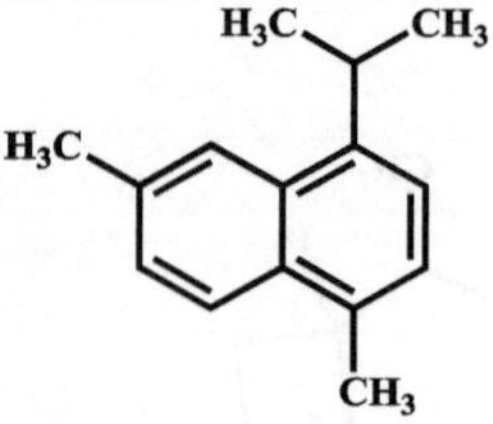

Cadalene

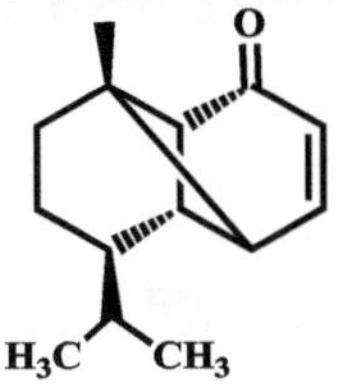

Mustakone

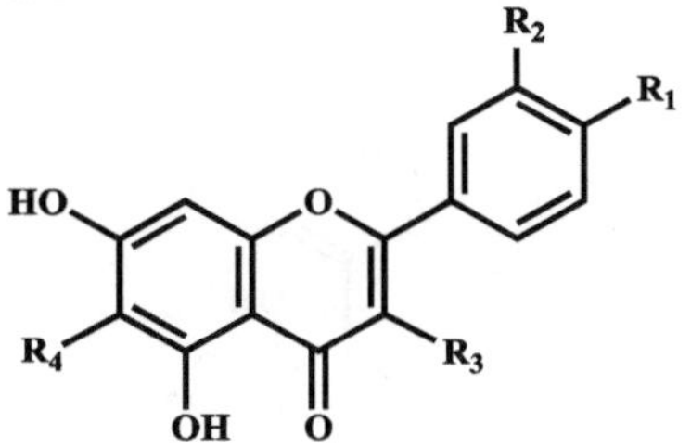

**Kaempferol: $R_1 = R_3 = OH$; $R_2 = R_4 = H$**
**Quercetin: $R_1 = R_2 = R_3 = OH$; $R_4 = H$**
**Pinoquercetin: $R_1 = R_2 = R_3 = OH$; $R_4 = CH_3$**

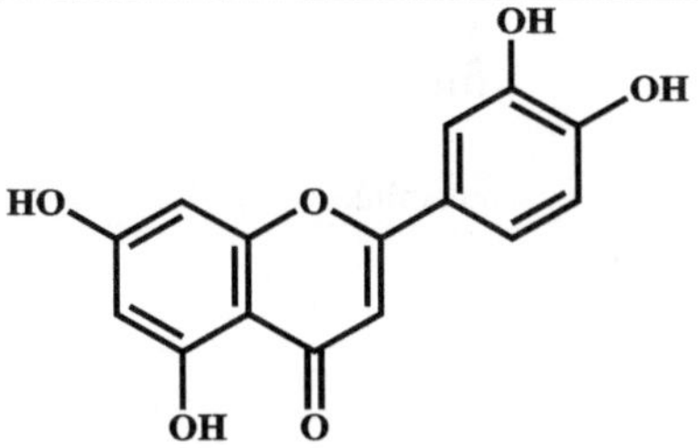

Luteolin

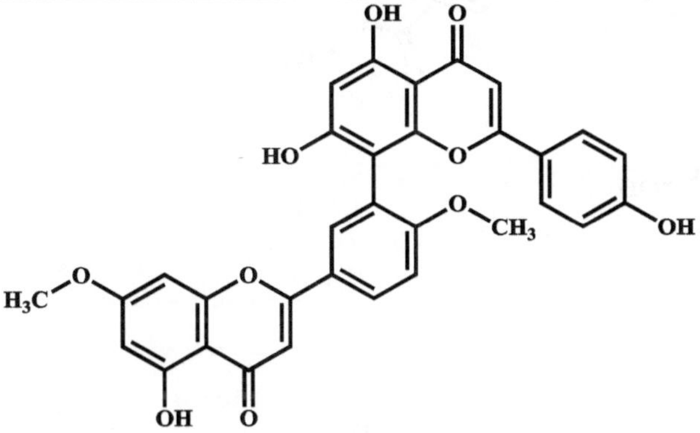

Ginkgetin

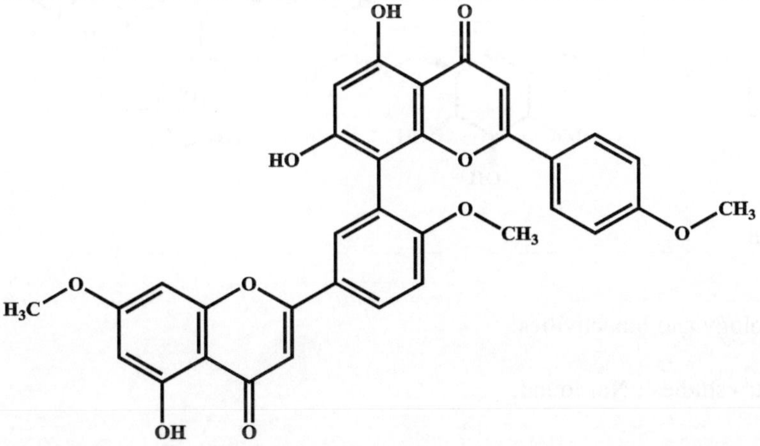

Sciadopitysin

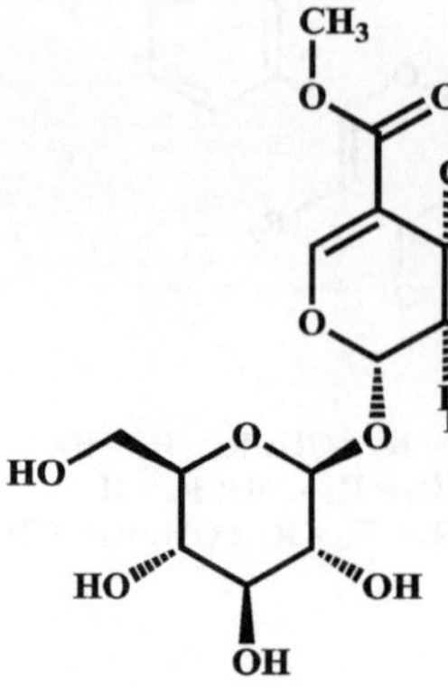

Ipolamiide

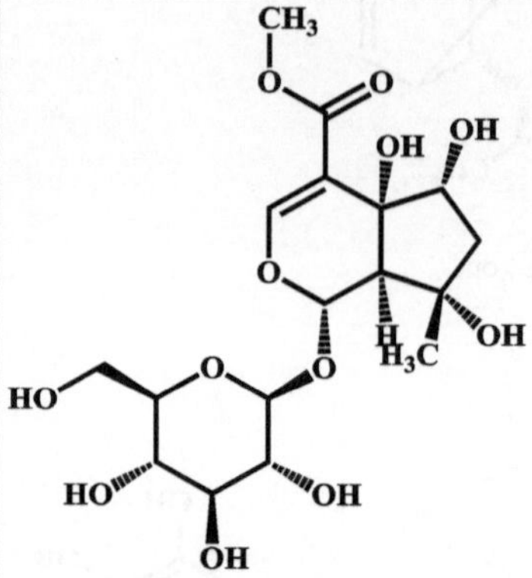

6β-Hydroxyipolamiide

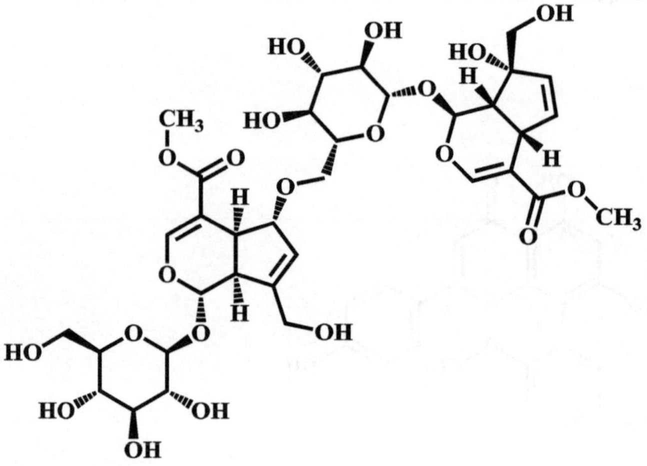

Rotunduside A

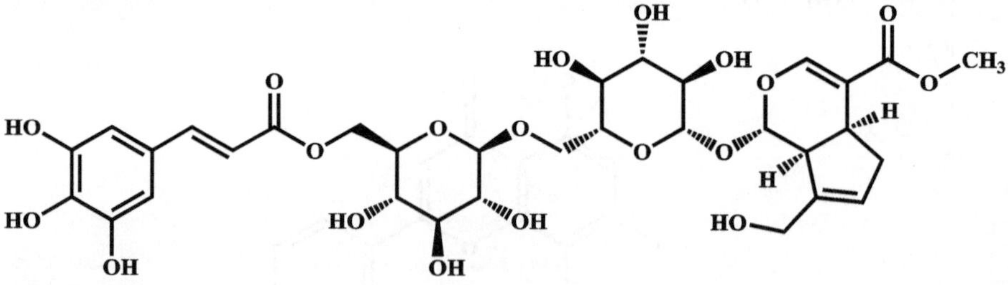

Rotunduside B

4) Pharmacology and bioactivities:

A) In vitro studies : Not found.

B) In vivo studies:
A study has demonstrated that the ethanolic extracts of the roots of *Cyperus rotundus* L. possess antidepressant activity, which may be related to inhibit the activity of monoamine oxidase A (MAO-A). The ethanolic extracts can significantly reduce the immobility time in the TST and FST. Cell membrane chromatography (CMC) showed that it has the similar retention behavior with fluoxetine. Moreover, the MAO-A half inhibition rate of the ethanolic extracts was 50.00 μg/mL; however, the fluoxetine's concentration was 346.89 μg/mL (Jia and Zou 2014).
In the despair mice models, rotunduside F showed significant antidepressant activity at the dosage of 50 mg/kg, which was close to the positive control fluoxetine (20 mg/kg). Treatment with rotunduside F at 50 mg/kg which significantly reduced duration of immobility time in FST and TST produced no significant difference in number of crossing activity of mice in OFT (Lin et al. 2015).

5) Indications and usage:
Soothe liver and regulate Qi, regulate menorrhea, and alleviate pain. Treat liver depression and qi stagnation syndrome, menoxenia, dysmenorrhea, and breast distending pain (Chinese Pharmacopoeia Commission 2010).
6) Precautions and side effects:
Not found.
7) Dosage:
A) Internal use:
6–10 g/day, decocted in water for oral administration, taken twice a day, generally for 5–7 days, modificated according to symptoms (Chinese Pharmacopoeia Commission 2010).
B) External use:
Not found.

## 7.18 References


Flora of China Editorial Committee. Flora of China. Beijing and St. Louis: Science Press and Missouri Botanical Garden Press. 2010; 23: 223.

Xu Y, Li DX, Ling TJ. Advances in Study on Chemical Constituents of Rhizomes of *Cyperus rotundus*. Chinese J Exp Traditional Med Formula. 2010;16(11): 214–18.

Sonwa MM, König WA. Chemical study of the essential oil of *Cyperus rotundus*. Phytochemistry. 2001;58 (5): 799–810.

Tam CU, Yang FQ, Zhang QW, Guan J, Li SP. Optimization and comparison of three methods for extraction of volatile compounds from *Cyperus rotundus* evaluated by gas chromatography-mass spectrometry. J Pharmaceut Biomed Analysis. 2007;44(2): 444–9.

Kilani, S, Ledauphin J, Bouhlel I, Sghaier, MB, Boubaker J, Skandrani I, Mosrati R, Ghedira K, Barillier D, Chekir-Ghedira L. Comparative study of *Cyperus rotundus* essential oil by a modified GC/MS analysis method. Evaluation of its antioxidant, cytotoxic, and apoptotic effects. Chem Biodivers. 2008;5(5): 729–42.

Nima ZAM, Jabier MS, Wagi R.I, Hussain HAAK. Extraction, identification and antibacterial activity of cyperus oil from Iraqi *C. rotundus*. Eng Technol. 2008;26(10): 1156–63.
Zhu M, Luk HH, Fung HS, Luk CT. Cytoprotective effects of *Cyperus rotundus* against ethanol induced gastric ulceration in rats. Phytother Res. 1997;11(5): 392–4.
Kilani S, Abdelwahed A, Ammar RB, Hayder N, Ghedira K, Chraief I, Hammami M, Chekir-Ghedira L. Chemical composition, antibacterial and antimutagenic activities of essential oil from (Tunisian) *Cyperus rotundus*. J Essential Oil Res. 2005;17(6): 695–700.
Tang W, Eisenbrand G. Handbook of Chinese medicinal plants: chemistry, pharmacology, toxicology, Vol 2. KGaA, Weinheim: Wiley-VCH. 2010.
Xu Y, Zhang HW, Wan XC, Zou ZM. Complete assignments of $^{1}$H and $^{13}$C NMR data for two new sesquiterpenes from *Cyperus rotundus* L. Magn Res Chem. 2009;47(6): 527–31.
Mohamed GA. Iridoids and other constituents from *Cyperus rotundus* L. rhizomes. Bulletin of Faculty of Pharmacy, Cairo University. 2015;53: 5–9.
Zhou Z, Zhang H. Phenolic and iridoid glycosides from the rhizomes of *Cyperus rotundus* L. Med Chem Res. 2013;22(10): 4830–5.
Lin SQ, Zhou ZL, Zhang HL, Yin WQ. Phenolic glycosides from the rhizomes of *Cyperus rotundus* and their antidepressant activity. J Korea Soc Appl Biol Chem. 2015;58(5): 685–91.
Jia HM, Zou ZM. Antidepressant effect evaluation of the ethanolic extract from the roots of *Cyperus rotundus* L. on cell membrane chromatography and different depression models. Eur J Integ Med. 2014;6(6): 742.
Chinese Pharmacopoeia Commission. Pharmacopoeia of People's Republic of China. Beijing: China Medical Science Press. 2010.

## 7.19 *Citrus aurantium* L.

### *7.19.1* Fructus Aurantii

#### 7.19.1.1 (Rutaceae)

1) Common names:
   Bitter orange, Zhi-qiao (Pinyin in Chinese).
2) Photograph of the species (Fig. 7.9)
3) Description of the species:

   A) Habitat and world distribution:
      Cultivated and sometimes naturalized in the south of the Qinling mountains (Flora of China Editorial Committee 2008).
   B) Morphology of the plant:
      Small trees. Branches with spines to ca. 8 cm. Petiole obovate, with 1–3 × 0.6–1.5 cm, base narrow; leaf blades dark green, thick. Inflorescences racemes, with few flowers or solitary flowers. Flowers hermaphrodite or male by more or less complete abortion of pistil; buds ellipsoid to subglobose. Calyx lobes 4 or 5. Petals with 2–3.5 mm in diameter. Stamens 20–25, typically basally connate into bundles. Fruit orange to reddish, globose to oblate, surface coarse; pericarp thick, sometimes difficult to remove; sarcocarp with 10–13 segments, acidic and sweet or sometimes bitter. Seeds numerous, large, with ridges; embryo(s) solitary to numerous; cotyledons milky white. Flowering in April–May and fruiting in September–December (Flora of China Editorial Committee 2008).

**Fig. 7.9** *Fructus Aurantii*

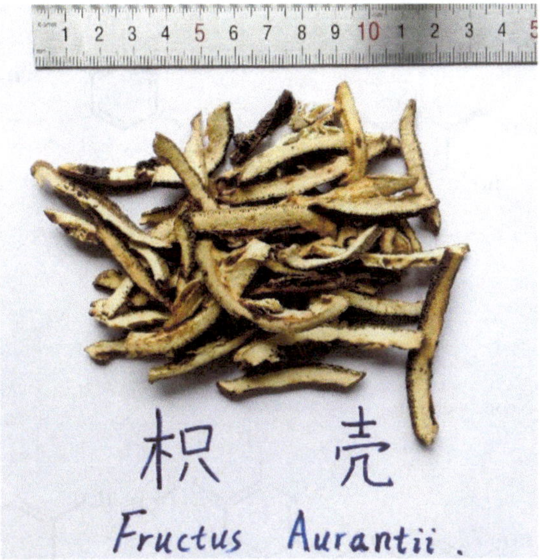


C) Medicinal parts:
Dry unripe fruit of *Citrus aurantium* L. has been used (Flora of China Editorial Committee 2008).

D) Chemical constituents:
*Fructus aurantii*'s chemical composition is diverse, mainly containing flavonoids, alkaloids, coumarins, essential oil and trace elements, and other compounds. **Flavonoids** (hesperidin, neohesperidin, naringin, narirutin, neoeriocitrin, neoponcirin, hesperetin, naringenin, nobiletin, tangeretin, 5-*O*-desmethylnobiletin, nornobiletin, sinensetin, 4-*O*-methylscutellarein, 4′,5,7,8-tetramethoxyflavone, 5-hydroxy-6,7,8,4'-tetramethoxyflavone) (Yan et al. 1994); **essential oil** (limonene account for 40.90 %, linalool account for 13.13 %, 2-undecanone account for 10.80 %, γ-terpinene account for 8.66 %, etc.) (Liao et al. 2004); **alkaloids and amines** (synephrine, *N*-methyl tyramine, octopamine, tyramine, hordenine, etc.) (Cristine 2005); **coumarins** (auraptene, umbelliferone, isoimperatorin, marmin acetonide, pranferin, isomerancin, bergaptene, meranzin, meranzin-O-glucoside, epoxybergamottin, etc.) (Chen and Lin 2004); **others** (quinic acid, citrusin I, ichangin-4-*O*-β-D-glucopyranoside, etc.) (Wu et al. 2015).

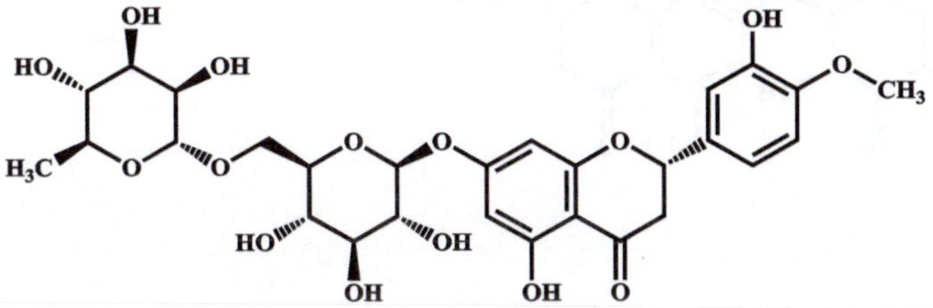

Hesperidin

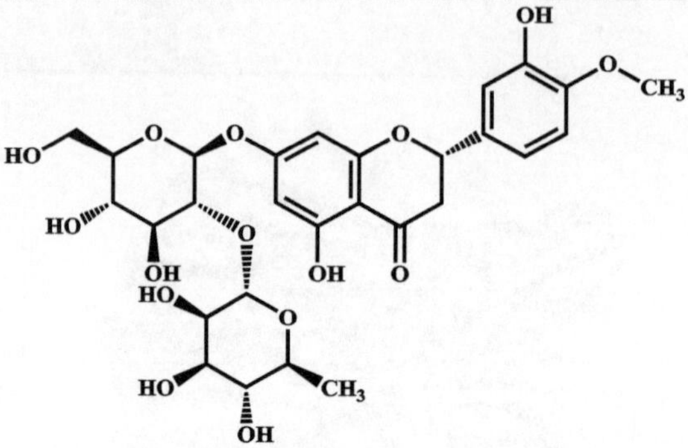

Neohesperidin

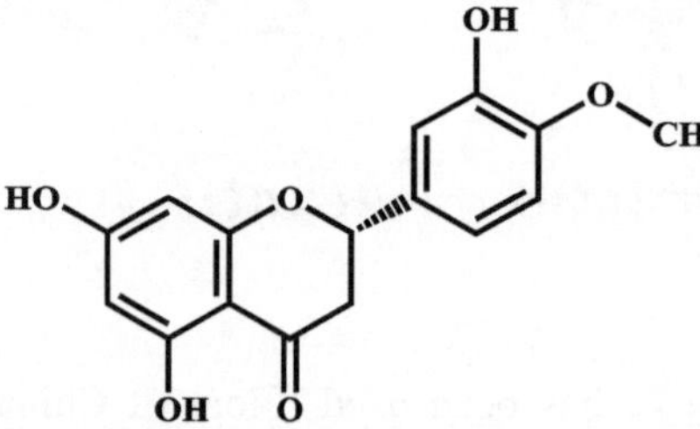

Hesperetin

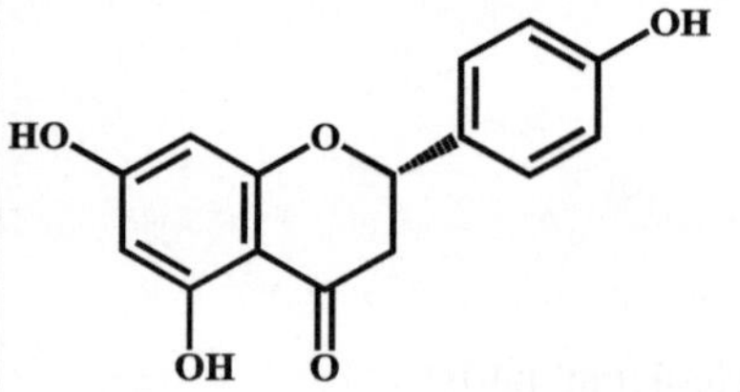

Naringenin

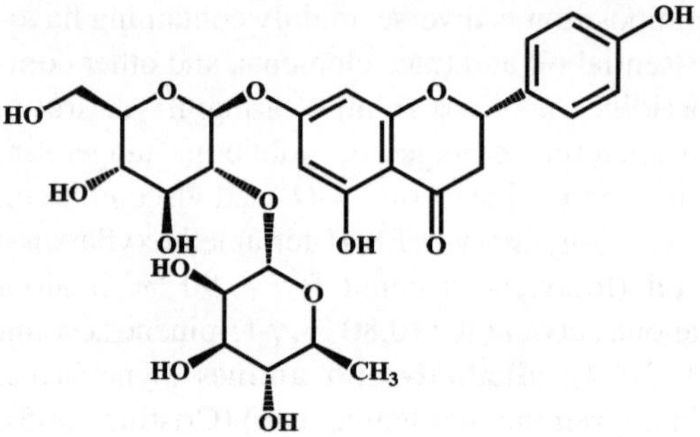

Naringin

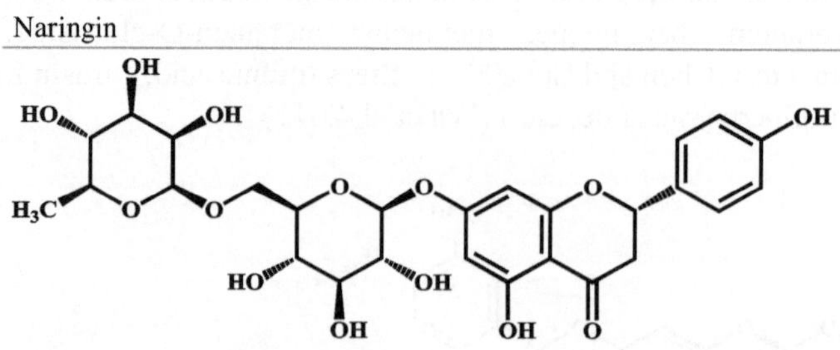

Narirutin

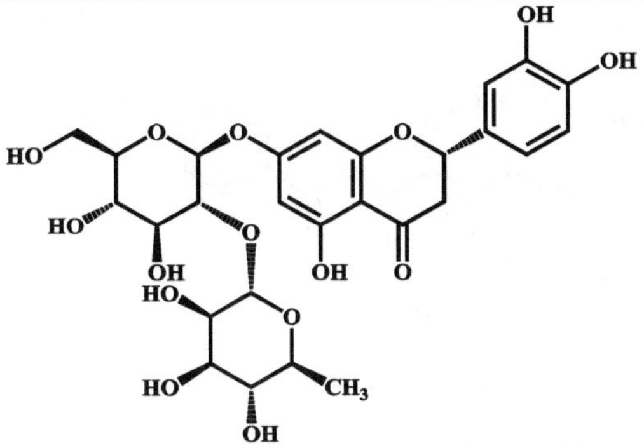


Neoeriocitrin

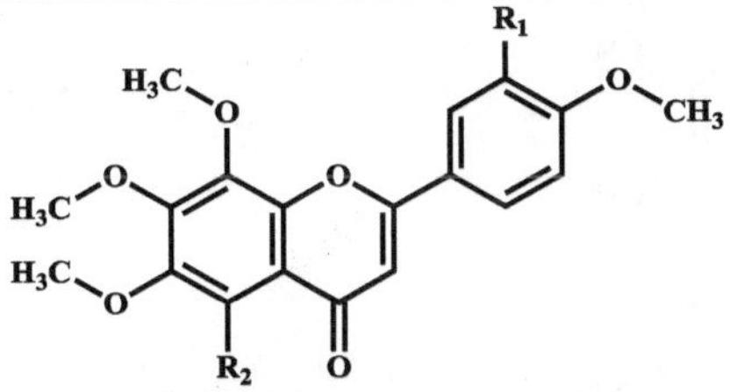


**Nobiletin: $R_1 = R_2 = OCH_3$**

**Tangeretin: $R_1 = H$; $R_2 = OCH_3$**

**5-*O*-Desmethylnobiletin: $R_1 = OCH_3$; $R_2 = OH$**

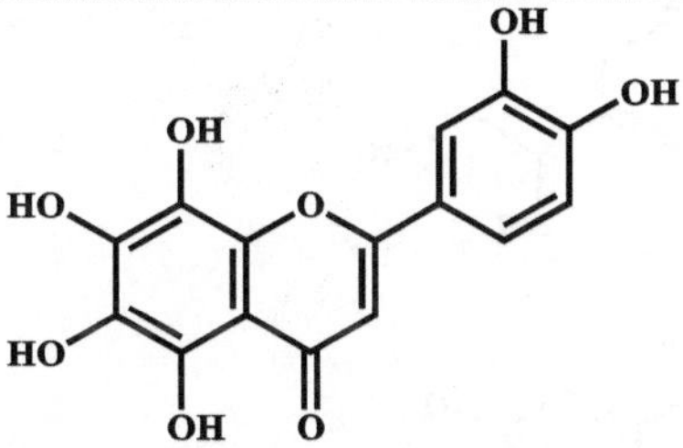


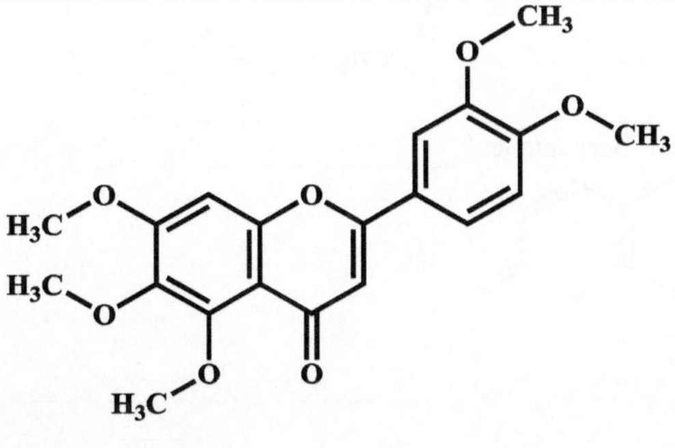


Sinensetin

| | |
|---|---|
| 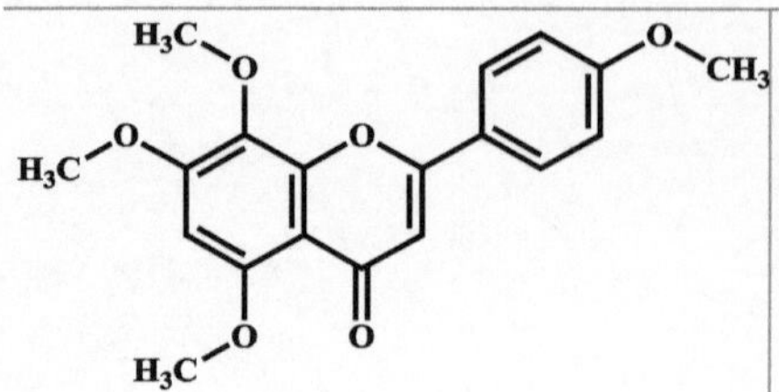<br>4′,5,7,8-Tetramethoxyflavone | 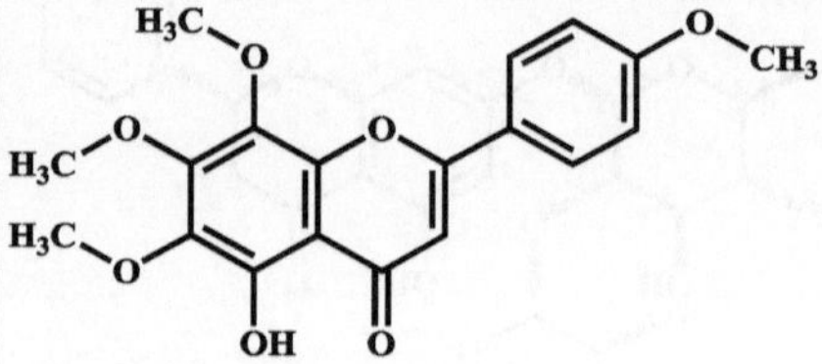<br>5-Hydroxy-6,7,8,4′-tetramethoxyflavone |
| 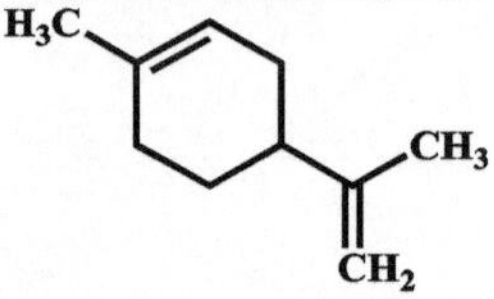<br>Limonene | 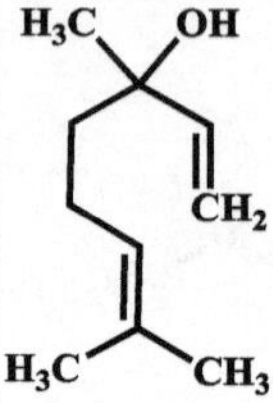<br>Linalool |
| 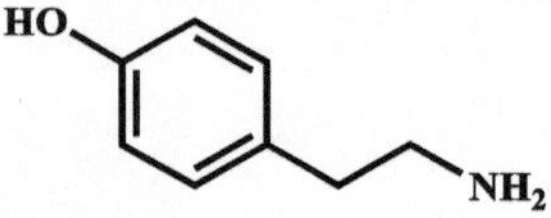<br>Tyramine | 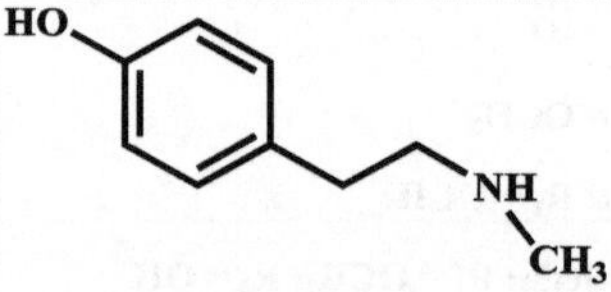<br>n-Methyl tyramine |
| 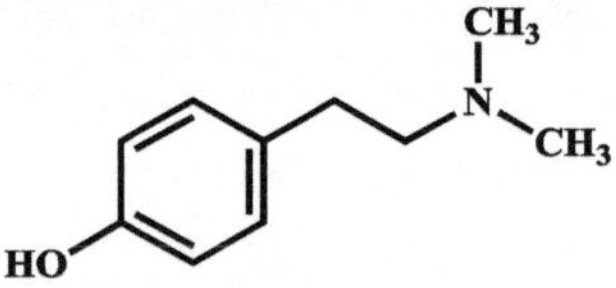<br>Hordenine | 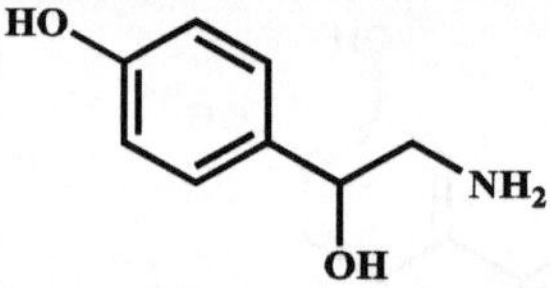<br>Octopamine |
| 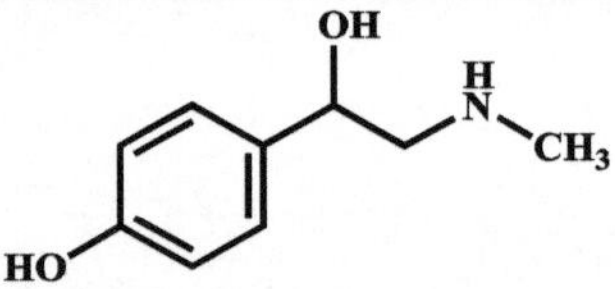<br>Synephrine | 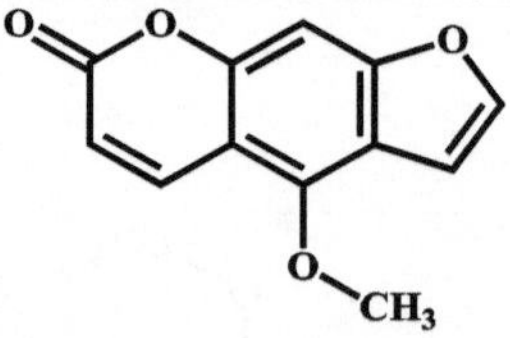<br>Bergaptene |
| 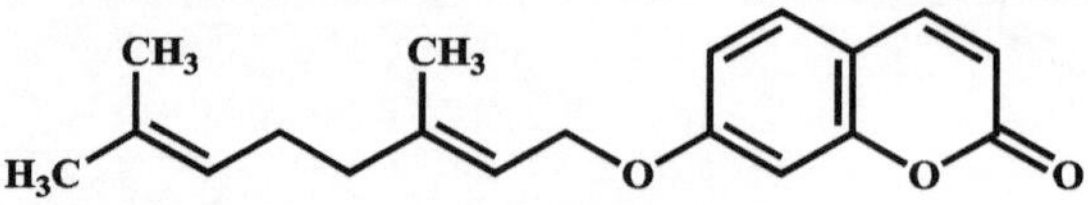<br>Auraptene | |

4) Pharmacology and bioactivities:

A) In vitro and in vivo:
A study including in vivo and in vitro results suggested that Zhi-Qiao has an antidepressant effect. Aqueous extract of *Citrus aurantium* L. significantly reduced the mice immobility time during the FST and the TST. MTT assays showed both aqueous extract and its major constituents (naringin, hesperidin, neohesperidin, and nobiletin) had neuroprotective effect on corticosterone-induced neurotoxicity in PC12 cells. Its four major constituents had the protection in a dose-dependent manner; the aqueous extract also showed protection at concentrations of 5 and 10 mg/L but had cytotoxicity at high concentrations (50, 100 mg/L) (Wu et al. 2015).

5) Indications and usage:
Promote qi circulation, alleviate middle energizer, and relieve distention. Treat food stagnation in the gastrointestine, diarrhea and dysentery of damp-heat type, chest impediment, thoracic accumulation, and chest and hypochondrium's pain due to qi stagnation (Chinese Pharmacopoeia Commission 2010).

6) Precautions and side effects:
Used with caution for pregnant women (Chinese Pharmacopoeia Commission 2010).

7) Dosage:

A) Internal use:
3–10 g/day, decocted in water for oral administration, taken twice a day, generally for 5–7 days, modificated according to symptoms (Chinese Pharmacopoeia Commission 2010).

B) External use:
Not found.

## 7.20 References

Flora of China Editorial Committee. Flora of China. Beijing and St. Louis: Science Press and Missouri Botanical Garden Press. 2008;11: 91–6.

Yan RN, Li F, Cai D. Determination of trace elements in Zhiqiao before and after processing. Guangdong Trace Element Sci. 1994;1(5): 50–1.

Liao F.X, Xin L.T, Chen H. Comparative analysis of constituents of volatile oil from *Fructus aurantii* Immaturus and Fructus aurantii. J Chongqing Univ. 2004;27(5): 38–40.

Cristine HNB. Hemodynamic effects of ephedra-free weight -los supplements in humans. Am J Med. 2005;118(9): 998–1003.

Chen ZhX, Lin L. Study of coumarins in Radix orange. *Chinese Herbal Medicine*, 2004;27(8): 577–8.

Wu M, Zhang H, Zhou C, Jia H, Ma Z, Zou Z. Identification of the chemical constituents in aqueous extract of Zhi-Qiao and evaluation of its antidepressant effect. Molecules. 2015;20(4): 6925–40.

Chinese Pharmacopoeia Commission. Pharmacopoeia of People's Republic of China. Beijing: China Medical Science Press. 2010.

## 7.21 *Ginkgo biloba* L.

### *7.21.1* Ginkgo Folium

#### 7.21.1.1 (Ginkgoaceae)

1) Common names:
   Ginkgo leaf, Yin-xing-ye (Pinyin in Chinese).
2) Photograph of the species (Fig. 7.10)
3) Description of the species:

   A) Habitat and world distribution:
      Scattered in broad-leaved forests and valleys on acidic, well-drained, yellow loess (pH = 5–5.5); found at 300–1100 m. Perhaps native in northwest Zhejiang (Tianmushan); widely and long cultivated below 2000 m in Anhui, Fujian, Gansu, Guizhou, Henan, Hebei, Hubei, Jiangsu, Jiangxi, Shaanxi, Shandong, Shanxi, Sichuan, and Yunnan of China. North Korea, Japan, Europe, and the United States have also cultivated it in garden (Flora of China Editorial Committee 1999).
   B) Morphology of the plant:
      Trees up to 40 m tall; trunk up to 4 m diameter at breast height; bark light gray or grayish brown, longitudinally fissured mainly on old trees; crown conical initially, finally broadly ovoid; long branchlets pale brownish yellow initially, finally gray, internodes with (1-) 1.5–4 cm; short branchlets blackish gray, with dense, irregularly elliptic leaf scars; winter buds yellowish

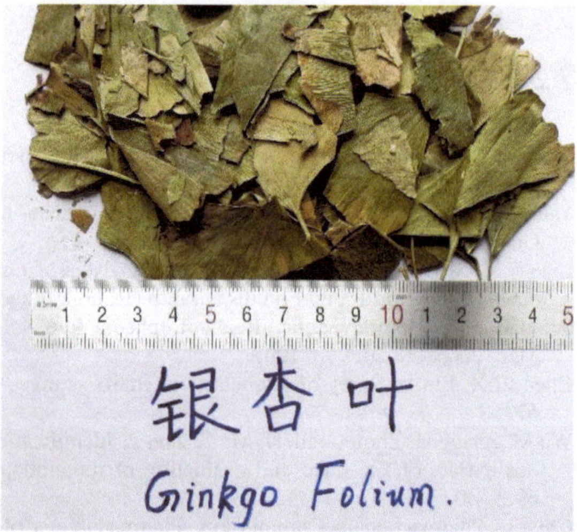


**Fig. 7.10** *Ginkgo Folium*

brown, ovate. Leaves with petiole with (3-)5–8(-10) cm; blade pale green, changing to bright yellow in autumn, with 13 × 8(-15) cm on young trees but typically with 5–8 cm wide, those on long branchlets divided by a deep, apical sinus into 2 lobes each further dissected, those on short branchlets with undulate distal and margin notched apex. Pollen cones ivory colored, 1.2–2.2 cm; pollen sacs with a boat shape, with widely gaping slit. Seeds elliptic, narrowly obovoid, ovoid, or subglobose, 2.5–3.5 × 1.6–2.2 cm; sarcotesta yellow, or orange-yellow glaucous, with rancid odor when ripen; sclerotesta white, with 2 or 3 longitudinal ridges; endotesta pale reddish brown. Pollination in March–April, seed maturity in September–October (Flora of China Editorial Committee 1999).

C) Medicinal parts:
The dry leaves of *Ginkgo biloba* L. has been used. Leaf extracts are available as supplements.

D) Chemical constituents:
**Terpene trilactones** (ginkgolides A, B, C, and J, bilobalide); **flavonoids** and **proanthocyanidins** (kaempferol, quercetin, isorhamnetin, kaempferol-3-*O*-(6″-O-rhamnosyl)glucoside, kaempferol-3-*O*-(6″-*O*,2″-*O*-di-rhamnosyl)glucoside, quercetin-3-*O*-(6″-*O*-rhamnosyl) glucoside,quercetin-3-*O*-(6″-*O*,2″-*O*-di-rhamnosyl)glucoside, kaempferol-3-*O*-(2″-*O*-(6″-*O*-p-coumaroyl)-glucosyl)rhamnoside, quercetin-3-*O*-(2″-*O*-(6″-*O*-p-coumaroyl)-glucosyl)rhamnoside, amentoflavone, bilobetin, 5′-methoxybilobetin, sequojaflavone, ginkgetin, isoginkgetin and sciadopitysin, procyanidin, prodelphinidin); **alkylphenols** (ginkgolic acids such as 6-alkyl salicylates, ginkgols, and bilobols), **carboxylic acids**; **polyprenols** (Smith and Luo 2004; van Beek 2002).

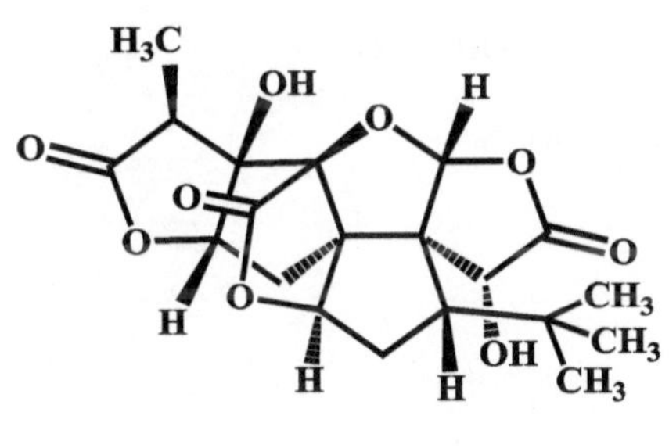

Ginkgolide A

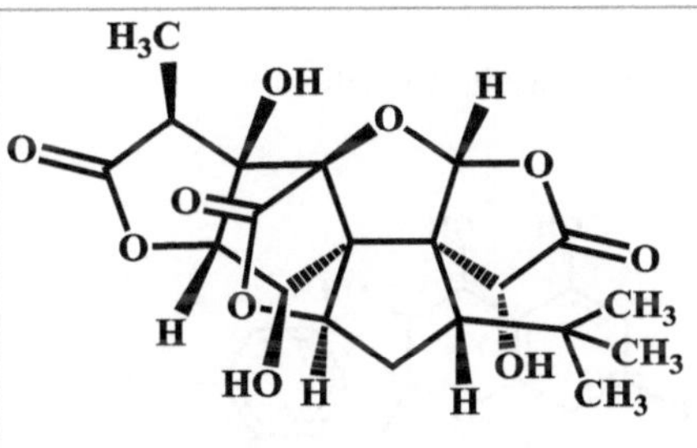

Ginkgolide B

| | |
|---|---|
| 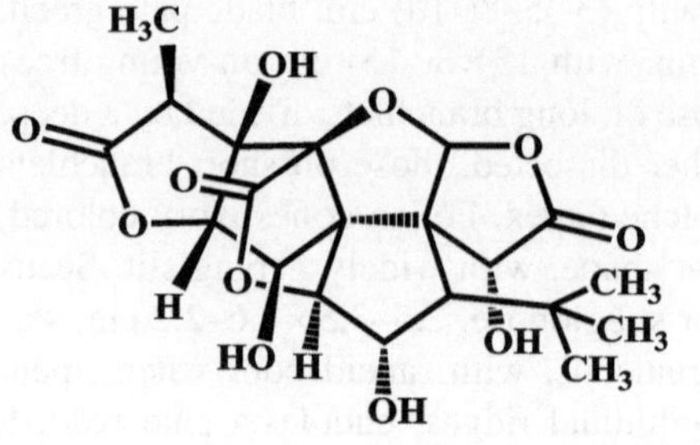 Ginkgolide C | 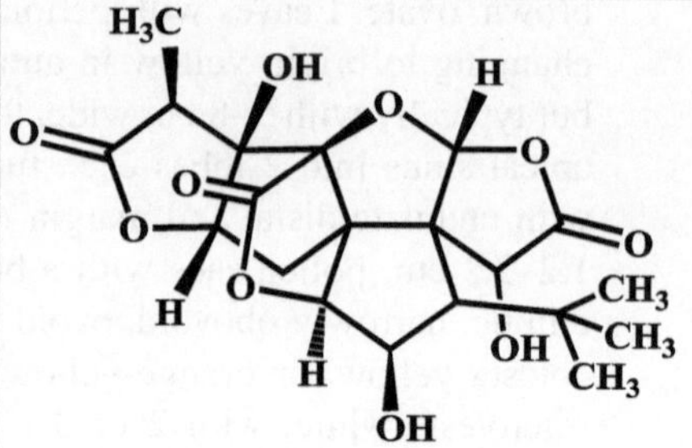 Ginkgolide J |
| 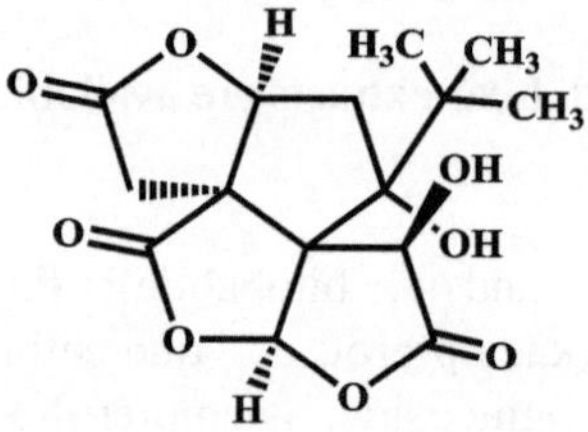 Bilobalide | 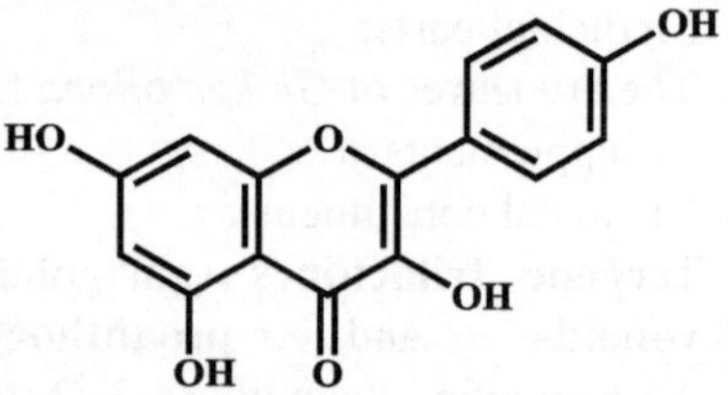 Kaempferol and its derivatives |
| 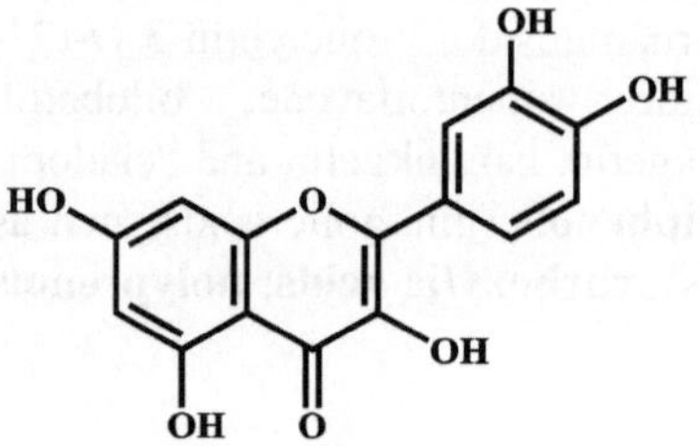 Quercetin and its derivatives | 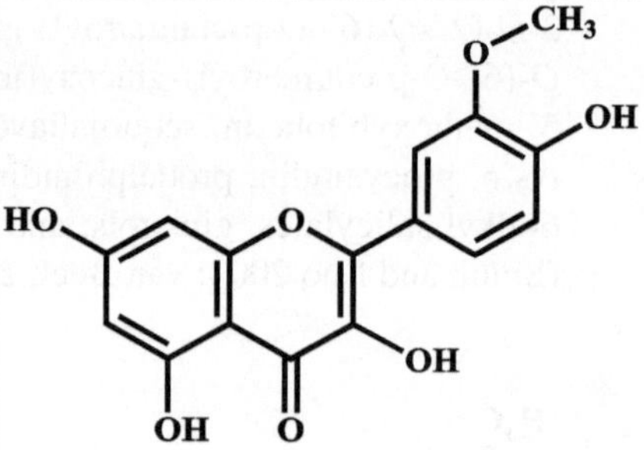 Isorhamnetin |
| 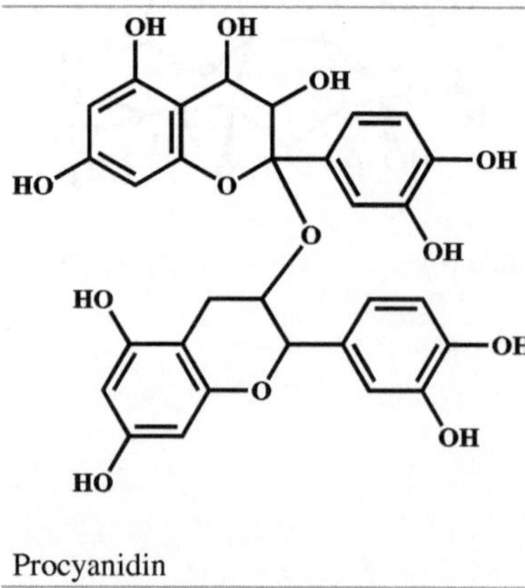 Procyanidin | |

4) Pharmacology and bioactivities:

   A) In vitro studies: Not found.
   B) In vivo studies:
      6-Alkyl salicylates are the bioactive components of the lipophilic extract of *Ginkgo biloba* leaves. Lipophilic extracts of *G. biloba* leaves at 50 and 100 mg/kg (p.o.) exhibited significant dose-dependent antidepressant activity as reflected by a reduction in the immobility time in FST and a reduced number of failures to escape from shock chamber to safe chamber (escape failure) compared with vehicle-treated animals in learned helplessness model of depression (Kalkunte et al. 2007). In another study, rats that were administered lipopolysaccharide (LPS) exhibited significant decreases in sucrose intake, food consumption, body weight, and DA contents in the nucleus accumbens. The result demonstrated that the extract of *G. biloba* leaves EGB 761 reversed depressive-like symptoms in rats which might be associated with its MAO-B inhibitor property. The study suggested that EGb 761 may provide a beneficial role in the treatment of anhedonia depressive-like behavior (Yeh et al. 2015).

5) Indications and usage:
   Activate blood and dissolve stasis, activate the meridians and relieve pain, astringe the lung for relieving cough, resolve turbidity, and reduce lipid. Treat obstruction of collaterals by blood stasis, choking sensation in chest and cardiac pain, apoplectic hemiplegia, cough with phlegm and asthma, and hyperlipidemia (Chinese Pharmacopoeia Commission 2010).
6) Precautions and side effects:
   Occasional adverse effects with excessive consumption of *G. biloba* leaves extract have been reported which include gastrointestinal disturbances, headaches, dizziness, excessive bleeding, allergic skin reactions, and occasional anaphylaxis-like reactions (only with intravenous administration) (De Smet 2002; Benjamin et al. 2001). The other components of *G. biloba* leaves extract are the ginkgolic acids (alkyl phenols), which are considered to be toxic. They are bilobol, cardanols, cardols, and ginkgol and are known to cause gastrointestinal and allergic reactions (McKenna et al. 2001).
7) Dosage:

   A) Internal use:
      9–12 g/day, decocted in water for oral administration or made into pill and powder, taken 2–3 times a day, generally for 5–7 days, modificated according to symptoms (Chinese Pharmacopoeia Commission 2010).
   B) External use:
      Proper dosage is for external application, pounded for applying or decocted for washing.

## 7.22 References

Flora of China Editorial Committee. (1999). Flora of China. Beijing and St. Louis: Science Press and Missouri Botanical Garden Press, Vol. 4, 8.
Smith JV, Luo Y. Studies on molecular mechanisms of *Ginkgo biloba* extract. Appl Microbiol Biotechnol. 2004;64(4): 465–72.
van Beek TA. Chemical analysis of *Ginkgo biloba* leaves and extracts. J Chromatograph A. 2002;967: 21–55.
Kalkunte SS, Singh AP, Chaves FC, Gianfagna TJ, Pundir VS, Jaiswal AK, Vorsa N, Sharma S. Antidepressant and antistress activity of GC-MS characterized lipophilic extracts of *Ginkgo biloba* leaves. Phytother Res. 2007;21(11): 1061–5.
Yeh KY, Shou SS, Lin YX, Chen CC, Chiang CY, Yeh CY. Effect of *Ginkgo biloba* extract on lipopolysaccharide-induced anhedonic depressive-like behavior in male rats. Phytother Res. 2015;29(2): 260–6.
Chinese Pharmacopoeia Commission. Pharmacopoeia of People's Republic of China. Beijing: China Medical Science Press. 2010.
De Smet PA. Herbal remedies. New Engl J Med. 2002;347(25): 2046–56.
Benjamin J, Muir T, Briggs K, Pentland B. A case of cerebral haemorrhage-can Ginkgo biloba be implicated. Postgrad Med J. 2001;77(904): 112–3.
McKenna DJ, Jones K, Hughes K. Efficacy, safety, and use of ginkgo biloba in clinical and preclinical applications. Alternat Ther Health Med. 2001;7(5): 70–86, 88–90.

## 7.23 *Panax ginseng* C. A. Mey.

### *7.23.1* Ginseng Radix et Rhizoma

#### 7.23.1.1 (Araliaceae)

1) Common names:
   *Panax ginseng*, Ren Shen (Pinyin in Chinese).
2) Photograph of the species (Fig. 7.11)
3) Description of the species:
   A) Habitat and world distribution:
      Mixed forests, deciduous broad-leaved forests. East of Heilongjiang, East of Jilin, East of Liaoning; recently cultivated in Hebei and Shanxi [other countries: Korea, East of Russia] (Flora of China Editorial Committee 2007).
   B) Morphology of the plant:
      Perennial herb, 30–60 cm tall. Rootstock usually with 1- or 2-fascicled roots, fusiform or cylindric. Leaves 3–6, verticillate at apex of stem, palmately compound; petiole base without stipule or stipule-like appendages; leaflets 3–5, membranous, abaxially glabrous, adaxially sparsely setose (trichomes ca. 1 mm), base broadly cuneate, margin densely serrulate, apex long acuminate; central leaflet elliptic to oblong-elliptic, 8–12 × 3–5 cm; lateral leaflets ovate to rhombic-ovate, 2–4 × 1.5–3 cm. Inflorescence, terminal solitary umbel with 30–50 flowers; peduncle with 15–30 cm, normally longer than petiole; pedicels with 0.8–1.5 cm. Ovary bicarpellate; styles 2, distinct. Fruit red, compressed-globose, with 4–5 × 6–7 mm; seeds nephroid, white (Flora of China Editorial Committee 2007).

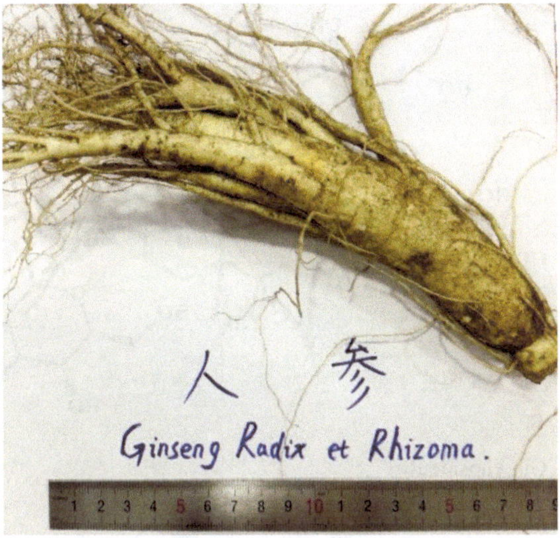


**Fig. 7.11** *Ginseng Radix et Rhizoma*

C) Medicinal parts:
The dry root and rhizome of the perennial herbaceous plant *Panax ginseng* C. A. Mey. are used.

D) Chemical constituents:
The bioactive ingredients in ginseng root include more than 60 ginsenosides (**triterpenoid saponins**), e.g., Rf, Rb1, Rb2, Rb3, Rc, Rd, Re, Rg1, Rg2, and Rg3, as well as **polysaccharides**, **fatty acids**, **oligopeptides,** and **polyacetylenic alcohols** (Qi et al. 2010).

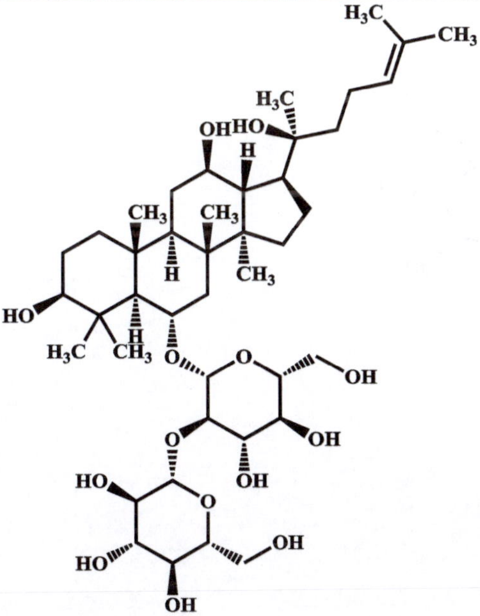

Ginsenoside Rf

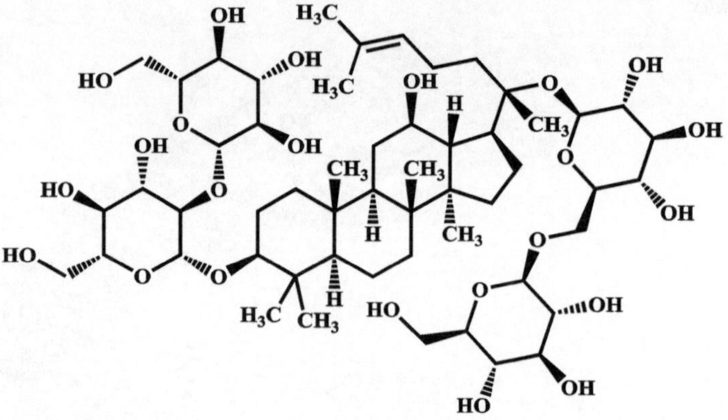

Ginsenoside Rb1

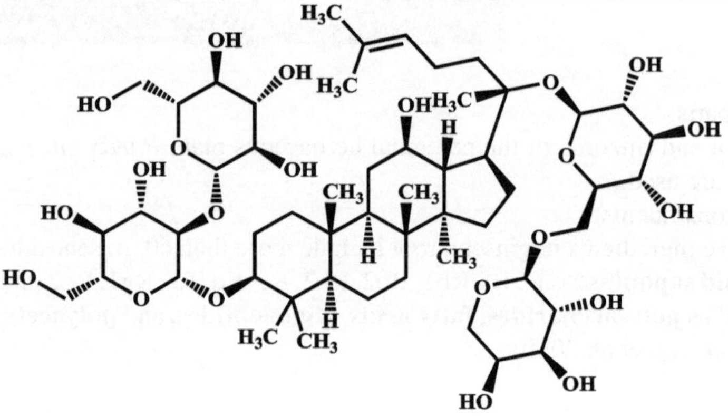

Ginsenoside Rb2

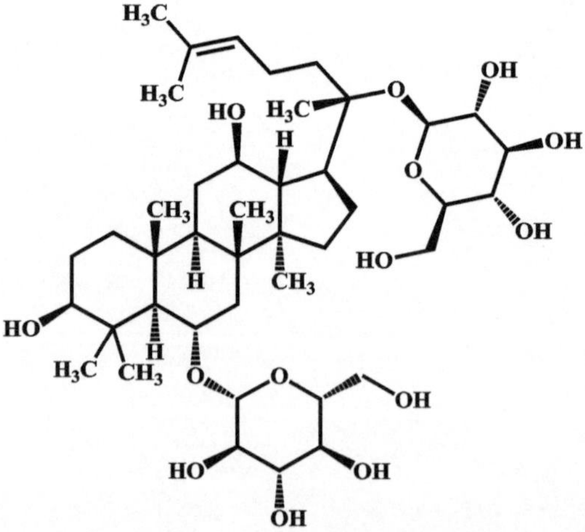

Ginsenoside Rg1

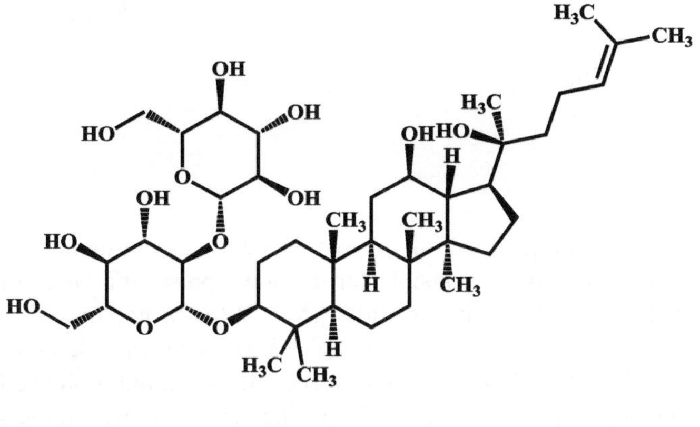

Ginsenoside Rg3

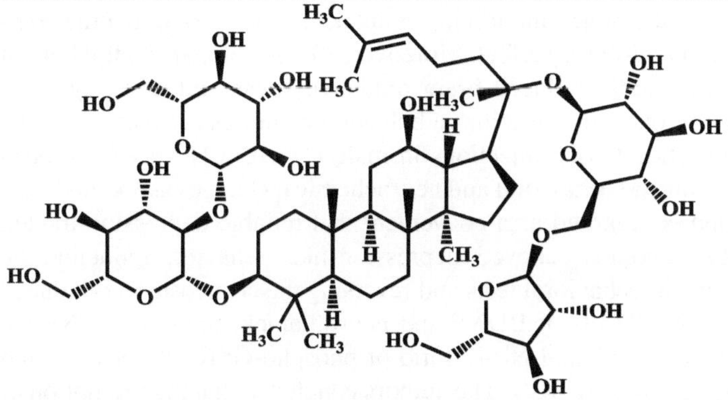

Ginsenoside Rc

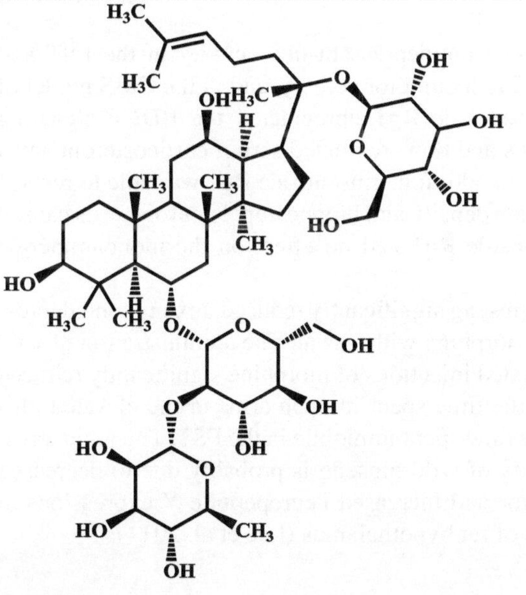

Ginsenoside Re

4) Pharmacology and bioactivities:

A) In vitro studies:
Not found

B) In vivo studies:
The metabolite 20(S)-protopanaxadiol, a post metabolism compound of ingested ginsenosides, may be responsible for the antidepressant activity of ginseng. Antidepressant-like activity of orally given 20(S)-protopanaxadiol was measured in animal tests including TST, FST, and rat olfactory bulbectomy depression model. In all those tests, this compound showed antidepressant-like activity as potent as fluoxetine. Bulbectomy animals treated with 20(S)-protopanaxadiol had higher levels of monoamine neurotransmitters in the brain, and in vitro reuptake assay showed that the compound had a mild inhibitory effect. Moreover, 20(S)-protopanaxadiol but not fluoxetine significantly reduced brain oxidative stress and downregulated serum corticosterone concentration in bulbectomy animals (Xu et al. 2010).

Repeated corticosterone injections in male C57BL/6 N mice have been reported to mimic the behavioral and neurochemical changes associated with depression and is regarded as a convenient and reliable depression model. Ginseng total saponins relieved depression-like behavior, reducing the immobility time in behavioral tests and reversing the corticosterone-induced changes in mRNA levels of BDNF and neurofilament light chain (NF-L), protein levels of BDNF and NF-L, ratio of phospho-GSK-3β/GSK-3β, and ratio of phospho-CREB/CREB. The authors concluded that the promotion of GSK-3β inhibitory phosphorylation which activates the CREB-BDNF signaling pathway may account for the antidepressant-like activity of ginseng total saponins (Chen et al. 2014).

Ginsenoside Rg1 exhibited antidepressant-like activity in the FST and TST in mice, without affecting locomotor activity, and in the CMS model of depression. Moreover, ginsenoside Rg1 upregulated the BDNF signaling pathway in the hippocampus and downregulated serum corticosterone level during the CMS procedure. In addition, ginsenoside Rg1 was able to reverse the decrease in dendritic spine density and hippocampal neurogenesis caused by CMS. However, ginsenoside Rg1 had no effect on the monoaminergic system (Jiang et al. 2012).

Administration of wild ginseng significantly reduced anxiety- and depression-like behaviors during morphine withdrawal. The administration of wild ginseng to rats prior to repeated injections of morphine significantly reduced the number of entries and the time spent in open arms in the elevated plus maze test and decreased the time spent immobile in the FST. The antidepressant and anti-anxiety activity of wild ginseng is probably due to decreased corticotropin-releasing factor and increased neuropeptide Y expressions in the paraventricular nucleus of rat hypothalamus (Lee et al. 2011).

5) Indications and usage:
Powerfully reinforce original qi, tonify zang-organ qi, promote the generation of body fluid to alleviate thirst and nourish blood, tranquilize the mind, and improve

the intelligence. Treat exhaustion syndrome due to qi deficiency, spleen qi and lung qi deficiency syndrome, both qi and yin deficiency syndrome, diabetes, blood deficiency syndrome, syndrome of both qi and blood deficiency, palpitation, insomnia, and amnesia. In addition, it can tonify kidney qi and can be indicated for impotence, seminal emission, and other diseases due to kidney qi deficiency (Chinese Pharmacopoeia Commission 2010).

6) Precautions and side effects:
   Antagonizing to Li-Lu (*Veratrum nigrum* L.) and Wu-Ling-Zhi (*Trogopterus Dung*); the warm nature of sun-dried raw Ren Shen is weaker than that of Hong Shen (*Red Ginseng*), which has been peeled, heated through steaming at a high temperature, and then dried or sun-dried (Chinese Pharmacopoeia Commission 2010).
7) Dosage:
   A) Internal use:
      5–10 g/day, decocted in water for oral administration, taken twice a day, generally for 5–7 days, modificated according to symptoms, 10–30 g can be used to rescue the exhaustion syndrome due to qi deficiency. Be smashed into powder, 0.5–1 g taken each time, twice a day. It can be decocted alone or mixed with other herbal tea by slow fire (Chinese Pharmacopoeia Commission 2010).
   B) External use:
      Not found.

## 7.24 References

Flora of China Editorial Committee. Flora of China. Beijing and St. Louis: Science Press and Missouri Botanical Garden Press. 2007;13: 489, 491.

Qi X, Ignatova S, Luo G, Liang Q, Jun FW, Wang Y, et al. Preparative isolation and purification of ginsenosides Rf, Re, Rd and Rb1 from the roots of *Panax ginseng* with a salt/containing solvent system and flow step-gradient by high performance counter-current chromatography coupled with an evaporative light scattering detector. J Chromatograph A. 2010;1217(13): 1995–2001.

Xu C, Teng J, Chen W, Ge Q, Yang Z, Yu C, et al. 20(S)-protopanaxadiol, an active *ginseng* metabolite, exhibits strong antidepressant-like effects in animal tests. *Prog Neuropsychopharmacol Biol Psychiatry*, 2010;34(8): 1402–11.

Chen L, Dai J, Wang Z, Zhang H, Huang Y, Zhao Y. The antidepressant effects of ginseng total saponins in male C57BL/6 N mice by enhancing hippocampal inhibitory phosphorylation of GSK-3beta. Phytother Res. 2014;28(7): 1102–06.

Jiang B, Xiong Z, Yang J, Wang W, Wang Y, Hu Z.L, et al. Antidepressant-like effects of ginsenoside Rg1 are due to activation of the BDNF signalling pathway and neurogenesis in the hippocampus. Br J Pharmacol. 2012;166(6): 1872–7.

Lee B, Kim H, Shim I, Lee H, Hahm DH. *Wild ginseng* attenuates anxiety- and depression-like behaviors during morphine withdrawal. J Microbiol Biotechnol. 2011;21(10): 1088–96.

Chinese Pharmacopoeia Commission. Pharmacopoeia of People's Republic of China. Beijing: China Medical Science Press. 2010.

## 7.25 *Hypericum perforatum* L.

### *7.25.1* Hyperici perforati herba

#### 7.25.1.1 (Guttiferae)

1) Common names:
   Common St. John wort herb with root, Eola weed, goat weed, Klamath weed, rosin rose, Tipton's weed, Guan-ye-lian-qiao, Guan-ye-jin-si-tao (Pinyin in Chinese).
2) Photograph of the species (Fig. 7.12)
3) Description of the species:
   A) Habitat and world distribution:
      Open woodlands, meadows, grasslands and steppes, riverbanks, stony and grassy slopes, roadsides, in dry or well-drained habitats; 100–2800 m. Gansu, Guizhou, Hebei, Henan, Hubei, Hunan, Jiangsu, Jiangxi, Shaanxi, Shandong, Shanxi, Sichuan, Xinjiang, and Yunnan [other countries: Northwest India, Kazakhstan, Kyrgyzstan, Mongolia, Russia, northwest Africa, Southwest to Central Asia, Atlantic islands, Europe (except extreme North); introduced into many parts of the world] (Flora of China Editorial Committee 2007).
   B) Morphology of the plant:
      Perennial herbs, 20–60(-100) cm tall, erect from creeping and rooting base; stems numerous to few, much branched especially distally. Stems 2-lined, with few black glands on lines. Leaves sessile to subsessile; blade more or

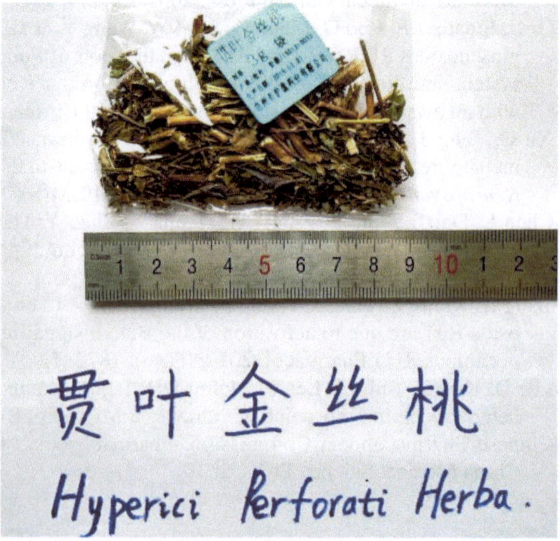


**Fig. 7.12** *Hyperici perforati herba*

less narrowly elliptic to more or less narrowly oblong or linear, with dimensions of 1-2.5 cm × 3–7(-15) mm; thickly papery, abaxially paler; laminar glands pale, scattered and sometimes black, few, dots; intramarginal glands black, spaced, interspersed with small dense pale ones; main lateral veins ca. 2-paired, tertiary reticulation lax or barely visible; base subcordate-amplexicaul to rather narrowly cuneate, margin entire, plane or more or less recurved, apex obtuse. Inflorescence with 3 to numerous flowers, from 1 to 3 nodes; flowering branches curved-ascending from up to 15 or occasionally more nodes below, the whole nearly flat-topped to broadly pyramidal or cylindric; bracts and bracteoles up to 4(-7) mm, narrowly lanceolate to linear, margin entire. Flowers with 1.5–2.5(-3) cm in diameter, stellate; buds narrowly ovoid, apex acute. Sepals free, with 3–4(-5) × 1–1.2 mm, erect in bud, recurved in fruit, narrowly oblong or lanceolate to linear, equal; laminar glands pale and often a few black, in 2(-4) rows, basal streaks to all dots; intramarginal glands absent or few, black; margin entire, apex acute to acuminate with tip somewhat glandular; veins 3(-5). Petals with golden yellow color, oblong to oblong-elliptic, (0.8-)1.2–1.5 cm × 5–6 mm, asymmetric; laminar glands black or pale, dots to lines, or frequently absent; intramarginal glands black or pale, distal, in sinuses when present; margin distally more or less crenate. Stamens 40–60, apparently tri-fascicled, longest with 6–8 mm. Ovary narrowly ovoid to ovoid-ellipsoid; styles 3, 4.5–6 mm, broadly to rather narrowly spreading. Capsule ovoid-conic to ovoid, 3–6.5 × 3–5 mm; valves with abaxial vittae and lateral yellowish, elongate or short vesicles. Seeds ca. 1 mm, dark brown; testa finely linear-foveolate. Flowering in June–September, fruiting in July–October (Flora of China Editorial Committee 2007).

C) Medicinal parts:
The dried aboveground part of *Hypericum perforatum* L. has been used.

D) Chemical constituents:
**Phloroglucinols** (hyperforin, adhyperforin) (Shan and Hu 2001); **naphtodianthrones** (hypericin, pseudohypericin, protohypericin, protopseudohypericin, cyclopseudohypericin, etc.) (Li et al. 2007); **flavonoids** (quercetin-3-*O*-galactoside, quercetin-3-*O*-rhamnoside, quercetin-3-*O*-glucoside, quercetin-3-*O*-rutinoside, etc.) (Li et al. 2007); **other compounds** (laurate, myristic acid, isovaleric acid, β-carotene, etc.) (Li et al. 2007).

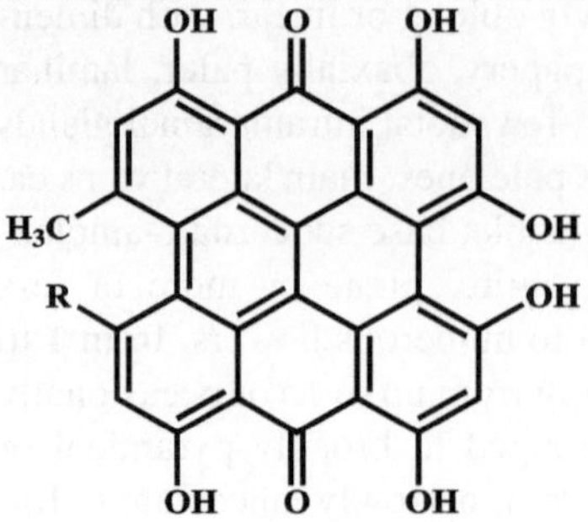

**Hypericin** **R = $CH_3$**

**Pseudohypericin R = $CH_2OH$**

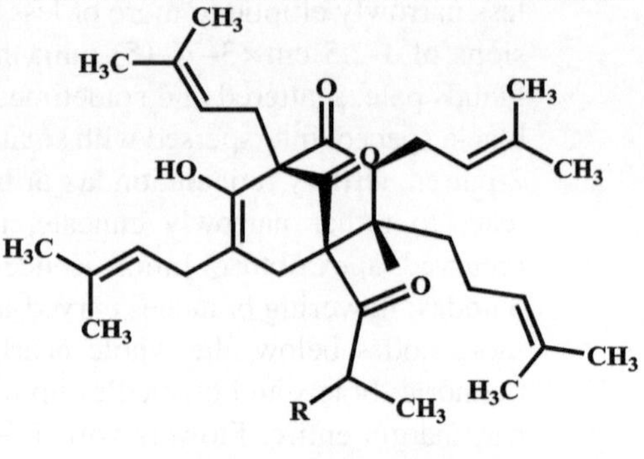

**Hyperforin R = $CH_3$**

**Adhyperforin R = $CH_2CH_3$**

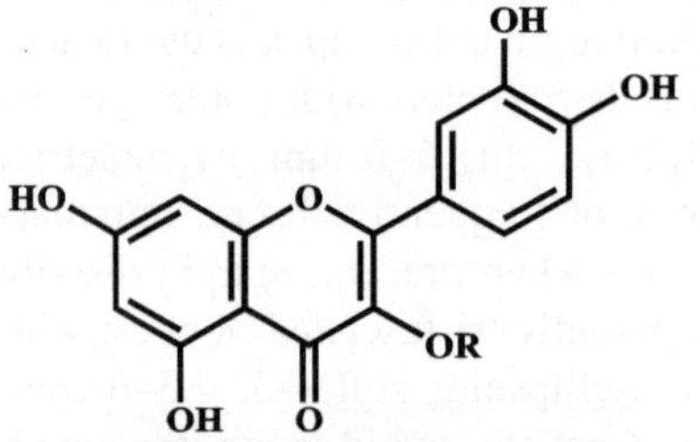

R = galactose: quercetin-3-*O*-galactoside
R = rhamnose: quercetin-3-*O*-rhamnoside
R = glucose: quercetin-3-*O*-glucoside
R = rutinose: quercetin-3-*O*-rutinoside

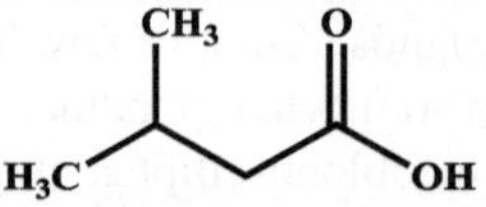

Isovaleric acid

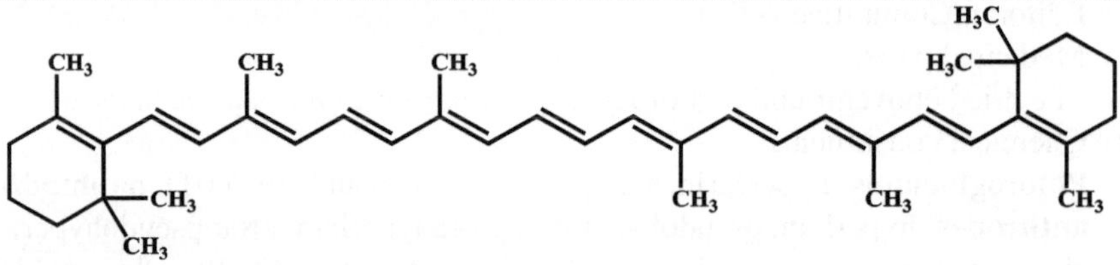

β-Carotene

4) Pharmacology and bioactivities:

A) In vitro studies:
Not found.

B) In vivo studies:
A result of a study indicates that *Hypericum extract* WS 5572 is an effective and well-tolerated drug for the treatment of mild to moderate major depressive disorder (Kalb et al. 2001). Another study provided evidence that *Hypericum* extract STW 3-VI in a once-daily dosing regimen may be an effective and well-tolerated option for patients with moderate depressive disorders (Uebelhack et al. (2004). *Hypericum* extracts are more effective than

placebo in people with major depression. They are similarly effective as standard antidepressants but with fewer side effects (Ernst 2009). A comparison of patterns of spontaneous adverse drug reaction reporting with St. John's wort and fluoxetine during the period 2000–2013 has been done. The result of the study demonstrated that herbal preparations can result in adverse drug reaction similar to those of prescription medications (Hoban et al. 2015).

5) Indications and usage:
Soothing liver qi stagnation, clearing heat and promoting diuresis, relieving swelling, and promoting lactation. Treat liver qi stagnation, gloomy and depressed mood, joint swelling and pain, acute mastitis and hypogalactia (Chinese Pharmacopoeia Commission 2010).
6) Precautions and side effects:
Similar to other commercially available SSRIs, such as fluoxetine (Hoban et al. 2015).
7) Dosage:
   A) Internal use:
   2–3 g/day, decocted in water for oral administration, taken twice a day, generally for 5–7 days, modificated according to symptoms (Chinese Pharmacopoeia Commission 2010).
   B) External use:
   Proper dosage is for external application, pounded for applying (Chinese Pharmacopoeia Commission 2010).

## 7.26 References

Flora of China Editorial Committee. Flora of China. Beijing and St. Louis: Science Press and Missouri Botanical Garden Press. 2007;13:5, 30, 31.

Shan MD, Hu LH. Three new hyperforin analogues from *Hypericun perforatum*. J Nat Product. 2001;64(1):127–34.

Li Y, Cao X. L, Fu P. Research progress on chemical constituents and extraction methods of *Hypericun perforatum* L. J Beijing Technol Business Univ Nat Sci Ed. 2007;25(6): 6–10.

Kalb R, Trautmann-Sponsel RD, Kieser M. Efficacy and tolerability of *hypericum* extract WS 5572 versus placebo in mildly to moderately depressed patients. A randomized double-blind multicenter clinical trial. Pharmacopsychiatry. 2001;34(3): 96–103.

Uebelhack R, Gruenwald J, Graubaum HJ, Busch R. Efficacy and tolerability of *Hypericum* extract STW 3-VI in patients with moderate depression: a double-blind, randomized, placebo-controlled clinical trial. Adv Ther. 2004;21(4): 265–75.

Ernst E. Review: *St John's wort* superior to placebo and similar to antidepressants for major depression but with fewer side effects. *Evid Based Ment Health*. 2009;12(3): 78.

Hoban CL, Byard RW, Musgrave IF. A comparison of patterns of spontaneous adverse drug reaction reporting with *St. John's Wort* and fluoxetine during the period 2000-2013. Clin Exp Pharmacol Physiol. 2015;42(7): 745–51.

Chinese Pharmacopoeia Commission. Pharmacopoeia of People's Republic of China. Beijing: China Medical Science Press. 2010.

## 7.27 *Nelumbo nucifera* Gaertner

### *7.27.1* Nelumbinis Semen

#### 7.27.1.1 (Nymphaeaceae)

1) Common names:
   Lotus seed, Lian-zi, Ou-shi, Shui-zhi-dan (Pinyin in Chinese).
2) Photograph of the species (Fig. 7.13)
3) Description of the species:
   A) Habitat and world distribution:
      Lakes, ponds, cultivated. Throughout China except Nei Mongol, Qinghai, and Xizang [other countries: Bhutan, India, Indonesia (Java), Japan, Korea, Malaysia, Myanmar, Nepal, New Guinea, Pakistan, Philippines, Russia (Far East), Sri Lanka, Thailand, Vietnam, Southwest Asia, Australia] (Flora of China Editorial Committee 2001).
   B) Morphology of the plant:
      Herbs perennial, aquatic. Petiole with 1-2 m, terete, fistulous, glabrous or papillae hard and scattered; leaf blade abaxially blue-green, orbicular, with 25–90 cm in diameter, papery, glabrous, glaucous, water-repellent, margin entire. Flowers with 10–23 cm in diameter; peduncles longer than petioles, glabrous or sparsely spinulate. Tepals, 5–10 × 3–5 cm, caducous, pink or white, oblong-elliptic to obovate. Stamens slightly longer than receptacle; filament slender; anther linear, 1–2 mm; connective appendage clavate, to

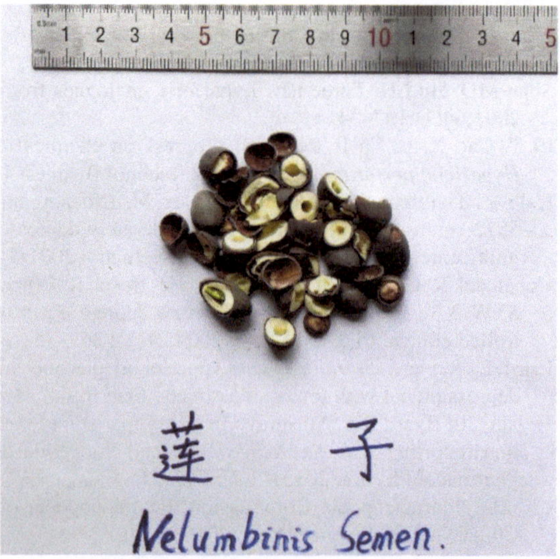


**Fig. 7.13** *Nelumbinis Semen*

7 mm, incurved. Receptacle accrescent, turbinate, 5–10 cm in diameter. Fruit oblong to ovoid, with 1.0–2.0×7–15 cm, glabrous; pericarp thick, hardened. Flowering in June–August (Flora of China Editorial Committee 2001).

C) Medicinal parts:
The ripe seeds of *Nelumbo nucifera* Gaertner has been used.

D) Chemical constituents:
**Alkaloids** (neferine, liensinine, isoliensinine, nelumborine A, nelumborine B, nelumboferine, (−)-roemerine, (−)-nuciferine, *N*-nornuciferine, *O*-nornuciferine, dauricine); **alkyl 4-hydroxybenzoates** (methyl-4-hydroxybenzoate, ethyl-4-hydroxybenzoate, propyl-4-hydroxybenzoate, and butyl-4-hydroxybenzoate) (Youn et al. 2010; Itoh et al. 2011).

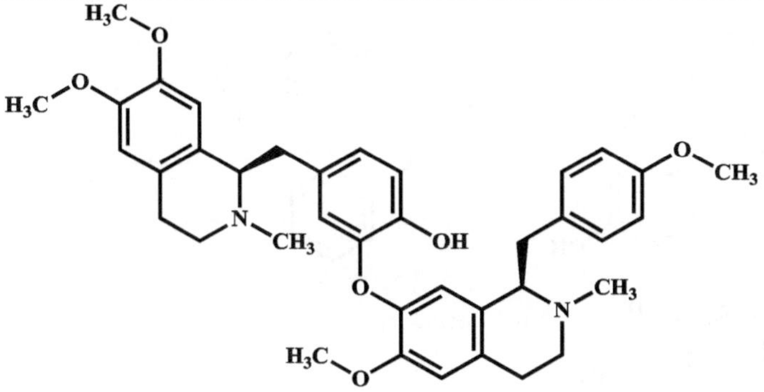


Neferine

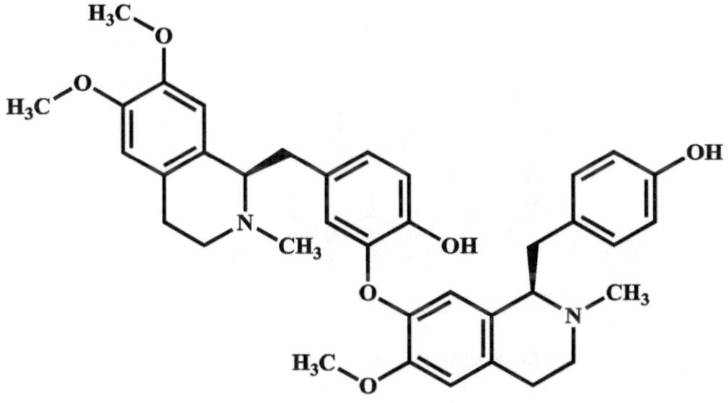


Liensinine

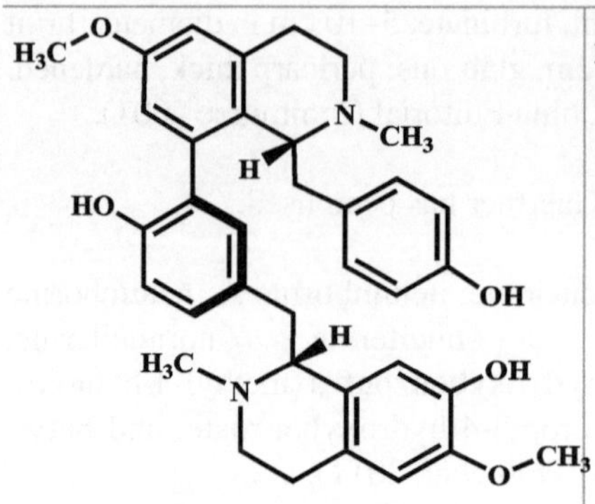

Nelumborine A

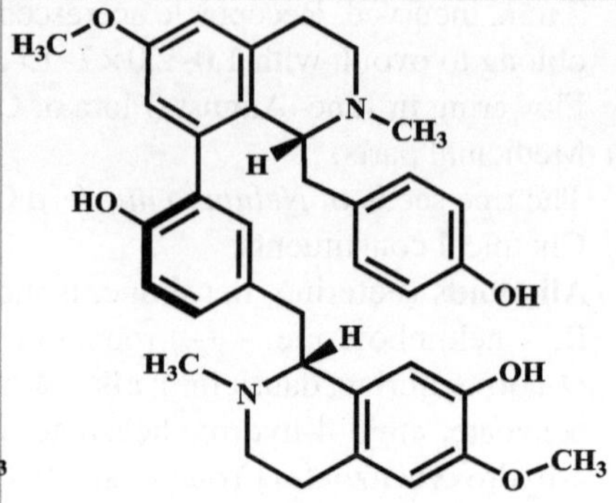

Nelumborine B

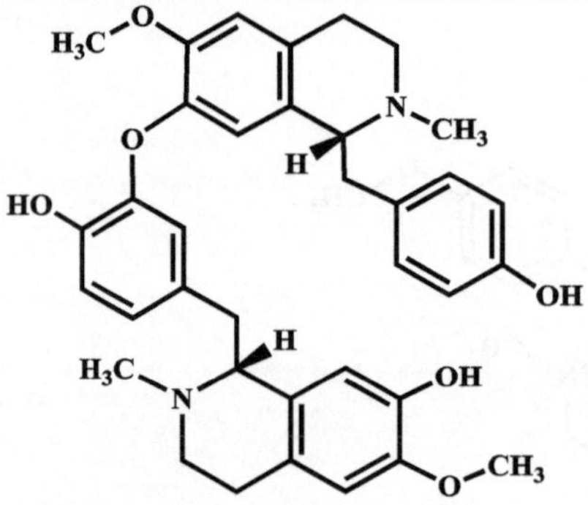

Nelumboferine

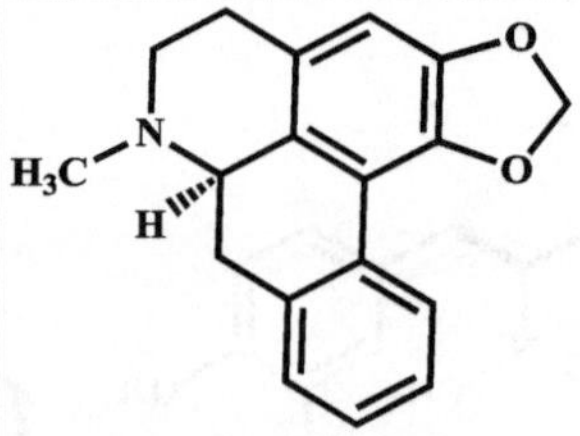

(−)-Roemerine

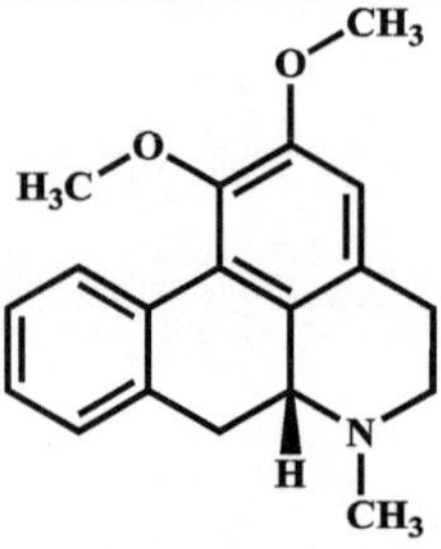

(−)-Nuciferine

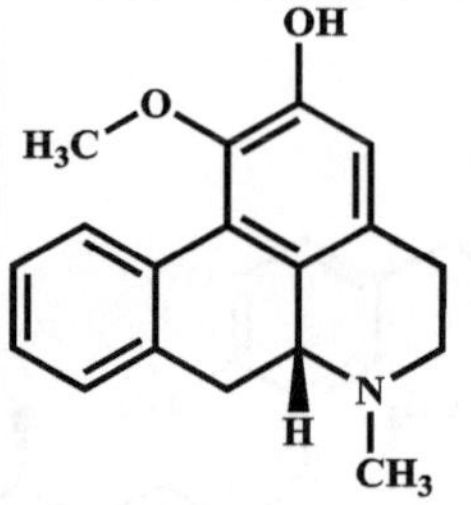

*O*-Nornuciferine

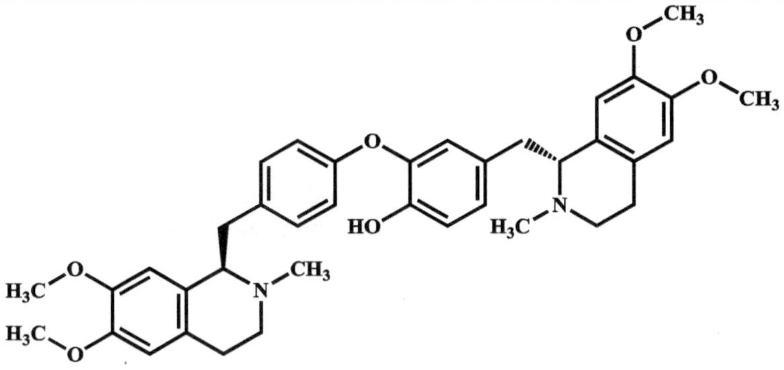


Dauricine

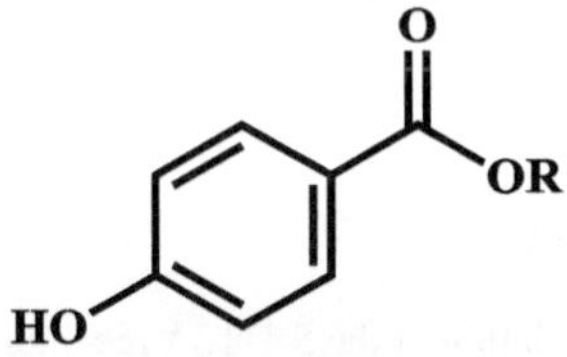


**Methyl-4-hydroxybenzoate: R = $CH_3$**
**Ethyl-4-hydroxybenzoate: R = $CH_2CH_3$**
**Propyl-4-hydroxybenzoate: R = $CH_2CH_2CH_3$**
**Butyl-4-hydroxybenzoate:R = $CH_2CH_2CH_2CH_3$**

4) Pharmacology and bioactivities:

A) In vitro studies:
Not found.

B) In vivo studies:
The effect of neferine on immobility in the FST was investigated in mice. The study implied neferine showed antidepressant-like effects and these effects are mediated by the $5\text{-HT}_1A$ receptor (Sugimoto et al. 2010).

In other study, the authors used a $5\text{-HT}_1A$ receptor binding assay, with a specific $5\text{-HT}_1A$ receptor agonist, 8-OH-DPAT (8-hydroxy-2-(di-n-propyl-amino) tetralin), to assess the mechanism of the antidepressant effect of *N. nucifera* on rats, and the effects were compared with two well-known antidepressants. *N. nucifera* treatment significantly reversed the decreased sucrose intake under chronic mild stress (CMS) as well as reversed the CMS-induced decrease in $5\text{-HT}_1A$ receptor binding in the frontal cortex, hippocampus, and hypothalamus of rat brains (Jang et al. 2004).

5) Indications and usage:
Tonify spleen to check diarrhea, tonify kidney and secure essence, stop leukorrhagia, nourish heart, and induce tranquilization. Treat diarrhea and dysentery due to spleen deficiency, seminal emission and spermatorrhea, leukorrhagia and palpitation, and insomnia (Chinese Pharmacopoeia Commission 2010).

6) Precautions and side effects:
   Not found.
7) Dosage:
   A) Internal use:
      10–15 g/day after being crushed, decocted in water for oral administration, taken twice a day, generally for 5–7 days, modificated according to symptoms. It is better to keep the seeds when treating restlessness due to disharmony between heart and kidney. It needs to be pounded and remove the core (Chinese Pharmacopoeia Commission 2010).
   B) External use:
      Not found.

## 7.28 References

Flora of China Editorial Committee. Flora of China. Beijing and St. Louis: Science Press and Missouri Botanical Garden Press, 2001; 6:116.

Youn UJ, Lee JH, Lee YJ, Nam JW, Bae H, Seo EK. Regulation of the 5-HT3A receptor-mediated current by alkyl 4-hydroxybenzoates isolated from the seeds of Nelumbo nucifera. Chem Biodivers. 2010;7(9): 2296–302.

Itoh A, Saitoh T, Tani K, Uchigaki M, Sugimoto Y, Yamada J, Nakajima H, Ohshiro H, Sun S, Tanahashi T. Bisbenzylisoquinoline Alkaloids from Nelumbo nucifera. Chem Pharmcal Bull (Tokyo). 2011;59(8): 947–51.

Sugimoto Y, Furutani S, Nishimura K, Itoh A, Tanahashi T, Nakajima H, Oshiro H, Sun S, Yamada J. Antidepressant-like effects of neferine in the forced swimming test involve the serotonin (1A) (5-$HT_1$A) receptor in mice. Eur J Pharmacol. 2010;634(1–3): 62–7.

Jang CG, Kang M, Cho JH, Lee SB, Kim H, Park S, Lee J, Park SK, Hong M, Shin MK, Shim IS, Bae H. Nelumbinis semen reverse's a decrease in 5-HTIA receptor binding induced by chronic mild stress, a depression-like symptom. Arch Pharm Res. 2004;27(10): 1065–72.

Chinese Pharmacopoeia Commission. Pharmacopoeia of People's Republic of China. Beijing: China Medical Science Press. 2010.

## 7.29 *Paeonia lactiflora* Pall.

### *7.29.1* Paeoniae Radix Alba

#### 7.29.1.1 (Ranunculaceae)

1) Common names:
   White peony root, Hangzhou white peony, Bai-shao (Pinyin in Chines).
2) Photograph of the species (Fig. 7.14)
3) Description of the species:
   A) Habitat and world distribution:
      Woods, grasslands; 400–2300 m. South of Gansu, Hebei, Heilongjiang, Jilin, Liaoning, Nei Mongol, South of Ningxia, Shaanxi, and Shanxi in

China, and Japan, Korea, Mongolia, and Russia (Far East, Siberia) (Flora of China Editorial Committee 2001).

B) Morphology of the plant:
Perennial herb, up to 70 cm tall. Roots thick, attenuate toward tip, up to 1.3 cm in diameter, stems glabrous. Proximal leaves biternate; all leaflets decurrent at base, terminal ones often bi- or tri-segmented; leaflets and segments up to 15, lanceolate or ovate lanceolate, 4.5–16 × 1.5–4.8 cm, occasionally pubescent abaxially, bristly along veins adaxially, margin white cartilaginous denticulate, apex acuminate. Flowers typically several per shoot, both terminal and axillary, sometimes only terminal one developed, single (in wild plants) or double (in cultivated plants), 8–13 cm wide. Bracts 4 or 5, lanceolate, unequal. Sepals 3 or 4, broadly ovate or suborbicular, with 1–1.5 × 1–1.7 cm. Petals 9–13, with 3.5–6 × 1.5–4.5 cm, white or pink (in wild plants), or varying in color (in cultivated plants), obovate. Filaments yellow, 0.7–1.2 cm; anthers yellow. Disc yellow, annular. Carpels 2–5, green or purple, glabrous or rarely tomentose. Follicles oblong-ellipsoid, 2.5–3 × 1.2–1.5 cm. Flowering in May–June (Flora of China Editorial Committee 2001).

C) Medicinal parts:
The dried root of *Paeonia lactiflora* Pall. has been used.

D) Chemical constituents:
**Monoterpenes and monoterpene glycosides** (paeoniflorin, oxypaeoniflorin, benzoylpaeoniflorin, albiflorin, albiflorin R1, benzoyloxypaeoniflorin, paeoniflorigenone, (Z)-(1S, 5R)-β-pinen-10-yl-β-vicianoside, lactiflorin, paeonilactones A, B, C, 6-*O*-β-D-glucosyl-lactinolide, paeonilactinone, 1-*O*-β-D-glucosyl-paeonisuffrone, lactinolide, 4-*O*-galloylalbiflorin, galloypaeoniflorin, 6-*O*-galloylalbiflorin) (Xiao-rong and Geng-yuan 2006; Jin et al. 2013); **triterpenoids** (palbinone, 11α,12α-epoxy-3β,23-dihydroxyolean-28,13β-olide,

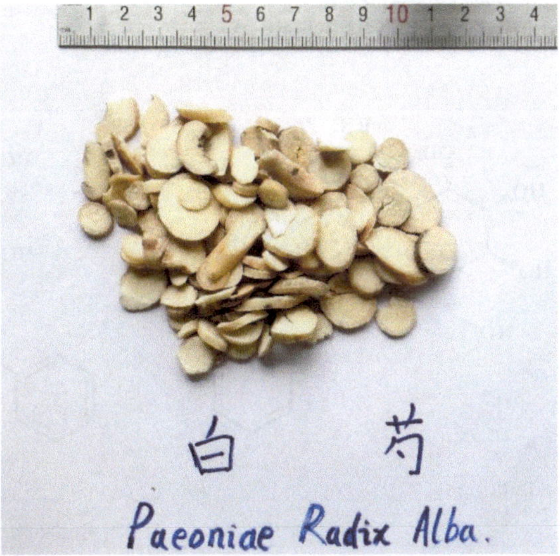


**Fig. 7.14** *Paeoniae Radix Alba*

3β-hydroxy-11α,12α-epoxy-olean-28-13β-olide, 3β-hydroxy-11-oxo-olean-12-en-28-oic acid, oleanolic acid, bederagenin, betulinic acid, 23-hydroxy-betulinic acid, 30-norhederagenin (Ikuta et al. 1995), 11α,12α-epoxy-3β, 23-dihydroxy-30-norolean-20(29)-en-28,13β-olide); **flavonoids** (kaempferol-3-*O*-glucoside, kaempferol-3,7-di-*O*-glucoside, catechin) (Kamiya et al. 1997); **volatile oil** (butyl alcohol, furfural, 3-hydroxypyridine, 2-furylmethyl ketone, benzaldehyde, 5-methyl-2-furfural, methyl benzoate, phenyl acetaldehyde, acetophenone, furfural alcohol, 4,7-dimethyl benzofuran,α,α–dimethylbenzyl alcohol, methyl salicylate, perilla alcohol, α-hydroxy acetophenone, ethyl salicylate, 2-methoxyphenol, benzyl alcohol, β-phenylethyl alcohol, 2,6-di-*tert*-butyl-4-methyl phenol, 2-acetyl pyrrole, phenol, o-cresol, nonyl phenol, *p*-cresol, *m*-cresol, *m*-*tert*-butyl phenol, *p*-methoxyacetophenone, thymol, carvacrol, benzoic acid, paeonol, 4-hydroxy-3-methoxy-acetophenone) (Mitsuo Miyazawa et al. 1984); **other compounds** (gallic acid, gallotannin, methylgallate, ethyl gallate, 6-O-galloyl-β-D-glucopyranose, pyrogallol, 1, 2, 3, 6-tetra-*O*-galloyl-β-D-glucose,1, 2, 3, 4, 6-pentagalloylglucose, di-(2-ethylhexyl) phthalate, sucrose, and β-sitosterol) (Xiao-rong and Geng-yuan 2006; Jin et al. 2013; Tan et al. 2010).

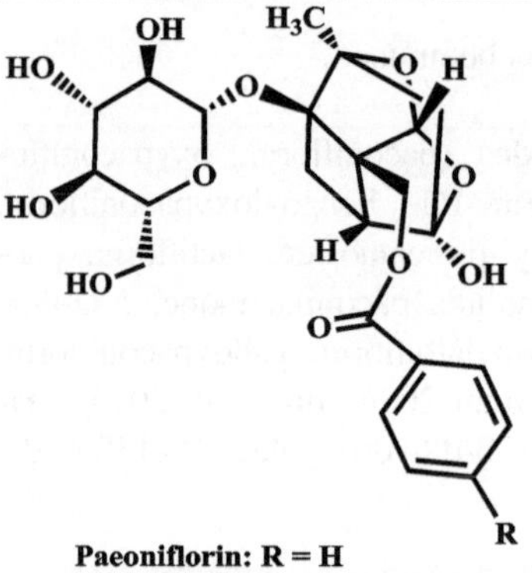

**Paeoniflorin: R = H**
**Oxypaeoniflorin: R = OH**

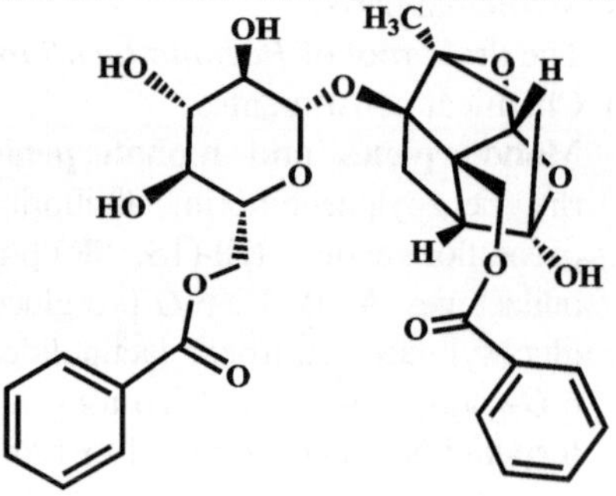

Benzoylpaeoniflorin

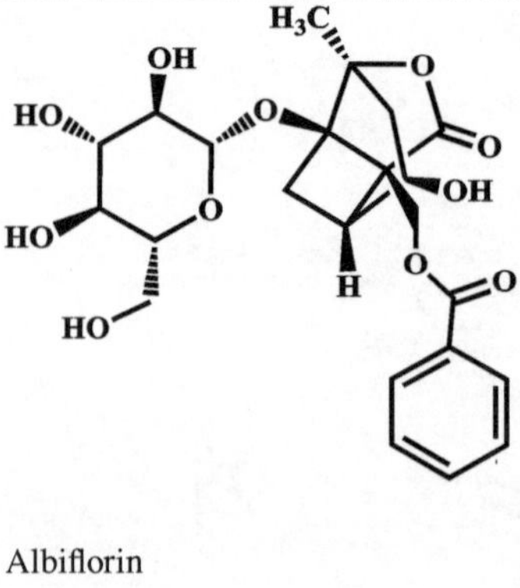

Albiflorin

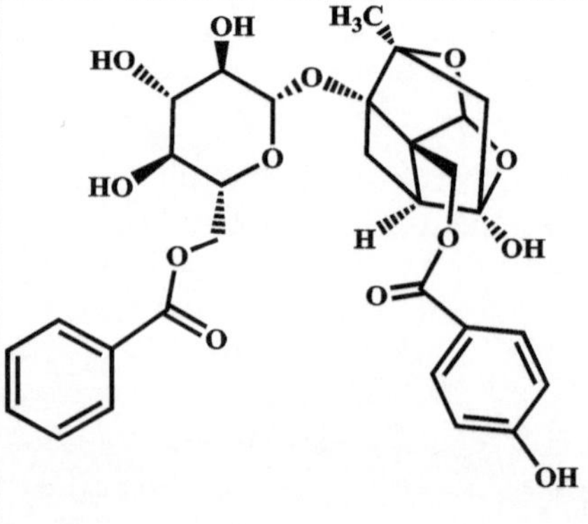

Benzoyloxypaeoniflorin

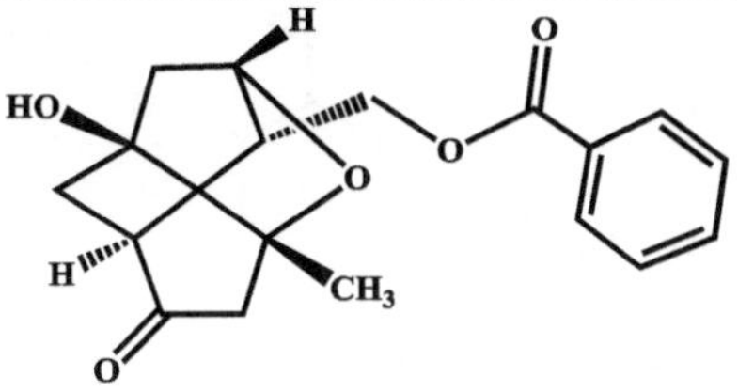

Paeoniflorigenone

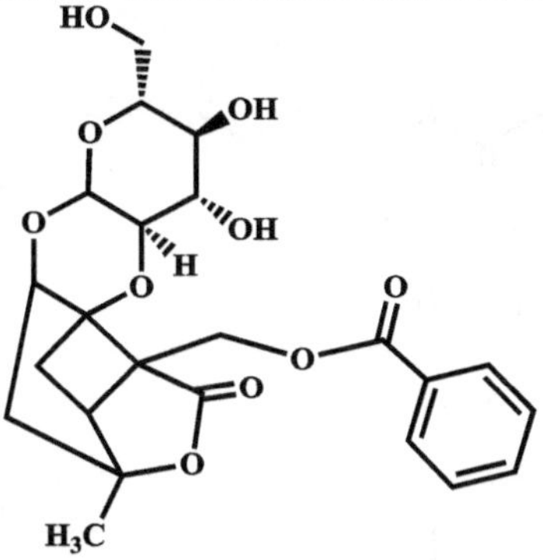

Lactiflorin

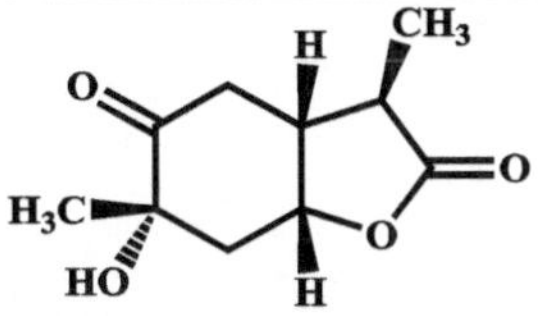

Paeonilactone A

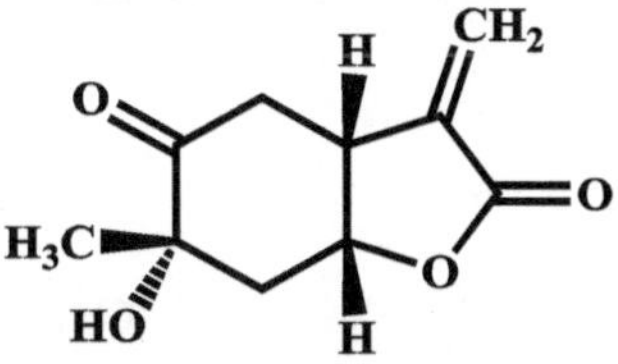

Paeonilactone B

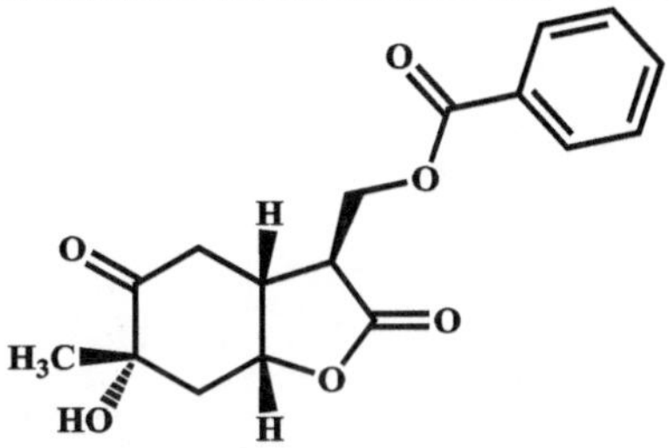

Paeonilactone C

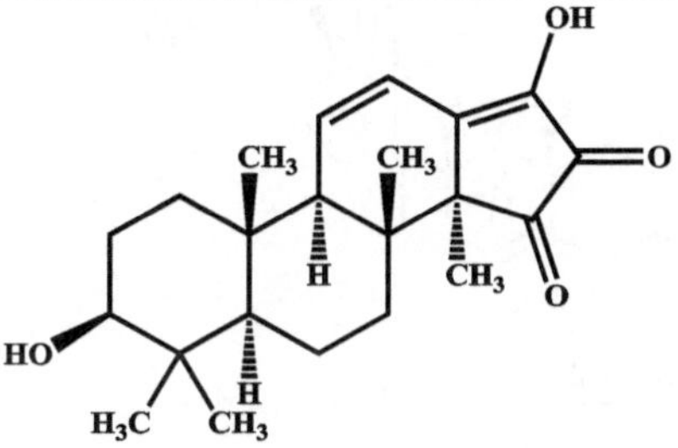

Palbinone

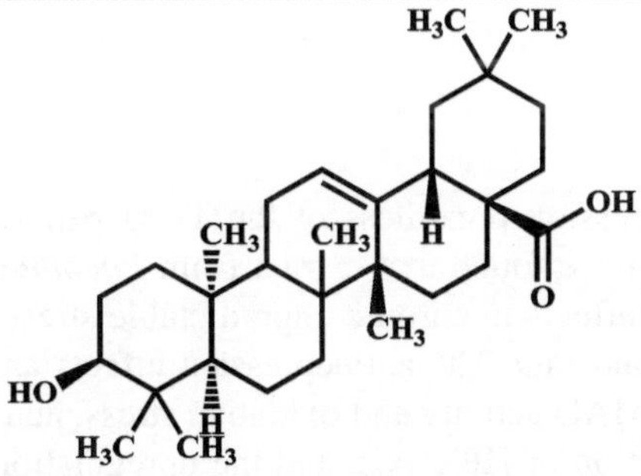

Oleanolic acid

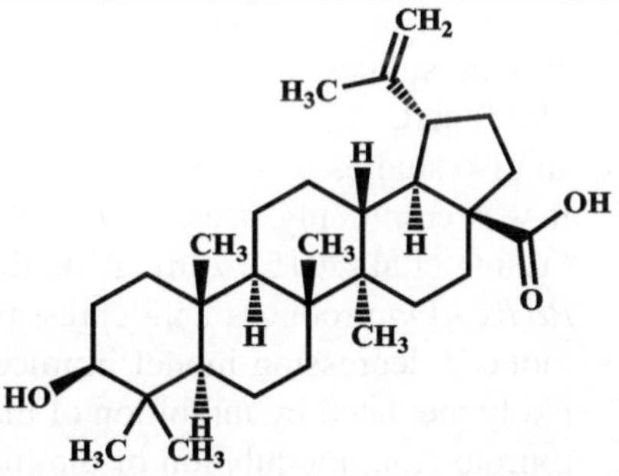

Betulinic acid

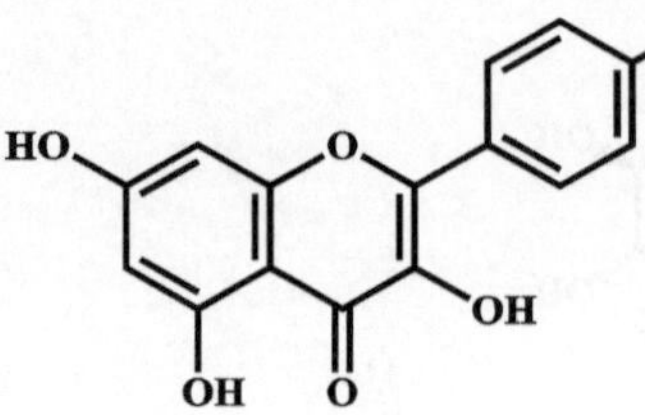

Kaempferol derivatives

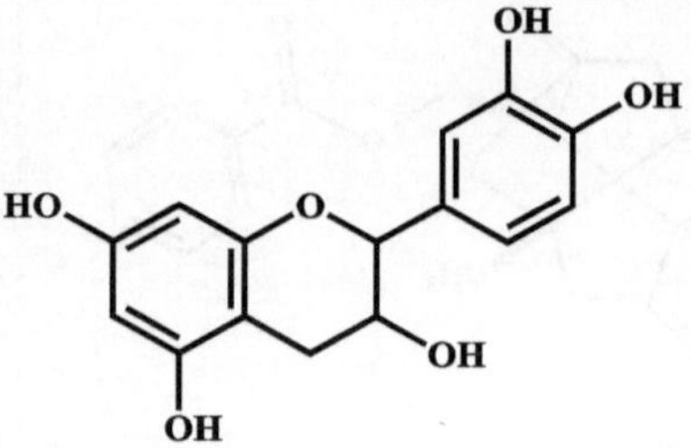

Catechin

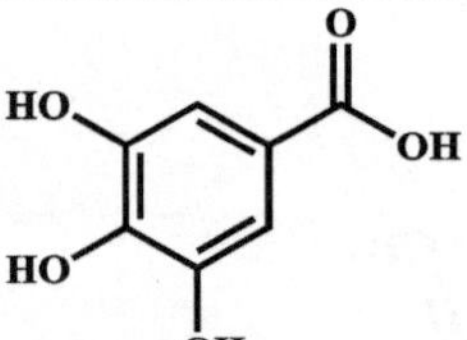

Gallic acid

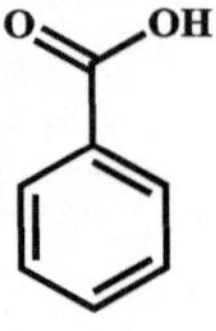

Benzoic acid

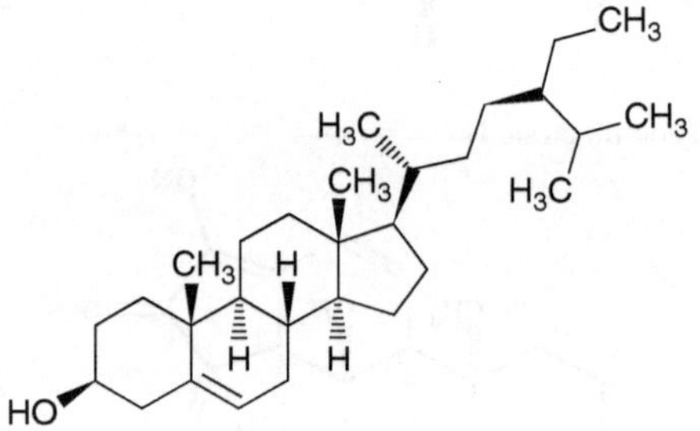

β-Sitosterol

4) Pharmacology and bioactivities:

A) In vitro studies:
Not found.

B) In vivo studies:
It was commonly used to treat depression regardless of the TCM pattern (Yeung et al. 2015). Animal studies have found that extract from *Paeoniae Radix Alba* produces antidepressant effects in chronic unpredictable stress-induced depression model in mice and rats. The antidepressant effects are likely mediated by inhibition of the MAO activity and oxidative stress, neuroprotection, modulation of the function of HPA axis, and the upregulation of neurotrophins (Mao et al. 2012; Mao et al. 2010). Studies have demonstrated that the total glycosides of peony exert antidepressant-like effects in animal models, with paeoniflorin being the main active glycoside of peony (Mao et al. 2008).

5) Indications and usage:
   Nourishing blood for regulating menstruation, astringing yin and suppressing sweating, softening liver to relieve pain, and calming liver yang. It can be used for blood deficiency chlorosis, irregular menstruation, spontaneous perspiration, night sweat, hypochondriac pain and abdominal pain, limbs spasm pain, headache, and dizziness (Chinese Pharmacopoeia Commission 2010).
6) Precautions and side effects:
   Should not be applied alone in yang deficiency syndrome. Should not be applied with Li-Lu (*Veratrum nigrum* L) (Zhang et al. 2014).
7) Dosage:
   A) Internal use:
      6–15 g/day, 15–30 g/day for large dosage, decocted in water for oral administration, taken twice a day, generally for 5–7 days, modificated according to symptoms. Raw material for pacifying liver and astringing Yin; stir-baked or stir-baked with alcohol for nourishing blood and regulating menstruation (Chinese Pharmacopoeia Commission 2010).
   B) External use:
      Not found.

## 7.30 References


Flora of China Editorial Committee. (2001). Flora of China. Beijing and St. Louis: Science Press and Missouri Botanical Garden Press, Vol. 6, 127.

Xiao-rong G, Geng-yuan T. Active principles of *Paeonia lactiflora* Pall. Chinese J New Drugs. 2006;15(6): 416–8.

Jin Y, Chen M. Tao J. Advances in studies on chemical constituents and pharmacological effects of *Paeonia lactiflora* Pall. Chinese J Pharmacol Toxicol. 2013;27(4): 745–50.

Ikuta A, Kamiya K, Satake T. Saiki Y. Triterpenoids from callus tissue cultures of Paeonia species. Phytochemistry 1995;38(5): 1203–7.

Kamiya K, Yoshioka K, Saiki Y, Ikuta A. Satake T. (1997). Triterpenoids and flavonoids from *Paeonia lactiflora. Phytochemistry*, 44(1), 141-144.

Mitsuo Miyazawa, Haruo Maruyama, Hiromu Kameoka. Essential oil constituents of "PAEONIAE RADIX" *Paeonia lactiflora* Pall. (P. albilora Pall.). Agric Biol Chem. 1984;48(11): 2847-2849.

Tan J, Zhao Q, Yang L, Shang Z, Du Z. Yan M. Chemical constituents in roots of *Paeonia lactiflora*. Chinese Trad Herb Drugs. 2010;41(08): 1245–8.

Yeung WF, Chung KF, Ng KY, Yu YM, Zhang SP, Ng BF, Ziea ET. Prescription of Chinese herbal medicine in pattern-based traditional Chinese medicine treatment for depression: a systematic review. Evid Based Complement Alternat Med. 2015;2015: 160189.

Mao QQ, Ip SP, Xian YF, Hu Z, Che CT. Anti-depressant-like effect of peony: a mini-review. Pharmaceut Biol. 2012;50(1): 72–7.

Mao QQ, Xian YF, Ip SP, Tsai SH. Che CT. Long-term treatment with peony glycosides reverses chronic unpredictable mild stress-induced depressive-like behavior via increasing expression of neurotrophins in rat brain. Behav Brain Res. 2010;210(2): 171–7.

Mao QQ, Ip SP, Tsai SH, Che CT. Antidepressant-like effect of peony glycosides in mice. J Ethnopharmacol. 2008;119(2): 272–5.

Chinese Pharmacopoeia Commission. Pharmacopoeia of People's Republic of China. Beijing: China Medical Science Press. 2010
Zhang X, Wang Y, Liang Q, Ma Z, Xiao C, Tan H, Gao Y. The correlation between chemical composition, as determined by UPLC-TOF-MS, and acute toxicity of *Veratrum nigrum* L. and *Radix paeoniae* alba. Evid Based Complement Alternat Med. 2014(2014). doi: 10.1155/2014/892797.

## 7.31 *Polygala tenuifolia Willd.*

### *7.31.1* Polygalae Radix

#### 7.31.1.1 (Polygalaceae)

1) Common names:
   Polygala root, Yuan-zhi (Pinyin in Chinese).
2) Photograph of the species (Fig. 7.15)
3) Description of the species:

   A) Habitat and world distribution:
      Prairies, grasslands, shrub forests, thickets on mountain slopes; (200-) 500–2300 m. Gansu, Hebei, Heilongjiang, Henan, Jiangsu, Jiangxi, Liaoning, Nei Mongol, Ningxia, Qinghai, Shaanxi, Shanxi, and Sichuan [other countries: Korea, Mongolia, Russia] (Flora of China Editorial Committee 2008).
   B) Morphology of the plant:
      Perennial herb, with 15–50 cm tall. Principal root up to 10 cm, stout, base fleshy. Stems much branched, cespitose, erect or inclined, ridged and sul-

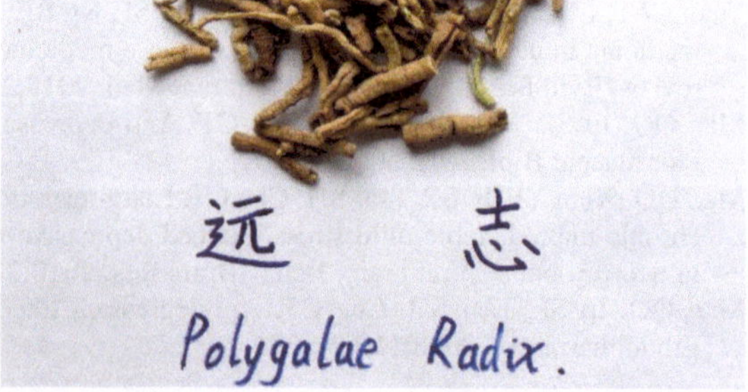


**Fig. 7.15** *Polygalae Radix*

cate, pubescent. Leaves subsessile; leaf blade with 10–30×0.5–1(-3) mm, linear to linear-lanceolate, papery, glabrous or very sporadically puberulent, midvein raised abaxially, impressed adaxially, lateral veins obscure, base cuneate, margin entire, recurved, apex acuminate. Racemes terminal, secund, 5–7 cm, slender, becoming slightly pendulous, sparsely few flowered; bracts 3, ca. 1 mm, caducous, lanceolate, apex acuminate. Sepals 5, persistent, glabrous; outer sepals 3, linear-lanceolate, ca. 2.5 mm, apex acute; inner sepals 2, petaloid, obovate or elliptic, with ca. 5×2.5 mm, midvein green, base unguiculate, margin membranous, violet, apex rounded, mucronate. Petals 3, connate in lower 1/3, purple; lateral petals obliquely oblong, ca. 4 mm, inside hairy at base; keel longer than lateral petals, apex with fimbriate appendages. Stamens 8; filaments forming an open staminal ciliate sheath split in upper 1/4 with 2 separate single filaments with anthers in middle, and on each side of these a united group of 3 filaments each carrying 3 sessile oblong anthers at tip. Ovary compressed-orbicular, apex retuse; style curved, apex trumpet-shaped; stigma in trumpet. Capsule orbicular, with ca. 4 mm in diameter, narrowly winged, eciliate, apex retuse. Seeds black, ovoid, ca. 2 mm in diameter, densely pubescent, with 2-lobed decurrent strophiole. Flowering and fruiting in May–September (Flora of China Editorial Committee 2008).

C) Medicinal parts:
The dry root of *Polygala tenuifolia* Willd. or *Polygala sibirica* L. has been used (Flora of China Editorial Committee 2008).

D) Chemical constituents:
**Xanthones** (lancerin, onjixanthones I and II, 1,3,7-trihydroxy-2-methoxyxanthone, 6-hydroxy-1,2,3,7-tetramethoxyxanthone, 1,2,3,6,7-pentamethoxyxanthone); **saccharide esters** (sibiricose A7, polygalatenoside A, polygalatenoside C, tenuifoliside C, 3′-*O*-3,4,5-trimethoxycinnamoyl-6-*O*-4-methoxybenzoyl sucrose); **flavonoids** (5,3′-dihydroxy-7,4′-dimethoxyflavonol-3-O-glucoside, rhamnetin-3-*O*-glucoside); **triterpenoids** (presenegenin, tenuifolin); **phenolic acid** (3,4,5-trimethoxycinnamic acid); **benzophenone glycoside** (sibiriphenone A) (Zhou et al. 2014).

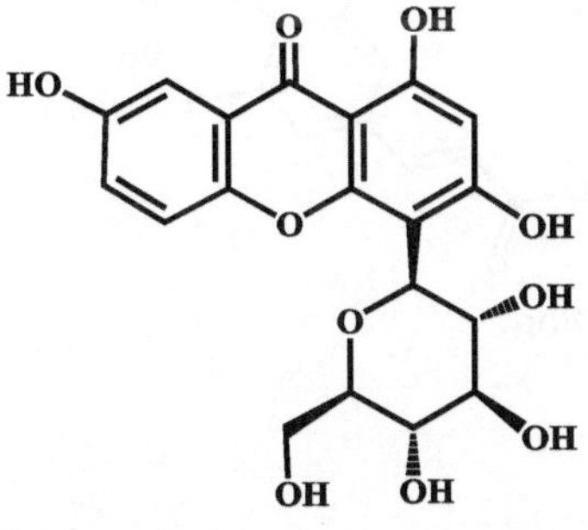

Lancerin

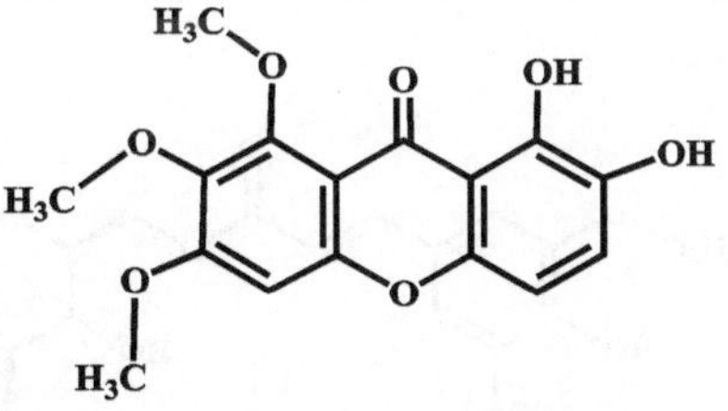

Onjixanthone I

| | |
|---|---|
| 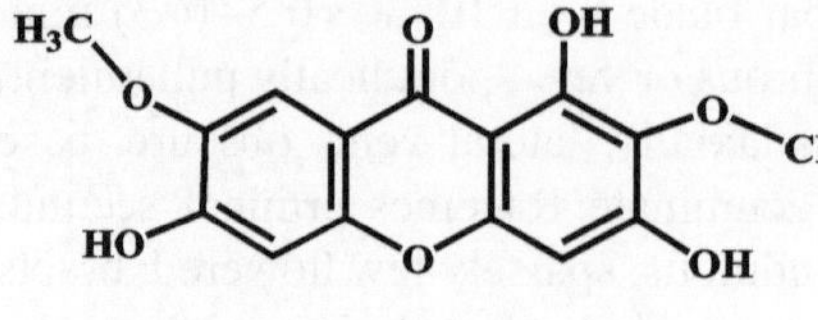 Onjixanthone II | 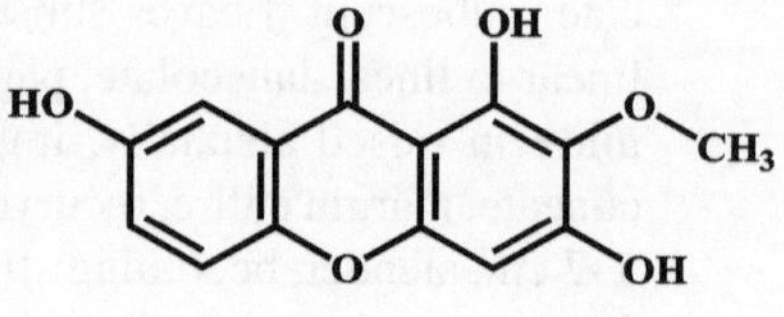 1,3,7-Trihydroxy-2-methoxyxanthone |
| 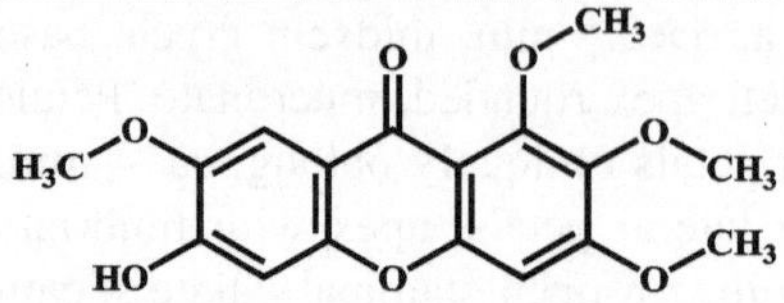 6-Hydroxy-1,2,3,7-tetramethoxyxanthone | 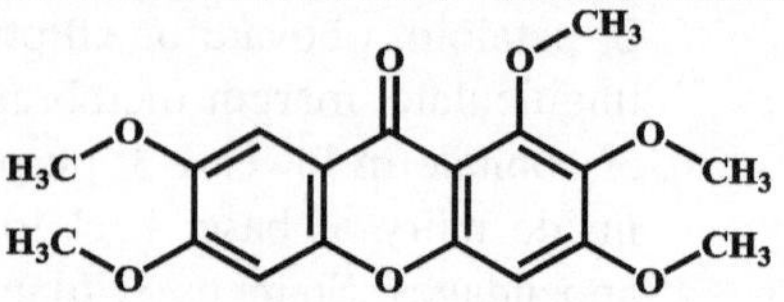 1,2,3,6,7-Pentamethoxyxanthone |
| 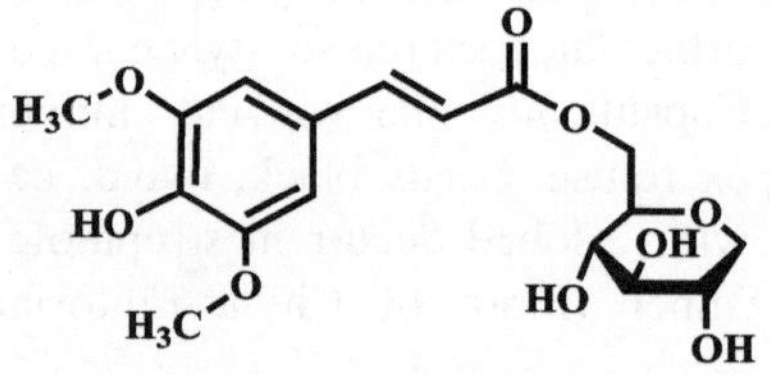 Sibiricose A7 | 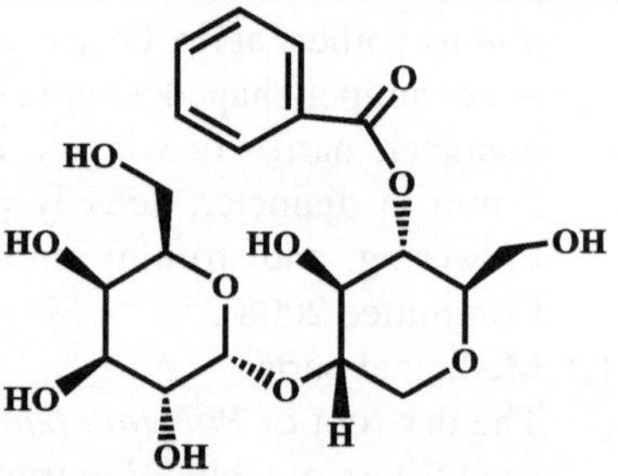 Polygalatenoside C |
| 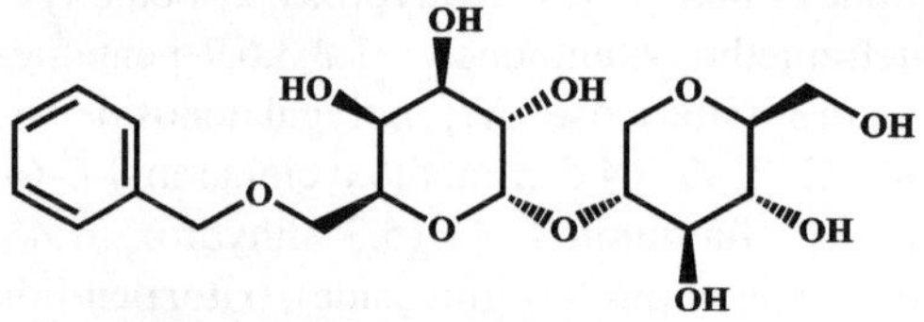 Polygalatenoside A | |
| 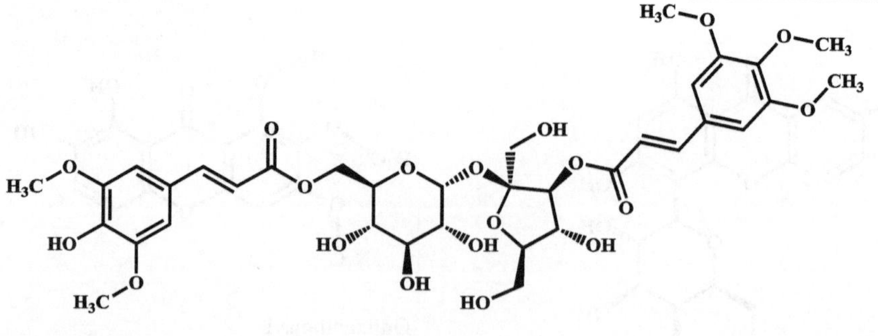 Tenuifoliside C | |

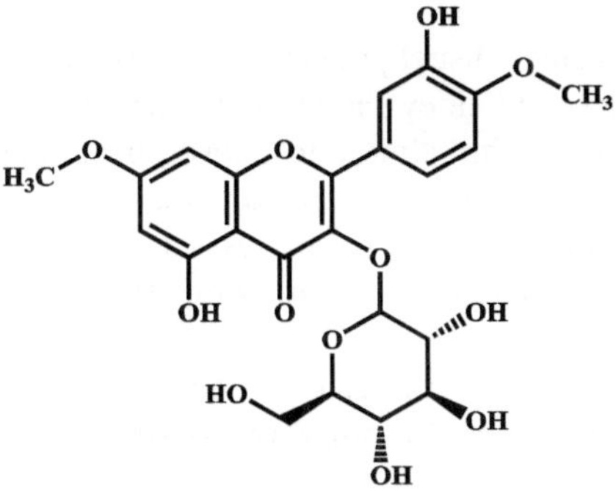

5,3′-Dihydroxy-7,4′-dimethoxyflavonol-3-O-glucoside

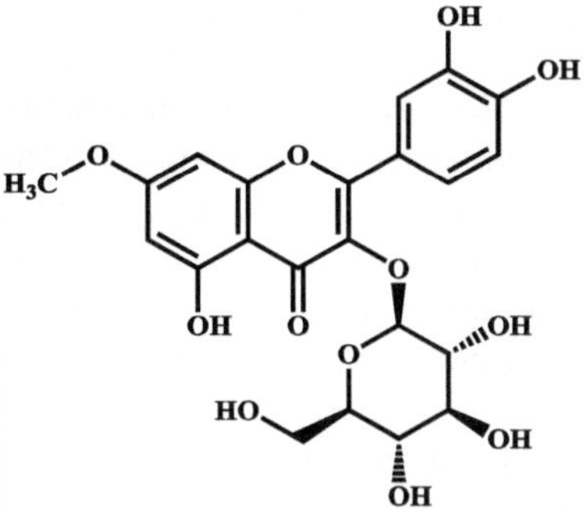

Rhamnetin-3-*O*-glucoside

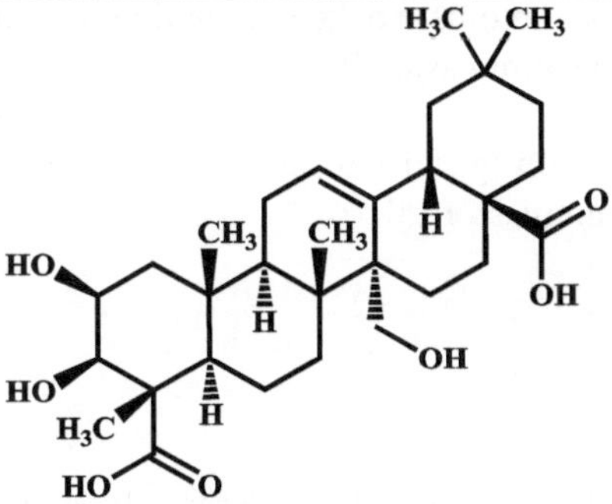

Presenegenin

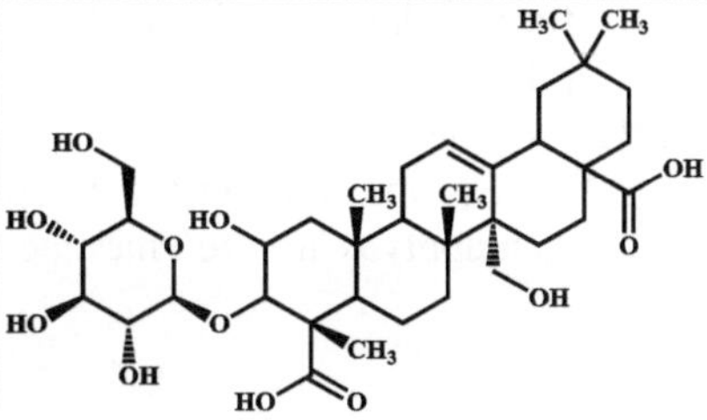

Tenuifolin

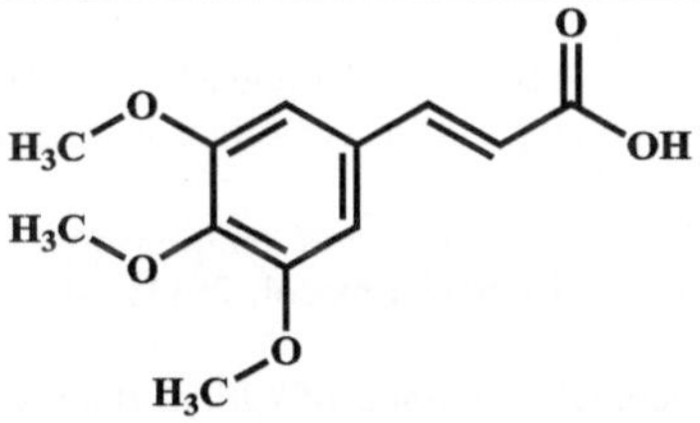

3,4,5-Trimethoxycinnamic acid

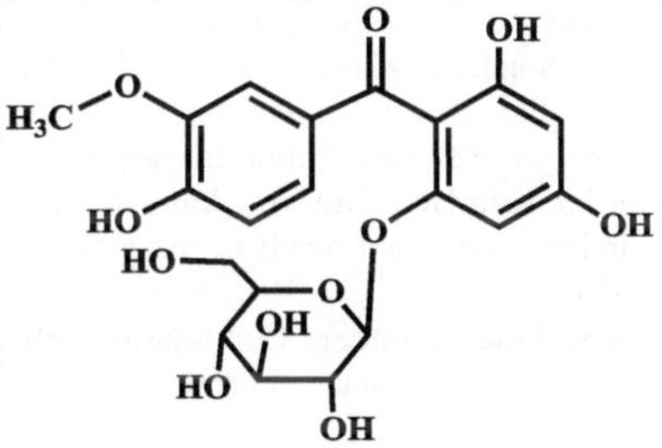

Sibiriphenone A

4) Pharmacology and bioactivities:

A) In vitro studies:
Not found.

B) In vivo studies:
*Radix Polygalae* extract exerts rapid-onset antidepressant effects by modulating glutamatergic synapses in brain circuits of depression (Shin et al. 2014).

*Radix Polygalae* has protective effects on dopaminergic neurons via its antioxidant and anti-apoptotic activity (Choi et al. 2011).

5) Indications and usage:
Tranquilize the mind, improve the intelligence, dispel phlegm for resuscitation, restore normal coordination between heart and kidney, cure abscess, and relieve swelling. Treat unsteadiness of heart-mind, fright palpitations, upset, insomnia, amnesia due to disharmony between heart and kidney, epilepsy and mania due to phlegm obstruction of heart, cough with profuse phlegm, abscess, deep-rooted carbuncle, and swollen sores (Chinese Pharmacopoeia Commission 2010).
6) Precautions and side effects:
Allergies, chest tightness, skin itching, gastritis, gastric ulcer patients, and pregnant women with caution (Chinese Pharmacopoeia Commission 2010).
7) Dosage:

A) Internal use:
3–10 g/day, decocted in water for oral administration, twice a day, generally for 5–7 days, modificated according to symptoms. Soaked in the liquor, into pill or powder (Chinese Pharmacopoeia Commission 2010).
B) External use:
Take proper dosage for external application, grind into powder with wine, then apply (Chinese Pharmacopoeia Commission 2010).

## 7.32 References


Flora of China Editorial Committee.Flora of China. Beijing and St. Louis: Science Press and Missouri Botanical Garden Press, Vol. 2008;11:142.

Zhou YH, Zhang SY, Guo Q, Chai XY, Jiang Y, Tu PF. Chemical investigation of the roots of *Polygala sibirica* L. Chinese J Nat Med. 2014;12(3): 225–8.

Shin IJ, Son SU, Park H, Kim Y, Park SH, Swanberg K, Shin JY, Ha SK, Cho Y, Bang SY, Lew JH, Cho SH, Maeng S. Preclinical evidence of rapid-onset antidepressant-like effect in *Radix Polygalae* extract. Public Library Sci One. 2014;9(2): e88617.

Choi JG, Kim HG, Kim MC, Yang WM, Huh Y, Kim SY, Oh MS. *Polygalae radix* inhibits toxin-induced neuronal death in the Parkinson's disease models. J Ethnopharmacol. 2011;134(2): 414–21.

Chinese Pharmacopoeia Commission. Pharmacopoeia of People's Republic of China. Beijing: China Medical Science Press. 2010.


## 7.33 *Polygonum multiflorum* (Thunberg) Haraldson

### *7.33.1* Polygoni Multiflori Caulis

#### 7.33.1.1 (Polygonaceae)

1) Common names:
Tuber fleece flower stem, Shou-wu-teng, Ye-jiao-teng (Pinyin in Chinese).
2) Photograph of the species (Fig. 7.16)

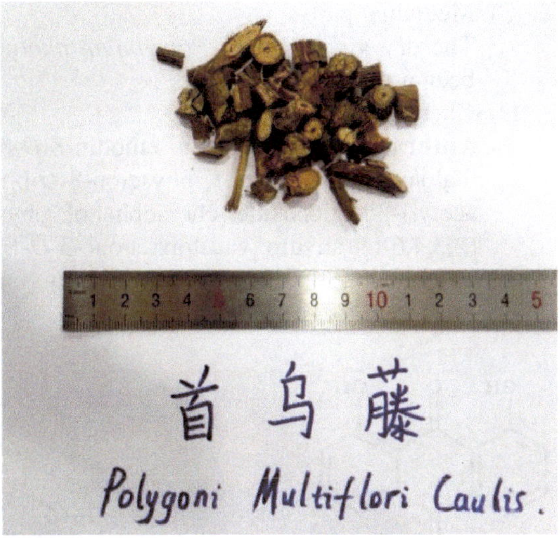


**Fig. 7.16** *Polygoni Multiflori Caulis*

3) Description of the species:

A) Habitat and world distribution:
Mountain slopes, rock crevices, thickets in valleys; 200–3000 m. Anhui, Fujian, Gansu, Guangdong, Guangxi, Guizhou, Hainan, Hebei, Heilongjiang, Henan, Hubei, Hunan, Jiangsu, Jiangxi, Jilin, Liaoning, Qinghai, Shaanxi, Shandong, Sichuan, Taiwan, Yunnan, and Zhejiang [other countries: Japan] (Flora of China Editorial Committee 2003).

B) Morphology of the plant:
Perennial herb. Root tuber black brown, narrowly elliptic, large, ligneous. Stems twining, with 2–4 cm, much branched, striate, glabrous, ligneous at base. Petiole with 1.5–3 cm; leaf blade ovate or narrowly ovate, 3–7 × 2–5 cm, glabrous in upper and lower surfaces or abaxially papillate along veins, base cordate or subcordate, margin entire, apex acuminate; ocrea 3–5 mm, membranous, oblique, glabrous. Inflorescence terminal or axillary, paniculate, spreading, with 10–20 cm; peduncle minutely papillate; bracts triangular-ovate, papillate, apex acute, each 2–4-flowered. Pedicel with 2–3 mm, slender, articulate at base, elongate in fruit. Perianth with white or greenish color; tepals elliptic, outer three larger, accrescent and winged on abaxial surface in fruit; wings decurrent below to pedicels on abaxial surface in fruit; perianth (including wings) in fruit nearly orbicular, 6–7 mm in diameter. Stamens 8; filaments dilated at base. Styles 3, very short; stigmas capitate. Achenes included in persistent perianth, black brown, shiny, ovoid, trigonous, 2.5–3 mm, smooth. Flowering in June–October, fruiting in July–November (Flora of China Editorial Committee 2003).

C) Medicinal parts:
The dry vine stem of *Polygonum multiflorum* (Thunberg) Haraldson has been used.

D) Chemical constituents:
**Anthraquinones** (emodin, emodin-8-*O*-β-D-glucoside, emodin-8-*O*-(6′-*O*-malonyl)-β-D-glucoside, physcion-8-*O*-β-D-glucoside, physcion-8-*O*-(6′-*O*-acetyl)-β-D-glucoside, chrysophanol, physcion) (Liang et al. 2009); **other** (2,3,4,6-tetrahydroxyacetophenone-3-*O*-B-D-glucopyranoside, β-sitosterol, 2,3,5,4′-tetrahydroxystilbene-2-*O*-β-D-glucoside, etc.) (Liang et al. 2009; Zhao et al. 2013).

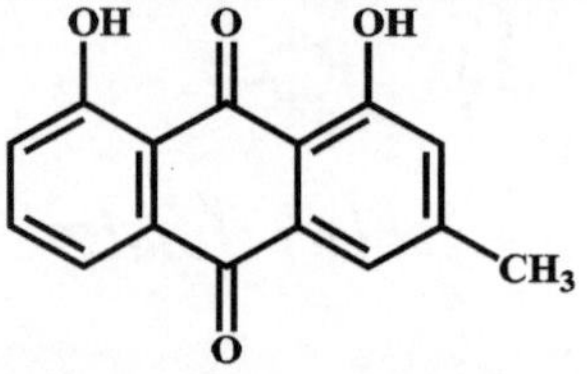
Chrysophanol

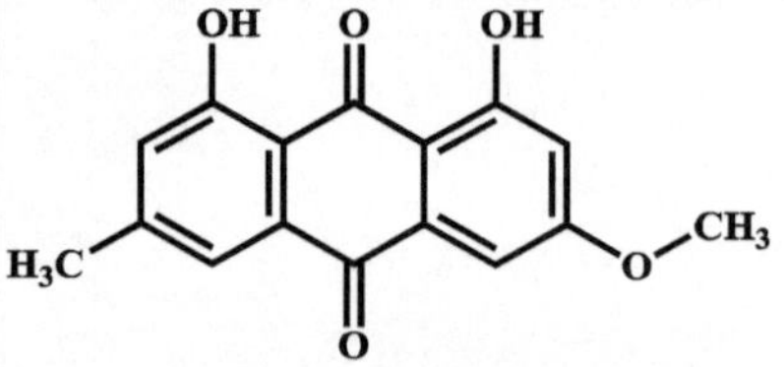
Physcion and its derivatives

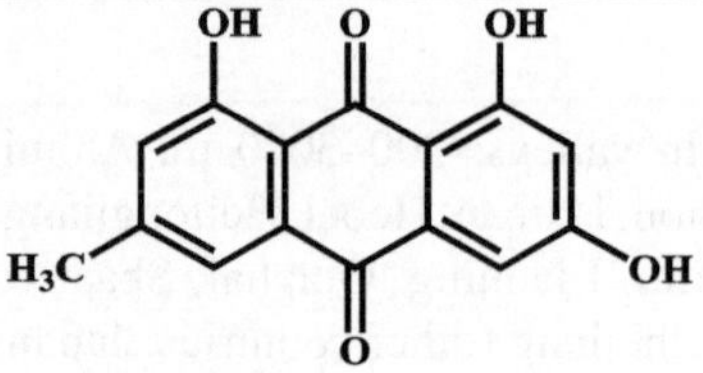
Emodin and its derivatives

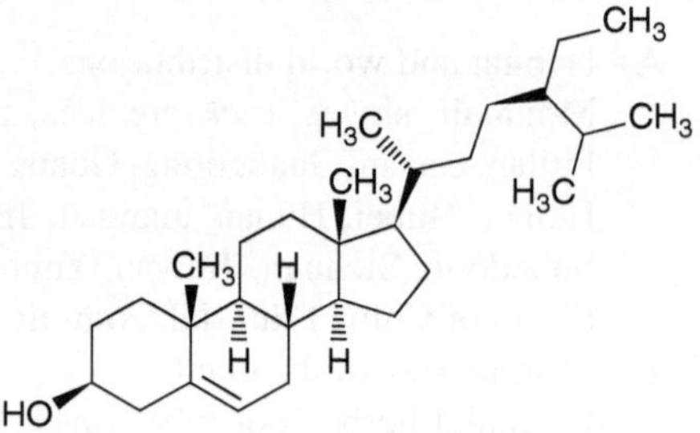
β-Sitosterol

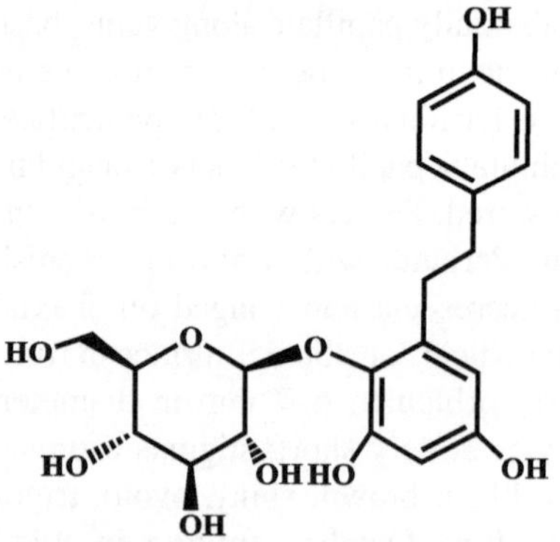
2,3,5,4′-Tetrahydroxystilbene-2-*O*-β-D-glucoside

4) Pharmacology and bioactivities:

   A) In vitro studies;
      Not found.
   B) In vivo studies:
      *Polygoni Multiflori Caulis* decoction was studied by the rolling cylinder method in mice and by the somnopolygraphic recording method in rats. The result from the rolling method showed that the decoction (9 g/kg) had synergistic action with the subthreshold hypnotic dose of sodium pentobarbital (20 mg/kg). The result from the somnopolygraphic recording experiment showed that the immediate hypnotic effect of an effective dose of the decoction (20 g/kg) was basically similar to that of diazepam (5 mg/kg) used under the same experimental condition, mainly an increase in the duration of slow wave sleep phase (SWS) and a decrease in the duration of paradoxical sleep phase (PS). By the end of a three-day session in dose of twice a day, the hypnotic effect appeared more prominent and the reduction of SWS latency became significant (Yang et al. 1990).

5) Indications and usage:
   Nourish the blood and induce tranquilization, dispel wind, and dredge collaterals. Used for insomnia due to restlessness of deficiency, dream-disturbed sleep, pain of body due to blood deficiency, wind-damp arthralgia, blood deficiency, and itching rash of skin (Chinese Pharmacopoeia Commission 2010).
6) Precautions and side effects:
   Not found.
7) Dosage:

   A) Internal use:
      9–15 g/day, decocted in water for oral dose, served twice a day for 7 days, modificated according to symptoms (Chinese Pharmacopoeia Commission 2010).
   B) External use:
      Take proper dose for washing with decoction.

## 7.34 References


Flora of China Editorial Committee. Flora of China. Beijing and St. Louis: Science Press and Missouri Botanical Garden Press. 2003;5: 317.

Liang Y, Tian WX, Ma XF. Chemical constituents of *Caulis Polygoni Multiflori*. *J Shenyang Pharmaceut Univ*. 2009;26(7): 536–8, 546.

Zhao Y, Kao CP, Chang YS, Ho YL. Quality assessment on *Polygoni Multiflori Caulis* using HPLC/UV/MS combined with principle component analysis. Chem Central J. 2013;7(1): 106.

Yang J, Xu J, Zhou L, Huang L, Yang S, Xiong T, Dong L. Sedative-hypnotic effect of caulis *Polygoni multiflori* decoction. Hua Xi Yi Ke Da Xue Bao, 1990;21(2): 175–7.

Chinese Pharmacopoeia Commission. Pharmacopoeia of People's Republic of China. Beijing, China Medical Science Press. 2010.

## 7.35 *Bupleurum chinense* de Candolle or *Bupleurum scorzonerifolium* Willdenow

### *7.35.1* Radix Bupleuri

#### 7.35.1.1 (Umbelliferae)

1) Common names:
   Chai-hu, Di-xun, Ci-hu, Shan-cai, Chai-cao, Ru-cao (Pinyin in Chinese).
2) Photograph of the species (Fig. 7.17)
3) Description of the species:

   A) Habitat and world distribution:
   ***Bupleurum chinense de*** **Candolle**: Grasslands, stream banks, sunny slopes, roadsides; 100–2700 m. Anhui, Gansu, Hebei, Heilongjiang, Henan, Hubei, Hunan, Jiangsu, Jiangxi, Jilin, Liaoning, Nei Mongol, Shaanxi, Shandong, Shanxi, and Zhejiang (Flora of China Editorial Committee 2005).

   ***Bupleurum scorzonerifolium*** **Willdenow**: Shrub forest margins, sunny mountain slopes, dry grasslands; 100–2300 m. Anhui, Gansu, Guangxi, Hebei, Jiangsu, Heilongjiang, Jilin, Liaoning, Nei Mongol, Shaanxi, Shandong, and Shanxi [other countries: Japan, Korea, Mongolia, Russia] (Flora of China Editorial Committee 2005).

   B) Morphology of the plant:
   ***Bupleurum chinense de*** **Candolle**: Perennial plants with 50–85 cm. Root stout, elongate, brown, woody, and frequently branched. Stem solitary or several, dichotomously much branched above, base without fibrous remnant sheaths. Basal leaves with 4–7 × 0.6–0.8 cm, oblanceolate or narrow-elliptic, base tapering into petioles, apex acuminate. Middle leaves with 4–12 × 0.6–1.8(-3) cm, broadly linear-lanceolate, 7–9-nerved, abaxially glaucous, apex apiculate. Apical leaves small. Umbels numerous, 2–6 cm across; peduncles slender, greatly spreading forming a large loose panicle; bracts 0 or 2–3, linear, 1–5 × 0.5–1 mm, 3-nerved; rays 3–8, very slender, 1–3 cm, unequal; bracteoles 5, lanceolate, 3–3.5 × 0.6–1 mm, shorter than flowers; umbellules 4–6 mm across, with 5–10 flowers. Petals bright yellow. Stylopodium low-conic, discoid, dark yellow. Fruit oblong, brown, ca. 3 × 2 mm; ribs prominent, narrowly winged, wings pale brown; vittae 3(-4) in each furrow, 4 on commissure. Flowering and fruiting in September–October (Flora of China Editorial Committee 2005).

   ***Bupleurum scorzonerifolium*** **Willdenow**: Perennial plants with 30–60 cm. Taproot stout, dark reddish brown, branched. Stems 1–3, flexuose, greatly dichotomously branched, base clothed with fibrous remnant sheaths. Basal leaves with 6–16 × 0.2–0.7 cm, linear, thick-papery, rigid, nerves 3–5, prominent abaxially, margin white cartilaginous, base slightly narrowed and clasp-

**Fig. 7.17** *Radix Bupleuri*

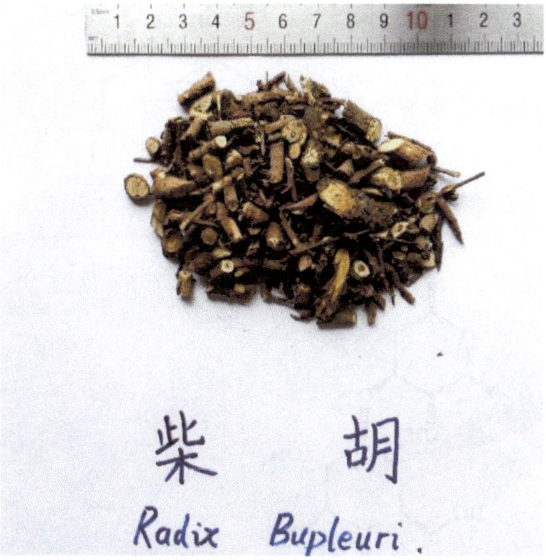


ing. Upper leaves small. Umbels numerous, 1.2–4 cm across; bracts 1–3, subulate, 0.5–4 × 0.2–0.6 mm, unequal, deciduous; rays (3-)4-6(-8), 1–2 cm, very slender, spreading; bracteoles 5, lanceolate, 2.5–4 × 0.5–1 mm, equaling or slightly exceeding umbellules; umbellules 2–5 mm across, with (6-)9-11(-15)flowers; pedicels 0.2–1 mm. Petals are yellow. Stylopodium low-conic, dark yellow. Fruit ellipsoid, dark brown, with 2.5–3 × 1.5–2 mm; ribs pale, prominent; vittae 5–6 in each furrow, 4–6 on commissure. Flowering and fruiting in July–September (Flora of China Editorial Committee 2005).

C) Medicinal parts:
The dry root of *Bupleurum chinense* de Candolle or *Bupleurum scorzonerifolium* Willdenow has been used. Separately named Bei-Chai-hu or Nan-Chai-hu.

D) Chemical constituents:
**Saponins** (saikosaponin-A, -B1, -B2, -C, and -D); **flavonoids** (quercetin, etc.); **coumarins** (scopoletin, scoparone, limettin, fraxetin, aesculetin, isoscoparome, angelicin); **polyacetylenes** (bupleurynol, (2Z,9Z)-pentadecadiene-4,6-diyene-1-ol, (2Z,8E,10E)-pentadecatriene-4,6-diyne-1-ol, (2Z,8Z,10E)-heptadecatriene-4,6-diyne-1-ol); **volatile compounds** (pentanal, *n*-hexanal, β-pinene, furfural, 2-amyl furan, dodecanal, etc.); **others** (polysaccharide, lignans, fatty acids and sterols, threonine, etc.) (Testai et al. 2005; Yu et al. 2014; Liu et al. 2015; Li et al. 2015; Xing et al. 2015).

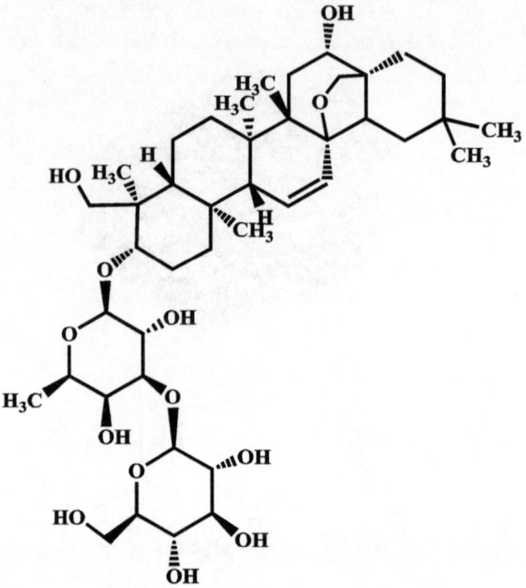

Saikosaponin-A

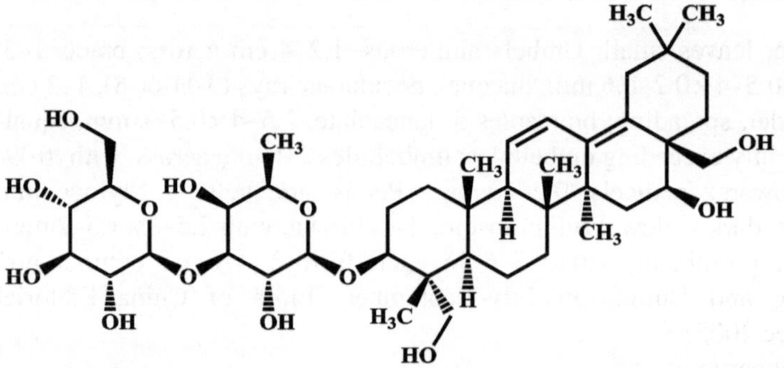

Saikosaponin-B1

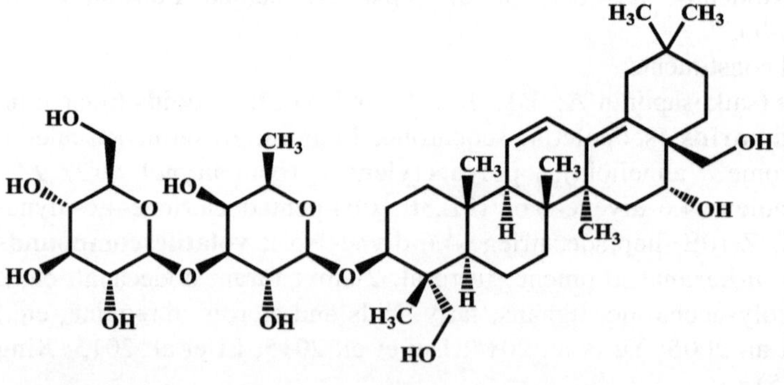

Saikosaponin-B2

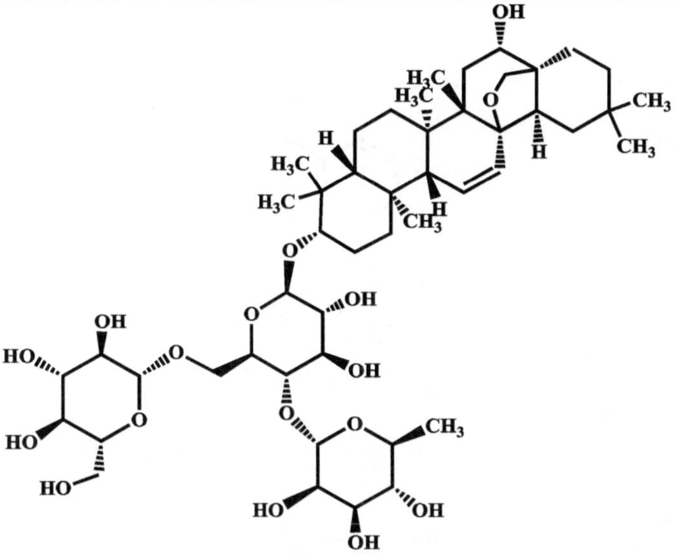

Saikosaponin-C

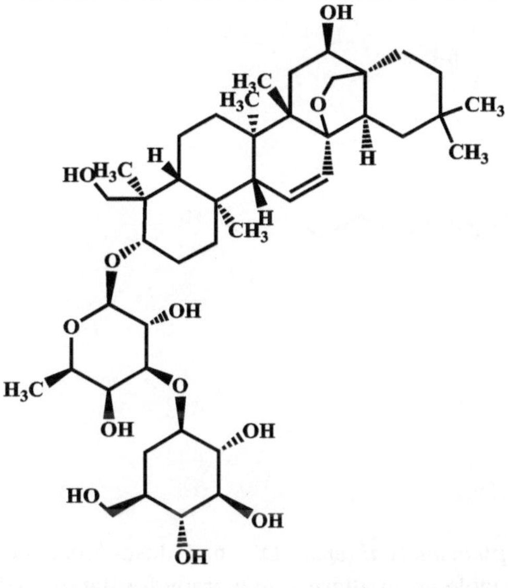

Saikosaponin-D

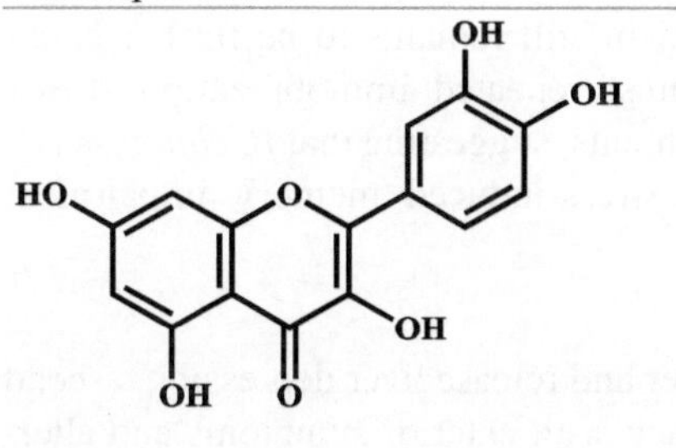

Quercetin

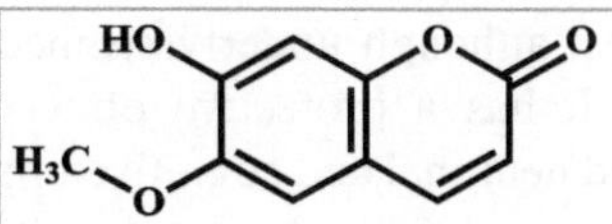

Scopoletin

| | |
|---|---|
| 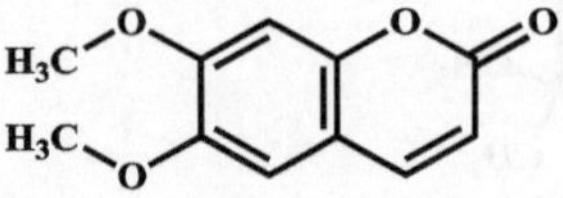 Scoparone | 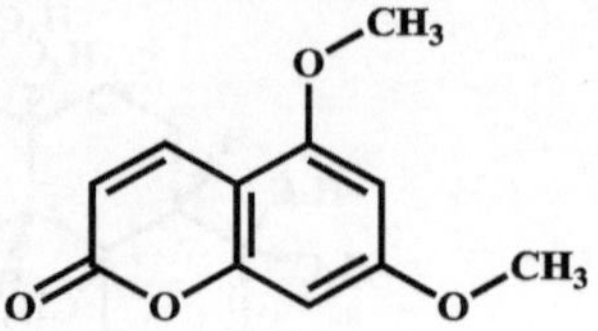 Limettin |
| 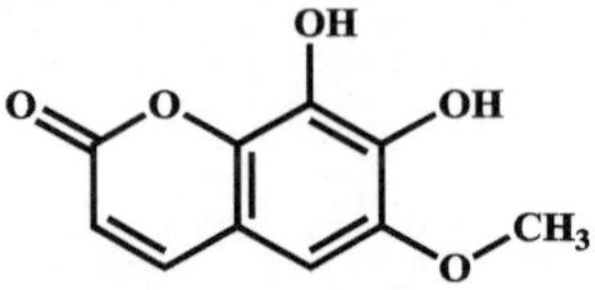 Fraxetin | 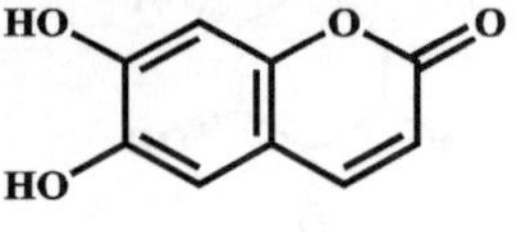 Aesculetin |
| 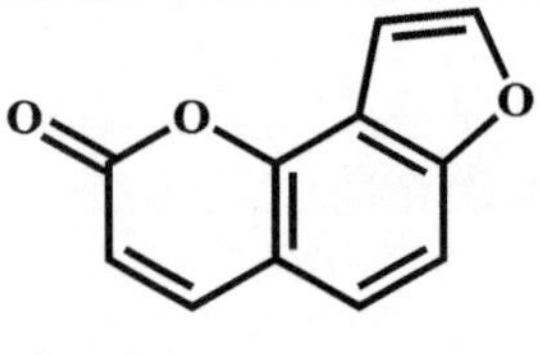 Angelicin | 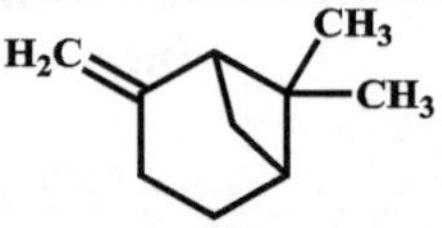 β-Pinene |
| 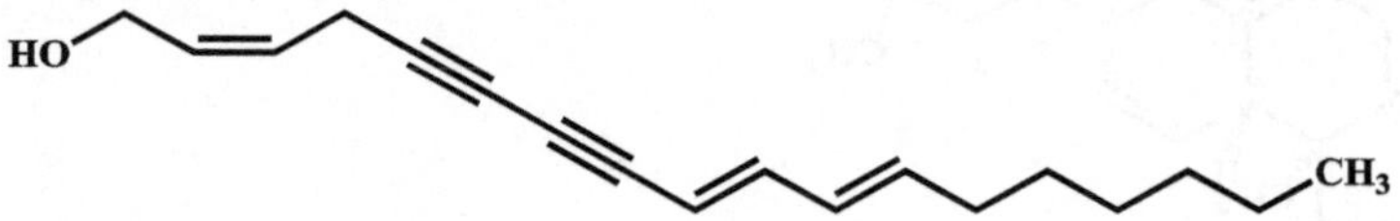 Bupleurynol | |

4) Pharmacology and bioactivities:

1) In vitro studies:
   Not found
2) In vivo studies:
   The methanol extract of *Bupleurum chinense* DC has dose-dependent antidepressant-like activity valuable as an alternative therapy for depression and the mechanism of action involves the serotonergic and noradrenergic systems, although underlying mechanism still remains to be further elucidated. It has a protective effect against repeated immobilization stress-induced neuronal and cognitive impairments, suggesting that *B. chinense* DC may be useful in the treatment of stress-induced memory impairment (Pennington et al. 2009).

5) Indications and usage:
   Release exterior and cure fever, soothe liver and release liver depression, ascend yang and qi. Treat external pathogens attack with exterior symptoms and alternating chills and fever due to Shao-yang syndrome. Treat the pain in chest and

hypochondrium, irregular menstruation due to the liver qi stagnation, and prolapse of the anus and uterine due to syndrome of sinking of middle qi (Chinese Pharmacopoeia Commission 2010). In addition, it also has the action of relieving fever and preventing malaria.

6) Precautions and side effects:
Hyperactivity of yang due to yin deficiency, hyperactivity of fire due to yin deficiency, endogenous liver wind, and ascent of *qi* movement must be cautious or inhibitory to use it (Chinese Pharmacopoeia Commission 2010).

7) Dosage:

   A) Internal use:
   3–10 g/day, decocted in water for oral administration, taken twice a day, generally for 3–5 days, modificated according to symptoms. For releasing exterior symptoms and fever, it should be used unprocessed in a slightly bigger dosage; for soothing the liver and easing depression, it should be processed with vinegar; for ascending yang, it can be used unprocessed or prepared with alcohol, in any way of preparation it should be used in a small dosage (Chinese Pharmacopoeia Commission 2010).

   B) External use:
   Not found.

## 7.36 References


Flora of China Editorial Committee. Flora of China. Beijing and St. Louis: Science Press and Missouri Botanical Garden Press. 2005;14: 70, 73.

Testai L, Silvio C, Ammar B, Luisa P, Vincenzo C, MartinottiE. Vasorelaxant effects of the chloroformic crude extract of *Bupleurum fruticosum* L. (Umbelliferae) roots on rat thoracic aorta. J Ethnopharmacol. 2005;96(1-2): 93–7.

Yu JQ, Deng AJ, Qin HL. Distinctive features of chemical composition of *Bupleurum chinense* applicable to original authentication. Analyt Methods. 2014;6: 1067–75

Liu J, Fang Y, Yang L, Qin X, Du G, Gao X. A qualitative, and quantitative determination and pharmacokinetic study of four polyacetylenes from *Radix Bupleuri* by UPLC-PDA-MS. J Pharmaceut Biomed Analysis. 2015;111, 257–65.

Li ZY, Sun HM, Xing J, Qin XM, Du GH. Chemical and biological comparison of raw and vinegar-baked *Radix Bupleuri*. J Ethnopharmacol. 2015;165: 20–28.

Xing J, Sun HM, Li ZY, Qin XM. Comparison of volatile components between raw and vinegar baked *Radix Bupleuri* by GC-MS based metabolic fingerprinting approach. Evid Based Complement Alternat Med. 2015;2015: 653791.

Pennington K, Focking M, McManus CA, Pariante CM, Dunn MJ, Cotter DR. A proteomic investigation of similarities between conventional and herbal antidepressant treatments. J Psychopharmacol. 2009;23(5): 520–30.

Chinese Pharmacopoeia Commission. Pharmacopoeia of People's Republic of China. Beijing: China Medical Science Press. 2010.

# 7.37 *Ligusticum chuanxiong* Hort.

## *7.37.1* Rhizoma Chuanxiong

### 7.37.1.1 (Umbelliferae)

1) Common names:
   *Ligusticum wallichii* Franchat, Szechuan lovage rhizome, Chuanxiong, Xiong-qiong, San-ju-qiong, Guan-xiong (Pinyin in Chinese).
2) Photograph of the species (Fig. 7.18)
3) Description of the species:

   A) Habitat and world distribution:
      This species prefers mild climate, abundant rainfall, ample sunshine, and more humid environment. Sichuan (Guan-xian), Gansu, Guizhou, Hubei, Jiangxi, Shaanxi, and Yunnan (Qiu et al. 1979).
   B) Morphology of the plant:
      Perennial herbs, 40–60 cm tall. Rhizome developed, irregular shape of the nodules, strong aroma. Stems erect, terete, with longitudinal stripes, upper multi-branches, lower internodes with large discoid. Lower stem leaves petiolate, 3–10 cm long handle, base expands into scabbard; contour of blade ovate triangular, 12–15 cm long, 10–15 cm wide, 3–4-ternate/pinnate, 4–5 pairs of pinnae, ovate lanceolate, 6–7 cm long, 5–6 cm wide, ultimate lobes linear-lanceolate shaped to long ovoid, 2–5, 1–2 mm wide, with small pointed; upper leaves gradually simplified. Synflorescence top or lateral; involucral bracts 3–6, linear, 0.5–2.5 cm long; calyx teeth agenesis; petals white, obovate to ovate, 1.5–2 mm, apex with folding small pointed; stylopo-

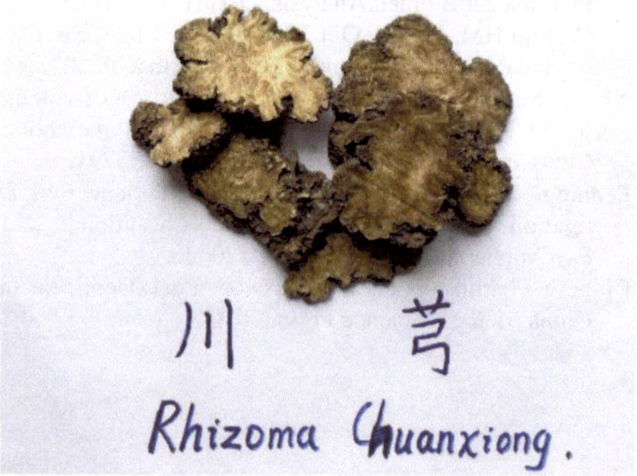


**Fig. 7.18** *Rhizoma Chuanxiong*

dium conic; styles 2, 2–3 mm long downward inflection. Fruitlets are flat on both sides, 2–3 mm long, ca. 1 mm wide; furrowed tubing 1–5 dorsal, lateral ribs groove tube 2–3, commissure vittae 6–8. Flowering July–August; Fruiting September–October (Qiu et al. 1979).

C) Medicinal parts:
The dry rhizome of *Ligusticum chuanxiong* Hort. has been used.

D) Chemical constituents:
**Essential oil** (phthalide lactones, sabinene, α-pinene, myrcene) (Qiu et al. 1979); **alkaloids and other nitrogen compounds** (1-β-ethylacrylate-7-carboxycarboline, chuanxiongzine, perlolyrine, anhydride, 1-acetyl-β-carboline, valine anhydride, trimethylamine, choline, uracil, adenine, adenosine, scopoletin) (Wu et al. 2011); **phenolic acids** (ferulic acid, 4-hydroxybenzoic acid, caffeic acid, vanillic acid, protocatechuic acid) (Wu et al. 2011); **phthalide lactones** (ligustilide, neoligustilide, 3-butylphthalide, butylidenephthalide, 4,7-dihydroxy-3-butylphthalide, senkyunolides A–S, ligustilide glycol, Z,Z´-6,6′-7,3′α-diligustilide, chuanxiongnolide A and chuanxiongnolide B, ligusticoside A, 4-hydroxy-3-butylphthalide, 3-butyl-3-hydroxyl-4,5-dihydrophthalide) (Wu et al. 2011); **other constituents** (1-hydroxy-1-(3-methoxy-4-hydroxyphenyl)-ethanol, sedanonic acid, chrysophanol, 4-hydroxy-3-methoxyphenylethylene, linoleic acid, palmitic acid, *n*-hexadecanoic acid, 5,5′-*O*-methyl furfural, 4-pentylcyclohex-3-ene-1α,2β-diol, spathulenol, β-sitosterol, sucrose, senkyunon, quercetin, augustic acid, xiongterpene, (4S)-*p*-menth-1-ene4,7-diol, aromadendrene-4α,10β-diol, aromadendrene-4β,10α-diol, apigenin, kaempferol-3-*O*-glucoside, alloaromadendrene-4β,10α,13,15-tetrol) (Rant al. 2011).

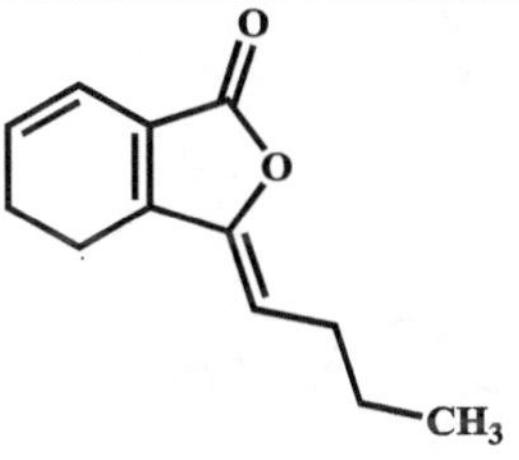


Ligustilide

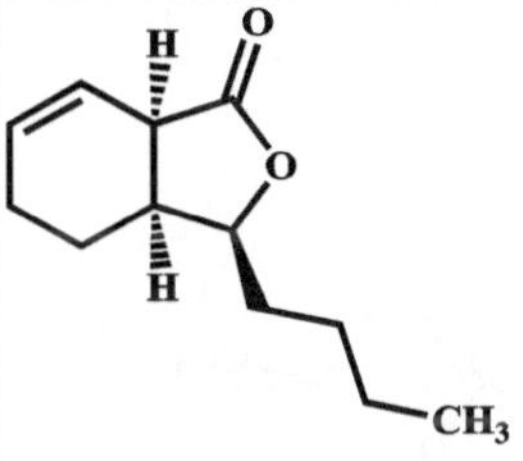


Cnidilide

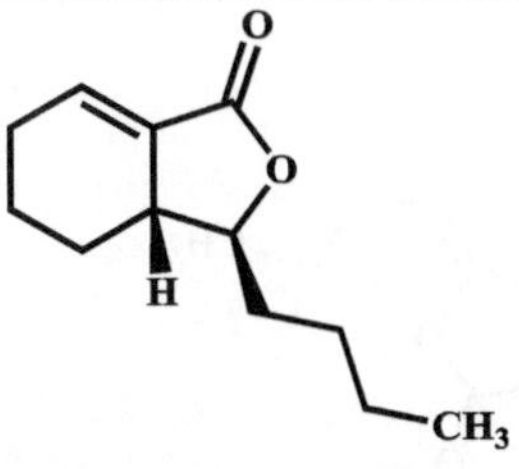


Neocnidilide

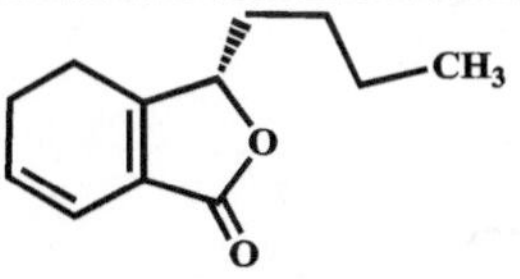


Senkyunolide A

| | |
|---|---|
| 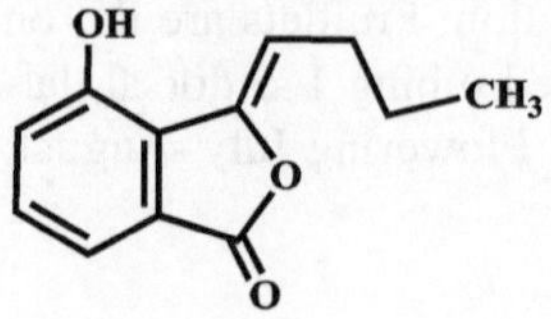<br>Senkyunolide B | 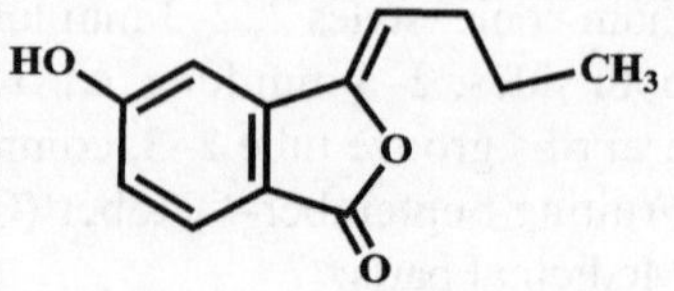<br>Senkyunolide C |
| 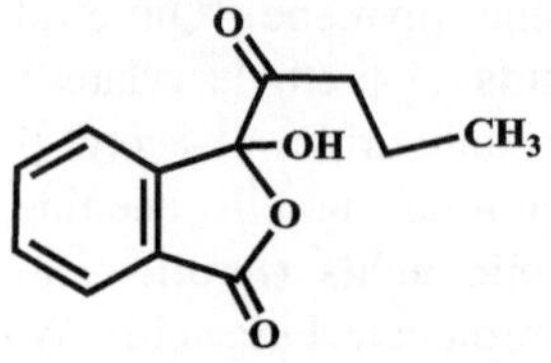<br>Senkyunolide D | 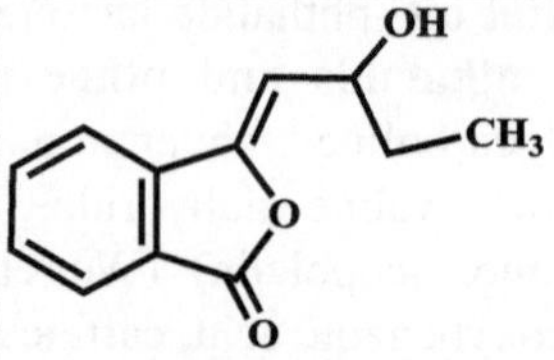<br>*Z*-Senkyunolide E |
| 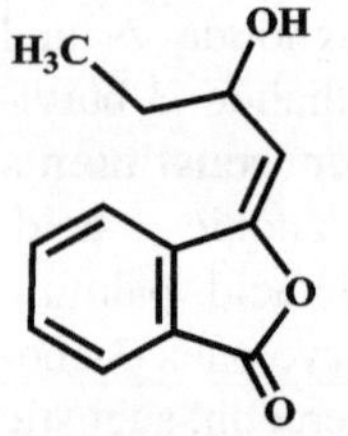<br>*E*-Senkyunolide F | 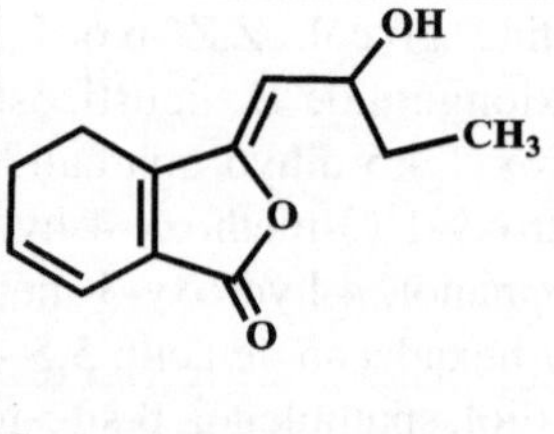<br>Senkyunolide F |
| 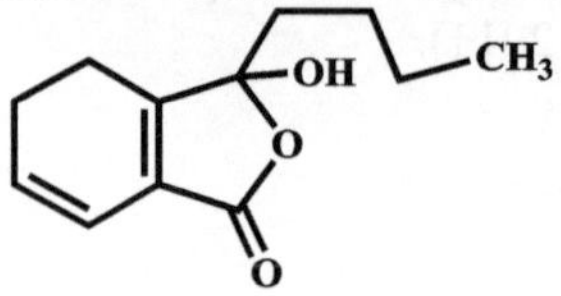<br>Senkyunolide G | 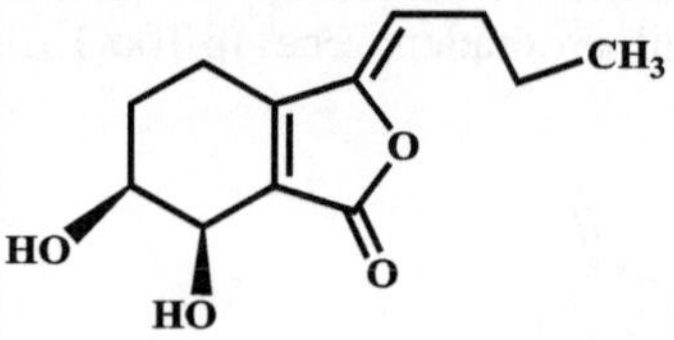<br>Senkyunolide H |
| 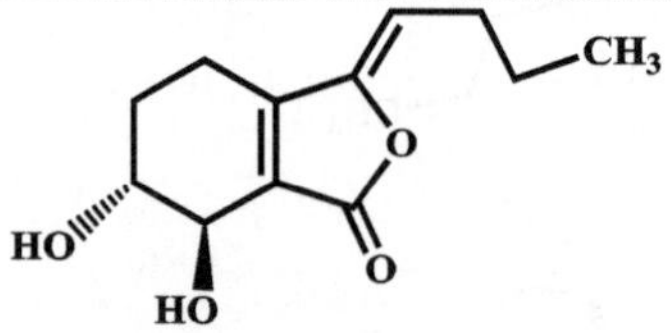<br>Senkyunolide I | 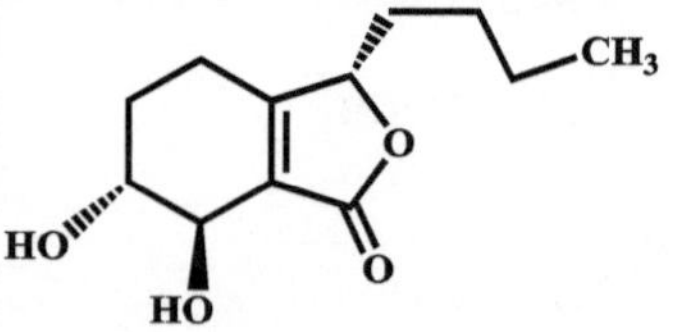<br>Senkyunolide J |
| 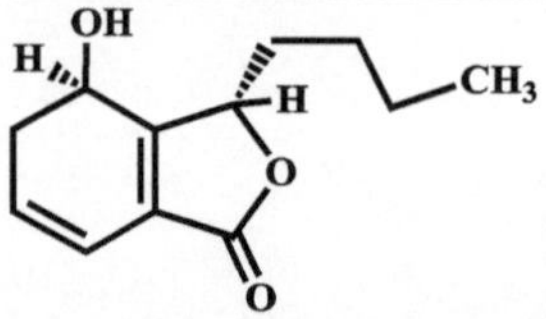<br>Senkyunolide K | 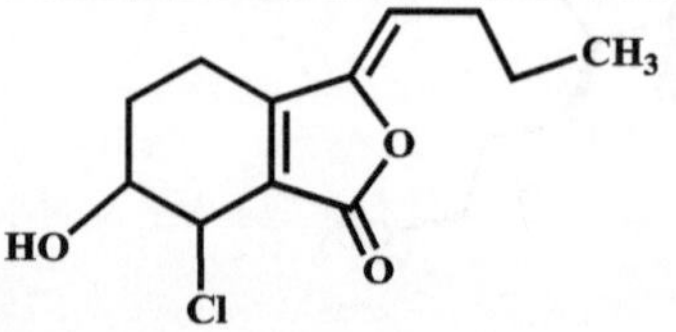<br>Senkyunolide L |

| | |
|---|---|
| 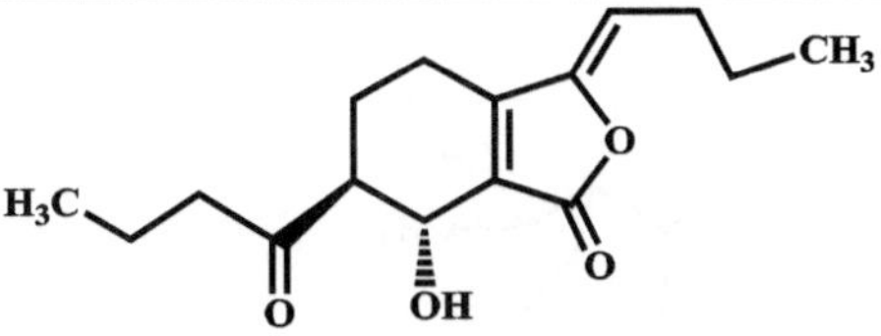<br>Senkyunolide M | 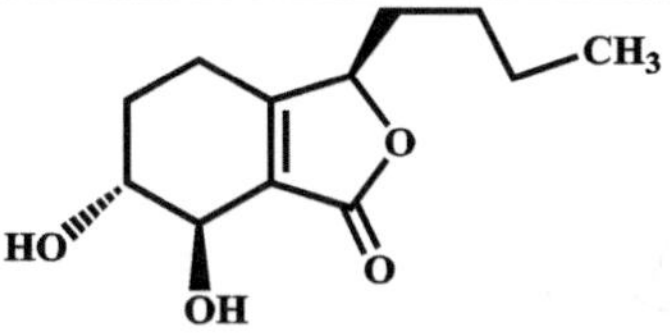<br>Senkyunolide N |
| 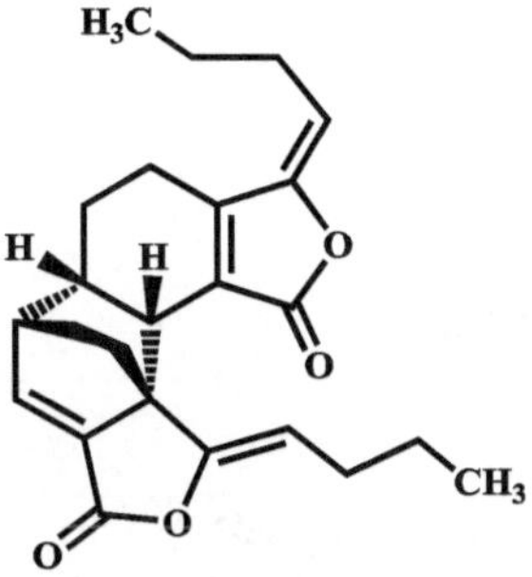<br>Senkyunolide O | 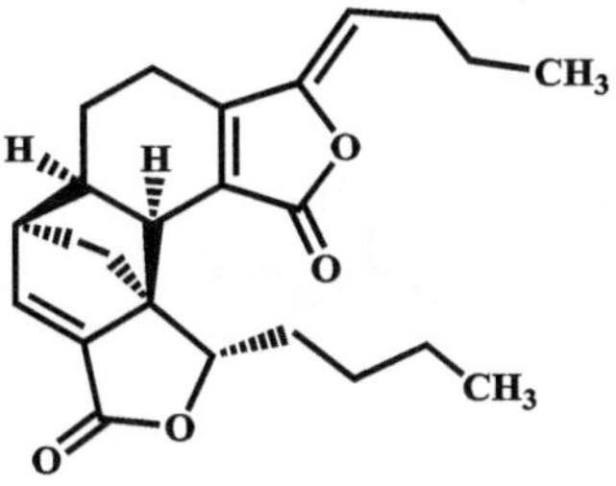<br>Senkyunolide P |
| 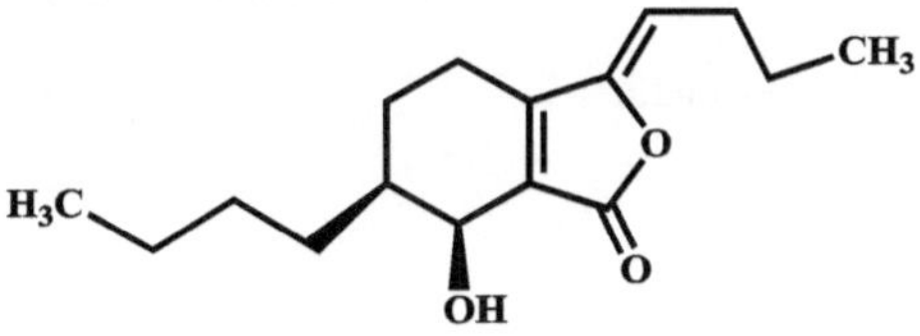<br>Senkyunolide Q | 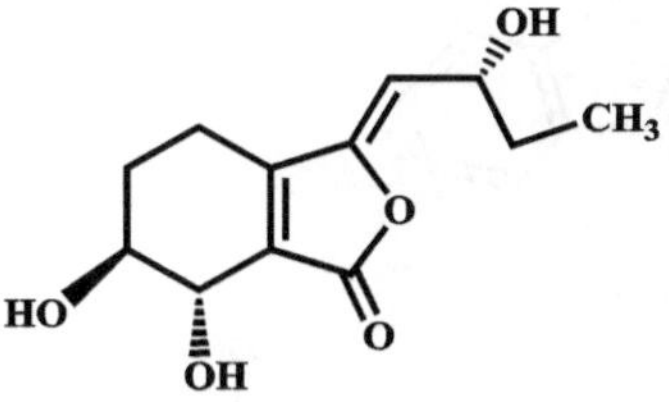<br>Senkyunolide R |
| 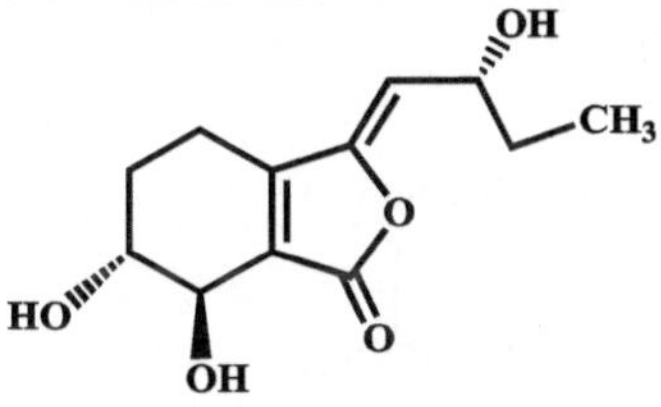<br>Senkyunolide S | 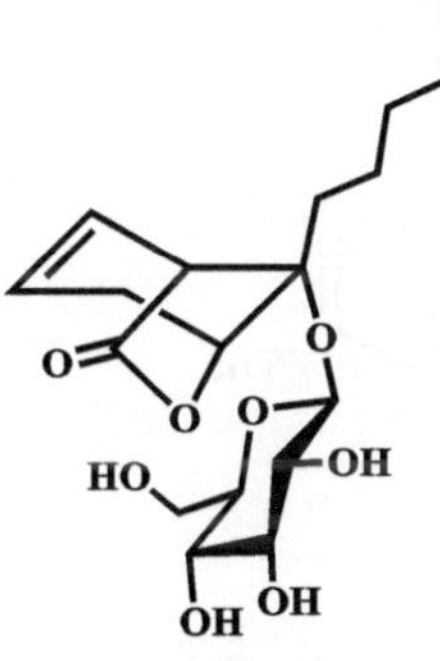<br>Ligusticoside A |
| 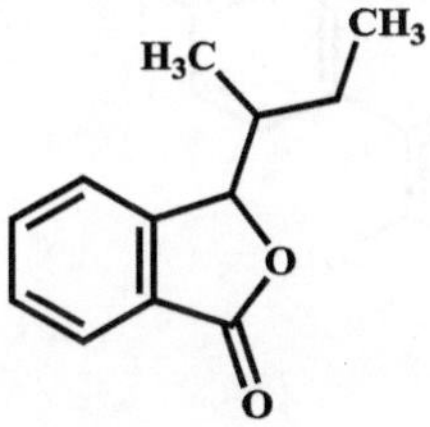<br>3-Butylphthalide | 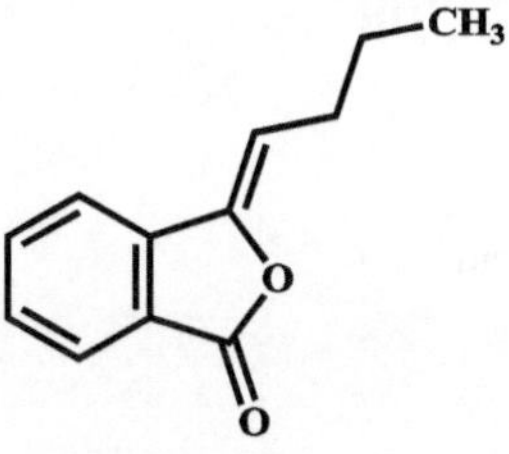<br>Butylidenephthalide |

| | |
|---|---|
| 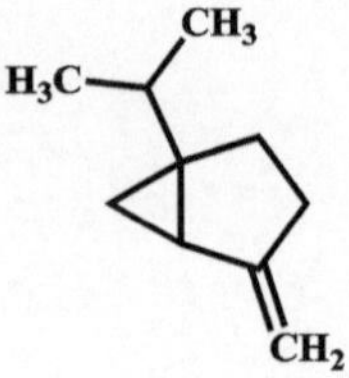<br>Sabinene | 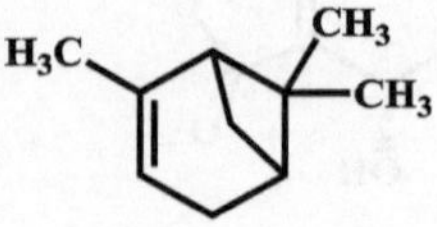<br>α-Pinene |
| 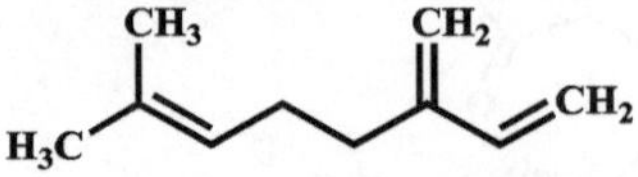<br>Myrcene | 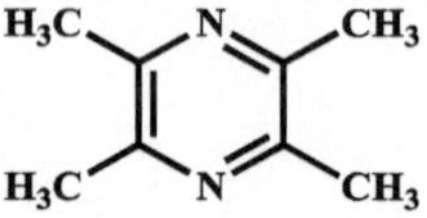<br>Chuanxiongzine |
| 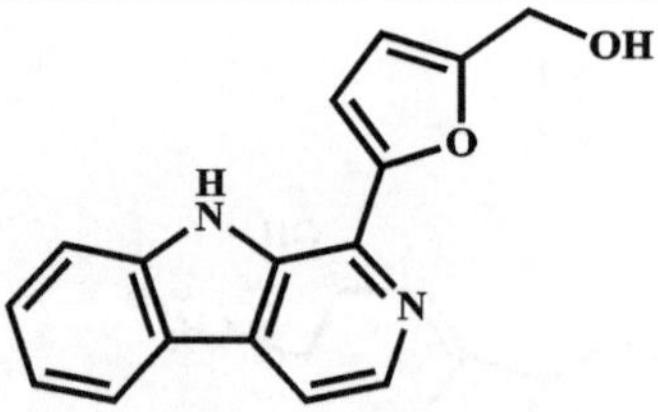<br>Perlolyrine | 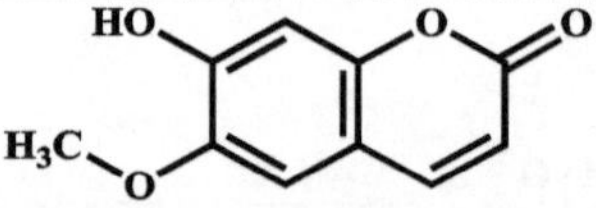<br>Scopoletin |
| 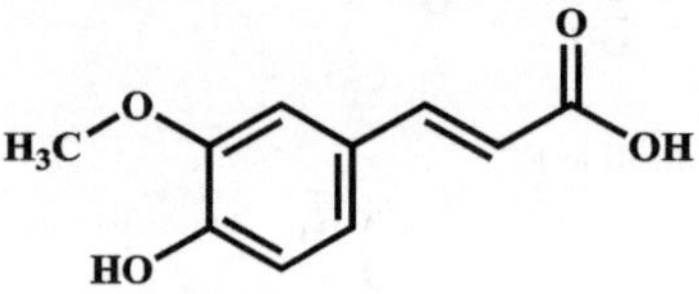<br>Ferulic acid | 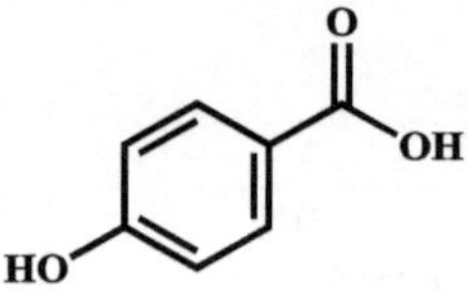<br>4-Hydroxybenzoic acid |
| 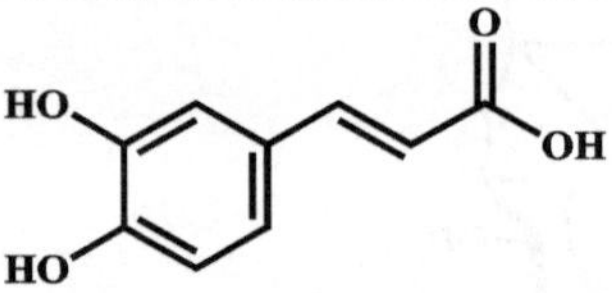<br>Caffeic acid | 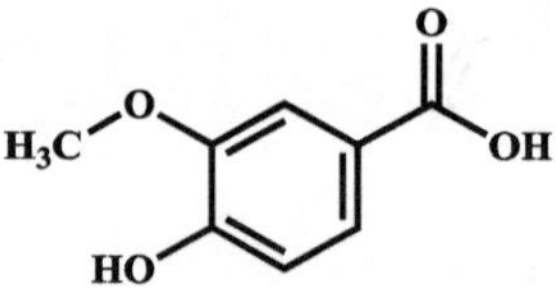<br>Vanillic acid |
| 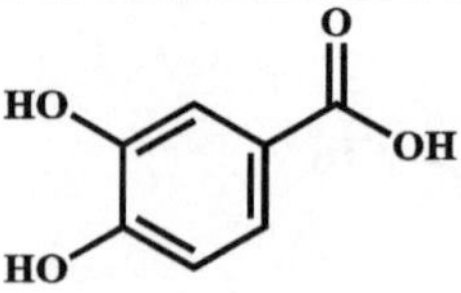<br>Protocatechuic acid | 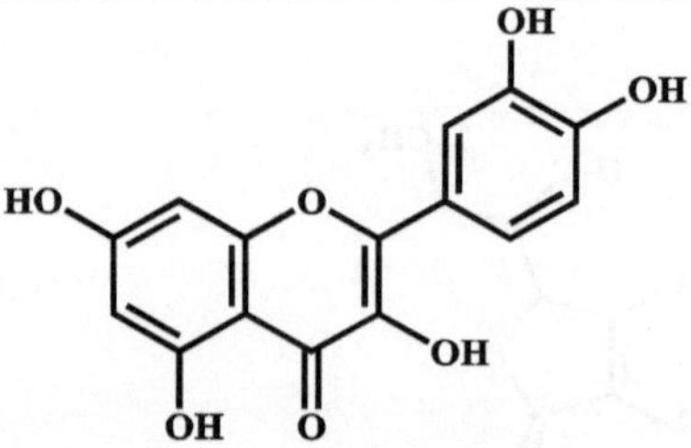<br>Quercetin |

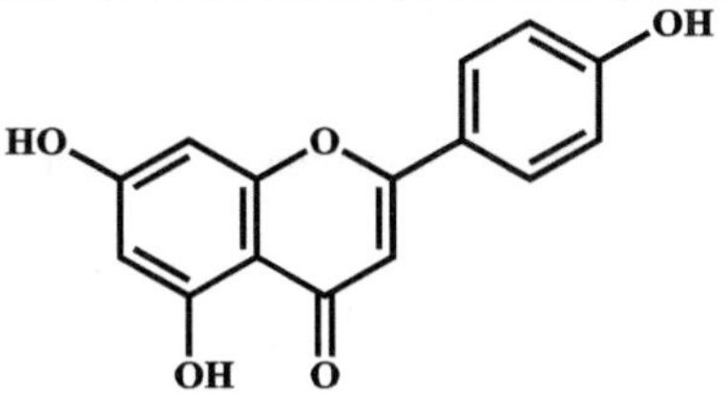


Apigenin

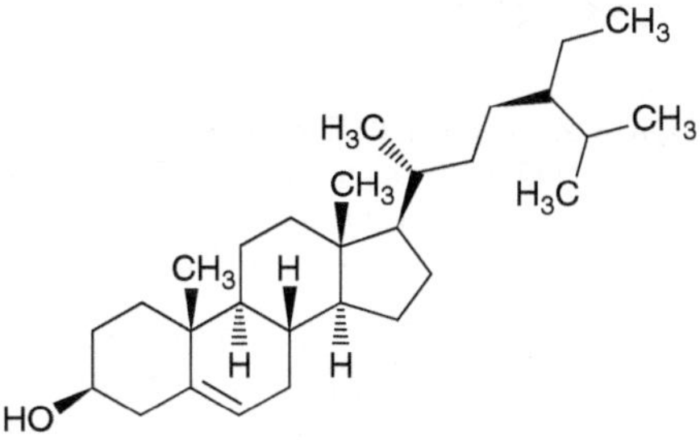


β-Sitosterol

4) Pharmacology and bioactivities:

   A) In vitro studies:
      Not found.
   B) In vivo studies:
      A study showed that the chuanxiongzine or 2,3,5,6-tetramethylpyrazine (TMP) can generate antidepressant effect on chronic mild stress (CMS)-induced rat model of depression. Long-term administration of TMP partially reversed the effects of CMS on the consumption of sucrose solution, the squares crossing and rearing in OFT, and the immobility time during FST in rats (Yu et al. 2012). Studies showed that *Rhizoma chuanxiong* was a component of the antidepression prescriptions, such as Yueju-Wan, Danggui-Shaoyao-San, Chai-hu–Shugan–San, and so on (Yu et al. 2012; Wei et al. 2008; Xue et al. 2011, 2013; Zhang et al. 2011).

5) Indications and usage:
   Activate blood circulation and move qi, expel wind, and alleviate pain. Treat for pain due to blood stasis and qi stagnation, such as pain of chest and hypochondria, stasis of heart vessel, angina pectoris, traumatic injuries and swelling pain, irregular menstruation, dysmenorrhea, amenorrhea, postpartum abdominal pain and so on, headache, and wind-damp arthralgia (Chinese Pharmacopoeia Commission 2010).
6) Precautions and side effects:
   It should be used with cautions for yin deficiency with effulgent fire because of its warm and dry properties, and be prohibited for pregnant women (Chinese Pharmacopoeia Commission 2010).
7) Dosage:

   A) Internal use:
      3–10 g/day, decocted in water for oral administration, twice a day, generally for 3–5 days, modificated according to symptoms (Chinese Pharmacopoeia Commission 2010).
   B) External use:
      Not found.

## 7.38 References

Qiu S, Zeng Y, Pan K, Tang Y, Xu J. On the nomenclature of the Chinese plant drug "Chuanxiong". Acta Phytotaxonom Sinica. 1979;17(2): 101–3.
Wu J.G, Ma L, Zhang SY, Zhu ZZ, Zhang H, Qin LP. Wei YJ. Essential oil from rhizomes of *Ligusticum chuanxiong* induces apoptosis in hypertrophic scar fibroblasts. Pharmaceut Biol. 2011;49(1): 86–93.
Ran X, Ma L, Peng C, Zhang H, Qin LP. *Ligusticum chuanxiong* Hort: a review of chemistry and pharmacology. Pharmaceut Biol. 2011;49(11): 1180–9.
Yu L, Jiang X, Zhang Y, Liao M, Ma R, Yu T. Antidepressant-like activity of tetramethylpyrazine measured by chronic experimental method in rat model of depression. Pharmacol Pharm. 2012;03(1), 52–7.
Wei XH, Cheng XM, Shen JS, Wang ZT. Antidepressant effect of *Yueju-Wan* ethanol extract and its fractions in mice models of despair. J Ethnopharmacol. 2008;117(2): 339–44.
Xue W, Zhou X, Yi N, Jiang L, Tao W, Wu R, Wang D, Jiang J, Ge X, Wang Y. *Yueju pill* rapidly induces antidepressant-like effects and acutely enhances BDNF expression in mouse brain. Evid Based Complement Alternat Med. 2013, 184367. http://doi.org/10.1155/2013/184367
Xu F, Peng D, Tao C, Yin D, Kou J, Zhu D, Yu B. Anti-depression effects of *Danggui-Shaoyao-San*, a fixed combination of traditional Chinese medicine, on depression model in mice and rats. Phytomed. 2011;18(13): 1130–6.
Zhang Y, Huang X, Wang Y, Xie Y, Qiu X, Ren P, Gao L, Zhou H, Zhang H. Qiao M. Ferulic acid-induced anti-depression and prokinetics similar to *Chaihu–Shugan–San* via polypharmacology. Brain Res Bull. 2011;86(3): 222–8.
Chinese Pharmacopoeia Commission. Pharmacopoeia of People's Republic of China. Beijing, China Medical Science Press. 2010.

## 7.39 *Curcuma longa* L.

### *7.39.1* Rhizoma Curcumae Longae

#### 7.39.1.1 (Zingiberaceae)

1) Common names:
   Yellow ginger, *Radix curcumae longae*, Turmeric, Jiang-huang, Bao-ding-xiang (Chinese Pinyin).
2) Photograph of the species (Fig. 7.19)
3) Description of the species:
   A) Habitat and world distribution:
      Adapt to a warm and humid climate and prefer the environment with adequate sunshine and abundant rainfall, hate cold and dry circumstance. Cultivated throughout Fujian, Guangdong, Guangxi, Sichuan, Taiwan, Xizang, and Yunnan [native origin unknown; cultivated throughout tropical Asia (Flora of China Editorial Committee 2000).
   B) Morphology of the plant:
      Plants with ca. 1 m tall. Rhizomes many branched, orange or bright yellow, cylindric, aromatic; roots tuberous at tip. Petiole with 20–45 cm; leaf blade green, oblong or elliptic, 30–45(-90) × 15–18 cm, glabrous, base attenuate,

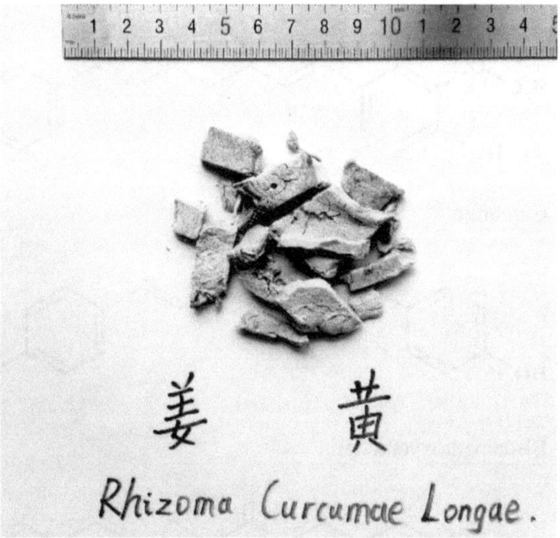


**Fig. 7.19** *Rhizoma Curcumae Longae*

apex shortly acuminate. Inflorescences terminal on pseudostems; peduncle with 12-20 cm; spike cylindric, with 12–18 × 4–9 cm; fertile bracts pale green, ovate or oblong, 3–5 cm, apex obtuse; coma bracts spreading, white and green, sometimes tinged reddish purple, apex acute. Calyx white, 0.8–1.2 cm, puberulent, apex unequally tri-toothed. Corolla pale yellow; tube to 3 cm; lobes deltoid, with 1–1.5 cm, central one larger, apex mucronate. Lateral staminodes shorter than labellum. Labellum, 1.2–2 cm, yellowish with central, yellow band, obovate. Anther spurred at base. Ovary sporadically hairy. Flowering in August (Flora of China Editorial Committee 2000).

C) Medicinal parts:
The dry root or rhizome of *Curcuma longa* L. has been used.

D) Chemical constituents:
**Curcumin and its derivatives** (curcumin, *p,p*'-dihydroxydicinnamoylmethane (bisdemethoxycurcumin), *p*-hydroxycinnamoylferuloylmethane (demethoxycurcumin) dihydrocurcumin); **sesquiterpenoids** (curlone, turmeronols A and B, germacrone-13-al, 4-hydroxybisabola-2,10-diene-9-one, procurcumadiol, curcumenone, dihydrocurdione, 4-methoxy-5-hydroxybisabola-2,10-diene-9-one, 2,5-dihydroxybisabola-3,10-diene, (4S,5S)-germacron-4,5-epoxide, α-turmerone, bisacurone, curcumenol, isoprocurcumenol, zedoaronediol, procurcumenol, eiprocurcumenol, 4,5-dihydroxybisabola-2, 10-diene); **acidic polysaccharides** (utonans A, B, C, D); **volatile oil** (α-turmerone, ar-turmerone, curcumene, germacrone, ar-curcumene, cineole, α-terpinene, curcumol, curzerenone, curdione, α-pinene, β-pinene, limonene, linalool, β-caryophyllene, borneol, etc.); **others compounds** (campesterol, stigmasterol, β-sitosterol, cholesterol; aliphatic acid and metallic elements (K, Na, Mg, Ca, Mn, Fe, Cu and Zn)) (Karioti et al. 2011; Chane-Ming et al. 2002).

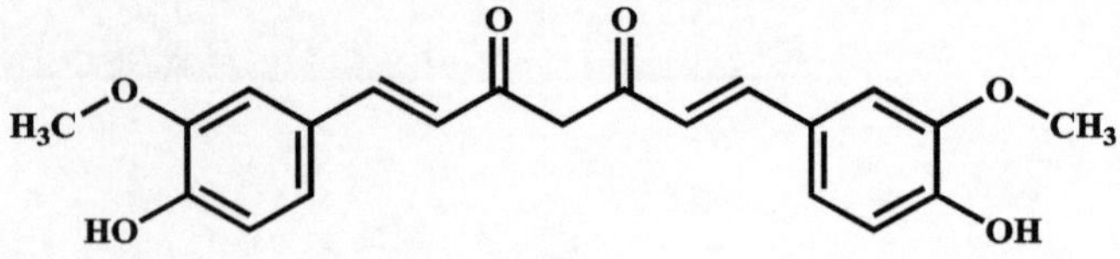

Curcumin

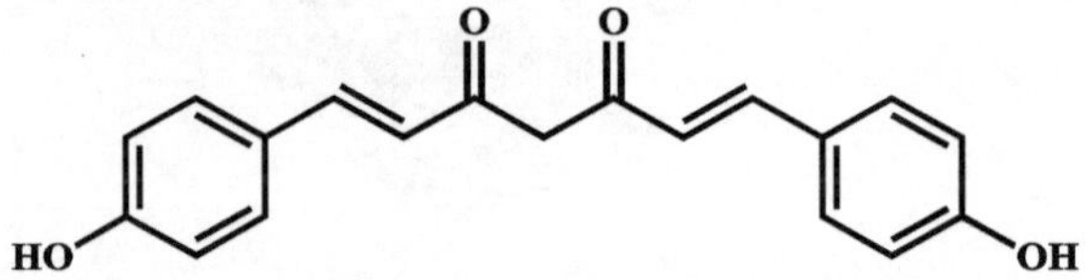

Bisdemethoxycurcumin

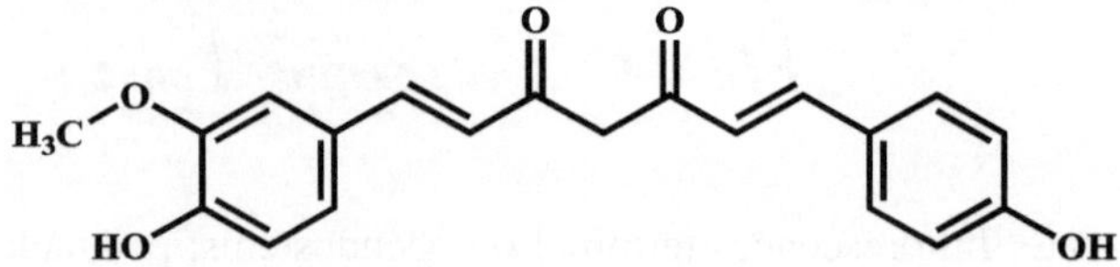

Demethoxycurcumin

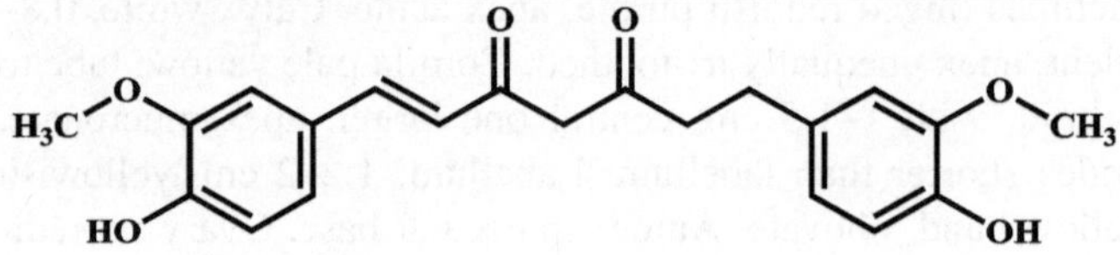

Dihydrocurcumin

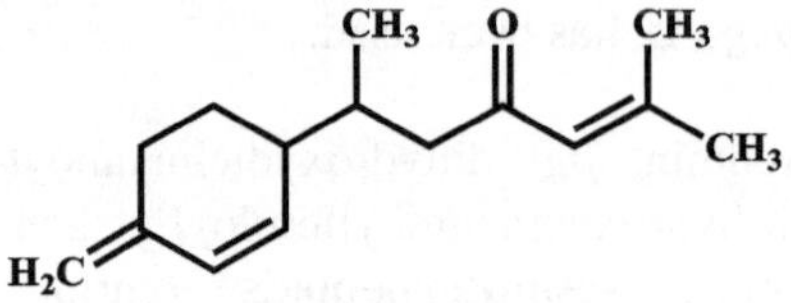

Curlone

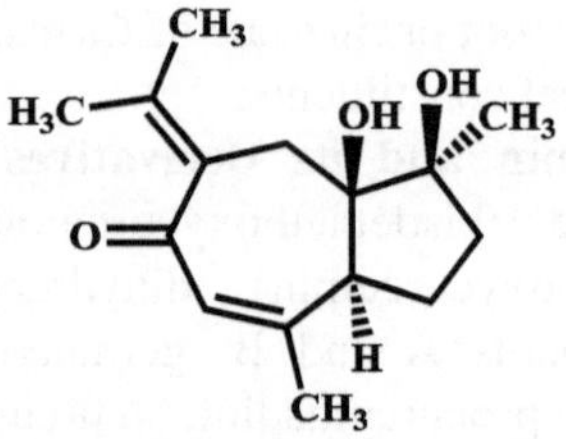

Procurcumadiol

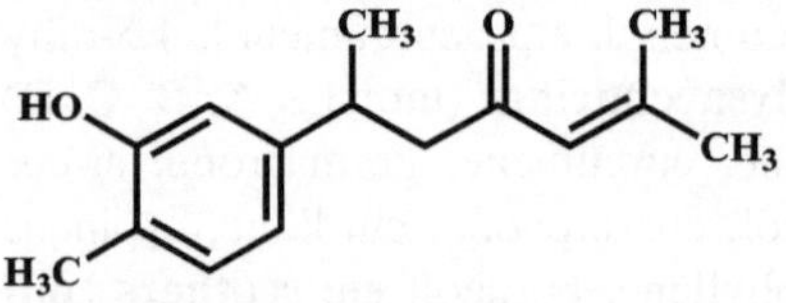

Turmeronol A

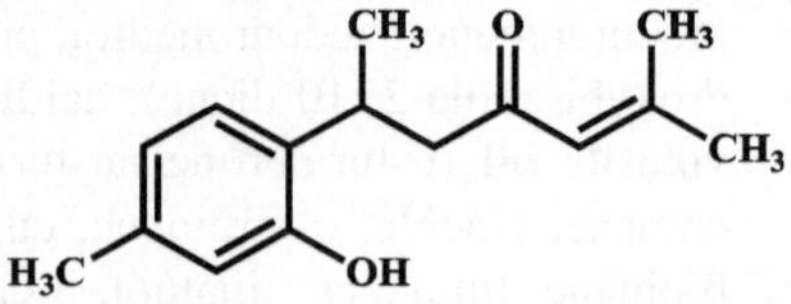

Turmeronol B

| | |
|---|---|
| 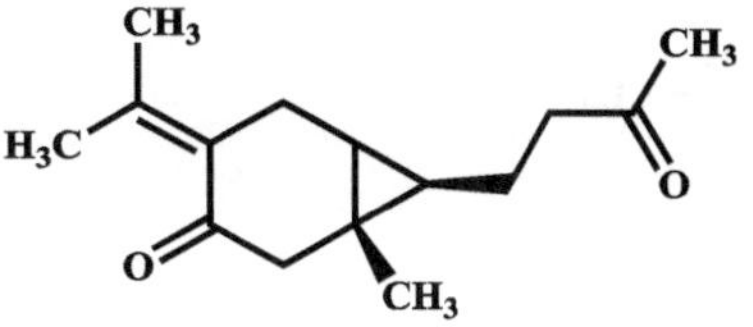 Curcumenone | 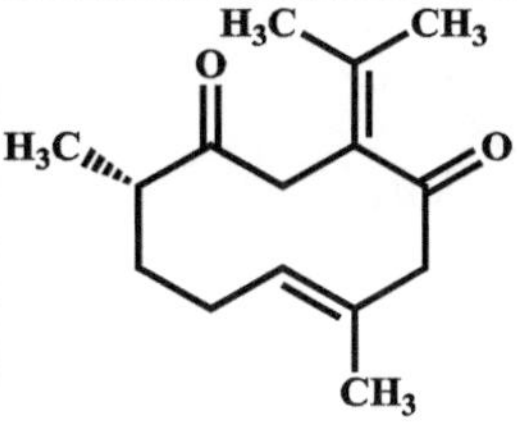 Dehydrocurdione |
| 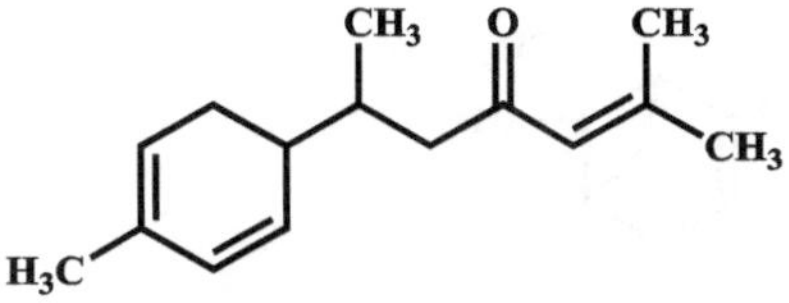 α-Turmerone | 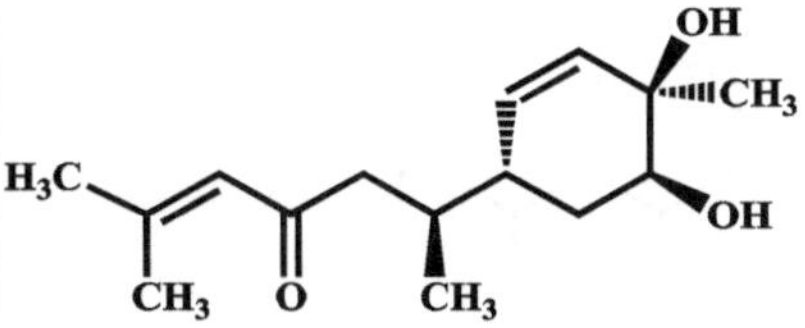 Bisacurone |
| 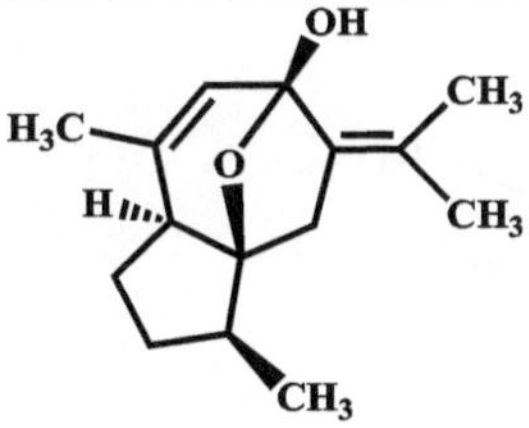 Curcumenol | 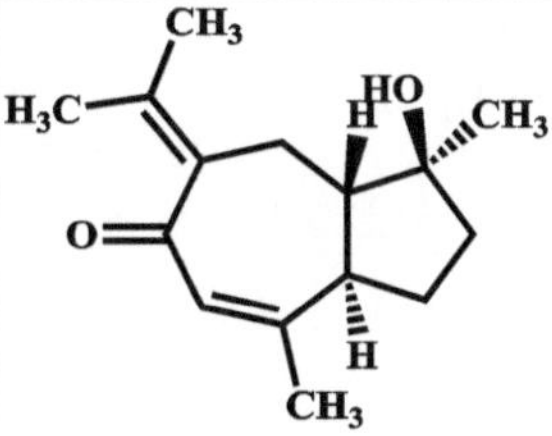 Procurcumenol |
| 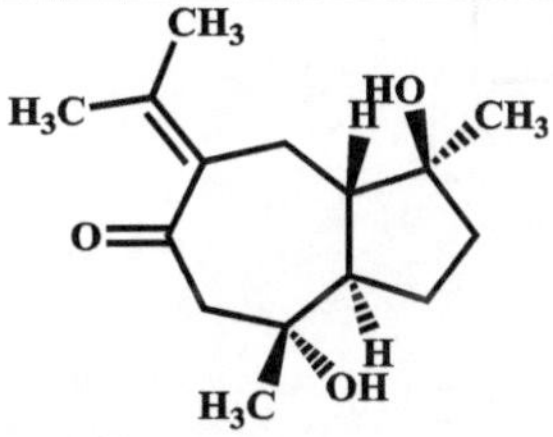 Zedoaronediol | 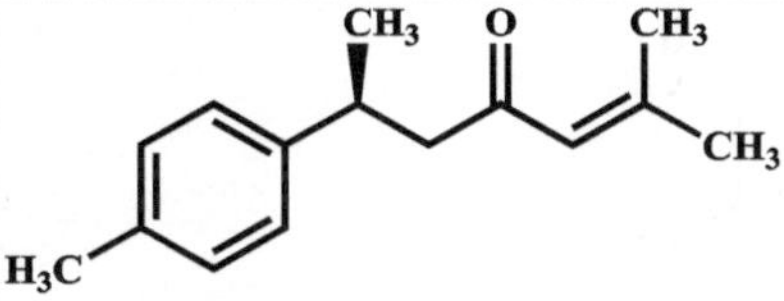 ar-Turmerone |
| 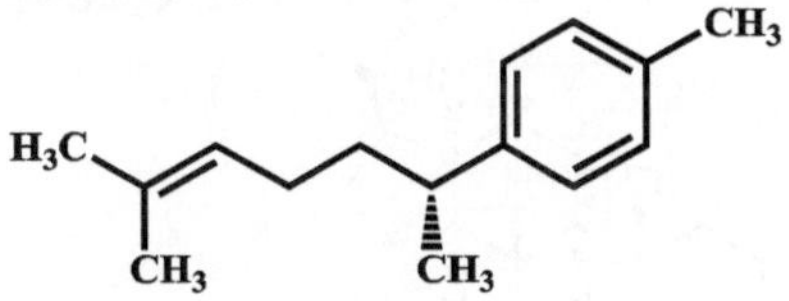 α-Curcumene | 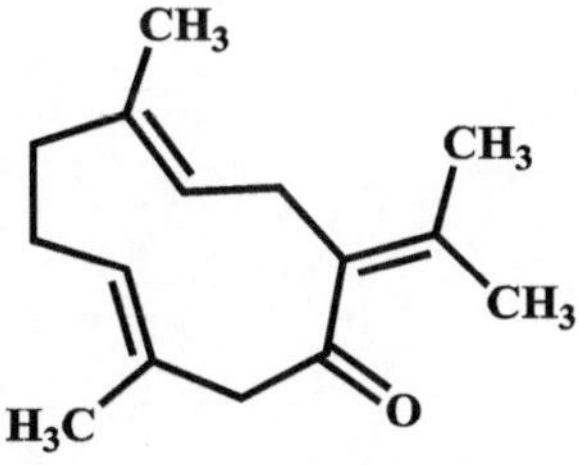 Germacrone |

| | |
|---|---|
| 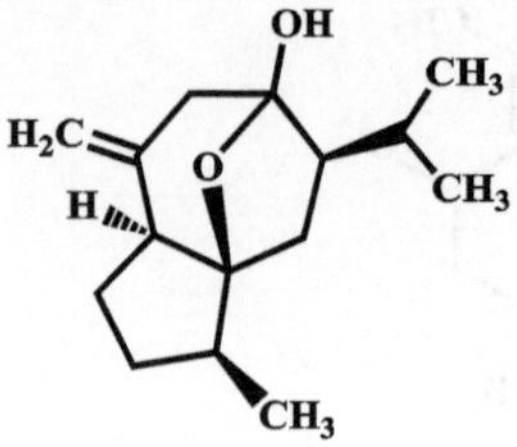<br>Curcumol | 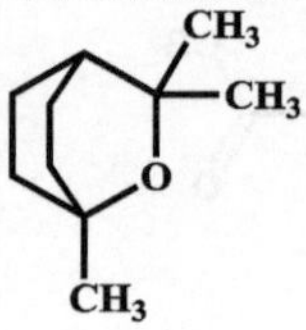<br>1,8-Cineole |
| 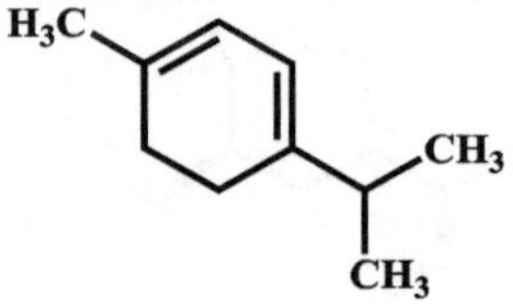<br>α-Terpinene | 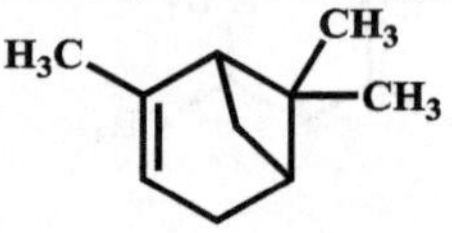<br>α-Pinene |
| 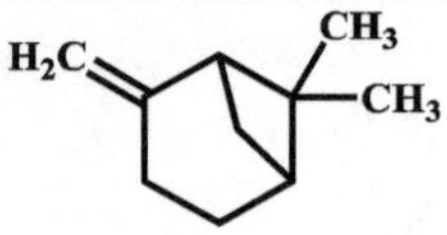<br>β-Pinene | 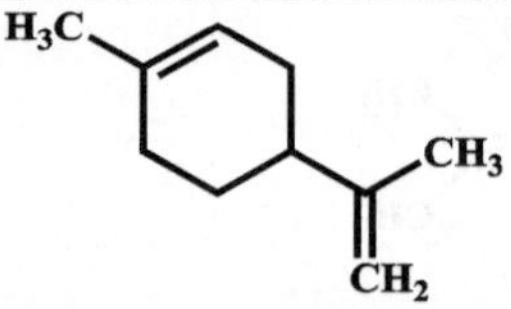<br>Limonene |
| 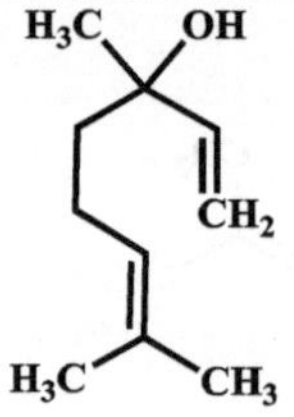<br>Linalool | 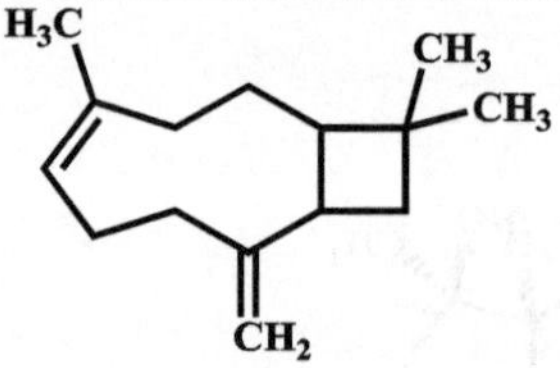<br>β-Caryophyllene |
| 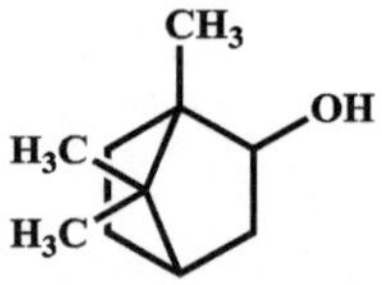<br>Borneol | 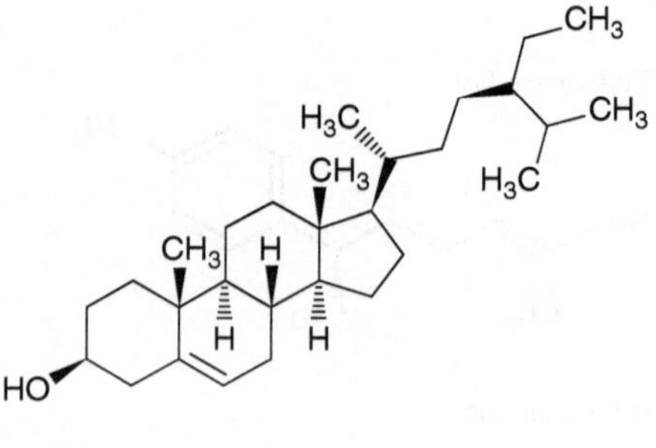<br>β-Sitosterol |

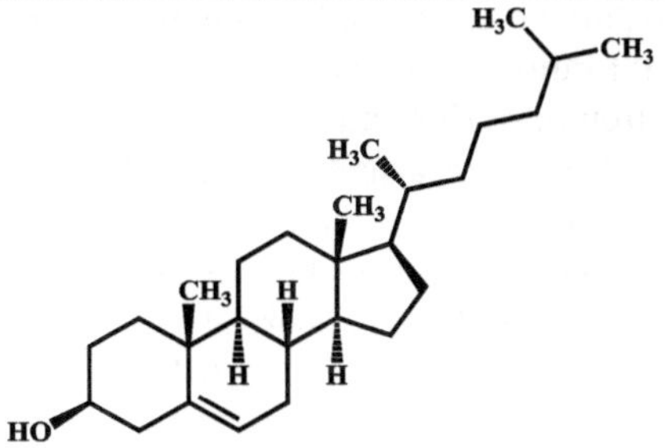

Cholesterol

4) Pharmacology and bioactivities:

A) In vitro studies:
Not found.

B) In vivo studies:
The antidepressant effects of *Curcuma longa* may be mediated by actions in the central monoaminergic neurotransmitter systems and its anti-inflammatory aptitude.

The extract at oral doses of 560 mg/kg for 14 days significantly decreased the duration of immobility in TST and FST in mice, indicating that the activity of *Curcuma longa* in antidepression may be mediated in part through MAO-A inhibition in mouse brain (Yu et al. 2002).

Curcumin administration to rats (or mice) reduced the immobility time in FST (or TST) and reversed the bilateral olfactory bulbectomy-induced behavioral abnormalities such as hyperactivity in OFT, as well as deficits in step-down passive avoidance. Curcumin also reversed depression-induced low levels of 5-HT, NA, 5-HIAA, and DOPAC in brain (Xu et al. 2005a, b). Chronic treatment with curcumin significantly reversed the CMS-induced behavioral abnormalities (reduced sucrose preference and decreased locomotor activity) in stressed rats. Additionally, curcumin effectively inhibited cytokine gene expression at both the mRNA and the protein level and reduced the activation of NF-κB (Jiang et al. 2013).

After one-week administration, turmerone produced antidepressant-like effects seemingly by increasing the monoamines level and decreasing the MAO-A activity. Turmerone (2.5, 5.0 mg/kg, p.o.) significantly reduced the immobility time of mice in both the FST and TST, and decreased the corticosterone level in the blood while it increased: the levels of 5-HT in cortex, striatum, hippocampus, and hypothalamus; the level of NE in striatum and hippocampus; the levels of MHPG and DOPAC in hypothalamus; the level of 5-HIAA in striatum; and the level of DA in striatum, hippocampus, and hypothalamus. Turmerone (2.5, 5.0 mg/kg) decreased the activity of MAO-A in the frontal cortex and hippocampus of mouse brain (Liao et al. 2013).

In a randomized controlled trial, 60 patients diagnosed with major depressive disorder were randomized in a 1:1:1 ratio for 6 weeks and received treatment with fluoxetine (20 mg) and curcumin (1000 mg) individually or in combination. The proportion of responders as measured by the HAM-D17 scale was higher in the combination group (77.8 %) than in the fluoxetine (64.7 %) and the curcumin (62.5 %) groups; however, these data were not statistically significant. The mean change in HAM-D17 score at the end of 6 weeks was comparable in all three groups (Sanmukhani et al. 2014).

5) Indications and usage:
   Activate blood and move qi, dredge meridians, and alleviate pain. Treat pain of the heart, chest, hypochondria, and abdomen; the irregular menstruation, dysmenorrhea, amenorrhea, and postpartum tormina due to qi stagnation and blood stasis; and arthralgia syndrome due to wind dampness (Chinese Pharmacopoeia Commission 2010).
6) Precautions and side effects:
   The blood deficiency without stagnation of qi and blood stasis must be cautious to use it (Chinese Pharmacopoeia Commission 2010).
7) Dosage:
   A) Internal use:
      3–10 g/day, decocted in water for oral administration, taken twice a day, generally for 5–7 days, modificated according to symptoms (Chinese Pharmacopoeia Commission 2010).
   B) External use:
      Proper dose for external application.

## 7.40 References


Flora of China Editorial Committee. Flora of China. Beijing and St. Louis: Science Press and Missouri Botanical Garden Press. 2000; 24: 361.

Karioti A, Fani E, Vincieri FF, Bilia AR. Analysis and stability of the constituents of *Curcuma longa* and *Harpagophytum procumbens* tinctures by HPLC-DAD and HPLC–ESI-MS. J Pharmaceut Biomed Analysis. 2011;55(3): 479–86.

Chane-Ming J, Chane-Ming J, Chalchat JC, Cabassu P. Chemical composition of essential oils from rhizomes, leaves and flowers of *Curcuma longa* L. from Reunion Island. J Essential Oil Res. 2002;14(4): 249–251.

Yu ZF, Kong LD, Chen Y. Antidepressant activity of aqueous extracts of *Curcuma longa* in mice. J Ethnopharmacol. 2002;83:161–5.

Xu Y, Ku BS, Yao HY, Lin YH, Ma X, Zhang YH, Li XJ. Antidepressant effects of curcumin in the forced swim test and olfactory bulbectomy models of depression in rats. Pharmacol Biochem Behav. 2005;82(1): 200–6.

Xu Y, Ku BS, Yao HY, Lin YH, Ma X, Zhang YH, Li XJ. The effects of curcumin on depressive-like behaviors in mice. Eur J Pharmacol. 2005;518(1): 40–6.

Jiang H, Wang Z, Wang YH, Xie K, Zhang QR, Luan QS, Chen WQ, Liu DX. Antidepressant-like effects of curcumin in chronic mild stress of rats: Involvement of its anti-inflammatory action. Prog Neuropsychopharmacol Biol Psychiatry. 2013;47 33–9.

Liao JC, Tsai JC, Liu CY, Huang HC, Wu LY, Peng WH. Antidepressant-like activity of turmerone in behavioral despair tests in mice. BMC Complement Alternat Med. 2013;13: 299.
Sanmukhani J, Satodia V, Trivedi J, Patel T, Tiwari D, Panchal B, Goel A, Tripathi CB. Efficacy and safety of curcumin in major depressive disorder: a randomized controlled trial. Phytother Res. 2014;28(4): 579–85.
Chinese Pharmacopoeia Commission. Pharmacopoeia of People's Republic of China. Beijing: China Medical Science Press. 2010.

## 7.41 *Rhodiola rosea L.*

### *7.41.1* Rhodiolae Rosea

#### 7.41.1.1 (Crassulaceae)

1) Common names:
   Rosenroot, golden root, Rhodiola, roseroot, Arctic root, Orpin rose, Rhodiole rougeâtre, Saoluomabuer (Tibetan name), Hong-jing-tian (Pinyin in Chinese).
2) Photograph of the species (Fig. 7.20)
3) Description of the species:
   A) Habitat and world distribution:
      Forested, grassy, or rocky slopes; 1800–2700 m. Central of Gansu, Hebei, Jilin, Shanxi, and Xinjiang [Japan, Kazakhstan, Korea, Mongolia, Russia; Europe, North America] (Flora of China Editorial Committee 2001).
   B) Morphology of the plant:
      Roots erect, robust. Caudex short. Caudex leaves scalelike. Flowering stems 10–30 cm. Stem leaves remote, sessile, oblong, elliptic-oblanceolate, or

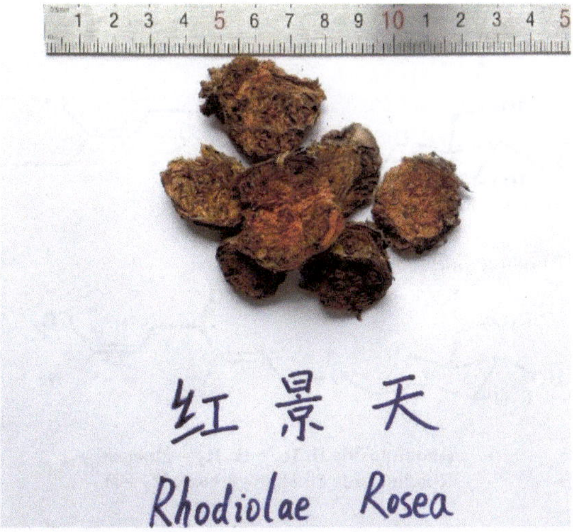


**Fig. 7.20** *Rhodiolae Rosea*

obovate, 0.7–3.5 × 0.5–1.8 cm, base subamplexicaul, margin entire, apically few dentate, or serrate, apex acute to acuminate. Inflorescences corymbiform or capitate, compact, ca. 2 × 3–6 cm, many flowered. Flowers unisexual, male ones unequally 4-merous. Sepals lanceolate-linear, ca. 1 mm, apex obtuse. Petals greenish yellow or yellow, linear-oblanceolate to oblong, ca. 3 mm, apex obtuse. Stamens 8, longer than petals. Nectar scales oblong, apically slightly narrowed, 1–1.5 × ca. 0.6 mm, apex emarginate. Carpels of female flowers erect. Styles recurved. Follicles lanceolate to linear-lanceolate, 6–8 mm, apical beak ca. 1 mm. Seeds lanceolate, ca. 2 mm, winged at one end. Flowering in April–August, fruiting in July–September (Flora of China Editorial Committee 2001).

C) Medicinal parts:
Dry root and rhizome of *Rhodiola rosea* L. have been used.

D) Chemical constituents:
The roots and rhizomes of *Rhodiola rosea* L. contain **essential oils** (the main classes: monoterpene hydrocarbons (25.40 %), monoterpene alcohols (23.61 %), and straight-chain aliphatic alcohols (37.54 %), e.g, *n*-decanol (30.38 %), geraniol (12.49 %) and 1,4-*p*-menthadien-7-ol (5.10 %), rosiridol, rosiridin, etc.), **glycosides** (alcoholic glycosides (rhodiolosides -A, -B, -C, -D, -E), cyanogenic glycosides (rhodiocyanoside A, lotaustralin), etc.), **organic acids** (oxalic, citric, malic, gallic, succinic acids), **flavonoids** (gossypetin, kaempferol, quercetin, herbacetin, etc.), **others** (phenylethanoids, phenylpropanoids, aryl glycosides, proanthocyanidins, and other gallic acid derivatives, fats, waxes, sterols, tannins and proteins, etc.) (Panossian et al. 2010; Petsalo et al. 2006).

*n*-Decanol

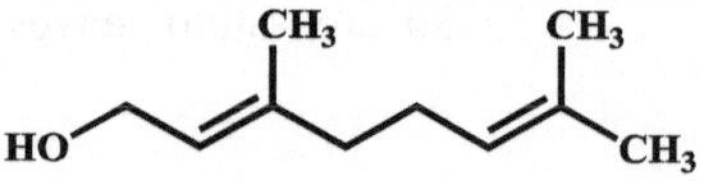


Geraniol

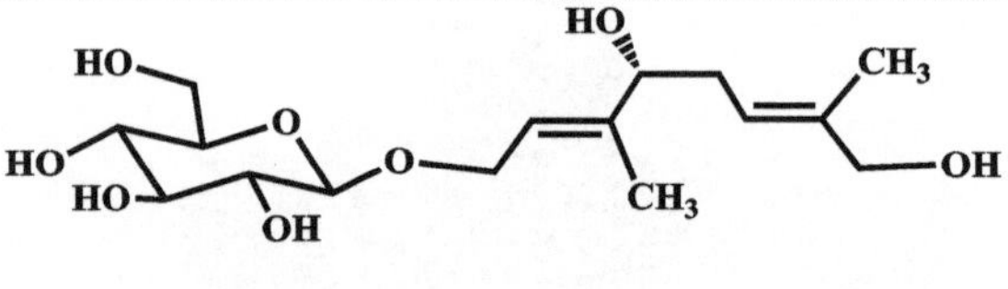


Rhodioloside A

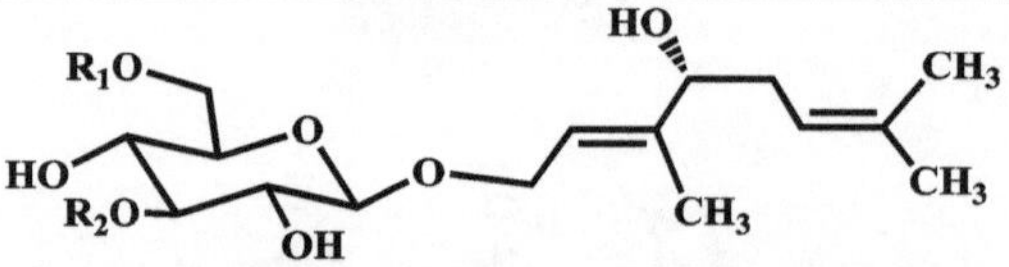


**Rhodioloside B: $R_1$ = H, $R_2$ = glucose**
**Rhodioloside C: $R_1$ = glucose, $R_2$ = H**

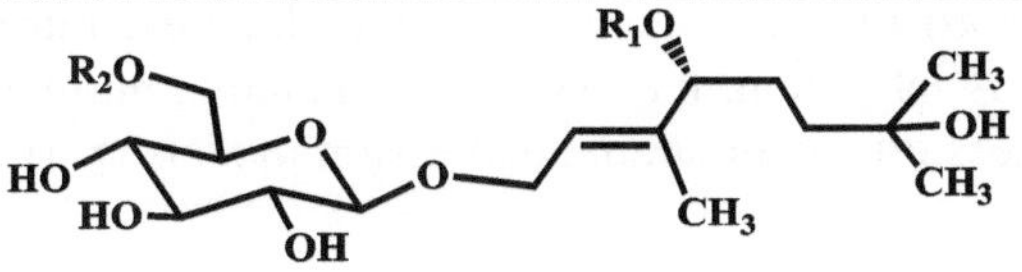

**Rhodioloside D: $R_1 = R_2 = H$**
**Rhodioloside E: $R_1 = H$, $R_2 = Ara$**

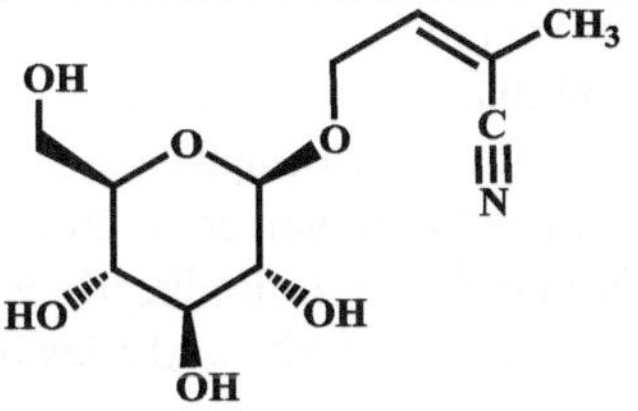

Rhodiocyanoside A

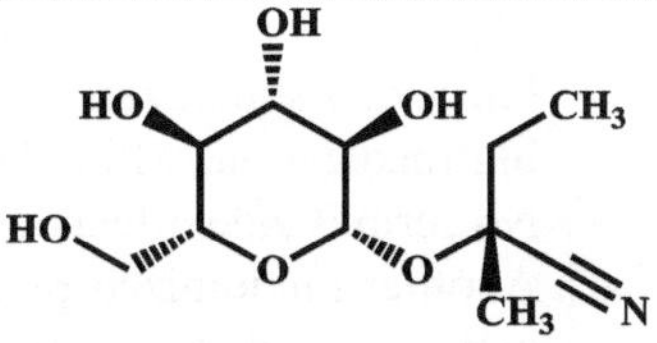

Lotaustralin

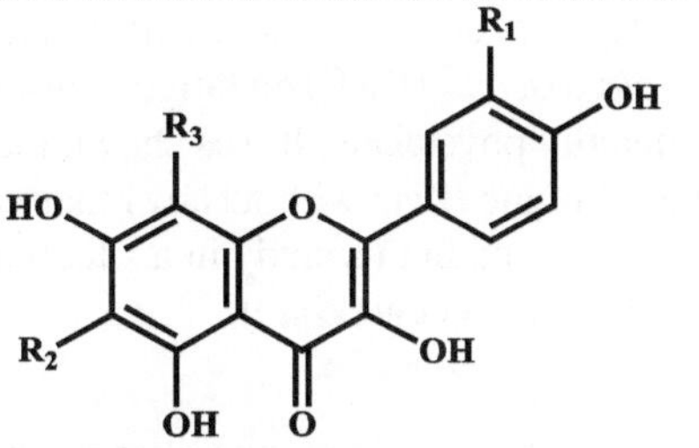

**Quercetin: $R_1 = OH$, $R_2 = R_3 = H$**
**Kaempferol: $R_1 = R_2 = R_3 = H$**
**Gossypetin: $R_1 = R_2 = OH$, $R_3 = H$**
**Herbacetin: $R_1 = R_2 = H$, $R_3 = OH$**

4) Pharmacology and bioactivities:

A) In vitro studies:
A study found that *R. rosea* supplementation could protect cultured cells against ultraviolet light, paraquat, and $H_2O_2$. However, it did not alter the levels of the major antioxidant defenses nor did it markedly activate the antioxidant response element or modulate heme-oxygenase-1 expression levels at relevant concentrations. In addition, *R. rosea* extract was not able to significantly degrade $H_2O_2$ in vitro. These results suggest that in human cultured cells *R. rosea* does not act as an antioxidant and that its mode of action cannot be sufficiently explained through a prooxidant and hormetic mechanism (Schriner et al. 2009).

B) In vivo studies:
**Antidepression**: *R. rosea* may be beneficial for the treatment of depression via known psychopharmacological actions including neuroendocrine

modulation (inhibition of cortisol, stress-induced protein kinases, nitric oxide), monoamine oxidase A inhibition, monoamine modulation, normalization of 5-HT, and anti-stress effects in animal depression models (Sarris et al. 2011).

A phase III clinical trial was carried out as a randomized double-blind placebo-controlled study with parallel groups over 6 weeks. 89 patients with mild to moderate depression were randomized to receive placebo or the extract SHR-5 at 340 or 680 mg/day over a 6-week period. For individuals in SHR-5 groups, overall depression, together with insomnia, emotional instability, and somatization, but not self-esteem, improved significantly following medication, while the placebo group did not show such improvements. No serious side effects were reported in any groups (Darbinyan et al. 2007). Another randomized placebo-controlled trial found *R. rosea* produced less antidepressant effect versus sertraline, but it resulted in significantly fewer adverse events and was better tolerated (Mao et al. 2015).

**Anti-fatigue effect**: A randomized, placebo-controlled, double-blind, cross-over study investigated the effect of repeated low-dose treatment with a standardized extract SHR-5 of rhizome *Rhodiola rosea* L. (RRE) on fatigue during night duty among a group of 56 young, healthy physicians. It was shown that *R. rosea* SHR-5 extract possesses a clear antifatigue effect without any reported adverse reactions or side effects, at the dosage used in this study in a situation of a moderate level of fatigue and stress (Darbinyan et al. 2000).

5) Indications and usage:
Tonifying qi and activating blood, clearing away the lung heat, and eliminating stasis to subdue swelling. Treat for leucorrhea, cough due to pneumonia, hematemesis, hemoptysis, traumatic injury, burn and scald for external application, deficiency of spleen qi, deficiency of lung yin and cough due to lung heat, neurosis, chest stuffiness, cardiodynia, hemiplegia, and altitude stress (State Administration of Traditional Chinese Medicine 1999).
6) Precautions and side effects:
The overdose of *R. rosea* may result in allergy, palpitations, stomach discomfort, headache, nausea, irritability, insomnia, fatigue, and unpleasant sensations (Iovieno et al. 2011).
7) Dosage:
   A) Internal use:
   3–6 g/day, decocted in water for oral administration, taken twice a day, generally for 5–7 days, modificated according to symptoms (State Administration of Traditional Chinese Medicine 1999).
   B) External use:
   For external use, make it into powder and use as paste or mash (State Administration of Traditional Chinese Medicine 1999).

## 7.42 References

Flora of China Editorial Committee. Flora of China. Beijing and St. Louis: Science Press and Missouri Botanical Garden Press, Vol. 2001;8:261.
Panossian A, Wikman G, Sarris J. Rosenroot (*Rhodiola rosea*): traditional use, chemical composition pharmacology and clinical efficacy. Phytomedicine. 2010;17(7): 481–93.
Petsalo A, Jalonen J, Tolonen A. Identification of flavonoids of *Rhodiola rosea* by liquid chromatography-tandem mass spectrometry. J Chromatograph A. 2006;1112(1):224–31.
Schriner SE, Avanesian A, Liu Y, Luesch H, Jafari M. Protection of human cultured cells against oxidative stress by *Rhodiola rosea* without activation of antioxidant defenses. Free Radic Biol Med. 2009;47(5): 577–84.
Sarris J, Panossian A, Schweitzer I, Stough C, Scholey A. Herbal medicine for depression, anxiety and insomnia: a review of psychopharmacology and clinical evidence. Eur Neuropsychopharmacol. 2011;21(12): 841–60.
Darbinyan V, Aslanyan G, Amroyan E, Gabrielyan E, Malmstrom C, Panossian A. Clinical trial of *Rhodiola rosea* L. extract SHR-5 in the treatment of mild to moderate depression. Nord J Psychiatry. 2007;61(5): 343–348.
Mao JJ, Xie SX, Zee J, Soeller I, Li QS, Rockwell K, et al. *Rhodiola rosea* versus sertraline for major depressive disorder: A randomized placebo-controlled trial. Phytomedicine. 2015;22(3): 394–399.
Darbinyan V, Kteyan A, Panossian A, Gabrielian E, Wikman G, Wagner H. *Rhodiola rosea* in stress induced fatigue--a double blind cross-over study of a standardized extract SHR-5 with a repeated low-dose regimen on the mental performance of healthy physicians during night duty. Phytomedicine. 2000;7(5):365–371.
State Administration of Traditional Chinese Medicine. Chinese Materia Medica Commission. Shanghai: Shanghai Scientific and Technical Publishers. 1999
Iovieno N, Dalton ED, Fava M, Mischoulon D. Second-tier natural antidepressants: review and critique. J Affect Disord. 2011;130:343–357.

## 7.43 *Valeriana officinalis* L.

### *7.43.1 Radix Valerianae*

#### 7.43.1.1 (Valerianaceae)

1) Common names:
   *Valerian root*, Xie-cao, Xiao-jiu-jia, Lu-zi-cao, Ba-di-ma, Qi-li-xiang, Xiang-cao, Mao-shi-cai (Pinyin in Chinese).
2) Photograph of the species (Fig. 7.21)
3) Description of the species:
   A) Habitat and world distribution:
      Grassy slopes, forests, by streams; below 2500 m, but up to 4000 m in Qinghai, Sichuan, and Xizang. Anhui, Chongqing, Gansu, Guizhou (Fanjingshan, Huangping, Kaili), Hebei, Henan, West of Hubei, West of Hunan (Sangzhi, Xinning), Jiangxi (Lu Shan), North and southeast of Nei Mongol, Qinghai, Shaanxi, Shandong (Changqing, Kunyu Shan, Taishan), Shanxi, Sichuan, Taiwan, Xizang (Jomda, Qamdo, Riwoqê), and Zhejiang

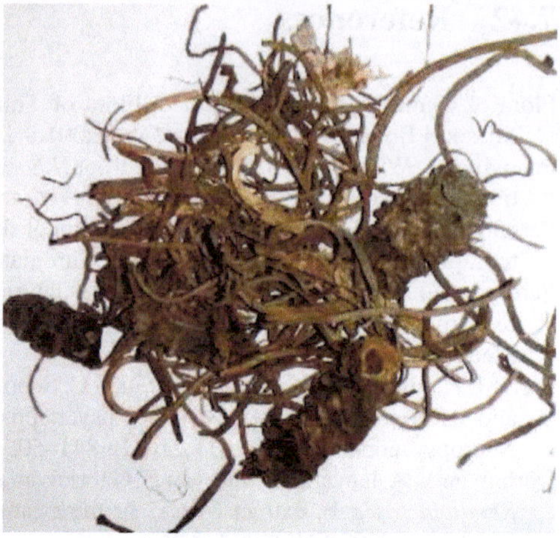

**Fig. 7.21** *Radix Valerianae*

(Tianmushan) [other countries: Japan, Russia, Europe] (Flora of China Editorial Committee 2000).

B) Morphology of the plant:

Perennial herb. Plants up to 150 cm tall, rarely stoloniferous. Rhizomes reduced; roots fascicled, clavate, 1.5–2.5 mm in diameter. Stems hirsute or hispidulous, particularly at nodes, rarely glabrous. Basal and lowermost cauline leaves usually wilted at anthesis; cauline leaves petiolate to sessile; petiole with 0–8 cm; blade 5–15 × 3–8 cm, ovate to broadly ovate, pinnatisect; segments 7–15, lanceolate or linear, glabrous or more or less hispidulous, base decurrent, margin entire or coarsely serrate; terminal and lateral segments similar in shape and size. Inflorescence corymbiform, large and compact even in fruit, or with 2 distinct, secondary lateral inflorescences; bracts leaflike below, small and linear-lanceolate otherwise; bracteoles elliptic-oblong, oblanceolate, or linear-lanceolate, margin scarious-ciliate, apex awned. Corolla with 4–6 mm; purplish red, pink, or white; funnelform; lobes elliptic. Stamens and style strongly exserted. Achenes narrowly ovoid, 4–5 mm, both sides glabrous or pubescent. Flowering in May–July, fruiting in June–October (Flora of China Editorial Committee 2000).

C) Medicinal parts:

The roots and rhizomes of *Valeriana officinalis* L. have been used.

D) Chemical constituents:

**Alkaloids** (actinidine, chatinine, valerianine, and valerine alpha-methyl pyrryl ketone) and **essential oil** (borneol, bornyl acetate, valeric acid, valerenic acid, valerenal, and valeranone). **Valepotriates** are esterified iridoid-monoterpenes. The principal diene valepotriates are valtrate, isovalerate, 7-desisovaleroyl-7-acetylvaltrate, and 7-homovaltrate, and the main monoene derivatives are isovaleroxyhydroxydidrovaltrate and didrovaltrate (Dennehy et al. 2005).

| | |
|---|---|
| 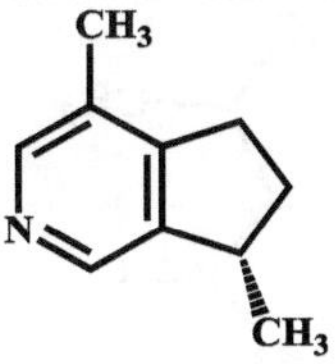 Actinidine | 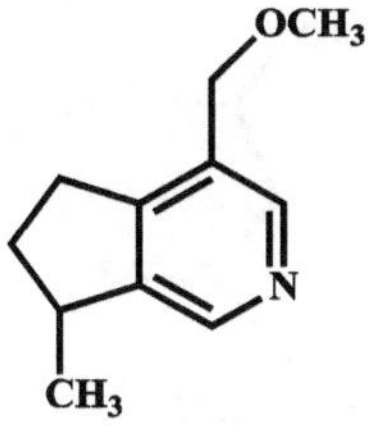 Valerianine |
| 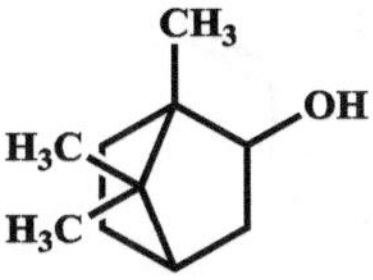 Borneol | 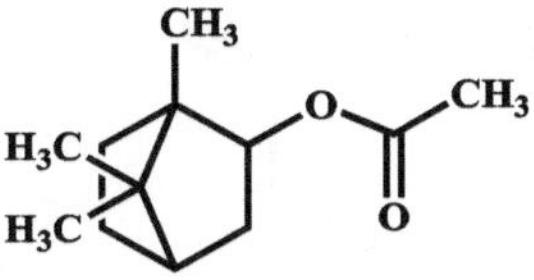 Bornyl acetate |
| 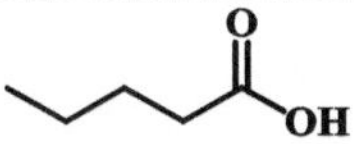 Valeric acid | 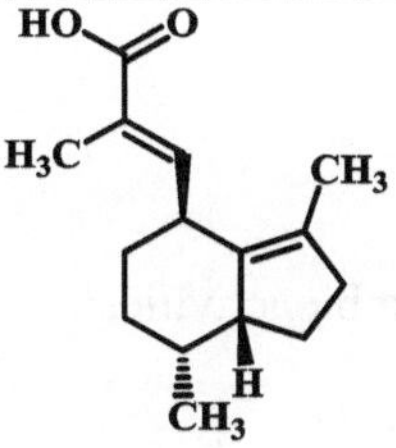 Valerenic acid |
| 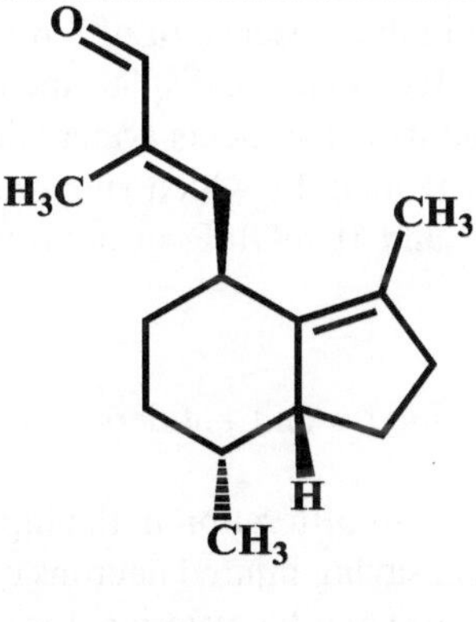 Valerenal | 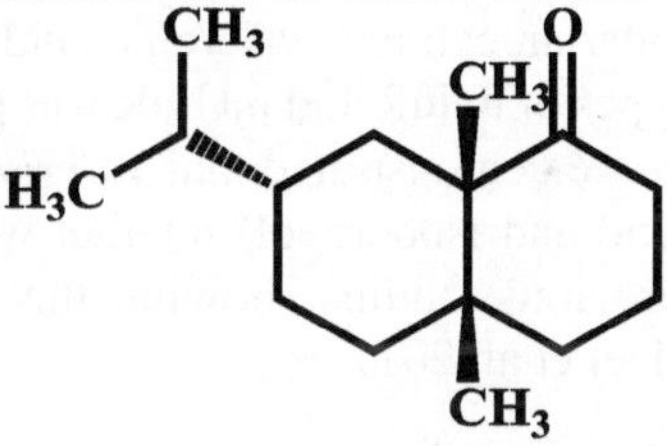 Valeranone |

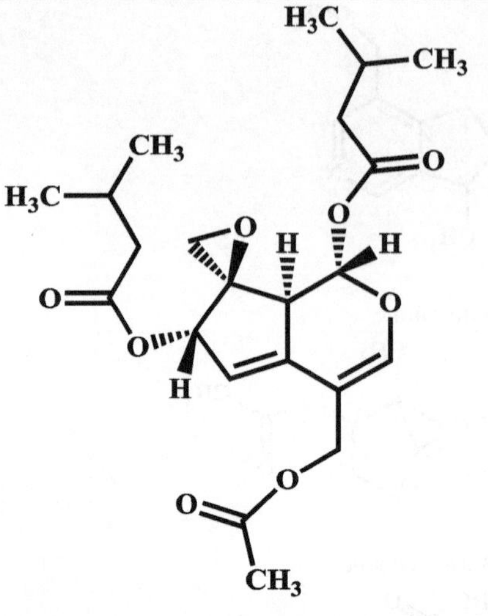

Valtrate

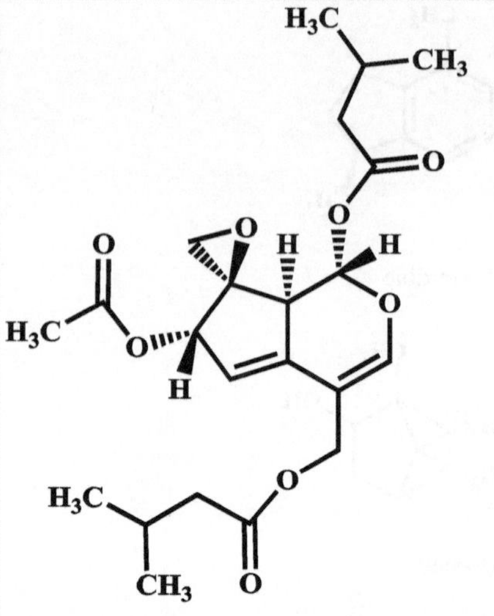

Isovaltrate

4) Pharmacology and bioactivities:

A) In vitro studies:

Many studies suggest that metabotropic glutamate receptors (mGluR) are involved in anxiety. One study using receptor-binding assays with aqueous valerian extracts, valerenic acid, and isoborneol in the presence of different types of iGluR and mGluR was performed. The [$^3$H] glutamate displacement curve demonstrated that *Valeriana officinalis* and its constituents (valerenic acid and isoborneol) interact with Group I and II mGluR, which supports previous studies showing the role of Group I and II mGluR in anxiety (Del et al. 2011).

B) In vivo studies:

In vivo studies show that valerian displays anxiolytic and antidepressant activities.

Valerian may promote the level of 5-HT and cell proliferation in the hippocampus of the depressive rats and plays a role in saving injured neurons of the hippocampus. After the administration of 100 mg/kg valerian for 3 weeks, the number of BrdU-positive cells, p-CREB-positive cells, and neurons in the hippocampus of the depressive rats were recovered to the normal status; 10 g/L sucrose intake was increased and recovered to that of normal control group of rats. The levels of 5-HT in the hippocampus in 100 mg/kg and 200 mg/kg valerian-treated groups were increased and recovered to normal level (Tang et al. 2008a, b). It also decreased the serum cortisol level and the number of caspase-3 cells in the hippocampus of depressive rats to normal level (Qin et al. 2009).

From different valerian extracts, only the patented special extract phytofin Valerian 368 showed activity in the FST indicating antidepressant properties and displayd anxiolytic (EPM test) but no sedative effect (locomotor activity and ether-induced anesthesia tests) thus contributing to the sleep-enhancing properties of valerian (Hattesohl et al. 2008).

Moreover, a randomized placebo-controlled pilot study showed that the valepotriates may have a potential anxiolytic effect on the psychic symptoms of generalized anxiety disorder (Andreatini et al. 2002).

5) Indications and usage:
Induce tranquilization, stop bleeding, and alleviate pain. Treat for irregular menstruation, traumatic injury, lumbago, neurasthenia, dyspepsia, insomnia, and anxiety-induced sleep disturbances (Ioachimescu et al. 2012).
6) Precautions and side effects:
It is contraindicated to weak people with yin deficiency. Large doses are known to cause withdrawal symptoms when stopped. Those with liver disease are advised not to use valerian (Cohen and Del 2008). It has been reported to cause headaches and night terrors in some individuals (Garges et al. 1998).
7) Dosage:
    A) Internal use:
    3–9 g/day, decocted in water for oral dose, or steep in wine, served twice a day for 3–5 days, modificated according to symptoms (State Administration of Traditional Chinese Medicine 1999).
    B) External use:
    Make it into powder and use as paste or mash for application (State Administration of Traditional Chinese Medicine 1999).

## 7.44 References

Flora of China Editorial Committee. Flora of China. Beijing and St. Louis: Science Press and Missouri Botanical Garden Press. 2000;667, 670.

Dennehy CE, Tsourounis C, Horn AJ. Dietary supplement-related adverse events reported to the California Poison Control System. *Am J Healthsyst Pharm.* 2005;62(14): 1476–82.

Del VLM, Ayala-Marin YM, Ortiz-Sanchez CM, Torres-Hernandez BA, Abdalla-Mukhaimer S, Ortiz JG. Selective interactions of *Valeriana officinalis* extracts and valerenic acid with [$^3$H] glutamate binding to rat synaptic membranes. *Evidence-Based Complementary and Alternative Medicine*: *eCAM*, 2011;2011:403591.

Tang JY, Zeng YS, Chen QG, Qin YJ, Chen SJ, Zhong ZQ. Effects of *Valerian* on the level of 5-hydroxytryptamine, cell proliferation and neurons in cerebral hippocampus of rats with depression induced by chronic mild stress. *Zhong Xi Yi Jie He Xue Bao.* 2008;6(3): 283–8.

Tang JY, Zeng YS, Chen QG, Qin YJ, Chen SJ, Zhong ZQ. Effects of *Valerian* on Weight and Behavior of Depressive Rats Induced by Chronic Mild Stress. J Sun Yat-Sen Univ Med Sci. 2008;29(5): 541–5.

Qin YJ, Lin YK, Zhou CC, Li Y, Zhong ZQ. Zeng YS. Effects of *Valerian* on serum cortisol level and caspase-3 positive cells of hippocampus in depressive rats induced by chronic mild stress. Anatomy Res. 2009;31(2): 88–93.

Hattesohl M, Feistel B, Sievers H, Lehnfeld R, Hegger M, Winterhoff H. Extracts of *Valeriana officinalis* L. s.l. show anxiolytic and antidepressant effects but neither sedative nor myorelaxant properties. Phytomedicine. 2008;15(1–2), 2–15.
Andreatini R, Sartori VA, Seabra ML, Leite JR. Effect of valepotriates (valerian extract) in generalized anxiety disorder: a randomized placebo-controlled pilot study. Phytother Res. 2002;16(7): 650–4.
Ioachimescu OC, El-Solh AA. Pharmacotherapy of insomnia. Expert Opin Pharmacother. 2012;13(9): 1243–60.
Cohen DL, Del TY. A case of valerian-associated hepatotoxicity. J Clin Gastroenterol. 2008;42(8): 961–2.
Garges HP, Varia I, Doraiswamy PM. Cardiac complications and delirium associated with valerian root withdrawal. JAMA. 1998;280(18): 1566–7.
State Administration of Traditional Chinese Medicine. Chinese Materia Medica Commission. Shanghai: Shanghai Scientific and Technical Publishers. 1999.

## 7.45 *Ziziphus jujuba* Mill. var. spinosa (Bunge) Hu ex H. F. Chow

### *7.45.1 Ziziphi Spinosae Semen*

#### 7.45.1.1 (Rhamnaceae)

1) Common names:
   Spine date seed, Suan-Zao-Ren, Zao-Ren (Pinyin in Chinese).
2) Photograph of the species (Fig. 7.22)

**Fig. 7.22** *Ziziphi Spinosae Semen*

3) Description of the species:

   A) Habitat and world distribution:
      Hills, sunny dry slopes, plains. It is mainly distributed in Anhui, Gansu, Hebei, Henan, Jiangsu, Liaoning, Nei Mongol, Ningxia, Shandong, Shaanxi, Shanxi, Xinjiang, partly in the Korea, and Russian (China flora editorial board 2007).
   B) Morphology of the plant:
      Plants often shrubs. Branches erect, not tortuous, with spines. Leaves small. Drupe subglobose or broadly oblong, small, 0.7–1.2 cm in diam.; mesocarp sour-tasting, thin; stone obtuse at both ends. Flowering in June–July, fruiting in August–September (China flora editorial board 2007).
   C) Medicinal parts:
      The dry ripe seed of *Ziziphus jujuba* Mill.var. *spinosa* (Bunge) Hu ex H. F. Chow has been used.
   D) Chemical constituents:
      **C-glycosyl flavone** (spinosin) (Zhang et al. 2014); **triterpene saponins** (jujuboside A and jujuboside B) (Zhang et al. 2014); **fatty acids** (e.g., 9-octadecenoic acid, 9,12-octadecadienoic acid, hexadecanoic acid, 9-eicosenoic acid, stearic acid, and arachidic acid) (Xie et al. 2012).

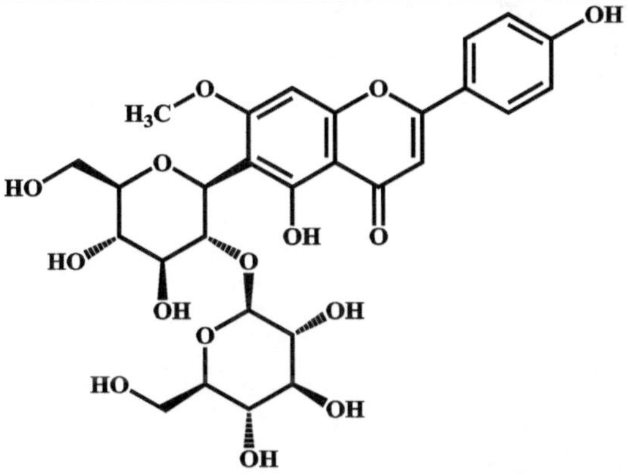


Spinosin

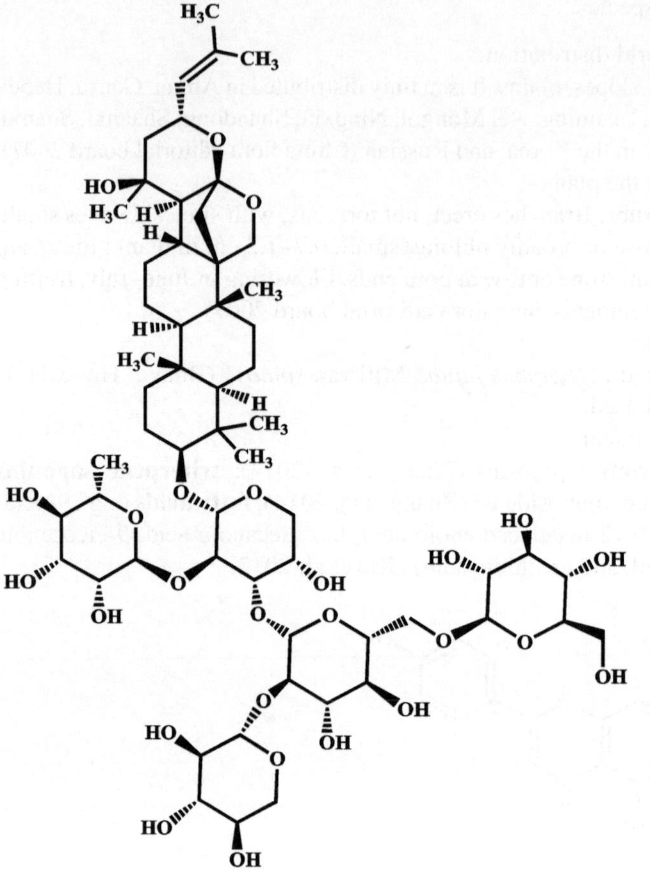


Jujuboside A

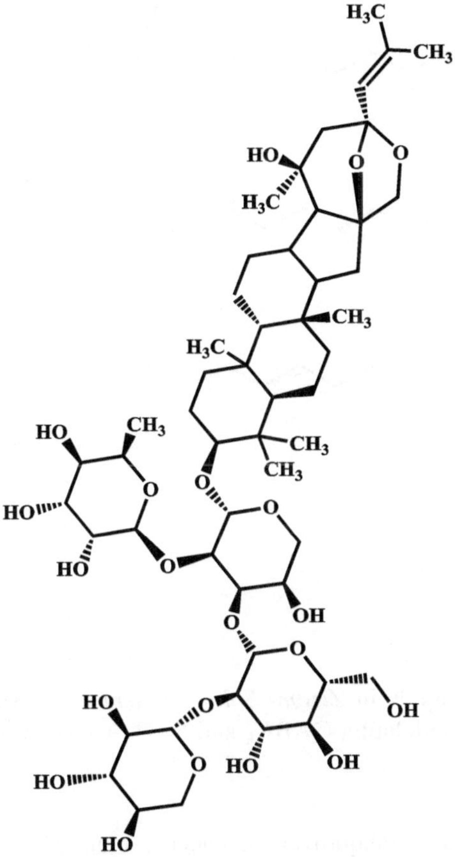


Jujuboside B

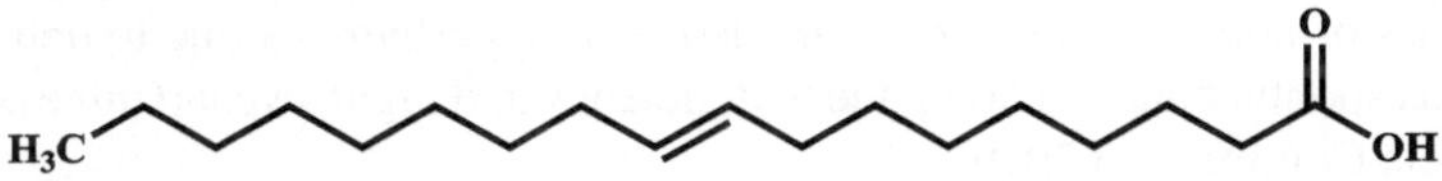


9-Octadecenoic acid

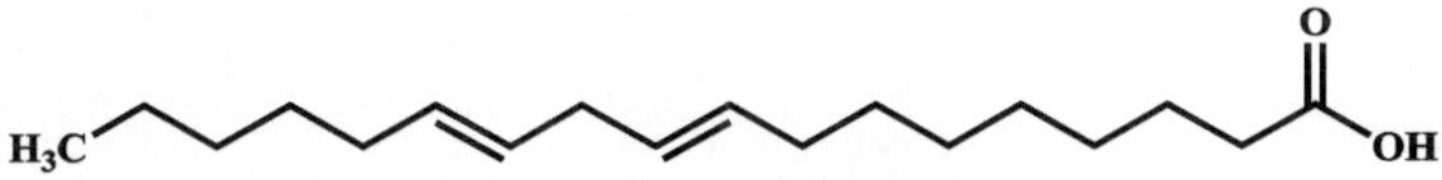


9,12-Octadecadienoic acid

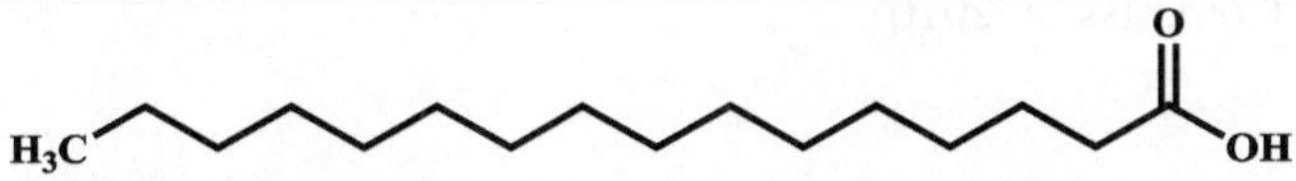


Hexadecanoic acid

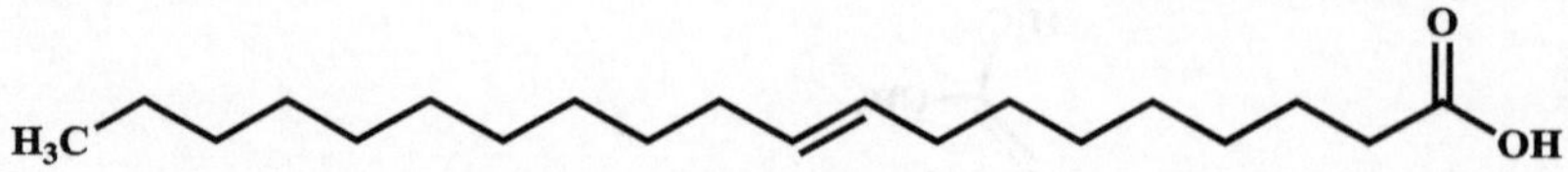

9-Eicosenoic acid

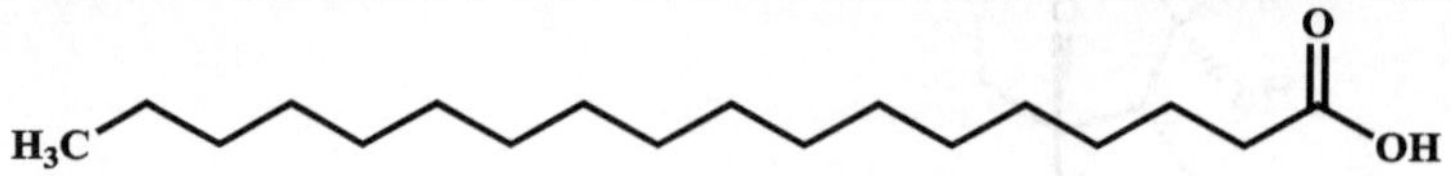

Stearic acid

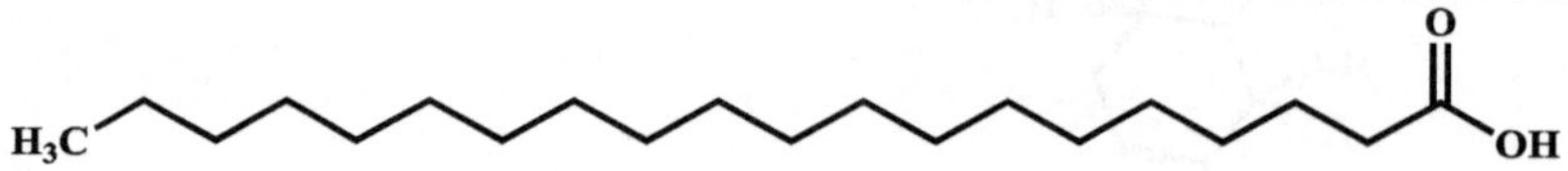

Arachidic acid

4) Pharmacology and bioactivities:

   A) In vitro studies:
      Not found.
   B) In vivo studies:
      Spinosin, one of the major flavonoids in *Ziziphi Spinosae Semen* (ZSS), exerts anxiolytic-like effects by modulating $GABA_A$ and $5\text{-}HT_1A$ receptors (Liu et al. 2015).

5) Indications and usage:
   Tonify heart and nourish liver, induce tranquilization, check sweating, and engender liquid. Treat syndrome of unsteadiness of heart-mind, such as palpitation, fearful throbbing, insomnia, and amnesia due to yin-blood deficiency and heart-mind loss of nourishment, and it can also treat hyperhidrosis due to deficiency, such as spontaneous sweating due to deficiency and night sweat (Chinese Pharmacopoeia Commission 2010).
6) Precautions and side effects:
   Not found.
7) Dosage:

   A) Internal use:
      9–15 g/day, decocted in water for oral dose, take powder 1.5–3 g/day, served 2–3 times a day for 7 days, modificated according to symptoms (Chinese Pharmacopoeia Commission 2010).
   B) External use:
      Not found.

## 7.46 References

China flora editorial board, Chinese academy of sciences. The flora of China. Science. 2007; 12: 119–23.

Zhang Y, Qiao L, Song M, Wang L, Xie J, Feng H. HPLC-ESI-MS/MS analysis of the water-soluble extract from *Ziziphi spinosae semen* and its ameliorating effect of learning and memory performance in mice. Pharmacognosy Magazine. 2014;10(40):509–16.

Xie J, Zhang Y, Wang L, Qi W, Zhang M. Composition of fatty oils from *semen ziziphi spinosae* and its cardiotonic effect on isolated toad hearts. Natural Product Res. 2012;26(5): 479–83.

Liu J, Zhai WM, Yang YX, Shi JL, Liu QT, Liu GL, et al. GABA and 5-HT systems are implicated in the anxiolytic-like effect of spinosin in mice. Pharmacol Biochem Behav. 2015;128: 41–9.

Chinese Pharmacopoeia Commission. Pharmacopoeia of People's Republic of China. Beijing, China Medical Science Press. 2010

# Chapter 8
# Antidepressant Medicinal Plants and Compounds Used in Traditional Medicines in North America

**Fabiola Domínguez, Eva González-Trujano, Juan Manuel Gallardo, and Sandra Orozco-Suárez**

**Abstract** The knowledge of medicinal plants is ancient and has transcended generations through tradition. It is considered that our ancestors obtained knowledge of these species after serving distinguish between eating and those who had any effect on the body, so from this they began to differentiate and select; thus, we have inherited an extensive knowledge of the use of plants to several diseases, including the nervous system. During the 1960s, research concerning medicinal plants that affect the central nervous system was particularly focused on remedies used by several different cultures. For example, Albert Hoffman discovered that ergot alkaloids are derivatives of lysergic acid, and the first total synthesis of ergotamine was performed in his laboratory in 1961. In 1975, the Mexican Institute for the Study of Medicinal Plants (IMEPLAM) was created. This institute was founded for the multidisciplinary study of the plants most widely utilized in Mexico to treat common illnesses. During the existence of IMEPLAM (1975–1980), numerous

F. Domínguez
Centro de Investigación Biomédica de Oriente, Instituto Mexicano del Seguro Social, Metepec, Puebla, Mexico
e-mail: irmafabiola@yahoo.com

E. González-Trujano
Dirección de Investigaciones en Neurociencias, Instituto Nacional de Psiquiatría. Ramón de la Fuente Muñiz, Ciudad de México, DF, Mexico
e-mail: evag@imp.edu.mx

J.M. Gallardo
Unidad de Investigación Médica en Enfermedades Nefrológicas, Hospital de Especialidades, Centro Médico Nacional Siglo XXI, Instituto Mexicano del Seguro Social, Ciudad de México, DF, Mexico
e-mail: jmgallardom@gmail.com

S. Orozco-Suárez (✉)
Unidad de Investigación Médica en Enfermedades Neurológicas, Hospital de Especialidades Centro Médico Nacional Siglo XXI, Instituto Mexicano del Seguro Social. Av. Cuauhtémoc 330 Col. Doctores, Ciudad de México, DF 06720, Mexico
e-mail: sorozco5@hotmail.com



C. Grosso (ed.), *Herbal Medicine in Depression*,
DOI 10.1007/978-3-319-14021-6_8

publications were produced, thus establishing the Institute as an icon in the research of medicinal plants in Latin America.

At the end of the twentieth century, a second stage of studies on medicinal plants and their therapeutic potential ushered in a new era of neurophytopharmaceuticals. Currently, new mechanism actions are being proposed to explain the pharmacological effects of these plant extracts on mental activity and cerebral functions. For example, *Ginkgo biloba*, *Hypericum perforatum*, *Sabal serrulata*, and *Crataegus oxyacantha* have been the subject of many research investigations in the past decade. The active chemicals in these plants belong to new and different groups of compounds (i.e., flavonoids and lactones that do not have psychedelic origins). In this chapter, we describe the traditional and commercial uses of some medicinal plants representative of Latin America based on animal models and some clinical studies.

**Keywords** Medicinal plants • Antidepressant compounds • Flavonoids

## 8.1 Introduction

Among the Aztecs, the three main spiritual centers were the head region (tonalli), the heart region (teyolia), and the region of the liver (ihiyotl). According to traditional Aztec beliefs, an accumulation of phlegm in the chest, for example, which could cause insanity, stupidity, or epilepsy by affecting diseases was sought by means of consultation with the spiritual world. The causes of, and remedies for, some diseases were sought by means of consultation with the spiritual world. Indeed, the Aztecs frequently used certain plants with hallucinogenic powers to change sensory perceptions and lead individuals to believe that they had traveled to other worlds, where they were among the gods and were able to obtain information about the cause of a disease and its remedies (Obiols and López 1975). The majority of plants that have been utilized as entheogens (i.e., substances that, upon ingestion, produce divine experiences) contain alkaloids that act on the nervous system in such a manner that the senses are altered and sensitized when confronted with distinct stimuli (Shultes and Hofmann 1982). The arrival of the Spaniards radically modified the medicine practices of the Aztecs and the use of medicinal plants. Diverse colonial documents, such as those of Martín de la Cruz, Juan Badiano, Bernardino de Sahagún, and Francisco Hernandez, provide examples of the use of medicinal plants from the viewpoint of the Aztecs in works such as *Libellus de Medicinalibus Indorum Herbis* (*Little Book of the Medicinal Herbs of the Indians*). Additional works describe the actions of Mexican medicinal plants and suggest their usefulness, such as *Historia de las cosas de la Nueva España* (*General History of the Things of the New Spain*) by Fray Bernardino de Sahagún (Viesca 1992, 1995).

*Libellus de Medicinalibus* was a manuscript completed in Mexico in 1552, which referred to native medicinal plants, the curative effects of which were indicated in Latin texts. This medicinal "herbarium" was written with the intention of showing the King of Spain the rich variety of medicinal plants that grew in the New World. An elderly Indian physician and a native of Tlatelolco, Martín de la Cruz, described the use of each of the plant and provided drawings of them, and Juan Badiano (a young Indian) translated the material from Nahuatl into Latin. Among the plants utilized for sleep disorders, tlazolpahtli is noteworthy. Its "supposed magical effect" was thought to cleanse the insomniac from sins and concerns. Reference was also made to mental diseases in which the sought-after effect from the plants, above all in the treatment of fear, was associated with modifications of the tonalli (a spiritual entity that Mexicans located anatomically in the cerebral cortex). Likewise, for the treatment of heart diseases, some of the plants in this herbarium were used to target another spiritual entity known as the teyolia, an organ that includes the heart and that the pre-Hispanic natives believed controlled mentality and higher mental faculties. Plants that were believed to act on the teyolia include *Talauma mexicana* or flor del corazón (heart flower) (Gimmel 2008). The association between spiritual effects and entities is part of a cosmovision because, according to pre-Hispanic Mexican thought, the diverse aspects of life interacted with others, and, thus, the body and the cosmos were believed to share structural and functional characteristics. According to this logic, because disease is the result of these interactions, treatments for diseases were directed toward restoring order, both in the body and in the cosmos (Bayardo et al. 2002). The use of medicinal plants by the ancient Mexican Indians was sustained by empirical observation, as well as by the use of magic. An example of this is the tropical vine *Mucuna sloanei*, a widely distributed plant in Mexico, but about which (to our knowledge) there are no specific studies. Despite the lack of empirical evidence, it is interesting to note that *Mucuna sloanei* was recorded as a medicinal plant whose seed was popularly known as deer's eye (ojo de venado). Among the Spaniards, it was used as an amulet to prevent the culturally complex condition causing unhealthy children known as evil eye (Martinez 1969). According to Mexican medical anthropology, evil eye was a filial syndrome that afflicted unearned infants and was clinically manifested in a loss of appetite, somnolence, and muscular hypotonia. In certain cases, it was accompanied by coma, respiratory failure, and death (Risser and Mazur 1995). According to popular tradition, the illness was provoked by cruel glances, particularly given by envious women unable to conceive a child. Although this condition was curable with the aid of several remedies, the deer eye (a *ripe Mucuna sloanei* seed attached to the child's wrist) was considered an efficacious amulet to protect children. A major progression in formal medical research occurred in 1888, when the National Medical Institute in Mexico was created by the order of President Porfirio Díaz. The objective of the Institute was to conduct studies of Mexican medicinal flora with the goal of incorporating medicinal plants into therapeutics at the national level. By 1915, the herbarium possessed 14,000 classified species and approximately 1000 chemical compounds that were obtained from plants (García 1981).

In 1975, the Mexican Institute for the Study of Medicinal Plants (IMEPLAM) was created. This institute was founded for the multidisciplinary study of the plants most widely utilized in Mexico to treat common illnesses. During the existence of IMEPLAM (1975–1980), numerous publications were produced, thus establishing the Institute as an icon in the research of medicinal plants in Latin America and reactivating this type of research (Lozoya 1976). At the same time, the Institute initiated the formation of the Medicinal Herbolarium, currently known as the Medicinal Herbolarium of the Mexican Institute of Social Security (IMSS), which is located at the Twenty-First Century National Medical Center in Mexico City. Its legacy comprises >120,000 specimens (Zolla 1980; Montes and Montes 2005). In 1980, the IMEPLAM became part of the IMSS Medical Research System. The IMEPLAM embodied the tendencies of the 1960s with respect to the increased research interest in natural plant drugs, including (a) the Indian *Rauwolfia* drugs and their derivatives, which were demonstrated to successfully treat various mental disorders and other diseases, and (b) the plant-derived drugs that induce psychotic symptoms akin to symptoms of mental illnesses. Indeed, interest in such psychosis-inducing drugs, which were traditionally used by healers and medicine men of "primitive" cultures, resulted in an extensive search for substances possessing hallucinogenic properties (Viesca et al. 1992). Thus, the first wave of renewed research interest in natural products in Europe and the USA began in the early 1960s and was closely linked to the discussion of ethnobotany in university lectures and the search for psychotropic for medicinal purposes (Duke 1995). The term psychotropic has been used in scholarly circles as a synonym for the popular term psychedelic. Psychotropic substances modify mental activity either through administration of a plant substance or a synthetic chemical. There are three types of psychotropic drugs: (a) sedatives, including narcotics, barbiturates, and other tranquilizers, many of which have been incorporated into modern psychiatric practice; (b) psychic stimulants, which induce changes in mood; and (c) psychic deviators or hallucinogens, including numerous plants containing substances such as lysergic acid, psilocybin, mescaline, bufotenine, harmine, and dimethyltryptamine (Szara 1967). During the 1960s, research concerning medicinal plants that affect the central nervous system was particularly focused on remedies used by several different cultures. Albert Hoffman discovered that ergot alkaloids are derivatives of lysergic acid, and the first total synthesis of ergotamine was performed in his laboratory in 1961. This discovery established that naturally occurring alkaloids, including those classified as clavine types, were of great value in therapy due to their oxytocic properties (Taber et al. 1963). Hoffman prepared lysergic acid diethylamide (LSD), which produces central excitation syndrome and may be of considerable importance in experimental psychiatry, neurophysiology, and other medical areas. This compound exhibited marked antagonism of serotonin and, thus, profoundly aided the study of the biological functions of serotonin, an important neurohumoral factor that has many-fold effects on the major structures and functions of organisms. For instance, serotonin increases permeability and elicits pain factors in patients with migraines (Hoffman 1979). The discovery of LSD sparked interest in psychotropic agents in general. This led Hoffman to examine the mushrooms of Mexico, which, according to ethnobotanical field studies, were reported to elicit psychic effects that are similar to those of LSD. A population of the ancient Indian mushroom, found in the remote mountains of southern Mexico, was discovered by Shultes and Shultes (1940) and was the first to suggest that the Aztec term "teonanácatl," which was associated with

Aztec rituals, was indeed a real mushroom. Wasson (1955) described the ceremonial use and the effects of the hallucinogenic mushrooms that were found in Oaxaca in southern Mexico and were later classified under the genus *Psilocybe*. Subsequently, Hoffman discovered the active compounds psilocybin and psilocin and was able to synthesize them in the laboratory (Wasson 1961). Shultes in 1940 also studied another enigmatic ritual plant of Mexico called "ololiuhqui" in Aztec (classified as *Turbina corymbosa*), which was considered the most important narcotic plant of the Aztecs (Guerra 1967). Pharmacologists found that "ololiuhqui" seeds contain d-lysergic acid amine (also called ergine) and lysergic acid methylcarbinolamide, both of which are hallucinogens that are closely related to LSD. Later, a new Mexican psychotropic plant became the topic of many research investigations.

*Lophophora williamsii*, the cactus called "peyote," was used by the Indians of the west coast of Mexico (the Huichol group) and by the legendary Tarahumara people of the Sierra. Its active constituent, mescaline (3,4,5-trimethoxyphenylalani ne), is related to norepinephrine. The discovery of mescaline led to a better understanding of the relationship between hallucinogenic compounds and neurotransmitters, as well as dopamine and their function in the brain (Bye 1979). Mescaline and norepinephrine have the same chemical base structure, and the psychotropic effects of mescaline depend on its interaction with cerebral norepinephrine receptors. Psilocybin and psilocin are derivatives of tryptamine, as is serotonin. Consequently, both psilocybin and psilocin interact with the serotonin system (Foster 2007). At the end of the twentieth century, a second stage of studies on medicinal plants and their therapeutic potential ushered in a new era of neurophytopharmaceuticals. Currently, new mechanism actions are being proposed to explain the pharmacological effects of these plant extracts on mental activity and cerebral functions. For example, *Ginkgo biloba*, *Hypericum perforatum*, *Sabal serrulata*, and *Crataegus oxyacantha* have been the subject of many research investigations in the past decade. The active chemicals in these plants belong to new and different groups of compounds (i.e., flavonoids and lactones) that do not have psychedelic origins (Fritze 2006; Bastianetto et al. 2007). The neuroprotective effects of flavonoids have demonstrated two principal mechanisms: oxidative stress and neuroinflammation (Limón et al. 2010).

**Search methods:** SCOPUS, PUBMED, Web of Knowledge, Information Etnobotánica del Herbario Medicinal del Instituto Mexicano del Seguro Social.

## 8.2 References


- Bastianetto S, Brouillette J, Quiri R. Neuroprotective effects of natural products: interaction with intracellular kinases, amyloid peptides and a possible role for transthyretin. Neurochem Res. 2007;32:1720–25.
- Bayardo SM, Malpica C, Berganza C. La enfermedad y la medicina en las culturas precolombinas de América; la cosmovisión náhuatl. Investigación en Salud 2002;3.

- Bye RA. Hallucinogenic plants of the Tarahumara. J Ethnopharmacol 1979; 1:23–48
- Duke JA. Commentary, novel psychotherapeutic drugs: a role for ethnobotany. Psychopharmacology 1995;31:177–84.
- Foster O. In: Triggle D, editor. Peyote and mescaline. Buffalo, NY, USA: State University of New York; 2007.
- Fritze J. Alternative and complementary medicine of psychiatric disorder. Eur Arch Psychiatr Clin Neurosci. 2006;256:461–3.
- García J. Historia de las instituciones de investigación en salud en América Latina, 1880–1930. Educación Medica y Salud. 1981;15:71–88.
- Gimmel M. Reading medicine in the codex de la Cruz Badiano. J Hist Ideas. 2008;69:169–92.
- Guerra F. Mexican phantastica—a study of the early ethnobotanical sources on hallucinogenic drugs. Br J Addiction Alcohol Other Drugs 1967;62:171–87.
- Hoffman A. How LSD originated. J Psychedelic Drugs. 1979;11:53–60.
- Limón D, Díaz A, Mendieta L, Luna F, Zenteno E, Guevara J. Los flavonoides: mecanismo de acción neuroprotección y efectos farmacológicos. In: Álvarez FG, Bustos J, editors. Mensaje Bioquímico vol XXXIV, México: UNAM; 2010. p. 143–54.
- Lozoya X. El Instituto Mexicano para el estudio de las plantas medicinales, A.C (IMEPLAM). In: Lozoya X, editor. Estado actual del conocimiento en plantas medicinales mexicanas. México: IMEPLAM; 1976. p. 243–8.
- Martínez M. In: Botas, editor. Las plantas medicinales de México. México; 1969.
- Obiols J, López M. El uso de drogas alucinógenas entre los aztecas y su relación con la medicina. Medicina e Historia 1975;49:7–26.
- Montes D, Montes F. La medicina alternativa y complementaria, una opción institucional. Rev San Militar Mex. 2005;59:385–8.
- Risser A, Mazur L. Use of folk remedies in a Hispanic population. Arch Pediatr Adolesc Med J. 1995;149:978–81.
- Shultes RE. Teonanacatl; the narcotic mushroom of the Aztecs. Am Anthropol. 1940;42:429–43.
- Shultes RE, Hofmann A. Plantas de los Dioses. Fondo de Cultura Económica. México DF; 1982.
- Szara S. The hallucinogenic drugs, curse or blessing? Am J Psychiatr 1967; 123:1513–18.
- Taber WA, Vining LC, Heacock RA. Clavine and lysergic acid alkaloids in varieties of morning glory. Phytochemistry 1963;2:65–70.
- Viesca C. El libellus y su contexto histórico. In: Kumate Jesús J, editor. Estudios Actuales sobre el *Libellus de medicinalibus indorum herbis*. México DF: Secretaría de Salud; 1992a. p. 49–85.
- Viesca C. Medicina prehispánica de México. México: Panorama; 1992b.

- Viesca C, Aranda A. Las alteraciones del sueño en el *Libellus de medicinalibus Indorum Herbis*, Estudios de la cultura Náhuatl, XXVI, México; 1996. p. 156.
- Viesca C, Ramos VM, Aranda A. Los tratamientos medicamentosos de las enfermedades mentales en la medicina Náhuatl prehispánica. In: Aceves P, editor. Tradiciones e intercambios científicos: Materia Médica, México: Farmacia y Medicina UAM; 2000. p. 27–44.
- Viesca AC, Sánchez G, Sánchez G. La materia médica en el *Libellus de Medicinalibus Indorum Herbis.* Revista de la Facultad de Medicina de la UNAM 2003;46:12–7.
- Wasson RG. The hallucinogenic fungi of Mexico: an inquiry into the origins of the religious idea among primitive peoples. Botanical Museum Leaflets, Harvard University; 1961. p. 137–62.
- Zolla C. Traditional medicine in Latin America, with particular reference to Mexico. J Ethnopharmacol. 1980;2:37–41.


## 8.3 *Hypericum perforatum* L

### *8.3.1 (Hypericaceae)*

1) **Common names:**
   Hipericon, St. John's wort, amber, amber touch-and-heal, chase-devil, demon chaser, goat weed, Hardhay, Klamath weed, Rosin rose, Saint Johannes wort, and tipton weed
2) **Photograph of the species (Fig. 8.1)**
3) **Description of the species:**
   A) **Habitat and world distribution:**
   St John's wort grows in temperate and subtropical regions of Europe, North America, and Asia. It can be found on uncultivated grounds, woods, hedges, roadsides, and meadows (Gilani 2005).
   B) **Morphology of the plant:**
   The genus *Hypericum* derives from the Greek words hyper (above) and eikon (picture), in reference to the plant's traditional use in warding off evil by hanging plants over a religious icon in the house during St. John's day. The species named *H. perforatum* refers to the presence of small oil glands in the leaves that look like windows, which can be seen when they are held against the light.

   St. John's wort is a perennial plant with widespread, creeping rhizomes. The stems are erect, branched in the upper section, and can grow up to 1 m high. Leaves are glabrous, opposite, sessile, and elliptical to ovate and grows 1.5–4 cm in length. They are yellow greenish, with transparent spots throughout the tissue and sometimes with a few black spots on the lower surface.

**Fig. 8.1** *Hypericum perforatum*. Source: Herbarium of medicinal plants, IMSS

Leaves exhibit obvious translucent spots when held up to the light, giving them a "perforated" appearance. Its five-petal flowers measure up to 2.5 cm across and are bright yellow with conspicuous black spots. The flowers appear in broad cymes at the ends of the upper branches, between late spring and early to midsummer. The sepals have glandular spots in the tissue. The numerous stamens are united at the base into three bundles. The pollen grains are ellipsoidal with 20–25 μm in diameter, with three pores and a smooth exile. A reddish/purple liquid is produced when flower buds or seed pods are crushed (Ciccarelli et al. 2001).

C) **Medicinal parts:**
The flowering tops of St. John's wort contain a share of approximately 30 % hypericins in the form of protopseudohypericin and protohypericin: buds (48 %), flowers (30 %), and leaves (17 %) (Poutaraud et al. 2001).

D) **Chemical constituents:**
**Anthraquinones** (hypericin and pseudohypericin), **prenylated phloroglucinol derivatives** (hyperforin, adhyperforin, furohyperforin), **flavonoids** (kaempferol, quercetin, luteolin, quercetin-3-*O*-glucoside, quercetin-3-*O*-rhamnoside, quercetin-3-*O*-galactoside), **phenolic acids** (caffeic acid, chlorogenic acid), **volatile oil**, **amino acids**, **vitamin C**, and **carotenoids** (Nahrstedt et al. 1997).

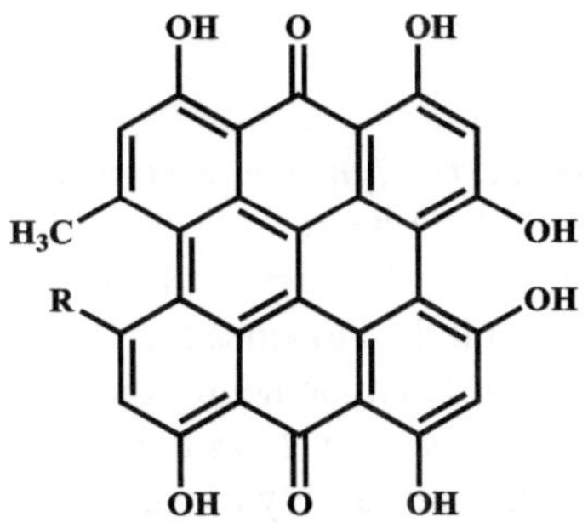

**Hypericin** R = $CH_3$

**Pseudohypericin** R = $CH_2OH$

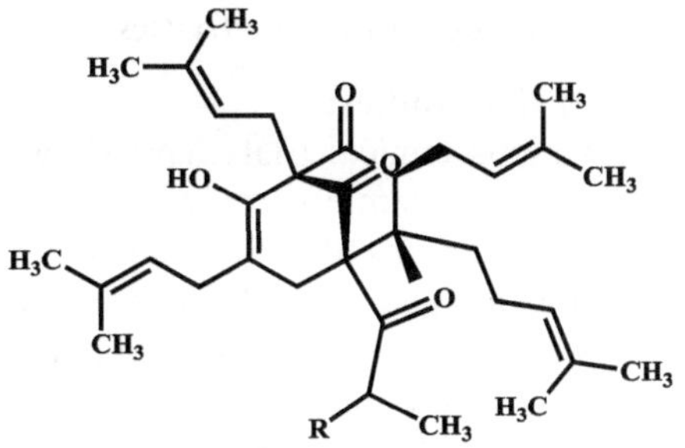

**Hyperforin** R = $CH_3$
**Adhyperforin** R = $CH_2CH_3$

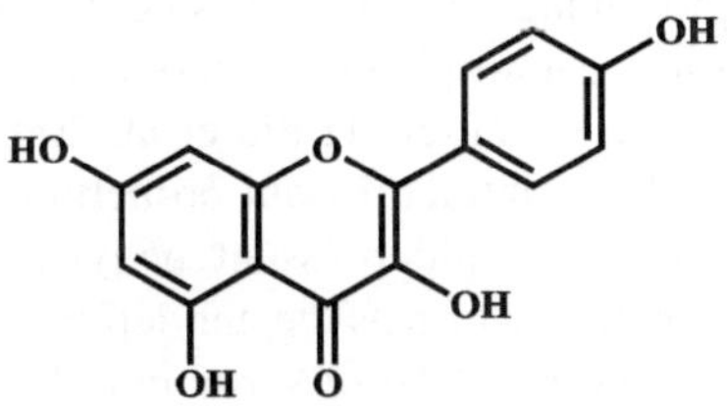

**Kaempferol**

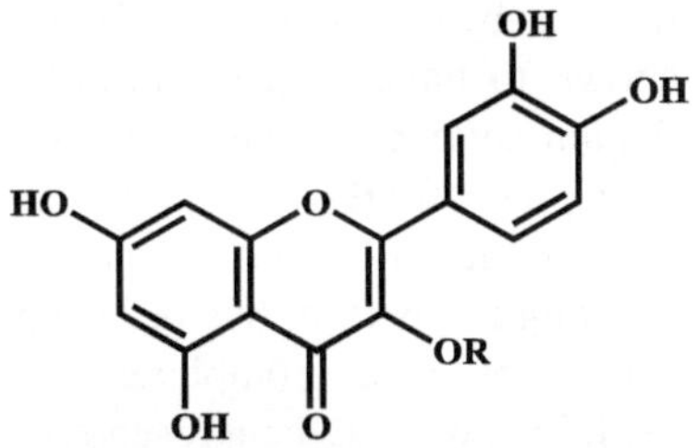

**Quercetin: R = H**

**R = glucose: quercetin-3-*O*-glucoside**

**Quercetin-3-*O*-rhamnoside: R = rhamnose**

**Quercetin-3-*O*-galactoside: R = galactose**

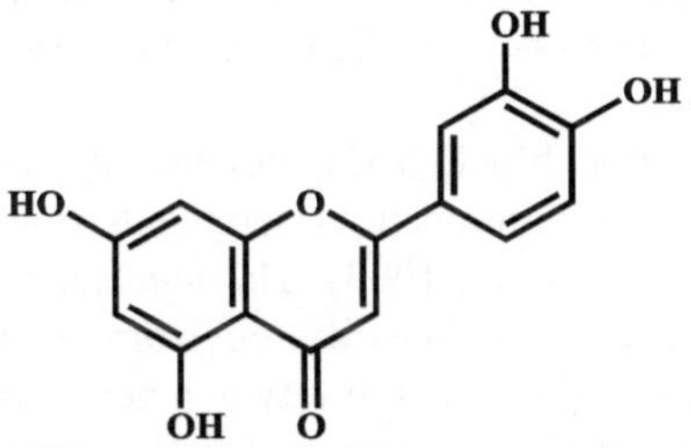

**Luteolin**

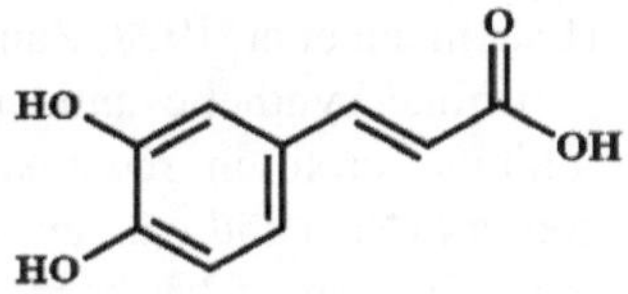

**Caffeic acid**

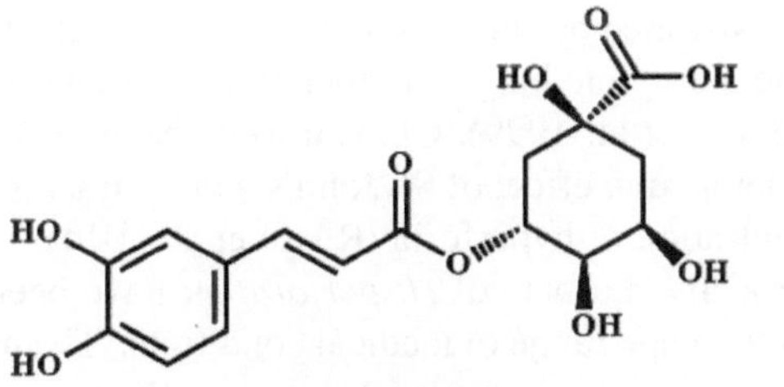

**Chlorogenic acid**

4) **Pharmacology and bioactivities:**

A) **In vivo studies:**

Various clinical trials have shown that *H. perforatum* has a comparable antidepressant efficacy as some currently used antidepressant drugs in the treatment of mild/moderate depression. Studies in vitro on rat brain mitochondria also found hypericin to strongly inhibit the enzymes MAO-A and MAO-B, which are involved in the degradation of amine neurotransmitters in vivo. Inhibiting their degradation increased their levels in the synapse. However, further studies determined that hypericin's ability to inhibit MAO was lower than was originally estimated. Moreover, the levels of hypericin necessary to obtain significant MAO inhibition were greater than those likely to be found in human brain tissue at normal doses. Hypericin has been shown to have a strong affinity for sigma ($\sigma$) receptors, which regulate dopamine levels. It also acts as a receptor antagonist at adenosine, benzodiazepine, $GABA_A$, $GABA_B$, and inositol triphosphate receptors, which regulate action potentials caused by neurotransmitters (Bilia et al. 2002). Although hypericin has been reported to have antidepressant properties, it cannot by itself completely account for the antidepressant activity of St. John's wort. Recent research focuses on hyperforin as the antidepression agent. Hyperforin is a potent inhibitor of the reuptake of serotonin, dopamine, noradrenaline, GABA, and L-glutamate from the synaptic cleft. The $IC_{50}$ values of approximately 0.05–0.1 μg/mL for neurotransmitters were reported in synaptosomal preparations. Blocking the reuptake of serotonin (5-HT) from the synaptic cleft alleviates symptoms of depression by allowing the serotonin to bind to 5-HT receptors and elicit a greater response (Laakmann et al. 1998; Zanoli 2004).

Unlike synthetic antidepressants that block 5-HT targets, hyperforin inhibits serotonin reuptake by elevating intracellular concentrations of sodium ($Na^+$) and calcium ($Ca^{2+}$) (Singer et al. 1999). The mechanism by which hyperforin has been shown to act is therefore different from that used by conventional antidepressants, perhaps pointing the way to a new class of antidepressants. Hyperforin also increases the number of 5-HT receptors (postsynaptic $5\text{-HT}_{1A}$ and $5\text{-HT}_{2A}$ receptors) with no changes in their affinity as demonstrated by studies on rat brains (Teufel-Mayer and Gleitz 1997), suggesting a possible long-term therapeutic benefit of St. John's wort treatment (De Vry et al. 1999). Clinical trials have also demonstrated that the level of therapeutic effect of St. John's wort extract is directly dependent on the concentration of hyperforin (Russo et al. 2014).

Additionally, extracts of *H. perforatum* have been traditionally recommended for a wide range of medical conditions. The most common modern-day use of St. John's wort is for depression. Studies have shown this species may be as effective as tricyclic antidepressants (TCAs) and selective serotonin reuptake inhibitor (SSRI) antidepressants for mild-to-moderate depression. The fraction of flavonoids in the extract of *H. perforatum* (39 %)

is structurally similar to that of MAO inhibitors. It suppresses the release of interleukin 6, a substance related to depression, as it modulates the release of cortisol (Laakmann et al. 1998).

Several meta-analyses of clinical studies have been conducted since the mid-1990s, perhaps the most comprehensive—and widely recognized—meta-analyses are the Cochrane reviews published by Linde et al. (2005). These reviews were performed through computerized searches of several databases, based on published and unpublished trials conducted to that effect. Each review had specific inclusion criteria, with the criteria evolving over time. For example, to be included in their 1998 review, trials had to be randomized, include patients with depressive disorders, compare preparations that included St. John's wort with placebos or other antidepressants, and include an objective clinical assessment of symptoms. The results showed that the therapeutic effect of St. John's wort in mild-to-moderate depression depends on its hyperforin content (Laakmann et al. 1998). Their 2005 review mandated that the trials had to be double-blind, administration of St. John's wort with other plant extracts had to be avoided, trials had an intervention of at least 4 weeks, and trials had to use only standard antidepressants (Linde et al. 2005).

Barnes (2001) noted that the trials comparing St. John's wort to synthetic antidepressants were criticized because the dosages of the latter were unrealistically low. A study conducted by the *Hypericum* Depression Trial Study Group was a double-blind multicenter investigation aimed at determining whether St. John's wort (LI-160) was useful in treating major depression. The study involved 340 adult outpatients at 12 academic and community psychiatric research clinics in the USA. Their study had the benefit of being three armed, involving comparisons against a control group receiving a placebo and a second receiving the SSRI sertraline. The proportion of patients showing full or partial response to St. John's wort was 47.6 %, which was actually lower than those receiving placebo (55.9 %) and sertraline (61.1 %). Perhaps most noteworthy was the finding that, statistically, response rates did not differ significantly between placebo and either St. John's wort or sertraline. The authors concluded that the findings failed to support any claims of efficacy of *H. perforatum* in treating moderately severe depression (Fava et al. 2005).

All the recent clinical trials mentioned in the literature suggest that *H. perforatum* is more tolerable than synthetic antidepressants, as it causes fewer side effects and also shows similar adverse reactions to those seen in placebo-controlled groups. This makes St. John's wort an attractive option for treating depression (Klemow et al. 2011; Chen et al. 2015).

5) **Indications and usage:**

St. John's wort extracts are used internally to treat low mood, depression, seasonal affective disorder, anxiety, and insomnia (particularly is associated with menopause). Within the EU products are licensed for treating mild-to-moderate forms of depression. Oil-based products are applied externally to treat minor wounds, bruises, burns, and swellings (Edwards et al. 2015).

The main active components are the hypericins (naphtodianthrones), which include hypericin, isohypericin, pseudohypericin, and photohypericin, and the hyperforins (prenylated phloroglucinols) including hyperforin, adhyperforin, and their derivatives are presents. Hyperforin and hypericin exert antidepressant activity in animal behavioral models, supporting the hypothesis that flavonoids are part of the constituents responsible for the therapeutic efficacy (Biloba 1999).

Considerable clinical data, including meta-analyses, have shown that St. John's wort is superior to placebo and as effective as synthetic antidepressants in treating certain types of depression. There is some evidence for the external use (Edwards et al. 2015).

6) **Precautions and side effects (Edwards et al. 2015):**
Extensive research supports the safe use of St. John's wort for a short duration (<3 months) at recommended doses in individuals that lack other medication intake. St. John's wort may cause anxiety, headache, muscle cramps, sweating, weakness, dry mouth, or skin irritation. When using St. John's wort, drugs metabolized by cytochrome P450 should be used with caution, as decreased drug effectiveness may occur.

St. John's wort may increase the risk of photosensitivity. Use cautiously in people with sensitive skin or those taking photosensitizing drugs.

St. John's wort may increase the risk of serotonin syndrome, a predictable consequence in antidepressant drugs interaction with excessive levels of serotonin which in turn produces a spectrum of specific symptoms including cognitive, autonomic, and somatic effects. St. John's wort should be used cautiously in people taking agents that increase the risk of serotonin syndrome.

Similarly, it may alter drug levels. Use cautiously in people taking agents for bacterial or fungal infections, agents for erectile dysfunction, anxiolytic agents, antihistamines, fertility agents, P-glycoprotein agents, pain relievers, or theophylline.

St. John's wort may cause mania or psychosis. Use cautiously in people with mental illnesses and those taking antipsychotics. It also has a high profile of interaction with many conventional drugs as it affects the pharmacokinetics of many drugs, by inducing cytochrome P450 (CYO) isozymes, such as CYP3A4, CYP2C19, CYP2C9, and the P-glycoprotein (P-gp) transporter. It inhibits the reuptake of 5-hydroxytryptamine (5-HT, serotonin). There are conflicting reports of interaction with theophylline. The main components responsible for the interactions are the hyperforins.

7) **Dosage:**

   A) **Internal use** (Russo et al. 2014):
   For anxiety, 900 mg of St. John's wort have to be taken orally twice daily for several weeks.

   For mild-to-moderate depression, 20–1800 mg St. John's wort have to be taken orally once to three times for 4–52 weeks. Extracts of St. John's wort used in studies included WS® 5570, WS 5572, WS 5573, ZE 117,

STW 3-VI, STW3, PM235, LoHyp-57, LI 160, Psychotonin® forte extract, and Hyperforat® and were generally standardized to contain 0.3 % hypericin and 2–5 % hyperforin. For severe depression, 900–1800 mg of St. John's wort (extracts LI 160 and WS® 5570) have to be taken orally daily for 8–12 weeks.

For irritable bowel syndrome, 450 mg of St. John's wort have to be taken twice daily for 12 weeks, without evidence of benefit.

For nerve pain, three 900 μg hypericin tablets are taken orally for two treatment periods of 5 weeks each.

For obsessive-compulsive disorder, 450–1800 mg (standardized to 0.3 % hypericin) are taken orally daily for 12 weeks.

For pain due to burning mouth syndrome, 300 mg capsules of St. John's wort (containing hypericin 0.31 % and hyperforin 3.0 %) have to be taken orally three times daily for 12 weeks.

For seasonal affective disorder (SAD), 900 mg and unspecified doses of St. John's wort (LI 160 and Kira®) have to be taken orally once to three times daily with or without light therapy for 4–8 weeks.

For smoking cessation, 300 mg St. John's wort (LI-160 extract) have to be taken orally once or twice daily for up to 3 months and a week.

For social phobia, 600–1800 mg St. John's wort have to be taken orally daily for 12 weeks.

For attention deficit hyperactivity disorder (ADHD), 300 mg of St. John's wort (standardized to 0.3 % hypericin) have to be used in children (under 18 years old) three times daily for 8 weeks, without evidence of benefit.

For children (under 18 years old) depression, 150–1800 mg St. John's wort are taken orally once to three times daily for up to 8 weeks.

St. John's wort may cause serious interactions with prescription drugs, herbs, or supplements. Therefore, people using any medications should consult their healthcare providers, including a pharmacist, prior to starting therapy.

## 8.4 References


- Barnes J, Anderson LA, Phillipson JD. St John's Wort (*Hypericum perforatum* L.): a review of its chemistry, pharmacology and clinical properties. J Pharm Pharmacol. 2001;53(5):583–600.
- Bilia AR, Gallori S, Vincieri FF. St. John's wort and depression: efficacy, safety and tolerability-an update. Life Sci. 2002;70(26):3077–96.
- Biloba G. Herbal remedies: adverse effects and drug interactions. Am Fam Physician. 1999;59(5):1239–44.
- Chen JA, Vijapura S, Papakostas GI, Parkin SR, Kim DJ, Cusin C, Baer L, Clain AJ, Fava M, Mischoulon D. Association between physician beliefs regarding

assigned treatment and clinical response: re-analysis of data from the *Hypericum* Depression Trial Study Group. Asia J Psychiatr. 2015;13:23–9.
- Ciccarelli D, Andreucci AC, Pagni AM. Translucent glands and secretory canals in *Hypericum perforatum* L.(Hypericaceae): morphological, anatomical and histochemical studies during the course of ontogenesis. Ann Bot. 2001;88(4):637–44.
- De Vry J, Maurel S, Schreiber R, De Beun R, Jentzsch KR. Comparison of hypericum extracts with imipramine and fluoxetine in animal models of depression and alcoholism. Eur Neuropsychopharmacol. 1999;9(6):461–8.
- Edwards SE, da Costa Rocha I, Williamson EM, Heinrich M. St. John's wort Hypericum perforatum L. Phytopharmacy. 2015;335.
- Fava M, Alpert J, Nierenberg AA, Mischoulon D, Otto MW, Zajecka J, Murck H, Rosenbaum JF. A double-blind, randomized trial of St John's wort, fluoxetine, and placebo in major depressive disorder. J Clin Psychopharmacol. 2005;25(5):441–7.
- Gilani AH, Rahmanb AU. Trends in ethnopharmacology. J Ethnopharmacol. 2005;100(1):43–9.
- Klemow KM, Bartlow A, Crawford J, Kocher N, Shah J, Ritsick M. 11 Medical attributes of St. John's Wort (*Hypericum perforatum*). Lester Packer, Ph.D., 2011;211.
- Laakmann G, Schüle C, Baghai T, Kieser M. St. John's worth in mild to moderate depression: the relevance of hyperforin for the clinical efficacy. Pharmacopsychiatry. 1998;31:54–9.
- Linde K, Berner M, Egger M, Mulrow C. St John's worth for depression Meta-analysis of randomised controlled trials. Br J Psychiatr. 2005;186(2):99–107.
- Linde K, Levente K, Gerta R, Jamil S, Schumann MD, Meissner K, Sigterman MD, Schneider K, Schneide A. Efficacy and acceptability of pharmacological treatments for depressive disorders in primary care: systematic review and network meta-analysis. Ann Fam Med. 2015;13:69–79.
- Nahrstedt A, Butterweck V. Biologically active and other chemical constituents of the herb of *Hypericum perforatum* L. Pharmacopsychiatry. 1997;30:129–34.
- Poutaraud A, Di Gregorio F, Tin VC, Girardin P. Effect of light on hypericins contents in fresh flowering top parts and in an extract of St. John's Wort (*Hypericum perforatum*). Planta Med. 2001;67(3):254–9.
- Russo E, Scicchitano F, Whalley BJ, Mazzitello C, Ciriaco M, Esposito S, Patané M, Upton R, Pugliese M, Chimirri S, Mammi M, Palleria C, De Sarro G. *Hypericum perforatum*: pharmacokinetic, mechanism of action, tolerability, and clinical drug–drug interactions. Phytother Res. 2014;28(5):643–55.
- Singer A, Wonnemann M, Müller WE. Hyperforin, a major antidepressant constituent of St. John's Wort, inhibits serotonin uptake by elevating free intracellular $Na^{+1}$. J Pharmacol Exp Therapeut. 1999;290(3):1363–8.
- Teufel-Mayer R, Gleitz J. Effects of long-term administration of *Hypericum* extracts on the affinity and density of the central serotonergic $5\text{-}HT_{1A}$ and $5\text{-}HT_{2A}$ receptors. Pharmacopsychiatry. 1997;30(Suppl 2):113-6.

**Fig. 8.2** *Magnolia dealbata* Zucc Source: Angélica Chávez Cortázar

- In: Ulbricht CE, Basch EM, editors. Natural standard herb and supplement reference: evidence-based clinical reviews. Mosby; 2005.
- Zanoli P. Role of hyperforin in the pharmacological activities of St. John's Wort. CNS Drug Rev. 2004;10(3):203–18.

## 8.5 *Magnolia dealbata* Zucc

### *8.5.1 (Magnoliaceae)*

1) **Common names:**
   Magnolia, eloxóchitl, and cowcumber
2) **Photograph of the species** (Fig. 8.2)
3) **Description of the species:**
   A) **Habitat and world distribution:**
      *Magnolia dealbata* Zucc is an endemic specie from Mexico, of limited distribution, with only six populations in the cloud forests located in the south-central region of this country (Gutiérrez et al. 2014). The species was thought to be extinct until 1977, when it was rediscovered. It survives in six populations, each with a few individuals, distributed among the cloud forest of south-central Mexico between 1200 and 1500 m above sea level (Gutierrez and Vovides 1997).
   B) **Morphology of the plant:**
      *M. dealbata*, a member of section Rytidospermum, is a deciduous tree species that grows to 25 m in height and has giant white flowers (20 cm) and leaves (80 cm). Each adult tree is able to produce several new stump sprouts

after they have been cut, but only one or two principal stems will survive (Sánchez and Pineda 2006). Seeds of *M. dealbata* are about 1.4 cm long and have a red sarcotesta or exocarp, rich in oils. The seeds are inside compound fruits (polyfollicles), 90×55 mm. The number of seeds per fruit is around 49 (Corral and Sánchez 2006).

C) **Medicinal parts:**
Bark and leaves (Gutiérrez et al. 2014).

D) **Chemical constituents:**
**Lignans** (honokiol and magnolol) (Dominguez et al. 2010).

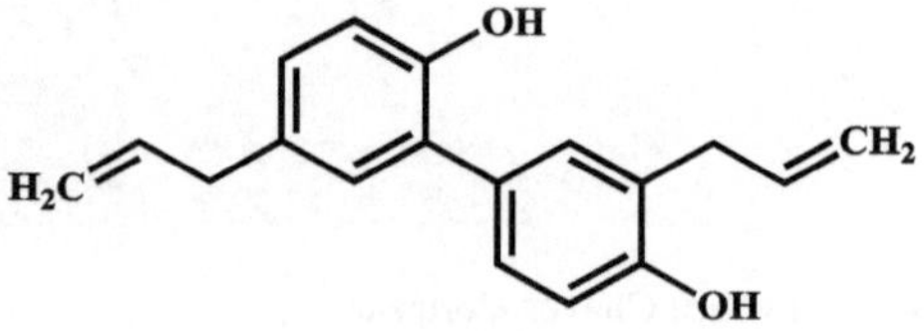


**Honokiol**

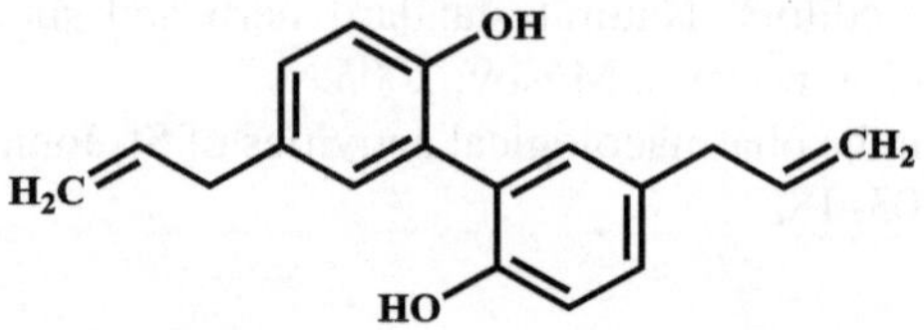


**Magnolol**

4) **Pharmacology and bioactivities:**

A) **In vivo studies:**
*Magnolia dealbata* possesses central nervous system (CNS) activity, such as reducing anxiety and tonic seizures and prevented mortality. Its active compounds honokiol and magnolol (lignans) are thought to be responsible for several of these pharmacological effects. These compounds have been identified as anxiolytic agents and as antioxidants that protect the brain tissue against lipid peroxidation (Martínez et al. 2006). The same way, honokiol and magnolol also exert a neuroprotective effect against focal cerebral ischemia–reperfusion injury, protect cortical neurons from chemical hypoxia, and attenuate heat stroke-induced neuronal damage (Kumar et al. 2015).

Honokiol and magnolol have been used to treat a variety of mental disorders including depression. Several studies have reported on the antidepressant-like effects of oral administration of the mixture of honokiol and magnolol in models of depression in rodents: forced swimming test (FST), tail suspension test (TST), and chronic mild stress (CMS) model (Xu et al. 2008). The mixture of honokiol and magnolol significantly decreased immobility time in the mouse FST and TST (without changing ambulatory or rearing behavior in the mouse open field test) and reversed CMS-induced reduction in sucrose consumption to prevent anhedonia in rats (Liu et al. 2015).

The mixture of honokiol and magnolol at 20 and 40 mg/kg significantly attenuated the CMS-induced decrease of 5-HT levels in frontal cortex, hippocampus, striatum, hypothalamus, and nucleus accumbens and markedly increased 5-HIAA levels in frontal cortex, striatum, and nucleus accumbens at 40 mg/kg and in frontal cortex at 20 mg/kg in the CMS rats (Xu et al. 2008). A subsequent reduction in 5-HIAA/5-HT ratio was found in hippocampus and nucleus accumbens in the CMS rats receiving this mixture. Furthermore, the mixture of honokiol and magnolol reduced the elevated corticosterone concentrations in serum to normalize the hypothalamic–pituitary–adrenal (HPA) hyperactivity in the CMS rats (Kalman et al. 2008). These results suggested that the mixture of honokiol and magnolol possesses potent antidepressant-like properties in behaviors involved in normalization of biochemical abnormalities in brain 5-HT and 5-HIAA, serum corticosterone levels, and platelet AC activity in the CMS rats (Li et al. 2013).

5) **Indications and usage:**
Magnolia bark extract is commercially available in the USA and throughout Europe. Dosage varies depending on the condition treated, with weight loss products containing magnolia bark extract available mostly in powder form. Several strengths are available for Relora, a patented blend marketed as "a natural stress management supplement" that includes magnolia bark extract (Talbott et al. 2013).

6) **Precautions and side effects:**
Honokiol and magnolol are not entirely without risks, although their limited empirical application to humans at therapeutic doses has limited the evaluation of their side effect profile thus far. There are potential risks that can be expected, including increased bleeding and potential neurotoxicity at high doses. Honokiol has been found to be a potent inhibitor of arterial thrombosis, so it may be advisable to avoid honokiol in coagulopathic patients or in those where bleeding or hemorrhage may be of concern (Park et al. 2011). These may include patients with hemorrhagic stroke, patients with hemorrhagic changes following ischemic stroke, patients on Coumadin or therapeutic Lovenox, and patients with clotting disorders such as hemophilia or von Willebrand's deficiency. While honokiol has been found to have neuroprotective effects at low doses, it has also been found to increase neuronal death in vitro at higher doses (100 μM applied directly to fetal cortical neurons) (Fukuyama et al. 2002). This suggests a need to further study its pharmacokinetic and pharmacodynamic parameters in humans before unintentionally administering a potentially toxic dose of this compound. Therefore, although the pharmacodynamics and pharmacokinetics of honokiol have now been well studied in rats, further studies need to be performed in humans before it can be widely administered (Sheng et al. 2014).

Drug–herb interactions are documented, and caution is advisable in patients self-medicating with magnolia bark extract. Magnolol may interact with acetaminophen. Additive adverse reactions with antiplatelet medications may occur in patients also self-medicating with magnolol. Magnolia bark extract and honokiol may interact with benzodiazepines. Magnolol may stimulate corticosterone secretion or increase steroid medication concentrations. Magnolia bark extract may cause excess sleepiness, vertigo, and dizziness in patients taking muscle relaxants (Kalman et al. 2008).

No dermatologic adverse effects have been documented with topical application of magnolol and honokiol. In one small clinical trial, a patient dropped out because of a number of adverse reactions, including heartburn, shaking hands, perilabial numbness, sexual dysfunction, and thyroid dysfunction (Shen et al. 2010).

7) **Dosage:**

Dosing recommendations include taking a 300 mg capsule of Relora two to three times per day. One small, randomized, controlled study treated patients with one 60 mg tablet of magnolia bark extract daily to help relieve menopausal symptoms (Talbott et al. 2013).

## 8.6 References


- Corral-Aguirre J, Sánchez-Velásquez LR. Seed ecology and germination treatments in *Magnolia dealbata*: an endangered species. Flora Morphol Distrib Funct Ecol Plants. 2006;201:227–32.
- Domínguez F, Chávez M, Garduño-Ramírez ML, Chávez-Avila VM, Mata M, Cruz-Sosa F. Honokiol and magnolol production by in vitro micropropagated plants of *Magnolia dealbata*, an endangered endemic Mexican species. Nat Prod Commun. 2010;5:235–40.
- Fukuyama Y, Nakade K, Minoshima Y, Yokoyama R, Zhai H, Mitsumoto Y. Neurotrophic activity of honokiol on the cultures of fetal rat cortical neurons. Bioorg Med Chem Lett. 2002;12(8):1163–6.
- Gutiérrez SLG, Chilpa RR, Jaime HB. Medicinal plants for the treatment of "nervios," anxiety, and depression in Mexican traditional medicine. Braz J Pharmacognosy. 2014;24:591–608.
- Gutierrez L, Vovides AP. An in situ study of *Magnolia dealbata* Zucc. In Veracruz State: an endangered endemic tree of Mexico. Biodivers Conserv. 1997;6(1):89–97.
- Kalman DS, Feldman S, Feldman R, Schwartz HI, Krieger DR, Garrison R. Effect of a proprietary Magnolia and *Phellodendron* extract on stress levels in healthy women: a pilot, double-blind, placebo-controlled clinical trial. Nutr J. 2008;7(11):1–6.
- Kumar GP, Anilakumar KR, Naveen S. Phytochemicals having neuroprotective properties from dietary sources and medicinal herbs. Pharmacognosy J. 2015;7(1).
- Martínez AL, Domínguez F, Orozco S, Chávez M, Salgado H, González M, González-Trujano ME. Neuropharmacological effects of an ethanol extract of the *Magnolia dealbata* Zucc. leaves in mice. J Ethnopharmacol. 2006;106(2), 250–5.

- Li L, Yang J, Ma SP, Qu R. Magnolol treatment reversed the glial pathology in an unpredictable chronic mild stress-induced rat model of depression. Eur J Pharmacol 2013;711(1):42–9.
- Liu Y, Wang D, Yang G, Shi Q, Feng F. Comparative pharmacokinetics and brain distribution of magnolol and honokiol after oral administration of Magnolia officinalis cortex extract and its compatibility with other herbal medicines in Zhi-Zi-Hou-Po Decoction to rats. Biomed Chrom. 2015; doi: 10.1002/bmc.3557.
- Park ES, Lim Y, Lee SH, Kwon BM, Yoo HS, Hong JT, Yun YP. Antiplatelet activity of obovatol, a biphenolic component of *Magnolia obovata*, in rat arterial thrombosis and rabbit platelet aggregation. J Atherosclerosis Thromb. 2011; 18(8):659–69.
- Sánchez-Velásquez LR, Pineda-López M, del R. Species diversity, structure and dynamics of two populations of an endangered species, *Magnolia dealbata* (Magnoliaceae). Revista de Biología Tropical. 2006;54:997–1002.
- Shen JL, Man KM, Huang PH, Chen WC, Chen DC, Cheng YW, ... Chen YH. Honokiol and magnolol as multifunctional antioxidative molecules for dermatologic disorders. Molecules. 2010;15(9):6452–65.
- Sheng Y, Xu J, Shi C, Li W, Xu H, Li N, Zhang XR. UPLC-MS/MS-ESI assay for simultaneous determination of magnolol and honokiol in rat plasma: application to pharmacokinetic study after administration emulsion of the isomer. J Ethnopharmacol. 2014;155(3):1568–74.
- Talbott S, Talbott J, Pugh M. Effect of Magnolia officinalis and Phellodendron amurense (Relora®) on cortisol and psychological mood state in subjects. Int J Sport Nutr. 2013;10(37).
- Xu Q, Yi L, Pan Y, Wang X, Li YC, Li JM, Kong LD. Antidepressant-like effects of the mixture of honokiol and magnolol from the barks of *Magnolia officinalis* in stressed rodents. Progr Neuro-Psychopharmacol Biol Psychiatr. 2008;32(3):715–25.


## 8.7 *Passiflora incarnata* L

### *8.7.1 (Passifloraceae)*

1) **Common names:**
   Wild passionflower, maypop, apricot vine, old field apricot, Holy Trinity flower, mayapple, molly pop, passion vine, pop apple, granadilla, maycock, maracoc, maracock, white sarsaparilla, Flor de la passion, and Pasionaria
2) **Photograph of the species (Fig. 8.3)**
3) **Description of the species:**
   A) **Habitat and world distribution:**
      The word *Passiflora* comes from the Latin word "passio" because in 1529, Spanish "conquistadores" described its flowers as symbols of the "passion of Christ" (Kinghorn 2001). *Passiflora incarnata* L., which originated in North

**Fig. 8.3** *Passiflora incarnata* L. Source: M en C. Alberto Hernández León

America, is the most common variety used in contemporary Western phytotherapy. This species, commonly known by the English name maypop, is native to the southeastern USA but is also cultivated in Europe, Asia, Africa, and Australia, both as an ornamental and as a medicinal plant (Dhawan et al. 2004).

B) **Morphology of the plant:**

*Passiflora incarnata* is an evergreen climber, rapidly growing up to 6 m. This plant is in leaf from December to January and in flower from June to July, and its seeds ripen from September to November. This plant possesses hermaphrodite flowers (possessing both male and female organs) which are pollinated by insects such as bees. *Passiflora incarnata* can grow in light (sandy), medium (loamy) or heavy (clay) soils, with a preference for well-drained soil, and it cannot grow in the shade due to the soil moisture in shady areas. Cotyledons are oval, thick and have a waxy appearance. The first true leaf is heart shaped and also glossy. Subsequent leaves have lobes (usually three, sometimes five) arising from a common point (palmately lobed leaves). Leaves are alternate along the stem, usually only slightly hairy. They have approximately 2 1/2–5 1/2 in. long and wide and occur on petioles. A pair of nectar-filled glands occurs at the base of the leaf blade and the junction of the petiole. Seedlings can occur but sprouts from the perennial rootstocks are much more common. Roots initially develop as a taproot but eventually develop a very deep perennial rootstock from which sprouts can emerge. Stems can trail along the ground or climb on other vegetation. Stems are usually only slightly hairy and may reach 6 1/2 feet in length. Solitary flowers arise from the area between stem and leaf petioles. Flowers range from 2 to 4 in. in length and are very attractive and light purple to lavender in color. Relatively large (1 1/2–3 in. long) berries that are green or yellowish green in color are produced. Berries contain many dark brown "dimpled" seed (reticulate seed) (Miroddi et al. 2013).

C) **Medicinal parts:**

Passionflower (*Passiflora incarnata*) was used traditionally in the Americas and later in Europe as a calming herb for anxiety, insomnia, seizures, and hysteria. It is still used today to treat anxiety and insomnia (Ernst 2001). Scientists believe passionflower works by increasing levels of a chemical called gamma-aminobutyric acid (GABA) in the brain. GABA lowers the activity of some brain cells, making you feel more relaxed. The effects of passionflower tend to be milder than valerian (*Valeriana officinalis*) or kava (*Piper methysticum*), other herbs used to treat anxiety (Elsas et al. 2010).

Spanish conquerors first learned of passionflower from pre-Colombian people who traditionally used it as a sedative to treat insomnia and nervousness. *Passiflora incarnata* or "passionflower" (flos passionis) acquired its name from the descriptions of the parts of its flower by Spanish missionaries in South America. It was known by the Spanish as "La Flor de las Cinco Llagas" or the "The Flower With The Five Wounds." "Passionis" refers to Christ's passion, representing various elements of the Crucifixion (Anagnostou 2005).

Passionflower was brought to Europe, where it became widely cultivated and was introduced to European folk medicine, becoming a popular traditional phytotherapeutic remedy as well as a homeopathic remedy for the relief of mild symptoms of mental stress, anxiety, and mild sleep disorder (Lakhan et al. 2010).

Today, passionflower is officially included in the national pharmacopeias of France, Germany, and Switzerland and is also monographed in the British Herbal Pharmacopoeia and the British Herbal Compendium (Blumenthal et al. 1998).

In European traditional medicine, passionflower has long been prescribed for various indications such as anxiety, nervousness, constipation, dispepsia, mild infections, and insomnia (Rotblatt et al. 2002).

Ground parts (flowers, leaves, and stems) of the passionflower are used for medicinal purposes with available forms in infusions, teas, liquid extracts, and tinctures (Movafegh et al. 2008).

D) **Chemical constituents:**

**Flavonoids** as luteolin, vitexin, isovitexin, apigenin, orientin, quercetin, chrysin, trisubstituted benzoflavone, and kaempferol (Dhawan et al. 2003). The extract from the aerial parts contains multiple bioactive metabolites such as flavonoids (like chrysin), amino acids (like GABA) (Lane et al. 2014), and harmala alkaloids, stimulants, and monoamine oxidase (MAO) inhibitors (Frye et al. 2007).

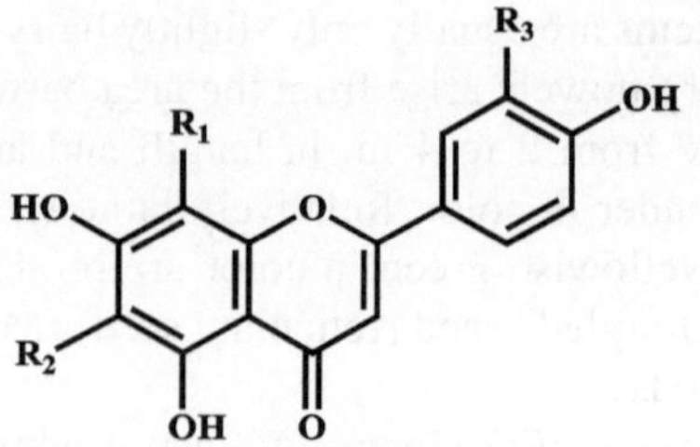

**Apigenin: $R_1$, $R_2$, $R_3$ = H**
**Vitexin: $R_1$ = glucose; $R_2$, $R_3$ = H**
**Isovitexin: $R_1$, $R_3$ = H; $R_2$ = glucose**
**Orientin: $R_1$ = glucose; $R_2$ = H; $R_3$ = OH**
**Luteolin: $R_1$, $R_2$ =H; $R_3$ = OH**

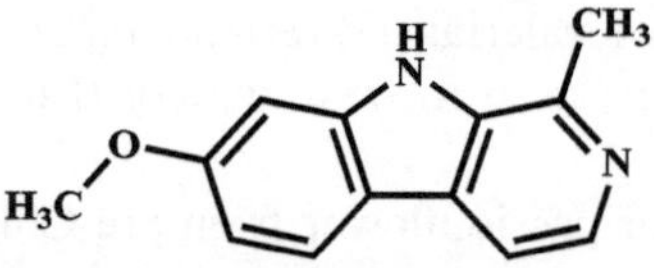

**Harmine**

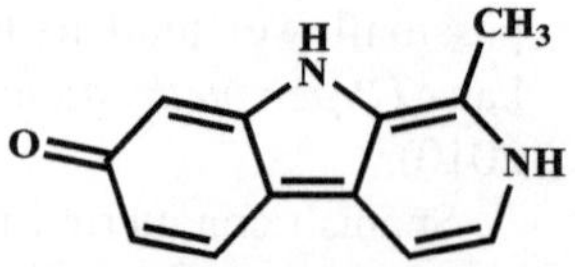

**Harmol**

4) **Pharmacology and bioactivities:**

A) **In vivo studies:**

Herbal medicines are popularly used worldwide and could be an option for treating anxiety if shown to be effective and safe. Passionflower extract is one of these cases. *P. incarnata* has been used to cure anxiety and insomnia since time immemorial. The anxiolytic effects of passionflower are well documented in mice (Dhawan et al. 2001, 2002). However, one of the problems with herbal supplements is that plant material contains thousands of phytochemicals, making it challenging to pinpoint the specific biochemicals responsible for the anxiolytic properties. In other words, although herbal remedies often produce positive results, identifying the active ingredients can be difficult. Therefore, users of herbal remedies may be consuming ineffective or possibly toxic substances in addition to the active, anxiolytic ingredients. To date, three human trials have documented the efficacy of passionflower as a treatment for anxiety-related disorders.

A fraction derived from the methanolic extract of *P. incarnata* has been observed to exhibit significant anxiolytic activity in mice using elevated plus maze (EPM) model of anxiety. The possibility of a phytoconstituent having benzoflavone nucleus as the basic moiety being responsible for the bioactivity of *P. incarnata* is highly anticipated. The potential anxiolytic effects of chrysin and the purported modulation of the benzodiazepine receptor on the $GABA_A$ receptor in laboratory rats have been tested. It has been hypothesized that chrysin decreases anxiety *via* interaction with the $GABA_A$ receptor in laboratory rats as measured by elevated plus maze

(EPM), corticosterone and catecholamine assays. In that study, each animal received an intraperitoneal injection of (1) vehicle (DMSO 4 %); (2) chrysin, 2 mg/kg; (3) midazolam, 1.5 mg/kg; or (4) flumazenil, 3 mg/kg, and chrysin, 2 mg/kg. The EPM was used to evaluate the behavioral component of anxiolysis, and catecholamine and corticosterone assays were examined to measure the neurohormonal effects of anxiety. No statistical difference was found among groups in catecholamine and corticosterone levels. The data suggested that chrysin may have anxiolytic properties similar to midazolam but to a lesser magnitude at the 2 mg/kg dose used in this study (Brown et al. 2007).

A study was designed to investigate the dual protective effect of the hydroethanolic extract of *Passiflora incarnata* in pentylenetetrazol (PTZ)-induced seizure and associated postictal depression (Singh et al. 2012). The animals, in different groups, were administered daily with varying doses of hydroethanolic extract of *P. incarnata* (150, 300, and 600 mg/kg; i.p.), diazepam (2 mg/kg; i.p.), and vehicle. On every fifth day, after PTZ treatment, seizure severity (score) was noted. The results led to the conclusions that the hydroethanolic extract of *P. incarnata* suppresses PTZ-induced seizures and ameliorates its associated postictal depression, which has been found to be get worsened with the standard antiepileptic drug, diazepam (Singh et al. 2012).

One double-blind, placebo-controlled study analyzed the difference in efficacy between oxazepam, a prescription benzodiazepine used to treat chronic anxiety symptoms, and passionflower in patients ($n=36$) who met the criteria for generalized anxiety disorder (GAD) (Akhondzadeh et al. 2001). The results showed no difference between the two anxiolytics with regard to the treatment of GAD, suggesting that passionflower is as effective as benzodiazepines in eliminating anxiety symptoms. Subjects from the passionflower group also reported lower job impairment performance than those in the benzodiazepine group; however, subjects in the benzodiazepine group reported a faster onset of symptom relief.

In an another example of a clinical study, 60 patients were randomized into two groups to receive either oral *P. incarnata* (500 mg, Passipy™ IranDarouk) ($n=30$) or placebo ($n=30$) as premedication, 90 min before surgery. A numerical rating scale (NRS) was used for each patient to assess anxiety and sedation before and 10, 30, 60, and 90 min after premedication. Psychomotor function was assessed with the Trieger dot test and the digit-symbol substitution test at arrival in the operating room, 30 and 90 min after tracheal extubation. The time interval between arrival in the postanesthesia care unit and discharge to home (discharge time) was recorded for each patient. The demographic characteristics of patients, American Society of Anesthesiologists (ASA) physical status, duration of surgery, basal NRS score, sedation at the present time intervals, and discharge time were similar in both groups. The NRS anxiety scores were significantly lower in the *Passiflora* group than in the control group ($p<0.001$). There were no significant differences in psychological variables in the postanesthesia care unit, and recovery of psychomotor function was comparable in both groups. In outpatient surgery, administration of oral *P. incarnata* as a premedication reduces anxiety without inducing sedation (Movafegh et al. 2008).

5) **Indications and usage:**

For generalized anxiety disorder (GAD).

6) **Precautions and side effects:**

Passionflower is likely safe for most people when taken orally in amounts normally found in food. It is possibly safe when taken short term (<2 months) as medicine or tea. It is possibly unsafe when taken by mouth in large amounts.

Passionflower can cause some side effects such as dizziness, confusion, irregular muscle action and coordination, altered consciousness, and inflamed blood vessels. There has also been a report of nausea, vomiting, drowsiness, a rapid heart rate, and abnormal heart rhythm in one person who took it (Capasso et al. 1995).

Pregnant women cannot take passionflower because it is unsafe. There are some chemicals in passionflower that might cause the uterus to contract. Not enough is known about the safety of taking passionflower during breastfeeding (Brinker et al. 1998).

Passionflower can affect the central nervous system. It might increase the effects of anesthesia and other medications on the brain during and after surgery. Stop taking passionflower at least 2 weeks before a scheduled surgery (Newall et al. 1996).

Passionflower might cause sleepiness and drowsiness. Taking passionflower along with sedative medications or with other herbs with sedative effects might cause too much sleepiness. Some sedative medications include pentobarbital (Nembutal), phenobarbital (Luminal), secobarbital (Seconal), clonazepam (Klonopin), lorazepam (Ativan), zolpidem (Ambien), and others. Some of the herbs and supplements with sedative effect include 5-HTP, *Acorus calamus* (calamus), California poppy (*Eschscholzia californica* Cham.), catnip (*Nepeta cataria* L.), hops (*Humulus lupulus* L.), Jamaican dogwood (*Piscidia piscipula* (L.) Sarg.), kava (*Piper methysticum* G. Forst.), St. John's wort (*Hypericum perforatum* L.), skullcap (*Scutellaria lateriflora*), valerian (*Valeriana edulis*), yerba mansa (*Anemopsis californica* (Nutt.) Hook. & Arn.), and others (Fugh et al. 1999).

7) **Dosage:**

A) **Internal use:**

For generalized anxiety disorder (GAD), 45 drops/day of passionflower liquid extract are taken orally. A specific tablet formulation 90 mg/day has also been used.

For reducing symptoms associated with narcotic drug withdrawal, 60 drops of passionflower liquid extract in combination with 0.8 mg of clonidine should be taken orally (Blumenthal et al. 1998).

B) **External use:**

There is not enough information to rate the safety of passionflower when applied to the skin.

## 8.8 References


- Akhondzadeh S, Naghavi HR, Vazirian M, Shayeganpour A, Rashidi H, Khani M. Passionflower in the treatment of generalized anxiety: a pilot double-blind randomized controlled trial with oxazepam. J Clin Pharm Therapeut. 2001;26:363–7.
- Anagnostou S.. Jesuits in Spanish America: contributions to the exploration of the American materia medica. Pharm Hist. 2005;3–17.
- In: Blumenthal M, Busse WR, Goldberg A, et al. editors. The Complete German Commission E Monographs. Boston, Mass: Integrative Medicine Communications; 1998. p. 179–80.
- Brinker F. Herb contraindications and drug interactions. 2nd ed. Sandy, Ore: Eclectic Medical; 1998. p. 109–10.
- Brown E, Hurd NS, McCall S, Ceremuga TE. Evaluation of the anxiolytic effects of chrysin, *a Passiflora incarnata* extract, in the laboratory rat. AANA J. 2007; 75(5):333–7.
- Capasso A, Pinto A. Experimental investigations of the synergistic-sedative effect of passiflora and kava. Acta Ther. 1995;21:127-114.
- Dhawan K, Dhawan S, Chhabra S. Attenuation of benzodiazepine dependence in mice by a tri-substituted benzoflavone moiety of *Passiflora incarnata* Linneaus: a non-habit forming anxiolytic. J Pharm Pharmaceut Sci. 2003;6(2):215–22.
- Dhawan K, Dhawan S, Sharma A. *Passiflora:* a review update. J Ethnopharmacol. 2004;94(1):1–23.
- Dhawan K, Kumar S, Sharma A. Anti-anxiety studies on extracts of *Passiflora incarnata* Linneaus. J Ethnopharmacol. 2001;78:165–70.
- Dhawan K, Kumar S, Sharma A. Comparative anxiolytic activity profile of various preparations of *Passiflora incarnata* Linneaus: a comment on medicinal plants' standardization. J Alternative Compl Med. 2002;8(3):283–91.
- Elsas SM, Rossi DJ, Raber J, White G, Seeley CA, Gregory WL, Mohr C, Pfankuch T, Soumyanath A. *Passiflora incarnata* L. (Passionflower) extracts elicit GABA currents in hippocampal neurons in vitro, and show anxiogenic and anticonvulsant effects in vivo, varying with extraction method. Phytomedicine. 2010;17:940–9.
- In: Ernst E, editor. Passionflower. The desktop guide to complementary and alternative medicine. Edinburgh: Mosby; 2001. p. 140–1.
- Frye A, Haustein C. Extraction, identification, and quantification of harmala alkaloids in three species of *Passiflora*. Am J Undergrad Res. 2007;27:19–26.
- Fugh-Berman A, Cott JM. Dietary supplements and natural products as psychotherapeutic agents. Psychosom Med. 1999;61(5):712–28.
- Kinghorn GR. Passion, stigma, and STI. Sex Transm Infect. 2001;77(5):370–5.
- Lakhan SE, Vieira KF. Nutritional and herbal supplements for anxiety and anxiety-related disorders: systematic review. Nutr J. 2010;9:42.
- Lane G. The effects of GABA, L-theanine, *Passiflora incarnata* and *Prunus cerasus* in reducing the symptoms of nervousness in individuals with anxiety and/or depression. J Nutr Therapeut. 2014;3(4):175–7.
- Miroddi M, Calapai G, Navarra M, Minciullo PL, Gangemi S. Passiflora incarnata L. Ethnopharmacology, clinical application, safety and evaluation of clinical trials. J Ethnopharmacol. 2013;150(3):791–804.

**Fig. 8.4** *Tilia americana* var. *mexicana* Source: Eva Aguirre-Hernández


- Movafegh A, Alizadeh R, Hajimohamadi F, Esfehani F, Nejatfar M. Preoperative oral *Passiflora incarnata* reduces anxiety in ambulatory surgery patients: a double-blind, placebo-controlled study. Anesth Analg. 2008;106(6):1728–32.
- Newall C, Anderson L, Phillipson J. Herbal medicines: a guide for health-care professionals. London, England: Pharmaceutical Press; 1996. p. 206–7.
- Rotblatt M, Ziment I. Evidence-based herbal medicine. Philadelphia, PA: Hanley & Belfus, Inc; 2002. p. 294–7.
- Singh B, Singh D, Goel RK. Dual protective effect of *Passiflora incarnata* in epilepsy and associated post-ictal depression. J Ethnopharmacol. 2012;139(1): 273–9.


## 8.9 *Tilia americana* var. *mexicana* (Schltdl.) Hardin

### *8.9.1 (Malvaceae)*

1) **Common names:**
   Flor de tila, tilo, tilio, jonote, sirimo, tila de bolita, tsirimu, and linden blossom
2) **Photograph of the species (Fig. 8.4)**
3) **Description of the species:**

   A) **Habitat and world distribution:**
   *Tilia americana* var. *mexicana* (Schltdl.). Hardin is considered endangered for the Mexican flora. This tree inhabits the canopy of the montane cloud forest (or lower montane forest), which is one of the types of vegetation covering less area and historically more disturbed in Mexico. It is distributed in 14 states of Mexico, from the northern states of Chihuahua and Coahuila to the southern states of Guerrero and Oaxaca. Even though a relatively large geographical distribution, the populations of this species are confined to the lower montane forest, which covers less than 1 % of Mexico (Hardin 1990).

B) **Morphology of the plant:**
*Tilia americana* var. *mexicana* is a tree up to 35 m in height and 1 m in diameter. The wood is white and soft, with strong fibers in the inner portion of the trunk. Winter buds are small, glabrous, and obtuse. Leaves are ovate-lanceolate, glabrous when mature, oblique at the base, and not cordate with ca. 10 cm long and ca. 6 cm wide; the apex is short and narrow, smooth on both sides, with 2–3 pairs of primary veins, and acutely serrate with gland-tipped teeth; stipules are deciduous; petiole is ca. 3 cm long; bracts are wide. Flowers are fragrant, yellow or white, and in cymes; peduncle is adnate at the base and regular or long; foliar bract is membranous; ovary is superior with five locales; stigma is penta-lobulated; stamens > 80, arranged in three to five anti-petal fascicles. Fruit is simple, dry, and indehiscent (nut) of ca. 0.40 cm wide; seed is one of ca. 0.2 cm wide and 0.3 cm long (Crow 1990).

C) **Medicinal parts:**
Inflorescences (Pérez-Ortega et al. 2008).

D) **Chemical constituents:**
**Flavonoids** (kaempferol derivatives, quercetin, quercetin-3-*O*-glucoside, quercetin-3-*O*-rutinoside, tiliroside) (Martinez et al. 2009), **sterols** (β-sitosterol), and **fatty acids** (Aguirre 2007a; Pérez et al. 2008; Cárdenas 2014).

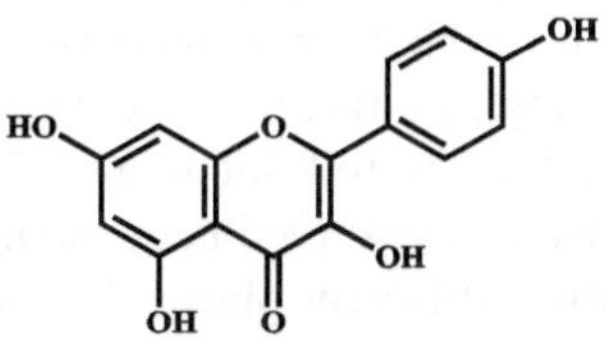

Kaempferol

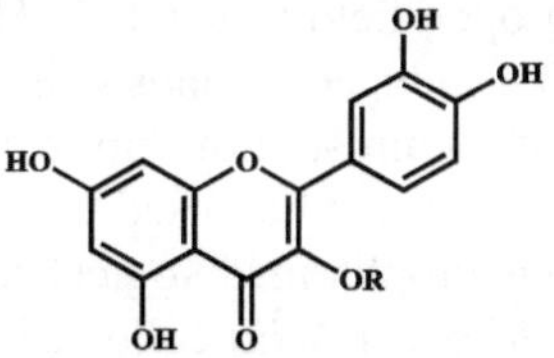

Quercetin: R = H

R = glucose: quercetin-3-O-glucoside

R = rutinose: quercetin-3-O-rutinoside

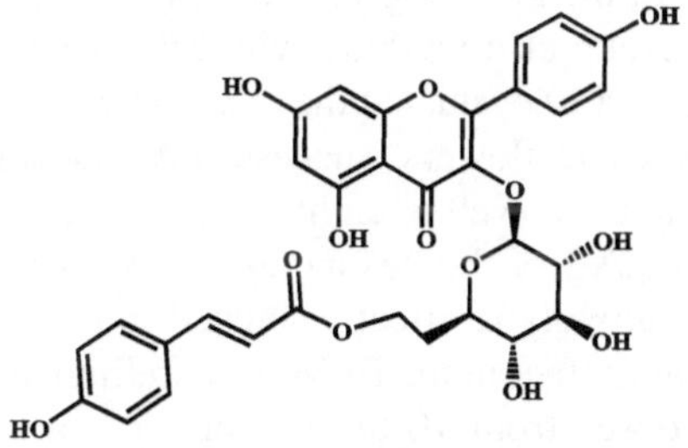

Tiliroside

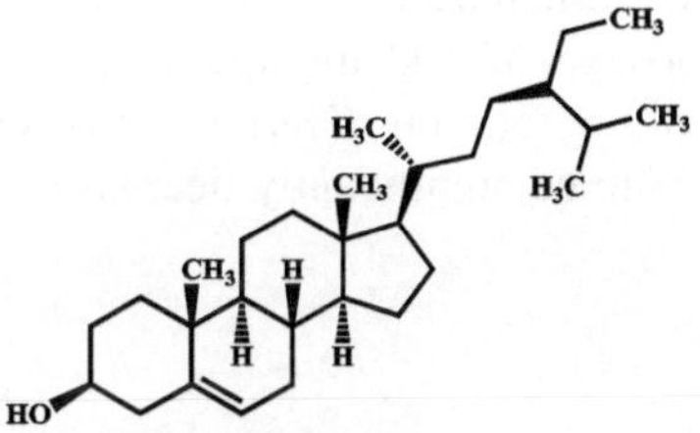

β-Sitosterol

4) **Pharmacology and bioactivities:**

A) **In vivo studies:**

The aerial parts of this species have been used in Mexican traditional medicine for their central nervous system depressant activities. The n-hexane (10 and 30 mg/kg, i.p.) and methanol (10–300 mg/kg, i.p.) extracts prepared from the inflorescences had an anxiolytic-like effect, and at higher doses it produced a decrease in the ambulatory activity (Aguirre-Hernández et al. 2007a). The anxiolytic activity was determined by elevated plus maze (EPM), hole-board test (HBT), and exploratory rearing. Acute toxicity was observed with a lower dose of methanol extract ($LD_{50}$=375 mg/kg) in comparison with the hexane extract ($LD_{50}$>2900 mg/kg). These researchers suggested that the hexane extract elicits a depressant action on the CNS, at least in part by the presence of β-sitosterol (1–10 mg/kg, i.p.), and some fatty acids that remain to be identified. A dose–response curve of β-sitosterol in the range 1–30 mg/kg doses indicated that this compound produced an anxiolytic-like action from 1 to 10 mg/kg and a sedative response when the dose was increased to 30 mg/kg; these effects resemble those produced by diazepam (0.1 mg/kg). To assess the anxiolytic-like response, methanol extracts of *Tilia* inflorescences (from 10 to 300 mg/kg, i.p.) were tested in mice using open field test (OFT), HBT, and EPM tests, as well as sodium pentobarbital-induced hypnosis. Since no toxicity was observed through this route of administration (up to 5000 mg/kg), therefore, it suggests that this plant is secure when used as tranquilizer in folk medicine. No difference in the neuropharmacological activity was observed between methanol extracts of Mexican *Tilia* collected in three different places. Nevertheless, quercetin and kaempferol aglycones were tested and showed anxiolytic-like response; therefore, the authors suggested that the pharmacological effect of *Tilia* inflorescences involved quercetin and kaempferol (10 mg/kg, i.p.), but it was independent of the kind of glycosides present in the samples.

Other research group also working with *Tilia americana* var. *mexicana* (Herrera-Ruiz et al. 2008) evaluated the anxiolytic effect of four extracts of the combined bracts and flowers, successively macerated with n-hexane, ethyl acetate, methanol, as well as an hydroalcoholic extract. These authors reported that 100 mg/kg of the hexane and ethyl acetate extracts via oral (p.o.) displayed no anxiolytic effect evaluated in the EPM test in mice. In contrast, a significant effect in the EPM was preliminary observed with the hexane extract at doses from 10 to 100 mg/kg, via intraperitoneal (i.p.), where b-sitosterol was found to be responsible, at least in part, of the anxiolytic activity (Aguirre-Hernández et al. 1997b). The methanol and hydroalcoholic extracts at dosage of 100 mg/kg, p.o., showed anxiolytic effect without motor activity alteration (Herrera-Ruiz et al. 2008), whereas ambulatory activity of mice significantly decreased after the same dosage administered by i.p. (Pérez-Ortega et al. 2008; Aguirre-Hernández et al. 2007a). The methanol extract was subjected to a bioassay-guided fractionation, which afforded a rich flavonoid mixture of quercetin and kaempferol derivatives with anxiolytic activity, identified as tiliroside, quercitrin, rutin, kaempferol, and quercetin (Herrera-Ruiz et al. 2008) as preliminary suggested (Pérez-Ortega et al. 2008) and confirmed by chromatographic analysis (Aguirre-Hernández et al. 2010). In these studies, not all the compounds were individually isolated and tested (Pérez-Ortega, et al. 2008; Herrera-Ruiz et al. 2008) except in the case of quercetin and kaempferol aglycones evaluated at 10 mg/kg, i.p. (Aguirre-Hernández et al. 2010).

Antinociceptive and anti-inflammatory activities of the *Tilia* inflorescences were also described by using inflammatory models in rats. In this study, the aqueous extract demonstrated significant activity tested at doses of 30 and 100 or 300 mg/kg, i.p., where the flavonoid quercetin was partially responsible of the antinociceptive effect of *Tilia* involving a serotonergic mechanism (Martínez et al. 2009) as also reported in the anxiolytic-like response in mice (Noguerón-Merino et al. 2015; Aguirre-Hernández et al. 2016). Preliminary anticonvulsant and antioxidant effects of the *Tilia* methanol extract at 100, 300, and 600 mg/kg, i.p. have been also reported (Cárdenas-Rodríguez et al. 2014).

5) **Indications and usage:**

Around the world, *Tilia* species have been recognized for their tranquilizing properties in traditional medicine (Argueta et al. 1994; Viola et al. 1994; Coleta et al. 2001) even as anticonvulsant (Tortoriello and Romero 1992). The infusion of inflorescences is also used as a remedy to treat inflammation and arthritic pain, among others kinds of affections associated with pain, such as enterocolitis, gastroenteritis, hepatic and nephritic cramps, menstrual cramps, and headache (Argueta et al. 1994).

Infusions of *Tilia* are typically prepared with water brought just to a boil and then poured over the inflorescences powder; it is covered and allowed to steep for 5 min. It can also be prepared in the drinking cup (by just pouring the heated water over the herb in the cup) or by dropping the powder of inflorescences into the pot which the water was heated in.

6) **Precautions and side effects:**

Although no recent clinical data has been published, the German Commission E monograph concluded that the linden flower is cardiotoxic.

Reports exist of specific toxicity such as contact urticaria, allergy from certain *Tilia* fruit oils in rats, seasonal pollinosis, organochlorine pesticide residues in linden-containing beverages, and occupational contact dermatitis with rhinoconjunctivitis from soft wood dust exposure (Mills et al. 2005).

7) **Dosage:**

A) **Internal use:**

Linden is available in several dosage forms. There are no recent clinical studies to support a specific dosage of linden. No more than 2–4 g/day of linden from teas or other preparations for internal use should be consumed. Similar dosage regimens can be found on various commercial websites (Mills et al. 2005).

B) **External use:**
Not for external use.

## 8.10 References


- Argueta A, Cano L, Rodarte M. Atlas de las plantas de la medicina tradicional mexicana I II y III. D.F. InstitutoNacional Indigenista Mexico. 1994;86: 175,1337,1355.
- Cárdenas-Rodríguez N, González-Trujano ME, Aguirre-Hernández E, Ruíz-García M, Sampieri A, 3rd, Coballase-Urrutia E, Carmona-Aparicio L. Anticonvulsant and antioxidant effects of *Tilia americana* var. mexicana and flavonoids constituents in the pentylenetetrazole-induced seizures. Oxid Med Cell Longev. 2014;329172.
- Coleta M, Campos MG, Cotrim MD, Cunha AP. Comparative evaluation of *Melissa officinalis* L., *Tilia europaea* L., *Passiflora edulis* Sims. and Hypericum perforatum L. in the elevated plus maze anxiety test. Pharmacopsychiatry. 2001;34:520–1.
- Herrera-Ruiz M, Román-Ramos R, Zamilpa A, Tortoriello J, Jiménez-Ferrer JE. Flavonoids from *Tilia americana* with anxiolytic activity in plus-maze test. J Ethnopharmacol. 2008;118(2):312–7.
- Martínez AL, González-Trujano ME, Aguirre-Hernández E, Moreno J, Soto-Hernández M, López-Muñoz FJ. Antinociceptive activity of *Tilia americana* var. mexicana inflorescences and quercetin in the formalin test and in an arthritic pain model in rats. Neuropharmacology. 2009;56(2):564–71.
- Mills S, Bone K. The essential guide to herbal safety. Elsevier Health Sci. 2005.
- Noguerón-Merino MC, Jiménez-Ferrer E, Román-Ramos R, Zamilpa A, Tortoriello J, Herrera-Ruiz M. Interactions of a standardized flavonoid fraction from *Tilia americana* with Serotonergic drugs in elevated plus maze. J Ethnopharmacol. 2015;164:319–27.
- Tortoriello J, Romero O. Plants used by Mexican traditional medicine with presumable sedative properties: an ethnobotanical approach. Arch Med Res. 1992; 23(3):111–16.
- Aguirre-Hernández E, González-Trujano ME, Terrazas T, Herrera-Santoyo J, Guevara-Fefer P. Anxiolytic and sedative-like effects of flavonoids from *Tilia americana* var. *mexicana*: GABAergic and serotonergic participation. Salud Mental. 2016;39(1):37–46.
- Viola H, Wolfman C, Levi de Stein M, Wasowski C, Peña C, Medina JH, Paladín AC. Isolation of pharmacologically active benzodiazepine receptor ligands from *Tilia tomentosa* (Tiliaceae). J Ethnopharmacol. 1994;44, 47–53.

**Fig. 8.5** *Valeriana edulis* Nutt. Source: Herbarium of medicinal plants, IMSS

## 8.11 *Valeriana edulis* Nutt. ex Torr. & A. Gray ssp. *procera* (Kunth) F.G. Mey

### *8.11.1 (Valerianaceae)*

1) **Common names:**
   Valeriana, Tuil bomol, Amantilla, Capon's tail, cat's love, common valerian, English valerian, fragrant valerian, and garden heliotrope.
2) **Photograph of the species (Fig. 8.5)**
3) **Description of the species:**

   A) **Habitat and world distribution:**
   The underground organs of several species of genus *Valeriana* are used by the traditional medicine of many cultures as mild sedative and tranquilizer, as a sleep-inducing aid. The genus is comprised of about 200 species and has been used medicinally before Christianity in the Mediterranean area, India, and China. In México, there is a native *Valeriana* species, *Valeriana edulis* ssp. *procera* (Rzedowski et al. 2003).

   B) **Morphology of the plant:**
   Herbaceous plant is erect, perennial, widely flaring, and tubular in shape; inflorescence has a long, often branched, dense cluster which blooms in May–June; leaves are opposite, thick, densely hairy, and 4″–12″ long; linear basal leaves are usually undivided or with two lower lobes; the smaller stem leaves are pinnately divided (http://wisflora.herbarium.wisc.edu/taxa/index.php?taxon=5326).

The part of the plant used medicinally is the root or rhizome. The rhizome is light grayfish brown, about the size of a finger joint, bearing many rootlets. The fresh root has no door, while the dried root smells distinctly unpleasant, akin to old gym socks, due to isovaleric acid. The plant itself is 50–150 cm tall with pinnate leaves and white or pink hermaphroditic flowers with three stamens; the stem is upright and without branches. It is sometimes used as a border in perennial gardens (Fleming 1998; Bisset 1994).

C) **Medicinal parts:**
The part of the plant used medicinally is the root or rhizome (Sharma et al. 2010).

D) **Chemical constituents:**
Iridoid valepotriates (0.5–2.0 %): valtrate, isovaltrate, dihydrovaltrate, valerosidate, and others. Volatile essential oil (0.2–02.8 %): bornyl isovalerate and bornyl acetate; valerenic, valeric, isovaleric, and acetoxyvalerenic acids; valerenal, valeranone, and cryptofaurinol; and other monoterpenes and sesquiterpenes. Alkaloids (0.01–0.05 %) valeranine, chatinine, alpha-methyl pyrrylketone, actinidine, skyanthine, and naphthyridylmethylketone. Lignans 8-hydroxypinoresinol (Thies 1968; Cropley et al. 2002).

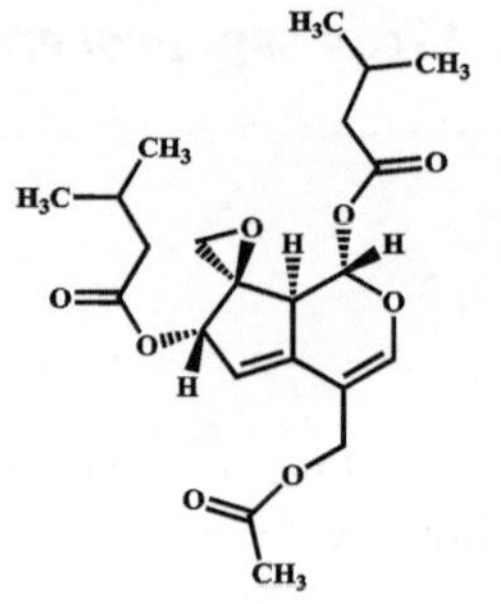
**Valtrate**

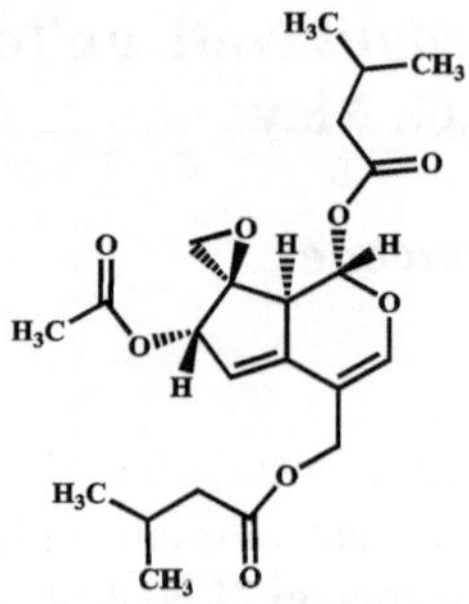
**Isovaltrate**

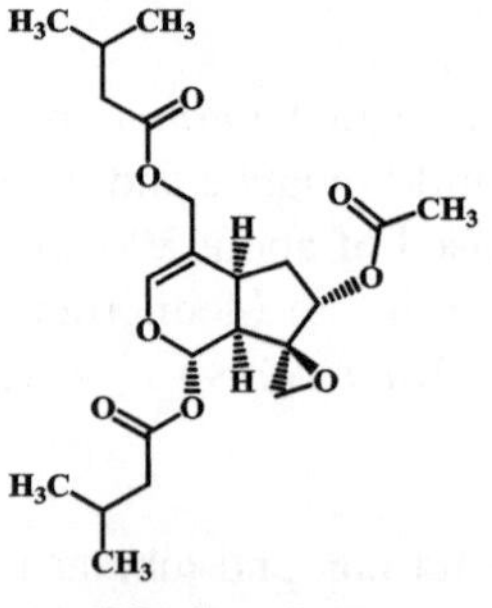
**Dihydrovaltrate**

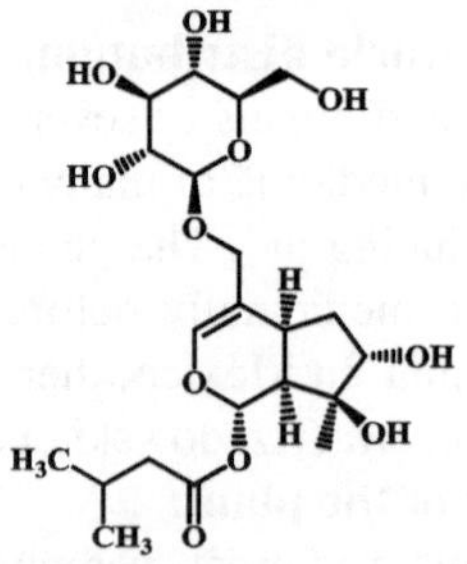
**Valerosidate**

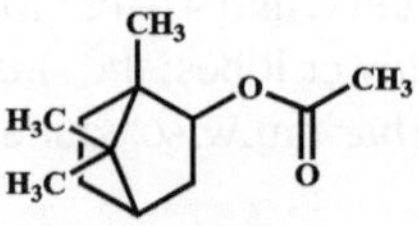
**Bornyl acetate**

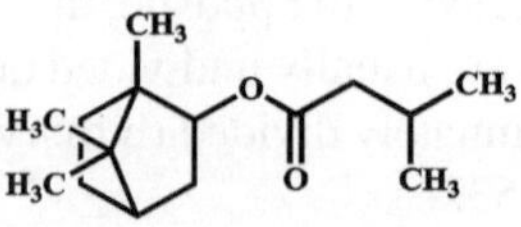
**Bornyl isovalerate**

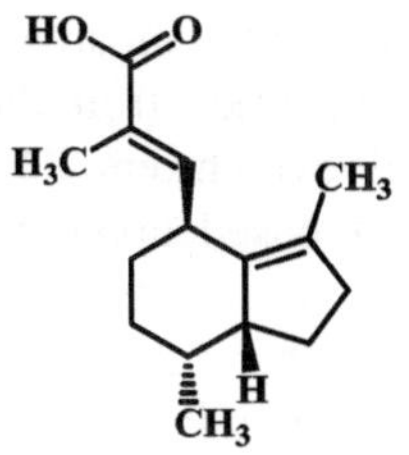


**Valerenic acid**

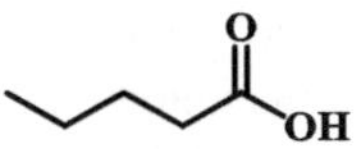


**Valeric acid**

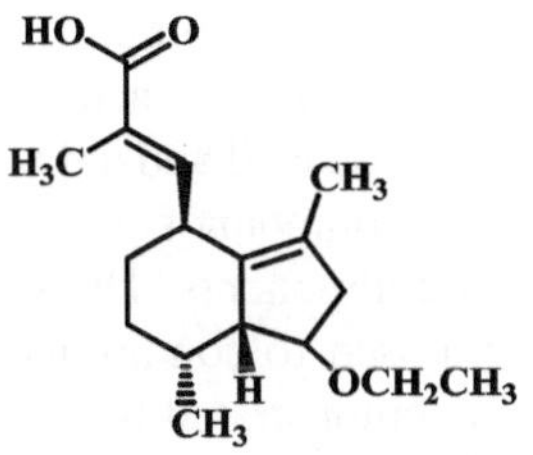


**Acetoxyvalerenic acid**

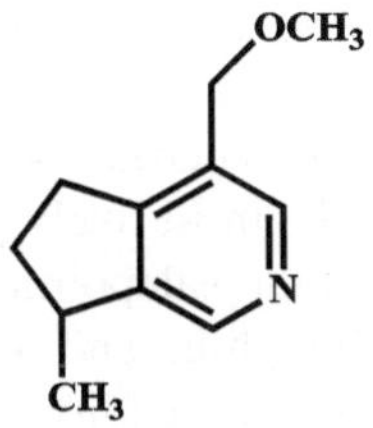


**Valerianine**

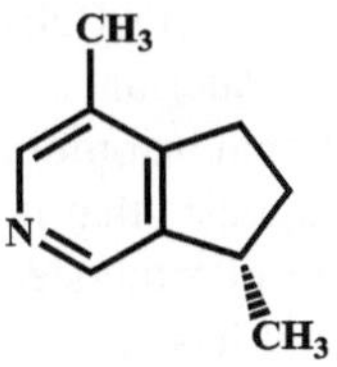


**Actidine**

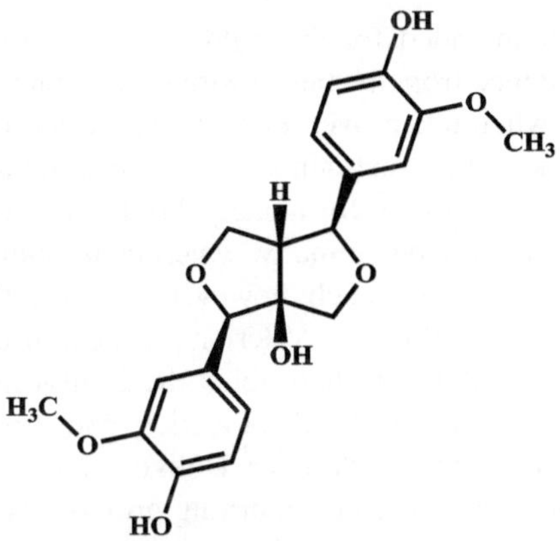


**8-hydroxypinoresinol**

4) **Pharmacology and bioactivities:**

   A) **In vitro studies:**

   The in vitro effect of aliquots from 14 commercially available single-entity and blended products containing valerian root on cytochrome P450 CYP3A4-mediated metabolism and P-glycoprotein transport has been determined with aqueous, ethanol, and acetonitrile extracts. Hydroxyvalerenic acid, acetoxyvalerenic acid, and valerenic acid content was analyzed, and a wide variation was found between samples and compared to the concentrations noted on the product labels. Valerian extracts from the products tested also exhibited a marked capacity to inhibit cytochrome P450 3A4-mediated metabolism and P-glycoprotein transport based upon the ATPase assay. There is wide variation between commercially available samples of valerian root. The findings from this study suggest that valerian root may have an initial inhibitory effect when taken with therapeutic products (Lefebvre et al. 2004).

B) **In vivo studies:**

Research into physiologic activity of individual components has demonstrated direct sedative effects (valepotriates, valeric acid) and interaction with neurotransmitters such as GABA (valeric acid and unknown fractions). The most likely pharmacologically active compounds in valerian are found in the valepotriates and the essential oil components. The valepotriates are lipophilic and highly unstable when exposed to conditions of heat, moisture, or acidity. Possible cytotoxicity is demonstrated by two valepotriates, valtrate and didrovaltrate, which has raised safety concerns. In vitro studies have shown that these compounds caused cell death in cultured rat hepatoma cells (Oliva et al. 2004).

Valerian is commonly recommended for the mild treatment for occasional insomnia. However, evidence from the best positive study on valerian suggests that it is only useful when taken over an extended period of time for chronic sleep disorders. Like other treatments used for insomnia, valerian has also been proposed as a treatment for anxiety, but there is no reliable evidence as yet that it is effective. Finally, valerian is sometimes suggested as a treatment for a nervous stomach; however, as of yet, there is no supporting scientific evidence for this use. Valerian has been used as a sleep aid for over 1000 years. Its ability to help relax the central nervous system, promote feelings of calm, decrease levels of anxiety and stress, and enhance sleep is known to millions the world over. Unlike some prescription sleep aids, valerian is not known to cause morning grogginess and is nonaddictive. The use of valerian as a sedative and antianxiety treatment has been reported for more than 2000 years. For example, in the second century AD, Galen recommended valerian as a treatment for insomnia (Houghton 1999).

Another study showed that valerian extract may be comparable to the effects of the prescription benzodiazepine drug oxazepam (Serax) for insomnia. Further research is necessary to confirm these results. Several clinical studies have shown that valerian is effective in the treatment of insomnia, most often by reducing sleep latency (Herrera et al. 2011).

The anxiolytic activity of ethanol extract (70 %) of Mexican valerian roots was assessed in mice using the exploratory rearing as behavioral model. The roots were obtained by micropropagation method, and extract showed anxiolytic and anticonvulsant activity at 100, 300, and 1000 mg/kg (i.p.). In addition, the extract decreased rotarod performance and traction force and prolonged pentobarbital-induced sleeping time at the highest dose (Oliva et al. 2004). The pharmacological effects have been attributed to their valepotriates. The anxiolytic activity of valepotriates has been tested in patients using a mixture of 0 % dihydrovaltrate, 15 % valtrate, and 5 % acevaltrate. Thirty-six patients with generalized anxiety disorder were randomized after a 2-week washout and assigned one of the following treatments for 4 weeks: (1) valepotriates (mean daily dose, 81.3 mg), (2) diazepam (mean daily dose, 6.5 mg), or (3) placebo. A parallel, double-blind, flexible-dose, placebo-controlled design was employed. The results suggested that the valepotriates may have a potential anxiolytic effect of the psychic symptoms of anxiety (Andreatini et al. 2002). On the other hand, it has been described that the valepotriates act in the brain through GABA receptors. Adenosine (A1) receptor interacts with GABA modulation (increased and decreased degradation of GABA) (Ugalde et al. 2005).

However, in vivo studies have shown no significant cytotoxic effects of valtrate on mouse bone marrow early progenitor cells. Generally, valepotriates have been described to impart spasmolytic or relaxing effects on the smooth muscle system, sedative properties, anticonvulsant activity, psychostimulant activity, and effectiveness in the treatment of behavior disorders. They have also been observed to antagonize the hypnotic effect of alcohol and to blunt the toxicity of ethanol in animals. It has been suggested that valepotriates interact synergistically with the essential oil constituents to produce nervous system activity (Dietz et al. 2005).

5) **Indications and usage:**

Valerian was an official remedy in the United States Pharmacopeia from 1820 to 1936. Valerian root's popularity as a sedative herb seems to be increasing along with the stresses of modern life. For at least 500 years, valerian root has been among the most popular remedies in the USA (Mowrey et al. 1986).

Restlessness and sleep-induction disturbances due to nervous conditions. The medicine is made from the root. Valerian is most commonly used for sleep disorders, especially the inability to sleep (insomnia). It is frequently combined with hops, lemon balm, or other herbs that also cause drowsiness. Some people who are trying to withdraw from the use of "sleeping pills" use valerian to help them sleep after they have tapered the dose of the sleeping pill (Sharma et al. 2010).

Valerian is also used for conditions connected to anxiety and psychological stress including nervous asthma, hysterical states, excitability, fear of illness (hypochondria), headaches, migraine, and stomach upset. Some people use valerian for depression, mild tremors, epilepsy, attention deficit hyperactivity disorder (ADHD), and chronic fatigue syndrome (CFS). Continuous use for several days, even up to 4 weeks, may be needed before an effect is noticeable (Rodenbeck et al. 1998).

6) **Precautions and side effects:**
Valerian can cause some side effects such as headache, excitability, uneasiness, and even insomnia in some people. A few people feel sluggish in the morning after taking valerian, especially at higher doses. It is best not to drive or operate dangerous machinery after taking valerian. The long-term safety of valerian is unknown. To avoid possible side effects when discontinuing valerian after long-term use, it is best to reduce the dose slowly over a week or two before stopping completely (Herrera et al. 2001).

There is not enough information about the safety of valerian during pregnancy or breastfeeding. Stay on the safe side and avoid use (Occhiuto et al. 2009).

Valerian slows down the central nervous system. Anesthesia and other medications used during surgery also affect the central nervous system. The combined effects might be harmful (Spinella 2006).

Alcohol interacts with valerian and can cause sleepiness and drowsiness. Valerian might also cause sleepiness and drowsiness. Taking large amounts of valerian along with alcohol might cause too much sleepiness (Givens et al. 2000).

Valerian interacts with sedative medications (Benzodiazepines) including alprazolam (Xanax), clonazepam (Klonopin), diazepam (Valium), lorazepam (Ativan), midazolam (Versed), temazepam (Restoril), triazolam (Halcion), and others, causing too much sleepiness (Bent et al. 2006).

Valerian might decrease how quickly the liver breaks down some medications. Taking valerian along with some medications that are broken down by the liver can increase the effects and side effects of some medications. Some medications changed by the liver include lovastatin (Mevacor), ketoconazole (Nizoral), itraconazole (Sporanox), fexofenadine (Allegra), triazolam (Halcion), and many others (Meolie et al. 2005).

7) **Dosage:**

A) **Internal use:**
A specific combination product, providing valerian root extract 160 mg and lemon balm leaf extract 80 mg (Euvegal forte, Schwabe Pharmaceuticals), has been tried to reduce symptoms of serious restlessness (dyssomnia) in children under the age of 12. Early results show it might be effective, but more research is needed.

Early research suggests that taking 600 mg of valerian for 7 days reduces blood pressure, heart rate, and feelings of pressure when under stress. Other research found that taking 100 mg before speaking in front of an audience reduces feelings of anxiety. Another study found that taking a combination product containing valerian and lemon balm might lower anxiety caused by stress at low doses but increase anxiety when taken in larger doses.

For inability to sleep (insomnia): 400–900 mg valerian extract up to 2 h before bedtime for as long as 28 days or valerian extract 120 mg, with lemon balm extract 80 mg three times daily for up to 30 days, or a combination product containing valerian extract 187 mg plus hops extract 41.9 mg per tablet, two tablets at bedtime for 28 days. Take valerian 30 min to 2 h before bedtime (http://www.webmd.com/vitamins-supplements/ingredientmono-870-valerian.aspx?activeingredientid=870&activeingredientname=valerian).

For mild insomnia: Adults (aged 18 or older):

Capsules: 300–1800 mg of valerian has been taken by mouth.

Aqueous or aqueous-ethanol extract: A dose of 1.5–3 g of herb has been taken by mouth 30–60 min before going to bed.

Tea: A dose of 1.5–3 g of valerian root steeped in 150 mL of boiling water for 5–10 min has been taken orally about 30–60 min before going to bed.

Children (younger than 18):

The dosing and safety of valerian have not been studied thoroughly in children, and valerian is therefore not recommended.

For sedation or stress reduction:

Aqueous or aqueous-ethanol extract: A dose of 100–600 mg taken by mouth before or after stressful events has been used.

Tea: A dose of 1.5–3 g of valerian root steeped in 150 mL of boiling water has been taken by mouth 5–10 min before or after stressful events. Experts in herbal preparations recommend that valerian products should be standardized to contain 0.8 % valerenic or valeric acid.

To relieve insomnia, one of the above dosages may be taken 30–45 min before bedtime. It may take 1–2 weeks of regular use before the herbal preparation takes effect. When giving valerian to children, recommended adult dosages should be adjusted in proportion to the child's weight. Most dosages of herbal products are calculated for an adult weighing 150 lb (70 kg). A child weighing 75 lb (35 kg) should therefore receive 1/2 the adult dose (Sharma et al. 2010).

B) **External use:**

Not for external use.

## 8.12 References


- Andreatini R, Sartori VA, Seabra ML, Leite JR. Effect of valepotriates (valerian extract) in generalized anxiety disorder: a randomized placebo-controlled pilot study. Phytother Res. 2002;16(7):650–4.
- Bent S, Padula A, Moore D, Patterson M, Mehling W. Valerian for sleep: a systematic review and meta-analysis. Am J Med. 2006;119(12):1005–12.
- Bissett NG. Herbal drugs and phytopharmaceuticals. Stuttgart: MedPharm CRC Pressl; 1994. p. 566.

- Castillo P, Zamilpa A, Márquez J, Hernández G, Lara M, Alvarez L. Comparative study of differentiation levels and valepotriate content of in vitro cultures and regenerated and wild plants of *valeriana edulis* ssp. *procera*. J Nat Prod. 2002; 65(4):573–75.
- Cropley M, Cave Z, Ellis J, Middleton RW. Effect of kava and valerian on human physiological and psychological responses to mental stress assessed under laboratory conditions.
- Dietz BM, Mahady GB, Pauli GF, Farnsworth NR. Valerian extract and valerenic acid are partial agonists of the 5-HT 5a receptor in vitro. Mol Brain Res. 2005;138(2), 191–7.
- Fleming T. PDR for herbal medicines. Montvale, NJ: Medical Economics Company, Inc.; 1998. p. 27.
- Givens M, Cupp MJ. Valerian. In: Toxicology and clinical pharmacology of herbal products. Humana Press; 2000. p. 53–66.
- Herrera-Arellano A, Luna-Villegas G, Cuevas-Uriostegui ML, Alvarez L, Vargas-Pineda G, Zamilpa-Alvarez A, Tortoriello J. Polysomnographic evaluation of the hypnotic effect of *Valeriana edulis* standardized extract in patients suffering from insomnia. Planta Med. 2001;67(08):695–9.
- Hoffman D. The complete illustrated holistic herbal. Rockport, MA: Element Books Inc.; 1996.
- Houghton PJ. The scientific basis for the reputed activity of Valerian. J Pharm Pharmacol. 1999;51:505–12.
- Klich R. Behavior disorders in childhood and their therapy. Med Welt. 1975;26:1251–4.
- Lefebvre T, Foster BC, Drouin CE, Krantis A, Livesey JF, Jordan SA. In vitro activity of commercial valerian root extracts against human cytochrome P450 3A4. J Pharm Pharm Sci. 2004;7(2):265–73.
- Meolie AL, Rosen C, Kristo D, Kohrman M, Gooneratne N, Aguillard RN, Fayle R, Troell R, Townsend D, Claman D, Hoban T, Mahowald M. Oral nonprescription treatment for insomnia: an evaluation of products with limited evidence. J Clin Sleep Med. 2005;1(2):173–87.
- Mowrey DB. The scientific validation of herbal medicine. New Canaan, Conn.: Keats Pub.; 1986. p. 316.
- Occhiuto F, Pino A, Palumbo DR, Samperi S, Pasquale R, Sturlese E, Circosta C. Relaxing effects of *Valeriana officinalis* extracts on isolated human non-pregnant uterine muscle. J Pharm Pharmacol. 2009;61(2):251–6.
- Oliva I, González-Trujano M, Arrieta J, Enciso-Rodríguez R, Navarrete A. Neuropharmacological profile of hydroalcohol extract of *Valeriana edulis* ssp. procera roots in mice. Phytother Res 2004;18(4):290–6.
- Rodenbeck A, Simen S, COhrs S et al. Alterations of the sleep stage structure as a feature of GABAergic effects of a valerian hop preparation in patients with psychophysiological insomnia. Somnologie. 1998;2:26–31.

**Fig. 8.6** *Galphimia glauca* Cav. Malpighiaceae Source: Dr. Jaime Tortoriello


- Rzedowski J, de Rzedowski GC. Dos especies nuevas de *Valeriana* (Valerianaceae) del centro de México. Acta Botánica Mexicana. 2003;(62):65–71.
- Sharma M, Jain UK, Patel A, Gupta N. A comprehensive pharmacognotic report on valerian. Int J Pharm Sci Res. 2010;1(7):6–40.
- Spinella M. Herbal medicines and sleep. In: Sleep and sleep disorders. New York, USA: Springer; 2006. p. 297–303.
- Thies PW. Linarin-isovalerianate, a currently unknown flavonoid from *Valeriana wallichii.* Planta Med. 1968;16:363–71. Phytother Res. 2002;16:23–7.
- Ugalde M, Reza V, González-Trujano M, Avula B, Khan IA, Navarrete A. Isobolographic analysis of the sedative interaction between six central nervous system depressant drugs and *Valeriana edulis* hydroalcoholic extract in mice. J Pharm Pharmacol. 2005;57(5):631–39. http://www.webmd.com/vitamins-supplements/ingredientmono-870-valerian.aspx?activeingredientid=870&active ingredientname=valerian.

## 8.13 *Galphimia glauca* Cav.

### *8.13.1 (Malpighiaceae)*

1) **Common names:**
   The *Galphimia* genus has a scientific synonym, *Thryallis*, while, depending on the locality, the following popular names are used to refer to this species: "corpionchi," "huachacata," "nacachata," "vacachata," "calderona amarilla," "flor de diciembre," "ramo de oro," and "hierba del piojo."
2) **Photograph of the species (Fig. 8.6)**
3) **Description of the species:**

   A) **Habitat and world distribution:**
   It is a shrub widely distributed throughout Mexico due to its resistance to environmental conditions. Native to central Mexico, this evergreen plant grows preferentially in the states of Guerrero, Michoacan, San Luis Potosí, Morelos, Chiapas, and Sonora at altitudes of below 2350 m (Rojas et al. 2005).
   B) **Morphology of the plant:**
   *Galphimia glauca* is a robust, evergreen shrub which grows to a height of between 1 and 1.8 m. The trunk is covered with reddish fibers; the leaves (ca. 5 cm long) are thin and leathery, are bluefish-grayish green in color, and are either elliptical, oval, or bluntly rounded off at their base. The leaves are alternate along the thin branches. *Galphimia* is characterized by flowers with yellow petals borne in racemes, either solitary or arranged in panicles, 2 cm in size, and have two outer leaves, five sepals and five petals, and ten anthers which are alternately short and long, and three styles. This shrub is fairly capable of adapting to drier climates and can even withstand mild frosts of up to −2 °C. The shrub blooms at different times of the year, depending on where it is planted (Sharma et al. 2012).

   The fruit is a schizocarp that breaks into three cocci. Each pebble-like coccus contains a globose seed with a dark brown, smooth seed coat. Cocci generally are 2.5 mm long in *G. glauca* (Anderson 2007).
   C) **Medicinal parts:**
   Leaves and aerial parts (Tortoriello et al. 1993).
   D) **Chemical constituents:**
   *Galphimia glauca* Cav. has demonstrated anxiolytic activity attributable to **nor-seco-triterpenes** denominated galphimines (galphimines A, B, E), the most active of which is galphimine-B (Tortoriello et al. 1999; Lara et al. 2005; Herrera-Arellano et al. 2012).

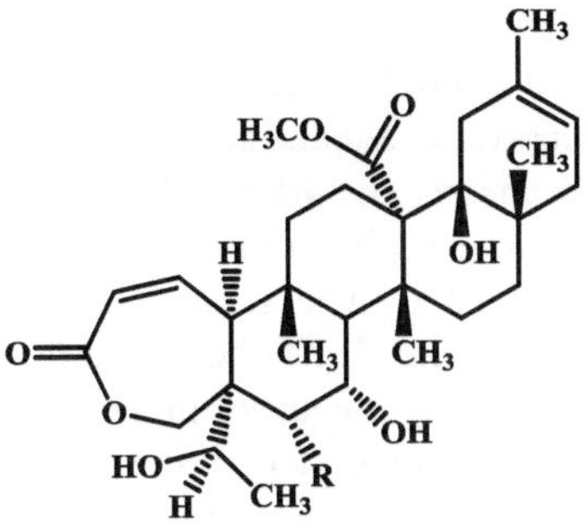

R = OH : Galphimine A
R = H : Galphimine B

4) **Pharmacology and bioactivities:**

A) **In vitro studies:**
Galphimine-B was evaluated using in vitro experimental models in which it was shown that the new compound administered produced a strong inhibition of the electrically induced contraction of guinea pig ileum. These effects suggested an interaction between this compound and the enteric and the CNS (Herrera et al. 2006).

B) **In vivo studies:**
In vivo, *Galphimia glauca* Cav. has demonstrated anxiolytic activity attributable to nor-seco-triterpenes denominated galphimines, the most active of which is galphimine-B. Galphimine-B inhibits ventral tegmental area dopaminergic neurons and interacts with the serotonergic system of the dorsal hippocampus. A previous clinical study administering a *G. glauca* herbal medicinal product for 4 weeks evidenced high percentages of therapeutic effectiveness and safety in patients with generalized anxiety disorder. Afterward, galphimine-A and galphimine-E were isolated and found to be active, although less potent than galphimine-B. Likewise, the standardized methanolic extract of *Galphimia glauca* administered thrice elicited anxiolytic-like effect when evaluated (Herrera-Arellano et al. 2012). In another study, it was observed that the structural feature responsible for the anxiolytic activity of the compounds was the presence of free hydroxyl groups at C-4, C-6, and C-7 and the double bond in the A ring (Tortoriello et al. 1993). In pharmaco-toxicological effects, the administration of the extracts for 56 days in mice did not show deaths, and the histopathologic analysis of different organs did not show any alterations or any changes in the biochemical parameters for liver functions, only a decrease of the spontaneous activity (Herrera-Ruiz et al. 2006, 2007).

The therapeutic effectiveness, safety, and the tolerability of a new standardized herbal medicinal product was next developed from the aqueous extract of *G. glauca* on patients with generalized anxiety, in a controlled randomized double-blind clinical trial. The result was important anxiolytic effectiveness, very similar to that observed with lorazepam (Herrera-Arellano et al. 2007). Later, this herbal product was evaluated administering it for 15 weeks to patients with generalized anxiety disorder and showed greater anxiolytic effectiveness than that obtained with lorazepam, with high percentages of therapeutic tolerability and safety. The mode of action of *G. glauca*, its main active compound, galphimine-B, has been demonstrated to selectively inhibit discharges of dopaminergic neurons in the ventral tegmental area and to interact with the serotonergic system of the dorsal hippocampus (Jiménez-Ferrer et al. 2011).

5) **Indications and usage:**

*Galphimia glauca* Cav. (Malpighiaceae) is used in Mexican traditional medicine as a sedative in the treatment of mental disorders. Sedative properties of a methanolic extract of the aerial parts of this plant have been established in animal trials, and an active compound, named galphimine-B, has already been isolated. This compound was submitted to neuropharmacological testing, where it was shown that galphimine-B had no significant effect as an anticonvulsant, while it exhibited a strong depressant activity on the nervous system (Tortoriello et al. 2011).

The safety of the *G. glauca* anxiolytic extract has been evaluated in preclinical toxicology assays using rodents under a chronic administration scheme. Determination of the possible toxic effects of this medicinal species was conducted with different extracts with the purpose of conciliating, on the one hand, the greatest efficacy of the extract with the least number of side effects or, in the best of cases, the absence of the latter (Aguilar-Santamaría et al. 2007). After oral administration of the 2.5 g/kg dose of the aqueous, methanolic, and ethanolic extracts of G. glauca in a 28-day daily scheme to Balb-C mice of both genders, an evaluation on the liver was performed, due to that administration via oral route ensures the passage of this phytopharmaceuticals through this organ by means of the portal circulation. Based on the anatomic position between the gastrointestinal tract and systemic circulation, the liver plays an important role in the metabolism of exogenous substances. The hepatotoxic effects produce tissue alterations accompanied by the liberation of the cellular contents into the plasma, such as the alanine aminotransferase enzyme (ALT), the aspartate aminotransferase enzyme (AST), and the alkaline phosphatase enzyme (ALP), which takes place after an acute inflammatory process and which is eventually related with the oxide-reduction equilibrium of the hepatocytes. The results of the serum analysis of these enzymes, in the different groups of mice with the extracts administered, show that there are no changes in the activity of these with respect to the control group, which suggests a lack of liver damage (Tortoriello et al. 2011).

6) **Precautions and side effects:**
   There is not enough information about the safety of *Galphimia* during pregnancy or breastfeeding (Aguilar et al. 2007).

   *G. glauca* slows down the central nervous system. Anesthesia and other medications used during surgery also affect the central nervous system. The combined effects might be harmful. Stop taking *Galphimia* at least 2 weeks before a scheduled surgery. Alcohol and sedative medications (benzodiazepines) may interact with *Galphimia* and can cause sleepiness and drowsiness (Herrera et al. 2007).

   Chronic administration of the extracts of *G. glauca* caused changes in the behavior of the mice. In special fashion, the methanolic extract produced piloerection, loss of equilibrium, and diminution in the straightening reflex, without modification of other signs, such as tearing, modification of the diameter of the pupil, respiratory movements, paralysis, and diminution of prehensile activity. Dehydration was also observed, as a consequence of treatment, which can be associated with a possible non-quantified diuretic effect, while it cannot be associated with diarrheic evacuations. These changes can be due to a potent depressor effect of the CNS produced by the prolonged administration time of the extracts at doses as high as 2.5 g/kg (Tortoriello et al. 2011).
7) **Dosage:**
   A) **Internal use:**
      The experimental group was treated with capsules containing the aqueous and the dry extract of *Galphimia glauca*, standardized in 0.350 mg of galphimine-B at a dose of one capsule every 12 h for four continuous weeks (Herrera et al. 2006).
   B) **External use:**
      Not for external use.

## 8.14 References


- Aguilar-Santamaría L, Ramírez G, Herrera-Arellano A, Zamilpa A, Jiménez JE, Alonso-Cortés D, Ledesmac N, Tortoriello J. Toxicological and cytotoxic evaluation of standardized extracts of *Galphimia glauca*. J Ethnopharmacol. 2007; 109(1):35–40.
- Anderson C. Revision of *Galphimia* (Malpighiaceae). Contributions from the University of Michigan Herbarium. 2007;25:1–82.
- Herrera-Arellano A, Jiménez-Ferrer E, Zamilpa A, Morales-Valdéz M, García-Valencia CE, Tortoriello J. Efficacy and tolerability of a standardized herbal product from *Galphimia glauca* on generalized anxiety disorder. A randomized, double-blind clinical trial controlled with lorazepam. Planta Med. 2007;73(08): 713–17.

**Fig. 8.7** *Annona muricata* L. Source: Fabiola Domínguez


- Herrera-Ruiz M, González-Cortazar M, Jiménez-Ferrer E, Zamilpa A, Álvarez L, Ramírez G, Tortoriello J. Anxiolytic effect of natural galphimines from *Galphimia glauca* and their chemical derivatives. J Nat Prod. 2006;69(1):59–61.
- Herrera-Ruiz M, Jiménez-Ferrer JE, De Lima TCM, Avilés-Montes D, Pérez-García D, González-Cortazar M, Tortoriello J. Anxiolytic and antidepressant-like activity of a standardized extract from *Galphimia glauca*. Phytomedicine. 2006;13(1):23–8.
- Herrera-Arellano A, Jiménez-Ferrer JE, Zamilpa A, García-Alonso G, Herrera-Alvarez S, Tortoriello J. Therapeutic effectiveness of *Galphimia glauca* vs. lorazepam in generalized anxiety disorder. A controlled 15-week clinical trial. Planta Med. 2012;78:1529–35.
- Jiménez-Ferrer E, Herrera-Ruiz M, Ramírez-García R, Herrera-Arellano A, Tortoriello J. Interaction of the natural anxiolytic Galphimine-B with serotonergic drugs on dorsal hippocampus in rats. J Ethnopharmacol. 2011;37(1):724–9.
- Lara-Ochoa F, Guillén-Torres A, Espinosa-Pérez G, Ortega-Hernández A. Conformational study of galphimines A and B. Spectrochim Acta A Mol Biomol Spectrosc. 2005;61(11):2677–86.
- Rojas G, Aranda E, Navarro V, Zamilpa A, Tortoriello J. In vitro propagation of *Galphimia glauca* and content of the sedative compound galphimine-B in wild and micropropagated plants. Planta Med. 2005;71(11):1076–8.
- Sharma A, Folch JL, Cardoso-Taketa A, Lorence A, Villarreal ML. DNA barcoding of the Mexican sedative and anxiolytic plant *Galphimia glauca*. J Ethnopharmacol 2012;144(2):371–8.
- Tortoriello J, Ortega A. Sedative effect of galphimine B, a nor-seco-triterpenoid from *Galphimia glauca*. Planta Med. 1993;59(05):398–400.

• Tortoriello J, Herrera-Arellano A, Herrera-Ruiz ML, Zamilpa A, González-Cortazar M, Jiménez-Ferrer JE. In: Kalinin VV, editor. New anxiolytic phytopharmaceutical elaborated with the standardized extract of *Galphimia glauca.* 2011. p. 185.

## 8.15 *Annona muricata* L.

### *8.15.1* (Annonaceae)

1) **Common names:**
   Soursop, graviola, sursop, and guanabana.
2) **Photograph of the species (Fig. 8.7)**
3) **Description of the species:**

   A) **Habitat and world distribution:**
   Graviola or soursop is a fruit native to tropical America whose tree grows spontaneously from southern Mexico to Brazil. It was one of the first plants to be exported to Europe after the discovery of America. Soursop is cultivated in many tropical areas in countries such as Angola, Brazil, Colombia, Costa Rica, Cuba, Jamaica, India, Mexico, Panama, Peru, the USA, Puerto Rico, Venezuela, and Asia (Pinto and Silva 1996).

   B) **Morphology of the plant:**
   *Annona muricata* L. is a samall evergreen tropical tree with low branching and bushy grows up to 7–9 meters in heigh; its bark long, 2.5–7.6 cm wide, leathery, obovate to elliptic, glossy on top, glabrous on underside, and simple; stipules are absent; the blade is oblanceolate, green on top, and paler and dull on under side with fine lateral nerves; it has a strong, pungent odor; petioles are short, 3–10 mm long. The flowers have three sepals and three tepals, the flowers sit directly at the trunk (cauliflory) or the branches (ramiflory).

   C) **Medicinal parts:**
   Leaves and soursop (Gajalakshmi et al. 2012).

   D) **Chemical constituents:**
   *Annona* contains sodium and potassium in abundance as well as vitamin C, fibers, starch, calcium, phosphorus, iron, protein, and essential amino acids. Marques and Farah (2009) identified phenolic acids (gallic acid, chlorogenic acid), fatty acids (amyl and geranyl caproic acids) (Bueso 1980; Pinto and Silva 1994), flavonoids, alkaloids, tannins, and saponins. **Alkaloids** isoquinoline derivatives annonaine, nornuciferine, and asimilobine. **Cyclohexapeptide** (annomuricatins A and C), **cyclopeptide** (annomuricatin), **acetogenins** (cohibins A and B; sabadelin; muricoreacin; murihexocins A, B, and C; murihexol; donhexocin; annonacin A; annohexocin).

Annonaceous acetogenins are a series of polyethers which contains either the adjacent or the nonadjacent tetrahydrofuran (THF) or tetrahydropyran (THP) ring and also an a,b-unsaturated c-lactone ring. They possess the most beneficial antitumor, cytotoxic, antimalarial, and antifeedant properties. Acetogenins gets interacted with the NADH–ubiquinone oxidoreductase (complex I) in mammalian and in the insect mitochondrial electron transport systems and/or with ubiquinone-linked NAD(P)H oxidase in the cancer cells cytoplasmic membranes (Hiroyuki et al. 2008).

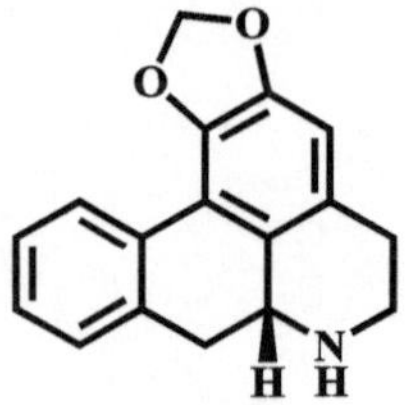

**Anonaine**

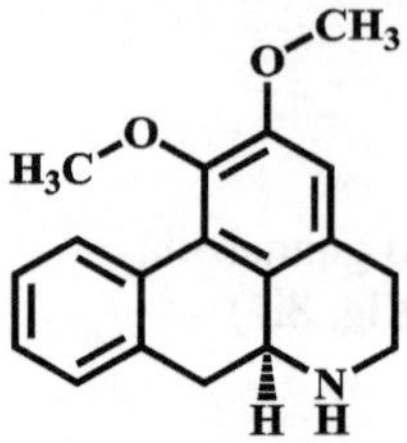

**Nornuciferine**

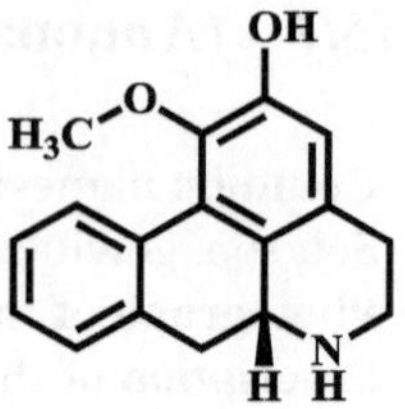

**Asimilobine**

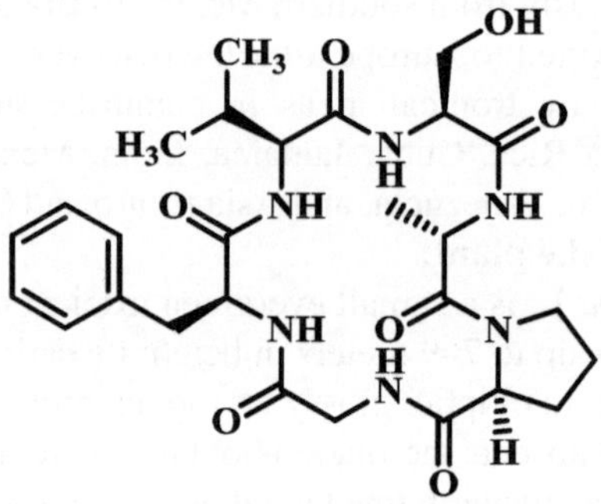

**Annomuricatin C**

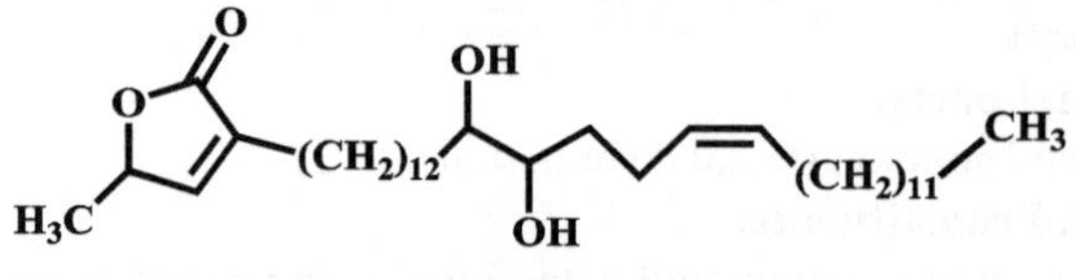

**Cohibin A**

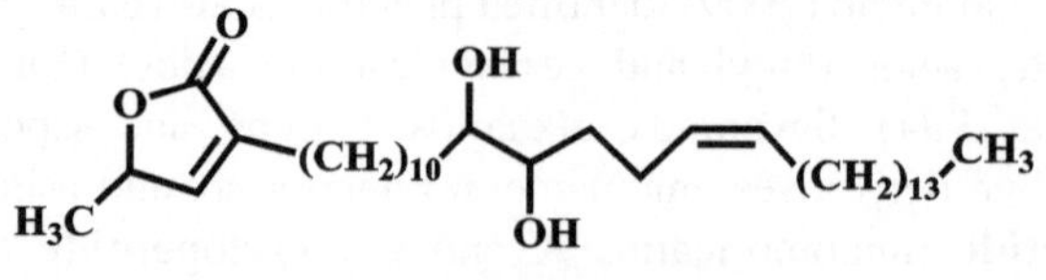

**Cohibin B**

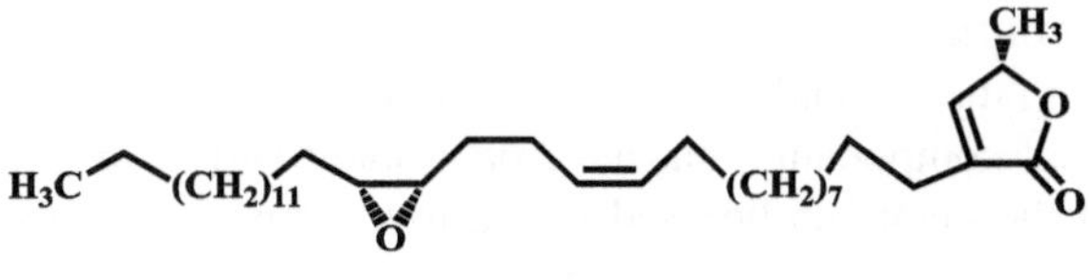

**Sabadelin**

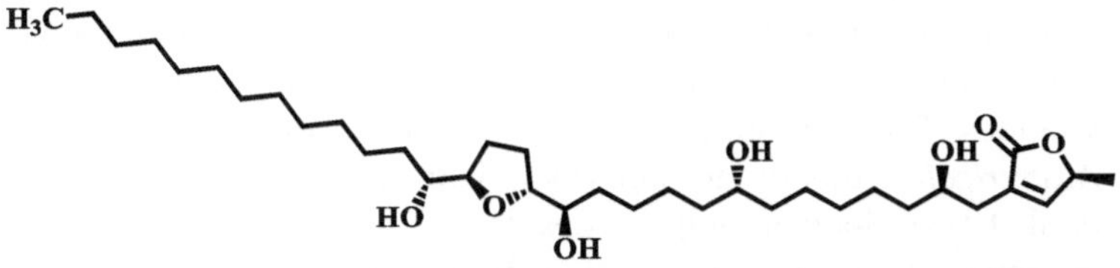

**Annonacin**

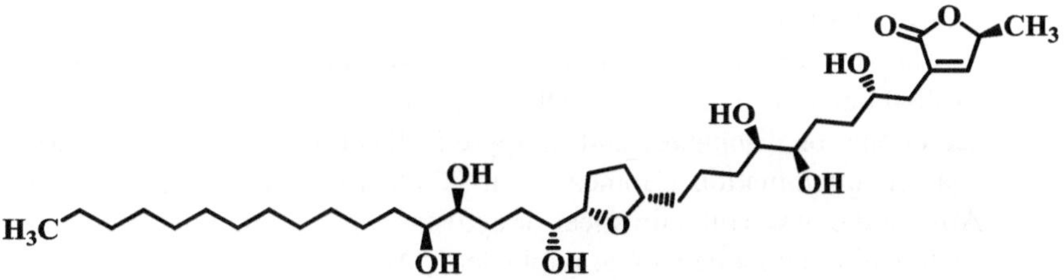

**Murihexocin A**

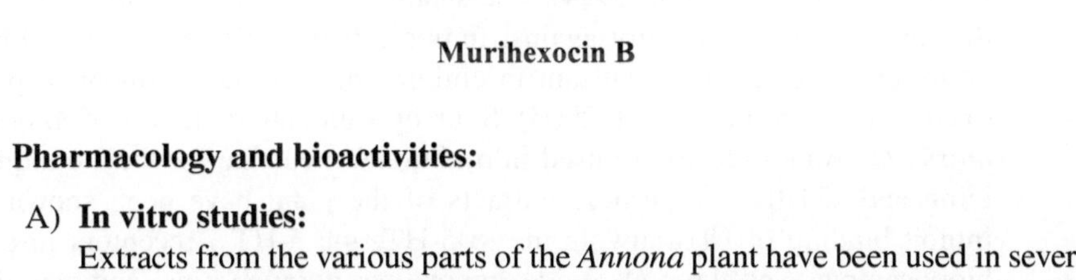

**Murihexocin B**

4) **Pharmacology and bioactivities:**

A) **In vitro studies:**

Extracts from the various parts of the *Annona* plant have been used in several trials to see their cytotoxic capacity in tumor cells; the main active ingredients were cytotoxic acetogenins muricine H, annomontacine, and annonacinone, which showed significant cytotoxic effect on human hepatoma cell Hep G2 and Hep 2,2,15 and in human tumor cell lines of prostatic carcinoma (PC-3) and pancreatic carcinoma (PACA-2) (Rodriguez et al. 2010).

B) **In vivo studies:**

Improving the use and marketing of the soursop is desirable because of the nutritional components that have been identified in the fruit. The various parts of the plant can be used for various purposes, for example, the fruit (juice) is used against dysentery, intestinal washings, diarrhea, fever, and congestion; the seed is used as vermifuge and anthelmintic; the bark and root have antispasmodic, hypotensive, and sedative activities; and the stem and leaf are anticancerigenous. *A. muricata* pulps are useful foods because they contain proteins, fatty acids, fiber, carbohydrates, minerals, and vitamins (Bueso 1980; Leal 1990; Lizana and Reginato 1990). However, *A. muricata* fruits do not contribute with many calories to the diet (kalil et al. 1979). Soursop pulp is considered to be aromatic and exotic and is consumed mostly after processing into cold beverages or sometimes fresh. The edible portion constitutes 67.5 % of total fruit. The characteristic flavor of this fruit is produced by amyl and geranyl caproic acids (Bueso 1980; Pinto and Silva 1994).

The soursop leaves are traditionally used in Brazil for liver problems (Solis-Fuentes et al. 2010), also they are used as suppurative (against mucus, secretions, or discharge) and antipyretic (Badrie and Schauss 2010) and against inflammation (Gomez et al. 2009; Vieira de Sousa et al. 2010). Among the powerful anticancer or cytostatic active ingredients, they found acetogenins contained in graviola leaf. Acetogenins of Annonaceae are waxy substances resulting from the combination of long-chain fatty acids (C32 or C34) with 2-propanol unit at carbon 2 to form a terminal lactone (the lactone is the beginning of the chain). As other acetogenins, the annonacin blocks complex I, which is responsible for converting NADH to $NAD^+$, and creates accumulation of protons through the inner mitochondrial membrane. Alcoholic leaf extracts have antispasmodic and relaxant activity on the smooth muscles; antiulcer activity against indomethacin induced ulcers and reduces the effect of stress on ulcer induction. These effects are produced by various compounds, including flavonoids, alkaloids, tannins, and saponins (Langason et al. 1994). The steam bark contains 4-entkaurenoids that have cytotoxic activity against tumor cell lines (Fatope et al. 1996). Other uses are as anticonvulsant in children and in the treatment of pain (You et al. 1995; Pinto et al. 2005). Soursop fruit and the leaves of *Annona muricata* (Annonaceae) are used in traditional medicine for their tranquilizing and sedative properties. Extracts of the plant have been shown to inhibit binding of [$^3$H]rauwolscine to 5-HTergic 5-$HT_{1A}$ receptors in calf hippocampus, and three alkaloids, annonaine, nornuciferine, and asimilobine, isolated from the fruit, have been shown to have $IC_{50}$ values of 3, 9, and 5 μM, respectively, although in ligand-binding studies it was not possible to determine whether interaction of these ligands with the receptor was agonistic or antagonistic. The inhibition of cAMP accumulation was tested in NIH-3T3 cells stably transfected with the 5-$HT_{1A}$ receptor from man.

None of the alkaloids showed antagonistic properties toward the $5\text{-HT}_{1A}$ receptors because in the antagonistic tests no influence on the forskolin-stimulated increase of cAMP level was detected. Full agonistic properties were measured for all three compounds; the inhibition constants ($K_i$) for three compounds were <10 μM. Inhibition of the binding of the radioligand to the $5\text{-HT}_{1A}$ receptor was observed in every ligand-binding assay performed with the alkaloids; the $K_i$ values for 1, 2, and 3 were in the μM range.

These results imply that the fruit of *Annona muricata* possesses antidepressive effects, possibly induced by these three compounds, and that in the past potent leads for the development of antidepressive therapeutics have not been used (Hasrat et al. 1997).

The effects of an ethanol extract of the mature leaves of *Annona muricata* (Annonaceae) were investigated on pentylenetetrazol (PTZ)-induced clonic–tonic seizures in mice. The latency of clonic PTZ seizures and the incidence of clonic and tonic episodes of PTZ seizures were analyzed. The results demonstrated that the plant extract significantly reduced the incidence of tonic PTZ seizures and mortality and lengthened the onset to clonic PTZ seizures. This study indicates that extract of the leaves of *A. muricata* contains active substance(s) that preferentially suppressed the tonic phase of PTZ-induced clonic–tonic seizure (N'gouemo et al. 1997).

5) **Indications and usage:**

An extract is being advocated for treatment of cancer, mainly on the Internet. Traditionally, the leaves are taken as an herbal tea for a wide variety of illnesses, including cancer, hypertension, diabetes, fever, indigestion, nervousness, palpitations, and skin diseases. It is also applied to kill ectoparasites.

There is experimental evidence supporting many of the traditional uses, and selective cytotoxicity has been observed in vitro in various cancer lines. There is no clinical evidence available.

Leaves, seeds, and bark are not recommended for internal use. The fruit is considered safe if the seeds are removed or remain whole when consumed (Edwards et al. 2015).

6) **Precautions and side effects:**

The benefits and side effects of graviola herbal supplements are not fully understood. Graviola may sometimes cause sleepiness, sedation, or relaxation. But an abnormally frequent atypical levodopa-unresponsive, akinetic–rigid syndrome with some similarity to Parkinson's disease was identified in the Caribbean island Guadalupe and was associated with the consumption of plants of the Annonaceae family, especially *Annona muricata*, suggesting a possible toxic etiology because it contains two groups of potential compounds, alkaloids and acetogenins. Both alkaloids and annonacin, the most abundant acetogenin, were toxic in vitro to dopaminergic and other neurons. Annonacin was toxic in nanomolar concentrations, whereas micromolar concentrations of alkaloids were needed; acetogenins are potent mitochondrial poisons like other parkinsonism-inducing compounds.

Annonacin are present in the fruit or aqueous extracts of the leaves *ofA. muricata* can cross the blood–brain barrier since it was detected in brain parenchyma of rats treated chronically with the molecule and induced neurodegeneration of basal ganglia in these animals, similar to that observed in atypical parkinsonism (Lannuzel et al. 2006).

Also, an interesting neurological syndrome, characterized by recurrent optic neuritis, cervical myelopathy from syringomyelia, paraparesis, amenorrhea–galactorrhea, and other endocrine problems, has been described among young black women in the French West Indies. The etiology remains unknown, but possible links with Devic's disease, acute disseminated encephalomyelitis, and neurotoxicity from quinolines in *Annona muricata* teas have been postulated (Román 1998).

*Annona muricata* may lower blood pressure. Caution is advised in patients taking medications that lower blood pressure may increase the amount of drowsiness caused by some drugs. Examples include benzodiazepines such as lorazepam (Ativan®) or diazepam (Valium®), barbiturates such as phenobarbital, narcotics such as codeine, some antidepressants, and alcohol. Caution is advised while driving or operating machinery. *A. muricata* may have additive effects with antibiotics, antidepressants, antifungals, anticancer agents, antiparasitic agents, antivirals, and vasodilators and may interfere with drugs used to treat Parkinson's disease (Edwards et al. 2015).

7) **Dosage:**

Not recommended so a dose cannot be given (Edwards et al. 2015)

## 8.16 References


- Adewole SO, Caxton-Martins EA. Morphological changes and hypoglycemic effects of *Annona muricata* Linn.(Annonaceae) leaf aqueous extract on pancreatic β-cells of streptozotocin-treated diabetic rats. Afr J Biomed Res. 2006;9(3).
- Beneto JR, Rodriguez AJ, Raman de Sandoval A. A soursop pulp extraction procedure. J Agr Univ Puert Rico. 1971;55:518–9.
- Bories C, Loiseau P, Cortes D, Myint SH, Hocquemiller R, Gayral P, Cave A, Laurens A. Antiparasitic activity of *Annona muricata* and *Annona cherimolia* seeds. Planta Med. 1991;57(5):434–6.
- Bueso CE. Soursoup, tamarind and chironja. In: In: Nagi S, Shaw PE, editors. Tropical and subtropical fruits—compositions, properties and uses. Westport, CT, USA: Avi Publishing Inc.; 1980. p. 375–87.
- Edwards SE, Rocha IDC, Williamson EM, Heinrich M. Phytopharmacy: an evidence-based guide to herbal medical products. 2015. p. 189–90.
- Gajalakshmi S, Vijayalakshmi S, Devi Rajeswari V. Phytochemical and pharmacological properties of Annona muricata: a review. Int J Pharm Pharmaceut Sci. 2012;4(2):3–6.

- Hasrat JA, De Bruyne T, De Backer JP, Vauquelin G, Vlietinck AJ. Isoquinoline derivatives isolated from the fruit of *Annona muricata* as 5-HTergic 5-HT1A receptor agonists in rats: unexploited antidepressive (lead) products. J Pharm Pharmacol. 1997;49:1145–9.
- Konno H, Okuno Y, Makabe H, Nosaka K, Onishi A, Abe Y, Sugimoto A, Akaji K. Total synthesis of cis-solamin A, a mono-tetrahydrofuran acetogenin isolated from *Annona muricata*. Tetrahedron Lett. 2008;49:782–5.
- Kalil AC, Lerner BR, Kuroba CH, Boog MCF, Oliveira MCM, Stefarini MLR, Sobrinho ONN, Lepper RM, Philippi ST, Faria Z. *Manual Básico de Nutriçao*. Estado de Sao Paulo, Sec.do Estado de Saude, Secao Nutricao. 1979;5, E: 11–3, 69–70.
- Langason RBF, Akunyili DN, Akubue PI. A preliminary study of the gastrointestinal effects of some Nigerian medicinal plants. Fitoterapia. 1994;65:235–41.
- Lannuzel A, Höglinger GU, Champy P, Michel PP, Hirsch EC, Ruberg M. Is atypical parkinsonism in the Caribbean caused by the consumption of Annonaceae?. In Parkinson's disease and related disorders. Vienna: Springer; 2006. p. 153–7.
- Leal F. Sugar apple. In: Tropical and subtropical fruits—compositions, properties and uses. In: Nagi S, Shaw PE, Wardowski WF, editors. Lake Alfred, FL, USA: Florida Science Source; 1990. p. 149–58.
- Lizana LA, Reginato G. Cherimoya. In: Nagi S, Shaw PE, Wardowski WF, editors. Fruits of tropical and subtropical origin: composition, properties and uses. Lake Alfred, FL, USA: Florida Science Source; 1990. p. 131–48.
- Morón Rodríguez F, Morón Pinedo D, Nodarse Rodríguez M. Assessment of scientific evidence recommending *Annona muricata* L. (soursop tree) for cancer prevention or treatment. Revista Cubana de Plantas Medicinales. 2010;15(3): 169–81.
- N'gouemo P, Koudogbo B, Tchivounda HP, Akono-Nguema C, Etoua MM. Effects of ethanol extract of *Annona muricata* on pentylenetetrazol-induced convulsive seizures in mice. Phytother Res. 1997;11(3):243–5.
- Orwa C. Agroforestry Database 4.0: a tree reference and selection guide. World Agroforestry Centre; 2010.
- Pinto AC, de Q, Silva EM. Graviola para exportacao: aspectos tecnicos da producao. Brasilia, Brasil: Embrapa/SPI; 1996.
- Román G. Tropical myeloneuropathies revisited. Curr Opin Neurol. 1998;11(5): 539–4.
- You M, Wickramaratne DBM, Silva GL, Chai H, Chagwedera TE, Famsworth NR, Cordell GA, Kinghom AD, Pezzuto JM. (–) Roemerine, an aprophine alkaloid from *Annona senegalensis* that reverses the multidrug resistance phenotype with cultured cells. J Nat Prod. 1995;58:598–604.
- Lannuzel A, Höglinger GU, Verhaeghe S, Gire L, Belson S, Escobar-Khondiker M, Poullain P, Oertel WH, Hirsch EC, Dubois B, Ruberg M. Atypical parkinsonism in Guadeloupe: a common risk factor for two closely related phenotypes? Brain. 2007;130:816–27.

# Chapter 9
# Antidepressant Plant Species from the Portuguese-Speaking African Countries (PALOP)

**Generosa Teixeira, Eurico S. Martins, and Luís Catarino**

**Abstract** The Portuguese-speaking African countries belong to the sub-Saharan region and are known by the acronym PALOP. As in other African countries, the traditional medicine is an important element in their cultural patrimony and the value of those ancestral practices is recognized.

The access to the biological resources from developing countries became easier but also the awareness to its preservation as well as the alert to the need of covering the insufficient scientific data on the safety, efficacy, and quality of traditional medicines.

We have done a research on the available information about medicinal plants commonly used for treatment of depression in the PALOP. The ethnobotanical data were firstly obtained through a review conducted on several works on the medicinal flora of these countries and their neighbors and information contained in the herbarium labels.

The plants with more reliable information as antidepressants were *Bacopa monnieri* (Scrophulariaceae), *Boophone disticha* (Amaryllidaceae), *Centella asiatica* (Apiaceae), *Cissampelos* genus (Menispermaceae), *Griffonia simplicifolia* (Fabaceae), *Mondia whitei* (Apocynaceae), *Palisota hirsuta* (Commelinaceae), *Securidaca longepedunculata* (Polygalaceae), and *Xysmalobium undulatum* (Apocynaceae). For these plants, we combined the data gathered on bioecology and uses with their chemical compounds and their reported pharmacological activity.

G. Teixeira (✉)
Faculdade de Farmácia, Centre for Ecology, Evolution and Environmental Changes, (Ce3C), Universidade de Lisboa, Av. Professor Gama Pinto, Lisbon 1649-003, Portugal
e-mail: gteixeira@ff.ulisboa.pt

E.S. Martins
JBT/IICT, Jardim Botânico Tropical, Instituto de Investigação Científica Tropical, Trav. Conde da Ribeira 9, Lisbon 1300-142, Portugal
e-mail: martins.eurico@gmail.com

L. Catarino
Faculdade de Ciências, Centre for Ecology, Evolution and Environmental Changes, (Ce3C), Universidade de Lisboa, Campo Grande, Lisbon 1749-016, Portugal
e-mail: lcatarino@iict.pt



C. Grosso (ed.), *Herbal Medicine in Depression*,
DOI 10.1007/978-3-319-14021-6_9

These seem to be plants with psychotropic effects, potentially active and to which modern pharmacology can appeal. Further studies are needed to completely elucidate the chemical groups responsible for the reported antidepressant effect, and additional research is necessary therefore to confirm the exact mechanisms involved in this central nervous system pathology.

**Keywords** Medicinal plants • Phytotherapy • Ethnopharmacology • Depression • Ethnobotany • Portuguese-speaking African countries (PALOP)

## 9.1 Introduction

### *9.1.1 The Portuguese-Speaking African Countries*

In Africa, there are five countries in which Portuguese is the official language: Angola, Cape Verde, Guinea-Bissau, Mozambique, and São Tome and Principe. All of them were Portuguese colonies and acquired their independence in 1974 and 1975.

The Portuguese-speaking countries in Africa are also known as PALOP—*Países Africanos de Língua Oficial Portuguesa* (formed in 1992)—and besides the official language, they also share a long common history and a strong cultural identity.

The PALOP occupy different geographic regions in the African continent: Cape Verde and Guinea-Bissau are located in Western Africa, Sao Tome and Principe and Angola in Middle (Central) Africa, and Mozambique in Eastern Africa (Fig. 9.1). Two of them, Cape Verde and Sao Tome and Principe, are small archipelagos, while others are among the largest, in the case of Angola, the 7th largest country in the African continent (UN Statistics Division 2006).

These countries show a huge cultural and ethnical diversity. In Angola, Guinea-Bissau, and Mozambique, a large number of ethnical groups coexist, each one with its beliefs, practices, and ways of nature appropriation. In Cape Verde and Guinea-Bissau, Creole emerged as the vehicular language, while, in the remaining PALOP, Portuguese is the vehicular language.

In terms of natural resources, these five countries show huge differences. Some are among the poorest countries in the world, showing a high reliance on foreign aid, and their economy is dependent on such assistance. To exacerbate this situation, after the independence, some experienced periods of strong political instability, as in Guinea-Bissau, and others were even devastated by for a very long time civil wars, as in Angola and Mozambique.

They show huge differences and some are among the poorest countries in the world, showing a high reliance on foreign aid and their economy is dependent on such assistance. To exacerbate this situation, after the independence, some experienced periods of strong political instability, as in Guinea-Bissau, and others were even devastated by very long time civil wars, as Angola and Mozambique.

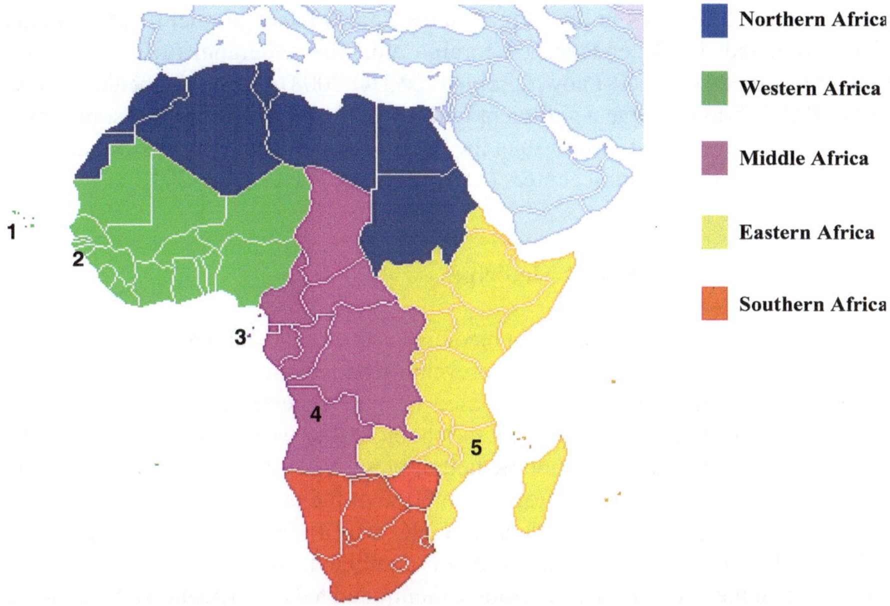


**Fig. 9.1** Geographical location of PALOP in Africa, according to the subdivisions of the UN (UN Statistics Division, 2006). 1, Cape Verde; 2, Guinea-Bissau; 3, Sao Tome and Principe; 4, Angola; 5, Mozambique

#### 9.1.1.1 Health Services

In most African cultures, the disease is an adversity imposed upon people by strange forces, and its cure is closely related with rituals and mystic practices, with the intervention of a native healer and so the therapy might not be attributed to the use of a medicine itself.

In spite of the progress made, as regards to basic care in most of African countries, the health status of the population continues to be a major challenge, since there is a high incidence of disease and the health-care system remains weak and fragmented (WHO 2009).

As with other sub-Saharan African countries, the PALOP show some sociodemographic asymmetries, high levels of poverty, poor schooling, and geographic population isolation, which do not contribute to the improvement of people's quality of life. Those asymmetries are in part due to migration flows and cause even deep social inequalities and serious infrastructural problems. In addition, they generate troubles on the inequality of access to the health services, also leading to the worsening of infectious disease transmission and to the absence of prevention and treatment of non-transmitted diseases (WHO 2008). Most of these countries have not yet met material conditions or know-how to improve the living status of their people, with basic sanitation, health, and education systems being yet under low implementation. The inadequacy of financial and human resources for those areas and the difficulties in the formation and maintenance of qualified technicians also compelled the dependence on foreign aid. The geographical distribution of human resources is a well-known problem (WHO 2008).In fact, the health systems of the PALOP do not have a sufficient number of health professionals, some showing a professional density lower than the African average not only in rural areas but also in the major population centers (Martins 2010; WHO 2008, 2010).

#### 9.1.1.2 African Traditional Medicine

According to the WHO, "traditional medicine is the total combination of knowledge and practices, used for the diagnosis, prevention and elimination of physical, mental and social diseases, which may be based on past experience and on observation handed down from generation to generation, on oral or written form" (WHO 2001).

Traditional medicine, or ethnomedicine, plays a central role in the health of a large part of the world's population, since 80 % of the population depends on traditional medicine for primary health care (WHO 2011a). In Africa, the culture, political instability, and socioeconomic problems combined with isolation, basic needs of survival, and the asymmetries previously mentioned led or perpetuated the underdevelopment of many regions. This compelled the population to deal with traditional medicine and local healers as the first and only source of health care (Agostinho and Silva 2013; Carvalho 2013; Diniz and Martins 2005). The traditional healers provide a popular and accessible service, especially in the rural areas across the African continent where the sanitary situation is characterized by a lack of qualified health workers, medicines, and equipment. So, the traditional medicine is an important element in the cultural patrimony and remains the main resource for the majority of people for treating health problems, particularly in rural areas (Romeiras et al. 2012; Rukangira 2001).

An African indigenous system of health is composed of three essential parts: the active medicinal substance, which may be plants, soil, or animals or animal parts; the spiritual forces, the gods and God; and the human spirit, which includes the healer and the patient. The functions of the traditional medicine practitioner are not limited to the diagnosis of diseases and the prescription of drugs. It is the medicine practitioner who provides the needed answers to the adversities imposed on the community by outside forces that are beyond the comprehension of ordinary people. These forces include curses, charms, evil spirits, aggrieved ancestors, witches, and the gods (Iwu 2014). It is the belief among most Africans that a herbal prescription from a medicine man could cure an indicated illness not necessarily because the plant has a demonstrable pharmacological activity per se, but because there is life-giving potency in every creature that the medicinal plants possess and also the acquired properties imparted to the plant through invocations, incantations, and rituals. The African medicinal agents consist of two main groups: (1) those using rituals, sacrifices, and other religious acts as part of the process of treatment of diseases and (2) those using plant extracts, herbs, seeds, roots, leaves, juices, liquids, powders, bones, minerals, and other substances that are supposed to have organic effects directly on the patient. Some of the procedures and rituals, and even the very plants used, may lack any overt value, but they are psychologically vital and play a great role in healing the sick and helping the afflicted.

After the Alma-Ata Conference, in 1978, the WHO suggested to the African governments the promotion and integration of traditional medicines in the primary health care, providing a source of affordable health care to local communities (WHO 1978). From these resolutions, the African traditional medicine began to achieve prominence, although it varies greatly from region to region (WHO 2001). In 2001 in Lusaka, the African Unity Organization declared the period 2001–2010 as the "Decade of African Traditional Medicine." During the presentation of the results of this first decade, it was renewed in the second "Decade of African Traditional Medicine" from 2011 to 2020 (WHO 2011b). In 2003, the African Union (ex-African Unity Organization), meeting in Mozambique, created the "African Traditional Medicine Day" to be celebrated on 31st of August, each year being devoted to a specific theme. Subsequently, some countries have instituted the "Traditional Medicine Week." In both cases, during the celebrations, there are several activities such as debates, symposiums, and cultural presentations, in order to enhance the traditional medicine (WHO 2011b).

In the course of these actions, traditional medicine has been recognized and also associated to a less dependence on foreign aid by reducing the reliance on expensive imported medicines. There were other achievements like an update on the knowledge about the African flora, with the emergence of research institutions in some African countries, though those activities were initially more related with the harvesting and taxonomy of plants and less with the information and study about their traditional uses. The guidelines for the protection of intellectual property rights and of the traditional medicine knowledge were established and the first steps toward the creation of an *African Herbal Pharmacopoeia* were given. With all this, the access to the biological resources from developing countries became easier but also the awareness to its preservation as well as the alert to the need of covering the insufficient scientific data on the safety, efficacy, and quality of traditional medicines (WHO 2011b). Indeed the problem begins in the fact that most of traditional medicinal plants are traded in local open markets, with few conditions, like lack of storage facilities and trading infrastructures which affects the final product quality (Stafford et al. 2005).

### 9.1.2 Depression: A Mental Disorder

The mental disorders are usually neglected, but they occur in all societies and affect hundreds of millions of people, women and men, at all stages of life, among rich and poor, both in urban and rural communities. If not treated, beyond the personal and family suffering, they lead to significant economic and social difficulties which affect society as a whole.

In most African societies, mental illness is seen as the clearest manifestation of the disturbed equilibrium of life. These diseases can be traced to a broken taboo, a contravened societal norm, or witchcraft. It is indeed futile to prescribe herbs and tranquilizers without first making a proper diagnosis of the causative factors responsible for the illness (Iwu 2014).

Despite the introduction of a number of antidepressant drugs with different pharmacological activities, depression continues to be a major medical problem. Even in industrialized countries, only a minority of people suffering from depression seeks or receives treatment. Part of the explanation lies in the symptoms themselves. Feelings of worthlessness, excessive guilt, and lack of motivation deter individuals from seeking help. In addition, such individuals are unlikely to appreciate the potential benefits of treatment. The integration of mental health services in primary care is the most viable way of ensuring that people have access to mental health care when they need it. Personal and social costs of the disease are very high. After anxiety, depression is the most common psychiatric disorder. It is estimated that about 10 % of people who consulted a doctor thinking that they have a physical problem actually suffer from depression. Those born in the last decades of the twentieth century seem to have a higher incidence of depression than previous generations (IOM 2010).

#### 9.1.2.1 Antidepressant Drugs and Medicinal Plants as Alternative Therapy

There are several types of currently available antidepressant drugs, such as selective serotonin reuptake inhibitors, tricyclic antidepressants, monoamine oxidase inhibitors, and psychostimulants. They all act on the nervous system and are often associated with several side effects. They have to be taken regularly for at least 4–6 weeks before they begin to work and are often associated with several side effects. The chances that any given antidepressant will work for a particular person are approximately 60–70 % (Nestler et al. 2002; Wong and Licinio 2001). Therefore, the identification of new therapeutic alternatives for the treatment of depression still becomes necessary.

The more traditional and less industrialized societies are based on the knowledge of the living world surrounding them and depend on the sustainable exploitation of the natural resources, for their food needs, housing, and medicine, among others. The importance of these ancestral practices is recognized (WHO 2011b) and its search also has substantially increased in Western countries, as it grows the demand for alternative methodologies against diseases, partly resulting from technological development and also of the complex way of life on those societies (WHO 2011b; Calixto 2005; Rates 2001).

Plants synthesize a variety of chemical compounds that are at the base of its medicinal and pharmaceutical importance. For a long time, the pharmaceutical industry uses raw materials of vegetable origin which were originally known and developed from the study of its traditional uses (Oliver-Bever 1986). These compounds can have a particular biological and pharmacological activity. In other cases, the isolated compounds from plants are used as chemical experimental models for the hemi-synthesis or complete synthesis of more effective and less toxic drugs, as well as many other applications in the pharmaceutical area. About 25 % of conventional medicines are originated directly or indirectly from plants (Cos et al. 2006). The available African natural resources are synonymous of abundant natural products, many of which possess known medicinal properties and as such its potential is enormous (Kuete 2013).

The therapeutic action of a plant depends on its chemical composition. It is important to ascertain if the plants from different geographical zones have the same chemical composition and so the same biological activity, as different ecological conditions might affect the biosynthetic pathways. In traditional medicine, local uses might differ for the same plant and this requires an accurate analysis of the associated biological and chemical data (Iwu 2014; Kuete 2013; Oliver-Bever 1986).

There are medicinal plants with psychotropic effects, potentially active and to which modern pharmacology can appeal, although the clinical use of medicines with antidepressant action obtained from plants is still limited (Iwu 2014; Kumar 2006; Zhang 2004).

## 9.2 Methods

Literature search was performed using electronic databases including ScienceDirect, Web of Science, Scopus, SciELO, PubMed, and Google Scholar. Keywords such as "alternative medicine, ethnomedicinal uses, pharmacological properties, mental disorders, depression, African plants, and antidepressant properties" were used. In addition, relevant information was also searched on the Internet, and publications containing original data and an adequate detailed description of methodology were considered, like academic dissertations, including Ph.D. and master's theses, and ethnobotanical works. We obtained free original articles where possible and also hand-searched for cross-referenced articles and books. Searches were restricted to English, French, and Portuguese language publications but no restrictions on year of publication were applied.

According to the literature survey, we have tried to combine the ethnomedicinal data collected about the plants with antidepressant activity with their chemical compounds and their reported pharmacological activities.

The search was done on the available information about medicinal plants commonly used for treatment of depression in the PALOP. As the explicit information on the traditional use for this condition is scarce, the ethnobotanical data were firstly obtained through a review conducted on several published works on the medicinal flora of these countries and secondly on the plants used in the neighboring sub-Saharan African countries and also belonging to the flora of the PALOP.

## 9.3 Conclusions

There is a consensus about plants as an important source of phytochemicals, potentially beneficial health agents. Those benefits are far from being fully exploited. Plants are used in different traditional medicine systems, but there is a vast amount of research still to be conducted to scientifically validate their traditional uses, that is, the search for the phytochemical constituents which are responsible for a certain biological activity. Many ethnomedicinal uses of plants from various sources are documented and compared, concentrating merely on the gross effects induced by herbal preparations. Relevant scientific studies on the plant's pharmacological properties and scientific literature for the reported constituents from the plant, which can scientifically validate the plant's traditional uses and pharmacological properties, are still out of focus. Ethnomedicinal studies are published every year without any discussion of the chemical constituents of the medicinal plants reported and scientific validations of their traditional medicinal uses. It is believed that a new approach on natural products can prove more valuable in not only identifying which diseases it may be beneficial but also contain associated chemical and biological data, which could lead to the discovery of new drugs. In this particular case, further studies are needed to completely elucidate the chemical groups responsible for the antidepressant effect of plant extracts, and further experiments are necessary therefore to confirm the exact mechanisms involved in this central nervous system pathology.

## 9.4 References


- Agostinho AB, Silva HL. Desafios da medicina tradicional africana no séc. XXI. Atas do Congresso Internacional Saber Tropical em Moçambique: História, Memória e Ciência. IICT—JBT. Lisboa; 2012. p. 24–6 outubro de 2012. ISBN 978-989-742-006-1. IICT, Lisboa, 2013.
- Calixto JB. Twenty-five years of research on medicinal plants in Latin America. A personal view. J Ethnopharmacol. 2005;100:131–4.
- Carvalho JCB. Estudo químico e biológico das espécies vegetais Cabo Verdianas, *Echium hypertropicum* Webb e *Echium stenosiphon* Webb subsp. *stenosiphon.* Niterói, Brasil: Dissertação de Mestrado Universidade Federal Fluminense; 2013.
- Cos P, Vlietinck AJ, Berghe DV, Maes L. Anti-infective potential of natural products: How to develop a stronger in vitro "proof-of-concept." J Ethnopharmacol. 2006;106:290–302.
- Diniz MA, Martins ES. Plantas medicinais na Guiné-Bissau. AfricaToday Fev. 2005;110–11.
- Iwu MM. Handbook of African medicinal plants. 2nd ed. CRC Press; 2014. ISBN-13: 978-1-4665-7198-3 (eBook—PDF).

- Kuete V. Medicinal plant research in Africa. Pharmacology and chemistry. Elsevier Inc.; 2013. ISBN: 978-0-12-405927-6. doi: http://dx.doi.org/10.1016/B978-0-12-405927-6.00001-1.
- Kumar V. Potential medicinal plants for CNS disorders: an overview. Phytother Res. 2006;20:1023–35.
- Martins MR. A Cooperação em saúde entre o Brasil e os PALOP (2003-2010). Dissertação de Doutoramento Faculdade de Ciências Económicas. Porto Alegre, Brasil: Universidade Federal do Rio Grande do Sul; 2010.
- Nestler EJ, Barrot M, DiLeonem RJ, Eisch AJ, Gold SJ, Monteggia LM. Neurobiology of depression. Neuron. 2002;34:13–25.
- Oliver-Bever B. Medicinal plants in tropical West Africa. Cambridge: Cambridge University Press; 1986. ISBN 978-0-521-10544-6.
- Rates SMK. Plants as sources of drugs. Toxicon. 2001;39:603–13.
- Romeiras MM, Duarte MC, Indjai B, Catarino L. Medicinal plants used to treat neurological disorders in West Africa: a case study with Guinea-Bissau flora. Am J Plant Sci. 2012;3:1028–36.
- Rukangira E. Medicinal Plants and Traditional Medicine in Africa: constraints and challenges. Sustain Dev Int. 2001;4:179–84.
- Stafford GI, Jager AK, van Staden J. Effect of storage on the chemical composition and biological activity of several popular South African medicinal plants. J Ethnopharmacol. 2005;97:107–15.
- UN Statistics Division. Composition of macro geographical regions, geographical sub-regions and selected economic and other groupings. 2006. In http://millenniumindicators.un.org/unsd/methods/m49/m49regin.htm. Accessed on Dec 2014.
- Wong ML, Licinio J. Research and treatment approaches to depression. Nat Rev Neurosci. 2001;2:343–51.
- WHO. *Primary health care*. Report of the international conference on primary health care. Alma-Ata, USSR. 6–12 Sept 1978. Geneva; 1978. (http://www.searo.who.int/entity/primary_health_care/documents/hfa_s_1.pdf).
- WHO. Promoting the role of traditional medicine in health systems: a strategy for the African region. Ouagadougou, Burkina Faso 28 August—02 September 2000. WHO Regional Office for Africa; 2001.
- WHO. The world health report 2008: primary health care now more than ever. Geneva; 2008. http://www.who.int/whr/2008/whr08_pr.pdf.
- WHO. First African Medicines Regulatory Authorities Conference Final Report. Addis Ababa, Ethiopia 31 October—03 November 2005. WHO Regional Office for Africa; 2009. ISBN: 978 929 023 1417.
- WHO. Análise dos recursos humanos da saúde (RHS) nos países africanos de língua oficial portuguesa (PALOP). In: Dussaultandi G, editor. Fronteira. Geneva: WHO; 2010. ISBN 978 92 4 859907 1.
- WHO. Health situation analysis in the African Region. Atlas of health statistics, 2011. Brazzaville: WHO Regional Office for Africa; 2011a.

- WHO. Década da medicina tradicional na Região Africana: Relatório dos Progressos. AFR-RC61-PR-2. Yamoussoukro, Côte d'Ivoire, 29 de Agosto—2 de Setembro. WHO Regional Office for Africa; 2011b.
- Zhang Z. Therapeutic effects of herbal extracts and constituents in animal models of psychiatric disorders. Life Sci. 2004;75:1659–99.


## 9.5 *Bacopa monnieri* (L.) Pennell

### 9.5.1 *Scrophulariaceae*

1) ***Synonyms*:**
   *Bacopa monnieria* (L.) Wettst.; *Gratiola monnieria* L.; *Lysimachia monnieri* L.; *Monniera africana* Pers.
2) ***Common names*:**
   Water hyssop (Eng.)
   Hysope d'eau (Fr.)
3) ***Photograph of the species (Fig. 9.2)***

**Fig. 9.2** *Bacopa monnieri* (L.) Pennell. Detail of leaves and flower (original photo taken by L. Allain)

4) ***Description of the species*:**

A) **Habitat, world distribution, and PALOP distribution:**

The genus *Bacopa* includes over 100 species of aquatic herbs distributed throughout the warmer regions of the tropics and subtropics (Philcox 1990). *B. monnieri* is native to Asia, mainly tropical Asia, and Africa and is a small sprawling herb common in fresh and brackish waters, found growing in wet pastures and margins of ponds and in rice fields as well. In some situations, it might be considered as an invasive species (http://www.issg.org/database/species/ecology.asp?si=1784; http://www.iucnredlist.org/details/164168/0).

In PALOP, it can be found in Mozambique (http://www.iucnredlist.org/details/164168/0).

B) **Morphology of the plant:**

Annual herb, creeping, glandular-punctate, glabrous, often rooting at the nodes, sometimes in flooded fields or swamps; stem usually simple, rarely branched. Leaves opposite, sessile, 10–15 × 5–7 mm, obovate to spatulate, obtuse at the apex, cuneate at the base, entire or rarely crenulate toward the apex, glandular-punctate. Flowers axillary, solitary, pedicellate, bracteate, hermaphrodite, irregular; pedicels up to 25 mm long at fruiting, always exceeding the leaves; bracts 2–3 mm long; calyx 5-partite, the lateral lobes narrow, oblong, the posterior one 4–6 × 3–4 mm, broadly ovate, the anterior ca. 5 × 2.5 mm, ovate; corolla 6–7.5 mm long, bilabiate, upper lip bilobed, lower trilobed; stamens 4, didynamous, included. Fruit, a capsule 2–2.5 mm long, ovoid; seeds small, numerous (Philcox 1990).

C) **Conservation status:**

There are no known conservation measures (http://www.iucnredlist.org/details/164168/0).

D) **Medicinal parts:**

Aerial parts.

E) **Chemical composition:**

*B. monnieri* contain **alkaloids** (brahmine, nicotinine, and herpestine) and **triterpenoid saponins** known as bacosides which comprise a group of 12 known analogs (Sivaramakrishna et al. 2006). Other saponins called bacopasides (I–XII), dammarane-type triterpenoid saponins also with biological interest, have been identified in the extract of the whole plant of *B. monnieri* (Garay et al. 2009). This extract also contained **betulinic acid, D-mannitol, Stigmastanol, $\beta$-sitosterol**, and **stigmasterol** (Al-Snafi 2013).

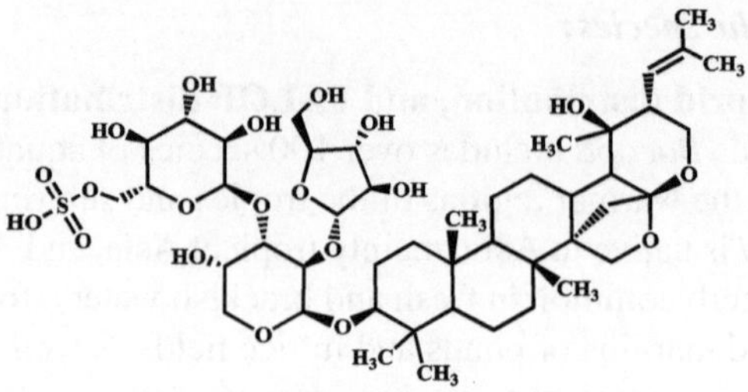

**Bacopaside I**

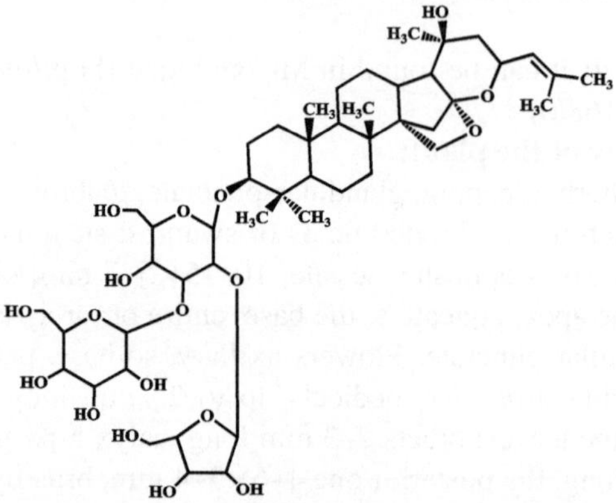

**Bacoside A3**

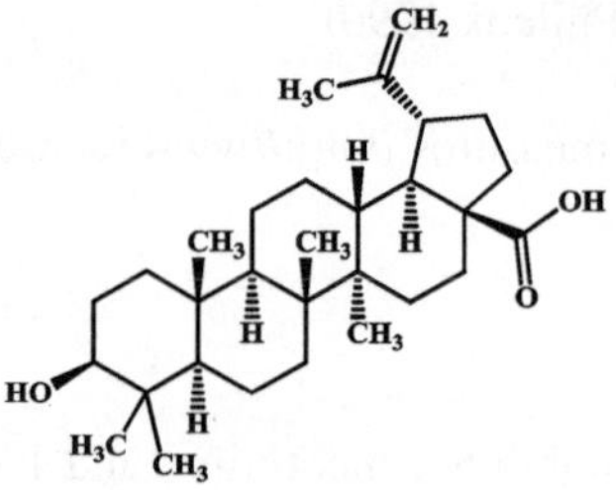

**Betulinic acid**

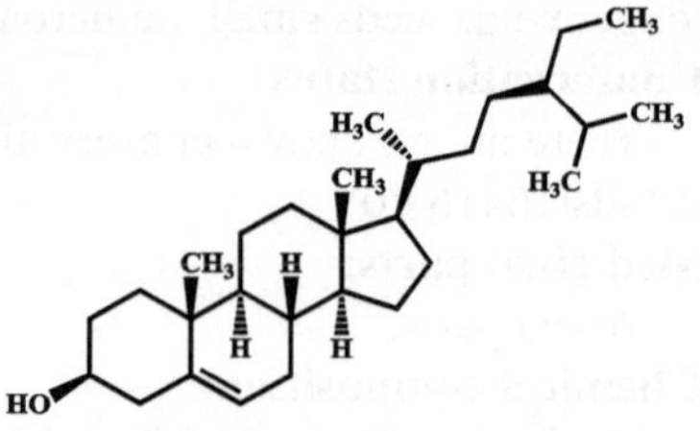

**beta-Sitosterol**

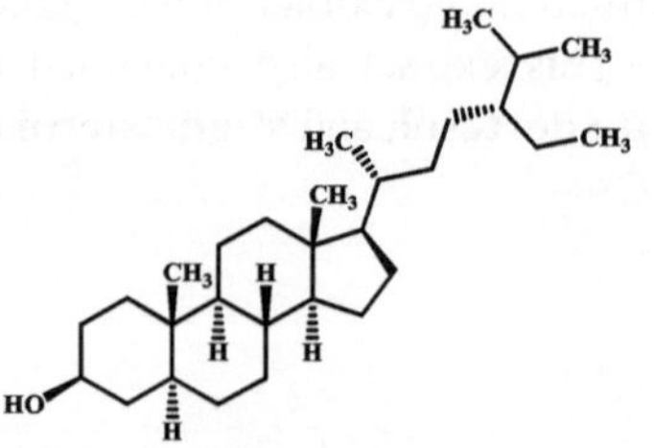

**Stigmastanol**

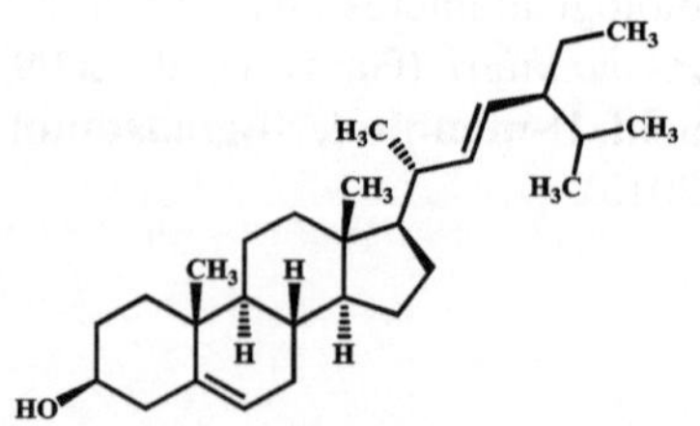

**Stigmasterol**

5) ***Major ethnomedicinal uses.***

There are references to the traditional use of *B. monnieri* for centuries (http://www.iucnredlist.org/details/164168/0; Al-Snafi 2013), to treat nervous disorders; as a brain tonic to enhance memory development, learning, and concentration; and to provide relief to patients with anxiety; it is also used for digestive complaints, for skin disorders, and as an antiepileptic, antipyretic, and analgesic. Other references mentioned its use as aphrodisiac and against snakebites (http://www.ars-grin.gov/cgi-bin/npgs/html/taxon.pl?102292).

6) ***Pharmacology and bioactivities*:**

The most important effects of the plant, plant extracts, and isolated triterpenoid saponins are on cognition and memory functions, their anxiolytic effects, and their role in management of convulsive disorders (Russo and Borrelli 2005; Kumar 2006). The extracts and isolated bacosides and bacopasides have been investigated extensively for their neuropharmacological effects. Singh et al. (1996) suggest an involvement of the GABAergic system in the mediation of CNS effects. The mechanism of action reported in various preclinical studies indicating a cognitive enhancing effect is still uncertain. Its multiple active constituents make its pharmacology complex (Russo and Borrelli 2005; Gomes et al. 2009). It is noteworthy that dammarane-type triterpenoid saponins are major constituents of a number of reputed herbal drugs, including ginseng (Russo and Borrelli 2005).

Later studies with some of the bacosides and bacopasides isolated from the whole plant of *B. monnieri* showed antidepressant activity when tested on common in vivo assays, like forced swimming test (FST) and tail suspension test (TST) in mice (Zhou et al. 2007). The methanolic extract from the whole plant of *B. monnieri* and different fractions were also evaluated for their antidepressant activity in the same in vivo rodent models of depression. These results showed that the methanolic extract, EtOAc fraction, and *n*-BuOH fraction significantly reduced the immobility times in mice after being administrated orally for five consecutive days and was comparable to that of imipramine, a tricyclic antidepressant (Sairam et al. 2002; Shen et al. 2009). A similar outcome was obtained with a hydroethanolic extraction and an *n*-butanolic extract of *B. monnieri* aerial parts (Abbas et al. 2011). These authors mentioned that the antidepressant effect was comparable to fluoxetine, a selective serotonin reuptake inhibitor. In another study, *n*-butanolic extract of *B. monnieri* potentiated yohimbine in mice. Yohimbine is an inhibitor of the presynaptic α-adrenergic receptors, and its potentiation induced by *B. monnieri* indicated the involvement of noradrenaline in the mediation of antidepressant actions of *B. monnieri* (Abbas et al. 2011).

7) ***Precautions and side effects*:**

According to Singh and Dharwan (*in* Al-Snafi 2013), contraindications and adverse effects of *B. monnieri* are not known.

8) ***Dosage*:**

According to the monograph of *B. monnieri* (Alternative Medicine Review 2004, *in* Al-Snafi, 2013), for *B. monnieri* extracts standardized to 20 % bacosides A and B, the dosage is 200–400 mg daily in divided doses for adults and 100–200 mg daily in divided doses for children.

## 9.6 References


- Singh HK, Dharwan BN. Neuropsychopharmacological effects of the Ayurvedic nootropic *Bacopa monnieri* Linn (Brahmi). Indian J Pharmacol. 1997;29: S359–65.
- Anonymous. Monograph, *Bacopa monnieri*. Alternative Med Rev. 2004;9:79–85.
- Abbas M, Subhan F, Rauf K, Khan M, Mohani S-H. The involvement of biogenic amines in the antidepressant effect of *Bacopa monnieri*. Pharmacologyonline. 2011;1:112–23.
- Al-Snafi AE. The pharmacology of *Bacopa monnieri*. A review. Int J Pharma Sci Res. 2013;4:154–9.
- Garay S, Mahato SB, Ohtani K, Yamasaki K. Dammarane triterpenoid saponins from Bacopa monnieri. Can J Chem. 2009;87:1230–4.
- Gomes NGM, Campos MG, Órfão JMC, Ribeiro CAF. Plants with neurobiological activity as potential targets for drug discovery. Progr Neuro- Psychopharmacol Biol Psychiatr. 2009;33:1372–89.
- Kumar V. Potential Medicinal Plants for CNS Disorders: an overview. Phytother Res. 2006;20:1023–35.
- Philcox D. Scrophulariaceae. In: Launert E, Pope GV, editors. Flora Zambeziaca 8(2). London: Flora Zambeziaca Managing Committee; 1990.
- Russo A, Borrelli F. Bacopa monnieri a reputed nootropic plant: an overview. Phytomed Int J Phytother Phytopharmacol. 2005;12:305–17. In http://www.thefreelibrary.com/Bacopa+monniera%2c+a+reputed+nootropic+plant%3a+an+overview.-a0133802203. Accessed on Mar 2015.
- Sairam K, Dorababu M, Goel RK, Bhattacharya SK. Antidepressant activity of standardized extract of Bacopa monnieri in experimental models of depression in rats. Phytomedicine. 2002;9:207–11.
- Shen Y-H, Zhou Y, Zhang C, Liu R-H, Su J, Liu X-H, Zhang W-D. Antidepressant effects of methanol extract and fractions of *Bacopa monnieri*. Pharmaceut Biol. 2009;47:340–3.
- Singh HK, Shanker G, Patnaik GK. Neuropharmacological and anti-stress effects of bacosides: a memory enhancer. Indian J Pharmacol. 1996;28:47.
- Sivaramakrishna C, Rao CV, Trimurtulu G, Vanisree M, Subbaraju GV. Triterpenoid glycosides from *Bacopa monnieri*. Phytochemistry. 2005;66: 2719–28.

- Zhou Y, Shen, Y-H, Zhang C, Su J, Liu R-H, Zhang W-D. Triterpene saponins from *Bacopa monnieri* and their antidepressant effects in two mice models. J Nat Prod. 2007;70:652–5.
- http://www.issg.org/database/species/ecology.asp?si=1784. Accessed Feb 2015.
- http://www.iucnredlist.org/details/164168/0. Accessed Feb 2015.


## 9.7 *Boophone disticha* (L.f.) Herb

### *9.7.1 (Amaryllidaceae)*

1) **Synonyms:**
   *Amaryllis disticha* L.f.
2) **Common names:**
   Moz: Tchindzondzonzo, (Tete).
   Windball, poison bulb, sore-eye flower (Eng.).
3) **Photograph of the species** (Fig. 9.3)

**Fig. 9.3** *Boophone disticha* (L.f.) Herb., inflorescence, in Mozambique (Torre, 13123, IICT Herbarium collection)

4) **Description of the species:**

A) **Habitat, world distribution, and PALOP distribution:**

Occurs in tropical and subtropical regions, from East Africa to South Africa. In PALOP, it can be found in Angola and Mozambique.

B) **Morphology of the plant:**

Bulbous plant with distichous leaves. Bulb up to 25 cm in diameter partially exposed above ground. Leaves developing after the flowers, sessile, up to 50 cm long and 1.5–4.2 cm wide, narrowly oblanceolate, prominently nerved, rigid, glaucous, the younger ones with a toothed cartilaginous margin. Scape 5–25 cm long, solid, with a large globose inflorescence subtended by two bracts 4–9 × 1–4 cm. Flowers up to 200, pink to purple, bisexual, regular, sweet-scented; pedicels 3–8.2 cm long at anthesis, up to 32 cm in fruit, angular; perianth with a 3–12-mm-long funnel-shaped tube and 6 identical segments 20–30 × 2–4 mm in two series of 3; stamens 6 with filaments 23–48 mm long and anthers yellow about 3 mm long; ovary inferior, 3-locular; style slender, exserted, up to 45 mm long. Fruit a 3-angled elongate capsule up to 4 cm long, with irregular dehiscence and a single pale fleshy globose seed (Zimudzi et al. 2008).

C) **Conservation status:**

*B. disticha* has been classified as "threatened" or "near threatened" because of unsustainable harvesting and as a consequence of large parts of its habitat have been destroyed for cultures (Williams et al. 2013). Sustainable harvesting is impossible since the plant is destroyed when bulbs are removed. As it is a slow-growing plant, its intensive culture is not attractive and discourages cultivators. In South Africa, the species is protected by environmental legislation and may not be harvested without permission (https://www.botanicalsociety.org.za/ProjectsAndActivities/Downloadable%20articles/FW07%20Boophone%20Gifbol.pdf; Williams et al. 2008).

D) **Medicinal parts:**

Bulbs and leaves.

E) **Chemical composition:**

The foremost compounds identified in bulbs and leaves are **eugenol**, an aromatic, volatile oil, smelling of cloves and having analgesic properties, and the **toxic alkaloids** buphandrin (=buphanidrine), buphanamine, buphenine, undulatine, buphanisine, and nerbowdine.

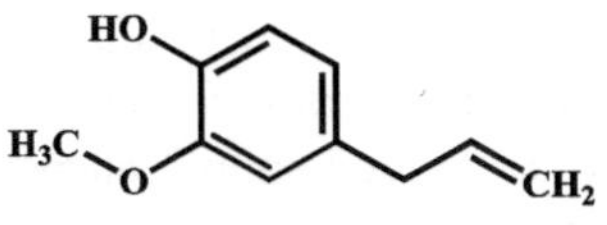

Eugenol

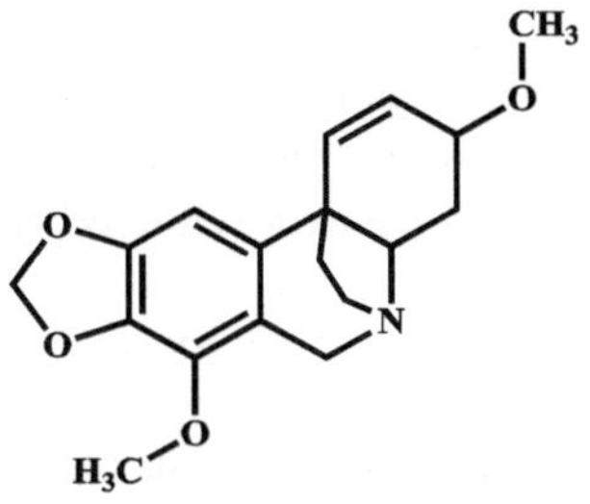

Buphandrin

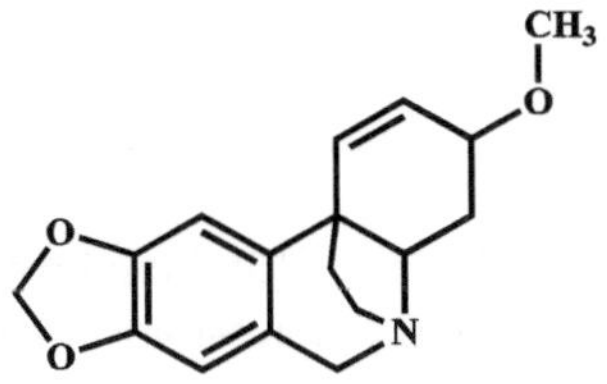

Buphanisine

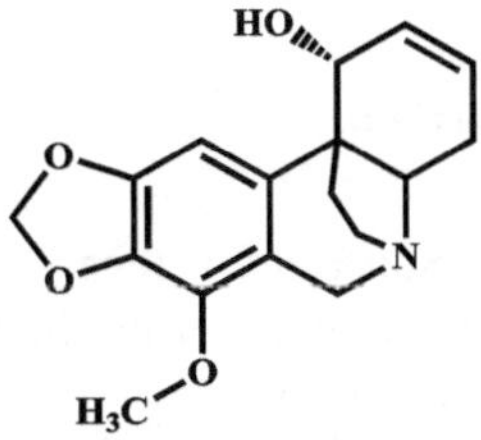

Buphanamine

5) **Major ethnomedicinal uses:**
   *B. disticha* bulb is very toxic and there are many documented deaths from its use, although it is easily found in local markets. Traditional healers use the bulb to treat pain and wounds, as the outer covering of the bulb is applied to boils and abscesses. Fresh leaves are used to stop bleeding of wounds and bulb infusions for anxiety, depression, and age-related dementia (Risa et al. 2004). There have been unsubstantiated reports of *B. disticha* being consumed recreationally for its hallucinogenic properties when taken in relatively small doses (http://www.plantzafrica.com/plantab/boophdist.htm).
6) **Pharmacological studies:**
   *B. disticha* showed high activity both in aqueous and ethanolic extracts of leaves and bulbs (Nielsen et al. 2004). Amaryllidaceae alkaloids are reported to exhibit several biological activities, including on the central nervous system (Lewis 1990; Elgorashi et al. 2006). Investigations on plants used to treat age-related dementia and debilitative mental disorders lead to the isolation of several compounds through bioassay-guided fractionations. In *B. disticha* bulb scale extracts, the identification of the alkaloids buphanamine, buphanadrine, buphanisine, and distichamine was achieved (Sandager et al. 2005; Neergaard et al. 2009; Nair and van Staden 2014). The first two alkaloids showed affinity for SERT (Sandager et al. 2005) and have a chemical structure, the benzo-1,3-dioxole moiety, similar to a number of the available clinically used selective serotonin reuptake inhibitor (SSRI) drugs, which could explain their activity (Stafford et al. 2008).

Pedersen et al. (2008) employed a rat model for depression to show that ethanolic extracts from *B. disticha* bulbs inhibited the serotonin transporter (SERT), noradrenaline transporter (NAT), as well as the dopamine transporter (DAT), all of which are believed to be involved in the pathophysiology of depression. The same bulb extracts were also investigated for in vivo antidepressant-like effects in animal models for depression that employed the forced swim test (FST) in both mice and rats and the tail suspension test (TST) in mice. The extracts exhibited antidepressant-like effects in those in vivo animal models (Pedersen et al. 2008).

## 9.8 References


- Elgorashi E, Stafford GI, Jäger AK, van Staden J. Inhibition of [3H] citalopram binding to the rat brain serotonin transporter by Amaryllidaceae alkaloids. Planta Med. 2006;72:470–3.
- Lewis JR. Amaryllidaceae alkaloids. Nat Prod Rep. 1990;9:183–91. In Elgorashi et al. 2006.
- Nair JJ, van Staden J. Traditional usage, phytochemistry and pharmacology of the South African medicinal plant *Boophone disticha* (L.f.) Herb. (Amaryllidaceae). J Ethnopharmacol. 2014;151:12–26.
- Neergaard J, Rasmussen HB, Stafford GI, Van Staden J, Jäger GI. Serotonin transporter affinity of (−)-loliolide, a monoterpene lactone from *Mondia whitei*. S Afr J Bot. 2010;76:593–6.
- Nielsen N, Sandager M, Stafford G, van Staden J, Jäger A. Screening of indigenous plants from South Africa for affinity to the serotonin reuptake transport protein. J Ethnopharmacol. 2004;94:159–63.
- Pedersen ME, Szewczyk B, Stachowicz K, Wieronska J, Andersen J, Stafford GI, van Staden J, Pilc A, Jäger KA. Effects of South African traditional medicine in animal models for depression. J Ethnopharmacol. 2008;119:542–8.
- Risa J, Risa A, Adsersen A, Gauguin B, Stafford GI, van Staden J, Jäger AK. Screening of plants used in southern Africa for epilepsy and convulsions in the GABAA-benzodiazepine receptor assay. J Ethnopharmacol. 2004;93:177–82.
- Sandager M, Nielsen N, Stafford GI, van Staden J, Jäger AK. Alkaloids from *Boophone disticha* with affinity to the serotonin transporter in rat brain. J Ethnopharmacol. 2005;98:367–70.
- Stafford G, Pedersen M, van Staden J, Jäger A. Review on plants with CNS-effects used in traditional South African medicine against mental diseases. J Ethnopharmacol. 2008;119:513–37.
- Williams VL, Raimondo D, Brueton VJ, Crouch NR, Cunningham AB, Scott-Shaw CR, Lötter M, Ngwenya AM. Boophone disticha (L.f.) Herb. National Assessment: Red List of South African Plants version 2014.1. 2008; In http://redlist.sanbi.org/species.php?species=2076-2. Accessed Mar 2015.

- Zimudzi C, Archer RH, Kwembeya EG, Nordal I. Amaryllidaceae, boophone. In: Timberlake JR, Martins ES, editors. Flora Zambesiaca 13 (part 1). 2008; 114–6.
- https://www.botanicalsociety.org.za/ProjectsAndActivities/Downloadable%20articles/FW07%20Boophone%20Gifbol.pdf. Accessed Mar 2015.
- http://www.plantzafrica.com/plantab/boophdist.htm. Accessed Jan 2015.


## 9.9 *Centella asiatica* (L.) Urb

### *9.9.1 (Apiaceae)*

1) **Synonyms:**
   *Hydrocotyle asiatica* L.; *H. thunbergiana* Spreng.; *H. pallida* DC.
2) **Common names:**
   STP: Olha d'atô.
   Angola: Mafoibale (muila).
   Asiatic or Indian pennywort, wild violet, and tiger herb (Eng.).
3) **Photograph of the species** (Fig. 9.4)
4) **Description of the species:**

   A) **Habitat, world distribution, and PALOP distribution:**
   Species occurring throughout tropical and subtropical regions of the world, native to Southeast Asian countries. In Africa, it is recorded from Central and Southern Africa. Among countries belonging to PALOP, this species can be found in Guinea-Bissau, Sao Tome and Principe, Angola, and Mozambique.

   B) **Morphology of the species:**
   Creeping perennial herb rooting at the nodes. Leaves alternate, simple, solitary or in groups of up to 5; petiole very variable, 2–20 cm long, glabrous or sometimes densely hairy when young; lamina 1–7 cm wide, reniform to almost circular with a deep basal sinus, margin crenate, pubescent or glabrous, with venation radiate and reticulate. Flowers 2–8 in subcapitate umbels, bisexual, subtending by two ovate bracts; peduncles 0.5–5 cm long, glabrous to hirsute; pedicels usually obsolete; calyx null; petals 5, dark crimson to greenish-white, sometimes tinged with purple, orbicular with a slender inflexed point; stamens 5, free; ovary inferior, 2-celled; styles 2, short, divergent. Fruit dry, laterally flattened, 3–3.5 mm long and 3.5–4 mm wide, orbicular to ellipsoid, deeply constricted at the commissure, dividing at maturity into two 1-seeded mericarps supported by an entire carpophore; mericarps with 5 primary ribs prominent when ripe, under which is present a small oil canal (Cannon 1978).

**Fig. 9.4** *Centella asiatica* (L.) Urb., from Mozambique (LISC Herbarium specimen, Torre 2602)

C) **Conservation status:**

The species is common throughout its regions. It has no known major threats and it is considered of least concern. No conservation actions are required. (http://www.iucnredlist.org/details/168725/0).

D) **Medicinal parts:**

Aerial parts.

F) **Chemical composition:**

Different chemical groups can be found in *C. asiatica*; standing out are large amounts of **triterpene saponosides** and their glycosides; the major ones are asiatic acid, madecassic acid, and asiaticosides and madecosides (Singh and Rastogi 1969; James and Dubery 2009; Hashim et al. 2011). **Monoterpene and sesquiterpene derivatives**, the most prominent being $\alpha$-pinene, $\beta$-pinene, and bisabolene, have also been identified in this species, as well as **steroids**, like stigmasterol and *β*-sitosterol; several **flavonoid derivatives**, like quercetin and kaempferol derivatives; **phenolic acids**; and **polysaccharides** (Jahan et al. 2012; Orhan 2012). According to the *African Herbal Pharmacopoeia* (2010), the chemical composition of the plant varies with its geographical distribution.

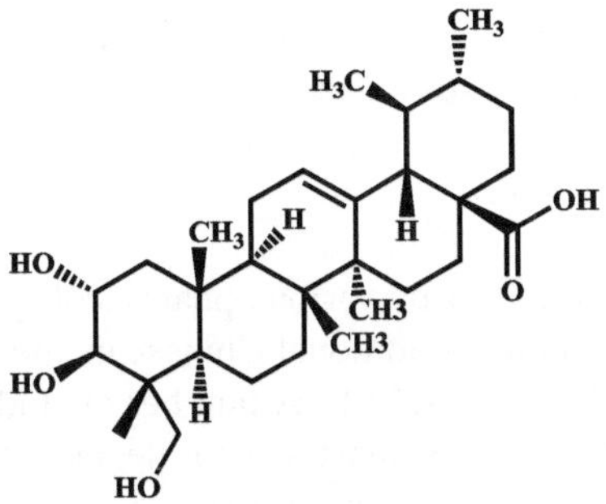

**Asiatic acid**

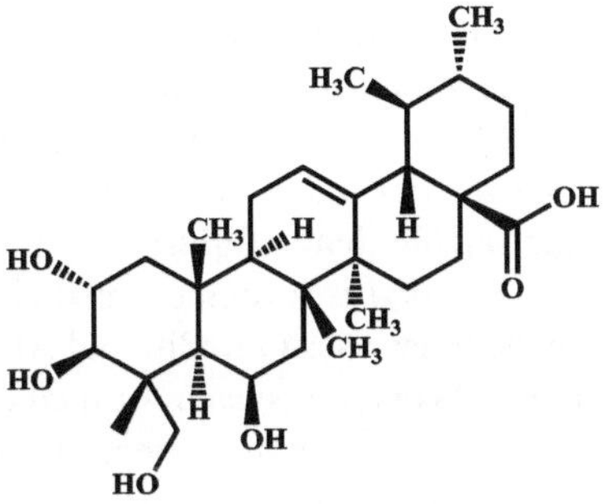

**Madecassic acid**

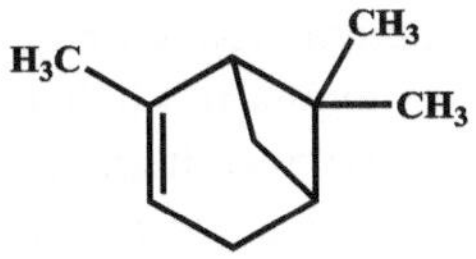

**α-Pinene**

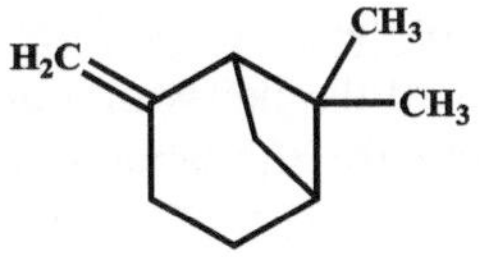

**β-Pinene**

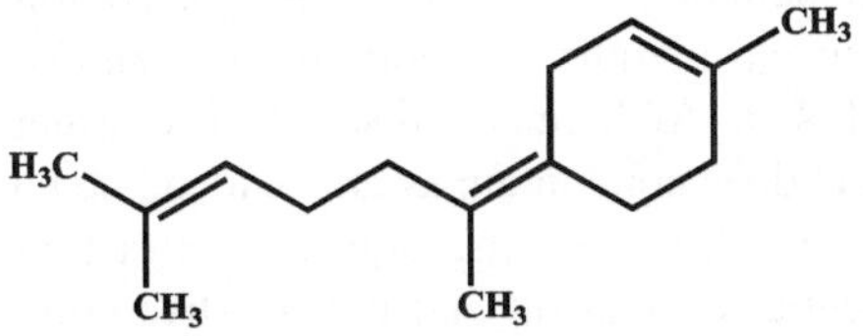

**Bisabolene**

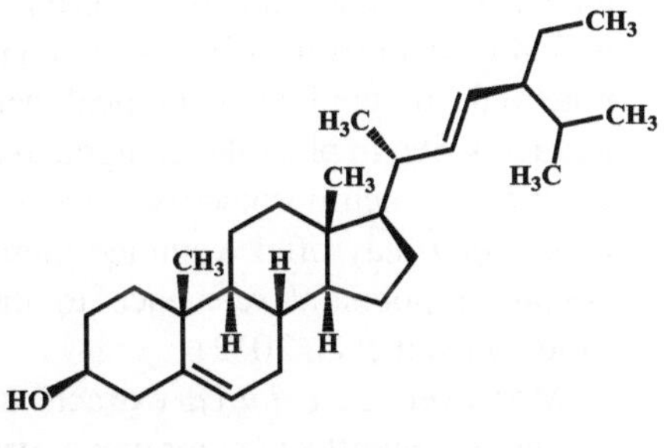

**Stigmasterol**

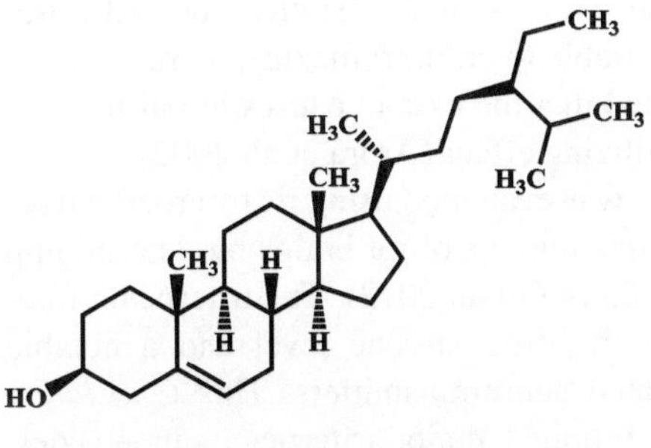

**Beta-Sitosterol**

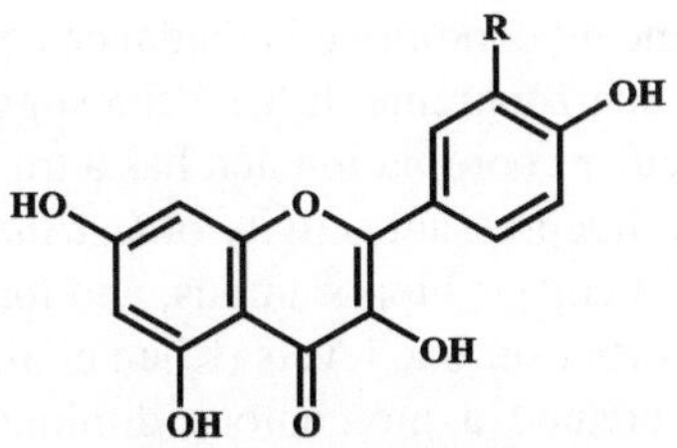

**R=OH: Quercetin**

**R=H: Kaempferol**

5) **Major ethnomedicinal uses:**

In Asia, *C. asiatica* is used as a vegetable, cooked or raw in salads and in curries. It has a long history of utilization in Ayurvedic and Chinese traditional medicines. The major ethnomedicinal uses of *C. asiatica* appear to be related to gastrointestinal disorders, skin wound healing and eczema, and memory enhancement or serving as nerve stimulant (Jahan et al. 2012). Among other uses, it is applied as a medicine for inflammation, cardiac tonic, depurative, febrifuge, hypotensive, nervine, sedative, astringent, stimulant, and cosmetic (Jahan et al. 2012; Orhan 2012). Monographs of the plant can be found in the *European Pharmacopoeia* and World Health Organization.

6) **Pharmacological studies:**

*C. asiatica* is a reputed medicinal plant due to its various pharmacological effects. It has uses in traditional African medicine, traditional Chinese medicine, and the Ayurvedic system of medicine (Jahan et al. 2012). A number of studies describe the protective effect of the plant extracts against several degenerative diseases of central nervous system, including Alzheimer's (Mukherjee et al. 2007) and Parkinson's diseases (Khan 2007; Awad et al. 2007). Soumyanath et al. (2012) used in vivo and in vitro standard tests to evaluate the cognitive effects of the aqueous extract of *C. asiatica*. These authors found that *C. asiatica* attenuated the behavioral abnormalities in a mouse model of Alzheimer's disease. In vitro, the same water extract protected some types of human neuroblastoma cells (SH-SY5Y cells and MC65) from toxicity induced by exogenously added and endogenously generated β-amyloid. The water extract did not show anticholinesterase activity or protect neurons from oxidative damage and glutamate toxicity, mechanisms of current Alzheimer's disease therapies. The combination of data from these in vivo and in vitro studies supports the *C. asiatica* potential for conferring clinical benefits in Alzheimer's disease. The water extract is rich in phenolic compounds and does not contain asiatic acid, a known *C. asiatica* neuroprotective triterpene. In this way, the aqueous extract of *C. asiatica* may offer a unique therapeutic mechanism and novel active compound of potential relevance to the treatment of this degenerative disease (Soumyanath et al. 2012).

Moreover, *C. asiatica* extracts have been found to have imipramine-like antidepressant effect (imipramine = tricyclic antidepressant) and significant anti-stress activity comparable to diazepam, a benzodiazepine (Sarma et al. 1998). In another study, trained rats treated with extracts of *C. asiatica* showed dose-dependent conditioned avoidance comparable to chlorpromazine, a drug used to treat schizophrenia. It has been suggested that the extract causes impairment of muscular coordination and has a tranquilizing effect (Arora et al. 2002).

Antidepressant activity of *C. asiatica* was evaluated using its triterpenic fraction in cortex, hippocampus, and thalamus regions of rat brains by determining the corticosterone levels (Chen et al. 2003 *in* Orhan 2012). The triterpenic fraction created a momentous diminution in corticosterone level and a notable increase in amount of monoamine-related neurotransmitters. This *C. asiatica* effect is generally attributed to the mentioned major triterpene saponosides, asiatic and madecassic acids, as well as to their heterosides. Liang et al. (2008) studied the potential antidepressant properties of the triterpene heteroside asiaticoside using valuable mice models to evaluate the antidepressant-like activity, and the results suggested that asiaticoside may have antidepressant-like action and the regulation of $\alpha$2-adrenergic receptor may be the major mechanism of this model, although the authors refer that further experiments evaluating the levels of noradrenaline and serotonin in different brain regions are still necessary to confirm this hypothesis.

7) **Precautions and side effects:**

*C. asiatica* has been used in traditional Indian Ayurvedic and Chinese medicines for decades. In pharmaceutical industries, it has shown good efficacy, performance, and safety, with low toxicity. According to WHO monographs (http://apps.who.int/medicinedocs/pdf/s2200e/s2200e.pdf), allergic contact dermatitis has been associated with topical application, but these reactions may be due to other ingredients in such preparations. Possible skin carcinogen in rodents after repeated topical application is also mentioned, but further experimentation is needed to substantiate this claim.

## 9.10 References


- African Herbal Pharmacopoeia. In: Brendler T, Eloff JN, Gurib-Faquim A, Phillips LD, editors. Graphic Press Ltd.; 2010. ISBN 978-99903-89-09-8.
- Arora D, Kumar M, Dubey SD. *Centella asiatica*—a review of its medicinal uses and pharmacological effects. J Nat Remedies. 2002;2:143–9.
- Awad R, Levac D, Cybulska P, Merali Z, Trudeau VL, Arnason JT. Effects of traditionally used anxiolytic botanicals on enzymes of the gamma-aminobutyric acid (GABA) system. Can J Physiol Pharmacol. 2007;85: 933–42.
- Cannon JFM. Umbelliferae, Centella. In: Launert E, editor. Flora Zambesiaca 4; 1978. p. 561–5.
- Chen Y, Han T, Qin L, Rui Y, Zheng H. Effect of total triterpenes from *Centella asiatica* on the depression behaviour and concentration of amino acid in forced swimming mice. J Chin Med Mater. 2003;26: 870–3.
- Arora D, Kumar M, Dubey SD. *Centella asiatica*—A review of its medicinal uses and pharmacological effects. J Nat Remedies. 2002;2:143–9.
- Howes MJR, Houghton P. Plants used in Chinese and Indian traditional medicine for improvement of memory and cognitive function. Pharmacol Biochem Behav. 2003;75:513–7.
- Hashim P, Sidek H, Helme M, Helan M, Sabery A, Palanisamy UD, Ilham M. Triterpene composition and bioactivities of *Centella asiatica*. Molecules. 2011;16:1310–22.
- Jahan R, Hossain S, Seraj S, Nasrin D, Khatun Z, Das PR, Islam T, Ahmed I, Rahmatullah M. *Centella asiatica* (L.) Urb.: ethnomedicinal uses and their scientific validations. Am-Eurasian J Sustain Agric. 2012;6:261–70.

- James JT, Dubery IA. Pentacyclic triterpenoids from the medicinal herb, *Centella asiatica* (L.) Urban. Molecules. 2009;14:3922–41.
- Liang X, Huang Y, Chen SW, Wang WJ, Xu N, Cui S, Liu XH, Zhang H, Liu YN, Liu S, Yang M, Dong Y. Antidepressant-like effect of asiaticoside in mice. Pharmacol Biochem Behav. 2008;89:444–9.
- Orhan IE. *Centella asiatica* (L.) Urban: from traditional medicine to modern medicine with neuroprotective potential. Evid Base Compl Alternative Med. 2012. doi: 10.1155/2012/946259.
- Singh B, Rastogi RP. A reinvestigation of the triterpenes of *Centella asiatica*. Phytochemistry 1969;8:917–21.
- Soumyanath A, Zhong YP, Henson E, Wadsworth T, Bishop J, Gold BG, Quinn JF. *Centella asiatica* extract improves behavioral deficits in a mouse model of Alzheimer's disease: investigation of a possible mechanism of action. Int J Alzheimer's Dis. 2012:381974. (http://www.iucnredlist.org/details/168725/0. Accessed Mar 2015.


## 9.11 *Cissampelos* Genus

### *9.11.1 (Menispermaceae)*

*Cissampelos mucronata* A. Rich., *Cissampelos owariensis* P. Beauv. ex DC., *Cissampelos pareira* L. var. *orbiculata* (D.C.) Miq.

1) **Synonyms:**

   *Cissampelos pareira* L. = *Cissampelos madagascariensis* Miers.

2) **Common names:**

   Angola: enhati, enyati, nhati (muila).

   GB: cauce-edjambaran (felupe); inétulo (mandinga); noferbalo, nopelebalo (fula); oreja di sandjo (criolo).

   Moz: nacadindi (macua); sissi (Tete); chipomba-folia (mateuè).

   Hairy heartleaf; heart-leaved vine (Eng.) (isto encontrei para C. mucronata in Fl Zimbab).

3) **Photograph of *Cissampelos mucronata*** (Fig. 9.5)
4) **Description of the genus:**

   A) **Habitat, world distribution, and PALOP distribution:**

      This genus is widely distributed throughout the tropics (Troupin, 1960). In PALOPs, some species can be found in Guinea-Bissau, Angola, and Mozambique (*C. mucronata* and *C. owariensis* (Semwal et al. 2014) and *C. pareira* var. *orbiculata* (Troupin 1960)).

   B) **Morphology of *C. mucronata*:**

      Twining liana with stem somewhat woody at the base and leafy stems puberulous or pubescent. Leaves alternate, simple, subpeltate, or clearly peltate; petiole up to 7 cm long, usually inserted 1–4 mm from the base of

**Fig. 9.5** *Cissampelos mucronata* A. Rich., from Guinea-Bissau, detail of inflorescences and leaves (original photos taken by L. Catarino)

the lamina, pubescent; lamina 2–12×2–12 cm, broadly cordate to ovate or orbicular, obtusely or emarginately mucronate or rarely acuminate at the apex, truncate to deeply cordate at the base, entire, membranous to papery, sparsely hairy above, puberulous to tomentose beneath, with 3–7 nerves from the base; stipules absent. Male flowers in solitary or paired axillary corymbose cymes up to 4 cm long, arranged in axils of bracts along an axis up to 10 cm long; sepals 4–5, ovate or obovate, 1.2–1.5×0.5 mm, keeled, hairy outside; petals united in a cup; stamens connate in a 4-locular synandrium. Female flowers in 5–9-flowered cymes arranged in axillary false racemes 5–10 cm long; bracts up to 1.5 cm in diameter, suborbicular to kidney-shaped, pubescent to tomentose; sepals usually 1, as in male flowers; petals 1, obtriangular to kidney shaped, ca. 1.5×2 mm; ovary superior, hairy with a thick 3-lobed style. Fruit a drupe 4–6×3–4 mm, curved, pubescent, 1-seeded; seed horseshoe-shaped tuberculate (Troupin 1960).

C) **Conservation status:**

Not threatened.

D) **Medicinal parts:**

Roots (tuberous roots) and leaves.

E) **Chemical composition:**

The **alkaloids** are the major chemical group reported in the genus *Cissampelos*. Most of the compounds were isolated from the roots of different species, including *C. pareira*, *C. owariensis*, and *C. mucronata*, and these include the **alkaloids** warifteine, methyl warifteine, berberine, hayatine, and hayatidin (Semwal et al. 2014).

In addition also **flavonoids** like quercetin and **saturated fatty acids**, **polysaccharides**, and **pectins** were isolated in roots and aerial parts of *Cissampelos* species (Semwal et al. 2014). Some authors remember that there are chemical variations in *Cissampelos*, and the plant organ, the developmental stage, the time of harvesting, and the geographical distribution must be taken into account for chemotaxonomic purposes (de Wet et al. 2011).

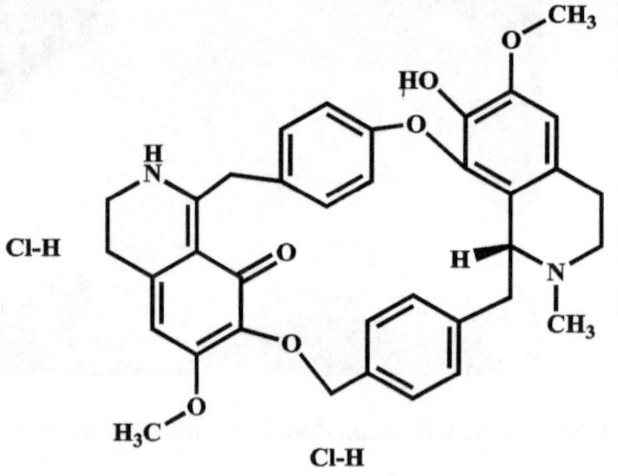

**Warifteine**

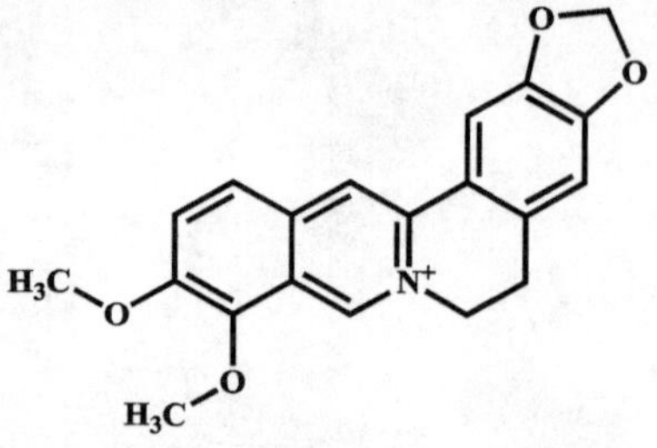

**Berberine**

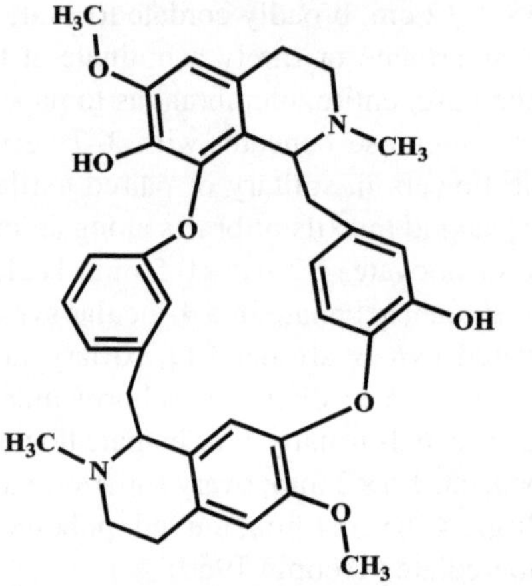

**Hayatine**

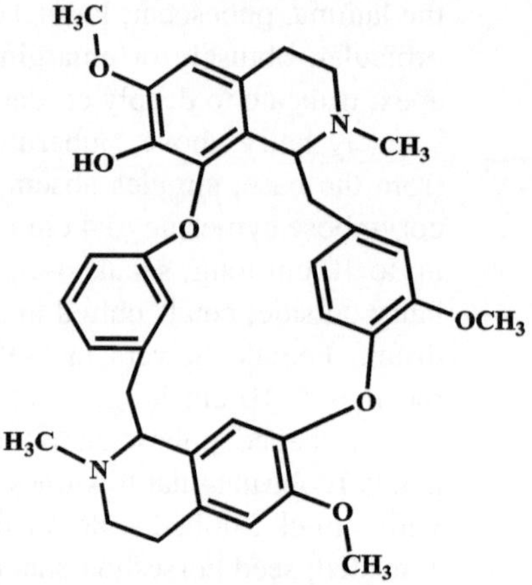

**Hayatidin**

5) **Major ethnomedicinal uses:**

*Cissampelos* species have a wide geographical distribution and vast traditional uses, being used because of both therapeutic and toxic properties. The roots of *C. owariensis* and *C. pareira* (which are often confused) are used as an emmenagogue, abortifacient, antipyretic, and diuretic (Oliver-Bever 1986). Roots and leaves are traditionally applied to asthma, cough, fever, arthritis, obesity, dysentery, snakebite, jaundice, blood pressure, and skin-related problems (Semwal et al. 2014). Iwu (2014) also mentions the use of aerial parts of *C. owariensis* against diarrhea, dysentery, colic, intestinal worms, and digestive complaints and as a tonic for amnesia and psychosis. In various parts of the world, some of *Cissampelos* species are used for arrow poison (Oliver-Bever 1986).

6) **Pharmacological studies:**

Some of the quinoline structure alkaloids isolated from different *Cissampelos* populations are known by its pronounced antipsychotic bioactivity and showed promising anti-allergic, immunosuppressive, antidepressant, anticancer, vasodilatory, and muscle-relaxant activities (Oliver-Bever 1986; Semwal et al. 2014). The total tertiary alkaloid fraction (containing warifteine) from *C. sympodialis* demonstrated an antidepressant effect in depression models, forced swimming test (FST), and reserpine test (Mendonca-Netto et al. 2008). The hydroalcoholic leaf extract of *C. sympodialis* (also containing warifteine) exhibited antidepressant effects together with antioxidant activity in the FST in mice (Zhang 2004). The ethanolic leaf extract was found to potentiate the toxicity of pentylenetetrazol, a convulsant drug, in mice. It also reduced the immobility period in the FST in mice and reversed the degree of catalepsy induced by reserpine in rats, an imipramine-like antidepressant effect (Almeida et al. 1998). The extracts of aerial parts of *C. owariensis* seem to have an effect against degenerative diseases of central nervous system, including Alzheimer's disease, as the methanolic extract showed activity against both acetylcholinesterase and butyrylcholinesterase (Elufioye et al. 2010). On the contrary, Hage et al. (2010), in order to select plants for the treatment of Alzheimer's disease, did not observe any effect on the production of β-amyloid peptide by n-hexane, dichloromethane, ethyl acetate, and water extracts of *C. owariensis*.

The ethanolic root extract of *C. mucronata* showed sedative activity in mice as it reduced ephedrine-induced spontaneous motor activity in rats and prolonged pentobarbitone sleeping time in mice (Akah et al. 2002).

Although these plants are used in different traditional medicine systems, there is a vast amount of research still to be conducted to validate their traditional use (Semwal et al. 2014).

## 9.12 References


- Akah PA, Nwafor SV, Okoli CO, Egbogha CU. Evaluation of the sedative properties of the ethanolic root extract of *Cissampelos mucronata*. Boll Chim Farm. 2002;141:243–6.
- Almeida RN, Navarro DS, DeAssis TS, DeMedeiros IA, Thomas G. Antidepressant effect of an ethanolic extract of the leaves of Cissampelos sympodialis in rats and mice. J Ethnopharmacol. 1998;63:247–52.
- de Wet H, Van Heerden FR, Van Wyk BE. Alkaloidal variation in Cissampelos capensis (Menispermaceae). Molecules 2011;16:3001–9.
- Elufioye TO, Obuotor EM, Sennuga AT, Agbedahunsi JM, Adesanya SA. Acetylcholinesterase and butyrylcholinesterase inhibitory activity of some selected Nigerian medicinal plants. Brazilian Journal of Pharmacognosy 2010;20:472–7.

- Hage, S., Kienlen-Campard, P., Octave, J. N., Quetin-Leclercq, J. 2010. In vitro screening on $\beta$-amyloid peptide production of plants used in traditional medicine for cognitive disorders. J Ethnopharmacol. 131: 585–91.
- Iwu M. M. 2014. Handbook of African Medicinal Plants. 2nd ed. CRC Press. International Standard Book Number-13: 978-1-4665-7198-3 (eBook—PDF).
- Mendonca-Netto, S., Varela, R., Fechine, M. F., Queiroga, M., Souto-Maior, F., Almeida, R., 2008. Antidepressant effects of total tertiary alkaloid fraction of Cissampelos sympodialis Eichler in rodents. Braz J Pharmacognosy. 18: 165–9.
- Nwosu MO. Herbs for mental disorders. Fitoterapia. 1999;70:58–63.
- Oliver-Bever B. Medicinal plants in tropical West Africa. Cambridge: Cambridge University Press; 1986. ISBN 978-0-521-10544-6.
- Semwal DK, Semwal RB, Vermaak I, Viljoen A. From arrow poison to herbal medicine—the ethnobotanical, phytochemical and pharmacological significance of Cissampelos (Menispermaceae). J Ethnopharmacol. 2014;155:1011–28.
- Troupin G. Menispermaceae. In: Exell AW, Wild H, editors. Flora Zambeziaca 1. London: Crown Agents for Oversea Governments and Administrations; 1960. p. 150–71.
- Zhang ZJ. The therapeutic effects of herbal extracts and constituents in animal models of psychiatric disorders. Life Sci. 2004;75:1.


## 9.13 *Griffonia simplicifolia* (Vahl ex DC.) Baill

### *9.13.1 (Fabaceae)*

1) **Synonyms:**
   *Bandeiraea simplicifolia* (Vahl ex DC.) Benth.; *Schotia simplicifolia* Vahl ex DC.
2) **Common local names:**
   Griffonia (Eng.)
3) **Photograph of the species** (Fig. 9.6)
4) **Description of the species:**
   A) **Habitat, world distribution, and PALOP distribution:**
      *G. simplicifolia* is native to West-Central Africa. It is usually found associated with termite hills and along the margins of primary forest and old farms (Iwu 2014). This species is found in Guinea-Bissau and Sao Tome and Principe (African Herbal Pharmacopoeia 2010).
   B) **Morphology of the species:**
      Liana or shrub with glabrous shining branches. Leaves alternate, simple, with petiole up to 1.5 cm long and lamina 6–12 × 3–6 cm, ovate, rounded to shortly acuminate at the apex, rounded to cordate at the base, entire, glabrous, 3-nerved from the base with reticulum prominent on both sides;

**Fig. 9.6** *Griffonia simplicifolia* (Vahl ex DC.) Baill., from Gabon (LISC Herbarium specimen, Le Testu 1926)

stipules minute, triangular. Flowers bisexual in axillary racemes 5–20 cm long; bracts and bracteoles minute, triangular, persistent; pedicels 3–4 mm long; calyx-tube 12–15 mm long, grayish tomentose, curved, gradually dilating and abruptly enlarged and ending in 5 obtuse lobes; petals 5, subequal, 10–12 mm long, elliptic, long-clawed, fleshy, sparsely short-hairy or glabrous, greenish; stamens 10, free with filaments up to 2 cm long, pubescent; ovary superior, ca. 4 mm long, stiped. Fruit an obliquely ovoid pod ca. 8×4 cm, leathery, blackish, with a stipe 1–1.5 cm long; seeds 1–4, orbicular, ca. 18×6×5 mm, glabrous (Aubréville 1968; Bosch 2008).

C) **Conservation status:**

*G. simplicifolia* is reported as a common species, but the high commercial value of its seeds is a serious threat to the reduction of populations. Sustainable harvesting needs to be implemented (Kumar et al. 2010) since cuttings and seed germination gave poor results (Iwu 2014).

D) **Medicinal parts:**

Roots, stems, and leaves.

E) **Chemical composition:**

In *G. simplicifolia* leaves, a **volatile oil** and **coumarins** have been isolated; a **cyanoglucoside** (lithospermoside = griffonin) was also isolated in this species roots (Kumar et al. 2010). Several **indole derivatives** were also isolated in the whole plant, among them stands out 5-hydroxy-L-tryptophan, known as 5-HTP (African Herbal Pharmacopoeia 2010), a natural occurring and highly absorbable type of aromatic amino acid and a direct precursor to serotonin. *G. simplicifolia* seeds have the richest 5-hydroxytryptophan content, but a similar composition is found in the seeds of the other species of *Griffonia*. *G. simplicifolia* seed extract is either a gray-white powder or pale brown crystals containing 95–98 % of 5-hydroxytryptophan (Kumar et al. 2010). **Lectins**, with insecticidal properties, were also isolated in *G. simplicifolia* seeds (Kumar et al. 2010).

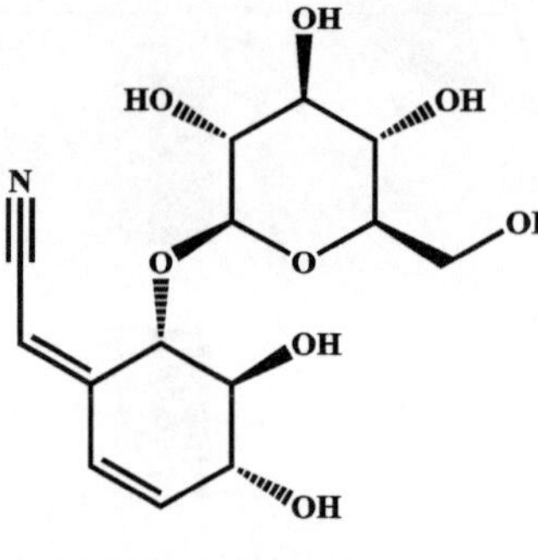


**Griffonin**

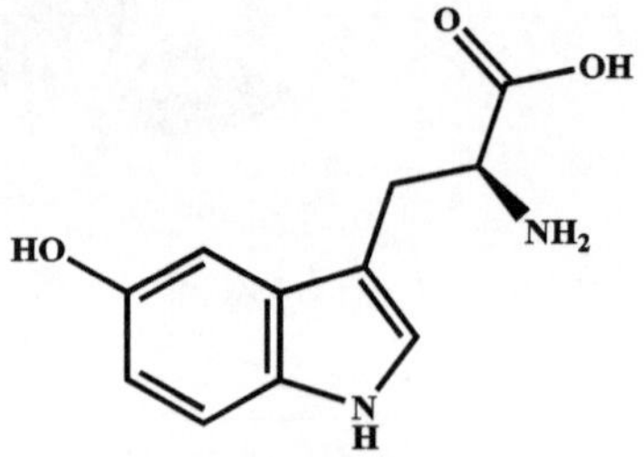


**5-Hydroxy-L-tryptophan**

5) **Major ethnomedicinal uses:**

The roots, stems, and leaves are used in chewing, a popular means of tooth cleaning, and aiding wound healing. A decoction of aerial parts is taken to stop vomiting; to treat gonorrhea, stomach problems, and congestion of the pelvis (African Herbal Pharmacopoeia 2010); and for mental disorders (Iwu 2014). Young leaves are eaten as an aphrodisiac (African Herbal Pharmacopoeia 2010). An extract from the powdered roots is used to treat sickle-cell anemia (African Herbal Pharmacopoeia 2010; Iwu 2014). The leaves are highly valued as animal feed and are said to stimulate reproduction (Iwu 2014).

6) **Pharmacological studies:**

*G. simplicifolia* is considered as a natural alternative for the antidepressant Prozac (fluoxetine, an SSRI), since 5-HTP is the precursor in the biosynthesis of the neurotransmitter serotonin within the central nervous system. 5-HTP crosses the blood-brain barrier (BBB) easily and is completely transformed into serotonin by 5-hydroxytryptophan decarboxylase, which is found in the nervous and liver tissues (Birdsall 1998; www.linnea-worldwide.com/download/5htp.pdf). Other neurotransmitters and CNS chemicals, such as melatonin, dopamine, norepinephrine, and β-endorphin, have also been shown to increase following oral administration of 5-HTP (Kumar et al. 2010). The use of 5-HTP has also been beneficial in the treatment of other low levels of serotonin situations, like in the treatment of fibromyalgia, insomnia, chronic headaches (Nicolodi and Sicuteri 1996; Birdsall 1998), and obesity (Cangiano et al. 1992).

Most studies on 5-HTP use in depression treatment were done 30–40 years ago, between the 1970s and 1990s, when there was a great interest in the serotonin hypothesis of depression (Iovieno et al. 2011). Kumar et al. (2010) mentioned studies following patients with different types of depression, unipolar and bipolar depression and severe depression. In the treatment of severe depression, when comparing the use of 5-HTP to selective serotonin reuptake inhibitor (SSRI) drugs, like fluoxetine, it was found that this indole derivative was at least as effective as the synthetic drugs, but with fewer side effects (Kumar et al. 2010).

After the introduction and approval of other SSRIs, the research on 5-HTP became less compelling to investigators, and since *G. simplicifolia* and 5-HTP are naturally occurring and cannot be patented as a pharmaceutical drug, industry has no profit incentive to market this natural compound (Kumar et al. 2010; Iovieno et al. 2011). So far there are no published reports comparing 5-HTP against newer antidepressants such as SSRIs (Turner et al. 2006; Iovieno et al. 2011). Results from the previous clinical studies and safety reports suggest that 5-HTP supplementation may have at least some antidepressant efficacy and generally good tolerability and may deserve to be evaluated in more rigorous studies in order to conclusively establish its effectiveness, both as monotherapy and as augmentation (Das et al. 2004; Turner et al. 2006; Iovieno et al. 2011).

7) **Precautions and side effects:**

Some patients initially may feel mild nausea after taking 5-HTP (African Herbal Pharmacopoeia 2010).

8) **Dosage:**

Initially 50 mg of 5-HTP 3 times a day, at mealtimes; after 2 weeks, the dose can duplicate, 100 mg keeping takes, i.e., three times a day, at mealtimes (African Herbal Pharmacopoeia 2010).

## 9.14 References


- African Herbal Pharmacopoeia. In: Brendler T, Eloff JN, Gurib-Faquim A, Phillips LD, editors. Graphic Press Ltd.; 2010. ISBN 978-99903-89-09-8.
- Aubréville A. Legumineuses—Sous-famille des Caesalpinioidées. Flore du Gabon n° 15. Paris: Muséum National d'Histoire Naturelle, Laboratoire de Phanérogamie; 1968.
- Birdsall TC. 5-Hydroxytryptophan: a clinically-effective serotonin precursor. Alternative Med Rev. 1998;3:271–80.
- Bosch CH. Griffonia simplicifolia (Vahl ex DC.) Baill. In: Schmelzer GH, Gurib-Fakim A, editors. Plant Resources of Tropical Africa 11(1). Medicinal plants 1. Wageningen, The Netherlands: PROTA Foundation; Leiden, The Netherlands: Backhuys Publishers, Wageningen, The Netherlands: CTA; 2008. p. 314–6.
- Cangiano C, Ceci F, Cascino A, Del Ben M, Laviano A, Muscariotoli M, Antonucci F. Rossi-Fanelli F. Eating behaviour and adherence to dietary prescriptions in obese adult subjects treated with 5-hydroxytryptophan. Am J Clin Nutr 1992;56:863–7.
- Das YT, Bagchi M, Bagchi D, Preuss HG. Safety of 5-hydroxy-L-tryptophan. Toxicol Lett. 2004;150:111–22.
- Iovieno N, Dalton ED, Fava M, Mischoulon D. Second-tier natural antidepressants: review and critique. J Affect Disord. 2011;130:343–57.
- Iwu MM. Handbook of African medicinal plants. 2nd ed. CRC Press; 2014. International Standard Book Number-13: 978-1-4665-7198-3 (eBook—PDF).
- L-5-Hydroxytryptophan *Griffonia* Seed Extract 99 %. www.linnea-worldwide.com/download/5htp.pdf.
- Kumar PS, Praveen T, Prakash JN, Jitendra B. A review on *Griffonia simplicifollia*—an ideal herbal antidepressant. Int J Pharm Life Sci. 2010;1:174–81.
- Nicolodi M, Sicuteri F. Fibromialgia and migraine, two faces of the same mechanism. Serotonin as the common clue for pathogenesis and therapy. Adv Exp Med Biol. 1996;398:373–9.
- Turner EH, Loftis JM, Blackwell AD. Serotonin a la carte: supplementation with the serotonin precursor 5-hydroxytryptophan. Pharmacol Therapeut 2006;109:325–38.


## 9.15 *Mondia whitei* (Hook. f.) Skeels

### *9.15.1 (Apocynaceae)*

1) **Synonyms:**

   *Chlorocodon whitei* Hook. f.

2) **Common names:**

   GB: Lacadje (Fula); ngaru (mandinga).

   Angola: mudondo.

   White's ginger, tonic root (Eng.).

**Fig. 9.7** *Mondia whitei* (Hook. f.) Skeels, flower, in Mozambique (Torre and Correia 17386, IICT Herbarium collection)

3) **Photograph of the species** (Fig. 9.7)
4) **Description of the species:**

   A) **Habitat, world distribution, and PALOP distribution:**
   Native to Africa, considered endemic in almost all African regions with the exception of the northern part of the continent, from sea level to 1800 m. It is found in the PALOP countries Guinea-Bissau, Angola, and Mozambique.

   B) **Morphology of the species:**
   Large liana with white latex and aromatic roots. Leaves opposite with petiole 2–7 cm long and lamina ovate to suborbicular, 10–20×6–14 cm, acuminate at apex, obtuse to cordate at base; interpetiolar stipules frill-like. Flowers with unpleasant odor, in lax axillary panicles, 10–20 in each panicle; peduncles 2–4 cm long; pedicels more or less 1 cm long; sepals 5, ovate, 2–3×1–2 mm, acute with fimbriate margin; corolla coriaceous, glabrous with tube 2–3 mm long, lobes 5, obliquely ovate, 9–11×4–6 mm, outside greenish, inside violet, maroon, mauve or reddish. Corona with 5 tri-segmented lobes, lateral ones flaplike, 1–2 mm long, greenish or creamy yellow, central one ligulate or corniculate, 5–8 mm long, darker colored. Fruit with two narrowly ovoid follicles 8–12×2–4 cm; seeds obliquely ovoid, 8–10 mm long with a 20–25-mm-long tuft of silky hairs. Species known from tropical Africa and Southern Africa, including Mozambique (Venter 2012).

   C) **Conservation status:**
   In West Africa, the plant is cultivated but in other regions it is seriously threatened by overharvesting (Iwu 2014).

D) **Medicinal parts:**
Roots, leaves, and flowers.

E) **Chemical composition:**
In *M. whitei* root and leaf extracts, **terpenes, flavonoids, coumarins, coumarinolignan**, as well as chlorinated coumarinolignan, **reducing sugars,** and **tannins** were detected. No alkaloids were found. The main compounds identified were both non-nitrogenous compounds: 2-hydroxy-4-methoxybenzaldehyde (Kubo and Kinst-Hori 1999) and monoterpene–lactone loliolide (Neergaard et al. 2010). Nutritional analysis indicated that *M. whitei* is rich in **minerals** and **vitamins** (Lamidi and Bourobou 2010; Iwu 2014).

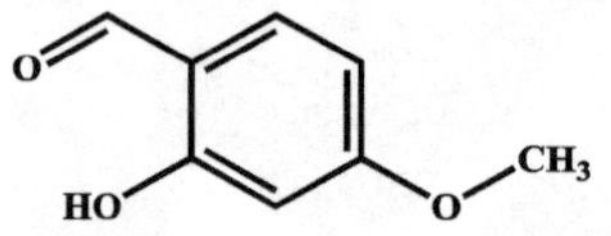

**2-Hydroxy-4-methoxybenzaldehyde**

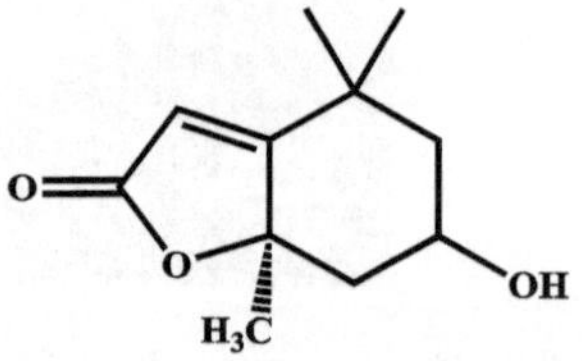

**(−) -Loliolide**

5) **Major ethnomedicinal uses:**
*M. whitei* root infusion is popularly used as aphrodisiac and in the treatment of male erectile dysfunction (African Herbal Pharmacopoeia 2010; Iwu 2014). It is also known to enhance cerebral and peripheral blood circulation, thus improving memory (http://www.infonet-biovision.org/default/ct/250/medicinal-plants). Leaf extracts are used in traditional medicine to treat nervous disorders (Neergaard et al. 2010).

6) **Pharmacological studies:**
In a screening of plant extracts for their affinity to the serotonin reuptake transporter protein, making use of an in vitro serotonin reuptake transporter (SERT) protein binding assay, according to Plenge et al. (1990), aqueous extracts of *M. whitei* leaves and flowers showed a mild affinity to SERT in rat brain, while the ethanolic leaf extract showed moderate antidepressant activity in vitro in the same SERT assay (Nielsen et al. 2004). Ethanolic leaf extracts of *M. whitei* were screened for affinity to the SERT in the [$^3$H]-citalopram-binding assay and for inhibitory effects on the SERT, the noradrenaline transporter (NAT), and the dopamine transporter (DAT) and showed antidepressant-like in vitro and in vivo effects (Pedersen et al. 2008). Further studies using a bioassay-guided fractionation lead to the isolation of the monoterpene–lactone loliolide (Neergaard et al. 2010), which showed affinity to SERT. Common antidepressant drugs showing SERT inhibitory activity contain an amine nitrogen atom and an aromatic ring. (−)-Loliolide, as a non-nitrogenous compound, seems to bind to SERT in a different way than standard nitrogen-containing serotonin reuptake inhibitors (Neergaard et al. 2010). According to these results, there seems to exist other success models of binding and blocking SERT, and further studies on functional SERT inhibition molecules are needed.

## 9.16 References


- African Herbal Pharmacopoeia. In: Brendler T, Eloff JN, Gurib-Faquim A, Phillips LD, editors. Graphic Press Ltd.; 2010. ISBN 978-99903-89-09-8.
- Iwu MM. Handbook of African medicinal plants. 2nd ed. CRC Press; 2014. International Standard Book Number-13: 978-1-4665-7198-3 (eBook—PDF).
- Kubo I, Kinst-Hori I. 2-Hydroxy-4-methoxybenzaldelyde: a potent tyrosinase inhibitor from African medicinal plants. Planta Med. 1999;65:19–22.
- Lamidi M, Bourobou H. Mondia whitei (Hook.f.) Skeels. In: Schmelzer GH, Gurib-Fakim A, editors. Prota 11(2): Medicinal plants/plantes médicinales 2. [CD-Rom]. Wageningen, The Netherlands: PROTA; 2010.
- Neergaard J, Rasmussen HB, Stafford GI, van Staden J, Jäger GI. Serotonin transporter affinity of (−)-loliolide, a monoterpene lactone from *Mondia whitei.* S Afr J Bot. 2010;76:593–6.
- Nielsen N, Sandager M, Stafford GI, van Staden J, Jäger A. Screening of indigenous plants from South Africa for affinity to the serotonin reuptake transport protein. J Ethnopharmacol. 2004;94:159–63.
- Pedersen ME, Szewczyk B, Stachowicz K, Wieronska J, Andersen J, Stafford GI, van Staden J, Pilc A, Jäger KA. Effects of South African traditional medicine in animal models for depression. J Ethnopharmacol. 2008;119:542–8.
- Plenge P, Mellerup ET, Nielsen M. Inhibitory and regulatory binding sites on the rat brain serotonin transporter: molecular weight of the [3H]paroxetine and [3H] citalopram binding proteins. Eur J Pharmacol Mol Pharmacol Sect. 1990; 189:129–34.
- Venter J. Mondia. In: Beentje HJ, editor. Flora of tropical East Africa, Apocynaceae (part 2). 2012. p. 143–6. (http://www.infonet-biovision.org/default/ct/250/medicinalplants. Mondia. Accessed Jan 2015.


## 9.17 *Palisota hirsuta* (Thunb.) K. Schum

### *9.17.1 (Commelinaceae)*

1) **Synonyms:**

   *Dracaena hirsuta* Thunb.; *Dianella triandra* Afzel.; *Palisota maclaudi*i Cornu.
2) **Common names:**

   GB: baru (fula); siquimbe (sosso); mabubé, m'bobé-n'kamasop (nalu).
   Eng (not found).
3) **Photograph of the species** (Fig. 9.8)
4) **Description of the species:**

   A) **Habitat, world distribution, and PALOP distribution:**

   Widespread in tropical Africa; common from Senegal to Zaire. It occurs as a weed on farmlands and in forest zones. It is also grown as an ornamental and might be used as a component of hedges. Found in Guinea-Bissau and Angola.

**Fig. 9.8** *Palisota hirsuta* (Thunb.) K. Schum., from Guinea-Bissau, detail of inflorescence and leaves (original photo taken by L. Catarino)

B) **Morphology of the species:**

Perennial herb 1–3 m high with several strong, simple, or little branched erect stems. Leaves opposite, sessile; sheath pubescent, the margin ciliate with about 5-mm-long reddish-brown hairs; lamina 15–25 × 4–9 cm, lanceolate to obovate, acute at the apex, attenuated to the base into a short pseudo-petiole, with long reddish-brown hairs at the margin, mainly to the base, parallel nerved. Flowers bisexual in a more or less cylindrical long panicle (sometimes 2–4 panicles) at the end of the stem, each branch or partial inflorescence with 1–3 cm long peduncle, perpendicular to the main axis, subtended by a small bract; pedicels 1–3 mm long; perianth of 6 subequal tepals, white sometimes tinged with purple; stamens 3 and 2–3 staminodes, hairy; ovary 3-celled, glabrous, each cell with 1–8 ovules. Fruit globular, indehiscent, fleshy, about 1 cm in diameter, red, turning black and shining at maturity; seeds 10–16, little compressed, almost smooth (vanden Berghen 1988).

C) **Conservation status:**

Not threatened.

D) **Medicinal parts:**

Roots and leaves.

E) **Chemical composition:**

In the leaves, the presence of **terpenoids**, **alkaloids**, **tannins**, and **flavonoids** was tested by simple qualitative and quantitative methods being the last two groups and the most dominant constituents (Woode et al. 2010). In the roots, two rare **phytoecdysones** (ecdysteroids), ecdysterone, and 20, 26-dihydroxyecdysone along with ***β*-stigmasterol** and **eicosanoic acid** were isolated. In the fruits, **carotenoids** are the main pigments. These are esterified with **saturated fatty acids**, mainly with lauric acid (Neuwinger 1996).

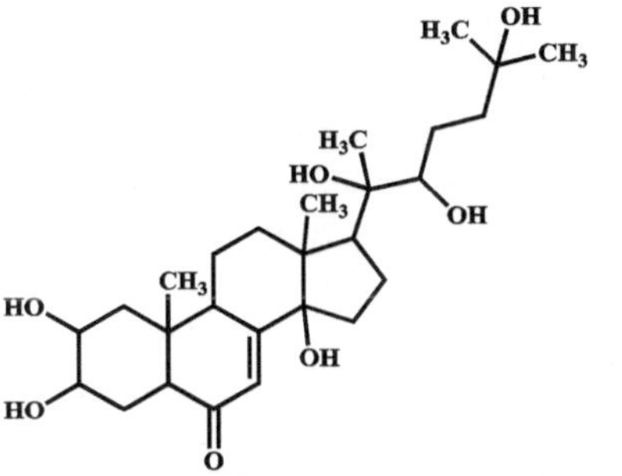

**Ecdysterone**

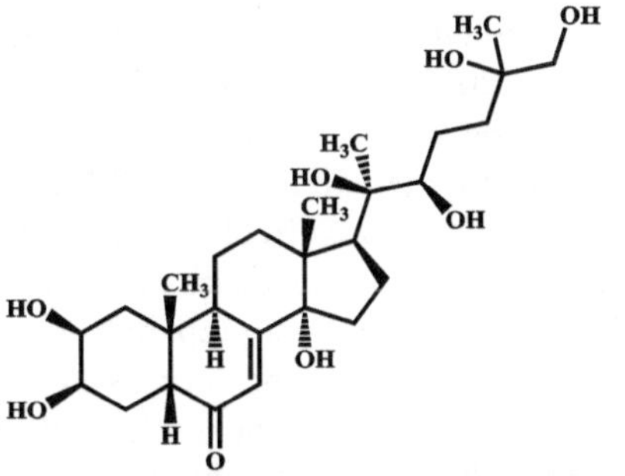

**20,26-Dihydroxyecdysone**

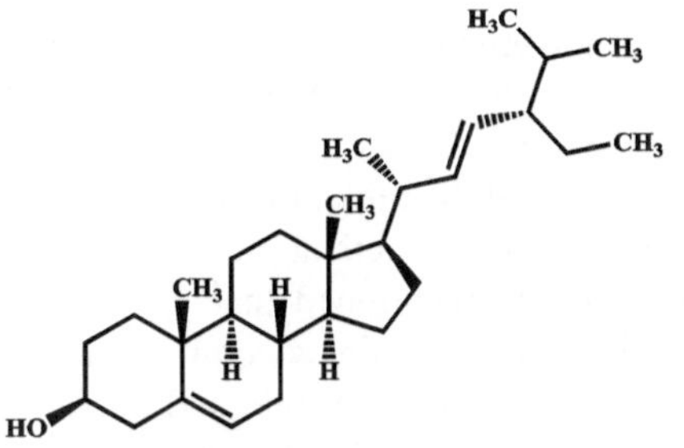

**Beta-Stigmasterol**

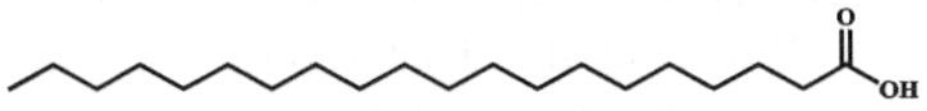

**Eicosanoic acid**

5) **Major ethnomedicinal uses:**

*P. hirsuta* is known for its analgesic and antiseptic effects. In many West African countries, a decoction of the twigs is used for gonorrhea; the roots for dysentery, anemia, and painful and inflammatory conditions (such as rheumatism); and the leaves as aphrodisiac (Benson et al. 2008) as well as in female infertility (Sarpong 2012). Whole plant is used as an analgesic and antiseptic (Mshana et al. 2000), antimicrobial (Anani et al. 2000; Iwu 2014), and antiviral (Anani et al. 2000). The root bark is employed as arrow poison and in the cicatrization of wounds (Neuwinger 1996). Dried leaves are smoked like tobacco for headaches and used in steam baths to treat malaria in southern Nigeria (Sarpong 2012).

6) **Pharmacological studies:**

The methanolic extract of the leaves of *P. hirsuta* showed antiviral activity against three test viruses, herpes simplex, Sindbis, and poliovirus (Anani et al. 2000), and the dominant activity in all cases was virucidal, that is, direct inactivation of virus particles (Hudson et al. 2000). The ethanolic root extract of *P. hirsuta* showed anti-inflammatory and antinociceptive/analgesic activities and a high antioxidant activity, which might explain in part the anti-inflammatory activity. These activities might be related with the compound 20-hydroxyecdysone, an ecdysteroid, isolated from the roots of *P. hirsuta*, which has proven to have a potent anti-inflammatory activity as well as antinociceptive action (Sarpong 2012).The ethanolic leaf extract of *P. hirsuta* also seems to have an hepatoprotective effect on carbon tetrachloride-induced hepatotoxicity on bioassays with rats (Imafidon et al. 2012). The hepatoprotective effects were dose dependent and more effective at higher doses. The presence of flavonoids, natural antioxidants, in the leaves plant might be responsible for that.

The ethanolic extract prepared from the roots of *P. hirsuta* exhibited an antipyretic effect in rats and an anti-inflammatory activity in the chick model of acute inflammation, comparable to the NSAID diclofenac, when administered preemptively and curatively (Boakye-Gyasi1 et al. 2008). According to these authors, the presence of flavonoids may contribute to the anti-inflammatory activity of the ethanolic root extract of *P. hirsuta.*

Woode et al. (2010) report on the effect of the ethanolic leaf extract of *P. hirsuta* in several common animal models used to study anxiety (light/dark box and elevated plus maze) and depression (forced swimming test and tail suspension test). The results suggest that the extract has anxiolytic and antidepressant-like effects in the models employed, possibly by GABAergic activation and/or effect on monoamine levels in the CNS. Those authors mentioned that it must be pointed out that the ethanolic leaf extract of *P. hirsute* contains several secondary metabolites, and therefore apart from the speculated GABAergic and dopaminergic involvement in the anxiolytic and antidepressant effects, other mechanisms and neurotransmitters may be involved, such as serotonergic and glutamatergic neurotransmissions. Further experiments may be necessary therefore to confirm the exact mechanism/s involved in these CNS effects.

## 9.18 References


- Anani K, Hudson JB, de Souza C, Alepagana K, Towers GHN. Investigation of medicinal plants of Togo for antiviral and antimicrobial activities. Pharmaceut Biol. 2000;38:40–5.
- Benson BB, Bekro YA, Mamyrbekova-Bekro JA, Coulibaly WK, Ehile EE. Assessment of sexual stimulant potential of total flavonoids extracted from leaves of *Palisota hirsuta* Thunb. K Schum (Commelinaceae). Eur J Sci Res. 2008;22:533–8.

- Boakye-Gyasi E, Woode E, Ainooson GK, Obiri DD, Ansah C, Duwejua M, Donkoh A. Anti-Inflammatory and antipyretic effects of an ethanolic extract of *Palisota hirsuta* K. Schum roots. Afr J Pharm Pharmacol. 2008;2:191–9.
- Berghen CV. Palisota. In: Jean Berhaut, Flore Illustrée du Sénégal, 1988;9: 137–8.
- Hudson JB, Anani K, Arnason JT, Gbeasor M. Further investigations on the antiviral activities of medicinal plants of Togo. Pharmaceut Biol. 2000;38:46–50.
- Imafidon KE, Lucky OO, Orhiere AV. Protective effect of ethanolic extract of *Palisota hirsuta* on CCl4 induced hepatotoxicity. Int Res J Pharmaceut. 2012; 2:143–7.
- Iwu MM. Handbook of African medicinal plants. 2nd ed. CRC Press; 2014. International Standard Book Number-13: 978-1-4665-7198-3 (eBook—PDF).
- Neuwinger HD. African ethnobotany: poisons and drugs, chemistry, pharmacology, toxicology. London: Chapman & Hall; 1996.
- https://books.google.pt/books?id=_j8ueEmakD0C&pg=PA360&lpg=PA360&dq=Palisota+distribution&source=bl&ots=ZA4TUdYDC4&sig=mpLeqws0gr9tuiVrK9S6Gl7LYZM&hl=en&sa=X&ei=4Iv4VI2CFIr3UNiigLgM&ved=0CBMQ6AEwAA#v=onepage&q=Palisota%20distribution&f=false. Accessed Jan 2015.
- Sarpong FM. Chemical Constituents, anti-inflammatory, anti-oxidant and antinociceptive activities of the roots of *Palisota hirsuta.* In: Schum K, (Commelinaceae). Thesis Doctor of Philosophy, Kumasi, Ghana: Faculty of Pharmacy and Pharmaceutical Sciences, Kwame Nkrumah University of Science & Technology; 2012. http://dspace.knust.edu.gh:8080/xmlui/handle/123456789/4627.
- Vanden Berghen C. Commelinacées. Flore Illustrée du Sénégal. 1988;9: 103–38.
- Woode E, Boakye-Gyasi E, Amidu N, Ansah C, Duwejua M. Anxiolytic and antidepressant effects of a leaf extract of *Palisota hirsuta* K. Schum (Commelinaceae) in mice. Int J Pharmacol. 2010;6:1–17.


## 9.19 *Securidaca longipedunculata* Fresen

### *9.19.1 (Polygalaceae)*

1) **Synonyms:**

   *Securidaca longipedunculata* var. *longipedunculata*; *Elsota longipedunculata* (Fresen.) Kuntze.

2) **Common names:**

   Angola: Muipanhoca (muila); munduku (muquilengue); muripanhoca (dimba); m'tchacha, muxaxa (muila); omutate (handa); utata (Cangombe); vint'hata (Cuito-Cuanavale).

**Fig. 9.9** *Securidaca longipedunculata* Fresen., from Guinea-Bissau, detail of fruits and inflorescences (original photos taken by L. Catarino)

GB: Mamampai (balanta); djutu (biafada); jurtú, jutù (creole); djúrô (fula); úli-élô (mandinga) (Catarino et al. 2006).

Moz: Boazi (Tete); colodi, colodji (Nampula); iosi (Namacurra); mupopo (Sofala); mupupo (Manica); naquehi, naquérè (Cabo Delgado, Zambezia); tsatse (Maputo) (de Koning 1993)

Violet tree (Eng).

3) **Photograph of the species** (Fig. 9.9)
4) **Description of the species:**

A) **Habitat, world distribution, and PALOP distribution:**

Species widespread in tropical and subtropical parts of Africa, in different climates and altitudes, in woodland and arid savanna soils (Ndou 2006; http://plants.jstor.org/compilation/Securidaca.longipedunculata). It is found in PALOP countries: Guinea-Bissau, Angola, and Mozambique.

B) **Morphology of the species:**

Shrub or small tree up to 6 m high, sometimes spiny. Leaves alternate, simple; petiole up to 5 mm long; lamina 1–5 × 0.5–1.8 cm, very variable in shape and size, from very narrowly elliptic to broadly oblong-elliptic, pubescent when young soon glabrescent. Flowers pink or purple, in terminal and axillary 3–5 cm long racemes , zygomorphic, sweet-scented; pedicels up to 14 mm long, pubescent; sepals 5, unequal, two lateral ones petaloid, 5–11 × 4–9 mm, suborbicular, two anterior ones up to 5 × 4.5 mm, broadly ovate, the posterior one up to 5 × 4 mm, ovate-acuminate with ciliate margins; corolla with 3 petals, the upper 2 up to 7.5 × 3.5 mm narrowly elliptic, ciliate at the base, the lowest (carina) keel-shaped, up to 10 mm long with a small lobed appendage 1 mm long near the apex. Stamens 8, the filaments united in staminal sheath ciliate on the upper margin. Fruit a samara, 3–5 × 0.8–2 cm, with an elliptic obliquely curved wing, sometimes with a second rudimentary wing and with a single rugulose or smooth seed 8–10 mm in diameter (Exell 1960).

C) **Conservation status:**

*S. longipedunculata* is a threatened and protected species; its roots are harvested and used as a medicine and that destroys the whole plant. Sustainable harvesting is difficult to implement in this model, but vegetative propagation of root cuttings is possible although difficult (Ndou 2006). Seeds germinate but seedlings are difficult to maintain (http://tropical.theferns.info/viewtropical.php?id=Securidaca+longipedunculata).

D) **Medicinal parts:**

Roots, root bark, and leaves.

E) **Chemical composition**

The distribution of the plant shows a variation in some features and chemical constituents, but the occurrence of **glycosides** (cardiac glycosides and glycosides of flavonoids and of saponins), **tannins**, **saponins**, and **alkaloids** was confirmed in different populations of Nigeria (Ajali and Chukwurah 2004; Junaid et al. 2008; Musa et al. 2013; Mongalo et al. 2015) and Cameroon (Moto et al. 2011).

The aqueous root extract of *S. longipedunculata* showed strong presence of saponins, alkaloids, and **volatile oil**, moderate presence of flavonoids and triterpenoids, and absence of steroids, tannins, and glycosides (Adebiyi et al. 2006). Another phytochemical analysis of *S. longipedunculata* root and root bark ethyl acetate, methanolic, and aqueous extracts confirms the existence of terpenes (methyl-2-hydroxybenzoate or methyl salicylate volatile oil), **steroids** (β-sitosterol), tannins, flavonoids (quercetin, rutin), **xanthone** (1,7-dihydroxy-4-methoxyxanthone), **phenolic acids** (*p*-coumaric acid, cinnamic acid, caffeic acid, and chlorogenic acid), and saponins (securidacaside A and securidacaside B, 3-*O*-β-D-glucopyranosylpresenegenin-28-O-β-D-apiofuranosyl-(1,3)-β-D-xylopyranosyl-(1,4)-[β-D-apiofuranosyl-(1,3)]-α-L-rhamnopyranosyl-(1,2)-{4-*O*-[(*E*)-3,4,5-trimethoxycinnamoyl]}-β-D-fucopyranosylester, and presenegenin) (Mongalo et al. 2015).

The essential oils of the root barks of *S. longipedunculata* were extracted by hydrodistillation with a Clevenger apparatus and analyzed by GC/MS and by $^{1}H$ and $^{13}C$ NMR spectra revealing methyl salicylate as the only compound (Nébié et al. 2004).

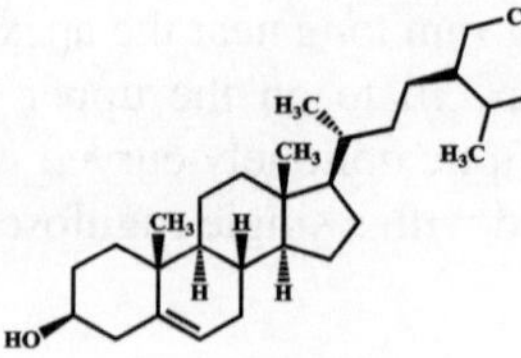

Beta-Sitosterol

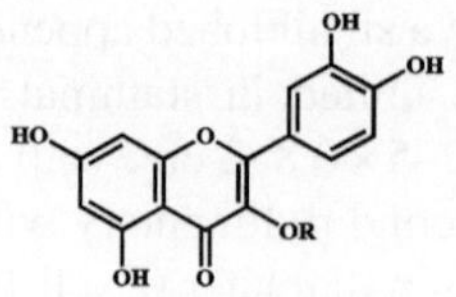

R=H: Quercetin
R=rutinose: rutin

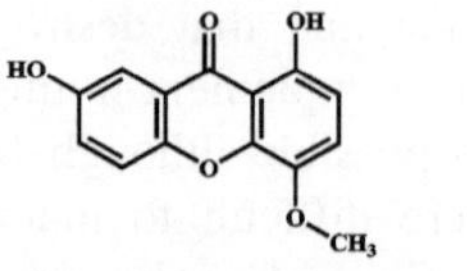

1,7-Dihydroxy-4-methoxyxanthone

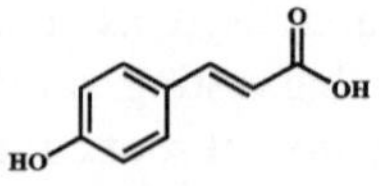

*p*-Coumaric acid

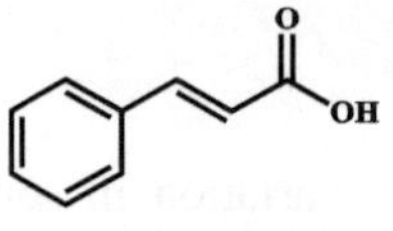

Cinnamic acid

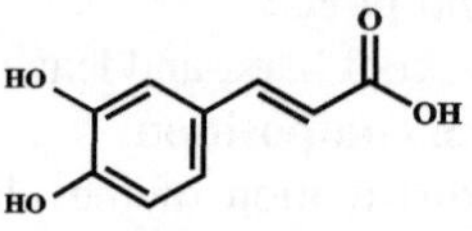

Caffeic acid

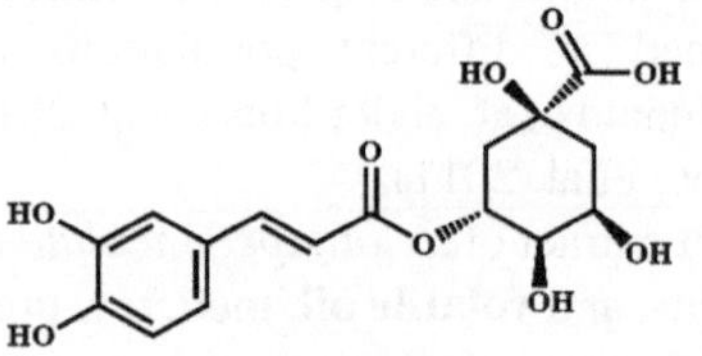

Chlorogenic acid

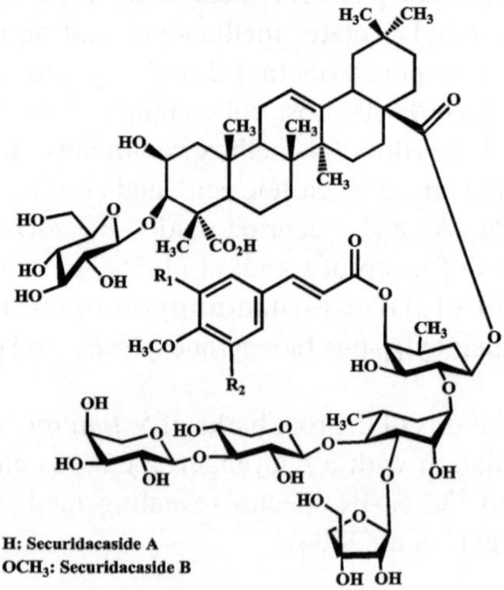

$R_1 = R_2 = H$: Securidacaside A
$R_1 = R_2 = OCH_3$: Securidacaside B

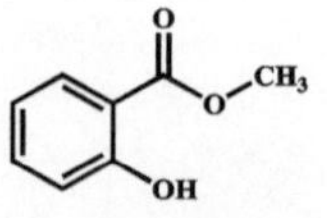

Methyl salicylate

5) **Major ethnomedicinal uses:**

*S. longipedunculata* is a very popular plant in the African medicine with a wide diversity of uses for its roots, although these might be poisonous. The leaves but above all the roots and root barks are taken orally either powdered or as infusions for treating various purposes highlighting epilepsy, insomnia, headache, inflammation, tuberculosis, malaria, diarrhea, abortion, infertility problems, venereal diseases (Ndou 2006; Moto et al. 2011; Mustapha 2013), and male erectile dysfunction as well as snakebites and used as arrow poison, natural pesticide to stored grains, and natural repellent (Ndou 2006). In some regions, the roots are used to treat people who are believed to be possessed by evil spirits (http://tropical.theferns.info/viewtropical.php?id=Securidaca+longipedunculata).

6) **Pharmacological studies:**

The methanolic extract of the roots and root barks of *S. longipedunculata* showed significantly high antimicrobial activity against pathogenic Gram + and Gram − bacteria and against multidrug-resistant bacteria (Junaid et al. 2008; Musa et al. 2013; Mongalo et al. 2015). Further studies are required in order to isolate the active compounds responsible for this antimicrobial activity, but the phytochemical screening of the extracts revealed that flavonoids, tannins, and saponins are present in the entire fractions, alkaloids were present in methanol and ethyl acetate in fractions only, and tannins are present in all the fractions except hexane fraction (Mongalo et al. 2015).

The aqueous extract of *S. longipedunculata* roots exhibited anticonvulsant, anxiolytic, and sedative activities in mice in a dose-dependent manner (Adeyemi et al. 2010; Moto et al. 2011), suggesting that the plant extract may be used in the management of convulsion and psychosis. The sedative properties could be related to the presence of some components in the extracts activating the benzodiazepine and/or GABA (gamma-aminobutyric acid neurotransmitter) sites in the $GABA_A$ receptor complex, and thus *S. longipedunculata* could be helpful in the treatment of insomnia and epilepsy in traditional medicine.

The aqueous root extract of *S. longipedunculata* was investigated for possible antinociceptive and central nervous system effects in mice using different assays, and results suggest that the extract possess antinociceptive and antidepressant-like effects with possible involvement of opioidergic pathways (Adebiyi et al. 2006). The tail-flick test is considered selective for opioid-like analgesic compound and the weak result of this test suggests that the extract of *S. longipedunculata* did not behave effectively as an opioid receptor agonist. It is possible that the extract presents a different central mechanism of action in this model (Adebiyi et al. 2006; Ojewole 2008). Concerning the exact mechanisms of anti-inflammatory effect of the aqueous root bark extract of *S. longipedunculata*, Ojewole (2008) suggests that, at least in part, it works by inhibiting the release, synthesis, and/or production of inflammatory mediators, including polypeptide kinins and prostaglandins, acting as the diclofenac.

On the other hand, the result of the forced swimming test (FST) clearly showed that the extract produced significant and dose-dependent antidepressant-like effect on the mice (Adebiyi et al. 2006). This effect was reversed by naloxone, an opioid antagonist, which might suggest the involvement of opioidergic pathway in the antidepressant action. Previous studies have shown that analgesia by antidepressant drugs can be inhibited by naloxone and enhanced by enkephalinase inhibitors, which are inhibitors of the enzymes that degrade endogenous enkephalin opioid peptides (Gray et al. 1998). Adebiyi et al. (2006) suggest that the extract might contain an antidepressant compound or the constituent of the extract may be simply behaving like an enkephalinase enzyme inhibitor. Once again further studies are needed to elucidate the chemical constituents responsible for the antidepressant effect of *S. longepedunculata* root extracts.

## 9.20 References


- Adebiyi RA, Elsa AT, Agaie BM, Etuk EU. Antinociceptive and antidepressant like effects of *Securidaca longepedunculata* root extract in mice. J Ethnopharmacol. 2006;107:234–9.
- Adeyemi O, Akindele AJ, Yemitan OK, Aigbe FR, Fagbo FI. Anticonvulsant, anxiolytic and sedative activities of the aqueous root extract of *Securidaca longepedunculata* Fresen. J Ethnopharmacol. 2010;130:191–5.
- Ajali U, Chukwurah BK. Antimicrobial activity of *Securidaca longipedunculata*. Phytomedicine. 2004;11:701–3.
- Catarino L, Martins E, Pinto-Basto MF, Diniz MA. Plantas vasculares e Briófitos da Guiné-Bissau. Lisboa: IICT/IPAD; 2006. 340 p.
- de Koning J. Checklist of vernacular plant names in Mozambique. Wageningen Agricultural University Papers 93. 1993;2.
- Exell AW. Polygalaceae. In: A.W. Exell & H. Wild, Flora Zambesiaca 1(part 1); 1960. p. 303–36.
- Gray AM, Spencer PS, Sewell RD. The involvement of the opioidergic system in the antinociceptive mechanism of action of antidepressant compounds. Br J Pharmacol. 1998;124:669–74. In: Adebiyi et al. 2006.
- Jayasekara TK, Stevenson PC, Hall DR, Belmain SR. Effect of volatile constituents from *Securidaca longepedunculata* on insect pests of stored grain. J Chem. Ecol. 2005;31:303–13.
- Junaid SA, Abubakar A, Ofodile AC, Olabode AO, Echeonwu GON, Okwori AJ, Adetunji JA. Evaluation of *Securidaca longipenduculata* leaf and root extracts for antimicrobial activities. Afr J Microbiol Res 2008;2:322–5.
- Mongalo NI, McGaw LJ, Finnie JF, van Staden J. *Securidaca longipedunculata* Fresen (Polygalaceae): a review of its ethnomedicinal uses, phytochemistry, pharmacological properties and toxicology. J Ethnopharmacol. 2015. http://dx.doi.org/10.1016/j.jep.2015.02.041i.
- Moto FCO, Mbafor JT, NgoBum E, Kouemou N, Kandeda AK, Talla E, Dimo T, Rakotonirira SV. Evaluation of the sedative and anticonvulsant properties of three Cameroonian plants. Afr J Tradit Complement Altern Med. 2011;8: 181–90.

- Musa AA, Oyewale AO, Ndukwe IG, Yakubu SE, Abdullahi MS. Phytochemical screening and antimicrobial activity of solvent fractions of *Securidaca longepedunculata* (Fresen) root bark methanol extract. J Chem Pharmaceut Res. 2013; 5:28–33.
- Mustapha AA. Ethno-medico-botanical uses of *Securidaca longepedunculata* Fresen (family-Polygalaceae) from Keffi Local Government, Nasarawa State, Nigeria. J Nat Remedies. 2013;13:133–7. www.jnronline.com.
- Nébié RHC, Yaméogo R, Bélanger A, Sib FS. Salicylate de méthyle, constituant unique de l'huile essentielle de l'écorce des racines de *Securidaca longepedunculata* du Burkina Faso. Comptes Rendues Chimie. 2004;7:1003–6.
- Ndou AP. Walter Sisulu National Botanical Garden. 2006. In http://www.plantzafrica.com/plantqrs/securidlong.htm. Accessed Jan 2015.
- Ojewole JAO. Analgesic, Anti-inflammatory and hypoglycaemic effects of *Securidaca longepedunculata* (Fresen.) [Polygalaceae] root-bark aqueous extract. Inflammopharmacology. 2008;16:174–81.


## 9.21 *Xysmalobium undulatum* (L.) W.T. Aiton

### *9.21.1 (Apocynaceae)*

1) ***Synonyms***:

   *Asclepias undulata* L.

2) ***Common names***:

   Angola: otyivatu (lunyaneka).

   Milk bush, uzara, wild cotton, wave-leaved xysmalobium (Eng.).

3) ***Photograph of the species (Fig. 9.10)***
4) ***Description of the species***:

   A) **Habitat, world distribution, and PALOP distribution:**

   *Xysmalobium* comprises 40–45 species known from tropical Africa and South Africa. *X. undulatum* is common in grassland areas and seasonally wet localities, from sea level up to 2000 m of altitude (Schmelzer 2011). It is found in Angola and Mozambique.

   B) **Morphology of the species:**

   Robust perennial herb with white latex and a thick tuberous rootstock from which grow one to several annual stout unbranched stems with spreading white hairs. Leaves opposite, hispid, with a very short petiole; lamina very variable in shape, linear to ovate, 8–20 × 0.5–8.5 cm, acute or acuminate at apex, rounded, truncate or cordate at base, undulate or crispate and often revolute at the scabrid margins; stipules absent. Flowers 12–26 in pubescent umbelliform inflorescences arising laterally from the axils of the upper leaves; peduncles robust 1.2–4.5 cm long; bracts linear 3–4 mm long; pedicels slender, 14–18 mm long; sepals 5, ovate or lanceolate, 4–8 mm long, pubescent; corolla greenish or yellowish, often tinged with brown or

**Fig. 9.10** *Xysmalobium undulatum* (L.) W.T. Aiton, from Angola (LISC Herbarium specimen, Mendes 2382)

pink outside divided almost to the base into 5 ovate lobes 5–10 × 2.5–6 mm, concave, acute and recurved at the apex, glabrous save in the upper ⅓ usually lanate or rarely glabrous or subglabrous. Corona very fleshy, without central cavity. Fruit a single follicle, 5–8 cm long and 1.5–2.5 cm wide, ovoid, often inflated, densely covered with soft hairy prickles; seeds flattened, ovate in outline with a tuft of long silky hairs (Goyder 2012).

C) **Conservation status:**

Not listed (African Herbal Pharmacopoeia 2010), but in some African regions, *X. undulatum* is officially protected because of overharvesting and habitat destruction. Tubers have been collected from the wild for local uses but export to the European market also happened. The tubers must be harvested at an early stage, as the older are thickened and deep-rooting. Seedlings should be transplanted by that time. Commercial plantations are currently made in small scale in South Africa (African Herbal Pharmacopoeia 2010; Schmelzer 2011).

D) **Medicinal parts:**

Tuberous roots.

E) **Chemical composition:**

The roots contain a mixture of **cardiac glycosides** (cardenolides) called uzarone or xysmalobin. The main compounds are uzarin (5.6 %), xysmalorin (1.5 %), and their isomers allo-uzarin (0.4 %) and allo-xysmalorin (0.1 %). Minor compounds include uzarigenin, ascleposide, coroglaucigenin, coroglaucigenin-3-*O*-β-glucoside, xysmalogenin, and pachygenol. From the seeds, the glycoside frugoside (0.5 %) was isolated (African Herbal Pharmacopoeia 2010; Schmelzer 2011).

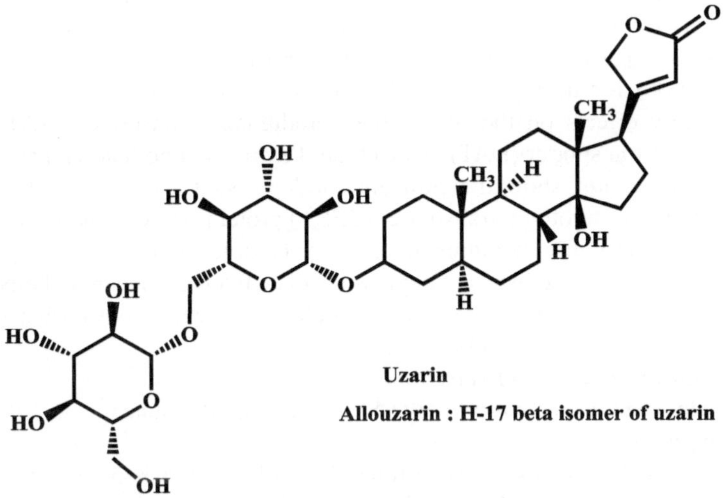

**Uzarin**

**Allouzarin : H-17 beta isomer of uzarin**

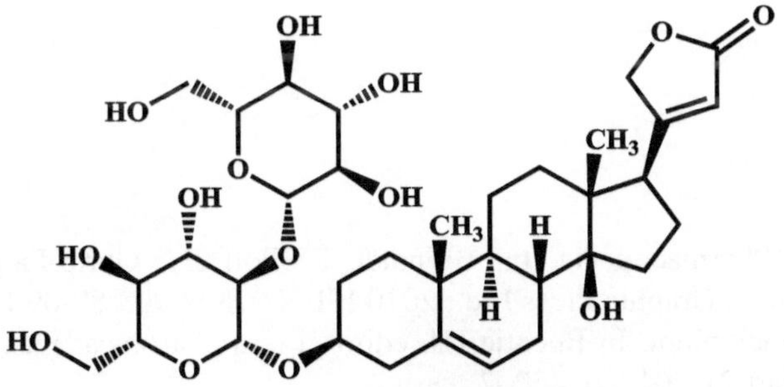

**Xysmalorin**

**Alloxysmalorin: H-17 beta isomer of xysmalorin**

5) ***Major ethnomedicinal uses*:**

The tubers of *X. undulatum* have traditional uses against acute diarrhea. A decoction or maceration of the pulverized tubercle is used against venereal diseases (African Herbal Pharmacopoeia 2010; Schmelzer 2011). The tubercle is also used as aphrodisiac and against stomach pain. The latexes are furthermore applied against snakebites (Schmelzer 2011).

6) ***Pharmacological studies*:**

The cardiac glycosides showed antidiarrheal and antispasmodic activity (African Herbal Pharmacopoeia 2010).

Several neurotransmitters are believed to be involved in the depression physiology, including serotonin, noradrenalin, and dopamine (Bang-Andersen 2006). It is assumed that depression is due to deficiency of one or another of these neurotransmitters (Walsh and Schwartz-Bloom 2005), although many other factors are believed to be involved. *X. undulatum* was screened, with other plant species as well, as a potential plant used for the treatment of depression, and ethanolic extracts from tubers and leaves showed high affinity to the serotonin reuptake transporter (SERT) (Nielsen et al. 2004). Other in vitro investigations (Pedersen et al. 2008), for affinity to the SERT in the [$^3$H]-citalopram-binding assay and for inhibitory effects on the SERT, the noradrenalin transporter (NAT), and the dopamine transporter (DAT) using ethanol extracts of both aerial parts and roots of *X. undulatum*, showed significant antidepressant activity by exhibiting high affinity to the serotonin transporter (SERT) protein but was unable to inhibit the uptake of serotonin, noradrenaline, and dopamine. In a further assay, the same ethanolic extracts showed antidepressant-like effects in one animal model (FST). This action might be attributed to xysmalorin and uzarin but further studies are needed (Pedersen et al. 2008).

7) ***Precautions and side effects*:**

Safe when used appropriately (African Herbal Pharmacopoeia 2010).

8) ***Dosage*:**

1g of herb contains 75 mg of total glycosides. Adults, 45–90 mg of total glycosides; children, 15–30 mg of total glycosides (African Herbal Pharmacopoeia 2010).

## 9.22 References


- African Herbal Pharmacopoeia. In: Brendler T, Eloff JN, Gurib-Faquim A, Phillips LD, editors. Graphic Press Ltd.; 2010 ISBN 978-99903-89-09-8.
- Goyder D. Xysmalobium. In Beentje HJ, editor. Flora of tropical East Africa. Apocynaceae (Part 2); 2012. p. 450–7.
- Moltzen EK, Bang-Andersen B. Serotonin reuptake inhibitors: the corner stone in treatment of depression for half a century—a medicinal chemistry survey. Curr Top Med Chem. 2006;6 1801–23.

- Nielsen N, Sandager M, Stafford GI, van Staden J, Jäger A. Screening of indigenous plants from South Africa for affinity to the serotonin reuptake transport protein. J Ethnopharmacol. 2004;94:159–63.
- Pedersen ME, Szewczyk B, Stachowicz K, Wieronska J, Andersen J, Stafford GI, van Staden J, Pilc A, Jäger KA. Effects of South African traditional medicine in animal models for depression. J Ethnopharmacol. 2008;119:542–8.
- Schmelzer GH. Xysmalobium undulatum (L.) W.T. Aiton. In: Schmelzer GH, Gurib-Fakim A, editors. Prota 11(2): Medicinal plants/plantes médicinales 2. [CD-Rom]. Wageningen, The Netherlands: PROTA; 2011. In http://database.prota.org/PROTAhtml/Xysmalobium%20undulatum En.htm (Accessed on January 2015).
- Stafford G, Pedersen M, van Staden J, Jäger A. Review on plants with CNS-effects used in traditional South African medicine against mental diseases. J Ethnopharmacol. 2008;119:513–37.
- Walsh CT, Schwartz-Bloom RD. Levine's pharmacology—drug actions and reactions. 7th ed. London: Taylor and Francis; 2005. ISBN 1-84214-255-0.

# Chapter 10
# Oceania: Antidepressant Medicinal Plants

**Ian Edwin Cock and Matthew J. Cheesman**

**Abstract** Despite having the smallest land mass of the World Wildlife Fund (WWF)-declared ecozones, Oceania is amongst the most diverse floral regions of the world. Geographically, the region consists of Australia and New Guinea as the largest land masses, as well as the islands of Melanesia, Micronesia and Polynesia. Due to the island nature of the region, the flora has developed in isolation in various climatic conditions within the region, resulting in an extremely high degree of endemism. Furthermore, the harsh climatic conditions in some regions have resulted in a wealth of unique phytochemicals not found in plants from other regions globally. Coupled with possibly the world's oldest continuous human inhabitation on the Australian mainland and a diversity of cultures in other Oceania regions, this has led to complex and sophisticated ethnopharmacological systems. Medicinal plants with unique properties have long been recognised by indigenous Oceania populations, and this lore has been passed from generation to generation. Whilst often not well recorded, there is a wealth of knowledge of the medicinal value of the regions' floral species for all types of therapeutic purposes. This chapter focuses on the plants of the region with known antidepressant uses and/or those plants which have phytochemistry consistent with antidepressant properties. This is by no means an exhaustive list, but instead serves to highlight some of the best known examples (e.g. kava-kava) and discuss examples of plants with established antidepressant mechanisms. For example, whilst we discuss the calmative properties of the Australian plant *Backhousia citriodora*, many other aromatic plants with similar essential oil components and thus similar therapeutic properties exist in the region and are not discussed here for the sake of brevity. Furthermore, despite the high degree of endemism of Oceania flora, several well-known species (e.g. *Areca catechu* L. and *Nelumbo nucifera* Gaertn.) have wide geographic ranges. Indeed, whilst

I.E. Cock (✉)
Environmental Futures Research Institute, Griffith University,
Nathan Campus, 170 Kessels Road, Nathan, QLD 4111, Australia

School of Natural Sciences, Griffith University,
Nathan Campus, 170 Kessels Road, Nathan, QLD 4111, Australia
e-mail: i.cock@griffith.edu.au

M.J. Cheesman
School of Biomedical Sciences, University of Queensland, St Lucia, QLD 4072, Australia
e-mail: m.cheesman@uq.edu.au



C. Grosso (ed.), *Herbal Medicine in Depression*,
DOI 10.1007/978-3-319-14021-6_10

native to Oceania, *A. catechu* is better known as a component of the pharmacopoeias of other regions (e.g. India). However, these species also make an important contribution to Oceanic antidepressant medicinal plants and are therefore discussed in this chapter.

**Keywords** Antidepressant • Australian plants • Melanesian plants • Micronesian plants • Polynesian plants • Complementary therapies

## 10.1 Geopolitical/Cultural Context

Oceania is a large region generally regarded as stretching from the Straits of Malacca to the west coast of the Americas. It is described as all the lands of the Pacific Ocean and comprises four main regions: Australia, Melanesia, Micronesia and Polynesia (Fig. 10.1). The Australian continent accounts for the majority of the land mass, consisting of approximately 7,686,850 km$^2$ (86 % of the total Oceania land area). New Guinea (consisting of Papua, West Papua and Papua New Guinea) accounts for a further 10 % of the total land mass, with the remainder of the land spread between the remaining 27 nations. Australia also contributes the majority of the population (approximately 23,000,000 or 56 % of the total Oceania population). East Timor (1,143,667; 3 %), Hawaii (1,360,301; 3 %), Maluku Islands (1,895,000; 5 %),

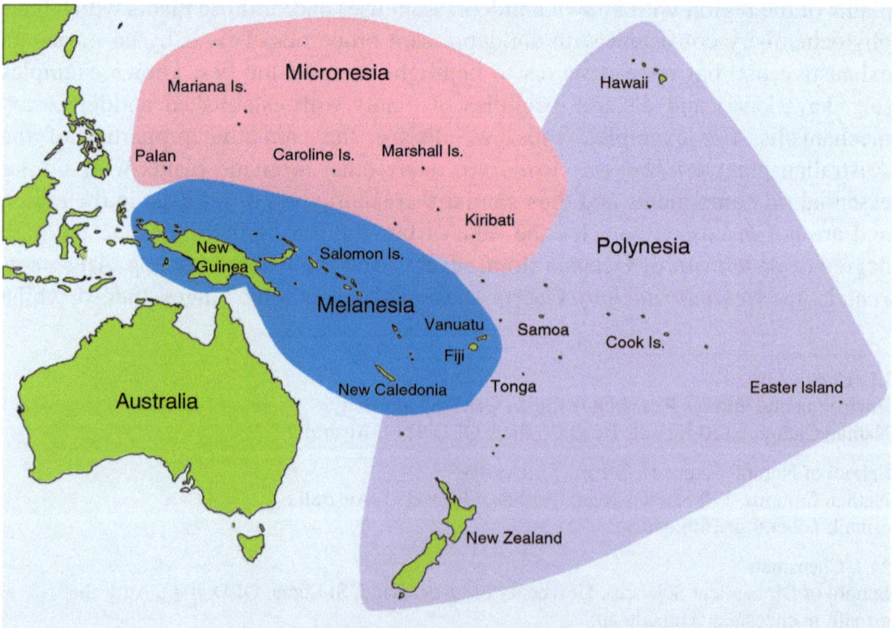


**Fig. 10.1** The Oceania region, highlighting Australia, Melanesia, Micronesia and Polynesia

New Zealand (4,465,900; 11 %), Papua (8 %) and Papua New Guinea (5,172,033; 13 %) also have significant populations, with lower populations in other regions (CIA World Fact Book 2015; United States Department of State 2015).

The region is culturally diverse, with a wide variety of traditional cultural groupings, as well as more recently arrived settlers. The region has some of the oldest continuous cultures in the world. Some studies estimate that the Australian Aborigines have lived continuously in Australia for as long as 80,000 years, making them possibly the oldest continuous culture in the world (Bowler et al. 2003). With such a long history, it is perhaps not surprising that the Aborigines developed a sophisticated and effective phyto-medical system and were able to treat most illnesses they encountered before European arrival (Cock 2011). At the other extreme, the Maori of New Zealand are a much more recent culture, arriving in New Zealand from other regions of Polynesia as recently as 700–800 years ago (McCormick 1939). Despite their recent history, the Maori people have also developed a plant-based medicinal system. It is also likely that they have brought traditional knowledge of similar plants with them when they arrived in New Zealand.

The cultural diversity of the region also accounts for the varied usage patterns of some therapeutic plants. *Piper methysticum* G. Forst (kava-kava) is now recognised for its sedative properties and is used therapeutically internationally (Cock 2015). However, in various Melanesian and Polynesian cultures, the consumption of kava-kava is social or ceremonial and may be drunk at formal gatherings and as a welcoming for visitors to the village. Whilst these ceremonies may have originated from a traditional understanding of the therapeutic properties of this plant, they have developed significance beyond medicinal for these cultures. Similarly, Australian Aborigine smoking ceremonies involving the burning of specific native plants to produce smoke were originally thought to be purely ceremonial. However, recent studies have demonstrated that the heating process may instead be required to create biologically active compounds (Sandgrove et al. 2014). Thus, whilst these practices took on a ceremonial aspect, it is possible that they originated from a traditional understanding of the medicinal properties of the local flora.

The Oceania region has diverse climatic and environmental conditions. The Australian continent ranges from arid/desert environments to tropical and subtropical regions with high rainfall. There are also significant temperate and equatorial regions. Much of Melanesia and Micronesia is also subtropical/tropical or equatorial. Polynesia has wide environmental diversity, with much of the region also being subtropical/tropical or equatorial. Much of New Zealand and other southern regions of Polynesia have a temperate climate, with several alpine regions. Such climatic diversity has resulted in high floral biodiversity. In many regions with harsh climates (e.g. the hot, arid inland regions of Australia), plant species have developed to survive in these environments, resulting in plants with unique secondary metabolite components. Furthermore, the isolation of the region from other regions has resulted in a high degree of endemism in floral species.

Local traditional medicine systems have developed for both physical and psychological complaints. As with other regions worldwide, depression is a significant medical issue in Oceania. Nowadays, much of the population relies on allopathic pharmaceuticals for treatment. There has recently been a revival in interest in plant-based remedies. However, due to the lack of written records of some Oceania cultural groups and the greater knowledge of plant species from other regions, much of the plant-based treatments rely on the better known treatments from Europe and Asia (e.g. St John's wort) (Cock 2015). Whilst usage of these remedies is relatively common in Oceania, they are exotic to the area and will not be considered in this chapter. A number of species that are either indigenous or endemic to Oceania are examined here. Some of those species (e.g. *Piper methysticum*) have well established antidepressant uses. Other species have received less study and are included because either they have effects which are likely to be therapeutic in individuals suffering from depression or they contain phytochemical components with established antidepressant properties.

## 10.2 Search Methods

Information was sourced using a variety of search engines including Google Scholar, PubMed and Scopus.

## 10.3 References


- Bowler JM, Johnston H, Olley JM, Prescott JR, Roberts RG, Shawcross W, Spooner NA. New ages for human occupation and climatic change at Lake Mungo, Australia. Nature 2003;421:837–40.
- CIA World Fact Book, Oceania (https://www.cia.gov/library/publications/the-world-factbook/wfbExt/region_aus.html). Accessed 10 Mar 2015.
- Cock IE. The safe usage of herbal medicines: counterindications, cross-reactivity and toxicity. Pharmacognosy Commun. 2015;5(1):2–50.
- Cock IE. Medicinal and aromatic plants—Australia. In: Ethnopharmacology, Encyclopedia of Life Support Systems (EOLSS). Developed under the auspices of UNESCO. Oxford, UK: EOLSS Publishers; 2011. http://www.eolss.net.
- McCormick EM, editor. The Maori. In Making of New Zealand. 1939;1(2): 8–9.
- Sandgrove N, Jones GL, Greatrex BW. Isolation and characterisation of (-)-genifuranal: The principal antimicrobial component in traditional smoking applications of *Eremophila longifolia* (Scrophulariaceae) by Australian aboriginal peoples. J Ethnopharmacol. 2014;154:758–66.
- United States Department of State. http://www.state.gov/misc/list/. Accessed 10 Mar 2015.

# 10.4 *Areca catechu* L.

## *10.4.1 (Arecaceae)*

1) **Synonyms:**
   *Areca faufel* Gaertn., *Areca hortensis* Lour., *Areca himalayana* Griff., *Areca nigra* Giseke., *Areca macrocarpa* Becc.
2) **Common names:**
   Betel nut, areca palm, betel palm, Indian nut, Pinang nut
3) **Photograph of the species** (Fig. 10.2).
4) **Description of the species:**

   A) **Habitat and world distribution:**
   *A. catechu* grows throughout much of Pacific Asia and East Africa. It is believed to have originated in the Philippines but has been naturalised widely in many areas, including New Guinea where it is widely grown.

   B) **Morphology of the species:**
   *A. catechu* is a single, slender trunk palm growing to 20 m in height and 20–30 cm in diameter. The palmate leaves are at 2–2.5 m long with broad leaflets and jagged tips. The flowers are white and sweet scented and occur clustered on a spike. These develop into orange-yellow fruit (betel nuts) approximately 2.5 cm in diameter with a fleshy pericarp and a fibrous mesocarp (Heatubun et al. 2012).

**Fig. 10.2** *A. catechu* tree and fruit. Photograph taken in Chiang Mai, Thailand in 2011 by Dr. Ian Cock

C) **Medicinal parts:**

The fleshy portion of the *A. catechu* nut is used for its stimulant effects. The pericarp is removed and chewed to induce euphoria and heightened alertness. It is usually administered as a preparation known as Paan, whereby the fleshy part of the nut is wrapped in a leaf (typically *Piper betle*), and often includes lime paste to bind the leaf and nut together (Franke et al. 2015).

D) **Chemical constituents:**

The *A. catechu* nut contains several **alkaloids** of which arecaidine and arecoline are considered to be responsible for the stimulant and intoxicating effects (Holdsworth et al. 1998; Amirkia and Heinrich 2014). The seed also contains **condensed tannins** called arecatannins (Yang et al. 2012).

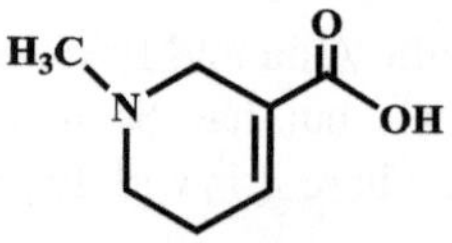


Arecaidine

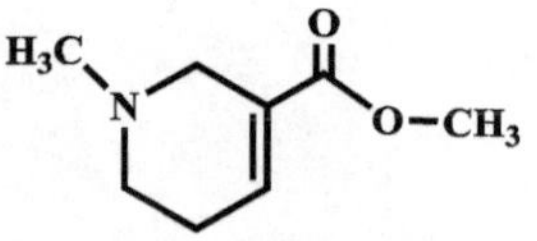


Arecoline

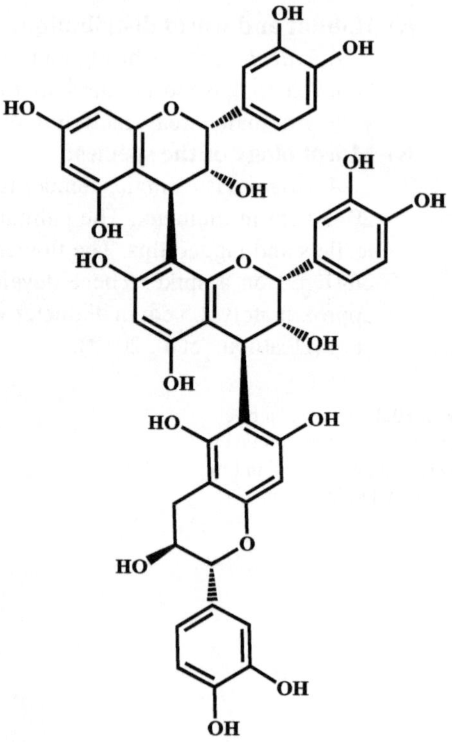


Arecatannin B1

5) **Pharmacology and bioactivities:**

A) **In vitro:**

Arecaidine and arecoline are believed to be the main bioactive compounds (Chu 2001). The pharmacology of arecoline has been more thoroughly investigated for its stimulatory effects. Arecoline is known to be a partial agonist of muscarinic acetylcholine receptors (Li et al. 2010) leading to parasympathetic effects (pupil constriction, bronchial constriction, etc.). It may also be an agonist of nicotinic acetylcholine receptors leading to its varied psychoactive properties. Whilst arecaidine is less well studied, its structural similarities indicate that it may have similar pharmacological properties.

B) **In vivo:**

Due to its muscarinic and nicotinic agonist effects, clinical studies have been conducted using these compounds. Healthy volunteers show increased mental alertness whilst intoxicated with betel quid although no improvements in cognitive performances were observed (Wyatt 1995; Osborne et al. 2011).

6) **Indications and usage:**

A) **Stimulatory activity:**

The chemical constituents of *A. catechu* include alkaloids that act as CNS stimulants and thus provide treatment options for the clinical manifestations of depression (Xiao et al. 2005). Whilst the precise mechanism(s) via which these alkaloids cause antidepressive outcomes in humans remains unknown, in rats it has been demonstrated that the ethanolic extracts of *A. catechu* can increase hippocampal serotonin and noradrenaline levels (Abbas et al. 2013) which may explain the mechanism of action of the *A. catechu* constituents.

B) **Alzheimer's disease:**

A definitive feature of Alzheimer's disease is a declining cognitive function. Due to the demonstrated clinical enhancements caused by *A. catechu* alkaloids, the compounds are now indicated as possible treatment regime for patients suffering from the disease. However, due to the availability of other more effective drugs, *A. catechu* compounds do not presently comprise first choice treatments.

7) **Precautions and side effects:**

*A. catechu* contains arecatannins which have been reported to lead to the carcinogenic effects arising from consumption of the nut (Jeng et al. 2001). A recent study has shown that chewing the nuts (or Paan) significantly increases the incidence of precancerous oral fibrosis (Merchant and Pitiphat 2015). Furthermore, consumption of tannins is believed to induce depression and/or exacerbate the illness, thus partially antagonising the positive therapeutic effects of the alkaloids.

8) **Dosage:**

*A. catechu* is generally consumed until the desired effect is reached, and thus dosages will vary between individuals. However, levels should be closely monitored as the alkaloids may reach toxic levels. For example, arecoline has been found to possess an $LD_{50}$ of approximately 100 mg/kg in mice when administered subcutaneously (Selvan et al. 1989).

## 10.5 References


- Abbas G, Naqvi S, Erum S, Ahmed S, Rahman AU, Dar A. Potential antidepressant activity of *Areca catechu* nut via elevation of serotonin and noradrenaline in the hippocampus of rats. Phytother Res. 2013;27:39–45.
- Amirkia V, Heinrich M. Alkaloids as drug leads—a predictive structural and biodiversity-based analysis. Planta Med. 2014;xlviii–liii.
- Chu N. Effects of betel chewing on the central and autonomic nervous systems. J Biomed Sci. 2001;8:229–36.
- Franke AA, Mendez AJ, Lai JF, Arat-Cabading C, Li X, Custer LJ. Composition of betel specific chemicals in saliva during betel chewing for the identification of biomarkers. Food Chem Toxicol. 2015. doi: 10.1016/j.fct.2015.03.012.
- Heatubun C, Dransfield J, Flynn T, Tjitrosoedirdjo S, Mogea J, Baker W. A monograph of the betel nut palms (Areca: Arecaceae) of East Malesia. Bot J Linn Soc. 2012;168:147–73.
- Holdsworth D, Jones R, Self R. Volatile alkaloids from *Areca catechu*. Phytochemistry. 1998;48:581–2.
- Jeng J, Chang M, Hahn L. Role of areca nut in betel quid-associated chemical carcinogenesis: current awareness and future perspectives. Oral Oncol. 2001;37: 477–92.
- Li C, Yang X, Tang W, Liu C, Xie, D. Arecoline excites the contraction of distal colonic smooth muscle strips in rats via the M3 receptor–extracellular Ca2+ influx—Ca2+ store release pathway. Can J Physiol Pharmacol. 2010;88: 439–47.
- Merchant A, Pitiphat W. Total, direct, and indirect effects of paan on oral cancer. Cancer Causes Control. 2015;26:487–91.
- Osborne P, Chou T, Shen T. Characterization of the psychological, physiological and EEG profile of acute betel quid intoxication in naïve subjects. PLoS One. 2011;6:e23874.
- Selvan RS, Venkateswaran KS, Rao AR. Influence of arecoline on immune system: I. Short term effects on general parameters and on the adrenal and lymphoid organs. Immunopharmacol Immunotoxicol. 1989;11:347–77.
- Wyatt T. Betel nut chewing and selected psychophysiological variables. Psychol Rep. 1996;79:451–63.
- Xiao JS, Zhang JJ, Huang CY, Huang HJ. Clinical trial of *Areca catechu* treatment of depression after stroke. J Math Med. 2005;18:444–5 (in Chinese).
- Yang WQ, Wang HC, Wang WJ, Wang Y, Zhang XQ, Ye WC. Chemical constituents from the fruits of *Areca catechu*. J Chin Med Mater. 2012;35:400–2 (Article in Chinese).

## 10.6 *Backhousia citriodora* F. Muell

### *10.6.1 (Myrtaceae)*

1) **Synonyms:**
   *B. citriodora* is distinct from, but sometimes confused with, *Eucalyptus staigeriana* F. Muell. ex Bailey (lemon ironbark)
2) **Common names:**
   Lemon myrtle, lemon-scented myrtle, sweet verbena tree, lemon-scented ironwood
3) **Photograph of the species** (Fig. 10.3).
4) **Description of the species:**

   A) **Habitat and world distribution:**
      *B. citriodora* is endemic to subtropical coastal rainforest regions of southern to central Queensland, although it can be cultivated in temperate climates and may also grow in colder regions if it is protected from frost when becoming established (Page and Olds 2004).
   B) **Morphology of the species:**
      *B. citriodora* is a large evergreen shrub to medium-sized tree that varies between 3 and 20 m in height. The aromatic leaves are lanceolate (5–12 cm long, 1.5–2.5 cm wide) and glossy green, with an entire margin. In summer,

**Fig. 10.3** *Backhousia citriodora* with flower clusters. Photograph was taken in 2009 in Brisbane, Australia, by Dr. Ian Cock

the tree becomes covered in clusters of small creamy-white flowers (5–7 mm in diameter). The calyx is persistent following petal fall (Page and Olds 2004). These develop into small dry indehiscent fruit which splits into two chambers as it ripens.

C) **Medicinal parts:**

Plantation trees are harvested by trimming. All aerial parts of *B. citriodora* are used for essential oil production. The dried leaf is also used as a culinary herb/functional food.

D) **Chemical constituents:**

*B. citriodora* is reputed to have the highest citral purity of any essential oil, typically containing 90–98 %, compared to *Citrus limonum* (lemon) which typically contains 3–10 % citral (Cock 2011). Citral is the collective name used to describe the isomeric aldehydes **neral** (α-citral) and **geranial** (β-citral). The oil also contains lesser, but still significant, amounts of **myrcene, linalool** and **cyclocitral** (Cock 2011).

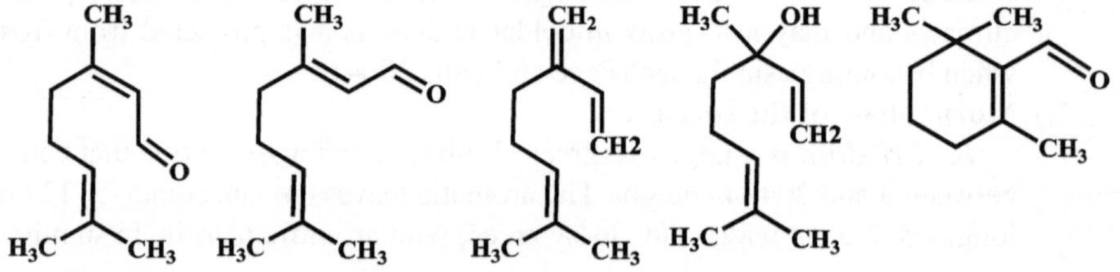


**Neral** **Geranial** **Myrcene** **Linalool** **Cyclocitral**

5) **Pharmacology and bioactivities:**

A) **In vitro:**

Due to its high citral content, *B. citriodora* has good antimicrobial properties. Numerous studies have reported the inhibitory effect of *B. citriodora* essential oil (Ryan et al. 2000; Wilkinson et al. 2003) and extracts (Cock 2008, 2013) against a wide variety of infective bacteria and fungi.

B) **In vivo:**

Anecdotal evidence from aroma therapists indicates that *B. citriodora* oil is effective in inducing relaxation, improving concentration, uplifting the individual emotionally and inducing soothing sleep (Webb 2000).

6) **Indications and usage:**

A) **Hypnotic and sedative:**

The oil is useful as a carmitive and sedative (Cock 2011). It may be either burnt as aromatherapy or applied and rubbed into the skin (Hayes and Markovic 2003). *B. citriodora* infusions are also reputed to have carmitive effects without adversely affecting concentration (Cock 2011).

B) **Antiseptic properties:**

*B. citriodora* has broad spectrum antibacterial (Cock 2008, 2013; Ryan et al. 2000; Wilkinson et al. 2003), anti-fungal (Wilkinson et al. 2003) and anti-viral activities (Burke et al. 2004; Webb 2000) and is therefore indicated for cold, influenza, bronchitis, indigestion and other GIT disorders, herpes and a wide variety of bacterial and fungal skin disorders (Cock 2011; Webb 2000).

7) **Precautions and side effects:**

- Avoid application on any inflammatory or allergic skin condition.
- Avoid undiluted application. The oil should be diluted with vegetable oil or other appropriate carrier.
- Avoid application on open or damaged skin.
- Where sensitive skin is expected, a skin patch test should initially be performed.

Due to its high citral content, the essential oil may cause skin irritation. There is no available data on skin sensitisation to citral so similar precautions to applying any aldehyde to the skin apply.

8) **Dosage:**

A) **Internal use:**

*B. citriodora* may be ingested, both as a food flavouring and as a functional food. A moderate amount of the dried leaf may be added to the food to the individual's taste. Alternatively, an infusion of the leaves is now commercially available and is becoming increasingly popular. If the oil is used for ingestion instead of the dried herb, one or two drops are sufficient due to the high citral content.

B) **External use:**

*B. citriodora* essential oil should not be applied undiluted to the skin (except directly on warts if their removal is the aim). The oil should be diluted to a 1 in 5 dilution (i.e. 1 part *B. citriodora* essential oil plus 4 parts carrier oil) for topical application. For aromatherapy and cleansing the environment of microbes, the oil may be burnt undiluted.

## 10.7 References


- Burke BE, Baillie JE, Olson RD. Essential oil of Australian lemon myrtle (*Backhousia citriodora*) in the treatment of molluscum contagiosum in children. Biomed Pharmacother. 2004;58(4):245–7.
- Cock IE. Antimicrobial activity of *Backhousia citriodora* (lemon myrtle) methanolic extracts. Pharmacognosy Commun. 2013;3(2):58–63.
- Cock IE. Medicinal and aromatic plants—Australia. In Ethnopharmacology, Encyclopedia of Life Support Systems (EOLSS). Developed under the auspices of UNESCO. Oxford, UK: EOLSS Publishers; 2011. http://www.eolss.net.

- Cock IE. Antibacterial activity of selected Australian native plant extracts. Internet J Microbiol. 2008;4(2).
- Hayes AJ, Markovic, B. Toxicity of Australian essential oil *Backhousia citriodora* (lemon myrtle). Part 2. Absorption and histopathology following application to human skin. Food Chem Toxicol. 2003;41(10):1409–16.
- Page S, Olds M, editors. Botanica: The illustrated A-Z of over 10,000 garden plants for Australian gardens and how to cultivate them. Australia: Random House; 2004.
- Ryan T, Cavanagh HMA, Wilkinson JM. Antimicrobial activity of *Backhousia citriodora* oil. Simply Essential. 2000;38:6–8.
- Webb MA. Bush sense. Australian essential oils and aromatic compounds. Adelaide, Australia: Griffin Press; 2000.
- Wilkinson JM, Hipwell M, Ryan T, Cavanagh HMA. Bioactivity of *Backhousia citriodora*: antibacterial and antifungal activity. J Agric Food Chem. 2003; 51(1):76–81.


## 10.8 *Duboisia hopwoodii* (F. Muell.) F. Muell

### *10.8.1 (Solanaceae)*

1) **Synonyms:**

   *Anthocercis hopwoodii* F. Muell

2) **Common names:**

   Pituri, pitchuri, pitcheri thorn apple, emu plant, camel poison, poison bush

3) **Photograph of the species** (Fig. 10.4).
4) **Description of the species:**

   A) **Habitat and world distribution:**

      *Duboisia hopwoodii* is native to the arid interior of Australia. It grows in deep sandy soils, commonly on sand dunes and sand plains. It may also grow in mallee or *Acacia* woodlands and is often found growing with *Tridia* spp., *Codonocarpus* spp. and *Callitris* spp. (Lassak and McCarthy 2011; Page and Olds 2004).

   B) **Morphology of the species:**

      *D. hopwoodii* is a medium-sized shrub which grows to 1–3 m tall and up to 3 m in diameter. It has smooth dark brown bark on young plants and newer growth, which becomes yellow brown and rough textured on older growth. The freshly cut heartwood is lemon yellow in colour, with a distinct vanilla aroma. Leaves are alternate, slightly fleshy and variable in size (usually 2–13 cm in length). They are generally narrow linear with straight margins tapered at both ends. White bell-shaped flowers occur in open groups at the end of branchlets. The globose berry (up to 6 mm in diameter) contains one or two seeds and is black when ripe (Lassak and McCarthy 2011; Page and Olds 2004).

**Fig. 10.4** *Duboisia hopwoodii* leaves, flowers and berries. Photograph was taken in central Northern Territory, Australia by Dr Ian Cock

C) **Medicinal parts:**

The leaves, flowers and twigs are used by Australian Aborigines as a tobacco substitute (Cock 2011). The plant material is dried, powdered and mixed with the ash of specific *Acacia* spp. (*A aneura, A. calcicola, A. coriacea, A. eutrophiolata, A. ligulata, A. pruinocarpa, A. beauverdiana*) and rolled into pituri balls which are generally chewed (or sometimes smoked).

D) **Chemical constituents:**

*D. hopwoodii* has been reported to contain significant levels of the **alkaloids** nicotine and nornicotine (Luanratana and Griffin 1982). Indeed, the total alkaloid content of fresh plant material has been reported to be as high as 25 % of dry weight.

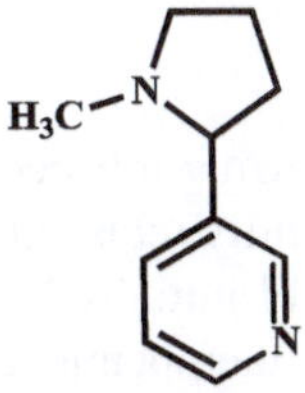


**Nicotine**

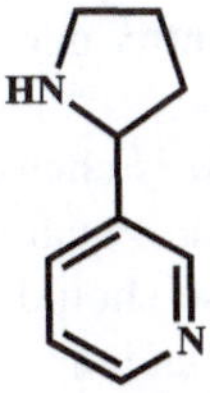


**Nornicotine**

5) **Pharmacology and bioactivities:**

A) **In vitro:**

Nicotine and nornicotine are nicotinic acetylcholine receptor (nAChR) agonists for most receptors (except the nAChRα9 receptors, where they act as antagonists) (Celie et al. 2004). Thus, nicotine can act as both a central nervous system (CNS) stimulant and a relaxant, which is believed to be responsible for its mood altering effects.

B) **In vivo:**

The nicotine is liberated from pituri balls by mouth acids and other alkaloids present in the ash. The ash also promotes rapid absorption of the nicotine through the mouth into the bloodstream (Latz 1995). Initially, the chewed *D. hopwoodii* mixture acts as a central nervous system stimulant, and early records indicate that it was used to excite participants before combat (Maiden 1889). Later, the effect is reversed and the user feels fatigued and sleepy. In small quantities, it may also act as an appetite suppressor. When smoked, *D. hopwoodii* leaves act as an anaesthetic and were used by Australian Aborigines in circumcision ceremonies.

6) **Indications and usage:**

A) **Stimulant:**

Due to the nicotine and nornicotine components of *D. hopwoodii,* it is useful as a stimulant in low doses (a couple of mg) (Lassak and McCarthy 2011; Cock 2011). It significantly enhances cognitive effects including fine motor skills, attention and memory.

B) **Anaesthetic:**

Nicotine and nornicotine both have analgesic effects (Lassak and McCarthy 2011; Cock 2011) and may therefore be useful in suppressing pain. However, as less toxic and/or addictive analgesic drugs are available, there is limited scope in modern medicine for the use of *D. hopwoodii* leaves for this purpose.

7) **Precautions and side effects:**

The $LD_{50}$ of nicotine is low, and 30–60 mg can be lethal in humans (Mayer 2014). Therefore, consumption of *D. hopwoodii,* as with any nicotine-containing formulation, must be carefully monitored. Despite this, it is unlikely that a person would overdose by either chewing or smoking *D. hopwoodii* due to the levels present and the quantity generally consumed.

8) **Dosage:**

As *D. hopwoodii* is chewed in a similar manner to chewing tobacco to provide a stimulant and/or analgesic effect, the minimum quantity required to elicit the desired response should be used. A low dosage should initially be trialled under supervision and use continued until the desired therapeutic effect is achieved. Usage should be discontinued if side effects/contraindications are noted.

## 10.9 References


- Celie PHN, van Rossum-Fikkert SE, van Dijk W, Brejc K, Smit AB. Nicotine and carbamylcholine binding to nicotinic acetylcholine receptors as studied in AChBP crystal structures. Neuron. 2004;41(6):907–14.
- Cock IE. Medicinal and aromatic plants—Australia. In: Ethnopharmacology, Encyclopedia of Life Support Systems (EOLSS). Developed under the auspices of UNESCO. Oxford, UK: EOLSS Publishers; 2011. http://www.eolss.net.
- Lassak EV, McCarthy T. Australian medicinal plants. A complete guide to identification and usage. Australia: New Holland Publishers; 2011.
- Latz PK. Bushfires and bushtucker. Aboriginal plant use in central Australia. Alice Springs, Australia: IAD Press; 1995.
- Luanratana O, Griffin WJ. Alkaloids of *Duboisia hopwoodii*. Phytochemistry. 1982;21(2):449–51.
- Maiden JH. The useful native plants of Australia. Sydney, Australia: Turner and Henderson; 1889.
- Mayer B. How much nicotine kills a human? Tracing back the generally accepted lethal dose to dubious self-experiments in the nineteenth century. Arch Toxicol. 2014;88(1):5–7.
- Page S, Olds M, editors. Botanica: the illustrated A-Z of over 10,000 garden plants for Australian gardens and how to cultivate them. Australia: Random House; 2004.


## 10.10 *Duboisia myoporoides* R.Br

### *10.10.1 (Solanaceae)*

1) **Synonyms:**
   None known
2) **Common names:**
   Soft corkwood, poison corkwood, cork tree, eye-opening tree, eye-plant, yellow basswood. Aboriginal names include onungunabie and ngmoo.
3) **Photograph of the species** (Fig. 10.5).
4) **Description of the species:**

   A) **Habitat and world distribution:**
   *D. myoporoides* grows in high rainfall coastal areas of eastern Australia on the margins of rainforests. Its range extends from the Sydney region northwards to the Cape York Peninsula in northern Queensland. It also grows in New Guinea and in New Caledonia (Lassak and McCarthy 2011; Page and Olds 2004).

**Fig. 10.5** *Duboisia myoporoides* leaves, flowers and fruit. Photograph was taken by Dr Ian Cock in Brisbane in 2011

B) **Morphology of the species:**

*D. myoporoides* is a large shrub/medium tree that may grow to 20 m in height. The bark is yellow brown to pale grey in colour and is thick and corky in texture. The leaves are alternate, obovate to narrowly elliptical in shape (4–15 cm long, 1–4 cm wide) and pale green in colour. White flowers with purple striations appear in spring in open bunches on the ends of branches. These develop into small, juicy black globular berries (4–8 mm diameter) containing several seeds during autumn (Lassak and McCarthy 2011; Page and Olds 2004).

C) **Medicinal parts:**

Whilst there is no clear evidence of medicinal usage by Australian Aborigines, the leaves are now used for the production of tropane alkaloids (especially scopolamine) (Cock 2011). Anecdotal accounts report that the tree sap and the leaves were used as a stupefacient for fishing (Lassak and McCarthy 2011). When added to a billabong, it would cause fish to float to the surface and also stun emus that drank the water, without poisoning the flesh.

D) **Chemical constituents:**

*D. myoporoides* leaves contain a number of medicinally important **alkaloids** including hyoscyamine, norhyoscyamine, scopolamine and nicotine (Foley 2006). The relative levels of these alkaloids (and thus the therapeutic properties) vary geographically. Hyoscyamine and norhyoscyamine are generally dominant in plants growing in cooler regions (e.g. Sydney and surrounds). In contrast, scopolamine is dominant in plants in warmer northern regions of Australia. Nicotine is the main alkaloid of plants growing in hot, tropical regions of New Guinea.

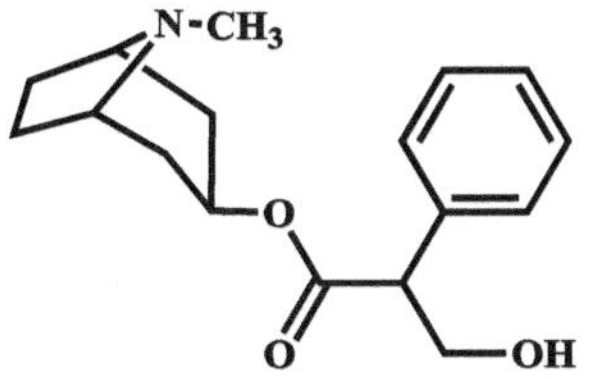


**Hyoscyamine**

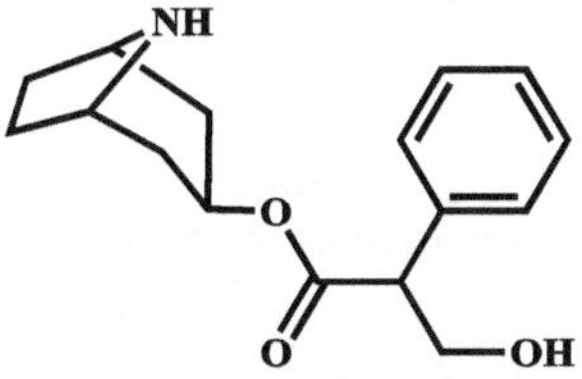


**Norhyoscyamine**

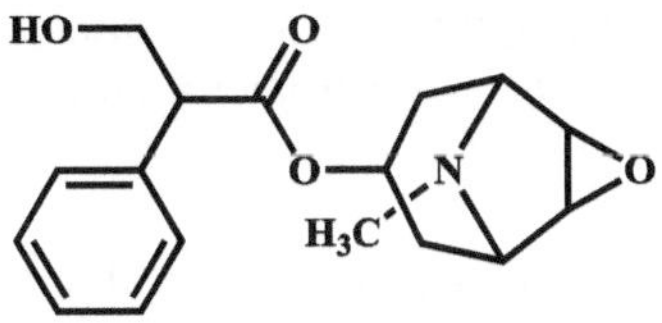


**Scopolamine**

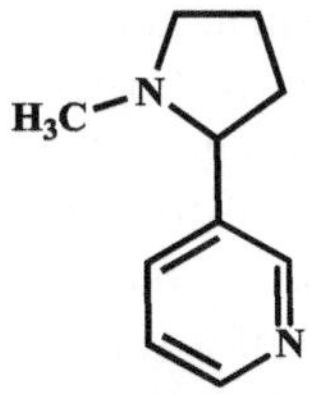


**Nicotine**

5) **Pharmacology and bioactivities:**

A) **In vitro:**

Hyoscyamine, norhyoscyamine (Furey and Devets 2006; Ricny et al. 2002) and scopolamine (Soejima and Noma 1984) are antagonists of the muscarinic acetylcholine receptors. They block acetylcholine binding to parasympathetic sites in many tissues including the central nervous system, cardiac tissue and the gastrointestinal tract, resulting in increased heart rate and cardiac output and lower blood pressure.

B) **In vivo:**

*D. myoporoides* leaf extracts induce similar effects to atropine, including paralysis of the vagus, stimulation of vasomotor centres at low doses but paralysis at higher doses, lower cardiac rate, increase blood pressure, accelerate respiration, inhibit saliva and sweat secretion and induce paralysis of the oculomotor nerve (Foley 2006). In humans, *D. myoporoides* leaf extracts have been reported to elicit sleepiness, delirium, increased pulse rate and respiration as well as hallucinations for up to 10 h (Lassak and McCarthy 2011; Cock 2011).

6) **Indications and usage:**

A) **Hypnotic and Sedative:**

Due to the tropane alkaloid components of *D. myoporoides,* it is useful as a sedative and hypnotic. In particular, scopolamine and hyoscyamine are both known to be potent sedatives and hypnotics (Lassak and McCarthy 2011; Cock 2011; Foley 2006).

B) **Ophthalmology:**

The mydriatic property of *D. myoporoides* extract was possibly the first therapeutic property recognised and exploited by modern allopathic medicine (Foley 2006). As early as 1878, hyoscyamine isolated from *D. myoporoides* was reported to have similar pupil dilatory activity as atropine, although the activity is more potent and its onset is more rapid and longer lasting.

C) **Treatment of Parkinson's disease:**

When *D. myoporoides* extracts (and the isolated components scopolamine and hyoscyamine) are used as sedatives, muscle relaxation precedes the onset of sleep (Foley 2006). This prompted the use of *D. myoporoides* products as early Parkinsonian treatments. Scopolamine is a successful treatment at doses as low as 0.2 mg, although its use was ultimately abandoned in favour of hyoscyamine due to its lower toxicity and the lower doses required (Foley 2006).

D) **Other disorders:**

*D. myoporoides* (and its isolated components) is also useful in the treatment of a variety of nervous disorders including epilepsy, Grave's disease and nystagmus (Foley 2006).

7) **Precautions and side effects:**

The same alkaloids which provide the therapeutic effects of *D. myoporoides* also render it toxic (Mayer 2014; Buckett and Haining 1965). Common side effects include dry mouth and throat, blurred vision and eye pain, dizziness, restlessness, arrhythmia and flushing (Foley 2006). An overdose may cause headache, nausea, vomiting, disorientation, hallucinations, short-term memory loss and coma in extreme cases. Intolerance to *D. myoporoides* alkaloids has also been frequently reported.

8) **Dosage:**

Several of the *D. myoporoides* alkaloids are toxic at relatively low doses. Hyoscyamine has an $LD_{50}$ of 375 mg/kg body weight (Buckett and Haining 1965), although toxic effects may be evident at doses as low as 0.2 mg (Foley 2006). Similarly, $LD_{50}$ values between 200 and 600 mg/kg have been reported for scopolamine (European Food Safety Authority 2008), and 30–60 mg nicotine can be lethal in humans (Mayer 2014). Therefore, consumption of *D. myoporoides* must be carefully monitored.

## 10.11 References


- Buckett WR, Haining CG. Some pharmacological studies on the optically active isomers of hyoscine and hyoscyamine. Br J Pharmacol. 1965;24:138–46.
- Cock IE. Medicinal and aromatic plants—Australia. In Ethnopharmacology, Encyclopedia of Life Support Systems (EOLSS). Developed under the auspices of UNESCO. Oxford, UK: EOLSS Publishers; 2011. http://www.eolss.net.

- European Food Safety Authority. Scientific opinion of the panel on contaminants in the food chain on a request from the European Commission on tropane alkaloids (from *Datura* sp.) as undesirable substances in animal feed. EFSA J. 2008;691:1–55.
- Foley P. *Duboisia myoporoides*: The medical career of a native Australian plant. Hist Record Aust Sci. 2006;17:31–69.
- Furey ML, Drevets WC. Antidepressant efficacy of the antimuscarinic drug scopolamine: a randomized, placebo-controlled clinical trial. Arch Gen Psychiatr. 2006;63(10):1121–9.
- Lassak EV, McCarthy T. Australian medicinal plants. A complete guide to identification and usage. Australia: New Holland Publishers; 2011.
- Mayer B. How much nicotine kills a human? Tracing back the generally accepted lethal dose to dubious self-experiments in the nineteenth century. Arch Toxicol. 2014;88(1):5–7.
- Page S, Olds M, editors. Botanica: the illustrated A-Z of over 10,000 garden plants for Australian gardens and how to cultivate them. Australia: Random House; 2004.
- Ricny J, Gualtieri F, Tucek S. Constitutive inhibitory action of muscarinic receptors on adenylyl cyclase in cardiac membranes and its stereospecific suppression by hyoscyamine. Physiol Res. 2002;51:131–7.
- Soejima M, Noma A. Mode of regulation of the Ach-sensitive K-channel by the muscarinic receptor in rabbit atrial cells. Pflügers Ach. 400(4):424–31


## 10.12 *Nelumbo nucifera* Gaertn

### *10.12.1 (Nelumbonaceae)*

1) **Synonyms:**
   *Nelumbium speciosum* Willd., *Nymphaea nelumbo* L.
2) **Common names:**
   Indian lotus, sacred lotus, bean of India, lotus
3) **Photograph of the species** (Fig. 10.6).
4) **Description of the species:**

   A) **Habitat and world distribution:**
      *N. nucifera* has a widespread distribution in Asia and Oceania, being native to tropical areas of Asia and northern Australia. It is a water plant which grows in rivers and water courses.
   B) **Morphology of the species:**
      *N. nucifera* grows in shallow water sources up to 3 m in depth in full sunlight and is anchored to the base of the water course through the root system whilst the leaves float on the water surface. The roots form a series

**Fig. 10.6** *N. nucifera* plant with a flower. Photograph was taken by Dr Ian Cock in Nanjing, China in June 2015

of long tubers (15–25 cm in length) that attach to the leaves via stems containing hollow air passages. Leaves are circular and large (up to 60 cm in diameter) with a large central attractive flower that may reach 20 cm in diameter. Flowers are either pink or white with striking central seed pods.

C) **Medicinal parts:**

All parts of the plant are used both as a food and as a therapeutic agent. The leaf, rhizome, seed and flower are used for the treatment of many types of ailments and can be used to treat depression, diabetes, inflammation and gastrointestinal, pulmonary and skin diseases. It is used in the treatment of cancer and has sedative properties (Mukherjee et al. 2009).

D) **Chemical constituents:**

*N. nucifera* contains interesting **flavonoids** including the flavonol miquelianin (=quercetin-3-*O*-glucuronide), as well as several **alkaloids** such as coclaurine, norcoclaurine, nuciferine and aporphine (Kashiwada et al. 2005). All of these compounds have been linked to various different antidepressive properties (Butterweck et al. 2000).

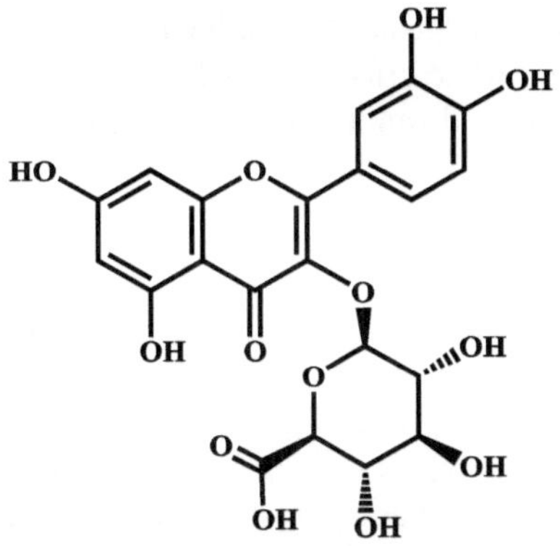

**Miquelianin**

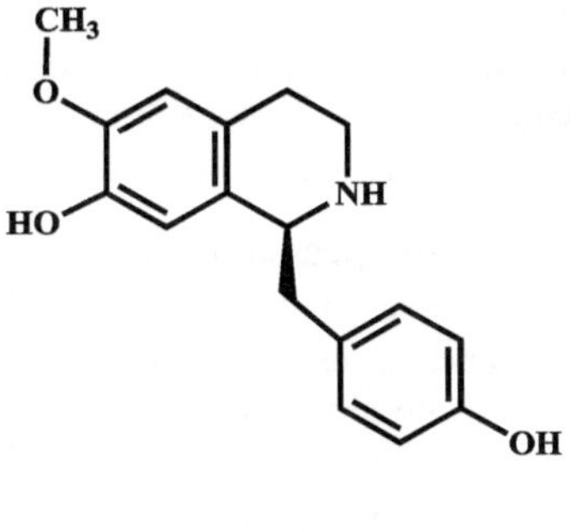

**Coclaurine**

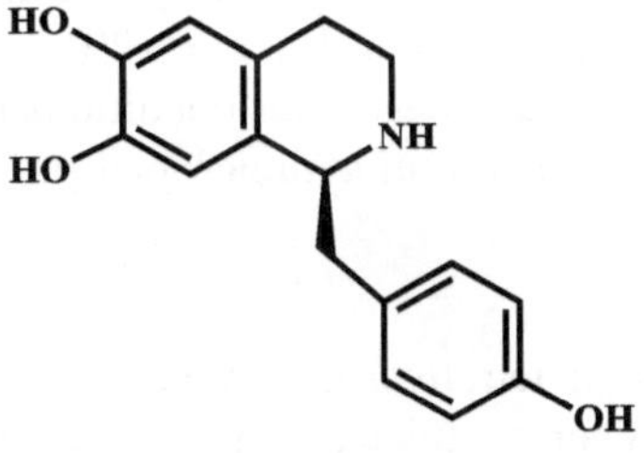

**Norcoclaurine**

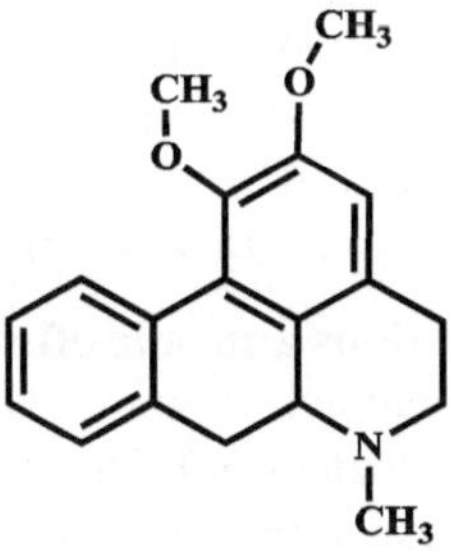

**Nuciferine**

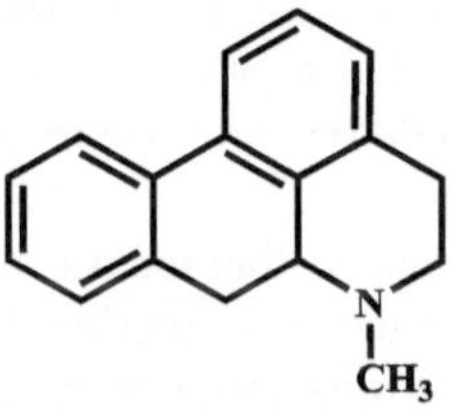

**Aporphine**

5) **Pharmacology and bioactivities:**

The sedative and antidepressant properties of *N. nucifera* are well established. Following absorption in the small intestine, miquelianin crosses the blood-brain barrier (BBB) (Juergenliemk et al. 2003) and acts on the central nervous system possibly by increasing the concentrations of serotonin (5-hydroxytryptamine; 5-HT) and noradrenaline in the synaptic cleft (Butterweck et al. 2002). The alkaloid components also possess antidepressant activity; coclaurine acts as a nicotinic acetylcholine receptor antagonist, whilst norcoclaurine is a $\beta_2$ adrenergic receptor antagonist (Exley et al. 2005). Nuciferine and aporphine are dopamine receptor antagonists and induce sedation and, in higher doses, catalepsy (Bhattacharya et al. 1978; Hedberg et al. 1996). They inhibit spontaneous motor activity and can potentiate morphine analgesia. Aporphine is also a partial $5\text{-}HT_{2A}$ receptor agonist and thus potentiates the effect of 5-HT, thereby having potential benefits as an antidepressant and anticonvulsant (Zetler 1988; Munusamy et al. 2013).

6) **Indications and usage:**

   A) **Antidepressant activity:**

   Due to its alkaloid and flavonoid composition, *N. nucifera* is useful as an antidepressant agent as well as having sedative and hypnotic properties. Some of the compounds act as $5\text{-}HT_{1A}$ receptor antagonists and partial agonists and thus assist in alleviating the symptoms of depression (Sugimoto et al. 2010). Furthermore, dopamine receptor antagonism provides the sedative effects.

   B) **Other therapeutic uses:**

   *N. nucifera* seed extracts have systemic anti-inflammatory activity (Liu et al. 2004, 2014) and block cell cycle progression (Liu et al. 2014). Thus, *N. nucifera* is indicated for inflammatory and cancerous conditions (Karki et al. 2012). *N. nucifera* also has broad spectrum antimicrobial properties (Sittiwet 2009; Knipping et al. 2012).

7) **Precautions and side effects:**

   *N. nucifera* contains significant levels of tetrahydroquinolines (coclaurine and norcoclaurine). This class of drug has been postulated as a possible cause of Parkinson's disease (Kobayashi et al. 2009), and thus caution should be taken with the therapeutic use of *N. nucifera.* Moreover, due to their β-adrenergic agonist activity on the trachea and heart muscle, deleterious respiratory and cardiovascular consequences may arise. In addition, the alkaloid component, nuciferine, has a relatively low $LD_{50}$ (<300 mg/kg) (Bhattacharya et al. 1978), and therefore, consumption should be carefully monitored.

8) **Dosage:**

   There is no available data on therapeutic dose range for *N. nucifera* plant material. It is generally ingested as a functional food for nutritional value and to aid health and well-being. Whilst some of the phytochemical components have reported toxicities, they are present in the plant in relatively low levels, meaning that the plant is generally safe to ingest in reasonable quantities. However, where the extract is used, the dose should be limited and administered under close supervision.

## 10.13 References


- Bhattacharya S, Bose R, Ghosh P, Tripathi V, Kay A, Dasgupta, B. Psychopharmacological studies on (-)-nuciferine and its Hofmann degradation product atherosperminine. Psychopharmacology. 1978;59:29–33.

- Butterweck V, Jürgenliemk G, Nahrstedt A, Winterhoff H. Flavonoids from *Hypericum perforatum* show antidepressant activity in the forced swimming test. Planta Med. 2000;66:3–6.
- Butterweck V, Nahrstedt A, Evans J, Hufeisen S, Rauser L, Savage J, Popadak B, Ernsberger P, Roth, B. In vitro receptor screening of pure constituents of St. John's wort reveals novel interactions with a number of GPCRs. Psychopharmacology (Berl). 2002;162:193–202.
- Exley R, Iturriaga-Vásquez P, Lukas R, Sher E, Cassels B, Bermudez, I. Evaluation of benzyltetrahydroisoquinolines as ligands for neuronal nicotinic acetylcholine receptors. Br J Pharmacol. 2005;146:15–24.
- Hedberg M, Linnanen T, Jansen J, Nordvall G, Hjorth S, Unelius l, Johansson A. ChemInform Abstract: 11-substituted (R)-aporphines: synthesis, pharmacology, and modeling of D2A and 5-HT1A receptor interactions. J Med Chem. 1996;39:3503–13.
- Juergenliemk G, Boje K, Huewel S, Lohmann C, Galla HJ, Nahrstedt A. In vitro studies indicate that miquelianin (quercetin 3-O-beta-D-glucuronopyranoside) is able to reach the CNS from the small intestine. Planta Med. 2003;69:1013–7.
- Karki R, Jung M, Kim K, Kim D. Inhibitory effect of *Nelumbo nucifera* (Gaertn.) on the development of atopic dermatitis-like skin lesions in NC/Nga mice. Evid Based Compl Alternative Med. 2012;2012:1–7.
- Kashiwada Y, Aoshima A, Ikeshiro Y, Chen YP, Furukawa H, Itoigawa M, Fujioka T, Mihashi K, Cosentino LM, Morris-Natschke SL, Lee KH. Anti-HIV benzylisoquinoline alkaloids and flavonoids from the leaves of *Nelumbo nucifera*, and structure-activity correlations with related alkaloids. Bioorg Med Chem. 2005;13:443–8.
- Knipping K, Garssen J, van't Land B. An evaluation of the inhibitory effects against rotavirus infection of edible plant extracts. Virol J. 2012;9:137.
- Kobayashi H, Fukuhara K, Tada-Oikawa S, Yada Y, Hiraku Y, Murata M, Oikawa S. The mechanisms of oxidative DNA damage and apoptosis induced by norsalsolinol, an endogenous tetrahydroisoquinoline derivative associated with Parkinson's disease. J Neurochem. 2009;108:397–407.
- Liu C, Tsai W, Lin Y, Liao J, Chen C, Kuo Y. The extracts from *Nelumbo nucifera* suppress cell cycle progression, cytokine genes expression, and cell proliferation in human peripheral blood mononuclear cells. Life Sci. 2004; 75:699–716.
- Liu SH, Lu TH, Su CC, Lay IS, Lin HY, Fang KM, Ho TJ, Chen KL, Su YC, Chiang WC, Chen YW. Lotus leaf (*Nelumbo nucifera*) and its active constituents prevent inflammatory responses in macrophages via JNK/NF-κB signaling pathway. Am J Chin Med. 2014;42:869–89.
- Mukherjee P, Mukherjee D, Maji A, Rai S, Heinrich M. The sacred lotus (*Nelumbo nucifera*)—phytochemical and therapeutic profile. J Pharm Pharmacol. 2009;61:407–22.

- Munusamy V, Yap B, Buckle M, Doughty S, Chung L. Structure-based identification of aporphines with selective 5-$HT_{2A}$ receptor-binding activity. Chem Biol Drug Des. 2013;81:250–6.
- Sittiwet C. Antimicrobial activity of essential oil from *Nelumbo nucifera* Gaertn. Pollen. Int J Pharmacol. 2009;5:98–100.
- Sugimoto Y, Furutani S, Nishimura K, Itoh A, Tanahashi T, Nakajima H, Oshiro H, Sun S, Yamada J. Antidepressant-like effects of neferine in the forced swimming test involve the serotonin1A (5-$HT_{1A}$) receptor in mice. Eur J Pharmacol. 2010;634:62–7.
- Zetler, G. Neuroleptic-like, anticonvulsant and antinociceptive effects of aporphine alkaloids: bulbocapnine, corytuberine, boldine and glaucine. Archives Internationales de Pharmacodynamie et de Thérapie. 1988;296:255–81.


## 10.14 *Piper methysticum* G. Forst

### *10.14.1 (Piperaceae)*

1) **Synonyms:**
   None known
2) **Common names:**
   kava-kava (Tongan), *'Awa* (Hawaiian), ava (Samoan), yaqona (Fijian), yangona, kava
3) **Photograph of the species** (Fig. 10.7).
4) **Description of the species:**

   A) **Habitat and world distribution:**
   *P. methysticum* is native to the western Pacific islands with the largest number grown in Polynesia and Melanesia. It prefers loose, well-drained soils and high rainfall (over 1000 mm per year), ideally growing in temperatures from 25 to 35 °C and high relative humidity. It generally grows well as an understory plant, as too much sunlight is harmful, particularly when the plant is becoming established.

   B) **Morphology of the plant:**
   *P. methysticum* is a vigorous and rapid growing medium to large plant, which produces large leaves with a dark green, glossy surface. The leaves are opposite and heart shaped, broadening at the base and having a pointed apex. They have prominent veins (palmate or pinnate) and grow to 15–20 cm in size, whilst the height of the entire plant reaches up to 4 m (Wagner et al. 1990). Its small flower spikes (green–yellow) are

**Fig. 10.7** Immature *P. methysticum* plant. Photograph was taken by Dr Ian Cock in Fiji in 2015

unisexual, although female flowers are rare and do not produce fruit via hand pollination. Cultivation is by propagation from stem or root cuttings.

C) **Medicinal parts:**

The roots of the plant are used to produce a drink that is reputed to have sedative, hypnotic and anaesthetic properties (Gounder 2006; Rychetnik and Madronio 2011).

D) **Chemical constituents:**

A total of 18 **kavalactones** have been identified in *P. methysticum* roots, a majority of which possess therapeutic properties. Of these, kavain, dihydrokavain, methysticin and dihydromethysticin, yangonin and desmethoxyyangonin are responsible for almost all of the pharmacological activities (Wang et al. 2015). It also contains minor constituents including **chalcones**, **flavokavins** A, B and C, as well as the toxic **alkaloid** pipermethystine (Shimoda et al. 2015).

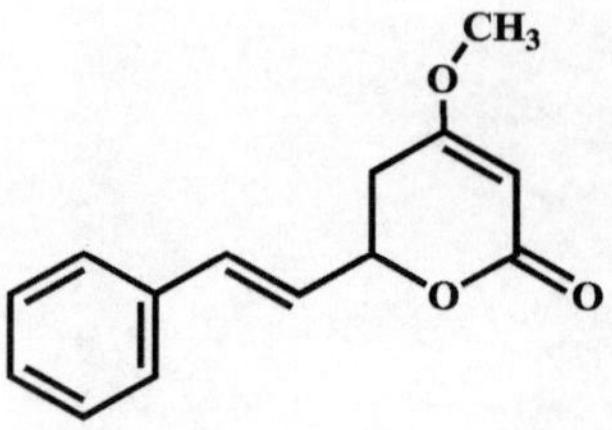


Kavain

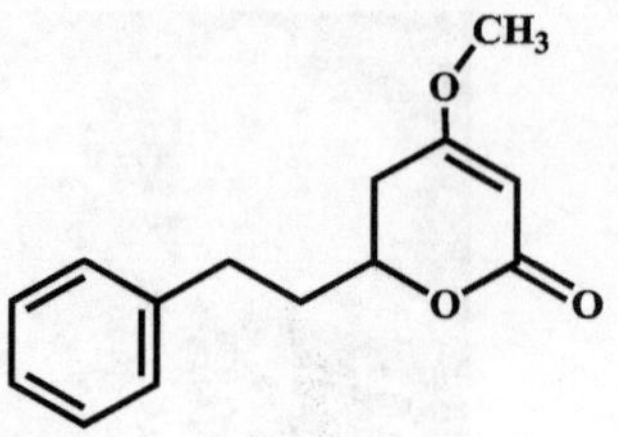


Dihydrokavain

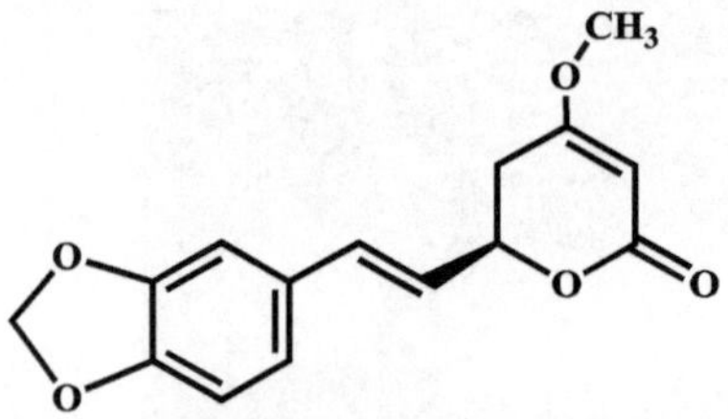


Methysticin

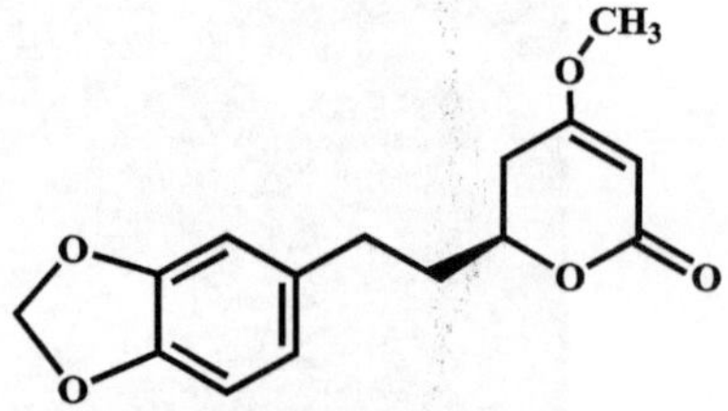


Dihydromethysticin

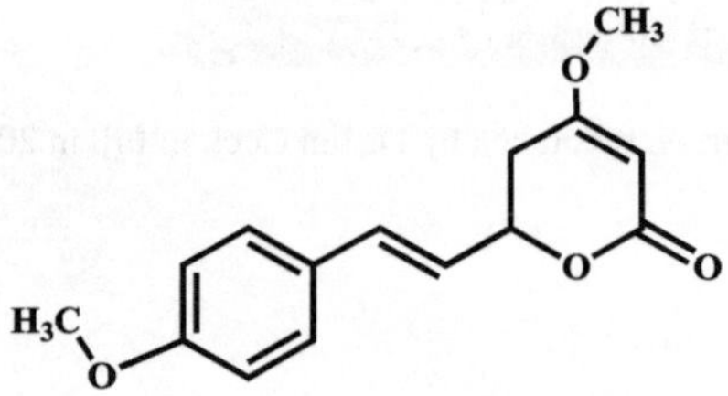


Yangonin

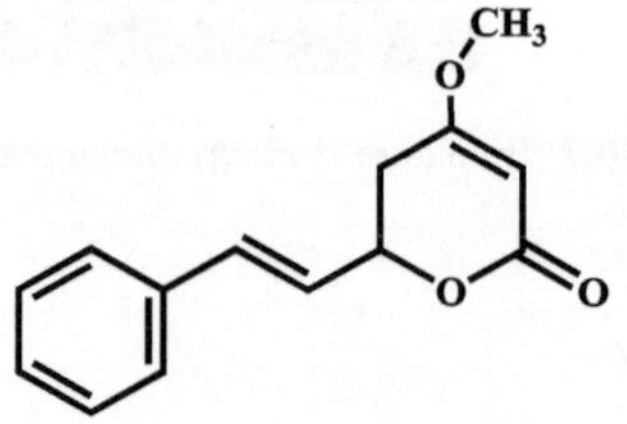


Desmethoxyyangonin

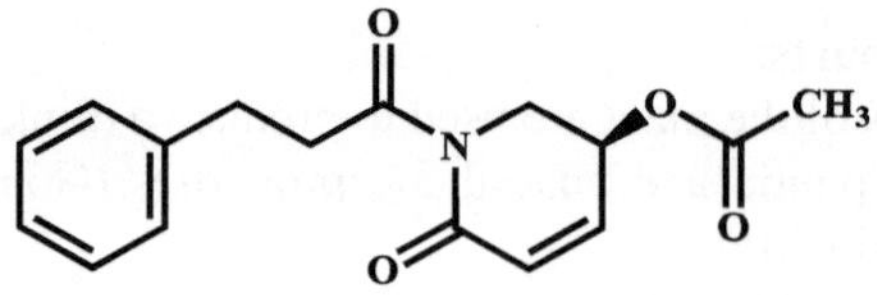


Pipermethystine

5) **Pharmacology and bioactivities:**

A) **In vitro:**

Kavain, dihydrokavain, methysticin, dihydromethysticin and yangonin potentiate $GABA_A$ receptor activity (Boonen and Häberlein 1998), whilst kavain and methysticin inhibit the reuptake of norepinephrine and dopamine and inhibit voltage-gated sodium and calcium channels (Dinh et al. 2001). All 6 bioactive kavalactones are irreversible inhibitors of monoamine oxidase B (Rowe et al. 2011). Furthermore, yangonin is known to be an agonist of the cannabinoid receptor type 1 ($CB_1$ receptor) (Ligresti et al. 2012). Kava extracts also have weak interactions with 5-hydroxytryptamine (5-HT) receptors and the benzodiazepine binding site of the $GABA_A$ receptor (Yuan et al. 2002; Wood 2003). Potentiation of the $GABA_A$ receptor may account for the reported therapeutic activities.

B) **In vivo:**

Kavalactones have been reported to elevate dopamine levels in the nucleus accumbens of the rat brain (Baum et al. 1998). Several studies have shown that kava extracts possess anxiolytic activity and thus may be an efficacious alternative to benzodiazepines and tricyclic antidepressants for the treatment of anxiety disorders (Wotjak 2003).

6) **Indications and usage:**

A) **Hypnotic and sedative:**

Due to the kavalactone components of *P. methysticum*, its use is indicated as a sedative and hypnotic (Singh and Singh 2002). In particular, the 6 aforementioned kavalactones appear to be the active pharmacological compounds responsible for these effects.

B) **Treatment of Parkinson's disease:**

Since *P. methysticum* extracts and its components inhibit the reuptake of dopamine, it has potential for treatment of Parkinson's disease. Studies have shown that coadministration of kava extract and levodopa enhances the efficacy of levodopa (Talati et al. 2009).

7) **Precautions and side effects:**

*P. methysticum* may cause serious (potentially fatal) hepatotoxicity due to pipermethystine and flavokavin B. However, several studies have failed to demonstrate the toxicity of these compounds, and it has recently been proposed that the hepatotoxicity may be due to contamination with aflatoxin arising from leaf mould (Teschke et al. 2011). This requires further study to ensure adequate quality control for commercial products. It should be noted that several adverse drug reactions may occur with the administration of kava extracts or compounds as they have been shown to induce the hepatic cytochrome P450 (CYP) 1A1 enzyme (Cock 2015; Li et al. 2001). Thus, careful monitoring of patients administered with drugs known to be CYP1A1 substrates, inhibitors or inducers is required. Other CYP enzymes, such as CYP3A4 and CYP2D6, may also be affected by kava (Cock 2015; Mathews 2005; Zou et al. 2004), thus causing complications in patients taking medications that interact with these drug-metabolising enzymes.

8) **Dosage:**

The chemical compositions of *P. methysticum* may vary between extracts, and thus dosages of the active compounds may fluctuate. Due to the hepatotoxicity of pipermethystine and flavokavin B (Rowe et al. 2011), and in particular the low $LD_{50}$ of these compounds, their consumption should therefore be carefully monitored.

## 10.15 References


- Baum SS, Hill R, Rommelspacher H. Effect of kava extract and individual kavapyrones on neurotransmitter levels in the nucleus accumbens of rats. Eur J Pharmacol. 1998;22(7):1105–20.
- Boonen G, Häberlein H. Influence of genuine kavapyrone enantiomers on the GABA A binding site. Planta Med. 1998;64:504–6.
- Cock IE. The safe usage of herbal medicines: counterindications, cross-reactivity and toxicity. Pharmacognosy Commun. 2015;5(1):2–50.
- Dinh L, Simmen U, Berger Bueter K, Bueter B, Lundstrom K, Schaffner W. Interaction of various *Piper methysticum* cultivars with CNS receptors in vitro. Planta Med. 2001;67:306–11.
- Gounder R. Kava consumption and its health effects. Pac Health Dialog. 2006; 13:131–5.
- Li Y, Mei H, Wu Q, Zhang S, Fang J, Shi L, Guo L. Methysticin and 7,8-dihydromethysticin are two major kavalactones in kava extract to Induce CYP1A1. Toxicol Sci. 2011;124:388–99.
- Ligresti A, Villano R, Allarà M, Ujváry I, Di Marzo V. Kavalactones and the endocannabinoid system: the plant-derived yangonin is a novel CB1 receptor ligand. Pharmacol Res. 2012;66:163–9.
- Mathews J. Pharmacokinetics and disposition of the kavalactone kawain: interaction with kava extract and kavalactones in vivo and in vitro. Drug Metabol Dispos. 2005;33:1555–63.
- Rowe A, Narlawar RW, Groundwater P, Ramzan I. Kavalactone pharmacophores for major cellular drug targets. Mini-Rev Med Chem. 2011;11:79–83.
- Rowe A, Zhang L, Ramzan I. Toxicokinetics of kava. Adv Pharmacol Sci. 2011; 2011:1–6.
- Rychetnik L, Madronio C. The health and social effects of drinking water-based infusions of kava: a review of the evidence. Drug Alcohol Rev. 2011;30:74–83.
- Shimoda L, Showman A, Baker J, Lange I, Koomoa D, Stokes A, Borris R, Turner H. Differential regulation of calcium signalling pathways by components of *Piper methysticum* ('Awa). Phytother Res. 2015; doi: 10.1002/ptr.5291.
- Singh Y, Singh N. Therapeutic potential of kava in the treatment of anxiety disorders. CNS Drugs. 2012;16:731–43.
- Talati R, Reinhart K, Baker W, White C, Coleman C. Pharmacologic treatment of advanced Parkinson's disease: a meta-analysis of COMT inhibitors and MAO-B inhibitors. Parkinsonism Relat Disord. 2009;15:500–5.
- Teschke R, Qiu S, Lebot V. Herbal hepatotoxicity by kava: Update on pipermethystine, flavokavin B, and mould hepatotoxins as primarily assumed culprits. Dig Liver Dis. 2011;43:676–81.
- Wagner W, Herbst D, Sohmer S. Manual of the flowering plants of Hawaii; 2012: [Honolulu]. Honolulu, USA: University of Hawaii Press; 1990.

- Wang J, Qu W, Bittenbender H, Li Q. Kavalactone content and chemotype of kava beverages prepared from roots and rhizomes of Isa and Mahakea varieties and extraction efficiency of kavalactones using different solvents. J Food Sci Technol. 2015;52:1164–9.
- Wood M. Therapeutic potential of 5-HT2C receptor antagonists in the treatment of anxiety disorders. Curr Drug Targets CNS Neurol Disord. 2003;2:383–7.
- Wotjak C. The endocannabinoid system of the brain: a potential therapeutic target for the treatment of anxiety disorders in traumatised patients? Pharmacopsychiatry. 2003;36–323.
- Yuan C, Dey L, Wang A, Mehendale S, Xie J, Aung H, Ang-Lee M. Kavalactones and dihydrokavain modulate GABAergic activity in a rat gastric-brainstem preparation. Planta Med. 2002;68: 1092–96.
- Zou L, Henderson G, Harkey M, Sakai Y, Li A. Effects of Kava (Kava-kava, 'Awa, Yaqona, *Piper methysticum*) on c-DNA-expressed cytochrome P450 enzymes and human cryopreserved hepatocytes. Phytomedicine. 2004;11:285–94.


## 10.16 *Piper novae-hollandiae* Miq

### *10.16.1 (Piperaceae)*

1) **Synonyms:**

   *Piper hederaceum* (Miq.) C.DC

2) **Common names:**

   Giant pepper vine, native pepper vine, climbing pepper, native pepper, curtain vine, Australian pepper vine, Mao-wararg (Aboriginal name)

3) **Photograph of the species** (Fig. 10.8).
4) **Description of the species:**

   A) **Habitat and world distribution:**

   *Piper novae-hollandiae* has a widespread distribution in the warm coastal rainforest regions of the eastern and northern Australia, occurring in New South Wales, Queensland and the Northern Territory (Lassak and McCarthy 2011; Page and Olds 2004).

   B) **Morphology of the plant:**

   *Piper novae-hollandiae* is a vigorous and rapid growing climbing plant resembling ivy, which grows over shrubs and trees. It climbs tree trunks using small adventitious roots. Once established, the stem becomes woody and forms branches up to 30 cm in diameter. The alternate leaves are broadly ovate (elliptical, broadened near the base and have a pointed apex), 6–12 cm long, 2–9 cm wide with a dark green glossy upper surface and a distinctly pale undersurface and have 5–7 noticeable veins (pinnate to palmate).

**Fig. 10.8** *Piper novae-hollandiae* vine. Photograph was taken in Toohey Forrest, Brisbane Australia in 2015 by Dr Ian Cock

Flowers are unisexual. Male flowers have cylindrical spikes 10–20 mm long with two or three stamens per flower. Female flower spikes are ovoid, 8–10 mm long. The fruit is an ovoid berry (5 mm long) which turns red as it ripens (Lassak and McCarthy 2011).

C) **Medicinal parts:**

*P. novae-hollandiae* leaves have historical uses as therapeutic agents. Australian Aborigines chewed the leaves as a stimulant and as a general tonic (Cock 2011). Reports also exist of the use of *P. novae-hollandiae* by Australian Aborigines for the treatment and cure of skin disorders and sexually transmitted diseases.

D) **Chemical constituents:**

The individual bioactive components are unknown, although several **alkaloids** including piperine, fagaramide, piperlonguminine and chavicine have been identified.

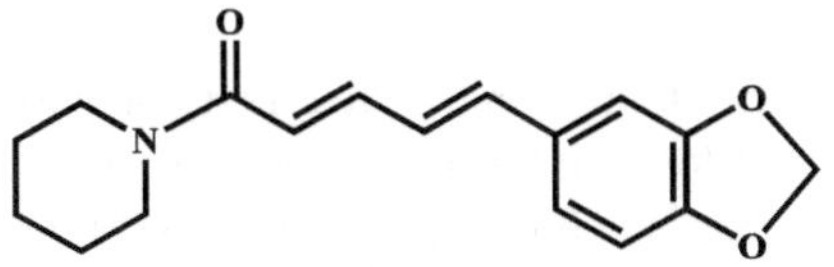

**Piperine**

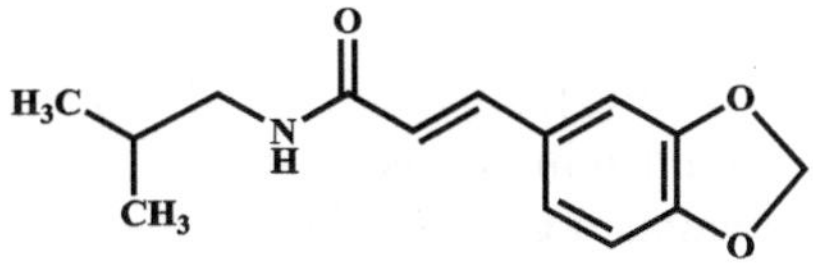

**Fagaramide**

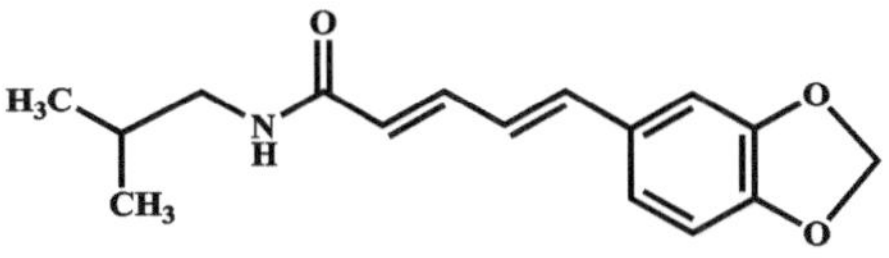

**Piperlonguminine**

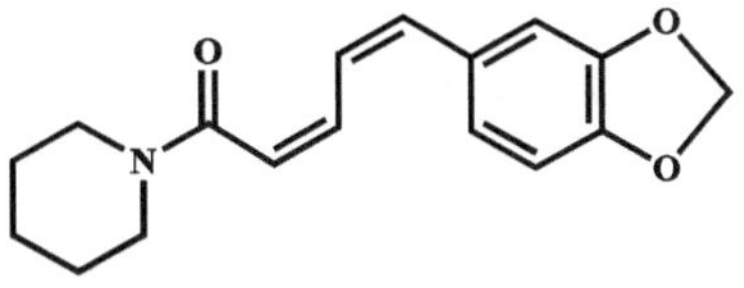

**Chavicine**

5) **Pharmacology and bioactivities:**

A) **In vitro:**

The sedative properties of several *Piper* species are well established. *Piper piscatorum* can induce changes in neuronal intracellular calcium concentration (McFerren et al. 2002). Ethanolic *Piper nigrum* extracts have significant anticonvulsant and sedative properties (Hu et al. 1996). *Piper methysticum* has similar effects, and pharmacological studies indicate that it may induce these effects by altering brain dopamine levels (Baum et al. 1998). However, there is a lack of published scientific studies on the therapeutic properties of *P. novae-hollandiae.* Of the bioactivities examined, the in vitro antimicrobial activity has been the most extensively reported (Cock 2011), although phytochemistry and pharmacological studies are lacking. Similarly, tumour inhibitory activity has also been reported for *P. novae-hollandiae* extracts (Loder et al. 1969), although pharmacological studies are required.

B) **In vivo:**

*P. novae-hollandiae* has traditional uses as a stimulant and general tonic (Maiden 1989). It also was chewed for sore gums and has been shown to contain an ether extractable component with numbing/analgesic activity (Webb 1948).

6) **Indications and usage:**

*P. novae-hollandiae* is an excellent stimulant and general tonic and may also act as a sedative at some doses (Cock 2011). It is useful for the treatment of gonorrhoea and mucous discharges (Webb 1948). It also was chewed for sore gums and has a numbing/analgesic effect (Webb 1948).

7) **Precautions and side effects:**

Rigorous pharmacological studies into *P. novae-hollandiae* are lacking. Toxicity studies are needed to establish safe dosage limits and to investigate side effects. Until these are thoroughly established, therapeutic usage should be avoided and/or closely monitored.

8) **Dosage:**

As the leaves are chewed to provide a stimulant or analgesic effect, the patient should initially use a minimal quantity under close supervision. Use may be continued until the desired therapeutic effect is achieved. Usage should be discontinued if side effects/contraindications are noted.

## 10.17 References


- Baum SS, Hill R, Rommelspacher H. Effect of kava extract and individual kavapyrones on neurotransmitter levels in the nucleus accumbens of rats. Eur J Pharmacol. 1998;22(7):1105–120.
- Cock IE. Medicinal and aromatic plants—Australia. In: Ethnopharmacology, Encyclopedia of Life Support Systems (EOLSS); Developed under the auspices of UNESCO. Oxford, UK: EOLSS Publishers; 2011. http://www.eolss.net.
- Hu SL, Ao P, Tan HG. Pharmacognostic studies on the roots of *Piper nigrum* L. II. Chemical and pharmacological studies. Acta Horticulturae. 1996;426: 175–8.
- Lassak EV, McCarthy T. Australian medicinal plants. A complete guide to identification and usage. Australia: New Holland Publishers; 2011.
- Loder JW, Moorhouse A, Russel GB. Tumour Inhibitory Plants. Amides of *Piper novae-Hollandiae* (Piperaceae). Aust J Chem. 1969;22:1531–38.
- Maiden JH. The useful native plants of Australia. Sydney, Australia: Turner and Henderson; 1889.
- McFerren MA, Cordova D, Rodriguez E. In vitro neuropharmacological evaluation of piperovatine, an isobutylamide from *Piper piscatorum* (Piperaceae). J Ethnopharmacol. 2002;83(3):201–7.
- Page S, Olds M, editors. Botanica: the illustrated A-Z of over 10,000 garden plants for Australian gardens and how to cultivate them. Australia: Random House; 2004.
- Webb LJ. Guide to the medicinal and poisonous plants of Queensland. CSIRO bulletin number 232. Melbourne, Australia: Government Printer; 1948.


## 10.18 *Tasmannia lanceolata* (Poir.) A.C.Sm

### *10.18.1 (Winteraceae)*

1) **Synonyms:**

*Drimys lanceolata* (Poir.) Baill., *Drimys aromatica* (R.Br.) F. Muell., *Tasmannia aromatica* R.Br., *Winterania lanceolata* Poir.

**Fig. 10.9** *Tasmannia lanceolata* leaves and berries. Photograph was taken in December 2014 in central Tasmania, Australia by Dr David Ruebhart and is reproduced here with the photographer's permission

2) **Common names:**

   Mountain pepper, mountain pepperberry, Tasmanian pepper

3) **Photograph of the species** (Fig. 10.9).
4) **Description of the species:**

   A) **Habitat and world distribution:**

   *T. lanceolata* is endemic to the woodlands and cool temperate rainforests of Tasmania and the south-eastern region of the Australian mainland (Cock 2013).

   B) **Morphology of the species:**

   *T. lanceolata* is a medium to large shrub that varies between 2 and 5 m in height. Individual plants are unisexual, having either male or female flowers. The stems, branches and twigs are red in colour. The aromatic leaves are lanceolate to narrowly elliptical in shape (4–12 cm long, 0.7–2.0 cm wide) with a distinctly pale undersurface. Small creamy-white unisexual flowers appear during the summer months. These develop into small fleshy black two-lobed berries (5–8 mm wide) during autumn (Cock 2013; Page and Olds 2004).

   C) **Medicinal parts:**

   *T. lanceolata* berries, leaves and bark have historical uses as food and as medicines (Cock 2013; Cock 2011). When the berry is air-dried, it forms a small, hard peppercorn which is suitable for milling or crushing. The berry has a pleasant spicy flavour and sharp aroma. *T. lanceolata* was used as a flavouring agent by Australian Aborigines and more recently by European settlers. Historically, the leaves have been used as a herb, and the berries have been used as a spice. Australian Aborigines also used *T. lanceolata* as a therapeutic agent to treat stomach disorders and as an emetic, as well as general usage as a tonic. Reports also exist of the use of *T. lanceolata* by Australian Aborigines for the treatment and cure of skin disorders, venereal diseases, colic and stomachache and as a quinine substitute (Cock 2011). Later, European colonists also recognised the therapeutic potential of *T. lanceolata*, and the bark was used as a common substitute for other herbal remedies (including those derived from the related South American Winteraceae species, *Drimys winteri* (winter bark)) to treat scurvy due to its high antioxidant activity (Konczak et al. 2010; Netzel et al. 2007).

D) **Chemical constituents:**

Volatile components account for the majority of the *T. lanceolata* phytochemical profile, accounting for as high as 6 % of the dry weight of the plant material (Menary et al. 2003). The **drimane sesquiterpene** polygoidal is the major component, accounting for up to 40 % of commercial *T. lanceolata* essential oil components (Menary et al. 2003). *T. lanceolata* also produces **phenylpropenes** including safrole and myristicin. Other **sesquiterpenoids**, including guaiol (4.36 %), calamenene (3.42 %), spathulenol (1.94 %), drimenol (1.91 %), cadina-1,4-diene (1.58 %), 5-hydroxycalamenene (1.47 %), bicyclogermacrene (1.15 %), α-cubebene (0.88 %), caryophyllene (0.87 %), α-copaene (0.48 %), cadalene (0.44 %), δ-cadinol (0.4 %), elemol (0.39 %), T-muurolol (0.39 %) and germacrene D are particularly abundant. Other sesquiterpenoids present in *T. lanceolata* essential oils include camphene (0.02 %), α-gurjunene (0.04 %) and viridiflorol (Menary et al. 2003). Many **monoterpenic compounds** are also present in significant levels in *T. lanceolata* with 1,8-cineole (0.77 %), α-pinene (0.86 %), β-pinene (0.38 %) and linalool (1.81 %) predominating (Menary et al. 2003). Other characteristic monoterpenes detected in the commercial *T. lanceolata* essential oils analysed in that study included sabinene, β- phellandrene, myrcene, terpinolene, α-terpineol, γ-terpinene, piperitone, limonene and cymene, although all of these were generally present at levels below 0.1 %. **Flavonoids** and **flavonoid glycosides** are also major components of *T. lanceolata*. These flavonoids include quercetin, quercetin-3-*O*-rutinoside (=rutin), cyanidin-3-*O*-glucoside and cyanidin-3-*O*-rutinoside. Recent studies have also reported the presence of **stilbenes** including the resveratrol glycoside piceid and several **combretastatins** (Cock et al. 2015).

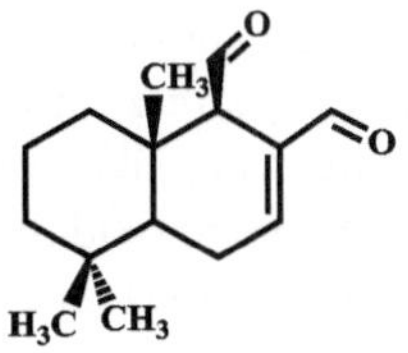

**Polygodial**

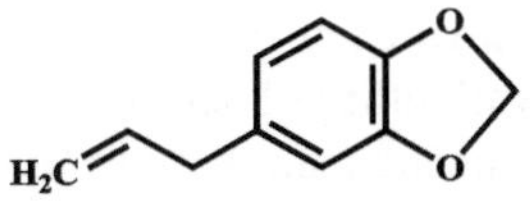

**Safrole**

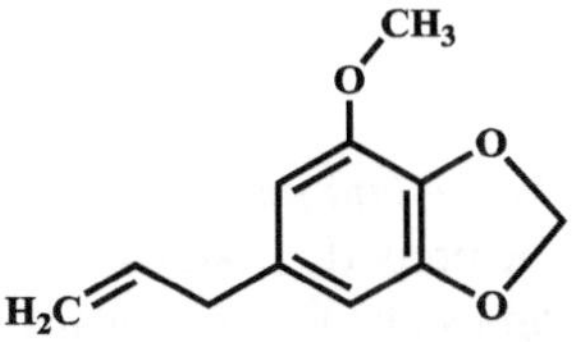

**Myristicin**

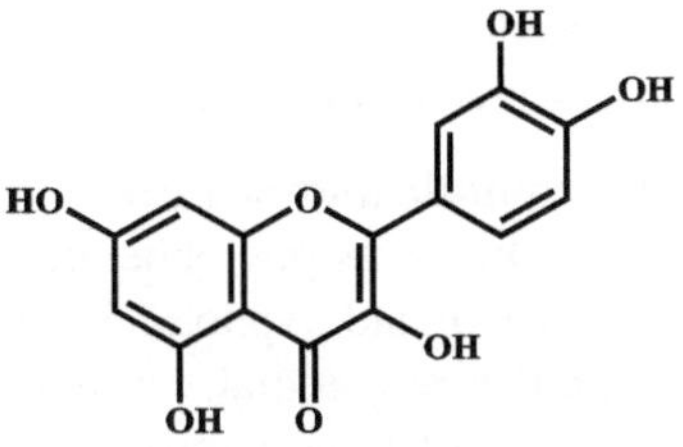

**Quercetin**

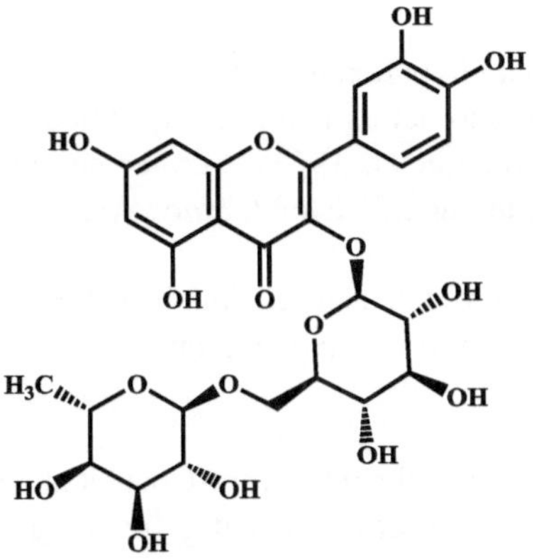

**Quercetin-3-*O*-rutinoside**

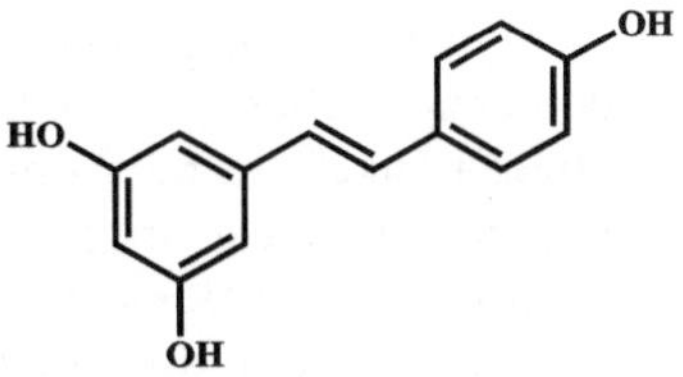

**Resveratrol**

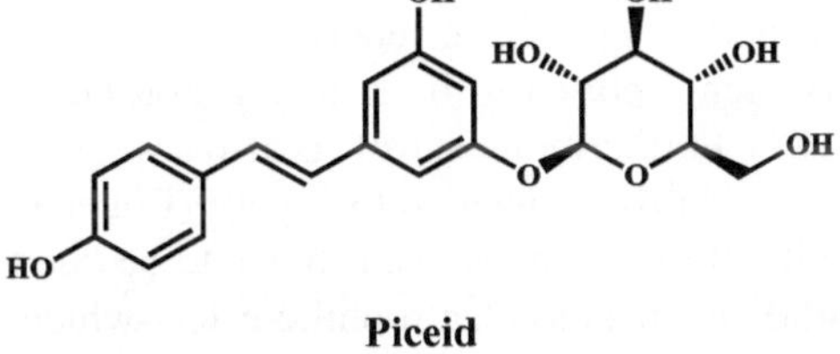

**Piceid**

5) **Pharmacology and bioactivities:**

Polygodial blocks TASK and TRESK potassium channels which result in mood modulatory effects (Beltrán et al. 2013). The stilbene resveratrol has antidepressant activity in vivo (Hurley et al. 2014), and it is likely that the glycosylated form (piceid) may have a similar activity. The flavonoids quercetin and rutin have been reported to have antidepressant effects, possibly by increasing the availability of serotonin and noradrenaline in the synaptic cleft although the exact mechanism is yet to be established (Anjaneyulu et al. 2003; Machado et al. 2008). Myristicin is a nonselective inhibitor of monoamine oxidase (MAO) and thus binds to both the A and B isoforms (Truitt et al. 1963), and a myristicin analogue has been shown to exert antidepressant effects in mice (Moreira et al. 2001). It is thus likely that myristicin also increases the availability of serotonin and noradrenaline in the synaptic cleft.

6) **Indications and usage:**

   A) **Hypnotic and Sedative:**

   Due to its phytochemical composition, *T. lanceolata* may be useful as a sedative and hypnotic, although this is yet to be properly evaluated. In particular, polygodial, myristicin, bioflavonoids and stilbene components have sedative/hypnotic properties (Anjaneyulu et al. 2003; Machado et al. 2008; Hurley et al. 2014; Beltrán et al. 2013; Truitt et al. 1963).

   B) **Stimulant:**

   Myristicin initially induces neuronal hyperexcitability followed by CNS depression. Thus *T. lanceolata* may also act as a stimulant, although this effect is likely to be short-lived (Truitt et al. 1963). Further studies are required to evaluate the potential stimulant effects of *T. lanceolata.*

   C) **Treatment of Parkinson's disease:**

   The presence of MAO-A/B inhibitors such as myristicin in *T. lanceolata* (Cock 2013) indicates that it may be useful in the treatment of Parkinson's disease.

   D) **Antiseptic:**

   *T. lanceolata* has broad spectrum antiseptic properties (Cock et al. 2015; Cock 2011; Winnett et al. 2014) and is therefore indicated for indigestion and other GIT disorders, urinary tract infections and a wide variety of bacterial and fungal skin disorders.

7) **Precautions and side effects:**

   *T. lanceolata* berries have been reported to contain significant levels of safrole, which has been reported to be mildly genotoxic and carcinogenic in rats (Miller et al. 1981). Furthermore, safrole is also a weak hepatotoxin and has been shown to induce oxidative damage to liver cells (Liu et al. 1999). The carcinogenicity and toxicity of safrole have been shown to be due to the conversion by rat cytochrome P450 enzymes to electrophilic esters which form covalent adducts with DNA (Miller and Miller 1983). However, it must be noted that these early carcinogenesis/toxicity studies were performed in rodent experimental systems. Parallel safrole metabolism studies in humans demonstrated that the carcinogenic metabolites present in rat urine were absent in humans (Strolin et al. 1977) and thus the carcinogenic activity of safrole may be milder or even non-existent for humans.

8) **Dosage:**

A) **Internal use:**

There is no available dosage information for *T. lanceolata* berry or leaves. Instead, it is generally ingested as a functional food to aid health and well-being. The main component, polygodial, is generally considered to be safe to ingest at relatively high doses, although its toxicological effects have not been thoroughly studied. Due to the possible negative effects of safrole, its content should be monitored and its consumption limited. However, its dosage is usually limited naturally by oral consumption due to the hot, spicy flavour and sharp aroma of polygodial. This limits the ingested dose of all other *T. lanceolata* phytochemicals, providing an inherent safety measure.

B) **External use:**

*T. lanceolata* essential oils may be applied externally to the skin as a general antimicrobial agent. No reports of negative side effects or toxicity were found for the external use of *T. lanceolata* oils.

## 10.19 References


- Anjaneyulu M, Chopra K, Kaur I. Antidepressant activity of quercetin, a bioflavonoid, in streptozotocin-induced diabetic mice. J Med Food. 2003;6(4):391–5.
- Beltrán LR, Dawid C, Beltrán M, Gisselmann G, Degenhardt K, Mathie K, Hofmann T, Hatt H. The pungent substances piperine, capsaicin, 6-gingerol and polygodial inhibit the human two-pore domain potassium channels TASK-1, TASK-3 and TRESK. Front Pharmacol. 2013;4:article 141. Doi: 10.3389/fphar.2013.00141.
- Cock IE, Winnett V, Sirdaarta J, Matthews B. The potential of selected Australian medicinal plants with anti-Proteus activity for the treatment and prevention of rheumatoid arthritis. Pharmacognosy Mag. 2015; 11(42 Suppl. 1): S190–S208
- Cock IE. The phytochemistry and chemotherapeutic potential of *Tasmannia lanceolata* (Tasmanian pepper): a review. Pharmacognosy Commun. 2013;3(4): 13–25.
- Cock IE. Medicinal and aromatic plants—Australia. In: Ethnopharmacology, Encyclopedia of Life Support Systems (EOLSS); Developed under the auspices of UNESCO. Oxford, UK: EOLSS Publishers; 2011. http://www.eolss.net.
- Hurley LL, Akinfiresoye L, Kalejaiye O, Tizabi Y. Antidepressant effects of resveratrol in an animal model of depression. Behav Brain Res. 2014;268:1–7.
- Konczak I, Zabaras D, Dunstan M. Antioxidant capacity and hydrophilic phytochemicals in commercially grown Australian fruits. Food Chem. 2010;123: 1048–54.
- Liu TY, Chen CC, Chen CL, Chi CW. Safrole-induced oxidative damage in the liver of Sprague-Dawley rats. Food Chem Toxicol. 1999;37:697–702.

- Machado DEG, Bettio LE, Cunha MP, Santos AR, Pizzolatti MG, Brighente IM, Rodrigues AL. Antidepressant-like effect of rutin isolated from the ethanolic extract from *Schinus molle* L. in mice: evidence for the involvement of the serotonergic and noradrenergic systems. Eur J Pharmacol. 2008;587:163–8.
- Menary RC, Dragar VA, Thomas S, Read CD. Mountain pepper extract. *Tasmannia lanceolata.* Quality stabilisation and registration. Rural Industries research and development Corporation (RIRDC) 2003; Publication number 02/148.
- Miller JA, Miller EC. The metabolic activation and nucleic acid adducts of naturally-occurring carcinogens. Recent results with ethyl carbamate and the spice flavors safrole and estragole. Br J Cancer. 1983;48:1–15.
- Miller EC, Sxanson AB, Phillips DH, Fletcher TL, Liem A. Structure-activity studies of the carcinogenicities in the mouse and rat of some naturally occurring and synthetic alkylbenzene derivatives related to safrole and estragole. Cancer Res. 1981;43:1124–34.
- Moreira DL, Souza PO, Kaplan MA, Pereira, NA, Cardoso GL, Guimarães EF. Effect of leaf essential oil from *Piper solmsianum* C.DC. in mice behaviour. Anais da Academia Brasileira de Ciências. 2001;73(1):33–7.
- Netzel M, Netzel G, Tian Q, Schwartz S, Konczak I. Native Australian fruits—a novel source of antioxidants for food. Innovat Food Sci Emerg Tech. 2007;8:339–46.
- Page S, Olds M, editors. Botanica: the illustrated A-Z of over 10,000 garden plants for Australian gardens and how to cultivate them. Australia: Random House; 2004.
- Strolin Benedetti M, Malnoe A, Louis Broillet A. Absorption, metabolism and excretion of safrole in the rat and man. Toxicology. 1977;7:69–83.
- Truitt EB, Duritz G, Ebersberger EM. Evidence of monoamine oxidase inhibition by myristicin and nutmeg. Exp Biol Med. 1963;112(3):647–50.
- Winnett V, Boyer H, Sirdaarta J, Cock IE. The potential of *Tasmannia lanceolata* as a natural preservative and medicinal agent: antimicrobial activity and toxicity. Pharmacognosy Commun. 2014;4(1):42–52.


## 10.20 *Terminalia ferdinandiana* Exell

### *10.20.1 (Combretaceae)*

1) **Synonyms:**

   *Myrobalanus edulis* Kuntze, *Terminalia edulis* F. Muell, *Terminalia latipes* subsp. *psilocarpa* Pedley, *Terminalia prostrata* Pedley
2) **Common names:**

   Kakadu plum, green plum, gubinge, billy goat plum, salty plum, wild plum
3) **Photograph of the species** (Fig. 10.10).

**Fig. 10.10** *Terminalia ferdinandiana* tree with fruit. Photograph was taken in 2014 near Darwin, Australia by Dr Ian Cock

4) **Description of the species:**

   A) **Habitat and world distribution:**

   *Terminalia ferdinandiana* is endemic to the tropical grasslands and subtropical woodlands of the Northern Territory and north western parts of Western Australia (Cunningham et al. 2009).

   B) **Morphology of the species:**

   *T. ferdinandiana* is a slender spreading tree which grows to 10 m high. It has rough greyish bark which is smooth in young trees but becomes flaky in older trees. The leaves are very large and broadly elliptical to ovate (15–22 cm long; 12–20 cm wide) with distinct veins. As with all *Terminalia*, the leaves are arranged at the ends (terminal) of the branches. The small creamy-white flowers form on flower spikes (to 20 cm in length) at the end of the dry season (September–November). These develop into small fleshy yellow-green fruits (sometimes with a reddish tinge) in the middle of the wet season (January–June). The fruit is smooth skinned, approximately 1.5–2 cm long and ovoid in shape, often with a short beak at the tip (Byrnes 1977).

C) **Medicinal parts:**

All parts of *T. ferdinandiana* have therapeutic value. Australian Aborigines used *T. ferdinandiana* leaves, bark and tree sap for the treatment and cure of a wide variety of bacterial and fungal skin disorders, venereal diseases, colic and stomachache (Cock 2011; Cock and Mohanty 2011). The fruit is used as a general tonic and may also be used to treat a wide variety of medical complaints including depression, gastrointestinal disorders, skin infections, inflammation and cancer (Mohanty and Cock 2012; Cock 2011). Recent studies have reported on the potent broad spectrum antibacterial and antirheumatic activities of both the fruit and leaves (Courtney et al. 2015; Sirdaarta et al. 2015; Cock and Mohanty 2011) and the anti-giardial activity of the fruit (Rayan et al. 2015).

D) **Chemical constituents:**

The phytochemistry of *T. ferdinandiana* is characterised by the high tannin content of the leaves and the high antioxidant content of fruit (Courtney et al. 2015; Mohanty and Cock 2012). Recent reports have identified several **tannins** including 4-galloylpyrogallol, ellagic acid dihydrate, trimethylellagic acid, chebulic acid, corilagin, punicalin, castalagin and chebulagic acid in *T. ferdinandiana* leaf extracts (Courtney et al. 2015). Less diversity and much lower levels of tannins are present in the fruit (Sirdaarta et al. 2015) with only relatively low levels of gallic acid and chebulic acid present.

A notable characteristic of the fruit is its exceptionally high antioxidant content (Konczak et al. 2010; Netzel et al. 2007). It has been reported that *T. ferdinandiana* fruit has the highest **ascorbic acid** levels of any plant in the world, with levels reported as high as 6 % of the recorded wet weight (Woods 1995; Miller et al. 1993). This is approximately 900 times higher (g/g) than the ascorbic acid content in blueberries (which were used as a standard). A number of **monoterpenoids** including isomyocorene, cineole (eucalyptol), cuminol, camphor and isomenthol have been reported in the fruit extract (Sirdaarta et al. 2015). The same study also identified several **flavonoids** as well as the aromatic esters and acids phthalane and eujavanoic acid. The presence of **stilbenes,** including the resveratrol glycoside piceid and several combretastatins, has also been reported (Sirdaarta et al. 2015).

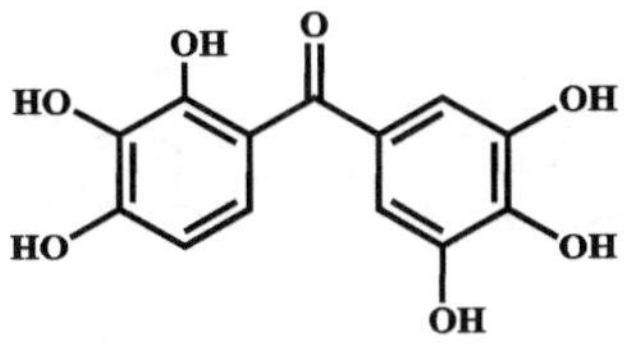


4-Galloylpyrogallol

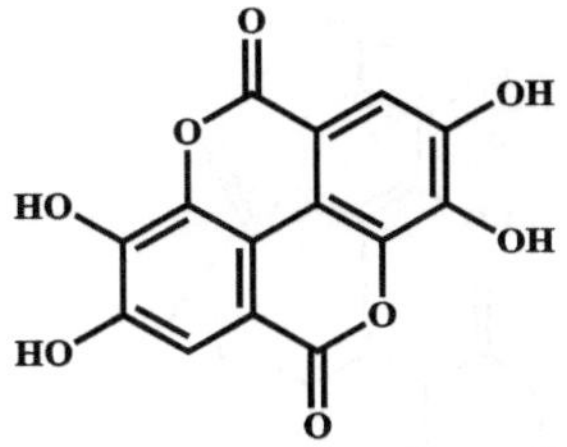


Ellagic acid

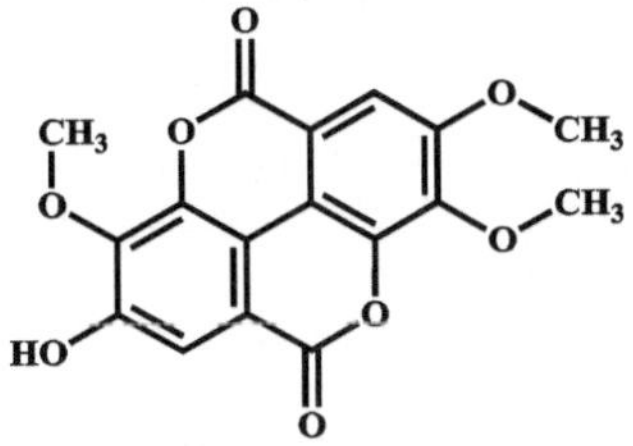


Trimethyl ellagic acid

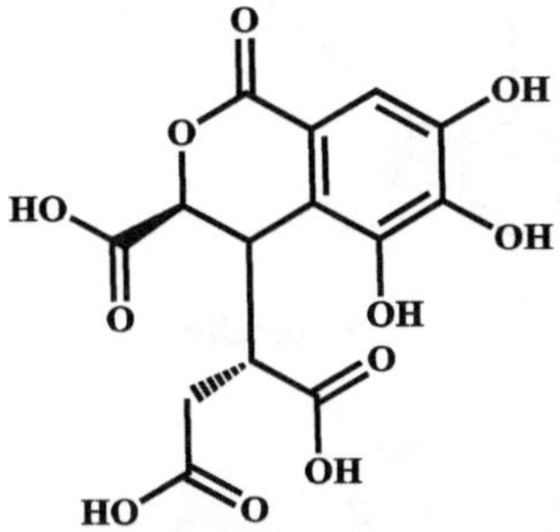


Chebulic acid

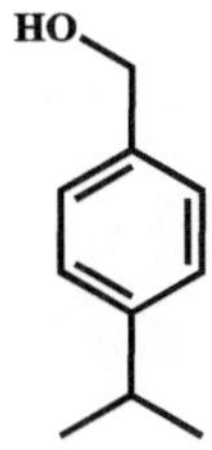


Cuminol

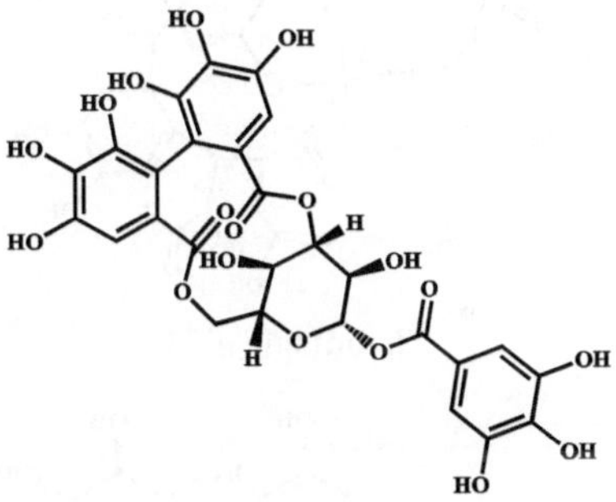

Corilagin

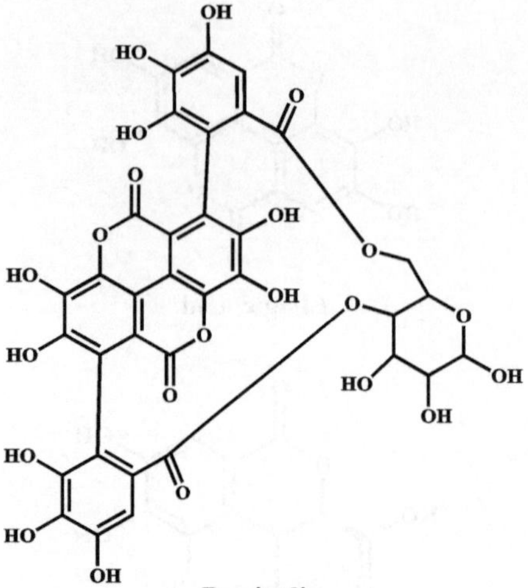

**Punicalin**

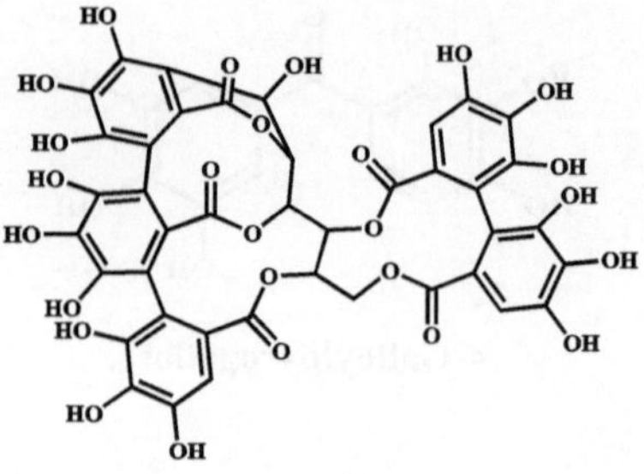

**Castalagin**

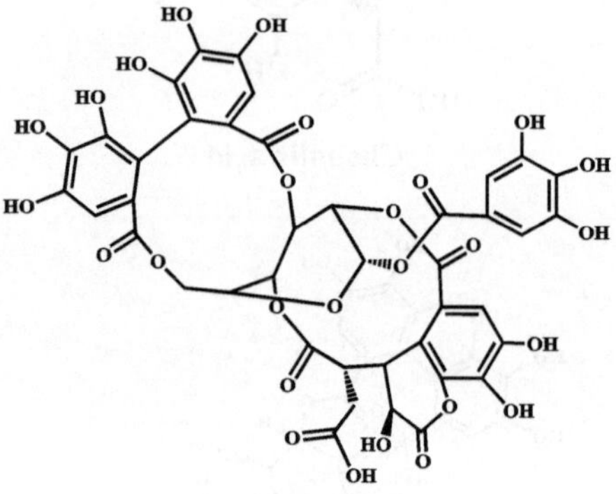

**Chebulagic acid**

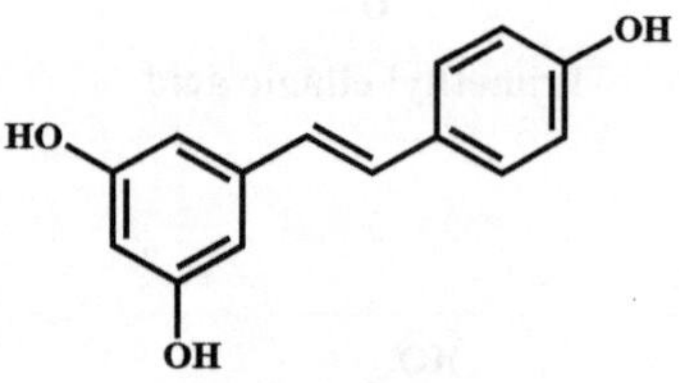

**Resveratrol**

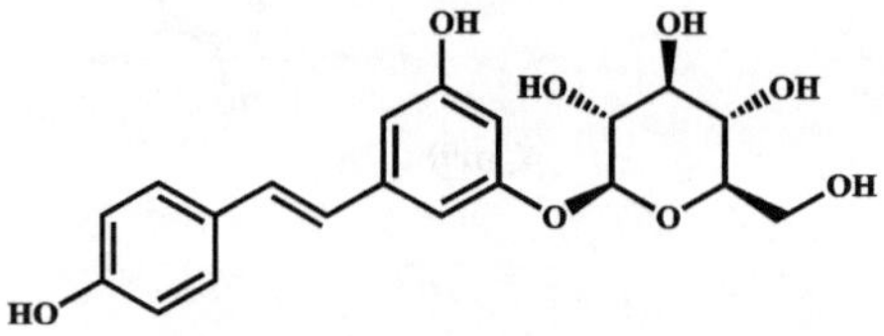

**Piceid**

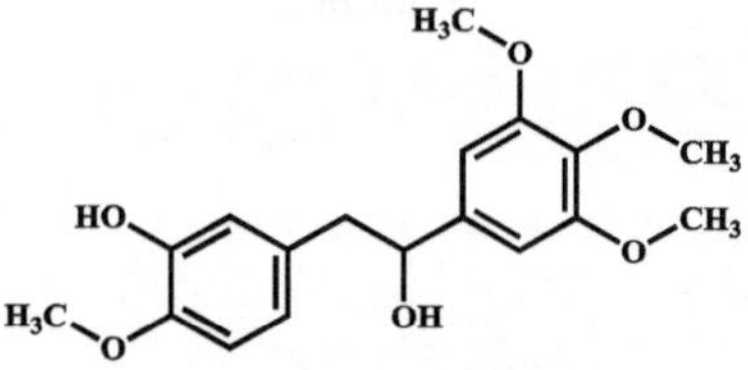

**Combretastatin**

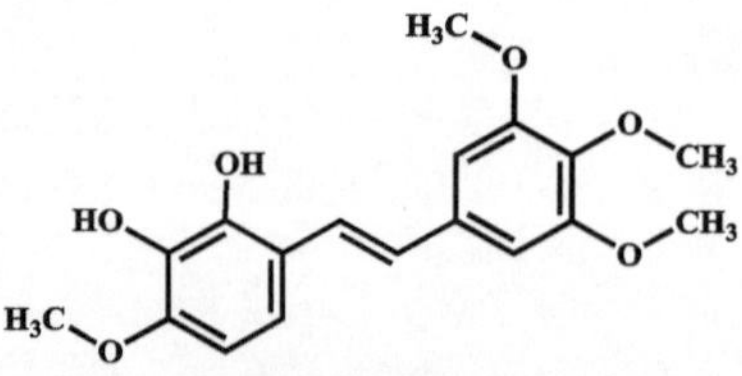

**Combretastatin A1**

5) **Pharmacology and bioactivities:**

*T. ferdinandiana* is characterised by its extremely high antioxidant activity (Konczak et al. 2010; Netzel et al. 2007). High levels of antioxidants can reverse the effects of anxiety and depression and have sedative/hypnotic effects (Gautron et al. 2012). High ascorbic acid levels (as are present in *T. ferdinandiana* fruit) are particularly effective at treating these conditions. Antioxidants may also act as serotonin reuptake inhibitors, thus potentiating the sedative effect (Khanzode et al. 2003). Stilbenes (such as piceid and the combretastatins) present in *T. ferdinandiana* fruit have sedative/hypnotic properties (Sirdaarta et al. 2015; Hurley et al. 2014). Flavonoids, which are present in high levels in *T. ferdinandiana* fruit, have been reported to have antidepressant effects, possibly by increasing the availability of serotonin and noradrenaline in the synaptic cleft (Anjaneyulu et al. 2003; Machado et al. 2008), although the mechanism is yet to be determined. Several monoterpenoids including cuminol, present in significant levels in *T. ferdinandiana* fruit, inhibit fibrillation of α-synuclein which, if aggregated, forms the inclusion bodies characteristic in Parkinson's disease (Morshedi and Aliakbari 2012).

6) **Indications and usage:**

A) **Hypnotic and sedative:**

Due to its phytochemical composition, *T. ferdinandiana* fruit may be useful as a sedative and hypnotic, although this is yet to be properly evaluated. In particular, ascorbic acid and the bioflavonoid and stilbene components have sedative/hypnotic properties (Anjaneyulu et al. 2003; Machado et al. 2008; Hurley et al. 2014).

B) **Antiseptic:**

*T. ferdinandiana* has broad spectrum antiseptic properties (Courtney et al. 2015; Sirdaarta et al. 2015; Cock 2011; Cock and Mohanty 2011) and is therefore indicated for indigestion and other GIT disorders, urinary tract infections and a wide variety of bacterial and fungal skin disorders.

C) **Inflammation and cancer:**

Recent studies have shown that *T. ferdinandiana* fruit (Sirdaatra et al. 2015) and leaf extracts (Courtney et al. 2015) inhibit the microbial triggers of several autoimmune inflammatory diseases including rheumatoid arthritis, ankylosing spondylitis and multiple sclerosis and may be useful in the prevention and treatment of these diseases. Extracts of the fruit can also directly block the activity of cyclooxygenase 2, possibly via the NF-κB, p44/42 mitogen-activated kinase and Akt pathways, resulting in a decreased release of the pro-inflammatory mediator prostaglandin $E_2$ (Tan et al. 2011a). *T. ferdinandiana* fruit extracts also have antiproliferative activity towards human leukaemia cells (HL-60) and colon adenocarcinoma cells (HT-29) (Tan et al. 2011b).

D) **Parkinson's disease:**

As cuminol inhibits fibrillation of α-synuclein (Morshedi and Aliakbari 2012), *T. ferdinandiana* fruit may be useful in the treatment of Parkinson's disease, although this is yet to be examined.

7) **Precautions and side effects:**

Consumption of tannins may induce depression (Girish et al. 2013). Therefore, the tannin-rich parts of the plant (leaves, bark, sap) should be avoided and only the *T. ferdinandiana* fruit used in the treatment of depression.

8) **Dosage:**

There is no available dosage information for *T. ferdinandiana* fruit, leaves, bark or sap. The fruit is generally ingested as a functional food to aid health and well-being (Cock 2011) and is generally considered to be safe to ingest at relatively high doses, although its toxicological effects have not been thoroughly studied.

## 10.21 References


- Anjaneyulu M, Chopra K, Kaur I. Antidepressant activity of quercetin, a bioflavonoid, in streptozotocin-induced diabetic mice. J Med Food. 2003;6(4):391–5.
- Byrnes NB. A revision of the Combretaceae in Australia. Brisbane, Australia: Queensland Herbarium; 1977.
- Cock IE. Medicinal and aromatic plants—Australia. In: Ethnopharmacology, Encyclopedia of Life Support Systems (EOLSS). Developed under the auspices of UNESCO. Oxford, UK: EOLSS Publishers; 2011. http://www.eolss.net.
- Cock IE, Mohanty S. Evaluation of the antibacterial activity and toxicity of *Terminalia ferdinandiana* fruit extracts. Pharmacognosy J. 2011;3(20):72–9.
- Courtney R, Sirdaarta J, Matthews B, Cock IE. Tannin components and inhibitory activity of Kakadu plum leaf extracts against microbial triggers of autoimmune inflammatory disorders. Pharmacognosy J. 2015;7(1):18–31.
- Cunningham AB, Garnett S, Gorman J, Courtenay K, Boehme D. Eco-Enterprises and *Terminalia ferdinandiana* : "Best Laid Plans" and Australian Policy Lessons. Econ Bot. 2009;63:16–28.
- Gautron M, Agrawal M, Gautam M, Sharma P, Gautam AS, Gautam S. Role of antioxidants in generalised anxiety disorder and depression. Indian J Psychiatr. 2012;54(3):244–7.
- Girish C, Raj, V, Arya J, Balakrishnan S. Involvement of the GABAergic system in the anxiolytic-like effect of the flavonoid ellagic acid in mice. Eur J Pharmacol. 2013;710:49–58.
- Hurley LL, Akinfiresoye L, Kalejaiye O, Tizabi Y. Antidepressant effects of resveratrol in an animal model of depression. Behav Brain Res. 2014;268:1–7.
- Khanzode SD, Dakhale GN, Khanzode SS, Saoji A, Palasodkar R. Oxidative damage and major depression: the potential antioxidant action of selective serotonin re-uptake inhibitors. Redox Rep. 2003;8(6):365–70.

- Konczak I, Zabaras D, Dunstan M. Antioxidant capacity and hydrophilic phytochemicals in commercially grown Australian fruits. Food Chem. 2010; 123:1048–54.
- Machado DEG, Bettio LE, Cunha MP, Santos AR, Pizzolatti MG, Brighente IM, Rodrigues AL. Antidepressant-like effect of rutin isolated from the ethanolic extract from *Schinus molle* L. in mice: evidence for the involvement of the serotonergic and noradrenergic systems. Eur J Pharmacol. 2008;587:163–8.
- Miller JB, James KW, Maggiore PM. Tables of composition of Australian Aboriginal foods. Acton, ACT, Australia: Aboriginal Studies Press; 1993. p. 256.
- Mohanty S, Cock IE. The chemotherapeutic potential of *Terminalia ferdinandiana*: phytochemistry and bioactivity. Pharmacognosy Rev. 2012;6(11):29–36.
- Morshedi D, Aliakbari F. The inhibitory effects of cuminaldehyde on amyloid fibrillation and cytotoxicity of alpha-synuclein. Modares J Med Sci Pathobiol. 2012;15(1):45–60.
- Netzel M, Netzel G, Tian Q, Schwartz S, Konczak I. Native Australian fruits—a novel source of antioxidants for food. Innovat Food Sci Emerg Tech. 2007;8:339–46.
- Rayan P, Matthews B, McDonnell PA, Cock IE. *Terminalia ferdinandiana* extracts as inhibitors of *Giardia duodenalis* proliferation: a new drug for the treatment of giardiasis. Parasitol Res. 2015. (In press).
- Sirdaarta J, Matthews B, White A, Cock IE. GC-MS and LC-MS analysis of Kakadu plum fruit extracts displaying inhibitory activity against microbial triggers of multiple sclerosis. Pharmacognosy Commun. 2015;5(2):100–15.
- Tan AC, Konczak I, Ramzan I, Zabaras D, Sze DMY. Potential antioxidant, anti-inflammatory, and proapoptotic anticancer activities of Kakadu plum and Illawarra plum polyphenolic fractions. Nutr Cancer. 2011a;63(7):1074–84.
- Tan AC, Konczak I, Ramzan I, Sze DMY. Australian native fruit polyphenols inhibit cell viability and induce apoptosis in human cancer cell lines. Nutr Cancer. 2011b;63(3):444–55.
- Woods B. A study of the intra-specific variations and commercial potential of *Terminalia fredinandiana* (the kakadu Plum). MSc thesis, Australia: Northern Territory University; 1995.

# Chapter 11
# Novel Drug Delivery Systems for Herbal Antidepressants

**Vandita Kakkar, Nancy Modgill, and Manoj Kumar**

**Abstract** Depression, a potentially life-threatening mood disorder, affects one in six persons in the United States, or approximately 17.6 million Americans each year. It is one of the top ten causes of morbidity and mortality worldwide as reported by the World Health Organization. Depressed patients are more likely to develop secondary diseases like type 2 diabetes and cardiovascular disease. Furthermore, depression is the third leading cause of global disease burden, accounting for 4.3 % of the total disability-adjusted life years. Predictions based on current exponential rise may assign it to be the leading cause of disease burden by the year 2030.

Despite advances in the treatment of depression with selective serotonin reuptake inhibitors (SSRIs) and serotonin and noradrenaline reuptake inhibitors (SNRIs), there continue to be several unmet clinical needs with respect to both the efficacy and the persisting side effects. Herbal drugs with evident antidepressant activity are purported to possess lower risk of precipitating side effects and can be explored as novel drug candidates. Though they find ample space to broaden their therapeutic segment at lab scale, their translation to clinics often has resulted in a failure. The latter is assigned to their compromised bioavailability across the blood–brain barrier (BBB) and attributed to their poor solubility and permeability, photodegradation and lower available systemic concentrations. All these concerns call for the development of novel drug delivery systems (NDDS), which include oral gastro-resistant formulations; transdermal delivery; nasal therapies; nanotechnology-based products, i.e. polymeric and lipidic nanoparticles; self-emulsifying novel drug delivery systems; liposomes; nanostructured lipid carriers; to nanodevices. The commercialisation of the herbal NDDS products especially the nanotech-based delivery system, is governed by stringent regulations, though a few products which fall into the category of modified conventional therapies may get approvals from FDA.

V. Kakkar (✉) • N. Modgill
University Institute of Pharmaceutical Sciences, Panjab University, Chandigarh 160014, India
e-mail: vanditakakkar@yahoo.co.in

M. Kumar
Advanced Drug Delivery Research, R&D, Pancea Biotec Limited, Lalru, Punjab 140501, India



C. Grosso (ed.), *Herbal Medicine in Depression*,
DOI 10.1007/978-3-319-14021-6_11

This chapter elaborates the need for developing the NDDS for herbal antidepressants (AD). Furthermore, an extensive review of the work done by the researchers around the globe and the constraints are also presented herewith. The regulations governing the approval of product for commercialisation and the future direction for these herbal antidepressants are also discussed.

**Keywords** Antidepressants • Bioavailability • Herbals • FDA • Novel delivery system

## 11.1 Introduction

### *11.1.1 Epidemiology*

Depression is one of the top ten causes of morbidity and mortality worldwide based on a survey by the World Health Organization (WHO 2012). It is a significant contributor to the global burden of disease and affects people in all communities across the world. A big strata of population (350 million people) is affected by this life-threatening disorder. A worldwide mental health survey conducted in 17 countries reported an incidence of one in 20 in a group of people (Mental and Day 2012). Depressive disorders which are often recurring start at a young age and reduce people's functioning. For these reasons, depression is the leading cause of disability worldwide in terms of total years lost due to disability. The burden of depression is 50 % higher for females (high-income and low- and middle-income countries) than males (Mathers et al. 2004). The demand for curbing depression and other mental health conditions is on the rise globally (Marcus et al. 2012). Further, research in developing countries suggests that maternal depression may pose a risk factor for poor growth in young children (Rahman et al. 2008). Maternal mental health in low-income countries may have a substantial influence on growth during childhood, with the effects of depression affecting not only this generation but also the next (Marcus et al. 2012). The latter indicates the influence of depression on growth of children not only for the present but also for the subsequent generations. Another serious concern associated with depression is the high suicidal rates, assigned to the occurrence of major depression disorder (MDD).

## 11.2 Management of Depression

Diagnosis and management of patients suffering from mild to moderate depression has been reliably taken up in a primary care centre by general physicians (Marcus et al. 2012). Unless specifically screened, depression may remain under-recognised and untreated in primary care/medical settings. Selective serotonin reuptake inhibitors

(SSRIs) are commonly the initial choice to treat depression in view of their ease of use, more tolerable side effects, superior safety profile and overall efficacy safety. Fluoxetine is the most economical and effective medication, which may be preferred in an otherwise healthy adult with depression; escitalopram is preferred in patients on multitherapy because of *low propensity to cause drug interactions*. More sedating SSRIs, e.g. paroxetine, may be preferred in patients with marked anxiety and insomnia. Sertraline is however extensively studied in patients with cardiac conditions and post-stroke depression. Generally, a lower dose of medications which are gradually increased over the usually recommended time period of 4–6 weeks is recommended (start low, go slow) for improvement in clinical symptoms (Pattanayak and Sagar 2014). *Though these serotonin inhibitors increase the synaptic level of serotonin, however, there is still a great variability in the response of the patients.*

This can be clearly revealed from the high rate of prescription filling for antidepressants (230 million prescriptions), making them one of the most-prescribed drugs in the United States. Despite all of these prescription drugs being taken, more than one in 20 Americans are depressed, according to the recent statistics report from the Centers for Disease Control and Prevention (CDC) (Mercola 2009). *These figures intensify the fact that the current therapeutic options of depression are far from ideal.* Most of the pharmacological interventions available today are not effective in more than 50 % cases and are often associated with a range of serious side effects and drug-drug/drug-food interactions (Baker et al. 2003; Meijer et al. 2004). Further, approximately 50 % of patients do not respond to initial first line antidepressant treatment, while approximately one third fail to achieve remission following several pharmacological interventions (O'Leary et al. 2015).

Unfortunately, these traditional allopathic medicines simply do not work any better than placebos, and there are many studies to support this fact. It is clear that more effective, rapid-acting antidepressants with novel mechanisms of action are thus required (O'Leary et al. 2015). The problems persisting with the current therapy (Fig. 11.1) warrant exploration of herbals for the treatment of depression.

## 11.3 Herbals to Rescue

According to the WHO, herbals are defined as the "medicines which use herbs, herbal materials, herbal preparations and finished herbal products, and contain active ingredients to alleviate suffering and maintain a healthy state". Herbs include the crude plant material such as leaves, flowers, fruit, seed, stems, wood, bark, roots, rhizomes or other plant parts, which may be entire, fragmented or powdered (WHO 2004). They contain active constituents referred to as the polyphenols which exhibit pluripharmacology (Wang and Xu 2012). Antioxidant and anti-inflammatory properties, in addition to their ability to prevent neurodegenerative diseases and cancer, have attracted several researchers to explore them as therapeutic agents (Scalbert et al. 2005).

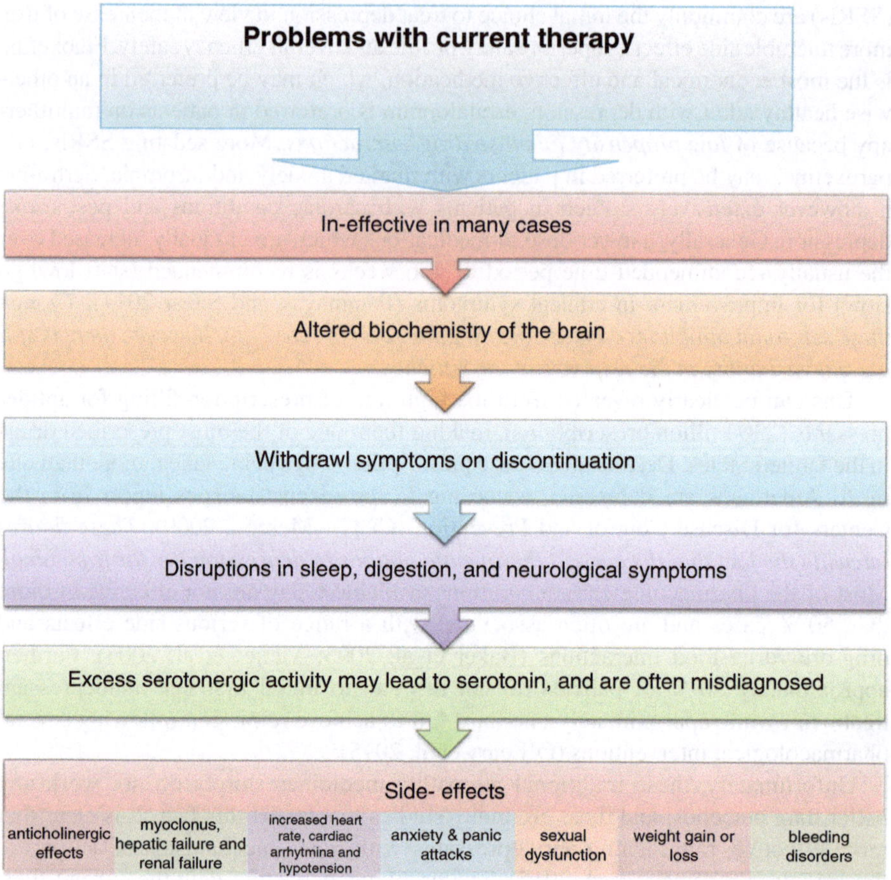


**Fig. 11.1** Various problems associated with current therapy

The active principles can be categorised into three main classes of compounds which include phenolics, isoprenoids and alkaloids. Polyphenolic nutraceuticals, by their pleiotropic effects, are believed to reduce the risk of several major diseases including cardiovascular, autoimmune and neurodegenerative disorder (Facchini 2001; Holstein and Hohl 2004). The major classes of polyphenols include phenolic acids, flavonoids, stilbenes and lignans (Manach et al. 2004). Besides, notable psychoactive constituents isolated from plants (usually containing alkaloids) such as cocaine from *Erythroxylum coca* Lam. (coca), morphine from *Papaver somniferum* L. (opium poppy) or arecoline from *Areca catechu* L. (betel nut), other less potent plants, such as *Hypericum perforatum* L. (St John's wort), have shown significant therapeutic activity over the last several decades in depression (Spinella 2001). Some of the phenolic compounds such as curcumin, low molecular proanthocyanidins, resveratrol, fisetin, piceatannol and ferulic acid, among others, have been used throughout Asia as traditional herbal medicines (Table 11.1). The diverse activities of these compounds and

**Table 11.1** List of herbal antidepressant drugs, with their active constituents mechanism of action and clinical relevance

| Herb | *Active constituents* | Mechanism of action | Clinical relevance |
|---|---|---|---|
| St. John's wort (capsule, tea or liquid extract) *or Hypericum perforatum* L. | *Hyperforin, hypericin and rutin* | Modulates monoamine transmission via Na+ channel<br>Inhibits reuptake of serotonin, dopamine and norepinephrine | Moderate to severe depression<br>Equally effective and sometimes more effective than conventional antidepressants |
| Saffron | Crocin, safranal | Inhibits synaptic serotonin reuptake (Hausenblas et al. 2013) | Performed well in treating mild to moderate depression in trials conducted in Iran in recent years<br>Delivered in 30 mg doses a day, it is as effective as Prozac and imipramine in the short term |
| *Echium amoenum* Fisch. & C.A.Mey. | Contains pyrrolizidine alkaloids | Increase of serotonin and dopamine levels in cerebrospinal fluid | Found superior to placebo in reducing depression on the Hamilton Depression Rating Scale (HAMD) at week four of a study with an effect size of 0.92 (Sayyah et al 2006) |
| *Lavandula* species | Linalool, linalyl acetate, terpinen-4-ol and camphor | Potentiates expression of $GABA_A$ receptors in cell culture, linalool inhibits glutamate binding in the brain<br>Linalyl acetate exerts a relaxing effect (Appleton 2012) | Oral supplementation containing a tincture of *L. angustifolia* with imipramine in the treatment of mild to moderate depression (Appleton 2012) |
| *Curcuma longa* L. | Curcumin, desmethoxycurcumin, bis-desmethoxycurcumin and tetrahydrocurcumin | Enhances serotonin level | The concomitant administration of curcumin and piperine showed a significantly enhanced level of serotonin<br>Synergistic effect when given jointly with antidepressants like fluoxetine, venlafaxine and bupropion |
| Grapes and red wine | Contain the polyphenolic compound *trans*-resveratrol (*trans*-3,5,4′-trihydroxystilbene) | Inhibitory effects of resveratrol on noradrenaline and 5-HT uptake activity<br>Reduced activity of MAO | Comparable results to fluoxetine in preventing symptoms of depression induced by reserpine, after subchronic ingestion for 20 days of *trans*-resveratrol (60 and 120 mg/kg body weight) |

(continued)

**Table 11.1** (continued)

| Herb | *Active constituents* | Mechanism of action | Clinical relevance |
|---|---|---|---|
| Extracts from mulberry extract | Containing cyanidin-3-*O*-beta-D--glucopyranoside | | Neuroprotective effect against the brain injury |
| Green tea | (–)-Epigallocatechin, (–)-epigallocatechin-3-gallate, (–)-epicatechin-3-gallate and (–)-epicatechin | Reduces the neuronal cell death in serum-starved cells and promote the neurite outgrowth | A clinical trial revealed higher consumption of tea decreased symptoms of depression in Japanese population (Zhu et al. 2012) |
| Fisetin | (3,3′,4′7′-Tetrahydroxyflavone) is a flavonoid which exists in many plants and foods, such as strawberries | Protects neuronal cells from oxidative stress and has a neurotrophic effect in improving the differentiation of PC-12 cells (Wang and Xu 2012) | |
| Ferulic acid | (4-Hydroxy-3-methoxycinnamic acid; FA) is an ample phenolic phytochemical found in plant components (Jerome et al. 2011) | Antioxidant; prevents neurons from Aß-induced cell death<br>It can increase the CREB phosphorylation (Wang and Xu 2012) | |

repeated demonstration to reduce the stress response, oxidative damage, inflammation and neuronal damage and act as neuroprotectants strongly suggest these compounds might have a significant impact on stress-induced depression and other affective disorders (Wang and Xu 2012).

Other nutrient cofactors which may alter the mood and neurotransmitter function include a combination of vitamin B referred to as B complex, vitamin C, vitamin $D_3$, essential fatty acids (omega-3's), L-theanine (amino acid), cysteine (amino acid) and zinc. Also, phytotherapeutics available from the Western herbal compendium compose of the St. John's wort, passion flower and valerian or from other ancient traditional paradigms, such as oriental medicinal herbs, are reported to possess a great potential to alleviate symptoms of depression (Pick 2015).

### *11.3.1 The State of Oxidant-Antioxidant Brain Balance*

The association of herbals/nutrients with the mental health has been evident since ages and has been highlighted by several researchers. A good nutrition is important for brain neurogenesis and plasticity so as to attain its functioning throughout the lifecycle (Parletta et al. 2013). Involvement of an oxidative imbalance has been reported in the pathogenesis of depression. Oxidative stress leads to the production of reactive oxygen species (ROS), which are unstable molecules with one or more unpaired electrons that initiate chain reactions which leads to damage of the tissue. Furthermore, these highly toxic molecules can result in impaired cellular lipid membranes and cell functions and oxidised proteins, DNA, RNA and finally cell death. The latter can further lead to a range of chronic degenerative diseases, including cardiovascular disease, cancer and premature ageing.

The brain being lipid rich and metabolically active with a modest level of antioxidant defence mechanisms is more susceptible to lipid peroxidation. This further decreases cell membrane fluidity and can damage membrane proteins; render receptors, enzymes and ion channels inactive; and ultimately break down membrane integrity. Other factors contributing to oxidation in brain include the reaction of neurotransmitters (dopamine, serotonin and norepinephrine) with oxygen, which leads to depletion in the levels of glutathione, the most abundant endogenous antioxidant in the brain (Guest and Grant 2012). This further releases iron and copper ions which cause damage to the brain in the forms that catalyse free radical activity. There is an increasing body of evidence to support the role of oxidative stress, together with inflammatory processes, in cognitive impairment, neurodegeneration and psychiatric disorders (Uttara et al. 2009).

Herbal products with phenolic compounds possess excellent antioxidant properties (Pandey and Rizvi 2009). Flavonoids which comprise *flavonols, flavones, flavanones, flavan-3-ols, chalcones, aurones and anthocyanins* act primarily by scavenging ROS and chelating ions and metals by breaking chains and preventing ROS formation. *The flavonoids act by direct scavenging of free radicals and assist antioxidant defence mechanisms within the cells via modulation of signalling cascades.* Other polyphenols include stilbenes (e.g. resveratrol in red grape skin), phenolic acids and lignans (in linseed). Recent studies have focused on nuclear factor erythroid 2 p45-related factor 2 (Nrf2), a transcription factor that is activated by phytochemicals and plays a key role in the expression of genes responsible for antioxidants such as glutathione and detoxification enzymes (Mattson 2008).

Although research has focused predominantly on antioxidant properties of dietary polyphenols from herbals, there is evidence that their beneficial effects in the brain extend to decreasing inflammation and numerous roles in neurological signalling pathways, e.g. *via* their role in expression of genes that encode antioxidant enzymes, neurotrophic factors and cytoprotective proteins—all of which contribute to neuronal stress adaptation, thereby decreasing neurodegeneration (Vauzour 2012).

### *11.3.2 Mechanisms Underlying the Anti-depressant Effects of Herbals*

Various mechanisms have been proposed to establish the antidepressant activities of the herbals. These include the monoamine (such as serotonin, dopamine and noradrenaline) reuptake inhibitors which enhance the binding and sensitisation of monoamine receptors, monoamine oxidase inhibitors and neuroendocrine modulators (Spinella 2001; Jerome et al. 2011; Kumar 2006). Other effects may include GABAergic effects, cytokine modulation (especially in depressive disorders with a comorbid inflammatory condition) and opioid and cannabinoid system effects (Spinella 2001). Though a multitude of biological effects of phytomedicine on reuptake and receptor binding of various monoamines, in addition to endocrine and psychoneuroimmunological modulation, have been reported (Butterweck and Schmidt 2007; Sarris et al. 2010), however, their antidepressant activity is still not clearly understood.

Some of the herbal medicines with mood-elevating effects (such as *Rhodiola rosea* L. and *Crocus sativus* L.) may also display anxiolytic effects due to dual modulation of neurological pathways (e.g. GABA, serotonin and noradrenaline systems). The latter may also be assigned to a “halo effect” which refers to the treatment of depression, resulting in reduction in anxiety (Roy-Byrne 1998). A clear example of halo effect was depicted by a clinical trial using *Piper methysticum* G. Forst (an established anxiolytic), which reported a significant reduction in depression in addition to its anxiolytic activity (Sarris et al. 2009).

Several theories have been proposed to support the mechanisms by which these herbal antidepressants are known to produce their effect. One of the hypotheses discussed in the previous section was the antioxidant hypothesis. Alternatively, an extract may contain compounds which do not directly affect the pathophysiological processes but may modify the absorption, distribution, metabolism and excretion of bioactive constituents or reduce their side effects (Williamson 2001). *One among these is their “adaptogenic” and “tonic” actions, which are posited to provide increased adaptation to exogenous stressors via complex effects on neurochemistry and the endocrine system* (Panossian and Wikman 2009).

Another hypothesis, i.e. the neurotrophic hypothesis, suggests that the stress-induced reduction of brain-derived neurotrophic factor (BDNF) reduction could, at least in part, induce the structural damage as well as the reduced neurogenesis, and most antidepressant treatment shares the effect of increasing BDNF and neurogenesis, which might be via downstream mechanisms (Panossian and Wikman 2009; Duman et al. 1997; Duman 2004). Neurotrophic factors are critical regulators of the formation and plasticity of neuronal neuritis. Studies have demonstrated that stress reduces BDNF expression which can be prevented by using antidepressants (Lee and Kim 2010).

Despite the great advances in our understanding of depression, little is known about the antidepressant mechanisms of many phytochemicals because their antidepressant effects always involve multiple complex mechanisms. Therefore, it is very important to identify the biomolecules and signalling network that can be specifically regulated by individual phytochemical. Moreover, there is a need to develop high-throughput assays to identify antidepressant phytochemicals. Further, in vivo and in vitro studies are warranted to reveal in more details about the antidepressant mechanisms of phytochemicals (Fowler et al. 2006).

### *11.3.3 Smart Drugs to Dirty Drugs: An Interplay*

"Smart drugs" also referred to as the "targeted drugs" or the "magic bullets" are losing their importance due to the involvement of complex mechanisms in various chronic disorders like cancer and neurodegeneration. This calls for a category of drugs referred to as the "dirty drugs", which bind to many different molecular targets or receptors in the body. The latter, however, may tend to cause several adverse reactions and unwanted effects while acting on the normal cells of the body.

Since decades, the general focus of the pharmaceutical industry has been to develop highly selective and targeted drugs. These, with their specificity, will help in attaining the optimal biological effect with reduced side effects. Drug specificity indeed has been encouraged and supported by regulatory agencies like the US Food and Drug Administration (FDA) and Health Canada (Pierce 2012). However, as the vast majority of chronic diseases are multifactorial with more than one signalling pathway and multiple proteins in the pathology, a failure with the targeted drug delivery has often resulted (Pires et al. 2012; Kukongviriyapan et al. 2012).

Herbal medicines are known to exhibit pleiotropism as a result of multicomponents attached to the mother nuclear structure. Conventional pharmacotherapeutic approaches often fall short for treating chronic conditions. Therefore, we need to delve into the identification of alternative nutraceutical agents that are more efficacious and safe while being flexible in their ability to treat a broader patient base.

Supporting the view, some researchers noted that conventional antidepressants are associated with numerous side effects, have a delay in onset, require higher therapeutic doses and yet are successful in treatment of only a small number of patients with major depressive disorder (Bredt et al. 2015). In comparison, many herbal nutraceuticals function in a pattern similar to antidepressants providing potential as alternative monotherapies, or as augmentative treatments, that are more tolerable and with similar or improved efficacy (Braithwaite et al. 2014).

Increasing popularity of these herbal therapeutics for curing a wide range of disorders calls for their incorporation in novel drug delivery systems (NDDS). NDDS for herbals particularly cure a disorder by making a target/receptor specific. It is a noteworthy point to be emphasised here that the modern era talks of transforming from smart to dirty drug concept. But if one analyses the depth of this fact, it can be sought that these dirty or pleiotropic molecules shall be delivered via smart novel drug delivery techniques. They are prophesised to deliver the encapsulated drug at the target site and at a predetermined rate so as to minimise the toxic effects. As a result, they increase the bioavailability of the incorported drug agent. *An interplay between the concept of "smart and dirty drugs" is thus hypothesised to establish the efficacy of NDDS products of herbal drugs.* Thus, this "pluri-target" concept which establishes a method of delivering a pluri-pharmacological or a multi-targeting drug *via* novel-targeted carrier system so as to control the release of the drug at defined site of action and prevent the chances of damage at the unwanted sites. The variety of novel herbal formulations like polymeric nanoparticles, nanocapsules, liposomes, phytosomes, nanoemulsions, microsphere, transfersomes and ethosomes has been reported using bioactive and plant extracts. The novel formulations are reported to have remarkable advantages over conventional formulations of plant actives and extracts which include enhancement of solubility, bioavailability, protection from toxicity, enhancement of pharmacological activity, enhancement of stability, improved tissue macrophages distribution, sustained delivery and protection from physical and chemical degradation (Saraf 2010).

### *11.3.4 Challenges with the Herbal Drugs*

Herbal drugs hold immense potential to contribute to the advancement of healthcare; however, major challenges must be overcome prior to their successful translation to therapeutics. These include their *inability to cross the blood–brain barrier (BBB), compromised bioavailability assigned to their poor aqueous solubility and permeability and their proneness to degradation via various pathways.* These are clearly depicted at various stages of their processing, upon administration and upon storage in Fig. 11.2.

Apart from the discussed issues associated with the herbals, the analytical validation for assay determinations of herbal formulations is a major hurdle to commercialisation of these products.

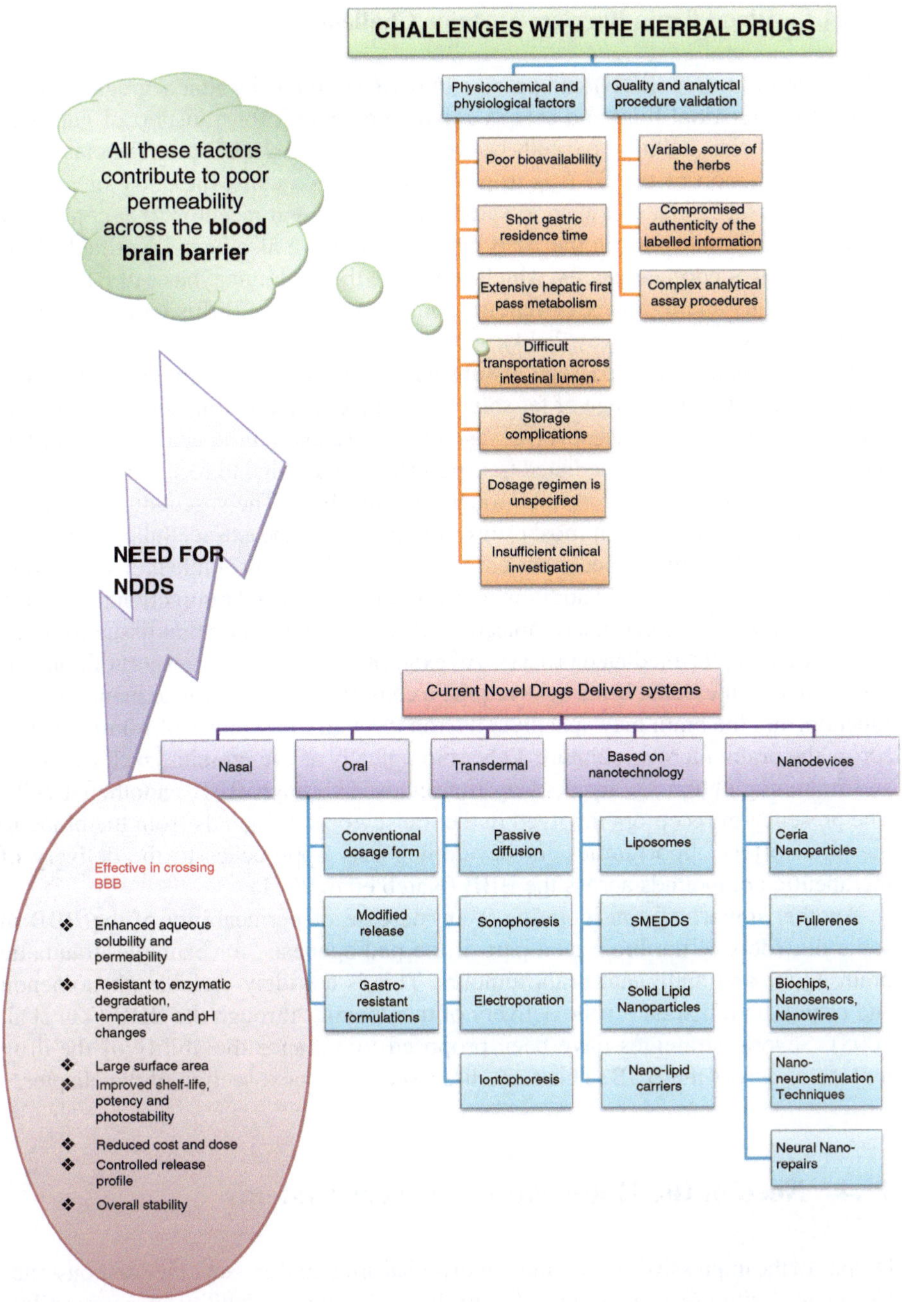


**Fig. 11.2** Need for novel drug delivery system for delivering antidepressants

#### 11.3.4.1 Blood Brain Barrier: A Major Challenge

Blood–brain barrier (BBB) is a passive impermeable barrier that segregates blood and brain interstitial fluid and acts as a dynamic conduit for transport of nutrients, peptides proteins and immune cells between the blood and the brain. The classically accepted function of the BBB is to protect the brain against the entry of noxious agents. Several carrier or transport systems, enzymes and receptors that control the penetration of molecules in the BBB endothelium have also been identified. Brain microvascular endothelial cells, which constitute the anatomical basis of BBB, form tight junctions due to a lack of fenestration and reduce the diffusion of molecules across the vessels (Upadhyay 2014).

The various modes of passage of substances across the BBB include (a) passive diffusion (allows the passage of fat-soluble substances and inhibits the water-soluble substances), (b) active transport (passage of glucose and amino acid occurs *via* this method) and (c) receptor-mediated transport (molecules bind to receptors on the surface of the brain and are escorted through (e.g. insulin)). The vasculature of normal BBB restricts the passage of most drugs from plasma to the extracellular space, with more than an 8-log difference in the entry rate of small, lipid-soluble molecules compared with large proteins. In addition to the unidirectional and bidirectional transport of small molecules, other macromolecules are able to enter the brain tissue from the blood by a receptor-mediated process. An example of this is the transport of transferrin (Tf) across the BBB. Brain cells require a constant supply of iron to maintain their function, and the brain may substitute its iron through transcytosis of iron-loaded Tf across the brain microvasculature. Other biologically active proteins, such as insulin and immunoglobulin G, are actively transcytosed through BBB endothelial cells. The presence of receptors involved in the transcytosis of ligands from the blood to the brain offers opportunities for developing new approaches to the delivery of therapeutic compounds across the BBB (Kateb et al. 2011).

Another important consideration is an increase of permeability of the BBB in some disorders of the brain as a part of the pathogenesis, for example, traumatic brain injury and malignant brain tumours. This is a widely variable phenomenon and cannot be relied on for the delivery of therapeutics through the BBB (Yin et al. 2008). Several strategies have been proposed to enhance the ability of the drug molecules across the BBB. These are discussed in the next section of the chapter.

## 11.4 Need of the Hour: Novel Delivery Systems

Despite of the impressive *in vitro* findings of herbal drugs and extracts, they demonstrates less or negligible *in vivo* activity due to their poor lipid solubility or permeability, resulting in poor absorption and hence poor bioavailability. With the advancement in the technology, NDDS opens the door towards the development of successful oral herbal products by protecting them from rapid elimination and degradation and enhancing solubility and permeation through gastrointestinal tract (GIT) membranes (Braithwaite et al. 2014).

Oral route, being non-invasive and most patient compliant, is considered to be the most acceptable and preferred route as it does not involve any special technique or complex instructions. However, this route is often influenced by dietary factors that may either enhance or impede its bioavailability. For example, lutein shows a dramatically increased absorption profile when consumed with a high fat meal, and lycopene absorption improves with concomitant intake of β-carotene (Alves-Rodrigues et al. 2004; Bernstein et al. 2001).

Novel herbal drug carriers cure particular diseases by targeting exactly the affected zone inside a patient's body and transporting the drug to that area (Dhiman et al. 2012). For the last one decade, many novel carriers such as liposomes, microspheres, nanoparticles, transfersomes, ethosomes, lipid based systems, etc. have been reported for successful modified delivery of various herbal drugs. Many herbal compounds including quercetin, genistein, naringin, sinomenine, piperine, glycyrrhizin and niaziridin have demonstrated capability to enhance the bioavailability (Kesarwani and Gupta 2013). Implantable devices that sense, stimulate, deliver and record from biological systems are being developed through microtechnology and nanotechnology. Transdermal delivery techniques, such as passive diffusion, sonophoresis, electroporation and iontophoresis, enhance the skin's permeability to drugs. Iontophoresis appears to be a promising and perhaps the most efficient assisted-delivery technique for future antidepressant transdermal delivery (Stevens et al. 2015).

### *11.4.1 Strategies Employed to Cross the BBB*

Various strategies employed for manipulating the BBB for drug delivery to the brain include osmotic and chemical opening of the BBB and the use of transport/carriers. Bypassing the BBB by an alternative route of delivery such as transnasal is an important route for delivering the drug (Abdelrahman et al. 2015). If targeted delivery to brain parenchyma is not the goal, alternative methods for crossing the blood–cerebrospinal fluid barrier may be considered or drugs may be introduced directly in the cerebrospinal fluid pathways by lumbar puncture (Welch and Hasbun 2010).

The upper limit of pore size in the BBB that enables passive flow of molecules across it is usually <1 nm; however, particles that have a diameter of several nm can also cross the BBB by carrier-mediated transport (De Jong and Borm 2008).

Ideal characteristics for delivering of therapeutics across the BBB include:

- Controlled release of the drug.
- It should not affect the integrity of the BBB.
- Carrier system used should be biodegradable and not toxic.
- Selective transport of drugs across the BBB.
- Systemic delivery should be targeted to the BBB and the site of intended action in the brain.
- Adequately achievable therapeutic concentrations in the brain.
- Maintenance of therapeutic concentrations for a sufficient duration of time for the desired efficacy.

#### 11.4.1.1 Nanotechnology-Based Drug Delivery Systems

Among the various approaches that are available, nanotechnology-based delivery methods provide the best prospects for crossing the BBB and enhancing the systemic drug concentration (Tiwari et al. 2012).

Nanoparticulate delivery may provide a significant advantage over currently used strategies as they can be transported across the BBB by carriers, also referred to as nanocarriers, without causing any damage to the BBB (De Jong and Borm 2008). Furthermore, this system allows a slow drug release of the drug in the brain, thus decreasing the peripheral toxicity. Various factors that influence the transport include type of polymer or surfactant used, nanoparticle (NP) size and the drug molecule. The various mechanisms involved for successful nanotechnological interventions include:

- Opening of the tight junctions between endothelial cells and enabling the drug to penetrate the BBB
- Transcytosis through the endothelial cell layer
- Endocytosis by endothelial cells and release of the drug inside the cell
- Inhibition of the transmembrane efflux systems (i.e. P-glycoprotein) by coating agents such as polysorbates

However, NPs may sometimes induce local toxic effects on the brain vasculature leading to a limited permeabilisation of the brain endothelial cells.

Work done by various researchers globally, in the development of nanotechnology-based delivery systems for herbal antidepressants, will be illustrated below.

(a) Self-microemulsifying drug delivery system (SMEDDS)
Self-microemulsifying drug delivery system (SMEDDS) is a system which can form fine oil-in-water droplets with a diameter size of less than 50 nm under mild agitation of the GIT without the dissolution process. Wu et al. (2015) reported that SMEDDS of AJS, an untitled novel drug synthesised based on the structure of cinnamamide, a Biopharmaceutics Classification System (BCS) class II drug, has a better action on the 5-hydroxytryptamine (5-HT) receptor attributed to the improved solubility and oral bioavailability (Wu et al. 2015).

(b) Solid Lipid Nanoparticles (SLNs)
SLNs constitute a colloidal delivery system, wherein the solid lipid is stabilised by a layer of surfactant and co-surfactant dispersed in an aqueous solution. A SLN is typically spherical with an average diameter between 10 and 1000 nm. The lipophilic features of the SLNs facilitate crossing the BBB to enter the brain by endocytosis. Several drugs have been incorporated into SLNs, which are potentially useful for the treatment of brain diseases. SLNs are deemed to be superior as compared to other delivery systems since there is no random fusion between the particles or with other membranes; the surface charge and molecular makeup can be easily modified with the possibility of multivalent attachment of small molecule ligands; the production procedures do not involve the use of organic solvent; they are stable over years and control the release of

the drug at the desired site of action. Kakkar and Kaur evaluated the antidepressant activity of the SLN of curcumin (a hydrophobic polyphenol) in the forced swim model of depression in mice. They reported that C-SLNs with their improved bioavailability and permeability revealed a higher antidepressant potential upon administration of a single and a much lower dose when compared to free curcumin (Kakkar and Kaur 2012).

(c) Liposomes

Liposome is a spherical vesicle composed of phospholipids, with at least one lipid bilayer, and used as a vehicle for administration of nutrients and pharmaceutical drugs (Torchilin 2011). Liposomes can be prepared by disrupting biological membranes (such as by sonication). Liposome properties vary substantially with lipid composition, size, surface charge and the method of preparation. Small liposomes, nanoliposomes, measure 25–50 nm in diameter. Though they are biocompatible, however, they possess high production costs. Liposomes can be used in strategies for therapeutic delivery across the BBB. Functionalisation of liposomes with ApoE-derived peptides facilitates cellular uptake and drug transport across a model of the BBB (Re et al. 2011).

Other nutraceuticals formulated within liposome delivery systems include *Echinacea* sp., glucosamine, *Piper methysticum* G. Forst., *Silybum marianum* (L.) Gaertn., *Hypericum perforatum* L., etc. (Braithwaite et al. 2014).

Antidepressant and cognitive effects of piperine-encapsulated liposomes (PL) in male Wistar rats have been reported. Oral piperine (5 mg/kg body weight/day) and intranasal PL (7.2 μg/day) were administered daily for 14 days to rats, which were subjected to forced swimming, Morris water maze and spontaneous motor behaviour tests. The quantitative analysis of the brain homogenates by HPLC indicated that orally or nasally delivered piperine is distributed to the hippocampus at a higher extent than the cortex and that the time to peak concentration of nasal PL was shorter than for the oral piperine. The authors concluded that intranasal PL showed significantly higher delivery of piperine, at a low dose, to exert its antidepressant and cognitive enhancing activities (Priprem et al. 2011).

Another study explored the effect of quercetin liposomes via nasal administration on the anxiety, depression-like activity and cognitive enhancing effect. They administered quercetin liposomes, containing 0.5 mg of quercetin, *via* intranasal route once daily continually for 4 weeks to male Wistar rats and reported the enhanced antidepressant activity of liposomal quercetin formulation, at all times of the treatment, in elevated plus maze test, forced swimming test and Morris water maze test, respectively (Wattanathorn et al. 2007). Other studies supporting the enhanced antidepressant effects of quercetin liposomes have also been reported at other dose levels (Priprem et al. 2008).

*It becomes imperative to highlight here that several NDDS are being explored not only for the herbal drugs but also the synthetic antidepressants for enhancing their bioavailability across the BBB. Strategies employed for enhancing the bioavailability of current synthetic antidepressants are discussed in the section below:*

(a) Polymeric nanoparticles

They have shown to be promising carriers for CNS drug delivery due to their potential in encapsulating drugs, hence protecting them from excretion and metabolism, and in delivering active agents across the BBB without inflicting any damage to the barrier (Rabanel et al. 2015). Polymeric NPs enable targeted drug delivery across the BBB with controlled release at the desired target. Another advantage of polymeric NPs is that they are biodegradable and NPs are constructed using approved raw materials which are relatively inexpensive and are easy to manufacture on largeside. However, some polymeric NPs, particularly those with a complex structure, are expensive. Some examples of polymeric NPs are chitosans, dendrimers, nanogels, poly(D,L-lactide-co-glycolide) (PLGA), poly(D,L-lactide) (PLA) and polybutylcyanoacrylate (PBCA). Polymeric NPs for delivery across the BBB should be biocompatible, nontoxic, nonthrombogenic and nonimmunogenic (Ma and Mumper 2013). A polysorbate 80 coating on the surface of PBCAs adsorbs ApoB and E, which facilitates their uptake by brain capillary endothelial cells via receptor-mediated endocytosis and assist their passage across the BBB (Re et al. 2011).

(b) Nanostructured lipid carriers (NLCs)

Nanostructured lipid carriers (NLCs) Consist of partially crystallised lipid particles with mean radii = 100 nm, dispersed in an aqueous phase containing emulsifier(s), as a potential delivery system. They possess some advantages over other colloidal carriers in terms of higher drug incorporation potential and stability and thus are chosen as excellent carrier systems for delivering herbal drugs. They enhance the bioavailability and stability of bioactive compounds, shelf life, consumer acceptability, functionality, nutritional value and safety of food systems and provide controlled release of encapsulated materials (Tamjidi et al. 2013).

NLCs have been identified as as a potential delivery system for drug agents apart from the natural molecules. Alam et al. (2012) reported an enhancement of eight times higher concentration of duloxetine (DLX) NLCs in the brain via intranasal administration when compared with the intravenous administration of DLX solution. Intranasal administration of nanostructured lipid carriers (NLCs) containing antidepressant drug circumvents the BBB and maintains the prolonged release at the site of action (Alam et al. 2012, 2014). The authors used pharmacoscintigraphy to determine the drug targeting potential (DTP), drug targeting efficiency (DTE) and biodistribution studies in different organs including the brain after administering duloxetine-loaded NLC (DLX-NLC). The biodistribution studies exhibited higher percentage of radioactivity/g for DLX-NLC formulations in the brain as compared with the DLX. They revealed higher DTP (86.80 %) and DTE (757.74 %) for DLX-NLC formulation and concluded the latter to have a better brain targeting efficiency than DLX solution (DTP = 65.12 %; DTE = 287.34 %) when administered intranasally.

The new antidepressant venlafaxine was incorporated into nanostructured lipid carrier (NLC) gel, for its intranasal delivery. Venlafaxine-loaded NLCs were prepared by emulsion solvent diffusion and evaporation method followed by ultrasonication. The particle size of optimised batch of NLCs was found to be 151.83 nm, and zeta potential was found to be –8.08 mV. Drug loading was found to be 34.33 %. A toxicological study carried out on sheep nasal mucosa showed no evidence of toxicity of the NLC-based gel. The authors reported NLC-based gel to show a rapid onset along with a prolonged duration of action as compared with pure drug solution and NLC dispersion (Joshi and Patravale 2006).

Apart from the above-cited approaches, several other novel drug delivery devices are being designed and evaluated. A brief summary of the work done in this area is discussed below.

(c) Gold nanoparticles

Researchers investigated the protective effect of *H. perforatum* gold nanoparticles' (nanohypericum-HPGNPs) role against restraint stress-induced behavioural and biochemical alterations in mice. HPE (200 mg/kg) and nanohypericum (20 mg/kg) treatment for 21 days significantly attenuated restraint stress-induced behavioural and oxidative damage (Prakash et al. 2010).

(d) Silver nanoparticles

A series of the six tricyclic antidepressants (imipramine, desipramine, clomipramine amitriptyline, nortriptyline and doxepin) known as a surface-active drug were adsorbed on the metal substrate in a form of silver nanoparticles (ca. 30 nm in diameter), and their interactions were examined in terms of surface-enhanced Raman spectroscopy (SERS). An analysis of SERS spectra revealed the dibenzopine moiety to be the primary site of the adsorption and was consistent with a model, in which the tricyclic ring was docked in the outer vestibule of biogenic amine transporters, whereas the dimethylaminopropyl side chain pointed to the substrate binding site. The authors clearly demonstrated potential of SERS technique in predicting a key functional group responsible for drug action (Jaworska 2014).

Aloe vera extract in NP, nanocurcumin, cumin NP and ayurvedic bhasma (a nano-preparation) has also been employed for enhanced bioavailability of these herbs (Sahu 2013).

## 11.5 Nanodevices

Treatment of common brain disorders at the molecular level, inside the specific cells, is now taken up through a nanodevice (Litt 2013).

This is a type of nanomedicine which is used for the treatment of depression (and other mental illnesses) and includes nanoelectronic biosensors (to help diagnose disease), nanocarriers (for superior drug delivery) and nanorobots (to modulate neurophysiological correlates of depression). Nanorobots could help correct genetic abnormalities that contribute to feelings of depression (and other conditions). Furthermore, nanorobots could guide doses of a drug to isolated regions of the brain, without affecting other (potentially non-problematic) regions.

Various studies on use of silica-based nanocarriers loaded with piracetam, a popular "racetam" nootropic, have been reported. The nanocarriers were able to transfer a greater amount of piracetam across the BBB than when taken orally (Jampilek et al. 2015).

Some nanomaterials, such as ceria NPs and water-soluble derivatives of buckminsterfullerene C60, are considered as neuroprotectives and are a unique class of NP compounds with potent antioxidant properties. Implantable devices such as biochips will also be in use to surpass the BBB in near future (Jain 2012).

(a) Nanosensors, nanowires and nano-neurostimulation techniques

Nanosensors are designed to interpret efficacy, potency and effects of certain drugs. Further, they may provide in-depth details regarding antidepressant (and psychotropic) mechanisms of action and provide complete patient-specific data.

Implant nanowires, which appear to be extremely thin and flexible, are used within the brain to read individualised signals of a depressed person. They however do not disrupt any brain tissue yet are able to collect data of neurotransmission and nerve signals within the brain.

Another innovation involves the stimulation of specific regions of the brain to alleviate the depressive symptoms. This is evidenced by repetitive transcranial magnetic stimulation (rTMS). rTMS in the left-prefrontal cortex is known to reduce depression. Furthermore, implanted electrodes or neural implants within the brain that stimulate the pleasure centres may also offset depressive symptoms.

(b) Neural nano-repairs

These are the preprogrammed nanorobots inserted into brains to repair the damage caused in the depressed patients naturally, or by the use of artificial nano-replacements, or carry stem cells to replace the damaged tissue (Jain 2012).

(c) Nanoscaled implants

They refer to the nanoscaled implants which mimic receptor proteins and can alter the neural pathways. Developed by Japanese researchers, these implants hold potential to increase the release of neurotransmitters responsible for both excitatory and inhibitory activity.

Other innovation in drug delivery includes the use of nanosomes, nanotubes and nanobubbles. It was also noted that nanoshells and dendrimers may provide additional therapeutic effects via enhanced potency and increased synergy (of combined treatments) (Jain 2012).

- *Nanosomes*: These are considered uniform liposomes that are manufactured with supercritical fluid technology.
- *Nanotubes*: These are tube-like structures consisting of graphite. They are regarded as being soluble and effective for transportation of some drugs.
- *Nanobubbles*: These are considered tiny (nano)bubbles that contain gas within a cavity- submerged in a liquid solution.

## 11.6 Intranasal Delivery

It has been shown in the literature from animal and human investigations that transport of exogenous materials directly from the nose to the brain is a potential route for bypassing the BBB (Illum 2000). This route involves the olfactory or trigeminal nerve systems which initiate in the brain and terminate in the nasal cavity at the olfactory neuroepithelium or respiratory epithelium, respectively. They are the only externally exposed portions of the CNS and therefore represent the most direct method of non-invasive entry into the brain. However, the quantities of drug administered nasally that have been shown to be transported directly from the nose to the brain are very low, normally less than 0.1 %, and hence the system is not currently used therapeutically, and no product is licensed specifically via this route (Illum 2004).

The strategy of applying drugs that are encapsulated into particulate vectors (such as synthetic NPs) to the olfactory epithelium could potentially improve the direct CNS delivery of drugs (Anton et al. 2012).

Thakkar et al. (2013) prepared and characterised mirtazapine microemulsion for intranasal delivery, to determine its brain drug delivery using pharmacokinetic studies, and assessed its performance pharmacodynamically for the antidepressant activity. Results showed that microemulsion containing Capmul MCM, Tween 80 and PEG 400 showed the highest solubilisation for mirtazapine. Mirtazapine microemulsion (MME; intranasal) and mirtazapine mucoadhesive microemulsion (MMME; intranasal) showed 11-folds and 14-folds higher Cmax (brain) and 18-folds and 20-folds higher AUC (brain) compared to MME (oral). Higher Cmax and AUC (brain) were observed for MMME (intranasal) compared to MME (intranasal) demonstrating the role of mucoadhesive agent. This may be attributed to longer residence time of the mucoadhesive microemulsion in the nasal cavity. Significant ($p<0.05$) reduction in assessed pharmacodynamic parameters was observed after intranasal administration of MMME against the control group. The authors demonstrated a more rapid and larger extent of transport of mirtazapine into the brain with intranasal MMME, which may prove useful in treating depression (Thakkar et al. 2013).

## 11.7 Transdermal Route

Drug delivery systems such as patches are more patient friendly and enable them to continue treatment for longer periods as a result of sustained delivery. Delivering medicine into the general circulation through the skin is seen as a desirable alternative to oral administration. In addition to increasing convenience, transdermal delivery can change the metabolism and bioavailability of compounds and their metabolites and thus alter the therapeutic index of a particular drug. Because of the reduced frequency of administration, compliance should be enhanced with transdermal delivery. Delivery of drug via a transdermal route also avoids the hostile environment of gastrointestinal tract, where drugs can be inactivated and absorption can vary depending on pH, food ingestion/interaction and other factors. In addition, oral medications may cause nausea because of local effects, and some cannot be taken if the patient is already nauseated. Transdermal route also avoids hepatic first-pass metabolism and associated side effects. Transdermal delivery offers convenience, especially notable in patches that require only once weekly application. It has been observed that many psychotropic drugs are non-compliant because of large and diverse array of adverse effects and drug adherence problems, and transdermal drug delivery may be useful in achieving patient compliance.

Transdermal drug delivery system (TDDS) of psychotropic drugs like haloperidol, imipramine, fluoxetine, selegiline and lithium has already been studied. Transdermal drug delivery system of selegiline, which is a monoamine oxidase (MAO) inhibitor, is approved by FDA with unique pharmacokinetic and pharmacodynamic properties (Jessen et al. 2008). This was developed to overcome limitations of orally administered MAOs, particularly dietary tyramine restrictions, and was found to be suitable in long-term treatment of major depressive disorder. Transdermal patches of sertraline hydrochloride have also been prepared and tested by using the polymer Eudragit RS-100, Eudragit RL-100, EC and dibutyl phthalate (Vijaya and Ruckmani 2011).

Dermal application may result in an accumulation of nutraceuticals in the skin and is an excellent alternative to the oral route, as it bypasses the GIT milieu and reduces hepatic and renal inactivation (Walle et al. 2004). However, some nutraceuticals, when applied topically to the skin, may become unstable on exposure to light or heat due to photodegradation. In addition to modified carrier systems available for nutraceutical formulations, a number of delivery aid mechanisms exist—like chemical penetration enhancers, iontophoresis, sonophoresis, etc.—that can be applied locally to assist in the transport of active principles to the skin and underlying tissues (Braithwaite et al. 2014).

## 11.8 Risks of Using Nanotechnology for Depression

Though the nano-based products provide ample advantages in delivering the herbal products for the treatment of depression, however, they are associated with few health and environmental concerns.

Firstly, there is dearth of reports on the clearance of the drug from the brain post-encapsulation in its nano form. Improper clearance may trigger inflammation and have other deleterious long-term consequences. The second issue relates to the environmental dilemma when using nanotechnology-based product for the treatment of depression. It was hypothesised in one of the reports that nanoparticles could make their way into wastewater streams, which may ultimately kill the useful bacteria of the ecosystems (MHDaily 2015).

The third serious concern depicts the ability of these nanoparticles to result in a cascade of unwanted immune reactions, possibly leading to sickness and/or death. Finally, the toxicity remains a concern for nanomaterials which could be toxic to an individual's CNS, ultimately damaging neurophysiology rather than imparting a healthy state.

## 11.9 FDA Regulations on Herbal Drugs

The law defines dietary supplements in part as products taken by mouth that contain a "dietary ingredient". Dietary ingredients include vitamins, minerals, amino acids and herbs or botanicals, as well as other substances that can be used to supplement the diet (www.fda.gov/downloads/ForConsumers/ConsumerUpdates/UCM050824.pdf).

Dietary supplements though have clean safety histories, but since they became widely available in 1994, the FDA and few researchers reported concerns pertaining to the use of some dietary supplements.

- The foremost issue was the contamination of herbs with germs, pesticides or toxic heavy metals.
- *Another problem associated was the falsifying information on the label.* Herbal supplements either do not mention their composition or contain either more or less than the amount specified, and in some cases, the ingredients are not listed on the label. This problem extends beyond the supplement producers/manufacturers and sellers. Some herbal suppliers (those who grow, harvest or sell the crops) may adulterate their crops with less expensive or more readily available plants. Moreover, the accidental contamination, when one plant grows in with others, and chances of flawed identity (when one plant looks like another) further worsen the authenticity of the product.

The FDA in its guidance documents has very aptly raised a few concerns regarding the dietary/herbal supplement. The laws, which are very relaxing, may entangle manufacturers to indulge in malpractices, leading to selling a low-quality product at much higher prices.

1. Federal law requires that every dietary supplement be labelled as such, either with the term "dietary supplement" or with a term that substitutes a description of the product's dietary ingredient(s) for the word "dietary" (e.g. "herbal supplement" or "calcium supplement"). Further, even more relaxed grounds are created when the "law does not require the manufacturer or seller to prove to FDA's satisfaction that the claim made on the label is accurate or truthful before it appears on the product". Moreover, the advertising procedure for these products, including ads broadcast on radio and television, falls under the jurisdiction of the Federal Trade Commission.

2. Law does not require dietary supplements to be proven safe to FDA's satisfaction before they are marketed. This requirement is one of the reasons which has led to the overflooding of market with the herbal products as such or the nano-converted form for enhancing their bioavailability. The manufacturers with this support from law are free to launch a nanoceutical (herbal product converted to a nano-based formulation) which may be otherwise has not been tested for its toxicity and safety data.

*The persisting role of FDA's control over these dietary products is far from ideal*, wherein its role begins once the product enters the marketplace. That is usually the agency's first opportunity to take action against a product that presents a significant or unreasonable risk of illness or injury or that is otherwise adulterated or misbranded. Safety monitoring responsibilities of FDA to report any side effects or any adverse effects reported by volunteers are taken up at this point of time. An important point of discussion comes when we say that "we are breathing in a nanotechnology era" and when a herbal drug is converted to a nanotechnology-based formulation and still is bearing on it a label of "modified dietary supplement". *How does the FDA regulate now? Time and again these questions are being raised at several stages, but the outcomes are mere discussions at various forums and publications in various guidance documents and journals.*

There are limitations to FDA oversight of claims in dietary supplement labelling and manufacturing. Despite all these issues, the FDA is not legally responsible for the safety of dietary supplements.

*Quality control and standardisation of herbal medicines:* Another concern in developing the herbals and its novel drug delivery system as therapeutics is the need to establish quality parameters for collection, handling, processing and production of herbal medicine as well as employ such parameters in ensuring the safety of the global herbal market. This involves adjusting the herbal drug preparation to a defined content of a constituent or a group of substances with known therapeutic activity by adding excipients or by mixing herbal drugs or herbal drug preparations.

Further, herbals made directly from crude plant material show substantial variation in composition, quality and therapeutic effects. Standardised extracts are high-quality extracts containing consistent levels of specified compounds, and they are subjected to rigorous quality controls during all phases of the growing, harvesting and manufacturing processes.

When the active principles are unknown, marker substances should be established for analytical purposes and standardisation. Marker substances are chemically defined constituents of herbal drugs that are important for the quality of the finished product. Ideally, the chemical markers chosen would also be the compounds that are responsible for the pharmacological effects in the body.

In the case of herbal drug preparations, the production and primary processing of the medicinal plant or herbal drug have a direct influence on the quality of the active pharmaceutical ingredients (APIs). Due to the inherent complexity of naturally growing medicinal plants and the limited availability of simple analytical techniques to identify and characterise the active constituents solely by chemical or biological means, there is a need for an adequate quality assurance system. This assurance is also required during cultivation, harvesting, primary processing, handling, storage, packaging and distribution. Deterioration and contamination through adulteration, especially microbial contamination, can occur at any one of these stages. It is extremely important to establish good agricultural, harvesting and manufacturing practices for herbal starting materials in order to minimise these undesirable factors. In this regard, producers, processors and traders of medicinal plants or herbal drugs have an obligation and a role to play. The manufacturers and suppliers of herbal products should adhere to quality control standards and good manufacturing practices.

## 11.10 Analytical Method Validation for Herbals: A Problem!

Herbals, which are mostly a combination of one or more constituents, often fail during a novel technology transfer due to the lack of standard analytical procedures for the analysis of its constituents. This is assigned to the current scenario, where no organisation or government body certifies an herb or supplement as being labelled correctly. The latter adds to the arduous task for developing a validated procedure as many of the unlisted constituents may interfere with the analytical procedure. Moreover, the information provided on the label is incomplete and often cannot be trusted to reveal the contents of the container. Studies on herbal products have revealed the chances of getting an authentic herbal product to be less than 50 %, the latter being assigned to the anomalous information on the label.

The word "standardised" on a product label does not assure of a quality product, since it does not have any legal definition. It is left to the discretion of consumers to assume the safety and efficacy of the product before becoming a guinea pig for the trial.

As referred and specified, the most pragmatic approach for quality control of herbal drugs thus would be to follow the published monographs in a pharmacopoeia (EMEA 2006; WHO 2003; FDA 2014). However, in cases where a monograph is not available, development and validation of analytical procedures have to be carried out by the manufacturer. The best strategy is to follow closely the pharmacopoeia definitions of identity, purity and content or assay. Valuable sources for general analytical procedures are included in the pharmacopoeias and in guidelines published by the WHO (2010). Furthermore, relevant information specifying the quantity of active ingredients, directions (dosage) and frequency of intake of the drug must be provided on the label to improve patient compliance.

## 11.11 Future Perspective and Conclusions

A plethora of evidence supports the health benefits and value of herbals in the treatment and prevention of depression. Depression, a life-threatening mood disorder, is on its peak to be the leading cause of disease burden by 2030. Lack of available effective therapeutic options, numerous side effects associated with the conventional antidepressants, their delayed onset, high dose levels for achieving efficacy and limited success rate have prompted the researchers to explore the folklore treasure so as to produce alternative agents that are more efficacious and safe, while being flexible in their ability to treat a broader patient base. Natural compounds due to their unique ability to promote the body's own natural healing process were used in preventative medicine but, now, are frequently prescribed and recommended as complementary or alternative treatments for ailments which involve complex mechanisms.

Though evidence supports the herbals to possess a reduced risk of side effects, their compromised bioavailability hampers their clinical translation as therapeutics.

Over the last few years, one of the prominent focus areas involves the development of the novel delivery methods to optimise dosage forms and formulations for improving bioavailability of promising herbal/natural drug candidates. The exponential growth in the novel drug delivery sector for herbals both in research and clinical applications is anticipated to result in the development of safe and effective treatment options in the next decade. Further, the issues of safety are expected to get resolved through thorough understanding of the mechanism of drug transport across the BBB. The latter shall aid in planning and execution of advanced knowledge-based therapies for delivering the herbal drugs via novel delivery mechanisms across the BBB. Several clinical trials are still warranted to establish the efficacy of existing as well as new drugs for neurological disorders.

The future emphasis of research will be on developing herbal drug delivery systems for the brain that can deliver adequate amounts of phytochemical in a controlled manner and in response to the requirements. Various techniques including the nano-based delivery systems, transdermal and intranasal, in addition to implantable devices such as biochips are on their way for delivering the drug precisely at the site of action. They can be developed as versatile tools for not only the diagnostic purpose but may act as therapeutic as well.

More to the above, maintaining a balance between stringent regulations for novel drug delivery systems especially the nano-based products and the oversight by the regulators for commercialising the herbal drugs or supplements is the need of the hour. Impediments to achieve a commercially viable novel carrier system for herbals are worsened with the quality of the drug and its source, in addition to the complexity involved in developing the validated analytical procedures. The above scenario heralds the potential of innovations for herbals and subsequently calls for a mutual partnership between the manufacturers and FDA to unveil an authentic composition so as to calm the herculean task of analytical method development for these novel drug delivery-based herbal products. Implementation of the latter will not only assist in improving the market viability of the herbal antidepressant-based NDDS product but also promulgate a feeling of faithfulness amid the consumers.

**Disclosure** The author reports no conflict of interest in this work.

## References

Abdelrahman FE et al. Investigating the cubosomal ability for transnasal brain targeting: In vitro optimization, ex vivo permeation and in vivo biodistribution. Int J Pharm. 2015;490(1–2):281–91. Available at: http://www.sciencedirect.com/science/article/pii/S0378517315004937.

Alam MI et al. Intranasal administration of nanostructured lipid carriers containing CNS acting drug: pharmacodynamic studies and estimation in blood and brain. J Psychiatr Res. 2012;46(9):1133–8.

Alves-Rodrigues A, Shao A. The science behind lutein. Toxicol Lett. 2004;150(1):57–83.

Anton N, Jakhmola A, Vandamme TF. Trojan microparticles for drug delivery. Pharmaceutics. 2012;4(1):1–25. Available at: http://www.ncbi.nlm.nih.gov/pmc/articles/PMC3834908/.

Appleton J. Lavender oil for anxiety and depression. Nat Med J. 2012;4(2).

Baker CB et al. Quantitative analysis of sponsorship bias in economic studies of antidepressants. Br J Psychiatry J Ment Sci. 2003;183:498–506.

Bernstein PS et al. Identification and quantitation of carotenoids and their metabolites in the tissues of the human eye. Exp Eye Res. 2001;72(3):215–23.

Braithwaite MC et al. Nutraceutical-based therapeutics and formulation strategies augmenting their efficiency to complement modern medicine: an overview. J Funct Foods. 2014;6:82–99.

Butterweck V, Schmidt M. St. John's wort: role of active compounds for its mechanism of action and efficacy. Wien Med Wochenschr. 2007;157(13-14):356–61.

Camfield DA, Sarris J, Berk M. Nutraceuticals in the treatment of obsessive compulsive disorder (OCD): a review of mechanistic and clinical evidence. Prog Neuro-Psychopharmacol Biol Psychiatry. 2011;35(4):887–95.

Dhiman A, Nanda A, Ahmad S. Novel herbal drug delivery system ( NHDDS ): the need of hour. IPCBEE. 2012;49:171–5.

Duman RS. Depression: a case of neuronal life and death? Biol Psychiatry. 2004;56(3):140–5.

Duman RS, Heninger GR, Nestler EJ. A molecular and cellular theory of depression. Arch Gen Psychiatry. 1997;54(7):597–606.

EMEA. Note for guidance specifications: test procedures and acceptance criteria for new drug substances and new drug products: chemical substances. Reproduction, 2(November 1999); 2006. p. 1–15. Available at: http://www.ncbi.nlm.nih.gov/pubmed/22167572.

Facchini PJ. Alkaloid biosynthesis in plants: biochemistry, cell biology, molecular regulation, and metabolic engineering applications. Annu Rev Plant Physiol Plant Mol Biol. 2001;52:29–66.

FDA. Guidance for Industry Analytical Procedures and Methods Validation for Drugs and Biologics Guidance for Industry Analytical Procedures and Methods Validation for Drugs and Biologics., (February), 2014; p. 18. Available at: http://www.fda.gov/downloads/Drugs/GuidanceCompliance RegulatoryInformation/Guidances/UCM386366.pdf [Accessed 25 Aug 2015].

Fowler A et al. A nonradioactive high-throughput/high-content assay for measurement of the human serotonin reuptake transporter function in vitro. J Biomol Screen. 2006;11(8):1027–34.

Guest JA, Grant RS. Effects of dietary derived antioxidants on the central nervous system. Int J Nutr Pharmacol Neurol Dis. 2012;2(3):195–7.

Hausenblas HA et al. Meta-analysis Saffron ( Crocus sativus L.) and major depressive disorder: a meta-analysis of randomized clinical trials. J Integr Med. 2013;11(6):377–83.

Holstein SA, Hohl RJ. Isoprenoids: remarkable diversity of form and function. Lipids. 2004;39(4):293–309.

Illum L. Is nose-to-brain transport of drugs in man a reality? J Pharm Pharmacol. 2004;56(1):3–17.

Illum L. Transport of drugs from the nasal cavity to the central nervous system. Eur J Pharm Sci. 2000;11(1):1–18.

Jain KK. Nanobiotechnology-based strategies for crossing the blood brain barrier. Nanomedicine. 2012;7(8):1225–33.

Jaworska A. A comparison between adsorption mechanism of tricyclic antidepressants on silver nanoparticles and binding modes on receptors. J Colloid Interface Sci. 2014;431:117–24.

Jerome S et al. Herbal medicine for depression, anxiety and insomnia: a review of psychopharmacology and clinical evidence. Eur Neuropsychopharmacol. 2011;21:841–60.

Jessen L, Kovalick LJ, Azzaro AJ. The selegiline transdermal system (Emsam): a therapeutic option for the treatment of major depressive disorder. Pharm Therapeut. 2008;33(4):212–46. Available at: http://www.ncbi.nlm.nih.gov/pmc/articles/PMC2730099/.

De Jong WH, Borm PJA. Drug delivery and nanoparticles: applications and hazards. Int J Nanomedicine. 2008;3(2):133–49. Available at: http://www.ncbi.nlm.nih.gov/pmc/articles/PMC2527668/.

Joshi M, Patravale V. Formulation and evaluation of nanostructured lipid carrier (NLC)-based gel of Valdecoxib. Drug Dev Ind Pharm. 2006;32(8):911–8.

Kakkar V, Kaur IP. Antidepressant activity of curcumin loaded solid lipid nanoparticles (C-SLNs) in mice. Am J Pharm Res. 2012;2(3).

Kateb B et al. Nanoplatforms for constructing new approaches to cancer treatment, imaging, and drug delivery: what should be the policy? NeuroImage. 2011;54 Suppl 1:S106–24. Available at: http://www.sciencedirect.com/science/article/pii/S1053811910001448 [Accessed 4 May 2015].

Kesarwani K, Gupta R. Bioavailability enhancers of herbal origin: An overview. Asian Pac J Trop Biomed. 2013;3(4):253–66.

Kukongviriyapan U et al. Preventive and therapeutic effects of quercetin on lipopolysaccharide-induced oxidative stress and vascular dysfunction in mice. Can J Physiol Pharmacol. 2012;90(10):1345–53.

Kumar V. Potential medicinal plants for CNS disorders: an overview. Phytother Res. 2006;20(12):1023–35.

Lee B-H, Kim Y-K. The roles of BDNF in the pathophysiology of major depression and in antidepressant treatment. Psychiatry Investig. 2010;7(4):231–5. Available at: http://www.ncbi.nlm.nih.gov/pmc/articles/PMC3022308/.

Litt B. Nanodevices to Treat Neurological Diseases. *Brain Research Foundation;* 2013. Available at: http://www.thebrf.org/Pages/Topic?id=1426 [Accessed 25 Aug 2015].

Ma P, Mumper RJ. Paclitaxel nano-delivery systems: a comprehensive review. Journal of nanomedicine nanotechnology. 2013;4(2):1000164. Available at: http://www.ncbi.nlm.nih.gov/pmc/articles/PMC3806207/.

Manach C et al. Polyphenols: food sources and bioavailability 1,2. Am J Clin Nutr. 2004;79: 727–47.

Marcus M et al. DEPRESSION: A Global Public Health Concern. *WHO Department of Mental Health and Substance Abuse*; 2012. p. 6–8. Available at: http://www.who.int/mental_health/management/depression/who_paper_depression_wfmh_2012.pdf [Accessed 28 Aug 2015].

Mathers C, Boerma T and Fat DM. The global burden of disease; 2004. p. 1–160. Available at: http://www.who.int/healthinfo/global_burden_disease/2004_report_update/en/ [Accessed 28 Aug 2015].

Mattson MP. Glutamate and neurotrophic factors in neuronal plasticity and disease. Ann N Y Acad Sci. 2008;1144:97–112. Available at: http://www.ncbi.nlm.nih.gov/pmc/articles/PMC2614307/.

Meijer WEE et al. Association of risk of abnormal bleeding with degree of serotonin reuptake inhibition by antidepressants. Arch Intern Med. 2004;164(21):2367–70.

Mercola J. 10 Antidepressant Alternatives Proven to Work; 2009. Available at: Mercola.com [Accessed 10 Aug 2015].

MHDaily. Nanotechnology for depression: the futuristic potential of nanopsychiatry. *Mental Health Blog*; 2015. Available at: http://mentalhealthdaily.com/2015/08/31/nanotechnology-for-depression-the-futuristic-potential-of-nanopsychiatry/ [Accessed 25 Sept 2015].

O'Leary O, Dinan T, Cryan J. Faster, better, stronger_ towards new antidepressant therapeutic strategies. Eur J Pharmacol. 2015;753:32–50.

Pandey KB, Rizvi SI. Plant polyphenols as dietary antioxidants in human health and disease. Oxidative Med Cell Longev. 2009;2(5):270–8. Available at: http://www.ncbi.nlm.nih.gov/pmc/articles/PMC2835915/.

Panossian A, Wikman G. Evidence-based efficacy of adaptogens in fatigue, and molecular mechanisms related to their stress-protective activity. Curr Clin Pharmacol. 2009;4(3):198–219.

Parletta N, Milte CM, Meyer BJ. Nutritional modulation of cognitive function and mental health. J Nutr Biochem. 2013;24(5):725–43.

Pattanayak RD, Sagar R. Depressive disorders in indian context: a review and clinical update for physicians. Journal of the Association of Physicians of India. 2014;62:827–32.

Pick M. Antidepressants and natural alternatives. Emotions, anxiety and mood. 2015. Available at: https://www.womentowomen.com/emotions-anxiety-mood/antidepressants-alternatives/.

Pierce G. Should we clean up the reputation of " dirty drugs "? Can J Physiol Pharmacol. 2012;1334(September):1333–4.

Pires AF et al. The role of endothelium in the vasorelaxant effects of the essential oil of Ocimum gratissimum in aorta and mesenteric vascular bed of rats. Can J Physiol Pharmacol. 2012;90(10):1380–5.

Prakash D et al. Effect of nanohypericum (Hypericum perforatum gold nanoparticles) treatment on restraint stressinduced behavioral and biochemical alteration in male albino mice. Pharmacognosy Res. 2010;2(6):330–4.

Priprem A et al. Antidepressant and cognitive activities of intranasal piperine-encapsulated liposomes. Adv Biosci Biotechnol. 2011;02(02):108–16.

Priprem A et al. Anxiety and cognitive effects of quercetin liposomes in rats. Nanomedicine. 2008;4(1):70–8.

Rabanel J-M et al. Effect of polymer architecture on curcumin encapsulation and release from PEGylated polymer nanoparticles: Toward a drug delivery nano-platform to the CNS. European journal of pharmaceutics and biopharmaceutics: official journal of Arbeitsgemeinschaft fur Pharmazeutische Verfahrenstechnik e.V; 2015.

Rahman A et al. The neglected "m" in MCH programmes-why mental health of mothers is important for child nutrition. Tropical Med Int Health. 2008;13(4):579–83.

Re F et al. Functionalization of liposomes with ApoE-derived peptides at different density affects cellular uptake and drug transport across a blood-brain barrier model. Nanomedicine. 2011;7(5):551–9.

Roy-Byrne P. Herbal Psychotropic Medications: Promises and Pitfalls. *Psychiatry;* 1998. Available at: http://www.jwatch.org/jp199811010000019/1998/11/01/herbal-psychotropic-medications-promises-and [Accessed 20 Aug 2015].

Sahu AN. Nanotechnology in herbal medicines and cosmetics. Int J Ayurveda Pharm. 2013;4(3):472–4.

Saraf S. Fitoterapia applications of novel drug delivery system for herbal formulations. Fitoterapia. 2010;81(7):680–9.

Sarris J et al. The Kava Anxiety Depression Spectrum Study (KADSS): a randomized, placebo-controlled crossover trial using an aqueous extract of Piper methysticum. Psychopharmacology. 2009;205(3):399–407.

Sarris J, Kavanagh DJ, Byrne G. Adjuvant use of nutritional and herbal medicines with antidepressants, mood stabilizers and benzodiazepines. J Psychiatr Res. 2010;44(1):32–41.

Sayyah M, Sayyah M, Kamalinejad M. A preliminary randomized double blind clinical trial on the efficacy of aqueous extract of *Echium amoenum* in the treatment of mild to moderate major depression. Prog Neuropsychopharmacol Biol Psychiatry. 2006;30(1):166–9.

Scalbert A et al. Dietary polyphenols and the prevention of diseases. Crit Rev Food Sci Nutr. 2005;45(4):287–306.

Spinella M. *The Psychopharmacology of Herbal Medicine Plant Drugs That Alter Mind Brain and Behavior by Marcello Spinella Information* 6586th ed., London, England: Mit Press; 2001. Available at: https://books.google.co.in/books?id=jZeaRiIFbhsC&dq=The+Psychopharmacology+of+Herbal+Medicine+Plant+Drugs+That+Alter+Mind+Brain+and+Behavior+by+Marcello+Spinella+Information&source=gbs_navlinks_s.

Stevens JR et al. The use of transdermal therapeutic systems in psychiatric care: a primer on patches. Psychosomatics. 2015;56(5):423–44. Available at: http://www.sciencedirect.com/science/article/pii/S0033318215000602.

Tamjidi F et al. Nanostructured lipid carriers ( NLC ): A potential delivery system for bioactive food molecules. Innovat Food Sci Emerg Tech. 2013;19:29–43.

Thakkar HP, Patel AA, Chauhan NP. Intranasal mucoadhesive microemulsion of mirtazapine: pharmacokinetic and pharmacodynamic studies. Asian J Pharmaceut. 2013;7(1):36–42.

Tiwari G et al. Drug delivery systems: an updated review. Int J Pharmaceut Investig. 2012;2(1):2–11. Available at: http://www.ncbi.nlm.nih.gov/pmc/articles/PMC3465154/.

Torchilin V. Multifunctional and stimuli-sensitive pharmaceutical nanocarriers. Eur J Pharm Biopharm. 2011;71(3):431–44.

Upadhyay RK. Drug delivery systems, CNS protection, and the blood brain barrier. BioMed Res Int. 2014;2014:869269. Available at: http://www.ncbi.nlm.nih.gov/pmc/articles/PMC4127280/.

Uttara B et al. Oxidative stress and neurodegenerative diseases: a review of upstream and downstream antioxidant therapeutic options. Curr Neuropharmacol. 2009;7(1):65–74. Available at: http://www.ncbi.nlm.nih.gov/pmc/articles/PMC2724665/.

Vauzour D. Dietary polyphenols as modulators of brain functions: biological actions and molecular mechanisms underpinning their beneficial effects. *Oxidative Medicine and Cellular Longevity*; 2012. p. 1–16.

Vijaya R, Ruckmani K. In vitro and In vivo characterization of the transdermal delivery of sertraline hydrochloride Films. DARU. 2011;19(6):424–32. Available at: http://www.ncbi.nlm.nih.gov/pmc/articles/PMC3436079/.

Walle T et al. High absorption but very low bioavailability of oral resveratrol in humans. Drug Metab Dispos. 2004;32(12):1377–82.

Wang R and Xu Y. Phytochemicals as antidepressants: the involvement of serotonin receptor function, stress resistance and neurogenesis. In: Kuang PH, Editor. Recent advances in theories and practice of Chinese medicine; 2012. p. 473–492. Available at: www.intechopen.com.

Wattanathorn J et al. quercetin.pdf. Am J Agr Biol Sci. 2007;2(1):31–5.

Welch H and Hasbun R. Chapter 3 - Lumbar puncture and cerebrospinal fluid analysis. In Vinken PJ and Bruyn GW, Editors. *Handbook of Clinical Neurology*. Department of Medicine, Section of Infectious Diseases, University of Texas Medical School, Houston, TX, USA: North-Holland Publishing Company; 2010. p. 31–49. Available at: http://www.sciencedirect.com/science/article/pii/S0072975209960031.

WFMH. Depression: A Global Crisis. *World Fedaration of Mental Health*; 2012. p. 1–32. Available at: http://www.who.int/mental_health/management/depression/wfmh_paper_depression_wmhd_2012.pdf [Accessed 12 Sept 2015].

WHO. Annex 2: The International Pharmacopoeia: revised concepts and future perspectives; 2003. 1(908): 22–25. Available at: http://www.who.int/medicines/areas/quality_safety/quality_assurance/PhIntRevisedConceptsFuturePerspectivesTRS908Annex2.pdf?ua=1 [Accessed 10 Oct 2015].

WHO. Good practices for pharmaceutical quality control laboratories; 2010. (957): 81–129. Available at: http://apps.who.int/prequal/info_general/documents/trs957/gpcl_trs957_annex1.pdf [Accessed 10 Oct 2015].

WHO. Suicide Prevention Across the Globe: Strengthening Protective Factors and Instilling Hope. International Association for Suicide Prevention; 2012. p. 1–3. Available at: www.iasp.info/wspd [Accessed 18 Oct 2015].

WHO. WHO guidelines on safety monitoring of herbal medicines in pharmacovigilance systems; 2004. p. 1–82. Available at: http://apps.who.int/medicinedocs/documents/s7148e/s7148e.pdf [Accessed 10 Oct 2015].

Williamson EM. Synergy and other interactions in phytomedicines. Phytomedicine. 2001;8(5):401–9.

Wu L et al. Research article a self-microemulsifying drug delivery system ( SMEDDS ) for a Novel Medicative Compound Against Depression: a Preparation and Bioavailability Study in Rats. *American Association of Pharmaceutical Scientists PharmSciTech*; 2015. (13).

Yin D et al. Increase in brain tumor permeability in glioma-bearing rats with nitric oxide donors. Clin Cancer Res. 2008;14(12):4002–9.

Zhu W et al. Green tea polyphenols produce antidepressant-like effects in adult mice. Pharmacol Res. 2012;65:74–80.

# Chapter 12
# Future Strategies for the Treatment of Depression

**Clara Grosso**

**Abstract** Despite almost 60 years of intensive research, depression still constitutes a huge economic and social burden worldwide. The reasons which led to this scenario as well as the strategies to overcome it are pointed out by several researchers from different fields of neuroscience and psychiatry: (1) low bioavailability and inefficacy of the current antidepressant drugs and the need for effective drug delivery systems, (2) the lack of appropriate biomarkers and the lack of a comprehensive knowledge of the genetic factors involved in the predisposition of depression, (3) the lack of suitable animal models to study treatment-resistant depression, and (4) the urgent need for a personalized medicine to treat depression patients.

This chapter aims, thus, at thinking about what we have learned from the past, what failed, and what has been achieved, in order to rethink the treatment of depression for the future.

**Keywords** Drug delivery systems • Pharmacogenomics • Personalized medicine • Treatment-resistant depression

## 12.1 What We Have Learned from the Past

Chapters 1 and 2 summarize all the theories of depression postulated since the 1950s and the mechanism of action of the current and past developed antidepressant drugs. Based on the monoamine theory, since the 1950s, several new pharmacological agents were approved for the treatment of people suffering from depression, such as the irreversible and nonselective monoamine oxidase (MAO) inhibitors **iproniazid** (introduced into medicine in 1958 but withdrawn in 1961 because of toxicity) (Zitrin and Thompson 1956; Crane 1957; Dooneief and Crane 1957; Loomer et al. 1957; Lundquist 1959), **phenelzine** (launched in 1961) (Dunlop 1959;

C. Grosso (✉)
REQUIMTE/LAQV, Laboratory of Pharmacognosy, Department of Chemistry, Faculty of Pharmacy, University of Porto, Porto, Portugal
e-mail: acfmtgrosso@ff.up.pt



C. Grosso (ed.), *Herbal Medicine in Depression*,
DOI 10.1007/978-3-319-14021-6_12

King 1959; Saunders et al. 1959; Rees and Davies 1961) and **tranylcypromine** (launched in 1960 but rarely used today) (Agin 1960; Aldridge and Oakley 1961; Lemere 1960); the tricyclic antidepressants (TCAs) **imipramine** (approved in 1959) (Azima 1959; Azima and Vispo 1958; Kuhn 1958; Lehmann et al. 1958; Mann and Macpherson 1959), **amitriptyline** (approved in 1961) (Barsa and Sauders 1961; Feldman 1961; Freed 1960; Vaisberg and Saunders 1961) and **clomipramine** (approved in 1964) (Blanchinet et al. 1970; Carlsson et al. 1969; Volmat et al. 1968); and the selective serotonin (5-HT) reuptake inhibitors (SSRIs) **zimelidine** (launched in 1982 but withdrawn one year after due to neurological side effects) (Benkert et al. 1977; Cox et al. 1978; Siwers et al. 1977) and **fluoxetine** (approved in 1987) (Fuller et al. 1974; Wong et al. 1974). Their introduction was followed by intensive studies of their putative mechanisms of action; then, in the following decades, the progressive development of new agents with different modes of action occurred. The novel treatments were safer, better tolerated, and had a broader therapeutic reach, but there was no profound gain in therapeutic efficacy compared to the first-generation psychotropic agents, leading to a sense of frustration. These new better tolerable agents included the reversible and selective MAO-A inhibitor **moclobemide** (not approved in the USA) (Da Prada et al. 1983; Grau 1983; Korn et al. 1986; Larsen and Mikkelsen 1984; Schlappi 1985; Wong et al. 1974), the SSRI **citalopram** (approved in 1998) (Christensen et al. 1977; Gottlieb et al. 1980; Hyttel 1977; Kragh-Sørensen et al. 1981; Overø 1978) and its *S*-enantiomer **escitalopram** (approved in 2002) (Montgomery et al. 2001; Owens et al. 2001; Sorbera et al. 2001b), the noradrenaline reuptake inhibitor (NARI) **reboxetine** (launched in 1997) (Cocchiara et al. 1991; Dostert et al. 1997; Riva 1985; Riva et al. 1989), and the combined serotonin and noradrenaline reuptake inhibitors (SNRIs) **venlafaxine** (approved in 1997) (Cantu et al. 1994; Khan et al. 1991; Pento 1988; Schweizer et al. 1988, 1991; Thomas Haskins et al. 1985) and **duloxetine** (approved in 2004) (Artigas 1995; Bymaster et al. 2001; Fuller et al. 1994; Goldstein et al. 2002; Kasahara et al. 1996; Kihara and Ikeda 1995; Martínez et al. 2001; Pitsikas 2000; Sorbera et al. 2000; Thor and Katofiasc 1995; Wong 1998; Wong et al. 1993).

Besides the increase of noradrenaline and serotonin in the synaptic cleft, other mechanisms of action based on monoamines have also been explored. **Bupropion**, introduced in 1985, is a dopamine reuptake inhibitor (DARI), but its effect was of modest potency (Castaner and Hopkins 1978; Cooper et al. 1980; Fann et al. 1978; Ferris et al. 1981; Findlay et al. 1981; Halaris et al. 1981; Katz and Sibel 1982; Soroko and Maxwell 1978; Soroko et al. 1977). The lack of enhanced clinical efficacy with SSRIs relative to TCAs encouraged the search for combining 5-HT reuptake inhibition with complementary properties. For instance, **vilazodone**, approved in 2011, is a 5-HT$_{1A}$ partial agonist/SSRI that more powerfully enhances extracellular 5-HT levels than pure SSRIs (Adamec et al. 2004; Bartoszyk et al. 1997; Dawson and Watson 2009; De Paulis 2007; Khan 2009; Laughren et al. 2011; Mealy and Bayés 2004; Murck et al. 2001; Rickels et al. 2009; Sorbera et al. 2001a; Treit et al. 2001). **Vortioxetine** was launched in 2013 and possesses 5-HT reuptake blocking activity plus multi-target actions at 5-HT$_{1A}$, 5-HT$_{1B}$, 5-HT$_3$, and 5-HT$_7$ receptors (Alvarez et al. 2012; Baldwin et al. 2012; Katona et al. 2012;

Mørk et al. 2012; Sanchez et al. 2015; Uldam et al. 2011). **Mirtazapine**, an atypical antidepressant that blocks both 5-$HT_{2C}$ and α2-AR receptors, was introduced in 1996 (Janssen 1995).

In an attempt to find more effective antidepressants, other theories of depression were explored. A 5-$HT_{2C}$ antagonism was synergistically joined to melatonergic agonism (involved in the circadian rhythm, see Chap. 1) in the first melatonergic antidepressant, **agomelatine** (approved in 2009) (Banasr et al. 2006; Barden et al. 2005; Bertaina-Anglade et al. 2006; Bień et al. 2002; Bourin et al. 2004; Den Boer et al. 2006; Fuchs et al. 2006; Hanoun et al. 2004; Kennedy 2007; Kupfer 2006; Millan et al. 2003, 2005; Papp et al. 2003; Pierre Olié and Kasper 2007; Pjrek et al. 2007; Tuma et al. 2005).

**Tianeptine**, an agent approved in 1988 by the EU for treating depression, was originally suggested to act by enhancing 5-HT reuptake (Mennini et al. 1987; Mocaer et al. 1988; Neuman 1979; Wilde and Benfield 1995), but more recent work suggests that it promotes synaptic plasticity by perhaps indirectly modulating glutamatergic transmission (see Chap. 1) (Hayley and Litteljohn 2013; Kasper and McEwen 2008; McEwen et al. 2010). More recently, there has been a revival of interest in the role of glutamate for management of depression due to the rapid-onset (hours) of antidepressant effect of a single dose of the noncompetitive NMDA receptor antagonist **ketamine** compared to the several weeks required for full efficacy of conventional antidepressants (Dutta et al. 2015; Yang et al. 2015).

Concerning the treatment of bipolar disorder, psychiatric patients started being administrated with **lithium** in 1949 (Cade 1949), and later studies in the early 1950s verified lithium's efficacy against mania. However, the acceptance of lithium was slow in psychiatry due to the lack of knowledge on the molecular mechanisms underlying its mood-stabilizing properties, and only in 1970 FDA finally approved it for the control of bipolar disorder. Alda (2015) and Oruch et al. (2014) reviewed the mechanisms of action of lithium as mood stabilizer, which include its influence on glutamatergic neurotransmission, up- or downregulation of the biosynthesis of neurotransmitters or their receptors, increase of serotonin release by neurons, inhibition of the enzyme glycogen synthase kinase-3 beta (GSK3β) involved in the circadian clock, inhibition of pAp (3′-5′ phosphoadenosine phosphate) phosphatases involved in RNA catabolism, involvement in neural plasticity through NMDA receptor/NO signaling, inhibition of the enzyme inositol monophosphatase, increase of 17-hydroxy-DHA formation in the brain, and finally, interaction with cAMP-mediated processes, thus blocking catecholamine receptors of the neurons.

The mood-stabilizing actions of **lamotrigine** (approved in 2003) and **carbamazepine** (approved 2005) seem to be complex. Lamotrigine is a glutamate release inhibitor approved for partial and tonic-clonic seizure and for bipolar disorder. Lamotrigine inhibits voltage-dependent $Na^+$ channels, $Ca^{2+}$ channels, and $K^+$ channels, thus decreasing glutamate release and increasing the AMPA receptor expression; it weakly inhibits serotonin reuptake and may exert neuroprotective effects (Lapidus et al. 2013). Carbamazepine stabilizes the inactive state of voltage-gated $Na^+$ channels and potentiates GABAergic transmission, by acting at various subunits of $GABA_A$ receptors (Millan et al. 2015b). **Valproic acid** also blocks $Na^+$ channels and increases cerebral levels of GABA (Millan et al. 2015b).

Therefore, after more than 60 years of intensive research and development, based mainly on animal models (see Chap. 4), human trials, and clinical practice, the ability to improve the lives of patients suffering from psychiatric and neurological disorders is still limited. One main factor contributing to this scenario was the relative disconnection of the big pharmaceutical companies from neuroscience, which was provoked by some important issues, such as (1) the difficult economic situation; (2) regulatory and pricing restrictions; (3) limited duration of patent protection; (4) costly, risky and prolonged clinical development; (5) competing demands for treating other important diseases like cancer, diabetes, and infective disorders; (6) promotion for the development of biological therapies such as antibodies and cell therapy which are of currently limited application to psychiatric disorders; and (7) the late recognition that sequencing the human genome will not lead to a rapid and direct solution to enrich the pipeline for psychiatric disorders (Millan et al. 2015a, b).

## 12.2 What We Are Learning with the Present and Dream for the Future

The decrease of the burden of depression will definitely englobe strategies such as (1) to find better and more efficient drug delivery systems for antidepressants and find new antidepressants with the same or different mode of actions, (2) to find the right animal models to test new antidepressant drugs, (3) to increase the knowledge on how the human genome influences the development of depression, and (4) to treat patients suffering from depression with a more personalized approach.

### *12.2.1 Find Better Drug Delivery Systems and Find New Antidepressants*

Thinking about the antidepressant drugs existing in the market, oral administration is perhaps a nonoptimal delivery system for most of them. Scientists are seeking for alternative drug delivery systems as a way to enhance the antidepressant response. Several alternative approaches to increase drug bioavailability were detailed in Chap. 11, ranging from oral gastroresistant formulations, transdermal delivery, nasal therapies and nanotechnology-based products. According to Kaminsky et al. (2015), based on the physicochemical characteristics of antidepressants and antipsychotic drugs, nasal, transdermal, and sublingual administration routes could be applied to enhance their effect.

Selegiline is a selective MAO-B inhibitor, but at high doses (≥30 mg/day) it lacks selectivity and inhibits MAO-A, thus displaying antidepressant activity (Gordon et al. 1999; Mann et al. 1989). A **selegiline** transdermal system was already approved by the FDA for depression treatment. It avoids hepatic first metabolism step, thus escaping from significant inhibition of gastrointestinal and hepatic MAO-A activity while still providing sufficient levels of selegiline in the brain to produce an antidepressant effect (Kaminsky et al. 2015; Lee and Chen 2007). For patients who experience dysphagia, oral disintegrating tablets (ODT) of **mirtazapine**, **risperidone** (an atypical antipsychotic), and **selegiline** are also on the market (Kaminsky et al. 2015).

On the other hand, although nanotechnology can help to increase drug bioavailability and pharmacokinetics, can provide technical assistance to in vivo imaging, and can model the human CNS (e.g., inorganic synapses and neural networks), we are still far from seeing a practical application in daily psychiatric practice due to the lack of safety studies (Fond et al. 2013). Nonetheless, several studies have been conducted to incorporate antidepressants in nanoparticles (Garcia et al. 2011; Haque et al. 2012, 2014; Jaworska and Malek 2014; Patel et al. 2012; Prakash et al. 2010; Tang et al. 2011; Varshosaz et al. 2015).

Finally, it cannot be forgotten that antidepressant-resistant patients which do not respond to the available antidepressant drugs may need new antidepressants more efficient and with less side effects which can act in the same targets or in different ones. Owing to the high diversity of terrestrial and marine organisms, natural products are some of the most successful source of drug leads for the treatment of many diseases and illnesses (David et al. 2015). Many medicinal plants, in particular, have been used for centuries in traditional medicine to treat depression, and they can be subjected to an intensive pharmacological research to find effective compounds or extracts with strong antidepressant effect. A list of antidepressant medicinal plants from different places of the world was compiled in Chaps. 5–10.

### *12.2.2 Find the Right Animal Models*

Animal models (see Chap. 4 for more information) and, with less extend of use but also important, cellular models (see Chap. 3 for more detailed information) have been of great value in understanding the cellular and neuroanatomical changes behind depression and in establishing in detail the mechanism of action of antidepressant drugs. However, Willner and Belzung (2015) recently formulated a very pertinent question: Are the animal models of depression in current use suitable for the development of treatments to overcome antidepressant resistance? Over the past two decades, research using animal models and clinical research have established that stress precipitates depression primarily by damaging the hippocampus and that antidepressants work by repairing this damage. The great limitation of this vision is that it assumes that stress-induced neurotoxicity is the problem that antidepressants need to overcome. However, there are many other factors that are known to increase vulnerability to depression, including a clinical history of depression (i.e., recurrent episodes of depression) and a variety of genetic, personality, or developmental (e.g., multiple loss events or childhood adversity) risk factors. These risk factors involve an increased sensitivity to stress, which means that in predisposing persons the first episode of depression is precipitated by very low levels of stress, compared with non-vulnerable persons in which high levels of stress are required to trigger the first depressive episode (Kendler et al. 2000, 2001, 2004; Willner and Belzung 2015).

Therefore, Willner and Belzung (2015) argued that the treatment-resistant depression may require a paradigm shift away from stress-based models to models of vulnerability to depression in which the conventional antidepressants are ineffective. According to the same authors, models of antidepressant-resistant depression should obey to three criteria: first, depression-like behavior should be elicited more readily in high-risk groups than in low-risk groups; second, the effectiveness of conventional antidepressants should be lower in high-risk than in low-risk groups; and third, that high-risk and low-risk groups should respond equally well to unconventional antidepressant treatments that act outside the hippocampus. Those which seems to be more promising are the chronic mild stress (CMS) and learned helplessness (LH) procedures applied to the Wistar-Kyoto (WKY) and congenital learned helplessness (cLH) rat strains; the high anxiety behavior (HAB) mouse strain and the CB1 cannabinoid receptor knockout; and the organic cation transporter 2 (OCT2) null mutant genetically modified mouse strains (Willner and Belzung 2015).

The congenital learned helplessness (cLH) strain exhibits a helpless phenotype even without prior exposure to inescapable stress and also exhibits anhedonia and/or anergia under a progressive ratio schedule, hyperactivity in the novel open field, but no spatial learning and memory deficits in the water maze test (Vollmayr et al. 2004). High anxiety-related behavior (HAB) rats are susceptible and vulnerable to stressor exposure and show signs of a hyperreactive hypothalamic-pituitary-adrenocortical (HPA) axis, thus resembling psychiatric patients (Landgraf and Wigger 2002). CB1 cannabinoid receptor-deficient mice present increased anxiety-like responses, higher sensitivity to develop depressive-like symptoms, decreased sensitivity to the rewarding properties of sucrose, cognitive impairments, changes in basal levels of corticosterone and increased corticosterone levels after stress exposure, deficit in BDNF neurotrophic factor in hippocampus under basal conditions and after the exposure to stress, and changes in the serotoninergic system (Valverde and Torrens 2012). At last, genetic deletion of OCT2 in mice turns them more sensitive to the action of unpredictable chronic mild stress (UCMS) and impairs HPA function (Courousse et al. 2015).

### *12.2.3 Increase the Knowledge on Human Genome*

Depression is highly heritable, with an estimation that 40–50 % of the risk for depression is genetically determined (Levinson 2006). Indeed, family and twin studies have clearly demonstrated the contribution of heritable factors to the aetiology of mood disorders (Bertelsen et al. 1977; Bierut et al. 1999; Kendler et al. 1994, 1999; Kendler and Prescott 1999; Li et al. 2008; Lyons et al. 1998; McGuffin et al. 2003; Zubenko et al. 2003). Understanding the genes involving in the pathogenesis of depression or those which contribute to the inefficacy of the antidepressant treatments is of great importance as a future strategy to fight depression.

Pharmacogenomic testing is clinically available to assist the selection of medication for treatment-resistant depression. Common tests include the cytochrome P450 (CYP) 2D6 and 2C19 enzymes, the serotonin transporter gene (5-HTTLPR), and the serotonin receptor genes (5-$HT_{2A}$ and 5-$HT_{2C}$).

Cytochrome P450 (CYP) enzymes are involved with the metabolism of most of medications, including antidepressants. The two major cytochrome P450 polymorphisms, regarding variable drug response, are those of CYP2C19 and CYP2D6. Both gene variations can cause abolished, reduced, normal, or augmented enzyme expression which is translated into the poor, intermediate, extensive, and ultrarapid metabolizer phenotypes (Ingelman-Sundberg et al. 2014). Identification of CYP2D6 or CYP2C19 poor metabolizers results in the recommendation of a lower dosage or avoidance of medications metabolized by CYP2D6 or CYP2C19 (Hamilton 2015). Moreover, the identification of homozygous for the short allele of 5-HTTLPR gene suggests more adverse effects, less response, or longer time to respond to SSRIs (Bijl et al. 2009; Kung and Li 2010; Stingl and Viviani 2015).

Besides the abovementioned genes, the number of interesting pharmacodynamic targets appears to be quite large, the most important of them seem to be genes coding for tryptophan hydroxylase, catechol-*O*-methyltransferase (COMT), MAO-A, norepinephrine transporter (NET), dopamine transporter (DAT), monoamine receptors (5-$HT_{1A}$, 5-$HT_{2A}$, 5-$H_{T2C}$, 5-$H_{T6}$, 5-$HT_{3A}$, 5-$HT_{3B}$, β1 adrenoceptor), dopamine (DA) receptors, glycogen synthase kinase-3 (GSK-3β), corticotropin-releasing hormone (CRH) receptor I (CRHR1), glucocorticoid receptor, angiotensin-converting enzyme, circadian locomotor output cycles kaput (CLOCK), nitric oxide synthase, interleukin (IL)-1β, and brain-derived neurotrophic factor (BDNF) (Nnadi et al. 2005; Porcelli et al. 2011).

Pharmacogenetic studies have not yet been successful, but there is a hope that combined data from genomics, proteomics, metabolomics, neuroimaging, and neuroendocrinology may eventually lead to effective personalized antidepressant treatments (Cuijpers 2014).

### 12.2.4 Personalized Medicine

Clinical trials have shown that, on average, antidepressants and psychotherapies have similar success in the first-line treatment of moderate depression and, on average, different antidepressants show equal or similar efficacy (Cuijpers 2014; Simon and Perlis 2010). So, why only about 33 % of the patients respond positively to the first-line treatment? The answer is quite simple: although these treatments are similarly effective, this does not mean that there are no differences within patients that led them to respond to a particular treatment in detriment of other treatments (Cuijpers 2014).

The need to match treatment with patient is recognized in Gordon Paul's famous question about psychotherapy: "What treatment, by whom, is most effective for this individual with that specific problem, and under which set of circumstances?" (Paul 1967). Personalized medicine promises to go beyond average effective treatments to identify the best treatment for each patient. In order to provide personalized medicine for depression, it is necessary to identify characteristics of individuals that reliably predict differences in benefits and/or adverse effects of alternative depression treatments, such as sociodemographic characteristics, clinical characteristics (such as symptom patterns or comorbidities), and biological markers (such as neuroimaging or genetic variation) (Cuijpers 2014; Simon and Perlis 2010).

In the past few years, several different approaches have attempted to develop personalized treatments. Preliminary research shows that pharmacotherapy is probably more effective than psychotherapy in patients with dysthymia and chronic depression and combined treatments are more effective than each treatment alone in older adults and mental health outpatients (Cuijpers et al. 2010).

The reason why a personalized approach is not already in practice is that depression is a very heterogeneous disorder, and many different effective treatments are available. If we wanted to show that one treatment is better for one patient with a specific characteristic, we would have to choose between the many treatments that may be more effective than others in that patient. This implies that the number of trials that should be performed is very large and with millions of participating patients. According to Cuijpers (Cuijpers 2014), only a small fraction of them have been conducted so far.

## 12.3 Conclusions

This chapter intended to summarize the challenges needed to be faced in order to improve treatments against depression and overcome antidepressant resistance. Progresses should be made toward a more personalized approach in which characteristics of each of the patients are taken into consideration. To select the appropriate biomarkers, the genetic background of patients should be considered, as well as other predisposing risk factors that could easily culminate in depressive episodes. More trials with more participants should be carried out in the future to establish the characteristics that make some patients more prone to be treated with a class of antidepressants in detriment of other antidepressant drugs.

## References


Adamec R, Bartoszyk GD, Burton P. Effects of systemic injections of vilazodone, a selective serotonin reuptake inhibitor and serotonin 1 A receptor agonist, on anxiety induced by predator stress in rats. Eur J Pharmacol. 2004;504:65–77.

Agin HV. Tranylcypromine in depression: a clinical report. Am J Psychiatr. 1960;117:150–1.

Alda M. Lithium in the treatment of bipolar disorder: pharmacology and pharmacogenetics. Mol Psychiatry. 2015;20:661–70.

Aldridge M, Oakley N. Side-effects of tranylcypromine. Lancet. 1961;278:932.

Alvarez E, Perez V, Dragheim M, Loft H, Artigas F. A double-blind, randomized, placebo-controlled, active reference study of Lu AA21004 in patients with major depressive disorder. Int J Neuropsychopharmacol. 2012;15:589–600.

Artigas F. Selective serotonin/noradrenal reuptake inhibitors (SNRIs). CNS Drugs. 1995; 4:79–89.

Azima H. Imipramine (Tofranil): a new drug for the depressed. Can Med Assoc J. 1959;80: 535–40.

Azima H, Vispo RH. Imipramine; a potent new anti-depressant compound. Am J Psychiatr. 1958; 115:245–6.

Baldwin DS, Loft H, Florea I. Lu AA21004, a multimodal psychotropic agent, in the prevention of relapse in adult patients with generalized anxiety disorder. Int Clin Psychopharmacol. 2012;27:197–207.

Banasr M, Soumier A, Hery M, Mocaër E, Daszuta A. Agomelatine, a new antidepressant, induces regional changes in hippocampal neurogenesis. Biol Psychiatry. 2006;59:1087–96.

Barden N, Shink E, Labbé M, Vacher R, Rochford J, Mocaër E. Antidepressant action of agomelatine (S 20098) in a transgenic mouse model. Prog Neuro-Psychopharmacol Biol Psychiatry. 2005;29:908–16.

Barsa JA, Sauders JC. Amitriptyline (Elavil), a new antidepressant. Am J Psychiatr. 1961;117: 739–40.

Bartoszyk GD, Hegenbart R, Ziegler H. EMD 68843, a serotonin reuptake inhibitor with selective presynaptic 5-HT(1A) receptor agonistic properties. Eur J Pharmacol. 1997;322:147–53.

Benkert O, Laakmann G, Ott L, Strauss A, Zimmer R. Effect of zimelidine (H 102/09) in depressive patients. Arzneimittelforschung. 1977;27:2421–3.

Bertaina-Anglade V, Drieu La Rochelle C, Boyer PA, Mocaër E. Antidepressant-like effects of agomelatine (S 20098) in the learned helplessness model. Behav Pharmacol. 2006;17:703–13.

Bertelsen A, Harvald B, Hauge M. A Danish twin study of manic-depressive disorders. Br J Psychiatry. 1977;130:330–51.

Bień E, Gruca P, Galoch Z, Papp M. Evidence for antidepressant- and anxiolytic-like activities of melatonin and agomelatine in animal models. Pharmacol Rep. 2002;54:185.

Bierut L, Heath AC, Bucholz KK, et al. Major depressive disorder in a community-based twin sample: are there different genetic and environmental contributions for men and women? Arch Gen Psychiatry. 1999;56:557–63.

Bijl MJ, Luijendijk HJ, van den Berg JF, Visser LE, van Schaik RHN, Hofman A, Vulto AG, van Gelder T, Tiemeier H, Stricker BH. Association between the CYP2D6*4 polymorphism and depression or anxiety in the elderly. Pharmacogenomics. 2009;10:541–7.

Blanchinet J, Léger JM, Vallat JN. Use of clomipramine in the treatment of depressive states. Bord Med. 1970;3:1163–4. passim.

Bourin M, Mocaër E, Porsolt R. Antidepressant-like activity of S 20098 (agomelatine) in the forced swimming test in rodents: involvement of melatonin and serotonin receptors. J Psychiatry Neurosci. 2004;29:126–33.

Bymaster FP, Dreshfield-Ahmad LJ, Threlkeld PG, Shaw JL, Thompson L, Nelson DL, Hemrick-Luecke SK, Wong DT. Comparative affinity of duloxetine and venlafaxine for serotonin and norepinephrine transporters in vitro and in vivo, human serotonin receptor subtypes, and other neuronal receptors. Neuropsychopharmacology. 2001;25:871–80.

Cade JF. Lithium salts in the treatment of psychotic excitement. Med J Aust. 1949;2:349–52.

Cantu TG, Korek JS, Romanoski AJ. Focus on venlafaxine: a new option for the treatment of depression. Hosp Formul. 1994;29:25–33.

Carlsson A, Corrodi H, Fuxe K, Hökfelt T. Effect of antidepressant drugs on the depletion of intraneuronal brain 5-hydroxytryptamine stores caused by 4-methyl-ϵ-ethyl-metatyramine. Eur J Pharmacol. 1969;5(4):357–66.

Castaner J, Hopkins SJ. Bupropion. Drugs Future. 1978;3:723–7.

Christensen AV, Fjalland B, Pedersen V, Danneskiold-Samsøe P, Svendsen O. Pharmacology of a new phthaline (Lu 10-171), with specific 5-HT uptake inhibiting properties. Eur J Pharmacol. 1977;41:153–62.

Cocchiara G, Battaglia R, Pevarello P, Benedetti MS. Comparison of the disposition and of the metabolic pattern of Reboxetine, a new antidepressant, in the rat, dog, monkey and man. Eur J Drug Metab Pharmacokinet. 1991;16:231–9.

Cooper BR, Hester TJ, Maxwell RA. Behavioral and biochemical effects of the antidepressant bupropion (Wellbutrin): evidence for selective blockade of dopamine uptake in vivo. J Pharmacol Exp Ther. 1980;215:127–34.

Courousse T, Bacq A, Belzung C, Guiard B, Balasse L, Louis F, Le Guisquet AM, Gardier AM, Schinkel AH, Giros B, Gautron S. Brain organic cation transporter 2 controls response and vulnerability to stress and GSK3βsignaling. Mol Psychiatry. 2015;20:889–900.

Cox J, Moore G, Evans L. Zimelidine: a new antidepressant? Prog Neuropsychopharmacol. 1978;2:379–84.

Crane GE. Iproniazid (marsilid) phosphate, a therapeutic agent for mental disorders and debilitating diseases. Psychiatr Res Rep. 1957;8:142–52.

Cuijpers P. Personalized treatment for functional outcome in depression. Medicographia. 2014; 36:476–81.

Cuijpers P, van Straten A, Schuurmans J, van Oppen P, Hollon SD, Andersson G. Psychotherapy for chronic major depression and dysthymia: a meta-analysis. Clin Psychol Rev. 2010;30:51–62.

Da Prada M, Kettler R, Keller HH, Haefely WE. Neurochemical effects in vitro and in vivo of the antidepressant Ro 11-1163, a specific and short-acting MAO-A inhibitor. Mod Probl Pharmacopsychiatry. 1983;19:231–45.

David B, Wolfender J-L, Dias DA. The pharmaceutical industry and natural products: historical status and new trends. Phytochem Rev. 2015;14(2):299–315.

Dawson LA, Watson JM. Vilazodone: a 5-HT1A receptor agonist/serotonin transporter inhibitor for the treatment of affective disorders. CNS Neurosci Ther. 2009;15:107–17.

De Paulis T. Drug evaluation: vilazodone—a combined SSRI and 5-HT1A partial agonist for the treatment of depression. IDrugs. 2007;10:193–201.

Den Boer JA, Bosker FJ, Meesters Y. Clinical efficacy of agomelatine in depression: the evidence. Int Clin Psychopharmacol. 2006;21:S21–4.

Dooneief AS, Crane GE. Iproniazid as adjunct in the treatment of debilitated patients with tuberculosis. N Y State J Med. 1957;57:3477–80.

Dostert P, Benedetti MS, Poggesi I. Review of the pharmacokinetics and metabolism of reboxetine, a selective noradrenaline reuptake inhibitor. Eur Neuropsychopharmacol. 1997;7:S23–35. http://dx.doi.org/10.1016/S0924-977X(97)00417-3.

Dunlop E. Depression: treatment of office patients with phenelzine (nardil). R I Med J. 1959; 42:656–7. passim.

Dutta A, McKie S, Deakin JFW. Ketamine and other potential glutamate antidepressants. Psychiatry Res. 2015;225:1–13.

Fann WE, Schroeder DH, Mehta NB. Clinical trial of bupropion HCl in treatment of depression. Curr Ther Res Clin Exp. 1978;23:222–9.

Feldman PE. Psychotherapy and chemotherapy (amitriptyline) of anergic states. Dis Nerv Syst. 1961;22(5Pt 2):27–31.

Ferris RM, Beaman OJ, Tang FLM, Russell A. Effects of bupropion, a new antidepressant, on CNS receptors. Pharmacologist. 1981;23(3):125.

Findlay JWA, Van Wyck Fleet J, Smith PG, Butz RF, Hinton ML, Blum MR, Schroeder DH. Pharmacokinetics of bupropion, a novel antidepressant agent, following oral administration to healthy subjects. Eur J Clin Pharmacol. 1981;21:127–35.

Fond G, Macgregor A, Miot S. Nanopsychiatry—the potential role of nanotechnologies in the future of psychiatry: a systematic review. Eur Neuropsychopharmacol. 2013;23:1067–71.

Freed H. On the parenteral use of amitriptyline (ElavilMerck): a preliminary report. Am J Psychiatr. 1960;117:455–6.

Fuchs E, Simon M, Schmelting B. Pharmacology of a new antidepressant: Benefit of the implication of the melatonergic system. Int Clin Psychopharmacol. 2006;21:S17–20.

Fuller RW, Hemrick-Luecke SK, Snoddy HD. Effects of duloxetine, an antidepressant drug candidate, on concentrations of monoamines and their metabolites in rats and mice. J Pharmacol Exp Ther. 1994;269:132–6.

Fuller RW, Perry KW, Molloy BB. Effect of an uptake inhibitor on serotonin metabolism in rat brain: studies with 3-(p-trifluoromethylphenoxy)-n-methyl-3-phenylpropylamine (Lilly 110140). Life Sci. 1974;15:1161–71.

Garcia X, Escribano E, Colom H, Domenech J, Queralt J. Tricyclic antidepressants-loaded biodegradable plga nanoparticles: in vitro characterization and in vivo analgesic and anti-allodynic effect. Curr Nanosci. 2011;7:345–53.
Goldstein DJ, Mallinckrodt C, Lu Y, Demitrack MA. Duloxetine in the treatment of major depressive disorder: a double-blind clinical trial. J Clin Psychiatr. 2002;63:225–31.
Gordon MN, Muller CD, Sherman KA, Morgan DG, Azzaro AJ, Wecker L. Oral versus transdermal selegiline: antidepressant-like activity in rats. Pharmacol Biochem Behav. 1999; 63(3):501–6.
Gottlieb P, Wandall T, Fredericson Overo K. Initial, clinical trial of a new, specific 5-HT reuptake inhibitor, citalopram (Lu 10-171). Acta Psychiatr Scand. 1980;62:236–44.
Grau M. Moclobemide. Drugs Future. 1983;8:14–126.
Halaris AE, Stern W, Harto-Truax N. Clinical efficacy of the new antidepressant bupropion (Wellbutrin®). Psychopharmacol Bull. 1981;17:140–2.
Hamilton SP. The promise of psychiatric pharmacogenomics. Biol Psychiatry. 2015; 77(1):29–35.
Hanoun N, Mocaër E, Boyer PA, Hamon M, Lanfumey L. Differential effects of the novel antidepressant agomelatine (S 20098) versus fluoxetine on 5-$HT_{1A}$ receptors in the rat brain. Neuropharmacology. 2004;47:515–26.
Haque S, Md S, Fazil M, Kumar M, Sahni JK, Ali J, Baboota S. Venlafaxine loaded chitosan NPs for brain targeting: pharmacokinetic and pharmacodynamic evaluation. Carbohydr Polym. 2012;89:72–9.
Haque S, Md S, Sahni JK, Ali J, Baboota S. Development and evaluation of brain targeted intranasal alginate nanoparticles for treatment of depression. J Psychiatr Res. 2014;48:1–12.
Hayley S, Litteljohn D. Neuroplasticity and the next wave of antidepressant strategies. Front Cell Neurosci. 2013;7.
Hyttel J. Neurochemical characterization of a new potent and selective serotonin uptake inhibitor: Lu 10-171. Psychopharmacology (Berl). 1977;51:225–33.
Ingelman-Sundberg M, Persson A, Jukic MM. Polymorphic expression of CYP2C19 and CYP2D6 in the developing and adult human brain causing variability in cognition, risk for depression and suicide: the search for the endogenous substrates. Pharmacogenomics. 2014;15(15):1841–4.
Janssen MJA. Mirtazepine (Remeron®). Pharm Weekbl. 1995;130:355–7.
Jaworska A, Malek K. A comparison between adsorption mechanism of tricyclic antidepressants on silver nanoparticles and binding modes on receptors. Surface-enhanced Raman spectroscopy studies. J Colloid Interface Sci. 2014;431:117–24.
Kaminsky BM, Bostwick JR, Guthrie SK. Alternate routes of administration of antidepressant and antipsychotic medications. Ann Pharmacother. 2015;49(7):808–17.
Kasahara T, Ishigooka J, Nagata E, Murasaki M, Miura S. Long-lasting inhibition of 5-HT uptake of platelets in subjects treated by duloxetine, a potential antidepressant. Jpn J Psychopharmacol. 1996;16:25–31.
Kasper S, McEwen BS. Neurobiological and clinical effects of the antidepressant tianeptine. CNS Drugs. 2008;22:15–26.
Katona C, Hansen T, Olsen CK. A randomized, double-blind, placebo-controlled, duloxetine-referenced, fixed-dose study comparing the efficacy and safety of Lu AA21004 in elderly patients with major depressive disorder. Int Clin Psychopharmacol. 2012;27:215–23.
Katz RJ, Sibel M. Animal model of depression: tests of three structurally and pharmacologically novel antidepressant compounds. Pharmacol Biochem Behav. 1982;16:973–7.
Kendler KS, Neale MC, Kessler RC, Heath AC, Lj E. The clinical characteristics of major depression as indices of the familial risk to illness. Br J Psychiatry. 1994;165:66–72.
Kendler KS, Gardner CO, Prescott CA. Clinical characteristics of major depression that predict risk of depression in relatives. Arch Gen Psychiatry. 1999;56:322–7.
Kendler KS, Kuhn JW, Prescott CA. Childhood sexual abuse, stressful life events and risk for major depression in women. Psychol Med. 2004;34:1475–82.
Kendler KS, Prescott CA. A population-based twin study of lifetime major depression in men and women. Arch Gen Psychiatry. 1999;56:39–44.

Kendler KS, Thornton LM, Gardner CO. Stressful life events and previous episodes in the etiology of major depression in women: an evaluation of the "kindling" hypothesis. Am J Psychiatr. 2000;157:1243–51.

Kendler KS, Thornton LM, Gardner CO. Genetic risk, number of previous depressive episodes, and stressful life events in predicting onset of major depression. Am J Psychiatr. 2001; 158:582–6.

Kennedy SH. Agomelatine: an antidepressant with a novel mechanism of action. Future Neurol. 2007;2:145–51.

Khan A. Vilazodone, a novel dual-acting serotonergic antidepressant for managing major depression. Expert Opin Investig Drugs. 2009;18:1753–64.

Khan A, Fabre LF, Rudolph R. Venlafaxine in depressed outpatients. Psychopharmacol Bull. 1991;27:141–4.

Kihara T, Ikeda M. Effects of duloxetine, a new serotonin and norepinephrine uptake inhibitor, on extracellular monoamine levels in rat frontal cortex. J Pharmacol Exp Therapeut. 1995;272: 177–83.

King PD. Phenelzine and ECT in the treatment of depressions. Am J Psychiatr. 1959;116:64–5.

Korn A, Eichler HG, Fischbach R, Gasic S. Moclobemide, a new reversible MAO inhibitor—interaction with tyramine and tricyclic antidepressants in healthy volunteers and depressive patients. Psychopharmacology (Berl). 1986;88:153–7.

Kragh-Sørensen P, Overø KF, Petersen OL, Jensen K, Parnas W. The kinetics of citalopram: single and multiple dose studies in man. Acta Pharmacol Toxicol. 1981;48:53–60.

Kuhn R. The treatment of depressive states with g 22355 (imipramine hydrochloride). Am J Psychiatr. 1958;115:459–64.

Kung S, Li X. The clinical use of pharmacogenomic testing in treatment-resistant depression. Prim Psychiatr. 2010;17:46–51.

Kupfer DJ. Depression and associated sleep disturbances: patient benefits with agomelatine. Eur Neuropsychopharmacol. 2006;16:S639–43.

Landgraf R, Wigger A. High vs low anxiety-related behavior rats: an animal model of extremes in trait anxiety. Behav Genet. 2002;32:301–14.

Lapidus KAB, Soleimani L, Murrough JW. Novel glutamatergic drugs for the treatment of mood disorders. Neuropsychiatr Dis Treat. 2013;9:1101–12.

Larsen JK, Mikkelsen PL. Moclobemide and clomipramine in the treatment of depression. A randomized clinical trial. Acta Psychiatr Scand. 1984;70:254–60.

Laughren TP, Gobburu J, Temple RJ, Unger EF, Bhattaram A, Dinh PV, Fossom L, Hung HMJ, Klimek V, Lee JE, Levin RL, Lindberg CY, Mathis M, Rosloff BN, Wang SJ, Wang Y, Yang P, Yu B, Zhang H, Zhang L, Zineh I. Vilazodone: clinical basis for the US Food and Drug Administration's approval of a new antidepressant. J Clin Psychiatry. 2011;72:1166–73.

Lee KC, Chen JJ. Transdermal selegiline for the treatment of major depressive disorder. Neuropsychiatr Dis Treat. 2007;3:527–37.

Lehmann HE, Cahn CH, De Verteuil RL. The treatment of depressive conditions with imipramine (G 22355). Can Psychiatr Assoc J. 1958;3:155–64.

Lemere F. Tranylcypromine (parnate) a new monoamine oxidase inhibitor. Am J Psychiatr. 1960;117:249.

Levinson DF. The genetics of depression: a review. Biol Psychiatry. 2006;60:84–92.

Li X, Sundquist J, Sundquist K. Age-specific familial risks of depression: a nation-wide epidemiological study from Sweden. J Psychiatr Res. 2008;42:808–14.

Loomer HP, Saunders JC, Kline NS. A clinical and pharmacodynamic evaluation of iproniazid as a psychic energizer. Psychiatr Res Rep. 1957;8:129–41.

Lundquist G. Treatment of endogenous depressions with iproniazid. Acta Psychiatr Scand. 1959;34:90.

Lyons MJ, Eisen SA, Goldberg J, et al. A registry-based twin study of depression in men. Arch Gen Psychiatry. 1998;55:468–72.

Mann AM, Macpherson AS. Clinical experience with imipramine (G22355) in the treatment of depression. Can Psychiatr Assoc J. 1959;4:38–47.

Mann JJ, Aarons SF, Wilner PJ, Keilp JG, Sweeney JA, Pearlstein T, Frances AJ, Kocsis JH, Brown RP. A controlled study of the antidepressant efficacy and side effects of (−)-deprenyl. Arch Gen Psychiatry. 1989;46:45–50.

Martínez J, Pérez S, Oficialdegui AM, Heras B, Orús L, Villanueva H, Palop JA, Roca J, Mourelle M, Bosch A, Del Castillo JC, Lasheras B, Tordera R, Del Río J, Monge A. New 3-[4-(aryl)piperazin-1-yl]-1-(benzo[b]thiophen-3-yl)propane derivatives with dual action at 5-$HT_{1A}$ serotonin receptors and serotonin transporter as a new class of antidepressants. Eur J Med Chem. 2001;36:55–61.

McEwen BS, Chattarji S, Diamond DM, Jay TM, Reagan LP, Svenningsson P, Fuchs E. The neurobiological properties of tianeptine (Stablon): from monoamine hypothesis to glutamatergic modulation. Mol Psychiatry. 2010;15:237–49.

McGuffin P, Rijsdijk F, Andrew M, Sham P, Katz R, Cardno A. THe heritability of bipolar affective disorder and the genetic relationship to unipolar depression. Arch Gen Psychiatry. 2003;60:497–502.

Mealy NE, Bayés M. Vilazodone hydrochloride. Drugs Future. 2004;29:976.

Mennini T, Mocaer E, Garattini S. Tianeptine, a selective enhancer of serotonin uptake in rat brain. Naunyn Schmiedebergs Arch Pharmacol. 1987;336:478–82.

Millan MJ, Brocco M, Gobert A, Dekeyne A. Anxiolytic properties of agomelatine, an antidepressant with melatoninergic and serotonergic properties: role of 5-HT2C receptor blockade. Psychopharmacology (Berl). 2005;177:1–12.

Millan MJ, Gobert A, Lejeune F, Dekeyne A, Newman-Tancredi A, Pasteau V, Rivet JM, Cussac D. The novel melatonin agonist agomelatine (S20098) is an antagonist at 5-hydroxytryptamine2C receptors, blockade of which enhances the activity of frontocortical dopaminergic and adrenergic pathways. J Pharmacol Exp Therapeut. 2003;306:954–64.

Millan MJ, Goodwin GM, Hamon M, Meyer-Lindenberg A, Ögren SO. 60 years of advances in neuropsychopharmacology for improving brain health, renewed hope for progress. Eur Neuropsychopharmacol. 2015a;25:591–8.

Millan MJ, Goodwin GM, Meyer-Lindenberg A, Ove Ögren S. Learning from the past and looking to the future: emerging perspectives for improving the treatment of psychiatric disorders. Eur Neuropsychopharmacol. 2015b;25:599–656.

Mocaer E, Rettori MC, Kamoun A. Pharmacological antidepressive effects and tianeptine-induced 5-HT uptake increase. Clin Neuropharmacol. 1988;11:S32–42.

Montgomery SA, Loft H, Sánchez C, Reines EH, Papp M. Escitalopram (S-enantiomer of citalopram): clinical efficacy and onset of action predicted from a rat model. Pharmacol Toxicol. 2001;88:282–6.

Mørk A, Pehrson A, Brennum LT, Møller Nielsen S, Zhong H, Lassen AB, Miller S, Westrich L, Boyle NJ, Sánchez C, Fischer CW, Liebenberg N, Wegener G, Bundgaard C, Hogg S, Bang-Andersen B, Bryan Stensbøl T. Pharmacological effects of Lu AA21004: a novel multimodal compound for the treatment of major depressive disorder. J Pharmacol Exp Therapeut. 2012;340:666–75.

Murck H, Frieboes RM, Antonijevic IA, Steiger A. Distinct temporal pattern of the effects of the combined serotonin-reuptake inhibitor and 5-HT1A agonist EMD 68843 on the sleep EEG in healthy men. Psychopharmacology (Berl). 2001;155:187–92.

Neuman M. S-1574. Drugs Future. 1979;4:522–4.

Nnadi CU, Goldberg JF, Malhotra AK. Pharmacogenetics in mood disorder. Curr Opin Psychiatry. 2005;18:33–9.

Oruch R, Elderbi MA, Khattab HA, Pryme IF, Lund A. Lithium: a review of pharmacology, clinical uses, and toxicity. Eur J Pharmacol. 2014;740:464–73.

Overø KF. Preliminary studies of the kinetics of citalopram in man. Eur J Clin Pharmacol. 1978;14:69–73.

Owens MJ, Knight DL, Nemeroff CB. Second-generation SSRIs: human monoamine transporter binding profile of escitalopram and R-fluoxetine. Biol Psychiatry. 2001;50:345–50.

Papp M, Gruca P, Boyer PA, Mocaër E. Effect of agomelatine in the chronic mild stress model of depression in the rat. Neuropsychopharmacology. 2003;28:694–703.

Patel K, Padhye S, Nagarsenker M. Duloxetine HCl lipid nanoparticles: preparation, characterization, and dosage form design. AAPS PharmSciTech. 2012;13:125–33.
Paul GL. Strategy of outcome research in psychotherapy. J Consult Psychol. 1967;31:109–18.
Pento JT. WY-45030. Drugs Future. 1988;13:839–40.
Pierre Olié J, Kasper S. Efficacy of agomelatine, a MT1/MT2 receptor agonist with 5-HT2C antagonistic properties, in major depressive disorder. Int J Neuropsychopharmacol. 2007;10:661–73.
Pitsikas N. Duloxetine Eli Lilly and Co. Curr Opin Investig Drugs. 2000;1:116–21.
Pjrek E, Winkler D, Konstantinidis A, Willeit M, Praschak-Rieder N, Kasper S. Agomelatine in the treatment of seasonal affective disorder. Psychopharmacology (Berl). 2007;190:575–9.
Porcelli S, Drago A, Fabbri C, Gibiino S, Calati R, Serretti A. Pharmacogenetics of antidepressant response. J Psychiatry Neurosci. 2011;36:87–113.
Prakash DJ, ArulKumar S, Sabesan M. Effect of nanohypericum (hypericum perforatum gold nanoparticles) treatment on restraint stressinduced behavioral and biochemical alteration in male albino mice. Pharmacognosy Res. 2010;2:330–4.
Rees L, Davies B. A controlled trial of phenelzine ("Nardil") in the treatment of severe depressive illness. J Ment Sci. 1961;107:560–6.
Rickels K, Athanasiou M, Reed C. Vilazodone, a novel, dual-acting antidepressant: current status, future promise and potential for individualized treatment of depression. Pers Med. 2009;6:217–24.
Riva F. FCE-20124. Drugs Future. 1985;10:905–6.
Riva M, Brunello N, Rovescalli AC, Galimberti R, Carfagna N, Carminati P, Pozzi O, Ricciardi S, Roncucci R, Rossi A, Racagni G. Effect of reboxetine, a new antidepressant drug, on the central noradrenergic system: behavioural and biochemical studies. J Drug Dev. 1989;1:243–53.
Sanchez C, Asin KE, Artigas F. Vortioxetine, a novel antidepressant with multimodal activity: review of preclinical and clinical data. Pharmacol Ther. 2015;145:43–7.
Saunders JC, Roukema RW, Kline NS, Bailey SDA. Clinical results with phenelzine. Am J Psychiatr. 1959;116:71–2.
Schlappi B. The lack of hepatotoxicity in the rat with the new and reversible MAO-A inhibitor moclobemide in contrast to iproniazid. Arzneimittelforschung. 1985;35:800–3.
Schweizer E, Clary C, Weise C, Rickels K. An open-label, dose-finding study of WY-45,030, a novel bicyclic antidepressant. Psychopharmacol Bull. 1988;24:195–7.
Schweizer E, Weise C, Clary C, Fox I, Rickels K. Placebo-controlled trial of venlafaxine for the treatment of major depression. J Clin Psychopharmacol. 1991;11:233–6.
Simon GE, Perlis RH. Personalized medicine for depression: can we match patients with treatments? Am J Psychiatr. 2010;167:1445–55.
Siwers B, Ringberger VA, Tuck JR, Sjoqvist F. Initial clinical trial based on biochemical methodology of zimelidine (a serotonin uptake inhibitor) in depressed patients. Clin Pharmacol Ther. 1977;21:194–200.
Sorbera LA, Castaner RM, Castaner J. Duloxetine oxalate: treatment of stress urinary incontinence antidepressant norepinephrine reuptake inhibitor 5-HT reuptake inhibitor. Drugs Future. 2000;25:907–16.
Sorbera LA, Rabasseda X, Silvestre J, Castañer J. Vilazodone hydrochloride. Drugs Future. 2001a;26:247–52.
Sorbera LA, Revel L, Martín L, Castañer J. Escitalopram oxalate. Antidepressant, 5-HT reuptake inhibitor. Drugs Future. 2001b;26:115–20.
Soroko FE, Maxwell RA (1978) The CNS pharmacology of bupropion HCl (wellbatrin ®), a novel antidepressant agent. Fed Proc 37:481.
Soroko FE, Mehta NB, Maxwell RA. Bupropion hydrochloride ((±) α-t-butylamino-3-chloropropiophenone HCl): a novel antidepressant agent. J Pharm Pharmacol. 1977;29:767–70.
Stingl J, Viviani R. Polymorphism in CYP2D6 and CYP2C19, members of the cytochrome P450 mixed-function oxidase system, in the metabolism of psychotropic drugs. J Intern Med. 2015;277:167–77.
Tang J, Slowing II, Huang Y, Trewyn BG, Hu J, Liu H, Lin VSY. Poly(lactic acid)-coated mesoporous silica nanosphere for controlled release of venlafaxine. J Colloid Interface Sci. 2011;360:488–96.

Thomas Haskins J, Moyer JA, Muth EA, Sigg EB. DMI, Wy-45,030, Wy-45,881 and ciramadol inhibit locus coeruleus neuronal activity. Eur J Pharmacol. 1985;115:139–46.

Thor KB, Katofiasc MA. Effects of duloxetine, a combined serotonin and norepinephrine reuptake inhibitor, on central neural control of lower urinary tract function in the chloralose-anesthetized female cat. J Pharmacol Exp Ther. 1995;274:1014–24.

Treit D, Degroot A, Kashluba S, Bartoszyk GD. Systemic EMD 68843 injections reduce anxiety in the shock-probe, but not the plus-maze test. Eur J Pharmacol. 2001;414:245–8.

Tuma J, Strubbe JH, Mocaër E, Koolhaas JM. Anxiolytic-like action of the antidepressant agomelatine (S 20098) after a social defeat requires the integrity of the SCN. Eur Neuropsychopharmacol. 2005;15:545–55.

Uldam HK, Juhl M, Pedersen H, Dalgaard L. Biosynthesis and identification of an N-oxide/N-glucuronide metabolite and first synthesis of an N-O-glucuronide metabolite of Lu AA21004. Drug Metab Dispos. 2011;39:2264–74.

Vaisberg M, Saunders JC. Amitriptyline in the treatment of depressive states (a pilot study). Dis Nerv Syst. 1961;22:334–8.

Valverde O, Torrens M. CB1 receptor-deficient mice as a model for depression. Neuroscience. 2012;204:193–206.

Varshosaz J, Zaki MR, Minaiyan M, Banoozadeh J. Preparation, optimization, and screening of the effect of processing variables on agar nanospheres loaded with bupropion HCl by a D-optimal design. BioMed Res Int. 2015;2015: Article ID 571816.

Vollmayr B, Bachteler D, Vengeliene V, Gass P, Spanagel R, Henn F. Rats with congenital learned helplessness respond less to sucrose but show no deficits in activity or learning. Behav Brain Res. 2004;150:217–21.

Volmat R, Allers G, Vittouris N. Clomipramine or anafranil. Apropos of treatment of 100 depressive states. Encéphale. 1968;57:116–42.

Wilde MI, Benfield P. Tianeptine: a review of its pharmacodynamic and pharmacokinetic properties, and therapeutic efficacy in depression and coexisting anxiety and depression. Drugs. 1995;49:411–39.

Willner P, Belzung C. Treatment-resistant depression: are animal models of depression fit for purpose? Psychopharmacology (Berl). 2015;232:3473–95.

Wong DT. Duloxetine (LY 248686): an inhibitor of serotonin and noradrenaline uptake and an antidepressant drug candidate. Expert Opin Investig Drugs. 1998;7:1691–9.

Wong DT, Bymaster FP, Horng JS, Molloy BB. 3-(p-trifluoromethylphenoxy)-N-methyl-3-phenylpropylamine (Lilly 110140): a specific inhibitor of serotonin uptake into synaptosomes of rat brain. J Pharmacol Exp Ther. 1974;193(3):804–11.

Wong DT, Bymaster FP, Mayle DA, Reid LR, Krushinski JH, Robertson DW. LY248686, a new inhibitor of serotonin and norepinephrine uptake. Neuropsychopharmacology. 1993;8:23–33.

Yang C, Shirayama Y, Zhang JC, Ren Q, Yao W, Ma M, Dong C, Hashimoto K. R-ketamine: a rapid-onset and sustained antidepressant without psychotomimetic side effects. Transl Psychiatry. 2015;5:e632.

Zitrin A, Thompson DS. EFfects of isonicotinic acid hydrazides on mental status of tuberculous patients. JAMA. 1956;161:204–10.

Zubenko GS, Maher B, Hughes HB, Zubenko WN, Stiffler JS, Kaplan BB, Marazita ML. Genome-wide linkage survey for genetic loci that influence the development of depressive disorders in families with recurrent, early-onset, major depression. Am J Med Genet B Neuropsychiatr Genet. 2003;123B:1–18.



# Index




C. Grosso (ed.), *Herbal Medicine in Depression*,
DOI 10.1007/978-3-319-14021-6

## D



## E



## F

**R**